Bock · Kostina · Phu · Rannacher (Eds.)
Modeling, Simulation and Optimization of Complex Processes

Hans Georg Bock · Ekaterina Kostina
Hoang Xuan Phu · Rolf Rannacher
Editors

Modeling, Simulation and Optimization of Complex Processes

Proceedings of the International Conference on High Performance Scientific Computing, March 10–14, 2003, Hanoi, Vietnam

With 231 Figures, and 34 Tables

Springer

Editors

Hans Georg Bock
Universität Heidelberg
Interdisziplinäres Zentrum
für Wissenschaftliches Rechnen (IWR)
Im Neuenheimer Feld 368
69120 Heidelberg, Germany
e-mail: bock@iwr.uni-heidelberg.de

Ekaterina Kostina
Universität Heidelberg
Interdisziplinäres Zentrum
für Wissenschaftliches Rechnen (IWR)
Im Neuenheimer Feld 368
69120 Heidelberg, Germany
e-mail: ekaterina.kostina@iwr.uni-heidelberg.de

Hoang Xuan Phu
Institute of Mathematics
Vietnamese Academy of Science
and Technology (VAST)
18 Hoang Quoc Viet Road
10307 Hanoi, Vietnam
e-mail: hxphu@math.ac.vn

Rolf Rannacher
Universität Heidelberg
Institut für Angewandte Mathematik
Im Neuenheimer Feld 294
68120 Heidelberg, Germany
e-mail: rannacher@iwr.uni-heidelberg.de

Library of Congress Control Number: 2004115281

Mathematics Subject Classification:
49-06, 60-06, 68-06, 70-06, 76-06, 85-06, 90-06, 93-06, 94-06

ISBN 3-540-23027-0 Springer Berlin Heidelberg New York

Springer is a part of Springer Science+Business Media

springeronline.com

Printed in Germany

Typesetting by the authors
Production: LE-TEX Jelonek, Schmidt & Vöckler GbR, Leipzig
Cover design: *design & production* GmbH, Heidelberg

Printed on acid-free paper 46/3142YL - 5 4 3 2 1 0

Preface

This volume contains a selection of papers referring to lectures presented at the International Conference on High Performance Scientific Computing held at the Hanoi Institute of Mathematics, Vietnamese Academy of Science and Technology (VAST), March 10–14, 2003. The conference has been organized by the Hanoi Institute of Mathematics, SFB 359 "Reactive Flows, Transport and Diffusion", Heidelberg, Ho Chi Minh City University of Technology and Interdisciplinary Center for Scientific Computing (IWR), Heidelberg.

High Performance Scientific Computing is an interdisciplinary area that combines many fields such as mathematics, computer science and scientific and engineering applications. It is a key high-technology for competitiveness in industrialized countries as well as for speeding up development in emerging countries. High performance scientific computing develops methods for computer aided simulation and optimization for systems and processes. In practical applications in industry and commerce, science and engineering, it helps to save resources, to avoid pollution, to reduce risks and costs, to improve product quality, to shorten development times or simply to operate systems better.

The conference had about 200 participants from countries all over the world. The scientific program consisted of more than 100 talks, 10 of them invited plenary talks given by internationally leading experts in the field. Topics were mathematical modelling, numerical simulation, methods for optimization and control, parallel computing, symbolic computing, software development, applications of scientific computing in physics, chemistry, biology and mechanics, environmental and hydrology problems, transport, logistics and site location, communication networks, production scheduling, industrial and commercial problems.

The submitted manuscripts have been carefully reviewed and 42 of the contributions have been selected for publication in this proceedings volume. We would like to thank all contributors and referees.

We would like also to use the opportunity to thank the sponsors whose support significantly contributed to the success of the conference: The German

Research Foundation (DFG) through SFB 359 "Reactive Flows, Transport and Diffusion"; Gottlieb Daimler- und Karl Benz-Stiftung; Deutscher Akademischer Austauschdienst (DAAD); The Abdus Salam International Centre for Theoretical Physics (ICTP); Hanoi Institute of Mathematics; Vietnamese Academy of Science and Technology (VAST); National Council for Natural Sciences of Vietnam; Key Project "Selected Problems of Optimization and Scientific Computing"; Mercedes-Benz Vietnam, Ho Chi Minh City.

Heidelberg, July 2004

Hans Georg Bock
Ekaterina Kostina
Hoang Xuan Phu
Rolf Rannacher

Contents

Constraint Retraction for Dynamic Constraint Satisfaction Problems over Disjoint Real Intervals

Duong Tuan Anh

Hochiminh City University of Technology, 268 Ly Thuong Kiet, Dist. 10, Hochiminh City, Vietnam
dtanh@dit.hcmut.edu.vn

Summary. In a dynamic constraint satisfaction problem (dynamic CSP),we can add a new constraint to the constraint network (a restriction) or delete an old constraint(a relaxation) at any time. Therefore, incrementality is crucial for solving a dynamic CSP since we do not want to resolve the whole constraint system from scratch whenever a restriction or a relaxation occurs. In this paper, we propose an algorithm that can handle incremental constraint retraction in dynamic CSPs over real intervals. Basing on the hierarchical arc-consistency technique for disjoint real intervals developed by G. Sidebottom and W.S. Havens, we extent the proposed algorithm to be the one dealing with constraint deletion in dynamic CSPs over disjoint real intervals. The extended algorithm makes incremental deletion of constraints over disjoint real intervals a feasible task that can be efficiently implemented.

1 Introduction

Arc-consistency is the most commonly used technique for solving constraint satisfaction problems over finite domains (Finite CSPs). The popularity of arc-consistency techniques is due to its simplicity and generality. Furthermore, basing on local propagation, arc-consistency techniques take advantage of the potential locality of typical constraint network.

It is shown that arc-consistency algorithms for CSPs over real intervals (ICSPs) have been developed through an approximation of the notion of arc-consistency for Finite CSPs [8]. These algorithms make use of Interval Arithmetic to compute the refined domains. However, all the techniques already developed for solving ICSPs deal only with the problems where there is one fixed set of constraints (also called *static ICSPs*).

Many real life applications involve reasoning in dynamic environments in which we can add a new constraint to the constraint system or remove a previously active constraint from the constraint system at any time. But the main difficulty in maintaining arc-consistency for dynamic CSPs is in the task involving with the deletion of a constraint: how to avoid resolving the remaining system from scratch when a formerly activated constraint is removed from the CSP.

In this paper, firstly we propose an algorithm that can handle incremental constraint retraction for dynamic CSPs over real intervals. The proposed algorithm follows the chain of dependencies among hyperarcs in constraint hypergraph and by doing so it updates only the part of the constraint hypergraph which is affected by the deletion, but maintains the rest of the hypergraph untouched. This algorithm is close in spirit to the one described in [5], however, their method is for finite domains and for constraint deletion at low-level - the constraints must be of basic form $X\ in\ r$ where X is a domain variable and r is a range. Our first algorithm for constraint deletion presented here can be viewed as a generalization of their method in order that it can handle user constraints over real intervals and of more general forms.

Basing on the hierarchical arc-consistency technique for disjoint real intervals developed by G. Sidebottom and W.S. Havens [10], we extend the first algorithm to be the one dealing with constraint retraction in dynamic CSPs over unions of disjoint real intervals. The key idea behind the extended algorithm is to employ the hierarchical data structure to represent a union of disjoint real intervals.

The rest of the paper is organized as follows. Section 2 give some necessary definitions and concepts. In the Section 3, we propose a method to deal with constraint deletion for dynamic CSPs over real intervals in which a reexecution from scratch can be avoided. Section 4 describes how to extend the method to handle constraint deletion for dynamic CSPs over unions of disjoint real intervals. Conclusion remarks are given in Section 5.

2 Background

A CSP is defined by a set of variables, each associated with a domain of candidate values and a set of constraints on subsets of the variables. A constraint specifies which values from the domains of its variables are compatible. A *solution* to the CSP is an assignment of values to all its variables which satisfies all the constraints.

Since this paper mainly deals with CSPs over real domains, some fundamental concepts of ICSPs will be explained briefly in this section.

2.1 CSPs over Real Intervals (ICSPs)

A CSP over real intervals, $(P = (V, D, S))$, is defined as a set of variables $V_1, \ldots, V_n$ taking their values respectively from a set D of continuous domains $D_1, \ldots, D_n$ and constrained by a set S of *constraints* $C_1, \ldots, C_m$. A domain is an interval of $\Re$.

A *state* of an ICSP at any time point is represented by the assignment of the form: $X_1 \leftarrow I_1, X_2 \leftarrow I_2, \ldots, X_i \leftarrow I_i, \ldots$ where each real interval I_i is the X_i's current domain. At the beginning, the ICSP under consideration has an *initial state* created from assigning the initial intervals to variables. These initial intervals are supplied by the user. During the process of local propagation, we are concerned with the current state of the ICSP. After applying the arc-consistency algorithm, if the ICSP becomes stable, its *terminal state* is also called the *interval solution* of the ICSP. Notice that the interval solution of an ICSP refers to a set of exact value solutions.

Example 2.1. Suppose we have a CSP over real intervals. The initial state is: $D_X = [1, 10], D_Y = [3, 8], D_Z = [2, 7], D_T = [-1000, 1000], D_U = [0, 15]$ and $D_V = [-20, 20]$ and the constraints are: $x + y = z(c1), y \leq x(c2), u = 2 * v(c3), x = 2 * t(c4)$

Let $var(c)$ denote the set of variables in constraint c. The constraint network of a CSP is described here as a *directed hypergraph* where variables are associated with nodes and each constraint c is associated with a set of *directed hyperarcs* of the form $< X, c >$ for each $X \in var(c)$. In this hypergraph, hyperarcs may connect one, two or more than two nodes. Given a directed hyperarc $< X, c >$ of a constraint hypergraph, X is called the *constrained variable* of the hyperarc. So each hyperarc here represents a *projection constraint*. For example, given the constraint c: X+ Y = Z, the hyperarcs $< X, c >$, $< Y, c >$ and $< Z, c >$ respectively represent the *projection constraints*: $X = Z - Y, Y = Z - X$ and $Z = X + Y$. Given a CSP, an arc consistency algorithm deletes inconsistent values from constrained variable domains.

Definition 1. *(Dependent on). Given $< X, c >, < Y, c' >$, two hyperarcs of the constraint system S, the hyperarc $< Y, c' >$ is dependent on $< X, c >$ iff $X \in var(c') \setminus \{Y\}$ and $c' \neq c$.* ■

Let denote $DH(X, c, S)$ the set of all hyperarcs in S dependent on $< X, c >$. In a local propagation, if the hyperarc $< X, c >$ has changed the domain of X, all the hyperarcs dependent on the hyperarc $< X, c >$ (i.e. the set $DH(X, c, S)$) will be activated to further the propagation.

Definition 2. *(Propagation graph) Given the constraint system S, we can build the directed graph in which each node represents a hyperarc in S and an arc with the direction from node h to node h' indicates that the hyperarc h'* ***dependent on*** *the hyperarc h. This directed graph is called* ***propagation graph****.* ■

Let c be a constraint. We write $ha(c)$ for the set of all hyperarcs of constraint c, and $DH(c)$ for the set of all hyperarcs dependent on any hyperarcs in $ha(c)$. That means $DH(c)$ consists of all the adjacent nodes of the nodes $ha(c)$ in the propagation graph. To explain the locality of constraint propagation in arc-consistency algorithm, we denote $DH^*(c)$ be the set of all hyperarcs that are reachable from any node in $ha(c)$ in the propagation graph.

A useful function for describing arc-consistency algorithms is *projection*, denoted Π, which maps a constraint c and a variable X in $var(c)$ to a subset of DX, the domain of X. $\Pi_X(c)$ is the set of values for X which is consistent with the constraint c and with all the domains of the other variables. Computing projections for some forms of numerical constraints is based on Interval Arithmetic [9].

To reduce the complexity of computing projections, there should be a restriction on the form of constraints. Each equality contains at most one function symbol and each inequality contains no function symbols. That is, constraints are of the form $'A1 = A2','A1 \leq A2','A1 + A2 = A3','A1.A2 = A3','A1 = sin(A2)'$, etc where A1, A2, and A3 are either real variables or real constants. In addition, constraints can be general linear equations like $'a_1X_1 + a_2X_2 + ... + a_nX_n = b'$.

2.2 Dynamic CSPs over Real Intervals

In a dynamic ICSP, we can add a new constraint to the constraint network (a restriction) or delete an old constraint (a relaxation) at any time.

A *state* of a dynamic ICSP after any change (constraint addition/removal) is represented by the assignment of the form: $X_1 \leftarrow I_1, X_2 \leftarrow I_2, \ldots X_i \leftarrow I_i, \ldots$where each real interval I_i is the X_i's current domain at that time point. If the tuple of intervals $< I_1, ..., I_n >$ is the terminal state of the current system which is arc-consistent, it is also called the *interval solution* of the dynamic ICSP at that time.

Dynamic ICSPs can have monotonic behaviour on adding a new constraint since the solution is the refinement of the larger earlier interval solution. However, dynamic ICSPs exhibit non-monotonic behaviour since deleting an formerly activated constraint can produce a quite different interval solution to the new ICSP.

2.3 Addition of a new constraint

Generally, the addition of constraints is easier to handle than the deletion of constraints. The procedure for addition of a new constraint c to the network S is given in Figure 1. It is a modification of the Davis's arc-consistency algorithm for ICSPs [3]. In the arc-consistency algorithm for ICSPs, first we put all the hyperarcs of all the constraints in the system (in any order) into the queue Q. But in the procedure *Add*, first we put the hyperarcs of the new constraint into the propagation queue Q instead. That means the domains of c's variables will be refined by the function *NARROW*. The changes in the domains of these variables will invoke some other constraints to be considered for interval propagation. This propagation goes on until the system becomes stable, i.e., there is no more domain to be narrowed.

Line 4 and 5 of the function *NARROW* specify that D_T is updated only if *NARROW* succeeds in refining it. Line 6 of procedure Add initializes Q to the set of hyperarcs of the new constraint. If $NARROW(T, c)$ refines D_T in line 10, then the queue Q is updated in line 11 by adding the set of hyperarcs dependent on the hyperarc $< T, c >$ into the queue Q so that these hyperarcs could be further revised.

```
   boolean function NARROW(T,c)
   begin
1    if ΠT(c) = ∅ then halt /* the original constraints were inconsistent */
2    else
3        CHANGED := ΠT(c) ⊂ DT;
4    if CHANGED then DT := ΠT(c);
5    return CHANGED
   end

   procedure Add(c,S)
   begin
6   Q:= A; /* A is set of hyperarcs of the added constraint c */
7   while Q not empty do
8     begin
9       dequeue any hyperarc(T,c) from Q;
10        if NARROW(T,c) then
11          Q := Q ∪ DH(T,c,S)
12     end
   end.
```

Fig. 1. The procedure for addition of a constraint

Proposition 1. *When a constraint c is added to a constraint system S, only the domains of the constrained variables of the hyperarcs in $DH^*(c) \cup ha(c)$ can be changed by Procedure* Add *and the domains of all the other variables remain the same.*

Proof. In the procedure Add, on the addition of the constraint c, only the hyperarcs in the set $DH^*(c) \cup ha(c)$ have the chance to be put into the propagation queue Q and only the constrained variables of these hyperarcs have the chance to be refined by the function NARROW. Therefore, only the domains of these variables can be changed. ■

Proposition 2. *When a constraint c is deleted from a constraint system S, only the domains of the constrained variables of the hyperarcs in $DH^*(c)$ can be changed by the deletion and the domains of all the other variables remain the same.*

Proof. Let denote $ha(S)$ be the set of all hyperarcs in a constraint system S. After deleting c from the set of constraint S, the propagation graph can be seen as consisting of two parts: $ha(S) \setminus DH^*(c)$ and $DH^*(c)$. Resolving the constraint system $S \setminus \{c\}$ from scratch means resetting all the variables in $S \setminus \{c\}$ to their respective initial domains and then applying local propagation through the whole network. In the local propagation, let do a specific way: we initialize the queue Q by putting first all the hyperarcs of $DH^*(c)$ and then all the hyperarcs in $ha(S) \setminus DH^*(c)$ into Q and during propagation, we do not activate the hyperarcs in $ha(S) \setminus DH^*(c)$ until all the hyperarcs of $DH^*(c)$ have been activated since the order of activating constraints does not affect the result. Doing so, the propagation works through the two subgraphs $DH^*(c)$ and $ha(S) \setminus DH^*(c)$ separately. By that way, the resetting and local propagation on the subgraph $ha(S) \setminus DH^*(c)$ would bring the domains of all variables in this part back to the same domains as they had before deleting c. In other words, the resetting and propagation on the subgraph $ha(S) \setminus DH^*(c)$ are unnecessary. So the deletion of c can affect only on the domains of the constrained variables in $DH^*(c)$. ■

3 Constraint Retraction for Dynamic CSPs over Real Intervals

This section describes the technique to deal with constraint deletion. It extends traditional arc-consistency algorithm to selectively delete from the constraint network a constraint and remove all its consequences on affected variables, but maintain the rest of the network untouched. Based on the Proposition 2.2, the constraint deletion consists of two fundamental tasks:

1. reset all the variables in the subgraph $DH^*(c)$ of the propagation graph to their initial domains,
2. apply the arc-consistency algorithm only for the subgraph $DH^*(c)$ after the change incurred by step (1).

Removing a formerly active constraint c from the system S is done by using the procedure *Delete* given in the Fig.2. Lines 1-9 in procedure *Delete* reset the domains of all variables in the subnetwork $DH^*(c)$ to their respective initial intervals. First, the queue Q is initialized to the set of all hyperarcs of the deleted constraint, i.e. $ha(c)$. The *while* loop in lines 3-9 is for visiting all the variables in the subgraph $DH^*(c)$. When a variable domain is reset to its initial interval, it will be marked.

Lines 10-18 perform the arc-consistency algorithm only for the subnetwork $DH^*(c)$ after the change incurred by resetting step. Lines 10-12 initialize the queue Q with all the hyperarcs that depends on each hyperarcs in $ha(c)$. The *while* loop in lines 13-18 propagate the domain changes from these neighbour constraints to their own neighbours and so on to go through the subgraph $DH^*(c)$. When the queue is empty, there are no more hyperarcs to reconsider for the propagating phase. The *while* loop in the propagating phase is exactly the same as the *while* loop in the arc-consistency algorithm for ICSPs.

Constraint deletion has a special feature. Whereas a restriction of a new constraint narrows the domains of a subset of variables in the network through propagation process, a constraint deletion is an *"resetting"* propagation which

```
procedure Delete(c,S)
begin
  /* Resetting */
  Q := ∅
  for each X ∈ var(c) do add <X,c> to the queue Q
  while Q not empty do
  begin
    dequeue the hyperarc(Y,c') from the queue Q
    mark Y; DY := DY^I; /* reset DX to its initial interval */
    for each (T,c") ∈ DH(Y,c',S\{c}) do
      if T is not marked then Add (T,c") to the queue Q
  end
  /* Propagating */
  Q := ∅
  for each (X,c) ∈ ha(c) do
      for each (Y,c') ∈ DH(X,c,S\{c}) do add (Y,c') to the queue Q
  while Q not empty do
  begin
    dequeue the hyperarc(Y,c') from the queue Q
    if NARROW(Y,c')then
        Add DH(Y,c',S\{c}) to the queue Q
  end
end.
```

Fig. 2. The procedure for deleting a constraint

makes the related domains larger. The enlarging propagation is performed through the network as long as there are still domains that should be enlarged.

An Example

Consider the CSP given in Example 2.1 in which the constraints are added into the system in the order $(c1, c2, c3, c4)$ and then c1 is removed from the system. The algorithm DACR will perform the sequence of computations given in the Table 1 to maintain the arc-consistency of the system. So after deleting c1: $z = x + y$, the interval solution of the constraint system becomes $D_X = [1, 10], D_Y = [3, 8], D_Z = [2, 7], D_T = [1.5, 5], D_U = [0, 15]$ and $D_V = [0, 7.5]$.

Now let resolve the constraint system as if we had only c2,c3,c4 from the beginning, the computation will be given as in Table 2. Again, we obtain the same result as in Table 1 with $D_X = [3, 10], D_Y = [3, 8], D_Z = [2, 7], D_U = [0, 15]$ and $D_V = [0, 7.5]$. However, the computational work in Table 4.2 is much more than the work after the line **Delete z = x+y** in Table 1 at the constraint deletion since the constraint $u = 2 * v$ still had been reconsidered.

Notice that the worst-case running time of the procedure *Delete* is slightly more than that of the procedure *Add* due to the overhead of resetting the domains of related variables to initial intervals.

Notice that the worst-case running time of the procedure *Delete* is slightly more than that of the procedure *Add* due to the overhead of resetting the domains of related variables to initial intervals.

Table 1.

Next hyperarc	New interval	Q
Add $z = x + y$		$z = x + y,$
		$y = z - x$
		$x = z - y$
$z = x + y$	$D_Z = ([1, 10] + [3, 8]) \cap [2, 7] = [4, 7]$	$y = z - x,$
		$x = z - y$
$y = z - x$	$D_Y = ([4, 7] - [1, 10]) \cap [3, 8] = [3, 6]$	$x = z - y$
$x = z - y$	$D_X = ([4, 7] - [3, 6]) \cap [1, 10] = [1, 4]$	
Add $y \leq x$		$y \leq x, x \geq y$
$y \leq x$	$D_Y \leq [1, 4] \Rightarrow D_Y = [3, 4]$	$x \geq y, z = x + y$
$x \geq y$	$D_X \geq [3, 4] \Rightarrow D_X = [3, 4]$	$z = x + y$
$z = x + y$	$D_Z = ([3, 4] + [3, 4]) \cap [4, 7] = [6, 7]$	
Add $u = 2v$		$u = 2v, v = u/2$
$u = 2v$	$D_U = (2 * [20, 20) \cap [0, 15] = [0, 15]$	$v = u/2$
$v = u/2$	$D_V = (0.5 * [0, 15]) \cap [20, 20] = [0, 7.5]$	
Add $x = 2t$		$x = 2t, t = x/2$
$x = 2t$	$D_X = (2 * [-1000, 1000) \cap [3, 4] = [3, 4]$	$t = x/2$
$t = x/2$	$D_T = (0.5 * [3, 4]) \cap [-1000, 1000] = [1.5, 2]$	
Delete $z = x + y$	$D_X = [1, 10], D_Y = [3, 8], D_Z = 2, 7]$	$y \leq x, x \geq y$
	$D_T = [-1000, 1000]$	$t = x/2, x = 2t$
$y \leq x$	$D_Y \leq [1, 10] \Rightarrow D_Y = [3, 8]$	$x \geq y, t = x/2$
		$x = 2t$
$x \geq y$	$D_X \geq [3, 8] \Rightarrow D_X = [3, 10]$	$t = x/2, x = 2t$
	$D_T = (0.5 * [3, 10]) \cap [-1000, 1000] = [1.5, 5]$	$x = 2t$
$x = 2t$	$D_X = (2 * [1.5, 5]) \cap [3, 10] = [3, 10]$	

Discussion

It is important to note that in our constraint retraction method, constraint dependencies are computed only when a constraint has to be removed. That means this method does not require any preparatory information recording during constraint addition as in DNAC4 algorithm [1] or DNAC6 algorithm [4] . These algorithms use *justifications*: for each value removal the applied responsible constraint is recorded. A generalization [7] of the information recording method relies upon the use of explanation sets (informally, a set of constraints that justifies a domain reduction).

Table 2.

Next Hyperarc	New interval	Q
Add $y \leq x$		$y \leq x$, $x \geq y$
$y \leq x$	$D_Y \leq [1,10] \Rightarrow D_Y = [3,8]$	$x \geq y$
$x \geq y$	$D_X \geq [3,8] \Rightarrow D_X = [3,10]$	
Add $u = 2v$		$u = 2v$, $v = u/2$
$u = 2v$	$D_U = (2 * [-20,20) \cap [0,15] = [0,15]$	$v = u/2$
$v = u/2$	$D_V = (0.5 * [0,15]) \cap [-20,20] = [0,7.5]$	
Add $x = 2t$		$x = 2t$, $t = x/2$
$x = 2t$	$D_X = (2 * [-1000,1000) \cap [3,10] = [3,10]$	$t = x/2$
$t = x/2$	$D_T = (0.5 * [3,10]) \cap [-1000,1000] = [1.5,5]$	

4 Constraint Retraction for Dynamic CSPs over Unions of Disjoint Real Intervals

This section represents how to adapt the method of constraint deletion for dynamic CSPs over real intervals to deal with incremental deletion of constraints over unions of disjoint real intervals.

4.1 Why Unions of Disjoint Real Intervals

We say that D_X, the domain of a variable X, is *convex* iff all the numeric values between $min(D_X)$ and $max(D_X)$ belong to D_X. Sometimes, the constraints of the ICSP are not of basic forms and therefore the variable domains are not convex. The variable domains have to be split into unions of disjoint real intervals due to three major reasons:

1. Definability of constraints. Certain values for the variables must be excluded by the definition of the constraints. For example, in the constraint $X = Y/Z$ all cases with $0 \in Z$ are prohibited.
2. Applicability of interval functions. Needed interval functions cannot always be defined or applied easily. For example, in the constraint $X^2 = Y$, the projection constraint for X should evaluated two cases $X \in [1,2]$ and $X \in [-2,-1]$ if $Y \in [1,4]$.
3. Disjunction of constraints. Disjunction of constraints can narrow variable domains into unions of intervals. For example, given $D_X = [1,10], D_Y = [3,8]$, the constraint $X \geq Y + 3 \vee Y \geq X + 3$ refines the $X's$ domain into $D_X = [1,5] \cup [6,10]$.

When domains are unions of real intervals, the computation of projection may split a union of disjoint real intervals into an ever larger set of intervals. This phenomenon is called *splitting problem.* Due to splitting problem, processing constraints over unions of disjoint real intervals becomes more complex [6]. An arc-consistency algorithm that is specially suitable for this case is the algorithm HACR, developed by Sidebottom and Havens [10] for the Echidna, a constraint logic programming language for disjoint real intervals.

4.2 Representation of Domains in HACR

HACR is an arc-consistency algorithm that can handle unions of disjoint real intervals. It represents directly a variable domain which is a union of disjoint real intervals by a binary tree whose node is labelled with a subinterval of the domain. The main idea behind HACR is that it considers a union of disjoint real intervals a hierarchical domain. The hierarchical domain is represented as a binary tree.

Let D_X be the domain of a variable X and Δ_X be the dynamic domain of X at any time of propagation process. The domains in the nodes of the tree for D_X are

$$D_X(k,s)(0 \leq k \leq m, 1 \leq s \leq 2^k)$$

where (k,s) specify a node in the tree. The integer k is the distance from the root and s is the number of the node at distance k from the root, counting from the left starting at 1.

The root domain $D_X(0,1)$ is exactly D_X. For $0 \leq k \leq m$, the children of (k,s) are $(k+1, 2s-1)$ and $(k+1, 2s)$ which satisfy:

$$D_X(k,s) = D_X(k+1, 2s-1) \cup D_X(k+1, 2s)$$

and

$$DX(k+1, 2s-1) \cap D_X(k+1, 2s) = \emptyset$$

The domain of each node in a binary tree represents a real interval. Nodes are associated with the lower and upper bound of the intervals they represent. The domains for two children of each node are the lower and upper halves of the parent domain. If $[x,y]$ is the interval of a node, the two intervals $[x, mid(x,y))$, $[mid(x,y), y]$ are associated with its left child and its right child respectively.

For a variable X, the relationship between the dynamic domain Δ_X and the binary tree for D_X is defined by the $markM_X(k,s)$ on node (k,s). The mark can be one of the three possible values:

1. '+' if $D_X(k,s) \subseteq \Delta_X$
2. '?' if $D_X(k,s) \not\subseteq \Delta_X$ and $D_X(k,s) \cap \Delta_X \neq \emptyset$
3. '×' if $D_X(k,s) \cap \Delta_X = \emptyset$

The data structure for the domain permits the arc-consistency to retain or eliminate whole subtrees as a unit, simply by manipulating the marks. At any time, the dynamic domain Δ_X of a variable X is the union of the domains of all nodes marked '+':

$$\Delta_X = \{D_X(k,s) \mid M_X(k,s) =' +'\}$$

4.3 The ReviseHACR Function

The *ReviseHACR* function, given in Fig. 3, plays the same role as the NAR-

```
boolean function ReviseHACR(T,c)
/* From [SH92]  */
begin
  for k= 0 to P_T do      /* P_T is the precision of variable T */
    for s = 1 to 2^k do
      TM_T(k,s) := 'x';
  for I ⊆ Π_T(c) do MarkTemp(D_T(1,0),I);
  Less := { D_T(k,s) | TM_T(k,s) < M_T(k,s) };   /* 'x' < '?' < '+' */
  for each D_T(k,s) ∈ Less do
    M_T(k,s) := TM_T(k,s);
  return Less ≠ Ø
end

procedure MarkTemp((D_T(k,s),I)
begin
  if (M_T(k,s) = 'x')∨(TM_T(k,s)='+')∨(I ∩ D_T(k,s) = Ø) then return
  else if (D_T(k,s) ⊆ I) ∨ (k=P_T) then TM_T(k,s) := '+'
  else
    begin
      TM_T(k,s) = '?';
      MarkTemp(D_T(k+1,2s-1),I); MarkTemp(D_T(k+1,2s),I);
    end
end.
```

Fig. 3. The ReviseHACR function of the algorithm HACR

ROW function in Procedure *Add*. *ReviseHACR* computes $\Pi_T(c)$ by generating a set of intervals whose union is approximately $\Pi_T(c)$. Conceptually, after computing the projection of a constraint c on the variable T using some interval function, *ReviseHACR(T,c)* performs the assignment of a new mark $M_T(k,s)$ to one of its possible values according to:

1. '+' if $D_X(k,s) \subseteq \Pi_T(c)$
2. '?' if $D_X(k,s) \not\subseteq \Pi_T(c)$ and $D_X(k,s) \cap \Pi_T(c) \neq \emptyset$
3. '×' if $D_X(k,s) \cap \Pi_T(c) = \emptyset$.

for each node $D_X(k,s)$ of D_T.

To keep the binary tree finite, HACR attaches a positive integer precision P to each variable X in the CSP. P is the maximum distance from the root to any node in the binary tree for Δ_X. When *ReviseHACR* determines that $D_X(k,s)$ should be refined, i.e. $M_X(k,s)$ should be set to '?' and its children analyzed, if the node (k,s) is at the precision limit ($k = P$) then $M_X(k,s)$ is left '+'. *ReviseHACR* can approximate the real domains by increasing the precision of variables as necessary. In other words, HACR can control the precision of the domain approximation. When assigning the precision P to a variable domain, we really approximate this infinite domain by a finite domain with 2^P discrete values. The precision P can be used as a parameter to control the convergence speed of the algorithm HACR in the same way as the bound width w does in the IP_2 algorithm in [8].

To discriminate the status of a domain before and after a refinement. HACR maintains a set of "temporary" marks associated with the nodes of the binary tree for D_T:

$$\{TM_{T'}(k,s) \mid 0 \le k \le P, 1 \le s \le 2^k\}$$

Complexity

The running time of HACR is proportional to the logarithm of domain size in the worst case [10]. But the domain here is considered as a finite domain with 2^P discrete values. Therefore, the running time of HACR is proportional to the precision P.

4.4 How to Adapt the Delete Procedure

It is straightforward to adapt the *Delete* procedure to deal with constraint retraction in dynamic CSPs over unions of disjoint real intervals. The adaptation requires two modifications on the *Delete* procedure:

1. The NARROW function used in the *Delete* procedure must be substituted with the ReviseHACR function.
2. In addition, the step resetting in the new procedure *Delete* becomes a non-trivial operation. The resetting a variable X to its initial domain adds values back to the current domain of X, thus changing the marks on some nodes that have been previously marked with '×'. For each constituent interval I in the initial domain, a depth-first traversal on the binary tree is required to put the marks back to the initial status. The procedure *Reset* for resetting a variable to its initial domain is given as in Fig. 4.

After the two modifications, we obtained the new *Delete* procedure that can deal with constraint retraction in dynamic CSPs over unions of disjoint real intervals. Obviously, this *Delete* procedure can support disjunctions of constraints.

```
procedure Reset(T,D_T^I)
begin
 /* D_T^I is the initial union of disjoint intervals for variable T */
 for k= 0 to P_T do
   for s = 1 to 2^k do
     M_T(k,s) := 'x';

 for each I ⊆ D_T^I do
     Reset-dom(D_T(0,1),I)
end

procedure Reset-dom(D_T(k,s),I)
begin
  if I ∩ D_T(k,s) = ∅ then M_T(k,s) := 'x';
  else if (D_T(k,s) ⊆ I) ∨ (k=P_T) then M_T(k,s) := '+';
  else
   begin
      M_T(k,s) = '?';
      Reset-dom(D_T(k+1,2s-1),I); Reset-dom(D_T(k+1,2s),I);
   end
end
```

Fig. 4. The procedure Reset for resetting a domain to its initial domain

5 Conclusions

This paper shows that it is possible to efficiently implement incremental constraint deletion in dynamic CSPs over real intervals as well as over unions of disjoint real intervals. The proposed algorithm follows the chain of dependencies among hyperarcs in constraint hypergraph and by doing so it updates only the part of the constraint hypergraph which is affected by the deletion, but maintains the rest of the hypergraph untouched. As for dealing with unions of disjoint real intervals, we employ a hierarchical data structure to represent a union of disjoint real intervals. The method helps to make an obviously complex task quite manageable.

Our future work includes improving the efficiency of the procedures *Delete* and its extention for unions of disjoint real intervals, and then applying the two algorithms in some real world scheduling applications that may involve with disjunctions of constraints.

References

[1] Bessiere, C.: Arc-consistency in Dynamic Constraint Satisfaction Problems. In: Proc. of AAAI'91, 221-226 (1991)

[2] Codognet, P., Diaz, D. and Rossi, P.: Constraint Retraction in FD, In: Proc. of 16th Conf. on Foundations of Software Technology and Theoretical Computer Science, Hyderabad, India, Dec., 168-179 (1996)

[3] Davis, E.: Constraint Propagation with Interval Labels. Artificial Intelligence, 32, 281-331 (1987)
[4] Debruyne, R.: Arc-consistency in Dynamic CSPs is no more Prohibitive. In: Proc. 8th Conference on Tools with Artificial Intelligence (TAI'96), 299-306 (1996)
[5] Geordet, Y, Codognet, P. and Rossi, F.: Constraint Retraction in clp(FD): Formal Framework and Performance Results. Constraints, an International Journal 4(1):5-42 (1999)
[6] Hyvonen, E.: Constraint Reasoning based on Interval Arithemetic: The Tolerance Propagation Approach. Artificial Intelligence, 58, 71-112 (1992)
[7] Jussien, N., Debruyne, R., and Boizumault, P.: Maintaining Arc-consistency within Dynamic Backtracking. In: Principles and Practice of Constraint Programming CP 2000, No. 1894, Lecture Notes in Computer Science, Springer-Verlag, 249-261 (2000).
[8] Lhomme, O.: Consistency Techniques for Numeric CSPs. In: Proc. IJCAI'93, 232-238 (1993)
[9] Moore, R.E.: Interval Analysis, Englewood Cliffs, New Jersey, Prentice-Hall (1966)
[10] Sidebottom, G., Havens, W.S.: Hierarchical Arc Consistency for Disjoint Real Intervals in Constraint Logic Programming. In: Proc. of Post-conference Workshop on Constraint Logic Programming Systems: Design and Applications, Washington D.C., USA, Nov., 120-150 (1992)

Computational Methods for Large Distributed Parameter Estimation Problems in 3D

Uri M. Ascher[1] and Eldad Haber[2]

[1] Department of Computer Science, University of British Columbia
Vancouver, BC, V6T 1Z4, Canada *
ascher@cs.ubc.ca

[2] Department of Mathematics and Computer Science, Emory University
Atlanta, GA, 30322, USA
haber@mathcs.emory.edu

Summary. This paper considers problems of distributed parameter estimation from data measurements on solutions of diffusive partial differential equations (PDEs). A nonlinear functional is minimized to approximately recover the sought parameter function (i.e., the model). This functional consists of a data fitting term, involving the solution of a finite volume or finite element discretization of the forward differential equation, and a Tikhonov-type regularization term, involving the discretization of a mix of model derivatives.

We develop methods for the resulting constrained optimization problem. The method directly addresses the discretized PDE system which defines a critical point of the Lagrangian. This system is strongly coupled when the regularization parameter is small.

We then apply such methods for electromagnetic data inversion in 3D, both in frequency and in time domains. This involves developing appropriate discretizations for the forward problems as well as handling very large systems of algebraic equations.

Finally, we explore the use of the so-called Huber's norm for the recovery of piecewise smooth model functions. Since our differential operators are compact, there are many different models that give rise to fields that fit practical data sets well.

1 Introduction

This paper mostly describes and surveys our efforts over the past several years to design efficient numerical techniques for electromagnetic data inversion in 3D. The full details are developed and presented in a series of papers including [15, 13, 1, 14, 2, 16], and we fall back on these sources as we try to paint the general picture.

* Supported in part under NSERC Research Grant 84306

In Section 2 we set up the scene for the problems considered, both the forward problems and those of data inversion for the recovery of a distributed parameter model. A nonlinear functional is minimized to approximately recover the sought parameter function (i.e., the model). This functional consists of a data fitting term, involving the solution of a finite volume or finite element discretization of the forward differential equation, and a Tikhonov-type regularization term, involving the discretization of a mix of model derivatives.

We develop methods for the resulting constrained optimization problem in Section 3. The methods directly address the discretized PDE system which defines a critical point of the Lagrangian. This system is strongly coupled when the regularization parameter is small.

In Section 4 we then apply such methods for electromagnetic data inversion in 3D, both in frequency and in time domains. This involves developing appropriate discretizations for the forward problems as well as handling very large systems of algebraic equations.

The last section of this paper, Section 5, is new. We consider the task of recovering a piecewise smooth model, i.e. one that admits discontinuities. We use "Huber's norm" for this and discuss the choice of its parameter. However, with the diffusive forward operators considered in this paper (and elsewhere) we must also cast some doubt on efforts which attempt to define the model "too uniquely" for realistic data. Indeed, we feel that often there are many scenarios that reasonably match the measured data in practice.

2 Forward problems and distributed parameter estimation

Several different applications give rise to a partial differential equation (PDE) of the form

$$\operatorname{div}(\sigma \nabla \mathrm{u}) = \mathrm{q}. \tag{1}$$

Here, the field $\mathrm{u}(\mathbf{x})$ satisfies the PDE plus boundary conditions for given sources $\mathrm{q} = \mathrm{q}(\mathbf{x})$ and a conductivity function $\sigma = \sigma(\mathbf{x})$. Applications include DC resistivity [25], magnetotelluric inversion [21], diffraction tomography [8], impedance tomography [6], oil reservoir simulation [9] and aquifer calibration [11]. Other applications of interest give rise to the more involved Maxwell's equations, e.g. written for the time harmonic case as

$$\nabla \times (\mu^{-1} \nabla \times \mathbf{E}) - \imath\omega\,\sigma\,\mathbf{E} = \imath\omega \mathbf{s}. \tag{2}$$

See, e.g., [18, 19, 15] and references therein. A typical domain Ω in 3D on which such a PDE is defined is depicted in Figure 1.

In a typical application we may have several PDEs such as (1) or (2) corresponding, e.g., to different frequencies and different sources and sinks.

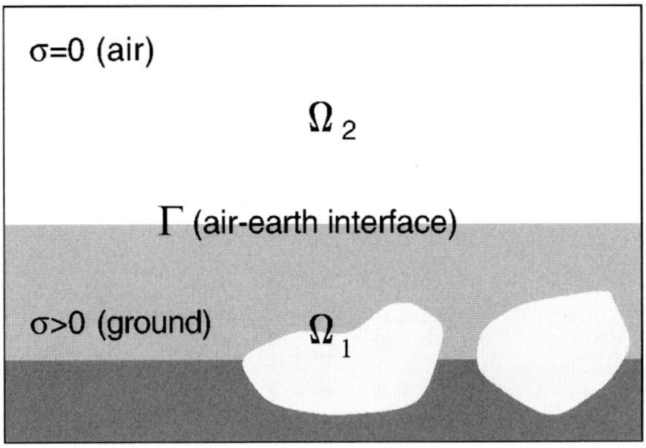

Fig. 1. A typical domain cross-section for a geophysical application

Whatever of these we have, we group them as well as the given boundary conditions into a unified notation

$$\mathcal{A}(\mathrm{m})\mathrm{u} = \mathrm{q}$$

where the *model* $\mathrm{m}(\mathbf{x})$ typically is a function of σ, e.g.,

$$\sigma(\mathbf{x}) = e^{\mathrm{m}(\mathbf{x})}, \ \ \mathbf{x} \in \Omega.$$

For the numerical solution of such a boundary value PDE we discretize it on a tensor product grid, not necessarily uniform, using a finite volume technique [13, 1]. We assume the material properties to be constant in each cell and call the resulting grid functions u, q and m. If needed, they are ordered into vectors. Thus, on a grid depicted in Figure 2 we obtain our discretized problem

$$A(m)u = q. \tag{3}$$

We assume further that the large and sparse matrix A is nonsingular. This assumption, as well as others regarding the type of grid and the use of the same grid for u and for m, are not strictly necessary, but they simplify both the exposition and the programming.

The *forward problem* is to find u satisfying (3) given m (and q, which is always assumed given).

The *inverse problem* is to recover the model m, given measurements b on the field u (and/or its derivatives, but let's do just u here) such that (3) holds.

However, it is well-known that while the forward problem is well-posed the inverse problem is not. Indeed, in practice for the available, noisy data typically there is no unique solution, i.e., there are many models m which yield a field u which is close to b to within the noise level, and moreover, such models m may vary wildly.

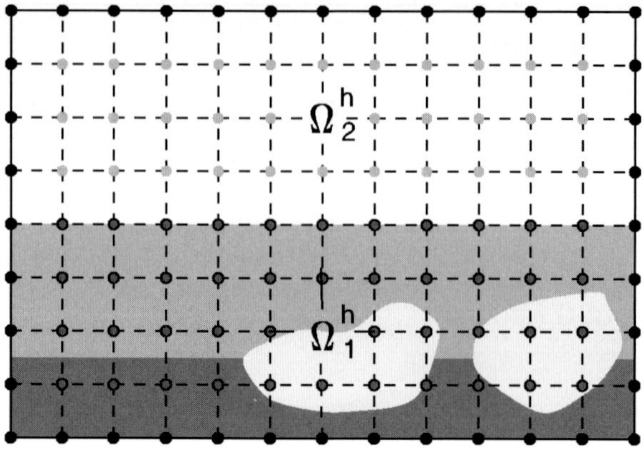

Fig. 2. The discretized domain: a cross-section of a 3D grid of cells.

Thus, we must add a priori information and isolate noise effects. Then the problem becomes to choose, from all the possible model solutions, the one closest to our a priori information. This Tikhonov-type regularization [26] leads to the optimization problem

$$\min_{m,u} \ \varphi = \frac{1}{2}\|Qu - b\|^2 + \beta R(m) \qquad (4)$$
$$\text{subject to} \qquad A(m)u - q = 0,$$

where Q is a matrix consisting of unit rows which projects to data locations, and $R(m)$ is a regularization term. The parameter $\beta > 0$ is the regularization parameter whose choice has been the subject of many a paper. Also, throughout this paper the notation $\|\cdot\|$ refers to the least squares norm.

For the regularization term, consider a same-grid discretization of

$$\mathrm{R(m)} = \int_\Omega \rho(|\nabla \mathrm{m}|) + \hat{\alpha}(\mathrm{m} - \mathrm{m}_{\mathbf{ref}}) \qquad (5)$$

where $\hat{\alpha}$ is a (typically very small, positive) parameter and $\mathrm{m}_{\mathrm{ref}}$ is a given reference function (typically, the half-space solution). We next present several options for the choice of the function ρ, but reserve most of the discussion to Section 5.

- The most popular least squares choice is

$$\rho(\tau) = \frac{1}{2}\tau^2,$$
$$\mathrm{R}'(\mathrm{m}) \leftarrow \mathrm{div}\, \nabla \mathrm{m}.$$

- In geophysics often depth is characteristically different from the horizontal direction. The weighted least squares choice is

$$R'(m) \leftarrow \operatorname{div} a\nabla m$$

for some diagonal matrix a.

- The total variation (TV) choice (e.g. [28, 10]) is

$$\rho(\tau) = \tau,$$
$$R'(m) \leftarrow \operatorname{div}\left(\frac{\nabla m}{|\nabla m|}\right)$$

- A mix of least squares and total variation is the so-called "Huber's norm",

$$\rho(\tau) = \begin{cases} \tau, & \tau \geq \gamma, \\ \tau^2/(2\gamma) + \gamma/2, & \tau < \gamma \end{cases}$$
$$R'(m) \leftarrow \operatorname{div}\left(\min\{\frac{1}{\gamma}, \frac{1}{|\nabla m|}\} \nabla m\right)$$

Whichever choice is made for ρ, there is a very large, nonlinear optimization problem to solve in (4). This problem is significantly harder to solve when total variation or Huber's norm are used, and it also gets harder the smaller β becomes. It is not unusual in our experience to solve problems of the form (4) for half a million unknowns, so care must be taken to do this efficiently: One cannot simply assume the existence of a general software package to do the job.

Fortunately, the matrices appearing in the necessary conditions for (4) are all very sparse and correspond to discretizations of elliptic PDEs, so we proceed below to exploit the special structure thus afforded.

3 Solving the optimization problem

The unconstrained approach

The most obvious first step in order to solve (4), taken by many, is to eliminate the field u using the forward problem. Thus we obtain a much smaller, although still large (and dense) unconstrained minimization problem

$$\min_m \varphi(m) = \frac{1}{2}\|QA(m)^{-1}q - b\|^2 + \beta R(m). \tag{6}$$

The *unconstrained approach* is to devise numerical methods for solving this unconstrained optimization problem.

Newton and Gauss-Newton methods are well-known. Iterations involve positive definite linear systems of form

$$(J^T J + \beta R'')\, \delta m = -p$$

where the vector p and the matrices J and R'' are evaluated based on a current iterate m and the next iterate is obtained as $m \leftarrow m + \tilde{\alpha}\,\delta m$, for a suitable step size $0 < \tilde{\alpha} \le 1$. The matrix J is the sensitivity matrix, and it will be reintroduced in more detail below.

The conventional wisdom is that it is good to have an unconstrained minimization problem. Indeed, the matrix $H_{red} = J^T J + \beta R''$ is positive definite. However, this matrix is full and designing good preconditioners for it is a difficult task. In the large, sparse context the superiority of the unconstrained approach may thus be challenged. Let us proceed to widen the scope.

The constrained approach

We continue to consider solving the constrained (4). Introducing the Lagrangian

$$\mathcal{L}(m,u,\lambda) = \frac{1}{2}\|Qu - b\|^2 + \beta R(m) + \lambda^T(A(m)u - q)$$

where λ is the vector (or more generally, a grid function like u) of Lagrange multipliers, we need to find an extremum (saddle) point of this Lagrangian.

This leads to the large system of nonlinear equations

$$\mathcal{L}_\lambda = Au - q = 0, \tag{7a}$$
$$\mathcal{L}_u = Q^T(Qu - b) + A^T\lambda = 0, \tag{7b}$$
$$\mathcal{L}_m = \beta R'(m) + G^T\lambda = 0; \quad \text{where } G = \frac{\partial A(m)u}{\partial m}. \tag{7c}$$

Note that all matrices appearing in (7), including the newly introduced G, are sparse.

For the numerical solution of (7) consider using a variant of Newton's method (e.g. Lagrange-Newton, SQP, Gauss-Newton; see for instance [20]). A Gauss-Newton instance could read

$$\begin{pmatrix} A & 0 & G \\ Q^TQ & A^T & 0 \\ 0 & G^T & \beta R'' \end{pmatrix} \begin{pmatrix} \delta u \\ \delta\lambda \\ \delta m \end{pmatrix} = - \begin{pmatrix} \mathcal{L}_\lambda \\ \mathcal{L}_u \\ \mathcal{L}_m \end{pmatrix}. \tag{8}$$

In (8) we have ordered the equations and the unknowns in a deliberate, unsymmetric fashion. Rather than highlighting the indefinite nature of the KKT matrix in its symmetric form we have placed on the main (block) diagonal the blocks corresponding to the highest derivative terms. For instance, if β were not small then the resulting matrix would have been block-diagonal dominant. (However, β *is* small and life is not that simple.)

As for the unconstrained case, the system (8) describes the equations to be solve at each iteration in order to find the correction vector for a current iterate (u, λ, m). We can distinguish the following *tasks* at each iteration:

1. Calculate the Gradients and Hessian.
2. Solve the large, sparse Hessian system.
3. Update the iteration.

Of these tasks the first is straightforward in the present context. The third has been the subject of many studies (involving methods such as damped Newton, trust region, directly updating λ, etc.) that need not be repeated here: see, e.g. [20] or other texts. We thus concentrate on the remaining, second task. There are two approaches here: the reduced Hessian approach and the simultaneous, all-at-once approach.

Reduced Hessian approach

Eliminating δu from the first block row of (8) (this involves solving the forward problem) and then $\delta\lambda$ from the second block row of (8) (this involves solving the adjoint of the forward problem) we obtain for δm a system of the form

$$H_{red}\ \delta m \equiv (J^T J + \beta R'')\ \delta m = -p \tag{9}$$

where

$$J = -QA^{-1}G \tag{10}$$

is the *sensitivity matrix*.

This sensitivity matrix is large and full for applications of the type considered here, so it is never evaluated or stored. To solve (9) we use a Preconditioned Conjugate Gradient method (e.g., [22]), i.e. the conjugate gradient method for

$$M^{-1}(J^T J + \beta R'')\delta m = -M^{-1}p,$$

where M is a preconditioner, e.g. $M = R''$. This conjugate gradient method requires only the evauation of matrix-vector products involving H_{red}.

Unfortunately, *evaluating* $H_{red}v$ for a given vector v is expensive!! It involves evaluating Jv and $J^T v$. While multiplying a vector by G or Q or their adjoints is fast, there is A^{-1} in (10) as well. The forward and adjoint problems must be solved for this purpose to a relatively high accuracy.

The system (9) is similar to that arising in the unconstrained approach. We now proceed to a different one, developed in [14, 2, 3, 4, 5].

Solving for the update direction simultaneously (all-at-once)

The following rationale is well-known: When the iterate for a nonlinear problem is far from the optimal solution, it is wasteful to solve the linearized problem to a high accuracy. Thus, one wants to balance accuracies of inner and outer iterations. This leads to inexact-Newton type methods; see, e.g., [20].

But now we apply the same rationale within the linear iteration: When the iterate for the update direction is far from the solution to the linear system, it is wasteful to eliminate some variables accurately in terms of others! Thus, one wants to balance accuracies inside the linear solver. This is where the unconstrained approach and the reduced Hessian approach described above fall over.

We are thus led to consider methods where δu, $\delta\lambda$ and δm are eliminated *simultaneously*. This approach is more natural also when considering the origin of the systems (7) and (8). These are, after all, discretizations of systems of PDEs, and the approach of eliminating some PDEs in terms of others instead of solving the given system as one is less usual.

We therefore consider solving the linear problem (8) at once. Unfortunately, the matrix is no longer symmetric or positive definite. Moreover, β is typically small, so the problem corresponds to a *strongly coupled* PDE system.

In [14] we proposed a preconditioned QMR method for the symmetrized system (8). The preconditioner consists of applying iterations towards the solution of the reduced problem (9). The catch is that this no longer has to be performed to a high accuracy. Biros & Ghattas had proposed an incredibly similar approach for a different application in [3, 4]. In the next section we demonstrate the performance of this method.

In [2] we proposed a multigrid method for the strongly coupled system (8). The resulting method is very nice and fast, but we have applied it only for the problem (1). The staggered discretization applied for the Maxwell equations in Section 4 complicates the multigrid technique significantly, so we chose the preconditioned QMR approach for the latter application.

4 3D electromagnetic data inversion in frequency and time domain

The inversion results reported in this section were obtained using the weighted least squares norm on $\nabla \mathrm{m}$ for the regularization functional R, where

$$\mathrm{m}(\mathbf{x}) = \ln \sigma(\mathbf{x}).$$

This transformation automatically takes care of the positivity constraint on σ, and it also reduces the contrast in the conductivity, which is reasonable for the mining exploration application and commensurate with the use of a least squares norm in R.

Maxwell's equations in frequency domain

The PDE system is written as

$$\nabla \times \mathbf{E} + \alpha\mu\mathbf{H} = \mathbf{s}_H \quad \text{in } \Omega,$$
$$\nabla \times \mathbf{H} - \hat{\sigma}\mathbf{E} = \mathbf{s}_E \quad \text{in } \Omega,$$
$$\mathbf{n} \times \mathbf{H} = 0 \quad \text{on } \partial\Omega,$$

where $\hat{\sigma} = \sigma + \alpha\epsilon$ and $\alpha = -\imath\omega$. The permeability $\mu(\mathbf{x})$ is assumed known also during data inversion and has potential (moderate) jumps only where $\sigma(\mathbf{x})$ has larger ones.

Typical parameter regimes in these applications satisfy $0 \leq \epsilon \ll 1$ and exclude high frequencies ω.

The $\nabla\times$ operator introduces a nontrivial nullspace. In [15] we have therefore reformulated these equations using the Helmholtz decomposition and a Coulomb gauge law,

$$\mathbf{E} = \mathbf{A} + \nabla\varphi$$
$$\operatorname{div}\mathbf{A} = 0.$$

Thus, $\mathbf{A}$ is the electric field induced by magnetic fluxes, and $\nabla\varphi$ is due to charge accumulation.

Next, we differentiate as in the pressure-Poisson equation in CFD (e.g. [24]) and stabilize, yielding the system

$$\nabla \times (\mu^{-1}\nabla \times \mathbf{A}) - \nabla(\mu^{-1}\operatorname{div}\mathbf{A}) + \alpha\widehat{\sigma}(\mathbf{A} + \nabla\varphi) = \alpha\mathbf{s}$$
$$\operatorname{div}\widehat{\sigma}(\mathbf{A} + \nabla\varphi) = \operatorname{div}\mathbf{s}.$$

For $\widehat{\sigma} = \sigma > 0$ this is an elliptic, diagonally dominant system.

The corresponding boundary conditions are

$$-(\nabla \times \mathbf{A}) \times \mathbf{n}\Big|_{\partial\Omega} = \mathbf{0},$$
$$\mathbf{A} \cdot \mathbf{n}\Big|_{\partial\Omega} = \frac{\partial\varphi}{\partial n}\Big|_{\partial\Omega} = 0,$$
$$\int_\Omega \varphi\, dV = 0.$$

See [13] for more.

A finite volume discretization is then applied on a staggered grid [15, 13]. This is summarized in Figure 3. The resulting system is written as

$$\begin{pmatrix} L_\mu + \alpha M_{\widehat{\sigma}} & \alpha M_{\widehat{\sigma}}\nabla_h \\ \nabla_h \cdot M_{\widehat{\sigma}} & \nabla_h \cdot M_{\widehat{\sigma}}\nabla_h \end{pmatrix} \begin{pmatrix} \mathbf{A} \\ \varphi \end{pmatrix} = \begin{pmatrix} \alpha\mathbf{s} \\ \nabla_h \cdot s \end{pmatrix}$$

In a multiple source/frequency experiment, we write the above as $A_k(m)u_k = q_k$. Then the forward problem is

$$A(m)u = \begin{pmatrix} A_1(m) & & & \\ & A_2(m) & & \\ & & \ddots & \\ & & & \ddots \\ & & & & A_s(m) \end{pmatrix} \begin{pmatrix} u_1 \\ u_2 \\ \vdots \\ \vdots \\ u_s \end{pmatrix} = \begin{pmatrix} q_1 \\ q_2 \\ \vdots \\ \vdots \\ q_s \end{pmatrix} = q. \qquad (11)$$

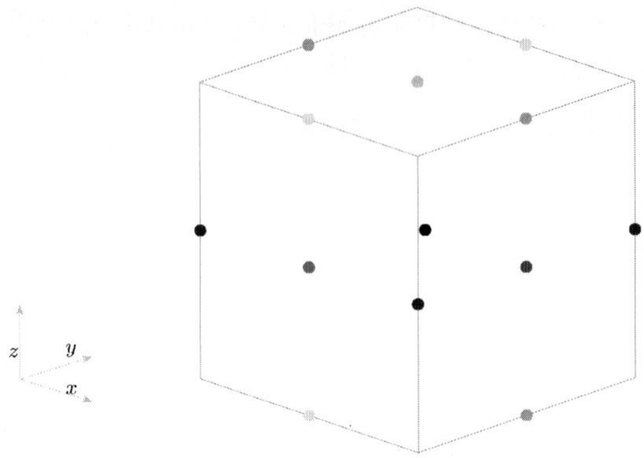

Fig. 3. Elements of a staggered, finite volume discretization. Here **A**, $\nabla\varphi$ and s are located at face centers (like **E**); φ and div **A** are at cell centers (like m); $\hat{\sigma}$ is obtained by harmonic averaging at face centers; and μ is obtained by arithmetic averaging at edge centers (like **H**).

Example 1

In [16] an experiment of an inversion of frequency domain data is described. It involves a transmitter and receiver geometry from an actual CSAMT field survey (from Penasquito, Mexico), although the conductivity model is synthesized.

The transmitter is a 1km grounded wire a few kilometers away from survey area – this is dealt with using a special procedure. The data is measured on 5 field components at frequencies 16, 64 and 512 Hz at 28 stations spaced $50m$ apart on each of 11 lines with linespacing of $100m$. This gives a total of 308 data locations and 4620 data values. The "true model" has two conductive bodies and one resistive body. This is used to generate "true data", which are then contaminated by Gaussian noise, 2% in amplitude and 2 degrees in phase. The $3350m \times 3000m \times 2000m$ volume is discretized into $64 \times 50 \times 30 = 96,000$ cells.

A simple continuation process in β is applied, where we start with a large β and decrease it until the data is deemed to be fitted sufficiently well. Results are accummulated in Table 1. The "misfit" reported here and later on is $\|Qu - b\|/\|b\|$, i.e., the relative l_2 norm of the predicted minus the observed data. The overall number of KKT iterations is rather reasonable here. Of the additional results displayed and discussed in [16] we only show Figure 4, to which we return in Section 5.

Table 1. Inverting electromagnetic data in the frequency domain.

$\beta = 100$	Final misfit $= 0.06$		
Nonlinear iter	KKT iter	$\|\|Au - q\|\|/\|\|q\|\|$	Rel-grad
1	4	$3e-2$	2e-1
2	4	$2e-4$	3e-2
3	3	$2e-6$	5e-4

$\beta = 1e0$	Final misfit $= 0.03$		
Nonlinear iter	KKT iter	$\|\|Au - q\|\|/\|\|q\|\|$	Rel-rad
1	8	$1e-6$	3e-3
2	6	$8e-7$	9e-4

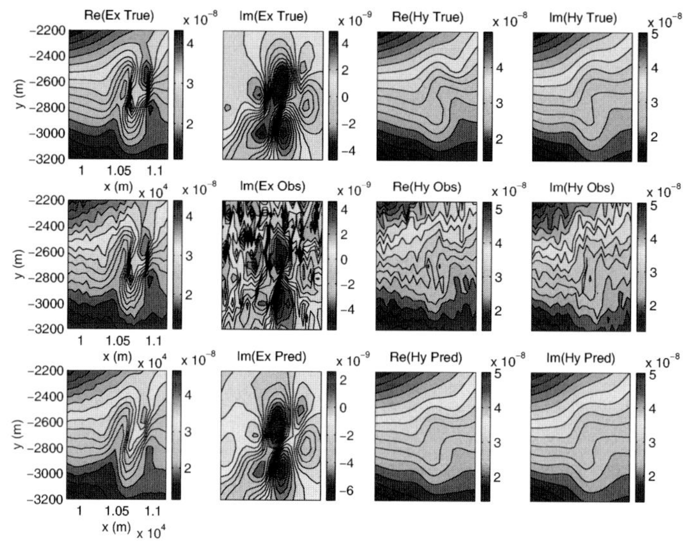

Fig. 4. The accurate E_x, H_y data for 512 Hz are shown in the top row. The error contaminated data are shown in the middle row and the bottom row displays the data predicted from the inverted model. The colour version of this figure can be found in Fig. A.1 on page 577.

Maxwell's equations in time domain

We continue to summarize [16]. The PDE system now reads

$$\nabla \times \mathbf{E} + \mu \frac{\partial \mathbf{H}}{\partial t} = \mathbf{0} \quad \text{in } \Omega,$$
$$\nabla \times \mathbf{H} - \sigma \mathbf{E} - \epsilon \frac{\partial \mathbf{E}}{\partial t} = \mathbf{s}_r(t) \quad \text{in } \Omega,$$
$$\mathbf{n} \times \mathbf{H} = 0 \quad \text{on } \partial\Omega.$$

For the choice of time discretization we note that (i) the relevant parameter regime over non-short time scales yields heavy dissipation, i.e., very stiff problems in time; (ii) the field is measured only beyond the initial, transient layer, so the details of this layer can be skipped; (iii) we cannot expect high accuracy, as sources are typically only continuous in time; and (iv) Lagrange multipliers (solution of the adjoint problem for a rough right hand side) are of low continuity.

All these together point at the unique choice of the backward Euler method for discretizing in time,

$$\alpha_n = (t_n - t_{n-1})^{-1}$$
$$\hat{\sigma}_n = \sigma + \alpha_n \epsilon$$

$$\begin{aligned} \nabla \times \mathbf{E}^n + \alpha_n \mu \mathbf{H}^n &= \alpha_n \mathbf{H}^{n-1} \equiv \mathbf{s}_H \quad \text{in } \Omega \\ \nabla \times \mathbf{H}^n - \hat{\sigma}_n \mathbf{E}^n &= \mathbf{s}_r^n - \alpha_n \epsilon \mathbf{E}^{n-1} \equiv \mathbf{s}_E \quad \text{in } \Omega \\ \mathbf{n} \times \mathbf{H}^n &= 0 \quad \text{on } \partial\Omega. \end{aligned}$$

Now we realize that the latter system has the same form as in the frequency domain. Thus, we apply the same treatment. We obtain

$$\begin{pmatrix} L_\mu + \alpha_n M_{\hat{\sigma}} & \alpha_n M_{\hat{\sigma}} \nabla_h & 0 \\ \nabla_h \cdot M_{\hat{\sigma}} & \nabla_h \cdot M_{\hat{\sigma}} \nabla_h & 0 \\ \alpha_n^{-1} M_\mu^{-1} \nabla_h \times & 0 & I \end{pmatrix} \begin{pmatrix} \mathbf{A}_n \\ \varphi_n \\ \mathbf{H}_n \end{pmatrix} = \begin{pmatrix} \alpha_n \mathbf{s} \\ \nabla_h \cdot s \\ \mathbf{H}_{n-1} \end{pmatrix}.$$

For the nth time step, we write the above as $B_n u_{n-1} + A_n(m) u_n = q_n$. Then forward problem for s time steps is

$$A(m)u = \begin{pmatrix} A_1(m) & & & \\ B_2 & A_2(m) & & \\ & \ddots & \ddots & \\ & & \ddots & \ddots \\ & & B_s & A_s(m) \end{pmatrix} \begin{pmatrix} u_1 \\ u_2 \\ \vdots \\ \vdots \\ u_s \end{pmatrix} = \begin{pmatrix} q_1 \\ q_2 \\ \vdots \\ \vdots \\ q_s \end{pmatrix} = q. \qquad (12)$$

Example 2

For an inversion of time domain data we considered in [16] a square loop $50m \times 50m$ just above earth's surface. Data is acquired at 20 depths in each of 4 boreholes surrounding a conductive body at 18 logarithmically spaced times between $10^{-4} - 10^{-1}$ *sec*. The "true model" is a conductive sphere of radius $15m$ buried in a uniform halfspace. The time discretization involves 32 steps equally spaced on a log-grid $10^{-7} - 10^{-1}$ *sec*. The inversion grid in space was $40 \times 40 \times 32$. The initial guess is a uniform halfspace equal to the true background conductivity.

Results are accummulated in Table 2 similarly to those for the previous example. The performance of our algorithm is again very satisfactory, as the total number of KKT iterations is only about three dozens.

Table 2. Inverting electromagnetic data in the time domain.

$\beta = 1e-1$	Final misfit $= 0.1$		
Nonlinear iter	KKT iter	$\|\|Au - q\|\|/\|\|q\|\|$	Rel-grad
1	2	$3e-3$	1e-2
2	3	$2e-4$	4e-3
3	2	$7e-6$	1e-3
4	2	$9e-7$	3e-4
$\beta = 1e-2$	Final misfit $= 0.04$		
Nonlinear iter	KKT iter	$\|\|Au - q\|\|/\|\|q\|\|$	Rel-grad
1	7	$4e-6$	2e-3
2	5	$6e-7$	7e-4
$\beta = 1e-3$	Final misfit $= 0.02$		
Nonlinear iter	KKT iter	$\|\|Au - q\|\|/\|\|q\|\|$	Rel-grad
1	8	$2e-6$	3e-3
2	7	$8e-7$	9e-4

5 Discontinuous solutions and Huber's norm

Recall that our argument leading to the minimization problem (4) has been that we introduce a priori information about smoothness of the model this way. But in many cases, including the examples of the previous section, our a priori knowledge is that the model probably contains jump discontinuities! So, in the regularization term

$$R(m) = \left[\int_{\Omega} \rho(|\nabla \mathrm{m}|) + \hat{\alpha}(\mathrm{m} - \mathrm{m}_{\mathbf{ref}})\right]_h$$

(where the subscript h implies that the integral has been discretized) we want to limit the effect of penalty through a jump discontinuity in m, as this should not be penalized for non-smoothness.

Huber's norm

Let us concentrate for a moment on the choice of norm in $R(m)$ that will allow discontinuities. For this we may assume that $u = m$, which leads one to the vast literature on image demoising, e.g. [23, 28]. Here we stick to the multiresolution view of exploring functions on a given grid as corresponding to a discretization of some limit process.

Note that:

- For $|\nabla \mathrm{m}| \to \infty$, $\int |\nabla \mathrm{m}|$ is integrable but $\int |\nabla \mathrm{m}|^{\mathbf{2}}$ is not. This suggests inadequacy of the least squares norm in the presence of discontinuities.
- For $|\operatorname{grad} \mathrm{m}| \to 0$, $\int |\nabla \mathrm{m}|$ yields problems when differentiating it to obtain necessary conditions, but $\int |\nabla \mathrm{m}|^{\mathbf{2}}$ does not.

These observations suggest to combine the two, which yields the so-called Huber's norm [17, 10, 23]

$$\rho(\tau) = \begin{cases} \tau, & \tau \geq \gamma, \\ \tau^2/(2\gamma) + \gamma/2, & \tau < \gamma \end{cases} \tag{13}$$

$$R'(\mathrm{m}) \leftarrow \mathrm{div}\left(\min\{\frac{1}{\gamma}, \frac{1}{|\nabla \mathrm{m}|}\} \nabla \mathrm{m}\right).$$

(Strictly speaking, "Huber's norm" is not a norm, because a multiplication of the argument by a constant may cause a switch, but this does not bother us here.)

Another way to handle the problem with the TV semi-norm is to modify the discretization $|\nabla \mathbf{m}|_{\mathbf{h}} = \sqrt{(\mathbf{D}_{+,\mathbf{x}}\mathbf{m})^2 + (\mathbf{D}_{+,\mathbf{y}}\mathbf{m})^2}$ into

$$|\nabla \mathbf{m}|_{\mathbf{h}} = \sqrt{(\mathbf{D}_{+,\mathbf{x}}\mathbf{m})^2 + (\mathbf{D}_{+,\mathbf{y}}\mathbf{m})^2 + \epsilon},$$

where the parameter $0 < \epsilon \ll 1$ ensures that $|\nabla \mathbf{m}|_{\mathbf{h}}^{-1}$ does not become too large. See, e.g., [28, 7]. If ϵ is too small then the resulting image tends to be blocky; if it's too large then discontinuities are again smeared, as in the least squares case. We have preferred to consider (13), even though it's rougher and harder to analyze, because the parameter γ appears more naturally.

Indeed, γ in (13) is interpreted as answering the question, "what is a jump"? i.e., it is the maximal size of a local change in $|\nabla \mathbf{m}|$ (no ϵ needed here) which is still interpreted as a smooth change in m on the scale of the current grid. This depends on the application. A lower value of γ favours the interpretation of a significant change in $|\nabla \mathbf{m}|$ as a jump, whereas a higher value favours the smooth interpretation.

We have found the following *automatic choice* to be particularly useful in practice. We set

$$\gamma = \frac{h}{|\Omega|_h} \left[\int_{\Omega} |\nabla \mathbf{m}|\right]_h. \tag{14}$$

Thus, γ depends on the solution, and indeed we adjust its value through the iteration in an obvious fashion.

Others choose this parameter using an expression from robust statistics involving medians of $|\nabla \mathbf{m}|$: see [23] and references therein. Also, in the image processing literature one often chooses to penalize even less through discontinuities than when using Huber's norm (e.g. using a Gaussian function or a Tukey biweight). However, this leads to non-convex functionals and local minima, which seems excessive in our more complex context where the forward problem is not simply the identity.

Consider solving the equations for the unconstrained approach (6) where R uses Huber's norm or the modified TV. If we set $QA(m)^{-1}q = m$ then an iterative method called *lagged diffusivity* suggests itself, whereby in the current iterate we set

$$R''\delta m \approx \operatorname{div}\left(\min\{\frac{1}{\gamma}, \frac{1}{|\nabla \mathbf{m}|}\}\ \nabla\delta\mathbf{m}\right), \tag{15}$$

with a similar expression for TV, and solve the resulting weighted least squares problem. This iteratively reweighted least squares (IRLS) algorithm has been analyzed in the image denoising context to show global convergence; see [28] and references therein. If γ is picked adaptively then we are short of a proof, although in practice no difficulty ever seems to arise.

Difficulties do arise, of course, when the forward model is (1) or (2), when even the least squares regularization functional can be testy. However, the lagged diffusivity approach can still be generalized. Indeed, for the nonlinear system (7) we use (15) for defining R'' in the linearization (8). The all-at-once methodology follows unchanged.

Recovering discontinuous surfaces from diffusive operators?

We have obtained good results using the techniques outlined in this section for examples in denoising and in SPECT tomography [12]. However, in the context of the present paper one wonders, is there really enough accurate data in applications to allow an honest identification of discontinuities using our diffusive forward operators?! Consider, for instance, Figure 4. Note how well the recovered field approximates the unpolluted data! Indeed, the solution of a PDE such as (1) or a low frequency instance of Maxwell's equations cannot tolerate high level noise – the forward problem is a low-pass filter. The recovered model fits the "true model" much less closely, and the differences are easily visible to the eye [16]. It would appear that many, rather different models for m could fit the data within the noise tolerance. Does it then make sense at all to seek a model with pinpointed discontinuities?

A full investigation of these questions is well beyond the scope of this paper. Here, to focus the discussion further, let us consider an example.

Example 3

This example is in $2D$, and $\Omega = [-1,1]^2$. To allow maximum possibility for recovery of discontinuities we consider the differential operator (1) in the form

$$\operatorname{div}(\mathrm{m}^{-1}\nabla \mathrm{u}) = \mathrm{q},$$

i.e. without the exponential transformation from σ to m, with natural boundary conditions, and assume (unrealistically) that data on u are available everywhere on a given grid.

The right hand side is chosen with source and sink,

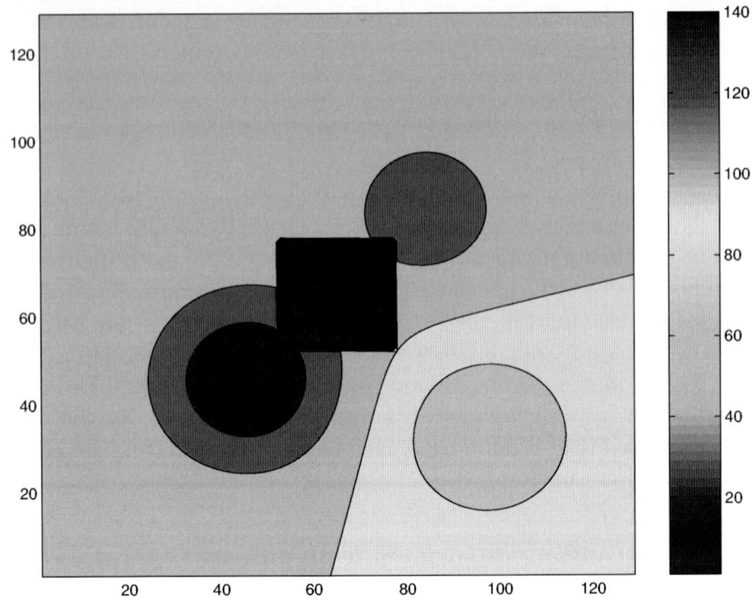

Fig. 5. Contour plot of the "true model" for Example 3. The colour version of this figure can be found in Fig. A.2 on page 578.

$$q = exp(-10((x + 0.6)^2 + (y + 0.6)^2)) - exp(-10((x - 0.6)^2 + (y - 0.6)^2)),$$

and the "true model", depicted in Figure 5, contains discontinuities. We use this true model to generate a field on a 129×129 cell-centered grid and contaminate this with 1% noise to yield the "observed data", b, depicted in Figure 6.

A standard finite volume discretization is applied to the forward problem (with harmonic averaging for m^{-1}). For the optimization problem we use a conjugate gradient solver with a multigrid preconditioner (see, e.g., [27]). Our multigrid preconditioner employs operator-induced, node-based prolongation. The advantages of using this preconditioner (over one based on simpler multigrid) will be discussed elsewhere. During the iteration we do not allow m to decrease below a pre-fixed, positive value. Convergence does not always come easy here, even though we do not use tight tolerances; but this is not our focus in this example.

Figure 7 displays the recovered model using least squares with $\beta = 10^{-5}$. The resulting misfit is 1.66×10^{-2}, which is not awfully far from the target of 1%. When reducing β to $\beta = 3 \times 10^{-6}$, Figure 8 is obtained and the misfit reduces to 1.50×10^{-2}. Reducing β to 10^{-6} produces a model where noise-related artifacts are apparent, so β should not be reduced further.

In comparison, our result using Huber's norm with $\beta = 10^{-5}$, which yields a final $\gamma = 4.6$ and a misfit 1.01×10^{-2}, is displayed in Figure 9.

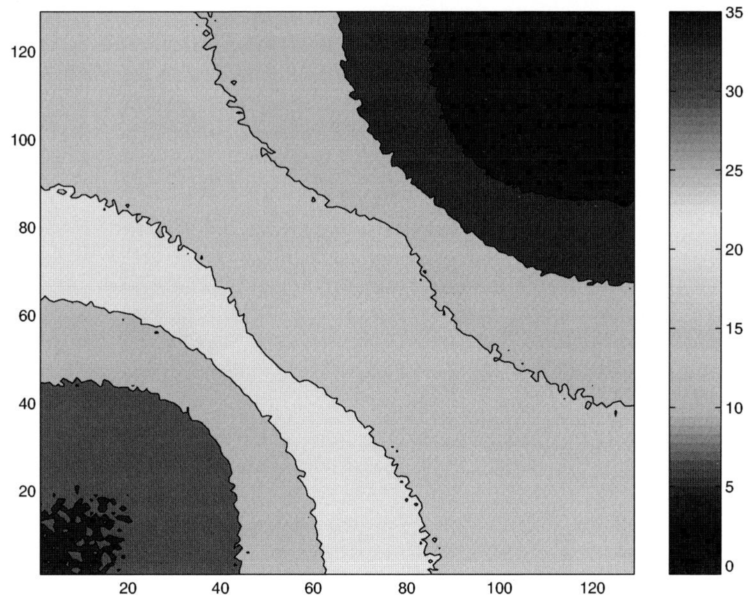

Fig. 6. Data for Example 3. The colour version of this figure can be found in Fig. A.3 on page 578.

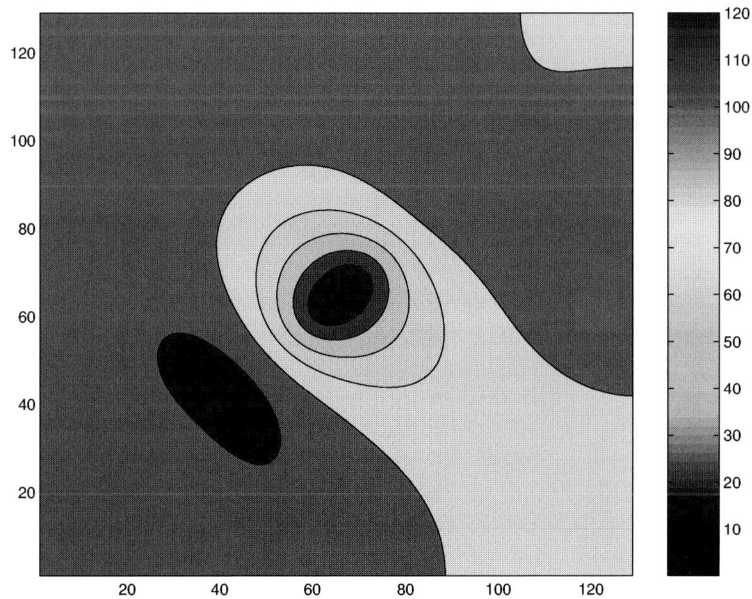

Fig. 7. Recovered model using least squares with $\beta = 10^{-5}$. The colour version of this figure can be found in Fig. A.4 on page 579.

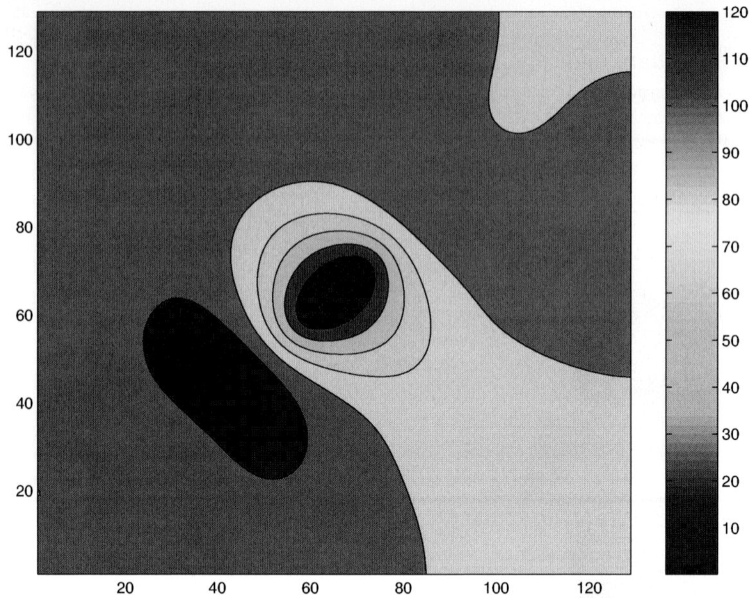

Fig. 8. Recovered model using least squares with $\beta = 3 \times 10^{-6}$. The colour version of this figure can be found in Fig. A.5 on page 579.

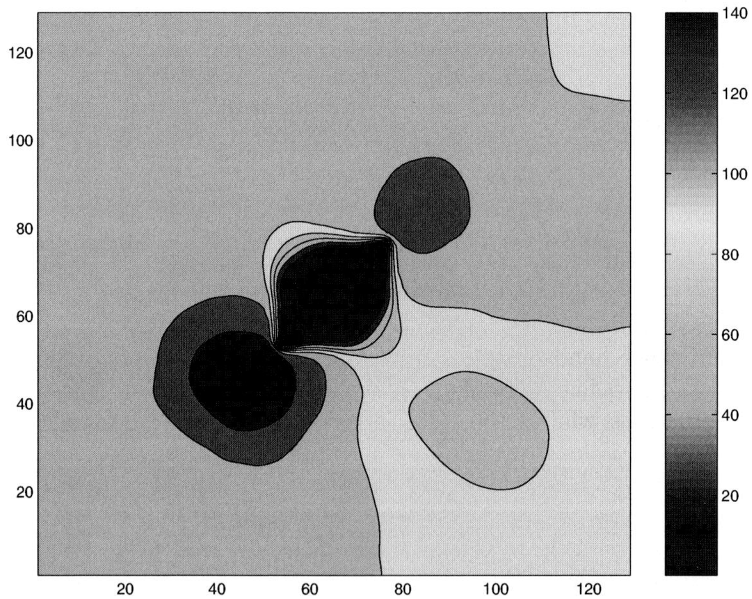

Fig. 9. Recovered model using Huber's norm with $\beta = 10^{-5}$. The colour version of this figure can be found in Fig. A.6 on page 580.

The reconstructed fields in all of these inversions are far smoother than the observed data and closer to the noiseless data (i.e. the exact discrete field) which is displayed in Figure 10. Recall our similar observation for Example 1.

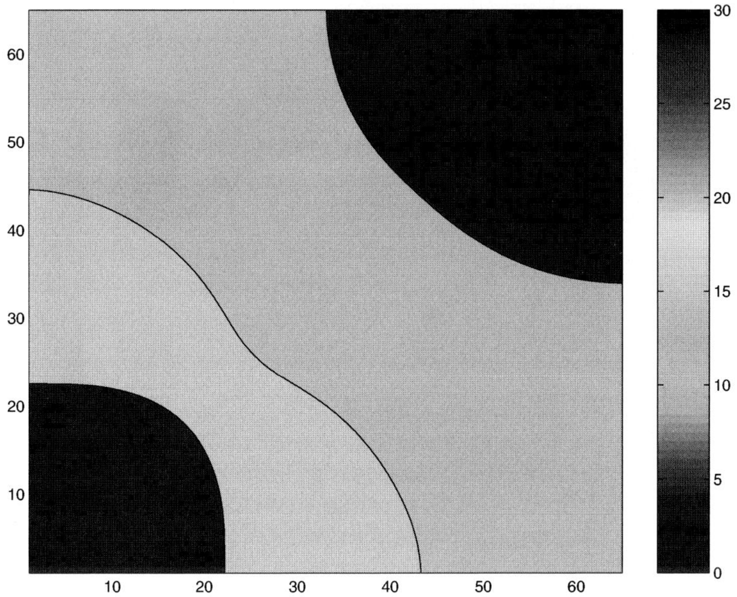

Fig. 10. The noiseless data, viz., the field corresponding to the "true model" using the applied discretization. The colour version of this figure can be found in Fig. A.7 on page 580.

The good news when comparing these figures is that the more careful Huber norm does yield better results, both in terms of misfit and in terms of closeness to the "true model".

On the other hand, the difference between a misfit of 1% and 1.5% is much too fine to define a cutting edge between a "good" and a "bad" model in realistic situations, where we do not know a "true model" either. One never has such a precise knowledge of the noise level (nor of its statistical distribution) in practice, and data are scarcely given everywhere - see Examples 1 and 2.

Of course, it can be said that, using the a priori information that the model contains discontinuities, we generate one such model which fits the data well enough and that is that. However, there are other models with discontinuities which fit the data well! For the present example we took the recovered model of Figure 8 and applied a simple thresholding procedure, whereby the range $[m_{\min}, m_{\max}]$ was divided into 5 equal subintervals, and then all values of m within each such subinterval were replaced by the midpoint value. The result is displayed in Figure 11. Clearly, the resulting model is rather far from the

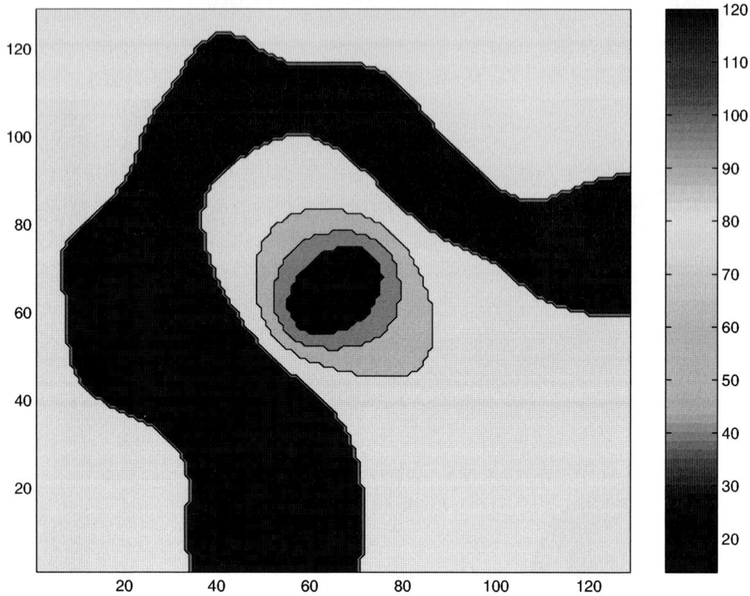

Fig. 11. The model of Figure 8 replaced by a piecewise constant approximation with 5 constant values. The colour version of this figure can be found in Fig. A.8 on page 581.

"true model" of Figure 5, and yet the misfit is a respectable 2.2×10^{-2}.

Our point with the above thresholding experiment is that in more realistic examples we may not know whether our recovered discontinuities are, even approximately, in the right place. Simply trusting that the misfit is small enough is insufficient for this purpose. (Considering the maximum norm of the predicted minus the observed fields proves insufficient as well.) It can then be argued that displaying a smooth blob, such as when using least squares regularization, is less committing than displaying a discontinuous solution (especially with only a few constant values), and as such is more commensurate with the actual information at hand. Nonetheless, the results using "Huber's norm" for this example are encouraging, especially if we disregard questions of cost in obtaining them.

References

[1] U. Ascher and E. Haber. Grid refinement and scaling for distributed parameter estimation problems. *Inverse Problems*, 17:571–590, 2001.

[2] U. Ascher and E. Haber. A multigrid method for distributed parameter estimation problems. *ETNA*, 2001.

[3] G. Biros and O. Ghattas. Parallel Newton-Krylov methods for PDE-constrained optimization. In *Proceedings of CS99, Portland Oregon*, 1999.

[4] G. Biros and O. Ghattas. Parallel preconditioners for KKT systems arising in optimal control of viscous incompressible flows. In *Proceedings of Parallel CFD*. North Holland, 1999. May 23-26, Williamsburg, VA.

[5] G. Biros and O. Ghattas. Parallel Lagrange-Newton-Krylov-Schur methods for PDE-constrained optimization Parts I,II. *Preprints*, 2000.

[6] L. Borcea, J. G. Berryman, and G. C. Papanicolaou. High-contrast impedance tomography. *Inverse Problems*, 12, 1996.

[7] T. F. Chan, G. H. Golub, and P. Mulet. A nonlinear primal-dual method for total variation-based image restoration. *SIAM Journal on Scientific Computing*, 20:1964–1977, 1999.

[8] A. J. Devaney. The limited-view problem in diffraction tomography. *Inverse Problems*, 5:510–523, 1989.

[9] R. Ewing (Ed.). *The mathematics of reservoir simulation.* SIAM, Philadelphia, 1983.

[10] C. Farquharson and D. Oldenburg. Non-linear inversion using general measures of data misfit and model structure. *Geophysics J.*, 134:213–227, 1998.

[11] S. Gomez, A. Perez, and R. Alvarez. Multiscale optimization for aquifer parameter identification with noisy data. In *Computational Methods in Water Resources XII, Vol. 2*, 1998.

[12] E. Haber. *Numerical Strategies for the Solution of Inverse Problems.* PhD thesis, University of British Columbia, 1997.

[13] E. Haber and U. Ascher. Fast finite volume simulation of 3D electromagnetic problems with highly discontinuous coefficients. *SIAM J. Scient. Comput.*, 22:1943–1961, 2001.

[14] E. Haber and U. Ascher. Preconditioned all-at-one methods for large, sparse parameter estimation problems. *Inverse Problems*, 17:1847–1864, 2001.

[15] E. Haber, U. Ascher, D. Aruliah, and D. Oldenburg. Fast simulation of 3D electromagnetic using potentials. *J. Comput. Phys.*, 163:150–171, 2000.

[16] E. Haber, U. Ascher, and D. Oldenburg. Inversion of 3D electromagntic data in frequency and time domain using an inexact all-at-once approach. 2002. Manuscript.

[17] P. J. Huber. Robust estimation of a location parameter. *Ann. Math. Stats.*, 35:73–101, 1964.

[18] G. Newman and D. Alumbaugh. Three-dimensional massively parallel electromagnetic inversion–i. theory. *Geophysical journal international*, 128:345–354, 1997.

[19] G. Newman and D. Alumbaugh. Three-dimensional massively parallel electromagnetic inversion–ii, analysis of a crosswell electromagnetic experiment. *Geophysical journal international*, 128:355–367, 1997.

[20] J. Nocedal and S. Wright. *Numerical Optimization.* New York: Springer, 1999.

[21] R. L. Parker. *Geophysical Inverse Theory.* Princeton University Press, Princeton NJ, 1994.

[22] Y. Saad. *Iterartive Methods for Sparse Linear Systems.* PWS Publishing Company, 1996.

[23] G. Sapiro. *Geometric Partial Differential Equations and Image Analysis.* Cambridge, 2001.

[24] D. Sidilkover and U. Ascher. A multigrid solver for the steady state navier-stokes equations using the pressure-poisson formulation. *Comp. Appl. Math. (SBMAC)*, 14:21–35, 1995.

[25] N.C. Smith and K. Vozoff. Two dimensional DC resistivity inversion for dipole dipole data. *IEEE Trans. on geoscience and remote sensing*, GE 22:21–28, 1984. Special issue on electromagnetic methods in applied geophysics.

[26] A.N. Tikhonov and V.Ya. Arsenin. *Methods for Solving Ill-posed Problems.* John Wiley and Sons, Inc., 1977.

[27] U. Trottenberg, C. Oosterlee, and A. Schuller. *Multigrid.* Academic Press, 2001.

[28] C. Vogel. *Computational Methods for Inverse Problem.* SIAM, Philadelphia, 2002.

Robust Parameter Estimation for Identifying Satellite Injection Orbits

Hans Georg Bock[1], Ekaterina Kostina[1], Johannes P. Schlöder[1], Gottlob Gienger[2], Siegmar Pallaschke[2], and Gerald Ziegler[2]

[1] Interdisciplinary Center for Scientific Computing, University of Heidelberg
Im Neuenheimer Feld 368, D-69120 Heidelberg, Germany
Bock@iwr.uni-heidelberg.de

[2] European Space Agency (ESA)
Robert-Bosch-Str. 5, D-64293 Darmstadt, Germany
Gottlob.Gienger@esa.int

Summary. Satellite tracking data typically consist of range, range rates, azimuth or elevation angles. The measurement times are usually clustered, not all states are measured and outliers may occur due to measurement errors. Moreover, range measurements are subject to ambiguities.

Due to the fact that the actual injection orbit of a satellite may significantly deviate from the planned one, e.g. if the launcher exhibits under-performance or malfunction, a fast and reliable the determination of initial orbits using the available data is particularly important.

Based on practical problems provided by European Space Agency (ESA) the paper first gives a mathematical formulation of orbit determination problems. An l_1 objective function is chosen in order to reduce the influence of outliers.

For the numerical treatment of orbit determination problems shooting strategies for estimation problems in nonlinear differential equations are described. Emphasis is put on the efficient treatment of the resulting large-scale nonlinear constrained weighted l_1 optimization problem.

The performance of the resulting codes is demonstrated using tracking data from ESA's Artemis mission.

1 Initial Orbit Determination Problems

When launching a satellite it may happen that the injection orbit it missed significantly due to launcher malfunction and/or underperformance. An example provides the launch of the Artemis satellite on 12. July 2001 where "... the Ariane 5 launcher had propelled the Artemis satellite into a transfer orbit ... with the apogee ... at only 17.000 km rather than the nominal 36.000 km" [10].

In practice it is of high importance to identify the actual orbit of a satellite as soon as possible in order to be able to predict the future trajectory and, if necessary, to perform correction maneuvers in order to eventually reach the operational mission orbit. The latter was done in the Artemis case. The recovery of the Artemis mission is described in [1].

The identification of the satellite orbits is based on measurements of the trajectory. Typically these data are collected at different ground stations. At each measurement point usually only some of the variables that determine the trajectory or functions of them are observed (e.g. range, range rate or azimuth and elevation angle). The measurement times are usually not evenly distributed over the ground station pass. Additionally the statistical quality of the measurements may differ from ground station to groundstation and there may be outliers e.g. due to ambiguity problems.

1.1 Model for Satellite Dynamics

The motion of satellites is described by a complex system of six nonlinear ordinary differential equations with initial values as unknown parameters

$$\dot{y}(t) = \begin{pmatrix} \dot{r}(t) \\ \dot{v}(t) \end{pmatrix} = \begin{pmatrix} v(t) \\ -\frac{GM_\oplus}{\|r\|^3} r(t) + \ \mathrm{pert}(r(t), v(t), t) \end{pmatrix} \tag{1}$$

$$y(t_0) = y_0 = \begin{pmatrix} r_0 v_0 \end{pmatrix} = p.$$

Here $y(t)$ is a vector containing Cartesian coordinates of the satellite and their velocities with respect to an inertial reference system, p is a vector of parameters to be estimated. The term $\mathrm{pert}(r(t), v(t), t)$ denotes perturbations that are due to external forces, e.g. gravitation of sun and moon, inhomogeneities of earth' gravitational field, air drag, solar radiation pressure etc. We refer the reader to [8] for the detailed description. We restrict ourselves to the described model. In a more general approach, the various perturbations may be estimated as well. In that case the parameter vector p should be augmented correspondingly. The methods that are described further can be applied to the general case as well.

1.2 Data and Measurement Functions

It is assumed, that at moments $t_i, i = 1, ..., k$, measurements η_{ij} of m_i observation functions b_j, $j = 1, ..., m_i$, of the state variables $y(t)$ and the parameters p are available

$$\eta_{ij} = b_j(y(t_i), p^{true}) + \varepsilon_{ij}, \qquad j = 1, ..., m_i, \qquad i = 1, ..., k,$$

which are subject to measurement errors ε_{ij}. Typically, these data contain measurements of range, range rate, azimuth and elevation angles. We again refer the reader to [8] for the detailed description of the functions $b_j(y(t), p)$.

According to the common approach, in order to determine the unknown parameters an optimization problem is solved. The possible constraints of this problem describe the specifics of the model (constraints on the initial and terminal states, constraints on parameters, etc). As the objective functional in the optimization problem, typically a norm of the measurement error is used. The type of the norm is motivated by the statistical distribution of the measurement error. If the errors are independent, normally distributed with zero mean and known variances ($N(0,\sigma_{ij}^2)$), minimizing a weighted least squares function

$$\min \sum_{i,j} (\eta_{ij} - b_j(y(t_i),p))^2/\sigma_{ij}^2$$

yields a maximum likelihood estimate. But in case of Laplace distribution l_1 estimation is appropriate:

$$\min \sum_{i,j} |\eta_{ij} - b_j(y(t_i),p)|/|\sigma_{ij}|$$

Further, it is well-known that l_1 parameter estimation possesses a very nice property, namely that the optimal solution is insensitive to outliers in the data. l_1 estimation is therefore used for robust parameter estimation [6].

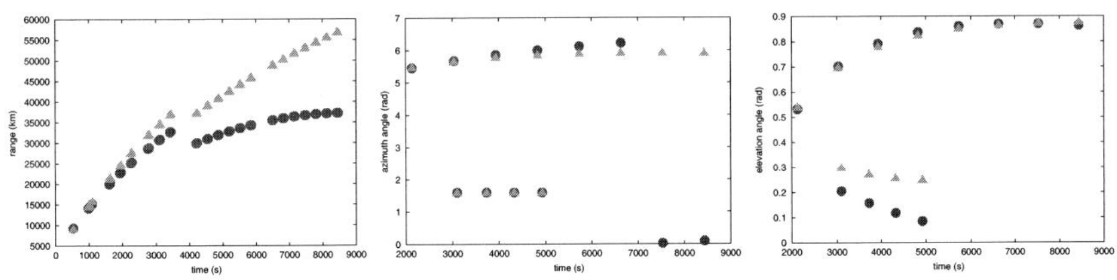

Fig. 1. Measurements (dots) vs model response before parameter estimation (triangles) (Artemis launch)

Fig. 1 shows the plots of measurements versus model response for the initial guess of the parameters. Obviously, there are two outliers in the measurements of the azimuth angle. Hence, for the orbit determination problem we choose the l_1 norm of the measurement error as the objective function.

1.3 l_1 Parameter Estimation Problem for Initial Orbit Determination

Mathematically the l_1 problem of parameter estimation for initial orbit determination can be written as follows:
Minimize the deviation of model response $b(y(t),p)$ to measurement values η such that satellite dynamics and initial conditions are fulfilled:

$$\min_{y,p} \sum_{j=1}^{l} \sum_{i=1}^{m_j} |\eta_{ij} - b_j(y(t_i), p)| / |\sigma_{ij}| \tag{2}$$

$$\text{s.t. } \dot{y}(t) = \begin{pmatrix} \dot{r}(t) \\ \dot{v}(t) \end{pmatrix} = \begin{pmatrix} v(t) \\ -\frac{GM_\oplus}{\|r\|^3} r(t) + \text{pert}(r(t), v(t), t) \end{pmatrix} := f(t, y(t), p)$$

$$y(t_0) = p.$$

2 Boundary Value Problem Methods

A typical solution approach to parameter estimation which is found very often in practice is the initial value or single shooting approach: the ODE system is repeatedly solved as an initial value problem, and unknown parameters including possibly initial values are iteratively improved by some optimization procedure.

In contrast to that, our numerical solution of the parameter estimation problem is based on the Boundary Value Problem (BVP) approach going back to [2]. The basic idea consists in parameterizing the dynamic equations (initial or boundary value problem) like a boundary value problem (e.g., by multiple shooting) and then performing simultaneously (in one iteration loop) the minimization of the cost function subject to the constraints given by the discretized boundary value problem. It has been shown [2, 3], that BVP methods (based on multiple shooting or collocation) are much more stable and efficient than the single shooting approach when solving parameter estimation problems.

2.1 Multiple Shooting

The scheme of multiple shooting consists in the following. First one chooses a suitable grid of multiple shooting nodes τ_j

$$t_0 = \tau_0 < \tau_1 < \ldots < \tau_m = t_l$$

covering the interval where measurements are given.

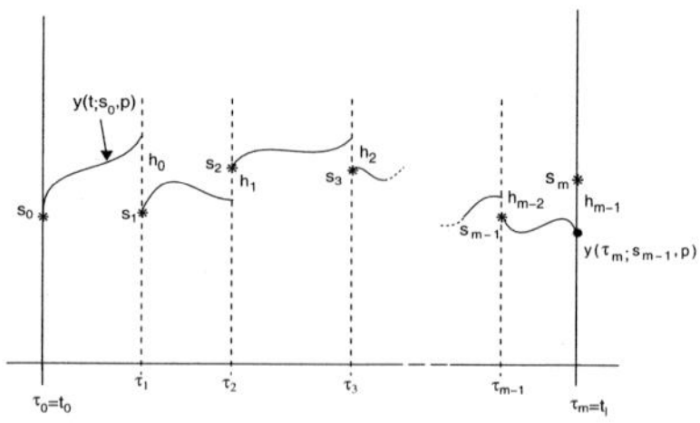

Fig. 2. Multiple shooting approach

At each grid point the values of the state variables s_j are chosen as additional unknowns and m ODE initial value problems

$$\dot{y} = f(t, y, p), \quad y(\tau_j) = s_j$$

are solved on each subinterval $I_j := [\tau_j, \tau_{j+1}]$ to yield a solution $y(t; s_j, p)$ for $t \in I_j$. Solutions of dynamic systems, generated by this procedure, are usually not continuous at τ_j. This has to be enforced by additional matching conditions. Inserting the computed values $y(t, s_j, p), \quad \tau_j \leq t \leq \tau_{j+1}$, into problem (2) one obtains a *constrained* optimization problem in the variables $(s, p) := (s_0, \ldots, s_m, p)$

$$\begin{aligned} &\min_{(s,p)} \ ||r_0(s, p)||_1, \qquad (3)\\ &\quad h_j(s_j, s_{j+1}, p) := y(\tau_{j+1}; s_j, p) - s_{j+1} = 0, \ j = 1, ..., m-1,\\ &\quad s_0 = p. \end{aligned}$$

For the case of the initial orbit determination, the parameters are the initial values for ODE, that is why one may eliminate the parameters from problem (3) obtaining the optimization problem only in variables s under matching conditions as equality constraints.

Multiple shooting possesses several advantages:

1. It is possible to include a priori information about the state variables, e.g. from the measurements or from expert knowledge, by a proper choice of initial guesses for the additional variables s_j. Thus, it can be ensured that the initial solutions $y(t; s_j, p)$ remain close to the observed data. It can be shown that this damps the influence of poor parameter guesses.
2. The adequate choice of initial guesses for the state variables (and the application of a Gauss-Newton method for the solution of the constrained least squares problem) typically avoids convergence to local minima with large residuals.
3. The scheme is numerically stable. The splitting of the integration interval limits error propagation and allows to solve parameter estimation problems even for unstable or chaotic systems.
4. The matching conditions induce a very specific BVP structure in the problem equations, which can be exploited in particular for parallelization.

2.2 l_1 Gauss-Newton Method

Parameterization of the BVP constraint yields a finite dimensional, possibly large scale, nonlinear constrained approximation problem (3) which can be formally written as

$$\min \ ||F_1(x)||_1 = \sum_{i=1}^{m_1} |F_{1i}(x)|, \ s.t. \ F_2(x) = 0. \qquad (4)$$

Note, that the equalities $F_2(x) = 0$ include the matching conditions induced by multiple shooting. We assume that the functions $F_i : D \subset R^n \to R^{m_i}, i = 1, 2,$ are twice continuously differentiable. The number of variables in problem (4) under consideration is equal to number of differential equations multiplied by the number of multiple shooting nodes plus the number of parameters.

To solve problem (4) we use a generalized Gauss-Newton method according to which a new iterate is (basically) generated by

$$x^{k+1} = x^k + t^k \Delta x^k, 0 < t_k \leq 1. \tag{5}$$

where the increment Δx is the solution of the following linear l_1 problem at $x = x^k$

$$\min_{\Delta x \in R^n} \quad ||F_1(x) + J_1(x)\Delta x||_1 = \sum_{i=1}^{m_1} |F_{1i}(x) + J_{1i}(x)\Delta x|, \tag{6}$$
$$\text{s.t.} \quad F_2(x) + J_2(x)\Delta x = 0.$$

Here $J_1(x)$ and $J_2(x)$ denote the Jacobians of $F_1(x)$ and $F_2(x)$ respectively, J_{1i} being the $i-$th row of the $J_1(x)$.

Choosing the step length t^k by means of classical line search methods based on the exact penalty function

$$T_1(x) := ||F_1(x)||_1 + \sum_{i=1}^{m_2} \alpha_i |F_{2i}(x)|$$

with sufficiently large weights $\alpha_i > 0, i = 1, ..., m_2$, ensures global convergence. However, it is well known that already in mildly ill-conditioned problems such a step size strategy may be very inefficient since it may produce very small step sizes. Therefore we use the "restrictive monotonicity test", see [3], [4], that has proved to be very effective in practical applications.

Note, that the generalized Gauss-Newton method (5) – (6) can be interpreted as a Sequential Linear Programming Method (SLP) since we ignore the information about second-order derivatives. As SLP method the generalized Gauss-Newton method has several advantages. First it does not use Hessians (second order derivative information) and the local linearized problems are linear programming ones. Under certain regularity assumptions at the solution the method shows quadratic rate of local convergence. There are problems, however, for which the Gauss-Newton method may have rather slow local convergence rate or may even fail. The reason is that the linearized model (6), which forms the basis of the Gauss-Newton method, is an inadequate representation of such nonlinear problems, since the second-order information cannot be ignored. Similar situations appear in applying the l_2 based Gauss-Newton method for parameter estimation problem with large residuals (see e.g. [3] for a detailed discussion). Using SQP-type methods for the nonlinear constrained l_1 approximation problem one could force convergence to a solution even in such a case. However, such solutions are undesirable in a certain sense. We can show that a solution of this type, even if it is a strict minimum of the nonlinear constrained l_1 problem (4), cannot be expected to be a continuous deformation of the "true" parameter values under perturbations caused by the measurement errors. Thus, slow local convergence of the full-step ($t^k \equiv 1$) Gauss-Newton indicates deficiencies in the model or lack of data and can be considered as an advantage of the method.

3 Solving the Linear l_1 Problem

At each iteration of a Gauss-Newton method a linear l_1 problem (6) has to be solved.

3.1 Condensing

The matrix $J(x) = \begin{pmatrix} J_1(x) \\ J_2(x) \end{pmatrix}$ of the l_1 problem under consideration shows the typical block structure of the BVP discretization which is induced by multiple shooting:

$$J(x) = \begin{pmatrix} D_1^0 & D_1^1 & \cdots & \cdots & D_1^m \\ G_0^l & G_0^r & & & \\ & G_1^l & \ddots & 0 & \\ & & \ddots & \ddots & \\ & 0 & & G_{m-1}^l & G_{m-1}^r \end{pmatrix}$$

Every block column corresponds to the derivatives with respect to the discretization variables in one subinterval. The block rows with G-matrices are the derivatives of the continuity conditions. The block rows with D-matrices correspond to the derivatives of the functions $F_1(x)$ of the cost functional of the nonlinear problem (4).

We use a fast, stable and efficient structure exploiting decomposition (see [2], [3]) to reduce a large linear l_1 problem to a linear constrained l_1 problem with smaller dimension. The number of variables in the resulting problem is in general case the number of parameters plus the number of differential equations. Since in the case of satellite orbit determination under consideration the parameters are initial values for ODE the resulting problem is a linear *unconstrained* problem with the number of variables equal the number of parameters, namely 6.

3.2 Solution of the Condensed Problem

In general the condensed problem is of the form

$$\min \; f(Y) = \sum_{i=1}^{M_1} |A_i^T Y + c_i|, \tag{7}$$
$$A_i^T Y + c_i = 0, \; i = M_1 + 1, ..., M_1 + M_2$$

in which the number M_1 of the components of the cost function is equal to the number m_1 of measurements. Note, that in the case of satellite injection orbit determination under consideration after condensing we get an unconstrained l_1 problem. The problem (7) is equivalent to the problem of linear programming with additional M_1 variables and $2 \times M_1$ inequality constraints

$$\begin{aligned} \min f_{LP}(Y) \; &= \sum_{i=1}^{M_1} \xi_i, \\ \text{s.t.} \; & \xi_i - A_i^T Y \geq c_i, i = 1, ..., M_1, \\ & \xi_i + A_i^T Y \geq -c_i, i = 1, ..., M_1, \\ & A_i^T Y + c_i = 0, \; i = M_1 + 1, ..., M_1 + M_2, \; \xi \geq 0, \end{aligned} \tag{8}$$

and can be solved by the simplex method of linear programming. However, it is more effective to solve the problem by the dual simplex method. Indeed, the dual problem for (8) and hence for (7) is an "ordinary" bounded-variable problem of linear programming

$$\begin{aligned} \min_{\lambda \in R^{M_1 + M_2}} \; & \varphi(\lambda) = c^T \lambda, \\ \text{s.t.} \; & A\lambda = 0, \; |\lambda_i| \leq 1, i = 1, ..., M_1. \end{aligned} \tag{9}$$

This kind of problems can be very effectively solved by the "long-step" dual simplex method [7], which takes into account the special structure of the inequality constraints in the linear programming problem (8).

It may happen that the matrix A in (7) does not have full rank and then the problem (7) may have an unbounded solution. This situation can be avoided by solving a regularized problem

$$\begin{aligned} \min \; f(Y) &= \sum_{i=1}^{M_1} |A_i^T Y + c_i|, \\ \text{s.t.} \; A_i^T Y + c_i &= 0, \; i = M_1 + 1, ..., M_1 + M_2, \; L \leq Y \leq U, \end{aligned} \tag{10}$$

which includes given lower L and upper U bounds on Y. For the solution of the problem (10) a modification of the "adaptive" method [5] can be effectively applied.

4 Numerical Results

In this section we present numerical results on solving a set of test problems. As a test set ten problems have been chosen as follows. The first one is the "original" problem of initial orbit determination for Artemis launch including the set of data and original guesses for the parameters provided by European Space Agency. For the remaining nine problems the original set of data has been taken whereas guesses for the parameters have been obtained from the original guess by random perturbations. The number of iterations necessary to solve the problems by the routine E04GBF, a Gauss-Newton method with relaxation, from the NAG Library [9] and by the software PARFIT-L1 for solving l_1 parameter estimation problems, incorporating the methods described in this paper, are presented in Table 1. PARFIT-L1 has been used in the single shooting mode. One can see that PARFIT-L1 solves all problems. In case E04GBF obtains the solution, PARFIT-L1 uses less iterations. Table 2 shows that the solutions obtained by l_2 estimator and l_1 estimator coincide up to three digits.

Table 1. Numerical results for solving test Artemis problems

	art^1	art^2	art^3	art^4	art^5	art^6	art^7	art^8	art^9	art^{10}
E04GBF	10	25	12	-	10	13	-	-	-	-
PARFIT-L1	9	8	8	6	8	10	10	7	12	7

Table 2. l_2 and l_1 solutions

	$p_1[km]$	$p_2[km]$	$p_3[km]$	$p_4[km/s]$	$p_5[km/s]$	$p_6[km/s]$
l_2-solution	5013.91	-7025.25	-263.102	7.89236	2.00387	-0.40286
l_1-solution	5013.99	-7025.28	-263.494	7.89226	2.00408	-0.40243

Fig. 3 shows the plots of measurements versus model response for the solution.

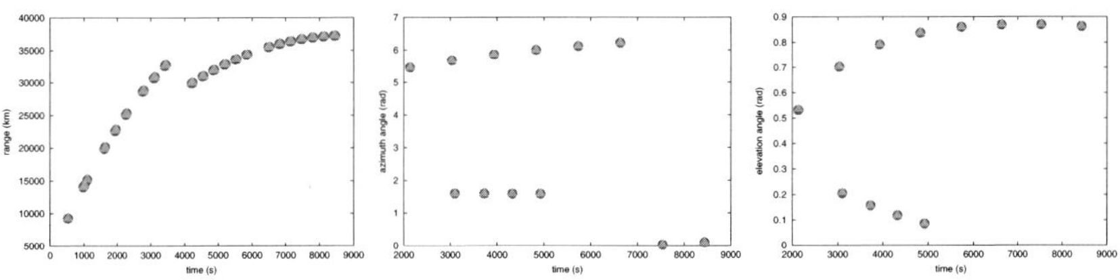

Fig. 3. Measurements (dots) vs model response after parameter estimation (triangles) (Artemis launch)

The numerical results show robustness and reliability of parameter estimation based on l_1 vs l_2 in case of data with outliers.

References

[1] G. Oppenhaeuser, A.G. Bird and L. van Holtz. Artemis - "A Lost Mission" on Course for a Full Recovery, ESA Bulletin No. 110, 9-16 (2002)

[2] H. G. Bock. *Numerical Treatment of Inverse Problems in Chemical Reaction Kinetics.* In: Ebert et al (eds.) Modelling of Chemical Reaction Systems, Springer Series in Chemical Physics **18**, Berlin Heidelberg (1981)

[3] H. G. Bock. *Randwertproblemmethoden zur Parameteridentifizierung in Systemen nichtlinearer Differentialgleichungen,* Bonner Mathematische Schriften **183**, Bonn (1987)

[4] H. G. Bock, E. Kostina, and J. P. Schlöder. On the Role of Natural Level Functions to Achieve Global Convergence for Damped Newton Methods. *System Modelling and Optimization. Methods, Theory and Applications,* M. Powell et al (Eds.), Kluwer, 51-74 (2000)

[5] R. Gabasov, F. M. Kirillova, E. A. Kostina. An adaptive method of solving l_1 extremal problems, *Computational Mathematics and Mathematical Physics,* vol. 38, 9, 1400 - 1411 (1998)

[6] P. J. Huber. *Robust Statistics,* John Wiley and Sons (1981)

[7] E. A. Kostina. The long step rule in the bounded-variable dual simplex method: numerical experiments, *Mathematical Methods of Operations Research,* 55, 3 (2002)

[8] O. Montenbruck, E. Gill. Satellite Orbits, Springer, Berlin Heidelberg New York (2000)

[9] NAG Fortran Library Manual, Mark 20, The Numerical Algorithms Group Ltd.

[10] Spaceflight now, Febr. 21 (2002)

On the Numerical Simulation of the Free Fall Problem

Sebastian Bönisch, Vincent Heuveline, and Rolf Rannacher

Numerical Analysis Group, IWR, University of Heidelberg
Im Neuenheimer Feld 294, D-69120 Heidelberg, Germany
sebastian.boenisch@iwr.uni-heidelberg.de
vincent.heuveline@iwr.uni-heidelberg.de
rolf.rannacher@iwr.uni-heidelberg.de

Summary. The numerical simulation of the free fall of a solid body in a viscous fluid is a challenging task since it requires computational domains which usually need to be several order of magnitude larger than the solid body in order to avoid the influence of artificial boundaries. Toward an optimal mesh design in that context, we propose a method based on the *weighted* a posteriori error estimation of the finite element approximation of the fluid/body motion. A key ingredient for the proposed approach is the reformulation of the conservation and kinetic equations in the solid frame as well as the implicit treatment of the hydrodynamic forces and torque acting on the solid body in the weak formulation. Informations given by the solution of an adequate dual problem allow to control the discretization error of given functionals. The analysis encompasses the control of the the hydrodynamic force and torque on the body. Numerical experiments for the two dimensional sedimentation problem validate the method.

Key words: Finite element method, a posteriori error estimation, free steady fall problem, fluid-structure coupling

1 Introduction

Over the last decades, the study of the motion of small particles in viscous liquid has been the object of intensive research activities in fluid mechanics. The investigation topics range from the theoretical mathematical analysis (existence and uniqueness proof) (see e.g. [13, 5, 14] and references therein) to the numerical simulation of the liquid-particle interaction (see e.g. [6, 8, 11, 15] and references therein). The present paper concentrates on the numerical simulation of the steady free fall of a unique solid body in a viscous flow. Many aspects related to this problem are still not well understood. Especially, the issue of the stability of *terminal states* in dependency with the body geometry and orientation needs to be addressed. We propose in that context a new weighted a posteriori error estimator in order to control the discretization error and to design adequate mesh leading to economical discretization for computing the physical quantities of interest (see [1] and references therein). These features are of great importance since the numerical simulation of the free fall of a solid body in a viscous fluid requires computational domains which are usually several orders of magnitude larger than the solid body in order to avoid the influence of artificial boundaries.

The weighted a posteriori error estimator relies on the resolution of an adequate dual problem which gives localized sensitivity factors with regard to the error measured by means of given functional. The key ingredients of the error estimator derivations in our context are the reformulation of the conservation and kinetic equations in the solid body frame as well as the implicit treatment of the hydrodynamic forces and torque acting on the body in the weak formulation. Our analysis encompasses the control of the free fall velocity, the orientation of the body, the hydrodynamic force and torque on the body.

The outline of the remainder of this paper is as follows. In section 2, we briefly derive the formulation of the stationary free fall problem. Section 3 deals with the weak formulation of the equations of the fluid-body motion and its discretization by means of the finite element method. Section 4 is dedicated to the derivation of an a posteriori error estimator for the hydrodynamic force and torque acting on the solid body. In section 5, numerical experiments for the two dimensional sedimentation problem are presented to validate the method.

2 Problem formulation

2.1 General formulation of the fluid/body interaction

We consider the free fall of a solid body $\mathcal{S} \subset \mathbb{R}^d$ $(d = 2, 3)$ in an incompressible liquid $\mathcal{L}$ filling the whole space $\mathcal{D} := \mathbb{R}^d \backslash \mathcal{S}$. The solid body $\mathcal{S}$ is assumed to be a bounded domain and the velocity of its mass center C (resp. its angular velocity) is denoted by $\mathcal{V}_C$ (resp. Ω) in the inertial frame $\mathcal{F}$. The region occupied by $\mathcal{S}$ at time t is described by $S(t)$ and the corresponding attached frame is denoted by $\mathcal{R}(t)$. In the inertial frame $\mathcal{F}$ the equations of conservation of momentum and mass of $\mathcal{L}$ in their non conservative form are given by

$$\left.\begin{aligned} \rho\frac{\partial \mathrm{v}}{\partial t} + \rho(\mathrm{v}\cdot\nabla)\mathrm{v} = \rho g + \nabla\cdot\mathcal{T}(\mathrm{v},\mathrm{p}) \\ \nabla\cdot\mathrm{v} = 0 \end{aligned}\right\} \quad \text{for} \quad (x,t) \in \bigcup_{t>0} [\mathbb{R}^d \backslash S(t)] \times \{t\}, \quad (1)$$

where ρ is the constant density of $\mathcal{L}$, v and p are the Eulerian velocity field and pressure associated with $\mathcal{L}$, $\mathcal{T}$ is the Cauchy stress tensor and ρg is the force of gravity which is assumed to be the only external force. We assume further a Navier-Stokes liquid model for which the Cauchy stress tensor is given by

$$\mathcal{T}(\mathrm{v},\mathrm{p}) := -\mathrm{p}\mathbf{1} + \mu(\nabla\mathrm{v} + (\nabla\mathrm{v})^T), \tag{2}$$

where μ is the shear viscosity. The boundary conditions are given by

$$\mathrm{v}(x,0) = 0, \qquad \lim_{|x|\to\infty} \mathrm{v}(x,t) = 0 \quad \text{for } x \in \mathbb{R}^d\backslash S(t) \tag{3}$$

$$\mathrm{v}(x,t) = \mathcal{V}_C(t) + \Omega(t) \times (x - x_C(t)) \ \text{ for } \ x \in \partial S(t). \tag{4}$$

The fluid/body coupling occurs through the Dirichlet boundary condition (4). It relies on the determination of the body motion which is obtained by requiring the balance of the linear and angular momentum:

$$\begin{cases} m_S \dot{\mathcal{V}}_C = m_S g - \displaystyle\int_{\partial S(t)} \mathcal{T}(\mathrm{v},\mathrm{p}) \cdot N \, d\sigma, \\ \frac{d(J_{S(t)}\cdot\Omega)}{dt} = - \displaystyle\int_{\partial S(t)} (x - x_C) \times [\mathcal{T}(\mathrm{v},\mathrm{p}) \cdot N] \, d\sigma, \end{cases} \tag{5}$$

where m_S is the mass of the body, N is the unit normal to $\partial S(t)$ oriented toward the body and J_S denotes the inertia tensor with respect to the mass center C. Further we assume $\mathcal{V}_C(0) = 0$, $\Omega(0) = 0$.

The straightforward formulation (1-5) has the disadvantage that the region occupied by the liquid $\mathcal{L}$ is time dependent. This can be avoided by reformulating these equations in the body frame $\mathcal{R}(\mathrm{t})$. If y denotes the position of a point P in the frame $\mathcal{R}(\mathrm{t})$ and x is the position of the same point in $\mathcal{F}$, we have

$$x = Q(t) \cdot y + x_C(t), \ \ Q(0) = \mathbf{1}, \quad x_C(0) = 0, \tag{6}$$

with Q an orthogonal linear transformation. Considering the transformation (6) one can reformulate the system of equations (1) in the following form

$$\left.\begin{array}{r} \rho\{\frac{\partial v}{\partial t} + ((v - V)\cdot\nabla)v + \omega \times v\} = \nabla \cdot T(v,p) + \rho G(t) \\ \nabla \cdot v = 0 \end{array}\right\} \tag{7}$$

for $(y,t) \in [\mathbb{R}^d\backslash S(0)] \times (0,\infty)$, where

$$v(y,t) := Q^T \cdot \mathrm{v}(Q\cdot y + x_C, t), \ p(y,t) := \mathrm{p}(Q\cdot y + x_C, t), \ G := Q^T \cdot g \tag{8}$$

$$V(y,t) := Q^T(\mathcal{V}_C + \Omega \times (Q\cdot y)), \ T(v,p) := Q^T \cdot \mathcal{T}(Q\cdot v, p)\cdot Q, \ \omega := Q^T \cdot \Omega. \tag{9}$$

The additional term $\omega \times v$ in the momentum equation $(7)_1$ corresponds to the *Coriolis force* induced by the frame transformation (6). Correspondingly the system equations (5) describing the motion of the body are transformed to

$$\begin{cases} m_S \dot{V}_C + m_S(\omega \times V_C) = m_S G(t) - \displaystyle\int_{\partial S} T(v,p) \cdot n \, d\sigma, \\ I_S \cdot \dot{\omega} + \omega \times (I_S \cdot \omega) = - \displaystyle\int_{\partial S} y \times [T(v,p) \cdot n] \, d\sigma, \\ \frac{dG}{dt} = G \times \omega, \end{cases} \tag{10}$$

where

$$V_C := Q^T \cdot \mathcal{V}_C, \quad n := Q^T \cdot N, \quad I_S := Q^T \cdot J_S \cdot Q, \quad \partial S := \partial S(0).$$

In order to keep compatible notations for both the two an three dimensional case, we assume for $d = 2$ that $\omega := (0, 0, \boldsymbol{\omega})$ and similarly $y \times [T \cdot n] = (0, 0, -y_2(T \cdot n)_1 + y_1(T \cdot n)_2)$. For $d = 2$, the equation $(10)_2$ reduces to a scalar equation.

In the body frame $\mathcal{R}$(t) the direction of the gravitational force G depends on the time t and becomes therefore an unknown to be resolved. The third additional equation of (10) provides the needed equation describing its variation. Its derivation relies on simple calculus related to the transformation (6). For more details regarding the overall derivation of these equations we refer to G.P. Galdi [5].

2.2 Formulation of the stationary free fall problem

The solid body $\mathcal{S}$ is said to undergo a *free steady fall* if the translational and angular velocity V_C and ω are constant and if the motion of the liquid $\mathcal{L}$ is stationary in the frame $\mathcal{R}(t)$. The study of such a configuration is of great interest since it corresponds to so called *terminate state* motions of sedimenting particles for which many questions still remain open: e.g. the number of possible terminal states for a given body geometry, the orientation of the solid body, the stability of the corresponding solution (see [5] and references therein). The free steady fall is thus obtained by requiring that v, p, V_C, ω and G are time independent. Comparing with (7-10), this leads to the following system of equations:

$$\left.\begin{aligned} \rho\{((v - V) \cdot \nabla)v + \omega \times v\} = \nabla \cdot T(v,p) + \rho G \\ \nabla \cdot v = 0 \end{aligned}\right\} \quad \text{for } \; y \in [\mathbb{R}^d \backslash S], \tag{11}$$

$$\lim_{|y| \to \infty} v(y) = 0 \tag{12}$$

$$v(y) = V(y) := V_C + \omega \times y \qquad \text{for } \; y \in \partial S \tag{13}$$

$$m_S(\omega \times V_C) = m_s G - \int_{\partial S} T(v,p) \cdot n \, d\sigma, \tag{14}$$

$$\omega \times (I_S \cdot \omega) = -\int_{\partial S} y \times [T(v,p) \cdot n] \, d\sigma, \tag{15}$$

$$G \times \omega = 0. \tag{16}$$

The system of equations (11-16) describes different class of free fall regimes and configurations which are outlined in [9]. They lead to different problem formulations. For the most general setup, we assume $\omega \neq 0$. Due to equation (16), this configuration can be attained only for $d = 3$. Further it imposes G parallel to ω. The free steady fall problem can then be stated as

Problem 1. Assume $d = 3$. Given ρ, $T = T(v, p)$, $|G| = |g|$, I_S and m_S, find v, p, V_C, ω, G whereas $G = |g||\omega|^{-1}\omega$ if $\omega \neq 0$, such that (11-15) holds.

3 Galerkin finite element discretization

For a domain $\Omega \subset \mathbb{R}^d$, let $L^2(\Omega)$ denote the Lebesgue space of square-integrable functions on Ω equipped with the inner product and norm

$$(f, g)_\Omega := \int_\Omega fg\; dx, \qquad ||f||_\Omega := (\int_\Omega |f|^2\; dx)^{\frac{1}{2}}.$$

Analogously, $L^2(\partial\Omega)$ denotes the space of square integrable functions defined on the boundary $\partial\Omega$. The L^2 functions with generalized (in the sense of distributions) first-order derivatives in $L^2(\Omega)$ form the Sobolev space H^1, while $H_0^1 = \{v \in H^1(\Omega), v|_{\partial\Omega} = 0\}$.

3.1 Variational formulation

The Galerkin finite element method starts from a variational formulation of the equations to be solved. We first consider the most general setup of problem 1 i.e. $\omega \neq 0$ and the related equations (11-15). The key ingredient for the derivation of a weak form of the equations (11-15) is an adequate choice of the velocity space allowing to eliminate the explicit formulation of the hydrodynamic force and torque on the solid body needed for the kinematic equations (14) and (15). This can be obtained by including the no-slip Dirichlet condition (13) in the velocity space:

$$\mathcal{H}_1(D) := \{(v, V, \omega) : v \in [H^1_{loc}(\overline{D})]^d, V \in \mathbb{R}^d, \omega \in \mathbb{R}^d, v = V + \omega \times y \text{ on } \partial S\} \tag{17}$$

where $D := \mathbb{R}^d \backslash S$. The pressure p which is defined modulo constants is assumed to lie in the space

$$L_0^2(D) := \left\{ q \in L^2(D) : \int_{D'} q = 0 \right\}. \tag{18}$$

where $D' \subset D$ bounded. For $u := \{(v, V_C, \omega), p\} \in \mathcal{H}_1(D) \times L_0^2(D)$ and $\varphi := \{(\varphi, \varphi_1, \varphi_2), q\} \in \mathcal{H}_1(D) \times L_0^2(D)$ we define the following semi-linear form

$$\begin{aligned} \mathcal{A}_1(u; \varphi) := {} & \rho(((v - (V_C + \omega \times y)) \cdot \nabla)v, \varphi)_D + (\omega \times v, \varphi)_D \\ & -(p, \nabla \cdot \varphi)_D + 2\mu \int_D D(v) : D(\varphi) - (\rho|g||\omega|^{-1}\omega, \varphi)_D \\ & -\varphi_1 \cdot [m_S(|g||\omega|^{-1}\omega - \omega \times V_C)] + \varphi_2 \cdot [\omega \times (I_S \cdot \omega)] \\ & -(\nabla \cdot v, q)_D, \end{aligned} \tag{19}$$

which is obtained by testing the equations (11) and (14-15) by $\varphi \in \mathcal{H}_1(D) \times L^2_0(D)$ and by integration by parts of the diffusive terms and the pressure gradient in $(11)_1$. $D(v)$ denotes the deformation tensor i.e. $D(v) := \frac{1}{2}(\nabla v + (\nabla v)^T)$. A weak form of problem 1 can therefore be formulated as

Problem V1. Find $u := \{(v, V_C, \omega), p\} \in \mathcal{H}_1(D) \times L^2_0(D)$ such that

$$\mathcal{A}_1(u; \varphi) = 0 \quad \forall \varphi \in \mathcal{H}_1(D) \times L^2_0(D). \tag{20}$$

The equation modeling the balance of the linear (resp. angular) momentum (14) (resp. (15)) can obviously be recovered by testing in (20) with the functions $\{(0, \varphi_1, 0), 0\}$ (resp. $\{(0, 0, \varphi_2), 0\}$).

Remark 1. The advantages of the formulation (20) rely on the fact that the force and torque on the solid body do not need to be computed explicitly. Numerical instabilities arising for the computation of these lower dimensional integrals can therefore be avoided (see [7, 12]).

3.2 Finite element discretization

First of all, the unbounded domain $D := \mathbb{R}^d \backslash S$ filled by the liquid $\mathcal{L}$ is replaced by a bounded domain $\Omega \subset \mathbb{R}^d \backslash S$. On the artificial boundary $\partial\Omega \backslash \partial S$ we prescribe homogeneous Dirichlet boundary conditions,

$$v(y) = 0 \qquad \text{for } y \in \partial\Omega \backslash \partial S.$$

In order to ensure that the impact of this simplification on the quantities of interest is smaller than the discretization error, we have to chose Ω large enough. For a detailed discussion of this issue we refer to [2, 16].

The discretization uses a conforming finite element space $W_1^h \subset \mathcal{H}_1(\Omega) \times L^2_0(\Omega)$ defined on a triangulation $\mathcal{T}_h = \{K\}$ consisting of quadrilateral or hexahedral cells K covering the domain $\overline{\Omega}$; the family of triangulations $\{\mathcal{T}_h\}_h$ is assumed to be quasi-uniform as $h \to 0$. For the trial and test spaces $W_1^h \subset \mathcal{H}_1(\Omega) \times L^2_0(\Omega)$ we consider the standard Hood-Taylor finite element [10] i.e.

$$\begin{aligned} W_1^h := \{((v, V, \omega), p) \in \left\{[C(\overline{\Omega})]^d \times \mathbb{R}^d \times \mathbb{R}^d\right\} \times C(\overline{\Omega}), \\ v|_K \in [Q_2]^d, p|_K \in Q_1, v|_{\partial S} = V + \omega \times y\right\}, \end{aligned}$$

where Q_r describes the space of isoparametric tensor-product polynomials of degree r (for a detailed description of this standard construction process see e.g. [3]). This choice for the trial and test functions has the advantage that it guarantees a stable approximation of the pressure since the uniform *Babuska-Brezzi* inf-sup stability condition is satisfied uniformly (see [4] and references therein). Compared to equal order function spaces for the pressure and the velocity, no additional stabilization terms are needed. Further, in order to facilitate local mesh refinement and coarsening, we allow the cells in the refinement zone to have nodes which lie on faces of neighboring cells (see figure 1). The degrees of freedom corresponding to such hanging nodes are eliminated by interpolation enforcing global conformity for the finite element functions. The discrete counterpart of problem (V1) reads

Problem V1′. Find $u_h := W_1^h$ such that

$$\mathcal{A}_1(u_h;\varphi_h) = 0 \quad \forall \varphi_h \in W_1^h. \tag{21}$$

4 A posteriori error control for the hydrodynamical force and torque

The implicit treatment of the hydrodynamical force and torque acting on the solid body $\mathcal{S}$ in terms of the natural boundary conditions (see section 3.1) allows to derive a specific a posteriori error control strategy. The proposed approach, inspired by the work of Giles et al. [7], takes great advantage of the special structure of the free steady fall problem and of the considered weak formulation leading to a natural derivation of error bounds for the hydrodynamical force and torque.

We consider the most general setup of problem 1 and define for $u := \{(v, V_C, \omega), p\} \in \mathcal{H}_1(D) \times L_0^2(D)$ the following weighted functional

$$J_\psi(u) := \int_{\partial S} [T(v,p)\cdot n]\cdot \psi \, d\sigma, \tag{22}$$

where $\psi := \psi_1 + \psi_2 \times y \in \mathbb{R}^3$ with $\psi_1, \psi_2 \in \mathbb{R}^3$. For $\psi = \psi_1$ (resp. $\psi = \psi_2 \times y$), the functional $J_\psi(u)$ corresponds obviously to the weighted hydrodynamical force (resp. hydrodynamical torque) since

$$J_{\psi_1}(u) = \psi_1 \cdot \int_{\partial S} [T(v,p)\cdot n] \, d\sigma \tag{23}$$

$$J_{\psi_2 \times y}(u) = \psi_2 \cdot \int_{\partial S} y \times [T(v,p)\cdot n] \, d\sigma. \tag{24}$$

Now, we define the following semi-linear form

$$\begin{aligned}\mathcal{A}(u;\varphi) := \rho(((v-(V_C+\omega\times y))\cdot\nabla)v,\varphi)_D + (\omega\times v,\varphi)_D \\ -(p,\nabla\cdot\varphi)_D + 2\mu\int_D D(v):D(\varphi) \\ -(\rho|g||\omega|^{-1}\omega,\varphi)_D - (\nabla\cdot v, q)_D,\end{aligned} \tag{25}$$

which, apart from the boundary terms φ_1 and φ_2, corresponds to the semi-linear form $\mathcal{A}_1(u;\varphi)$. Further we define the following velocity space

$$\mathcal{H}_1^\psi(D) := \mathcal{H}_1(D) \cap \{(v,V,\omega):\ \nabla\cdot v = 0 \ \text{ in } \ \Omega,\ V=\psi_1,\ \omega=\psi_2\}. \tag{26}$$

Then following lemma holds (see [9]):

Lemma 1. *Under sufficient regularity assumptions for the solution u of problem* V1*, we have*

$$J_\psi(u) = \mathcal{A}(u; w) \quad \forall w \in \mathcal{H}_1^\psi(D) \times L_0^2(D). \tag{27}$$

The discrete counterpart of $\mathcal{H}_1^\psi(D) \times L_0^2(D)$ is defined as

$$W_1^{\psi,h} := W_1^h \cap \{((v, V, \omega), p) : \; V = \psi_1, \; \omega = \psi_2\}\,.$$

Let $u_h \in W_1^h$ be the solution of the discrete problem V1$'$. One can easily show that the functional

$$\tilde{J}_\psi(u_h) := \mathcal{A}(u_h; w) \quad \forall w \in W_1^{\psi,h}, \tag{28}$$

is well defined since $\mathcal{A}(u_h; w)$ depends uniquely on the boundary value ψ of w. It is of importance to notice that in general

$$\tilde{J}_\psi(u_h) \neq J_\psi(u_h).$$

As shown in [7], the functional $\tilde{J}_\psi(u_h)$, rather than $J_\psi(u_h)$ is the appropriate approximation of $J_\psi(u)$. From now on, our purpose is then to derive error bounds for $J_\psi(u_h) - \tilde{J}_\psi(u_h)$. In order to derive an error representation for the error $J_\psi(u_h) - \tilde{J}_\psi(u_h)$, we define the following *linearized dual problem*:

Problem 4. Find $z := \left\{(z^v, z^{V_C}, z^\omega), z^p\right\} \in \mathcal{H}_1^\psi(D) \times L_0^2(D)$ such that

$$L(u, u_h; z, \varphi) = 0 \quad \forall \varphi \in \mathcal{H}_1^{\psi=0}(D) \times L_0^2(D). \tag{29}$$

Here, $L(u, u_h; z, \varphi)$ is assumed to be a bilinear form in z and φ chosen such that the following equality holds

$$L(u, u_h; z, u - u_h) = \mathcal{A}(u; z) - \mathcal{A}(u_h, z) \quad \forall z \in \mathcal{H}_1(D) \times L_0^2(D), \tag{30}$$

where u (resp. u_h) describes the solution of problem V1 (resp. V1$'$).
Due to the special nature of the nonlinear terms in $\mathcal{A}(\cdot; \cdot)$, $L(u, u_h; \cdot, \cdot)$ can be defined explicitly (see [9] for more details). Using these preliminaries we are now able to derive the needed error representation of $J_\psi(u_h) - \tilde{J}_\psi(u_h)$.

Proposition 1. *Let z be the solution of problem 4. Further let $\Pi : \mathcal{H}_1^\psi(D) \times L_0^2(D) \to W_1^{\psi,h}$ be some interpolation operator. We then have*

$$J_\psi(u_h) - \tilde{J}_\psi(u_h) = \mathcal{A}(u_h, z - \Pi z). \tag{31}$$

Proof. See [9].

Remark 2. The error representation (31) allows not only to control separately the hydrodynamical force and torque but also a weighted combination of both quantities. This can be done by an adequate definition of the weights ψ_1 and ψ_2 of the trace $\psi = \psi_1 + \psi_2 \times y$ in (27) and (28), respectively. The dual solution z depends on ψ exclusively through the enforcement of the Dirichlet boundary condition $z^v|_{\partial S} = \psi$.

5 Numerical experiments

We consider the free fall of a rectangular body of length $l = 6.10^{-2}m$ and width $L = 1.10^{-2}$ in a viscous fluid. The shear viscosity (resp. the density) is assumed to be $\mu = 0.1$ (resp. $\rho = 1$). Our numerical simulations lead to both horizontal and vertical position as terminal state. The vertical fall is however an instable terminal state and will not be further considered in the following (see figure 1). We assume homogeneous Dirichlet boundary conditions for the velocity on the outer boundary $\partial\Omega\backslash\partial S$ of the computational domain Ω (see [2] for more details). The large size needed for the computational domains imposes a careful mesh design. The a posteriori error estimates developed in section 4 allow to tackle this problem since it allows to construct economical meshes oriented toward the computation of a given parameter. Adaptive grids based on the a posteriori estimator (31) are plotted in figure 1.

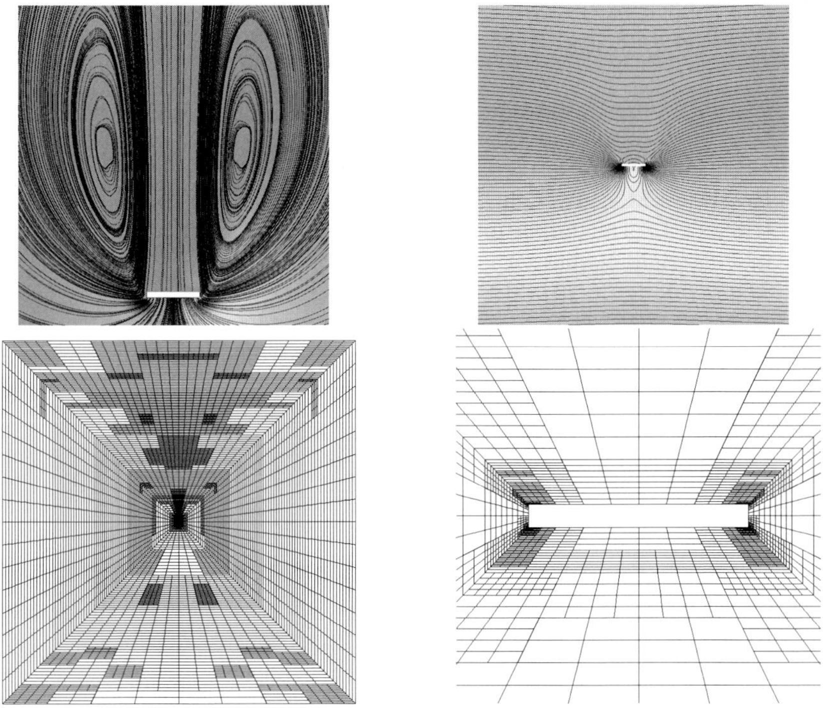

Fig. 1. (top left) Streamlines around the falling body for $\nu = 0.1$; (top right) pressure isolines; (bottom left and right) adaptive mesh obtains by means of (31) on a domain with diameter $D = 800$ and corresponding zoom around the body. The colour version of this figure can be found in Fig. A.11 on page 582.

References

[1] R. Becker and R. Rannacher. An optimal control approach to a posteriori error estimation in finite element methods. 1–102. A. Iserles, 2001.

[2] S. Bönisch, V. Heuveline, and P. Wittwer. Adaptive boundary conditions for exterior flow problems. Technical Report 2003-02, SFB 359, Universität Heidelberg, 2003.

[3] S.C. Brenner and R.L. Scott. *The mathematical theory of finite element methods.* Springer, Berlin-Heidelberg-New York, 1994.

[4] F. Brezzi and R. Falk. Stability of higher-order Hood-Taylor methods. *SIAM J. Numer. Anal.*, 28(3):581–590, 1991.

[5] G.P. Galdi. *On the motion of a rigid body in a viscous liquid: A mathematical analysis with applications.* Handbook of Mathematical Fluid Mechanics, S. Friedlander and D. Serre Eds, Elsevier, 2001.

[6] G.P. Galdi and A. Vaidya. Translational steady fall of symmetric bodies in a Navier-Stokes liquid, with application to particle sedimentation. *J. Math. Fluid Mech.*, 3:183–211, 2001.

[7] M. Giles, M. Larson, M. Levenstam, and E. Süli. Adaptive error control for finite element approximations of the lift and drag coefficients in viscous flow. Technical Report NA-97/06, Oxford University Computing Laboratory, 1997.

[8] R. Glowinski, T.W. Pan, T.I. Hesla, D.D. Joseph, and J. Periaux. A distributed Lagrange multiplier/fictitious domain method for flow around moving rigid bodies: Application to particulate flow. *Int. J. Numer. Meth. Fluids*, 30:1043–1066, 1999.

[9] V. Heuveline and R. Rannacher. The steady free fall problem. part 1: Numerical computation using adaptive finite element. *in preparation.*

[10] P. Hood and C. Taylor. A numerical solution of the navier-stokes equations using the finite element techniques. *Comp. and Fluids*, 1:73–100, 1973.

[11] H.H. Hu. Direct simulation of flows of solid-liquid mixtures. *Int. J. Multiphase Flow*, 22:335–352, 1996.

[12] H.H. Hu, D.D. Joseph, and M.J. Crochet. Direct simulation of fluid particle motions. *Theor. Comp. Fluid Dyn.*, 3:285–306, 1992.

[13] S. Nevcasová. Asymptotic properties of the steady fall of a body in viscous fluids. Technical Report 149/2002, Academy of Sciences of the Czech Republic, Mathematical Institute, 2002.

[14] D. Serre. Chute libre d'un solide dans un fluide visqueux incompressible. existence. *Jap. J. Appl. Math*, 4(1):99–110, 1987.

[15] S.O. Unverdi and G. Tryggvason. Computations of multi-fluid flows. *Physica D*, 60:70–83, 1992.

[16] P. Wittwer. On the structure of stationary solutions of the Navier-Stokes equations. *Commun. Math. Phys.*, 226:455–474, 2002.

Searching the Web: a Semantics-Based Approach

Tru H. Cao, Ta H. D. Nguyen, and Tran C. T. Qui

Faculty of Information Technology, Ho Chi Minh City University of Technology
268 Ly Thuong Kiet St., Dist. 10, Ho Chi Min City, Vietnam
tru@dit.hcmut.edu.vn

Summary. Current search engines such as Google are mainly keyword-based, whereby a query is represented by a set of keywords. Such a query language is not expressive enough to allow users to present the subject of the web documents that they want to find. Consequently, one often receives many useless results when searching the web. This paper proposes a semantics-based approach to web search engines in order to increase their precision. Our assumption is that the subjects of the documents that one wants to search for can be expressed by a set of concepts and relations between them. We propose to use conceptual graphs to represent both user queries and document descriptions, on the basis of an ontology built up for a particular domain. In order to reduce the computational cost, documents are first filtered to provide only those that contain the concepts and relations in the query. Graph matching is then performed to return relevant documents. A prototype of the proposed system is also presented for demonstration.

1 Introduction

Search engines have become indispensable tools to find useful information over the extremely large virtual pool of data over the World Wide Web. However, traditional keyword-based search engines both vary in recall (the rate of relevant web pages retrieved above relevant ones) and offer a bad precision (the rate of relevant web pages retrieved above the total retrieved). The main reason for these is that a set of keywords cannot exactly and fully describe what users want to search for.

To address this issue, we begin with answering the important question of how to express what a user wants in searching for documents. Using keyword is a way to do, but it is obviously inadequate. Our assumption is that the subject of the documents that one wants to search for can be expressed by a set of concepts and relations between them.

For example, suppose that a user is searching for research papers about "Application of Internet Computing to High Performance Scientific Computing". That subject has two concepts, namely, "Internet Computing" and "High

Performance Scientific Computing", and one relation "Applied To" between them. As such, it is different from the subject "Application of High Performance Scientific Computing to Internet Computing" and, thus, the two queries with these two subjects are expected to have different documents returned as their answers. We note that, for current keyword-based search engines, one cannot express the difference between the two subjects, whose queries comprise the same set of keywords, namely, "Internet Computing", "High Performance Scientific Computing" and "Application".

The next question is which formalism can be used to represent such a query. Conceptual graphs (CGs) ([5], [7], [18]), which are based on semantic networks and Peirce's existential graphs, appear to be a suitable language for search queries. Conceptual graphs have been used for solving problems in several areas such as, but not limited to, natural language processing, knowledge acquisition and management, database design and interface, and information systems. It was proposed to be a normative conceptual schema language by the ANSI standards committee on Information Resource Dictionary Systems, and a knowledge representation language in conceptual models by the ANSI standards committee on Information Processing Systems ([1], [2], [19]).

Our idea is that, for semantics-based searching, each document is annotated with a description summarizing its contents and represented by a conceptual graph (cf. [6]). A query is also represented by a conceptual graph, which will be matched to description graphs of documents to find the needed ones. However, in order to reduce the computation cost, documents are first filtered to provide only those that contain the concepts and relations in the query. Graph matching is then performed to return relevant documents.

The system requires an ontology of concepts and relations on an application domain. At this first step, we limit the study in the domain of scientific publications only, and more specifically in this paper our attention is focused on research papers in Artificial Intelligence (AI). However, the framework and methodology could be applied to other domains of web documents as well. Also, the results could be adapted for systems using other representation languages such as RDF graphs, which can be mapped to and from CGs (cf. [4]).

The paper is organized as follows. Section 2 summarizes the basic notions of conceptual graphs. Section 3 presents our development of an ontology of concepts and relations for research papers in Artificial Intelligence. Section 4 demonstrates a prototype of semantic search engines that we have implemented. Finally, Section 5 concludes the paper and suggests future research.

2 Conceptual Graphs

2.1 Syntax

A conceptual graph is a bipartite graph of *concept* vertices alternate with (conceptual) *relation* vertices, where edges connect relation vertices to concept vertices. Each concept vertex, drawn as a box and labelled by a pair of a *concept type* and a *concept referent*, represents an entity whose type and referent are respectively defined by the concept type and the concept referent in the pair. Each relation vertex, drawn as a circle and labelled by a *relation type*, represents a relation of the entities represented by the concept vertices connected to it. For brevity, we may call a concept or relation vertex a concept or relation, respectively. Concepts connected to a relation are called *neighbour concepts* of the relation. Each edge is labelled by a positive integer and, in practice, may be directed just for readability.

For example, the CG in Figure 1 expresses "Fuzzy is a theory. GA is an algorithm. There is a problem. Fuzzy is applied to GA. GA solves the problem", or briefly, "Fuzzy theory is applied to genetic algorithms to solve a problem". In a textual format, concepts and relations can be respectively written in square and round brackets as follows:

`[Theory:Fuzzy]`—1—`(applied_to)`—2—
`[Algorithm:GA]`—1—`(solve)`—2—`[Problem:*]`

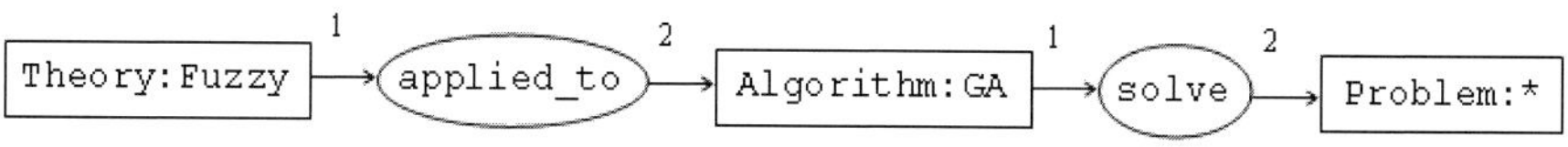

Fig. 1. An example CG

In this example, `[Theory: Fuzzy]`, `[Algorithm: GA]`, and `[Problem:*]` are concepts with `Theory`, `Algorithm`, and `Problem` being concept types, whereas `(applied_to)` and `(solve)` are relations with `applied_to` and `solve` being relation types. The referents `Fuzzy` and `GA` of the concepts `[Theory: Fuzzy]` and `[Algorithm: GA]` are *individual markers*. The referent `*` of the concept `[Problem:*]` is the *generic marker* referring to an unspecified entity.

Corresponding to the notion of sorts in order-sorted predicate logic, concept types are partially ordered by the concept subtype order. This order can be regarded as an information or specificity order in the sense that, given concept types t_1 and t_2 where t_2 is a concept subtype of t_1, a fact "Object x is of type t_2" is more informative than "Object x is of type t_1". So we write $t_1 \leq_\iota t_2$ to denote that t_2 is a concept subtype of t_1. For example, one may have `Method` $\leq_\iota$ `Algorithm`. Relation types can also be partially ordered, and the neighbour concept types of each relation are defined in its *signature*. For example, with the CG in Figure 1, one may have `applied_to` as a subtype of a relation type `act_on`, written as `act_on` $\leq_\iota$ `applied_to`. For a partial order on concept referents, which are basically individual markers and the generic marker only, it is simply that, for every individual marker i, $* \leq_\iota i$, and all individual markers are pairwise incomparable.

2.2 CG Projection

CG projection is an important operation on CGs. A projection maps a CG to another more or equally specific one, by mapping each vertex of the former to a vertex of the latter that has a more or equally specific label. The label (t_1, m_1) of a concept is said to be more or equally specific than the label (t_2, m_2) of another concept if and only if $t_2 \leq_\iota t_1$ and $m_2 \leq_\iota m_1$. The label t_1 of a relation is said to be more or equally specific than the label t_2 of another relation if and only if $t_2 \leq_\iota t_1$. The mapping must also preserve the adjacency of the neighbour concepts of a relation, that is, if a relation r_2 of type t_2 of arity n is mapped to a relation r_1 of type t_1 of the same arity then, for every i from 1 to n, the neighbour concept connected to r_2 by the edge labelled i must be mapped to the neighbour concept connected to r_1 by the edge labelled i. Figure 2 illustrates a CG projection from G to H.

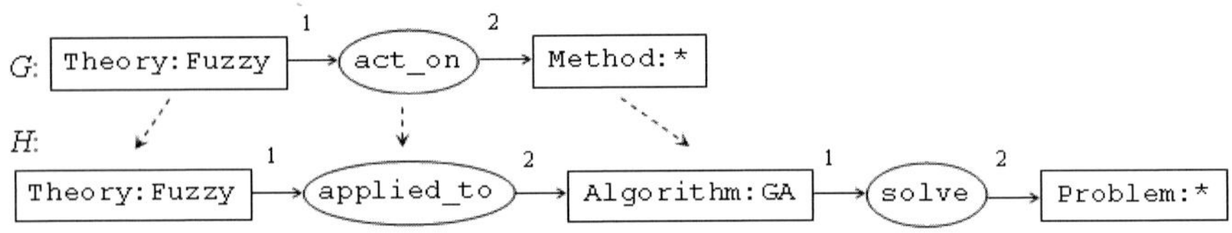

Fig. 2. A CG projection

CG projection is reflexive and transitive. That is, every CG has a projection to itself and, given CGs G, H and F, if G has a projection to H and H has a projection to F, then G has a projection to F. As such, CG projection defines a preorder on CGs, which can also be regarded as an information or specificity order, whereby if G has a projection to H, then H is more informative or more specific than G. Logically, if a CG G has a projection to a CG H, then H implies G.

With user queries and document descriptions represented by CGs, searching can be performed by projection of a query graph to description graphs. A document is relevant to the query if there is a projection from the query graph to its description graph. As shown in [14], there is a polynomial algorithm to find a projection from a tree CG to another CG.

3 Ontology for AI Research Papers

3.1 Building Methodology

A conceptual graph has no meaning if its concepts and relations are not linked to context through a semantic network. Sowa described some elements of this semantic network in [18]. After that, Mugnier and Chein integrated those elements in a *support* of a specific domain ([15]). In short, a support includes a lattice of concept types, a lattice of relation types, and the signatures of these relation types in a domain of discourse.

That notion of a support is a simple form of a domain-specific ontology, which is "a catalog of the types of things that are assumed to exist in a domain of interest" ([20]). Such an ontology provides basic information about the meaning and relationships of concepts and relations on a specific domain. Therefore, to develop a semantic search engine for the domain of AI research papers, an ontology for this domain has to be built.

For semantics-based applications, in particular for sharing knowledge on the web, ontology has increasingly become the quest and focus of significant research effort. Nevertheless, the important problem of how to build an ontology is still a difficult task. The main reasons for that difficulty are not only that there are many unstructured and ambiguous concepts and relations in a domain, but also that this task is closely related to terminology, philosophy and requires much knowledge and experience of domain experts.

Although many attempts have been developed to propose a method for building an ontology ([9], [10], [11], [21], [22]), none of those methods are complete and detailed enough ([8]). By studying and combining different methods, and through experience on building our own ontology, we propose a new method to build up a domain-specific ontology. Basically, our method is based on the outline given by Uschold and Grüninger in [22]. However, we integrate techniques from other methods, clarify some important points and propose a different approach when building a concept hierarchy. The main points are as follows:

1. Identifying requirements and scope of ontology.
   a) With the application: as noted in [11], the roles of the ontology in the application should be identified regarding why it is being built, what its intended uses are, and how it is used by the application and to what extent.
   b) With the ontology:
      i. Identifying the requirements and scope of the ontology based on the above results.
      ii. Proposing testing methods to guarantee the conformity of the ontology with the application's purposes.
      iii. Carefully surveying the application domain to pick out the most common and important features.
2. Building the concept hierarchy:
   a) This process includes two alternating tasks, namely, finding concepts and organizing these concepts into a hierarchy by their subtype (or "is-a-kind-of") relation. The two most important concerns in this process are rationality and consistency.
   b) In [22], the authors showed the disadvantages of the top-down and the bottom-up approaches, and proposed to use the middle-out ap-

proach starting from the most important concepts first. However, as we have experienced, to decide which concepts are the most important is difficult. Therefore, we propose to use both the top-down and the bottom-up approaches simultaneously.

c) During the construction process, classifying or grouping the concepts, one should often use testing methods mentioned above to verify if the ontology satisfies its requirements and scope.

3. Finding relations and their signatures: on the basis of the constructed concept hierarchy, examine every pair of concepts to find all possible relations between them that are suitable to the requirements and scope of the ontology.
4. Coding, testing, modifying and maintaining ontology.

3.2 Concepts and Relations for AI Papers

Using the above method, we have built an ontology for AI research papers to be employed by a semantic search engine. The concept hierarchy and the set of relations with their signatures are illustrated in the Appendix.

4 Semantic Search Engine Prototype

We have implemented a prototype of semantic search engines, where both query and documents are represented by conceptual graphs. The document collection used in this experiment is from the CiteSeer electronic library (http://citeseer.nj.nec.com/cs). For testing data, we have used the ontology presented above to represent the main contents of over fifty AI papers about Game Playing, Machine Learning and Fuzzy Theory. The results have shown that our ontology can represent the subjects of many papers in the AI domain. Using CG projection, the semantic search engine matches a query graph to paper graphs to find out relevant ones.

As for most other search engines, our proposed system architecture is a centralized crawler-indexer one (cf. [3]). The distinguishing features are the semantic search engine and the ontology of concepts and relations on a domain of discourse. The architecture can be adapted for different domains by changing only this ontology part.

The semantic searching process to answer a user query has two stages. In the first stage, the stored documents are filtered to provide only those that contain the concepts and relations in the query. In the second stage, the query graph is matched to the description graphs of the filtered documents to return relevant answers.

Experiment 4.1: Suppose that a user want to search for surveying papers about the Alpha-Beta algorithm. This query can be expressed by the following CG:

`[Survey]—1—(about)—2—[Algorithm:AlphaBeta]`

It will match with the following description CGs and the corresponding papers will be returned as answers:

1. *"Survey about the Alpha-Beta algorithm"*
   `[Survey]—1—(about)—2—[Algorithm:AlphaBeta]`
2. *"General survey about the Alpha-Beta algorithm"*
   `[GeneralSurvey]—1—(about)—2—[Algorithm:AlphaBeta]`
3. *"Detailed survey about the Alpha-Beta algorithm"*
   `[DetailedSurvey]—1—(about)—2—[Algorithm:AlphaBeta]`
4. *"Comparative survey about similarity between the Alpha-Beta and Mini-Max algorithms"*

```
[ComparativeSurvey: Similarity]
                 {—1—(about)—2—[Algorithm:AlphaBeta];
                  —1—(about)—2—[Algorithm:MiniMax];
                 }
```

In the last CG, the branches separated by ; and enclosed in {} are to denote that they are connected to the same concept `[ComparativeSurvey: Similarity]`. This experiment shows that the concepts in a returned paper are not necessarily the same as the respective ones in the query, but can be just their sub-concepts. For instance, `[GeneralSurvey]`, `[DetailedSurvey]`, and `[ComparativeSurvey: Similarity]` are sub-concepts of `[Survey]`.

Experiment 4.2: Suppose that a user want to search for papers about application of fuzzy theory to genetic algorithms. This query can be expressed by the following CG:

`[Theory:Fuzzy]—1—(applied_to)—2—[Algorithm:GA]`

After the filtering stage the papers of the following CGs whose concepts and relations include those in the query CG will be returned:

1. *"Survey about application of fuzzy theory to genetic algorithms"*

```
[Survey]—1—(about)—2—
        [Algorithm:GA]—2—(applied_to)—1—[Theory:Fuzzy]
```

2. *"Application of fuzzy theory to genetic algorithms for chess game playing"*

```
[GamePlaying:Chess]—2—(solve)—1—
        [Algorithm:GA]—2—(applied_to)—1—[Theory:Fuzzy]
```

3. *"Application of fuzzy theory to genetic algorithms for computer vision"*

```
[ComputerVision]—2—(solve)—1—
        [Algorithm:GA]—2—(applied_to)—1—[Theory:Fuzzy]
```

4. *"Comparative survey about using genetic algorithms and application of fuzzy theory to wavelet algorithms for computer vision"*

```
[ComparativeSurvey]
      {—1—(about)—2—[Algorithm: GA]—1—(solve)—2—
                                      [ComputerVision];
       —1—(about)—2—[Algorithm: Wavelet]—
```

```
                        {—1—(solve)—2—[ComputerVision];
                         —2—(applied_to)—1—[Theory: Fuzzy];
                        };
          }
```

5. *"Survey about application of genetic algorithms to fuzzy theory"*
```
   [Survey]—1—(about)—2—
            [Theory:Fuzzy]—2—(applied_to)—1—[Algorithm:GA]
```

In the matching stage, only the first three description CGs can match with the query CG. Meanwhile, a keyword-based search engine would also return the papers of the last two CGs as they contains the keywords "Fuzzy Theory" and "Genetic Algorithms" in the query.

5 Conclusion

We have proposed a new approach to web search engines in order to increase their precision by matching the meaning of a user query to those of web documents, rather than simply matching their keywords as in current search engines. Here the meaning of a query/document is expressed via the concepts it contains and the included relations between them. We have proposed to use conceptual graphs for representing this kind of semantics, as they are also based on concepts and conceptual relations.

An architecture of semantic search systems has been described, where user queries and document descriptions are represented by conceptual graphs. A searching process for a query has two stages, namely, concept-relation filtering and graph matching. In the first stage, only the documents whose description graphs contain the concepts and relations in the query are selected. Then, in the second stage, using CG projection the query graph will be matched to those description graphs to return relevant documents.

We have used this approach for development of a semantic search engine on scientific publications, and in particular research papers in Artificial Intelligence. We have proposed a mixed top-down and bottom-up method to build up an ontology, i.e., a hierarchy of concepts and relation signatures, for this domain. A prototype of the system has been implemented, showing that the system returns more precise answers than a traditional keyword-based system.

Towards a full-scale semantic search engine, there are two main problems to be solved. The first is how to organize a database of CGs over a large volume of scientific publications so that CGs can be retrieved and CG projection performed efficiently. The second is how to learn a description CG of a paper automatically. Regarding the first problem, the work [13], which used CGs to represent knowledge, has shown that using CG projection for searching is feasible. Regarding the second problem, in [24], machine learning techniques were applied to map link grammar structures of sentences to CGs in the domain of clothes descriptions, showing promising results.

Meanwhile, we have used the Naïve Bayes technique to automatically extract concept types in the proposed ontology from 500 AI research paper abstracts, with the accuracy varying from 55% to 65% ([23]). This performance could be improved by using a larger training data set. We are also investigating other advanced machine learning techniques such as support vector machines. The next step will be to research techniques for extraction of concept referents and relations.

References

[1] ANSI Report No. X3H4/93-196, IRDS Conceptual Schema (1993).
[2] ANSI Report No. X3T2/95-019r2, Information Interchange and Interpretation (1995).
[3] Baeza-Yates, R. and Ribeiro-Neto, B.: Modern Information Retrieval. Addison-Wesley (1999).
[4] Berners-Lee, T.: Conceptual Graphs and the Semantic Web. First draft (2001).
[5] Cao, T.H.: Fuzzy Conceptual Graphs: A Language for Computational Intelligence Approaching Human Expression and Reasoning. In: Sincak, P. et al. (eds.): The State of the Art in Computational Intelligence. Physica-Verlag (2000) 114-120.
[6] Cao, T.H.: Fuzzy Conceptual Graphs for the Semantic Web. Invited to the Berkeley Initiative in Soft Computing International Workshop on Fuzzy Logic and the Internet (2001).
[7] Chein, M., Mugnier, M.L.: Conceptual Graphs are also Graphs. Research Report No. 95003, Laboratoire d'Informatique, de Robotique et de Micro-électronique de Montpellier (1995).
[8] Fernández-López, M.: Overview of Methodologies for Building Ontologies. In Proceedings of the IJCAI'99 Workshop on Ontologies and Problem-Solving Methods: Lessons Learned and Future Trends. CEUR Publications (1999).
[9] Fernández-López, M., Gómez-Pórez, A., Juristo, N.: Methontology: from Ontological Art towards Ontological Engineering. In Proceedings of the AAAI'97 Workshop on Ontological Engineering, Spring Symposium Series, Stanford, USA (1997).
[10] Gómez-Pérez, A., Fernández, M., de Vincente, A.J.: Towards a Method to Conceptualize Domain Ontologies. In Proceedings of the ECAI'96 Workshop on Ontological Engineering (1996) 41-51.
[11] Grüninger, M., Fox, M.S.: Methodology for the Design and Evaluation of Ontologies. In Proceedings of the IJCAI'95 Workshop on Basic Ontological Issues in Knowledge Sharing, Montreal, Canada (1995).
[12] Jones, D., Bench-Capon, T., Visser, P.: Methodologies for Ontology Development. In Proceedings of IT&KNOWS Conference of the 15th IFIP World Computer Congress. Chapman-Hall (1998) 20-35.

[13] Martin, P., Elkund, P.: Embedding Knowledge in Web Documents. In Proceddings of the 8th International World Wide Web Conference (1999).
[14] Mugnier, M.L.: On Generalization/Specification for Conceptual Graphs. J. Expt. Theor. Artif. Intell. 7 (1995) 325-344.
[15] Mugnier M.L, Chein, M.: Conceptual Graphs Fundamental Notions. Revue d'Intelligence Artificielle **6** (1992) 365-406.
[16] Natalya, F.N.: Knowledge Representation for Intelligent Information Retrieval in Experimental Sciences. PhD Thesis, Northeastern University, Boston, MA (1997).
[17] Nicola, G.: Formal Ontology and Information System. In Proceedings of FOIS'98, Trento, Italy (1998).
[18] Sowa, J.F.: Conceptual Structures - Information Processing in Mind and Machine. Addison-Wesley Publishing Company (1984).
[19] Sowa, J.F.: Conceptual Graphs: Draft Proposed American National Standard. In: Tepfenhart, W., Cyre, W. (eds.): Conceptual Structures: Standards and Practices. Lecture Notes in Artificial Intelligence, Vol. 1640. Springer-Verlag (1999) 1-65.
[20] Sowa, J.F.: Knowledge Representation: Logical, Philosophical, and Computational Foundations. Brooks Cole Publishing Co. (2000).
[21] Uschold, M.: Building Ontologies: towards a Unified Methodology. In Proceedings of the 16th Annual Conference of the British Computer Society Specialist Group on Expert Systems (1996).
[22] Uschold, M. & Grüninger, M.: Ontologies: Principles, Methods and Applications. Knowledge Engineering Review **11** (1996).
[23] Vu, D.Q. & Huynh, T.N.: Automatic Concept Types Extraction from AI Research Papers. BEng Thesis, Ho Chi Minh City University of Technology (2003).
[24] Zhang, L. & Yu, Y.: Learning to Generate CGs from Domain Specific Sentences. In: Delugach, H.S., Stumme, G. (eds.): Conceptual Structures: Broadening the Base. Lecture Notes in Artificial Intelligence, Vol. 2120. Springer-Verlag (2001) 44-57.

A Appendix: An Ontology for AI Research Papers

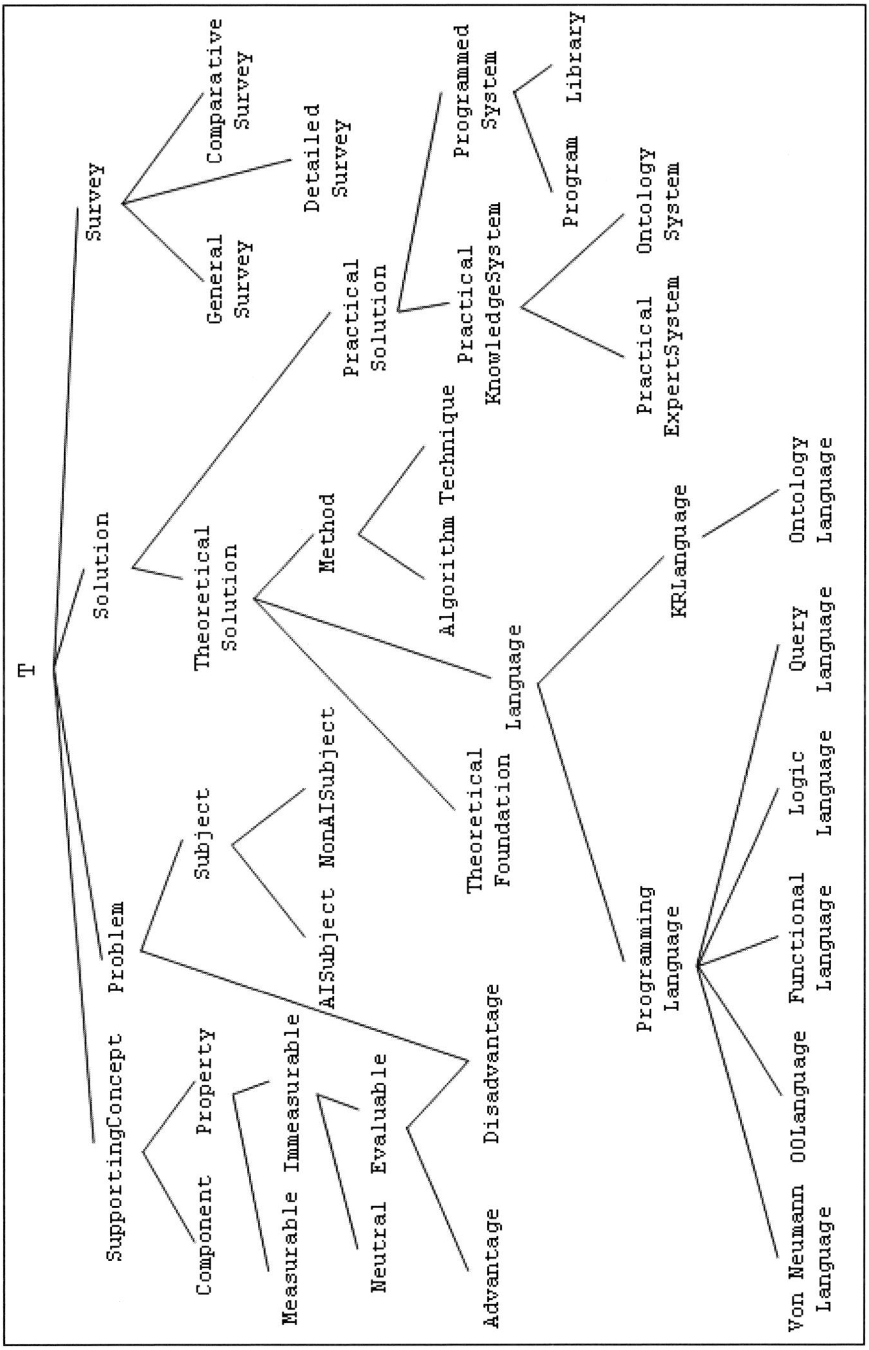

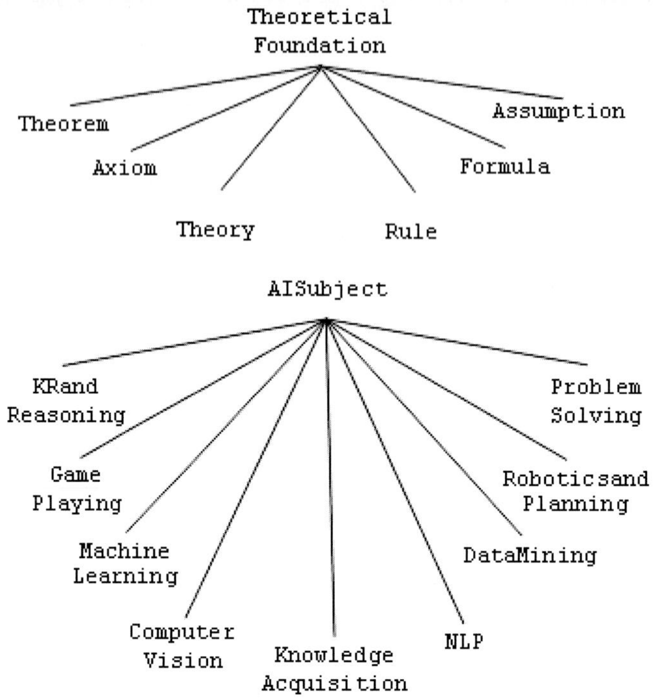

The set of relations and their signatures:

about	[Survey]—1—(about)—2—[T]
solve	[Solution]—1—(solve)—2—[Problem]
of	[SupportingConcept]—1—[of]—2—[Solution]
prove	[Solution]—1—(prove)—2—[Immeasurable]
measure	[Solution]—1—(measure)—2—[Measurable]
use	[Solution]—1—(use)—2—[Solution]
build	[Method]—1—(build)—2—[Solution]
applied_to	[Solution]—1—(applied_to)—2—[Solution]

Adaptive Computation with Perfectly Matched Layers for the Wave Scattering by Periodic Structures

Zhiming Chen[1] and Haijun Wu[2]

[1] Institute of Computational Mathematics, Chinese Academy of Sciences Beijing 100080, China.*
zmchen@lsec.cc.ac.cn
[2] College of Mathematics, Jilin University, Changchun 130012, China
whj@lsec.cc.ac.cn

1 Introduction

A posteriori error estimates are computable quantities in terms of the discrete solution and data that measure the actual discrete errors without the knowledge of exact solutions. They are essential in designing algorithms for mesh modification which equi-distribute the computational effort and optimize the computation. Recent studies (cf. e.g. [11], [7]) indicate that for appropriately designed adaptive finite element procedures, the meshes and the associated numerical complexity are quasi-optimal in the sense that the finite element discretization error is proportional to $N^{-1/2}$ in terms of the energy norm and proportional to N^{-1} in terms of the maximum norm, where N is the number of degrees of freedom of the underlying mesh. Therefore, in order to achieve an optimal solution method for elliptic problems, it is imperative to study efficient algorithms for solving the linear system of equations arising from the adaptive finite element discretization of elliptic problems.

The first objective of this note is to report our recent efforts in proving the uniform convergence of the multigrid V-cycle algorithm with local Gauss-Seidel relaxation performed only on new nodes and their immediate neighbors under the practical condition that the mesh refinements are carried out by the "newest vertex bisection" algorithm developed in [2] and [10]. This refinement algorithm has been widely used in the adaptive finite element community (see e.g. [11], [7]) and has been implemented in the package ALBERT [12].

Our main objective of this note is to consider the prediction of the scattered modes that arise when an electromagnetic wave is incident on some periodic structure which has many important applications in micro-optics. The media

* Partially supported by China NSF under grant 10025102 and by China MOST under grant G1999032802.

are assumed to be nonmagnetic and the magnetic permeability μ is constant everywhere. Then the electromagnetic fields in the whole space are governed by the following time harmonic (time dependence $e^{-\mathrm{i}\omega t}$) Maxwell equations

$$\nabla \times \mathbf{E} - \mathbf{i}\omega\mu\mathbf{H} = 0, \tag{1}$$

$$\nabla \times \mathbf{H} + \mathbf{i}\omega\varepsilon\mathbf{E} = 0. \tag{2}$$

Here $\mathbf{E}$ and $\mathbf{H}$ are the electric and the magnetic field vectors, respectively. The physical structure is described by the dielectric coefficient $\varepsilon(x), x = (x_1, x_2, x_3)$. We restrict ourselves to the two-dimensional setting (1D grating problem), the medium and the grating surface are assumed to be constant in x_2 direction. We assume the dielectric coefficient $\varepsilon(x) = \varepsilon(x_1, x_3)$ is periodic in x_1 direction with period $L > 0$

$$\varepsilon(x_1 + nL, x_3) = \varepsilon(x_1, x_3) \;\; \forall\, x_1, x_3 \in \mathbb{R}, \;\; n \text{ integer} .$$

The dielectric coefficient $\varepsilon(x)$ may be complex. We assume $\Im\varepsilon(x) \geq 0$ and $\Re\varepsilon(x) > 0$ whenever $\Im\varepsilon(x) = 0$. It is natural to assume that ε is constant away from a region $\{(x_1, x_3) : b_2 < x_3 < b_1\}$ which includes the structure, that is, there exist constants ε_1 and ε_2 such that

$$\varepsilon(x_1, x_3) = \varepsilon_1 \;\text{ in }\; \Omega_1 = \{(x_1, x_3) : x_3 \geq b_1\},$$
$$\varepsilon(x_1, x_3) = \varepsilon_2 \;\text{ in }\; \Omega_2 = \{(x_1, x_3) : x_3 \leq b_2\}.$$

In practical applications, we have $\varepsilon_1 > 0$ but ε_2 may be complex according to the substrate material used in Ω_2. Depending on the direction and polarization of the incident plane wave, the Maxwell equations can be simplified by considering the two fundamental polarizations: the transverse electric (TE) polarization and the transverse magnetic (TM) polarization. In the TE case, the electric field $\mathbf{E}$ is parallel to x_2 axis: $\mathbf{E} = (0, u, 0)^T \in \mathbb{R}^3$, where $u = u(x_1, x_3)$ satisfies the Helmholtz equation

$$\Delta u + k^2(x)u = 0 \quad \text{in} \quad \mathbb{R}^2. \tag{3}$$

Here $k^2(x) = \omega^2\varepsilon(x)\mu$ is the magnitude of the wave vector. Similarly, in the TM case, the magnetic field $\mathbf{H}$ is parallel to the x_2 axis: $\mathbf{H} = (0, u, 0)^T \in \mathbb{R}^3$, where $u = u(x_1, x_3)$ satisfies the equation:

$$\operatorname{div}\left(\frac{1}{k^2(x)}\nabla u\right) + u = 0 \;\text{ in }\; \mathbb{R}^2. \tag{4}$$

Our purpose is to develop efficient numerical methods solving the 1D grating problem for both the TE (3) and the TM (4) polarizations. In doing so, the first difficulty is to truncate the domain into an bounded computational domain. The finite element method based on variational formulation on the bounded domain Ω with periodic condition in x_1 direction and the transparent boundary condition on the top and bottom boundaries has the problem that the infinite series in the definition of the quasi-differential operator must be truncated. The second difficulty is the singularity of the solutions. Usually, the grating surface is piecewise smooth and across the surface the dielectric coefficient $\varepsilon(x)$ is discontinuous. Thus the solution of (4) will have singularities which slow down the finite element convergence when using uniform mesh refinements. Even in the TE case (3), when there are lossy materials beneath the grating surface, the transmitted waves decay exponentially. This makes the uniform mesh refinements uneconomical.

Here we explore the possibility of applying the recently introduced perfectly matched layer (PML) technique [4] to deal with the difficulty on truncating the unbounded domain. In practical applications involving PML method, there is a judicial compromise between a thin layer which requires a rapid variation of the artificial material property and a thick layer which requires more grid points and hence more computer time and more storage [9]. We propose to use *a posteriori* error analysis to design efficient adaptive method with error control which adaptively determines the finite element meshes and the PML parameters such as the thickness of the PML region and the medium property inside the region. Moreover, the derived *a posteriori* error estimate has the nice feature of exponentially decay in terms of the distance to the computational domain. This property leads to coarse mesh size away from the computational domain and thus makes the total computational cost insensitive to the thickness of the PML absorbing layer.

2 Uniform Convergence of Multigrid V-cycle on Adaptively Refined Finite Element Meshes

Let Ω be a bounded polygonal domain in R^2 with possibly reentrant corners. Consider the variational problem of finding $u \in H_0^1(\Omega)$ such that

$$A(u,v) = F(v) \quad \forall v \in H_0^1(\Omega), \tag{5}$$

where $F \in H^{-1}(\Omega)$, the dual space of $H_0^1(\Omega)$, and

$$A(u,v) = \int_\Omega [p(x)\nabla u \cdot \nabla v + r(x)uv]\, dx \quad \forall u,v \in H_0^1(\Omega).$$

We assume that $p \in C^1(\overline{\Omega}), r \in C^0(\overline{\Omega}), p(x) > 0$ on $\overline{\Omega}$, and $r(x) \geq 0$ on $\overline{\Omega}$.

Multigrid methods have been established as among the most efficient solvers for discretized elliptic problems. Let $\mathcal{M}_j$, $0 \leq j \leq J$, be a sequence of nested finite element meshes of Ω and $X_j \subset H_0^1(\Omega)$ the piecewise linear finite element space over $\mathcal{M}_j$ with dimension n_j. The distinct feature of applying multigrid methods on adaptively refined finite element meshes is that the number of nodes of the mesh $\mathcal{M}_j$ may not grow exponentially with respect to the number of the mesh refinements j. Thus the number of operations used for the multigrid method in which the relaxation is performed on all nodes can be as bad as $O(n_j^2)$ [10]. To reduce the computational cost, various local relaxation schemes are proposed in applying multigrid methods on adaptively refined finite element meshes ([1], [10], [6]). Numerical experiments in [10] strongly suggest that performing relaxation only at new nodes and their immediate neighboring nodes can guarantee uniform convergence of the multigrid methods for discrete elliptic problems with smooth coefficients. Here and henceforth, the immediate neighboring nodes of new nodes are referred to those of old nodes whose finite element basis functions are changed during the refinement procedure (see (6) below).

A considerable convergence theory for the multigrid methods has been developed to justify their use in the literature (cf. e.g. [5], [15] and the references therein). The common ingredient in the multigrid convergence analysis is the so-called "regularity and approximation" condition which depends crucially on the uniform refinement condition of the underlying finite element meshes. Convergence of multigrid methods with local relaxation is considered in [6], [5], and [15] under special conditions which are usually not satisfied in adaptive finite element mesh refinement strategies.

Our objective of this section is to report our efforts [14] in proving the uniform convergence of the multigrid V-cycle algorithm with local Gauss-Seidel relaxation performed only on new nodes and their immediate neighbors under the practical condition that the mesh refinements are carried out by the "newest vertex bisection" algorithm developed in [2] and [10]. This leads to an algorithm of optimal complexity for solving the linear system of equations resulting from the discretization of (5) by adaptive finite element methods. Here "optimal" means that one step of multigrid iteration can reduce the norm of the error of the approximate solution of the linear system by a factor that is bounded away from 1 independent of N, the size of the linear system, while using only $O(N)$ operations.

Denote by $\mathbb{N}_j$ the collection of interior nodes of $\mathcal{M}_j$. For any node $z \in \mathbb{N}_j$, we use the notation φ_j^z to represent the associated nodal finite element basis function of X_j which takes the value 1 at the node z and the value 0 at all other nodes. Let $\widetilde{\mathcal{N}}_j$ be the set of nodes on which local Gauss-Seidel relaxation are carried out:

$$\widetilde{\mathcal{N}}_j = \left\{ z \in \mathbb{N}_j : z \text{ is a new node or } z \in \mathbb{N}_{j-1} \text{but } \varphi_j^z \neq \varphi_{j-1}^z \right\}. \tag{6}$$

For convenience , we denote by $\widetilde{\mathcal{N}}_j = \left\{ x_j^k, k = 1, \cdots, \tilde{n}_j \right\}$ and $\varphi_j^k = \varphi_j^{x_j^k}$, the nodal finite element basis function corresponding to x_j^k.

For any $0 \leq j \leq J$, we define $A_j : X_j \to X_j$ by

$$(A_j w, v) = A(w, v), \quad \forall w, v \in X_j,$$

where the pairing $(\cdot, \cdot)$ is the inner product in $L^2(\Omega)$. We also define the orthogonal projections $Q_j, P_j : X_J \to X_j$ by

$$(Q_j w, v) = (w, v), \quad A(P_j w, v) = A(w, v), \quad \forall v \in X_j, \ \ \forall w \in X_J.$$

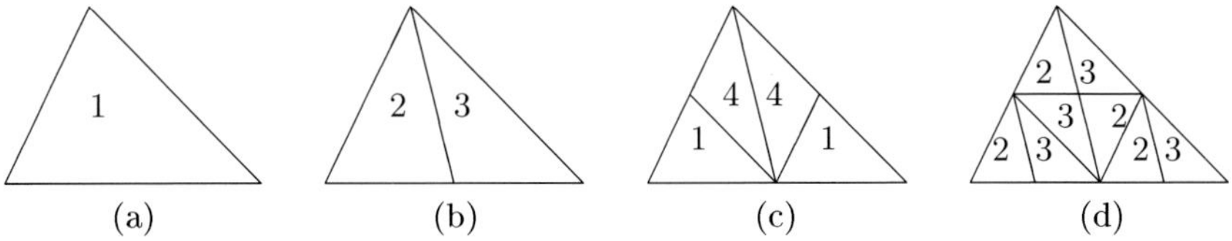

Fig. 1. Four similarity classes of triangles generated by "newest vertex bisection"

The standard V-cycle multigrid algorithm solves the system $A_J u_J = f_J$ by the iterative method

$$u_J^{(m+1)} = u_J^{(m)} + B_J(f_J - A_J u_J^{(m)}).$$

The operators $B_j : X_j \to X_j, 0 \le j \le J$ are recursively defined as follows:

Algorithm 1 (V-cycle) Let $B_0 = A_0^{-1}$. For $j > 0$ and $g \in X_j$, we define $B_j g = v^3$.

1. Pre-smoothing: $v^1 = R_j g$,
2. Correction: $v^2 = v^1 + B_{j-1} Q_{j-1}(g - A_j v^1)$,
3. Post-smoothing: $v^3 = v^2 + R_j^t(g - A_j v^2)$.

Let $P_j^k : X_J \to X_j^k := \text{span}\{\varphi_j^k\}$ defined by

$$A(P_j^k w, \varphi_j^k) = A(w, \varphi_j^k) \quad \forall w \in X_J.$$

The smoother R_j which performs Gauss-Seidel relaxation only at new nodes and their immediate neighbors is given by

$$R_j = \Big(I - \prod_{k=1}^{\tilde{n}_j}(I - P_j^k)\Big) A_j^{-1}.$$

We now briefly recall the "newest vertex bisection" algorithm for the mesh refinements. A detailed description of the algorithm can be found in [2], [10] or [12]. The "newest vertex bisection" algorithm consists of two steps: (1) the marked triangles are bisected by the edge opposite to the newest vertex finite number of times (the newest vertex of an element in the initial mesh is the vertex opposite to the longest edge), (2) all triangles with "hanging nodes" are bisected by the edge opposite to the newest vertex, this process is repeated until there are no hanging nodes. It is easy to check that this algorithm generates a sequence of meshes that all the descendants of an original triangle fall into four similarity classes indicated in Figure 1.

The following theorem is proved in [14].

Theorem 1. *Suppose the meshes $\mathcal{M}_j, 0 \le j \le J$, are obtained by the "newest vertex bisection" algorithm and satisfy the assumption that each element $K \in$*

$\mathcal{M}_j$ is obtained by refining some element $K' \in \mathcal{M}_{j-1}$ finite number of times so that $h_{K'} \le Ch_K$. Then there exists a constant $\delta < 1$ independent of the meshes $\mathcal{M}_j$ and J such that

$$\|I - B_J A_J\|_A < \delta.$$

The proof depends on an identity in [16] for estimating the norm of the product of non-expansive operators and two key ingredients which we now briefly describe.

For any $x_j^k \in \widetilde{\mathcal{N}}_j$, let φ_j^k be the corresponding finite element basis function in X_j, $\mathcal{E}_j^k$ be the collection of all edges of $\mathcal{M}_j$ included in the closure of the support of φ_j^k, and h_j^k be the length of the *shortest* edge of $\mathcal{M}_j$ with one vertex at x_j^k. The first key ingredient in our analysis is the following estimates which characterize the relationship of local mesh sizes

$$\sum_{i=j+1}^{J} \sum_{x_i^l \in \widetilde{\mathcal{N}}_i, x_i^l \in E_j^k} \left(h_i^l / h_j^k\right)^{3/2} \le C \quad \text{for any } x_j^k \in \widetilde{\mathcal{N}}_j,$$

$$\sum_{j=1}^{i-1} \sum_{x_j^k \in \widetilde{\mathcal{N}}_j, x_i^l \in E_j^k} \left(h_i^l / h_j^k\right)^{1/2} \le C \quad \text{for any } x_i^l \in \widetilde{\mathcal{N}}_i.$$

The second ingredient is the following property of the Scott-Zhang interpolation operator $\Pi_j : X_J \to X_j$

$$\sum_{j=1}^{J} \sum_{z \in \widetilde{\mathcal{N}}_j} |(\Pi_j v - \Pi_{j-1} v)(z)|^2 \le CA(v,v) \quad \forall v \in X_J.$$

This estimate is proved based on establishing appropriate connection between the adaptive meshes and uniformly refined meshes which extends similar idea in the analysis in [6], [5] and [15] but contains important differences.

3 The Wave Scattering and PML Technique

Now we turn to the wave scattering problem (3). Let $u_{\rm I} = e^{{\bf i}\alpha x_1 - {\bf i}\beta x_3}$ be the incoming plane wave which is incident upon the grating surface from the top, where $\alpha = k_1 \sin\theta, \beta = k_1 \cos\theta$, and $-\pi/2 < \theta < \pi/2$ is the incident angle. We are interested in quasi-periodic solutions u, that is, solutions u of (3) such that $u_\alpha = ue^{-{\bf i}\alpha x_1}$ are periodic in x_1 with period $L > 0$.

Denote by $\Gamma_j = \{(x_1, x_3) : 0 < x_1 < L, x_3 = b_j\}, j = 1, 2$. We wish to reduce the problem to the bounded domain

$$\Omega = \{(x_1, x_3) : 0 < x_1 < L \text{ and } b_2 < x_3 < b_1\}.$$

The radiation condition for the diffraction problem insists that u is composed of bounded outgoing plane waves in Ω_1 and Ω_2, plus the incident wave u_{I} in Ω_1.

For each integer n, let $\alpha_n = 2\pi n/L$. For any integer $n \in Z$ and $j = 1, 2$, we define

$$\beta_j^n = \beta_j^n(\alpha) = \begin{cases} \left(k_j^2 - (\alpha_n + \alpha)^2\right)^{1/2} & \text{if } k_j^2 \geq (\alpha_n + \alpha)^2 \\ \mathbf{i}\left((\alpha_n + \alpha)^2 - k_j^2\right)^{1/2} & \text{if } k_j^2 < (\alpha_n + \alpha)^2 . \end{cases}$$

We assume that $k_j^2 \neq (\alpha_n + \alpha)^2$ for all $n \in Z, j = 1, 2$. The assumption that only bounded outgoing plane waves except u_{I} exist in Ω_1 implies the following Rayleigh expansion in Ω_1:

$$u = u_{\mathrm{I}} + \sum_{n \in Z} A_1^n e^{\mathbf{i}(\alpha_n + \alpha)x_1 + \mathbf{i}\beta_1^n x_3} \;, \quad x \in \Omega_1 . \tag{7}$$

Similarly, we have the following Rayleigh expansion in Ω_2:

$$u = \sum_{n \in Z} A_2^n e^{\mathbf{i}(\alpha_n + \alpha)x_1 - \mathbf{i}\beta_2^n x_3} \;, \quad x \in \Omega_2 . \tag{8}$$

For any quasi-periodic function f which has the expansion $f = \sum_{n \in Z} f^{(n)} e^{\mathbf{i}(\alpha_n + \alpha)x_1}$, the following Dirichlet to Neumann Operator T_j is introduced in [3]

$$(T_j f)(x_1) = \sum_{n \in Z} \mathbf{i}\beta_j^n f^{(n)} e^{\mathbf{i}(\alpha_n + \alpha)x_1} \;, \quad 0 < x_1 < L \;,\; j = 1, 2 . \tag{9}$$

With this notation in mind, simple calculation shows that the Rayleigh expansion u in $\Omega_j, j = 1, 2$, defined in (7) and (8) satisfies respectively the following relations

$$\frac{\partial(u - u_{\mathrm{I}})}{\partial \nu} - T_1(u - u_{\mathrm{I}}) = 0 \;\text{ on }\; \Gamma_1, \qquad \frac{\partial u}{\partial \nu} - T_2 u = 0 \;\text{ on }\; \Gamma_2, \tag{10}$$

where ν stands for the unit outer normal to $\partial\Omega$. These are the transparent boundary conditions used in [3]. A variational formulation for the 1D grating problem (3) using the boundary conditions (10) can be defined and the existence of a unique solution is proved for all but a sequence of countable frequencies ω_j with $|\omega_j| \to +\infty$. Further uniqueness results can be obtained for any frequency ω if the dielectric coefficient $\varepsilon(x)$ has non-zero imaginary part in some subdomains in Ω. Here we shall not elaborate on this issue and assume in the following that the problem (3), (10) has a unique solution.

Now we turn to the introduction of absorbing PML layers. We surround our computational domain Ω with two PML layers of thickness δ_1 and δ_2 in Ω_1 and Ω_2 respectively. The specially designed model medium in the PML layers should basically be so chosen that the wave either never reaches its external boundary or the amplitude of the reflected wave is so small that it does not essentially contaminate the solution in Ω. Let $s(x_3) = s_1(x_3) + \mathbf{i}s_2(x_3)$ be the model medium property which satisfies

$$s_1, s_2 \in C(\mathbb{R}), s_1 \geq 1, s_2 \geq 0, \text{ and } s(x_3) = 1 \text{ for } b_2 \leq x_3 \leq b_1. \tag{11}$$

Here we remark that in contrast to the original PML condition which takes $s_1 \equiv 1$ in the PML region, we allow a variable s_1 in order to attenuate both the outgoing and evanescent waves there. The advantage of this extension makes our method insensitive to the distance of the PML region to the structure. Following the general idea in designing PML absorbing layers, we introduce the PML region

$$\Omega_1^{\rm PML} = \{(x_1, x_3) : 0 < x_1 < L \text{ and } b_1 < x_3 < b_1 + \delta_1\},$$
$$\Omega_2^{\rm PML} = \{(x_1, x_3) : 0 < x_1 < L \text{ and } b_2 - \delta_2 < x_3 < b_2\},$$

and the PML differential operator

$$\mathcal{L} := \frac{\partial}{\partial x_1}\left(s(x_3)\frac{\partial}{\partial x_1}\right) + \frac{\partial}{\partial x_3}\left(\frac{1}{s(x_3)}\frac{\partial}{\partial x_3}\right) + k^2(x)s(x_3).$$

The PML equations in the PML region are

$$\mathcal{L}(\hat{u} - u_{\rm I}) = 0 \quad \text{in } \Omega_1^{\rm PML}, \tag{12}$$
$$\mathcal{L}\hat{u} = 0 \quad \text{in } \Omega_2^{\rm PML}. \tag{13}$$

The equation satisfied by the PML solution $\hat{u}$ in the domain Ω is the original Helmholtz equation $\Delta\hat{u} + k^2(x)\hat{u} = 0$. Let $D = \{(x_1, x_3) : 0 < x_1 < L, b_2 - \delta_2 < x_3 < b_1 + \delta_1\}$. Due to the assumption (11) we can now formulate the PML model which we are going to solve in this paper

$$\mathcal{L}\hat{u} = -g \text{ in } D \tag{14}$$

with the quasi-periodic boundary condition $\hat{u}(0, x_3) = e^{-\mathbf{i}\alpha L}\hat{u}(L, x_3)$ for $b_2 - \delta_2 < x_3 < b_1 + \delta_1$ and the Dirichlet condition $\hat{u} = u_{\rm I}$ on $\Gamma_1^{\rm PML} = \{(x_1, x_3) : 0 < x_1 < L, x_3 = b_1 + \delta_1\}$, $\hat{u} = 0$ on $\Gamma_2^{\rm PML} = \{(x_1, x_3) : 0 < x_1 < L, x_3 = b_2 - \delta_2\}$. Here the source function

$$g = \begin{cases} -\mathcal{L}u_{\rm I} & \text{in } \Omega_1^{\rm PML}, \\ 0 & \text{elsewhere}. \end{cases}$$

To prove the existence and uniqueness of the above problem and derive an error estimate between $\hat{u}$ and u, we first find an equivalent formulation of the PML problem in the domain Ω. Similar to the argument leading to the Rayleigh expansion, we deduce from (12) that

$$\hat{u} = u_{\rm I} + \sum_{n \in Z}\left(A_1^n e^{\mathbf{i}\beta_1^n \int_{b_1}^{x_3} s(\tau)d\tau} + B_1^n e^{-\mathbf{i}\beta_1^n \int_{b_1}^{x_3} s(\tau)d\tau}\right) e^{\mathbf{i}(\alpha_n + \alpha)x_1} \text{ in } \Omega_1^{\rm PML}.$$

If we write $\hat{u}(x_1, b_1) = u_{\rm I}(x_1, b_1) + \sum_{n\in Z} \hat{u}_\alpha^{(n)}(b_1)e^{\mathbf{i}(\alpha_n+\alpha)x_1}$ on Γ_1, then the constants A_1^n, B_1^n can be uniquely determined by the additional boundary condition $\hat{u} = u_{\rm I}$ on $\Gamma_1^{\rm PML}$ through following equations

$$A_1^n + B_1^n = \hat{u}_\alpha^n(b_1)$$
$$A_1^n e^{\mathbf{i}\beta_1^n \int_{b_1}^{b_1+\delta_1} s(\tau)d\tau} + B_1^n e^{-\mathbf{i}\beta_1^n \int_{b_1}^{b_1+\delta_1} s(\tau)d\tau} = 0.$$

Thus we conclude that

$$\hat{u} = u_{\rm I} + \sum_{n\in Z} \frac{\zeta_1^n(x_3)}{\zeta_1^n(b_1)} \hat{u}_\alpha^{(n)}(b_1)e^{\mathbf{i}(\alpha_n+\alpha)x_1} \quad \text{in} \quad \Omega_1^{\rm PML}, \tag{15}$$

where $\zeta_1^n(x_3) = e^{-\mathbf{i}\beta_1^n \int_{x_3}^{b_1+\delta_1} s(\tau)d\tau} - e^{\mathbf{i}\beta_1^n \int_{x_3}^{b_1+\delta_1} s(\tau)d\tau}$. Similarly, we deduce from (13) that

$$\hat{u} = \sum_{n\in Z} \frac{\zeta_2^n(x_3)}{\zeta_2^n(b_2)} \hat{u}_\alpha^{(n)}(b_2)e^{\mathbf{i}(\alpha_n+\alpha)x_1} \quad \text{in} \quad \Omega_2^{\rm PML}, \tag{16}$$

where $\zeta_2^n(x_3) = e^{-\mathbf{i}\beta_2^n \int_{b_2-\delta_2}^{x_3} s(\tau)d\tau} - e^{\mathbf{i}\beta_2^n \int_{b_2-\delta_2}^{x_3} s(\tau)d\tau}$. Similar to (9), for any quasi-periodic function f which has the expansion $f = \sum_{n\in Z} f^{(n)} e^{\mathbf{i}(\alpha_n+\alpha)x_1}$, we define the following Dirichlet to Neumann operator $T_j^{\rm PML}$:

$$\left(T_j^{\rm PML} f\right)(x_1) = \sum_{n\in Z} \mathbf{i}\beta_j^n \coth(-\mathbf{i}\beta_j^n \sigma_j) f^{(n)} e^{\mathbf{i}(\alpha_n+\alpha)x_1}, \tag{17}$$

where $\coth(\tau) = \frac{e^\tau + e^{-\tau}}{e^\tau - e^{-\tau}}$, and

$$\sigma_1 = \int_{b_1}^{b_1+\delta_1} s(\tau)d\tau, \quad \sigma_2 = \int_{b_2-\delta_2}^{b_2} s(\tau)d\tau. \tag{18}$$

Then we know easily from (15), (16) that

$$\frac{\partial(\hat{u} - u_{\rm I})}{\partial\nu} - T_1^{\rm PML}(\hat{u} - u_{\rm I}) = 0 \ \text{ on } \ \Gamma_1, \quad \frac{\partial \hat{u}}{\partial\nu} - T_2^{\rm PML}\hat{u} = 0 \ \text{ on } \ \Gamma_2. \tag{19}$$

Let $\Delta_j^n = |k_j^2 - (\alpha_n + \alpha)^2|^{1/2}$ and $U_j = \{n : k_j^2 > (\alpha_n + \alpha)^2\}$, $j = 1, 2$. Then we have $\beta_j^n = \Delta_j^n$ for $n \in U_j$ and $\beta_j^n = \mathbf{i}\Delta_j^n$ for $n \notin U_j$. Let

$$\Delta_j^- = \min\{\Delta_j^n : n \in U_j\}, \quad \Delta_j^+ = \min\{\Delta_j^n : n \notin U_j\}.$$

The following lemma whose proof can be found in [8] plays the key role in the analysis.

Lemma 1. *For any φ, ψ such that $\varphi_\alpha = \varphi e^{-\mathbf{i}\alpha x_1}, \psi_\alpha = \psi e^{-\mathbf{i}\alpha x_1}$ are periodic in x_1 with period L, we have*

$$\left| \int_{\Gamma_j} (T_j \varphi - T_j^{\mathrm{PML}} \varphi) \bar{\psi} dx_1 \right| \le M_j \mathbf{x} \varphi L^2(\Gamma_j) \mathbf{x} \psi L^2(\Gamma_j),$$

where $M_j = \max \left(\frac{2\Delta_j^-}{e^{2\sigma_j^I \Delta_j^-} - 1}, \frac{2\Delta_j^+}{e^{2\sigma_j^R \Delta_j^+} - 1} \right)$, σ_j^R, σ_j^I *are the real and imaginary part of* σ_j *defined in* (18), *that is,* $\sigma_j = \sigma_j^R + \mathbf{i}\sigma_j^I$.

We note that the constant M_j approaches to zero exponentially as the PML parameters σ_j^R, σ_j^I tend to infinity. From the definition (18) we know that σ_j^R, σ_j^I can be calculated by the medium property $s(x_3)$, which is usually taken as power function

$$s(x_3) = \begin{cases} 1 + \sigma_1^m \left(\frac{x_3 - b_1}{\delta_1} \right)^m & \text{if } x_3 \ge b_1 \\ 1 + \sigma_2^m \left(\frac{b_2 - x_3}{\delta_2} \right)^m & \text{if } x_3 \le b_2 \end{cases}, \quad m \ge 1.$$

Thus we have

$$\sigma_j^R = \left(1 + \frac{\Re \sigma_j^m}{m+1} \right) \delta_j, \quad \sigma_j^I = \frac{\Im \sigma_j^m}{m+1} \delta_j. \tag{20}$$

It is obvious that either enlarge the thichness δ_j of the PML layers or enlarge the medium parameters $\Re \sigma_j^m$ and $\Im \sigma_j^m$ will reduce the PML approximation error.

The following theorem which gives an error estimate for the PML approximation can be proved easily from Lemma 1.

Theorem 2. *For sufficiently small constants* M_1, M_2, *the PML problem has a unique solution* $\hat{u}$. *Moreover, we have the following error estimate*

$$\begin{aligned} |||u - \hat{u}|||_\Omega := & \sup_{0 \ne \psi \in H^1(\Omega)} \frac{|b(u - \hat{u}, \psi)|}{\mathbf{x} \psi H^1(\Omega)} \\ & \le \hat{C} M_1 \| \hat{u} - u_{\mathrm{I}} \|_{L^2(\Gamma_1)} + \hat{C} M_2 \| \hat{u} \|_{L^2(\Gamma_2)} \end{aligned}$$

with $\hat{C} = \sqrt{1 + (b_2 - b_1)^{-1}}$.

4 Sharp *a posteriori* Error Estimates

Define

$$X(D) = \{ w \in H^1(D) : w_\alpha = w e^{-\mathrm{i}\alpha x_1} \text{is periodic in } x_1 \text{ with period } L \}$$

and introduce the sesquilinear form $a_D : X(D) \times X(D) \to \mathbf{C}$ as

$$a_D(\varphi, \psi) = \int_G \left(s(x_3) \frac{\partial \varphi}{\partial x_1} \frac{\partial \bar{\psi}}{\partial x_1} + \frac{1}{s(x_3)} \frac{\partial \varphi}{\partial x_3} \frac{\partial \bar{\psi}}{\partial x_3} - k^2(x) s(x_3) \varphi \bar{\psi} \right) dx.$$

Denote by $X_0(D) = \{w \in X(D), w = 0 \text{ on } \Gamma_1^{\rm PML} \cup \Gamma_2^{\rm PML}\}$. Then the weak formulation of the PML model (14) reads as follows: Find $\hat{u} \in X(D)$ such that $\hat{u} = u_{\rm I}$ on $\Gamma_1^{\rm PML}$, $\hat{u} = 0$ on $\Gamma_2^{\rm PML}$ and

$$a_D(\hat{u}, \psi) = \int_D g\bar{\psi}dx \qquad \forall\, \psi \in X_0(D). \tag{21}$$

Let $\mathcal{M}_h$ be a regular triangulation of the domain D. Remember that any triangle $T \in \mathcal{M}_h$ is considered as closed. We assume any element T must be completely included in $\overline{\Omega_1^{\rm PML}}$, $\overline{\Omega_2^{\rm PML}}$ or $\overline{\Omega}$. To define finite element space whose functions are quasi-periodic in x_1 direction, we also require that if $(0, z)$ is a node on the left boundary, then (L, z) is also a node on the right boundary, and vice versa. Let $V_h(D) \subset X(D)$ be the conforming linear finite element space and $V_h^0(D) = V_h(D) \bigcap X_0(D)$. Denote by $I_h : C(\bar{D}) \to V_h(D)$ the standard finite element interpolation operator.

The finite element approximation to the PML problem (21) reads as following: Find $\hat{u}_h \in V_h(D)$ such that $\hat{u}_h = I_h u_{\rm I}$ on $\Gamma_1^{\rm PML}$, $\hat{u}_h = 0$ on $\Gamma_2^{\rm PML}$ and

$$a_D(\hat{u}_h, \psi_h) = \int_D g\bar{\psi}_h dx \qquad \forall\, \psi_h \in V_h^0(D). \tag{22}$$

Let

$$A(x) = \begin{pmatrix} A_{11} & 0 \\ 0 & A_{22} \end{pmatrix} = \begin{pmatrix} s(x_3) & 0 \\ 0 & 1/s(x_3) \end{pmatrix}, \quad B(x) = k^2(x)s(x_3).$$

Then the definitions of $\mathcal{L}$ and a_D can be rewritten as

$$\mathcal{L} = \text{div}\,(A(x)\nabla) + B(x),$$
$$a_D(\varphi, \psi) = \textstyle\int_D \left(A(x)\nabla\varphi\nabla\bar{\psi} - B(x)\varphi\bar{\psi}\right) dx.$$

For any $T \in \mathcal{M}_h$, we denote by h_T its diameter. Let $\mathcal{B}_h$ denote the set of all sides that do not lie on $\Gamma_j^{\rm PML}$, $j = 1, 2$. For any $e \in \mathcal{B}_h$, h_e stands for its length. For any $T \in \mathcal{M}_h$, we introduce the residual

$$R_T := \mathcal{L}\hat{u}_h|_T + g|_T = \begin{cases} \mathcal{L}(\hat{u}_h|_T - u_{\rm I}|_T) \text{ if } T \subset \overline{\Omega_1^{\rm PML}}, \\ \mathcal{L}\hat{u}_h|_T \text{ otherwise}. \end{cases}$$

For any interior side $e \in \mathcal{B}_h$ which is the common side of T_1 and $T_2 \in \mathcal{M}_h$, we define the jump residual across e

$$J_e = (A\nabla\hat{u}_h|_{T_1} - A\nabla\hat{u}_h|_{T_2}) \cdot \nu_e,$$

using the convention that the unit normal vector ν_e to e points from T_2 to T_1. Denote by $\Gamma_{\rm left} = \{(x_1, x_3) : x_1 = 0, b_2 - \delta_2 < x_3 < b_1 + \delta_1\}$ and $\Gamma_{\rm right} = \{(x_1, x_3) : x_1 = L, b_2 - \delta_2 < x_3 < b_1 + \delta_1\}$. If $e = \Gamma_{\rm left} \cap \partial T$ for some element

$T \in \mathcal{M}_h$ and e' the corresponding side on Γ_{right} which is also a side of some element T', then we define the jump residual

$$J_e = A_{11}\left[\frac{\partial}{\partial x_1}(\hat{u}_h|_T) - e^{-\mathrm{i}\alpha L}\cdot\frac{\partial}{\partial x_1}(\hat{u}_h|_{T'})\right],$$
$$J_{e'} = A_{11}\left[e^{\mathrm{i}\alpha L}\cdot\frac{\partial}{\partial x_1}(\hat{u}_h|_T) - \frac{\partial}{\partial x_1}(\hat{u}_h|_{T'})\right].$$

For any $T \in \mathcal{M}_h$, denote by η_T the local error estimator which is defined as follows

$$\eta_T = \max_{x\in\tilde{T}} \rho(x_3)\cdot\left[h_T\|R_T\|_{L^2(T)} + \left(\frac{1}{2}\sum_{e\subset T} h_e\| J_e \|^2_{L^2(e)}\right)^{1/2}\right],$$

where $\tilde{T}$ is the union of all elements having non-empety intersection with T, and

$$\rho(x_3) = \begin{cases} |s(x_3)|e^{-R_j(x_3)} \text{ if } x \in \overline{\Omega_j^{\text{PML}}}, \\ 1 \text{ if } x \in \Omega. \end{cases}$$

with $R_j(x_3)$ $(j = 1,2)$ being defined by

$$R_1(x_3) = \min\left(\Delta_1^-\int_{b_1}^{x_3} s_2(\tau)d\tau, \Delta_1^+\int_{b_1}^{x_3} s_1(\tau)d\tau\right), \quad x_3 \geq b_1,$$

$$R_2(x_3) = \min\left(\Delta_2^-\int_{x_3}^{b_2} s_2(\tau)d\tau, \Delta_2^+\int_{x_3}^{b_2} s_1(\tau)d\tau\right), \quad x_3 \leq b_2.$$

Theorem 3. *There exists a constant $C > 0$ depending only on the minimum angle of the mesh $\mathcal{M}_h$, such that the following a posteriori error estimate is valid*

$$|||u - \hat{u}_h|||_\Omega \leq \hat{C}M_1\|\hat{u}_h - u_{\mathrm{I}}\|_{L^2(\Gamma_1)} + \hat{C}M_2\|\hat{u}_h\|_{L^2(\Gamma_2)}$$
$$+\hat{C}M_3\| I_h u_{\mathrm{I}} - u_{\mathrm{I}} \|_{L^2(\Gamma_1^{\text{PML}})} + C\left(\sum_{T\in\mathcal{M}_h}\eta_T^2\right)^{1/2},$$

where the constants $\hat{C}$ and $M_j (j = 1,2)$ are defined in Theorem 2 and Lemma 1 respectively, and

$$M_3 = \max\left(\frac{2\Delta_1^- e^{-\Delta_1^-\sigma_1^I}}{1 - e^{-2\Delta_1^-\sigma_1^I}}, \frac{2\Delta_1^+ e^{-\Delta_1^+\sigma_1^R}}{1 - e^{-2\Delta_1^+\sigma_1^R}}\right).$$

The proof of this theorem can be found in [8]. We notice that when the PML parameters σ_j^R and σ_j^I tend to infinity, the constants M_j decay exponentially. The important exponentially decay factors $e^{-R_j(x_3)}$ in the PML region Ω_j^{PML} allow us to take thicker PML layers without introducing unnecessary fine meshes away from the computational domain. Recall that thicker PML layers allow smaller PML medium property, which enhances numerical stability.

5 Numerical Results

The implementation of the adaptive algorithm in this section is based on the PDE toolbox of MATLAB. We use the *a posteriori* error analogue to that in Theorem 3 in the TM case to determine the PML parameters. According to the discussion in Section 3, we choose the PML medium property as the power function and thus we need only to specify the thickness δ_j of the layers and the medium parameters σ_j^m (see (20)). Recall from Theorem 3 that the *a posteriori* error estimate consists of two parts: the PML error $E_{\rm PML}$ and the finite element discretization error $E_{\rm FEM}$, where

$$E_{\rm PML} = M_1 \| \hat{u}_h - u_{\rm I} \|_{L^2(\Gamma_1)} + M_2 \| \hat{u}_h \|_{L^2(\Gamma_2)}, \tag{23}$$

$$E_{\rm FEM} = M_3 \| \hat{u}_h - u_{\rm I} \|_{L^2(\Gamma_1^{\rm PML})} + \Big(\sum_{T \in \mathcal{M}_h} \eta_T^2 \Big)^{1/2}. \tag{24}$$

$E_{\rm PML}$ and $E_{\rm FEM}$ should be changed accordingly in the TM case. In our implementation we first choose δ_j and σ_j^m such that $M_j L^{1/2} \le 10^{-8}$, which makes the PML error negligible compared with the finite element discretization errors. Once the PML region and the medium property are fixed, we use the standard finite element adaptive strategy to modify the mesh according to the *a posteriori* error estimate (24). For any $T \in \mathcal{M}_h$, we define the local *a posteriori* error estimator as follows:

$$\tilde{\eta}_T = \eta_T + M_3 \| I_h u_{\rm I} - u_{\rm I} \|_{L^2(\Gamma_1^{\rm PML} \cap \partial T)}.$$

Now we describe the adaptive algorithm we used in the paper.

Algorithm 2 Given tolerance $\texttt{TOL} > 0$. Let $m = 2, \delta_1 = \delta_2 = \delta$.

- Choose δ and σ_j^m such that $M_j L^{1/2} \le 10^{-8}$ for $j = 1, 2$;
- Set the computational domain $D = \Omega_2^{\rm PML} \cup \Gamma_2 \cup \Omega \cup \Gamma_1 \cup \Omega_1^{\rm PML}$ and generate an initial mesh $\mathcal{M}_h$ over D;
- While $E_{\rm FEM} > \texttt{TOL}$ do
  - refine the mesh $\mathcal{M}_h$ according to the following strategy

    if $\tilde{\eta}_T > \frac{1}{2} \max_{T \in \mathcal{M}_h} \tilde{\eta}_T$, refine the element $T \in \mathcal{M}_h$

  - solve the discrete problem (22) on $\mathcal{M}_h$
  - compute error estimators on $\mathcal{M}_h$

  end while

Now we report a numerical example to illustrate the performance of our adaptive algorithm. We only document the value δ of the thickness of the PML layers. The medium parameters σ_j^m are determined accordingly through the relation $M_j L^{1/2} \le 10^{-8}$ for $j = 1, 2$. We normalize the space variables

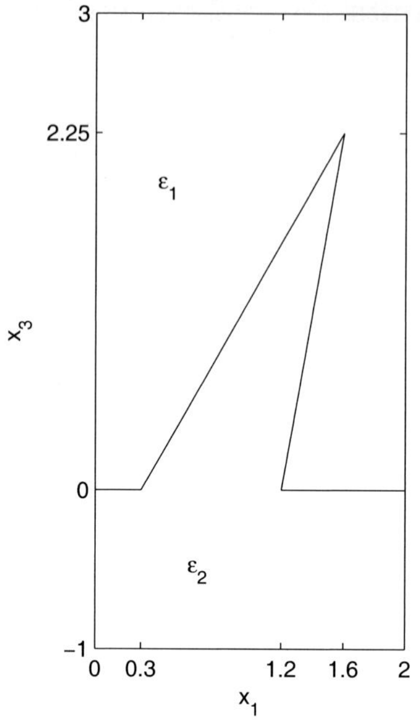

Fig. 2. Geometry of the domain.

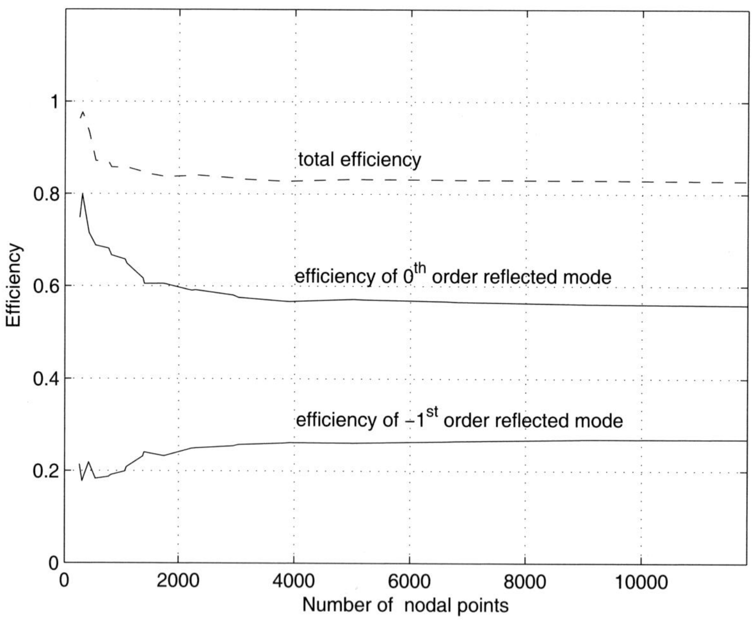

Fig. 3. Grating efficiency.

so that $\mu = 1$. We also scale the error estimator by a factor 0.15 as in PDE toolbox of MATLAB.

The example concerns the TM polarization on a grating surface with a sharp angle indicated in Figure 2. We choose $\varepsilon_1 = 1, \varepsilon_2 = (0.22 + 6.71\mathbf{i})^2$, $\theta = \pi/6$, $\omega = \pi$ and $L = 2$. There are two reflected out-going waves. The grating efficiency of the reflected waves as well as the total grating efficiency are displayed in Figure 3. Figure 4 shows the mesh and the amplitude of the associated solution after 17 adaptive iterations when the grating efficiency is stabilized. The mesh has 3905 nodes and the *a posteriori* error estimate over which is 0.7475. The initial *a posteriori* error estimate is 3.4165. This example shows clearly the ability of the proposed method to capture the singularities of the problem. The meshes near the upper PML boundary is rather coarse, as a result of the exponential decay factor in our *a posteriori* error estimator.

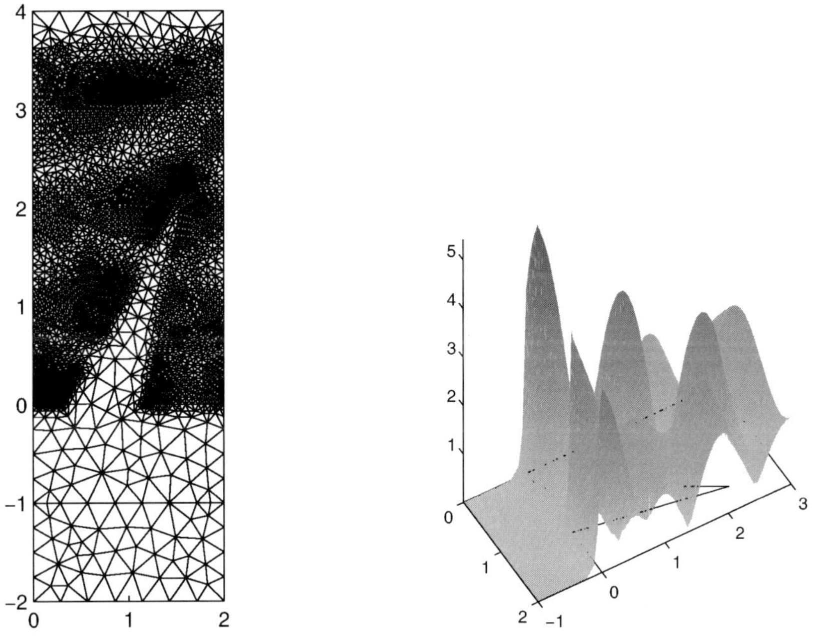

Fig. 4. The mesh and the surface plot of the amplitude of the associated solution after 17 adaptive iterations. The mesh has 3905 nodes.

References

[1] Bai, D., Brandt, A.: Local mesh refinement multilevel techniques. SIAM J. Sci. Stat. Comput., **8** , 109–134 (1987)

[2] Bänsch, E.: Local mesh refinement in 2 and 3 dimensions. Impact of Computing in Science and Engineering, **3**, 181–191 (1991)

[3] Bao, G., Dobson, D.C., Cox, J.A.: Mathematical studies in rigorous grating theory. J. Opt. Soc. Am. A, **12**, 1029–1042 (1995)

[4] Berenger, J.P.: A perfectly matched layer for the absorption of electromagnetic waves. J. Comput. Physics, **114**, 185–200 (1994)

[5] Bramble, J.H.: Multigrid Methods. Pitman Research Notes in Mathematical Sciences, **294**, Longman, Essex, (1993)

[6] Bramble, J.H., Pasicak, J.E.: New estimates for multigrid algorithms including the V-cycle. Math. Comp., **60**, 447–471 (1993)

[7] Chen, Z., Dai, S.: On the efficiency of adaptive finite element methods for elliptic problems with discontinuous coefficients. SIAM J. Sci. Comput., **24**, 443–462 (2002)

[8] Chen, Z., Wu, H.: An adaptive finite element method with perfectly matched absorbing layers for the wave scattering by periodic structures. SIAM J. Numer. Anal. (to appear)

[9] Collino, F., Monk, P.B.: Optimizing the perfectly matched layer. Comput. Methods Appl. Mech. Engrg., **164**, 157–171 (1998)

[10] Mitchell, W.F.: Optimal multilevel iterative methods for adaptive grids. SIAM J. Sci. Stat. Comput., **13**, 146–167 (1992)

[11] Morin, P., Nochetto, R.H., Siebert, K.G.: Data oscillation and convergence of adaptive FEM. SIAM J. Numer. Anal., **38**, 466–488 (2000)

[12] Schmidt, A. Siebert, K.G.: ALBERT: An adaptive hierarchical finite element toolbox. IAM, University of Freiburg (2000) `http://www.mathematik.uni-freiburg.de/IAM/Research/projectsdz/albert.`

[13] Scott, L.R., Zhang, S.: Finite element interpolation of nonsmooth functions satisfying boundary conditions. Math. Comp., **54**, 483–493 (1990)

[14] Wu, H., Chen, Z.: Uniform convergence of multigrid V-cycle on adaptively refined finite element meshes for second order elliptic problems. *submitted.*

[15] Xu, J.: An introduction to multigrid convergence theory. Lecture Notes for Winter School on Iterative Methods in Scientific Computing and their Applications, December 14–20 (1995)

[16] Xu, J., Zikatanov, L.: The method of alternating projections and the method of subspace corrections in Hilbert space. J. Amer. Math. Soc., **15**, 573–397 (2002)

Simulation and Optimization of Crawling Robots

Felix L. Chernousko

Institute for Problems in Mechanics of the Russian Academy of Sciences
pr. Vernadskogo 101-1, Moscow 119526, Russia
chern@ipmnet.ru

Summary. Crawling and climbing robots are complex mechatronic systems with many degrees of freedom which can move along various surfaces. Several types of these mobile robots are considered, namely: wall-climbing robots; tube-crawling robots; snake-like multilink mechanisms. The overall design of the robots and their specific features are described. The wall-climbing robots equipped with pneumatic suckers can move along vertical surfaces and perform different operations such as cleaning, inspection, painting, cutting, welding, etc. The tube-crawling robots move inside tubes and can be used for monitoring and repair of pipelines. The multilink mechanisms can perform snake-like locomotions along a horizontal plane. The motions of the robots occur under the influence of dry friction and control torques created by the actuators installed at the joints of the robots. Regular gaits of the robots are proposed, and their kinematics and dynamics are analyzed. Results of the computer simulation as well as experimental data are discussed. Optimization of the design and gaits of the robots is carried out. As a result, optimal geometrical and mechanical parameters are found which correspond to the maximal speed of the robot motion.

Key words: Robotics, Wall-climbing robots, Tube-crawling robots, Snake-like locomotions, Simulation, Dynamics, Multibody systems, Optimization.

1 Introduction

During the last decades, various types of mobile robots have been developed for different purposes arising in industry, construction, medicine, for operations in hazardous environments, for anti-terrorist and military purposes, etc. Mobile robots are based on different concepts of mechanical motion: they can use wheels, legs, tracks, suckers, etc.

In this paper, we restrict ourselves with several types of mobile robots which were developed and/or investigated in the Institute for Problems in Mechanics of the Russian Academy of Sciences (IPMech).

First, we consider wall-climbing robots equipped with pneumatic suckers. Several kinds of these robots were developed in the IPMech since 1984 [1–4].

Second, we analyze statics, kinematics, and dynamics of the multi-legged tube-crawling robot developed by F.Pfeiffer and his colleagues at the Technical University of Munich [5].

Then we consider snake-like mechanisms which can move along a horizontal surface [6–10].

For all these robotic systems, we briefly describe their structures and principles of motion. Regular gaits of the robots are proposed and analyzed. The dry friction forces play an important role for all types of the crawling robots considered in the paper. Computer simulation of the robot motions and their optimization are carried out. Optimal geometrical and mechanical parameters of the robots as well as optimal parameters of their gaits are obtained.

2 Wall-Climbing Robots

In many applications it is necessary for a robot to move along surfaces with a great slope, along vertical planes (walls) and horizontal planes from below (ceilings). Such climbing robots can be used for a number of operations (cleaning, painting, inspection, cutting, welding, etc.) in construction, maintenance, and repair of buildings and structures. These robots can be also useful in the shipbuilding industry, for fire fighting and other applications.

Climbing robots have special grippers (feet) that fix the robot to the surface along which it moves. These grippers may be magnetic or pneumatic [1–3, 11–13]. For microrobots, nano-tubes can be used. Magnetic grippers can be used only for surfaces made of ferromagnetic material. Vacuum, or pneumatic, grippers are more universal and can serve for any relatively smooth surface of metal, tiles, ceramics, concrete, plastic, wood, for painted surfaces, etc. The main requirement that the climbing robot must satisfy is a reliable contact with the surface: at each instant of time, the robot must be firmly attached to the surface.

Several versions of wall-climbing robots (WCR) with vacuum grippers have been developed in the IPMech since 1984 [1–4]. The general principles of the design of WCR were elaborated. The WCR consists of transport, technological, and control modules (Fig. 1).

The transport module includes two platforms (1 and 2) that can move with respect to each other. On the platform, ejector-type vacuum grippers (3) are mounted which ensure reliable contact with the wall.

The motion of the robot occurs as follows. At any instant, one of the platforms, say 1, is attached to the wall by means of its grippers, while the grippers of platform 2 do not touch the wall. Platform 2 is moved with respect to platform 1 by means of pneumatic drives. Platform 2 stops, its grippers approach the wall and are sucked onto it due to vacuum created between the grippers and the wall. After that, the vacuum in the grippers of platform 1 is removed, these grippers are disconnected from the wall and moved closer to the platform. Now platform 1 moves with respect to platform 2 which is

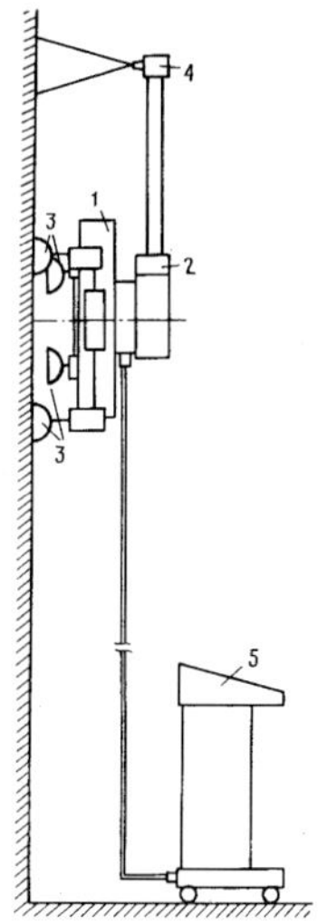

Fig. 1. Wall-climbing robot

fixed to the wall, and the process is repeated. A special mechanism equipped with an electric drive makes it possible to change the direction of motion by rotating one of the platforms about the axis perpendicular to the wall.

The compressed air for pneumatic drives and grippers is supplied to the WCR through the air hose from a compressor usually located on the ground. The compressor can be replaced by a cylinder with compressed air mounted on the robot.

To enhance the reliability of contact of the WCR with the wall, special ejector-type vacuum grippers were designed. The gripper consists of the central metallic part, with small pins on the contact surface, and a special cover made of a soft elastic material. The edge of the cover closely fits the wall and does not permit the atmosphere air to flow inside the sucker. The number of grippers varies from 8 to 24 for different types of WCR.

The technological module 4 (Fig. 1) depends on the task and include manipulators, sensors, and special equipment. The wall-climbing robots can carry out such technological operations as cleaning by means of special brushes and vacuum cleaners, painting, cutting, welding, inspection, they can carry TV cameras and different sensors. The robot equipped with a module for cutting is shown in Fig. 2. This robot can carry up to 120 kg and was successfully tested in fire-fighting conditions and at low winter temperatures.

The control module 5 (Fig. 1) is placed on the ground and is connected with the robot by a cable. The cable can be replaced by a wireless connection. The control unit receives information from various sensors installed on the robot. This information indicates the positions of different parts of the WCR, pressures, forces, etc. The signals from the technological equipment are also

Fig. 2. Robot for cutting

sent to the control module. This module includes a controller and a personal computer; an operator can supervise the control process. The control module is supplied with algorithms and programs that process the data obtained from the sensor and implement the feedback control both of the transport and technological modules.

Fig. 3. Robot system for cleaning internal walls

The WCR can be a part of more complex systems. For example, the robotic system for cleaning the internal walls of nuclear power plants was developed at the IPMech. This system shown in Fig. 3 includes a mobile wheeled robot for horizontal motions and a wall-climbing robot.

The reliability of contact of the WCR with the wall is the most important. To analyze this requirement, let us consider the equilibrium of the robot attached to the plane by means of n vacuum grippers (Fig. 4). Let the reference frame $Oxyz$ be connected to the plane, and the robot is situated in the half-space $z \geq 0$. Denote by $\mathbf{r}_i(x_i,\, y_i,\, 0)$ the positions of the centers of the grippers, by $\mathbf{F}_i(X_i,\, Y_i,\, N_i)$ the reaction force acting upon the ith gripper, and

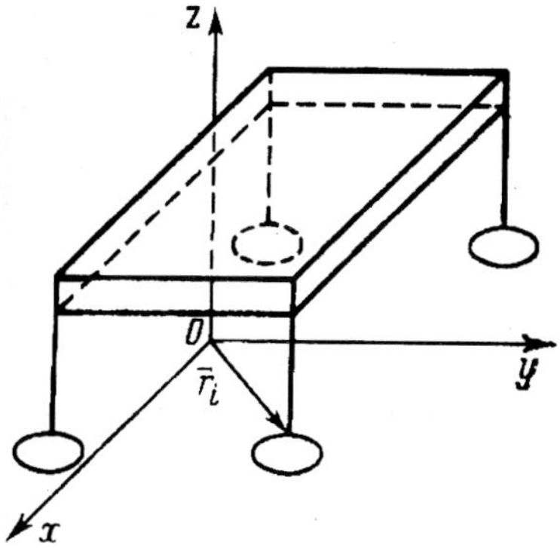

Fig. 4. Robot attached to the plane

by $\Phi_i = (p - p_i)S$ the vacuum force acting upon the ith gripper, $i = 1, \ldots, n$. Here, p is the atmospheric pressure, p_i is the pressure under the ith gripper, and S is the area of the gripper.

Denote by $\mathbf{R}(R_x, R_y, R_z)$ the net external active force acting upon the robot (e.g. the weight, tension of cables, etc.), and by $\mathbf{M}(M_x, M_y, M_z)$ the net torque of external active forces.

The friction forces must obey Coulomb's law

$$(X_i^2 + Y_i^2)^{1/2} \le f N_i, \quad i = 1, \ldots, n, \tag{1}$$

where f is the coefficient of friction, and $N_i \ge 0$ is the normal reaction at the ith gripper.

Six equations of equilibrium of the robot can be divided into two groups

$$\begin{gathered} R_z + \sum (N_i - \Phi_i) = 0, \quad M_x + \sum y_i (N_i - \Phi_i) = 0, \\ M_y - \sum x_i (N_i - \Phi_i) = 0, \end{gathered} \tag{2}$$

$$R_x + \sum X_i = 0, \quad R_y + \sum Y_i = 0, \quad M_z + \sum (x_i Y_i - y_i X_i) = 0. \tag{3}$$

Here, all sums are taken for all $i = 1, \ldots, n$.

The robot will be in equilibrium, if X_i, Y_i, and N_i satisfying equations (2), (3) and conditions (1) exist for all $i = 1, \ldots, n$. The first group of equations (2) can be regarded as conditions of non-separation: if they are satisfied with $N_i \ge 0$, then the robot will not lose contact with the plane $z = 0$. The second group of equations (3) are non-slipping conditions. If there exist X_i and Y_i satisfying (3) and inequalities (1) for $i = 1, \ldots, n$, then the robot will not slide along the plane $z = 0$.

Let us define K as a point in the Oxy plane with the coordinates

$$x_* = \frac{\sum x_i \Phi_i + M_y}{\sum \Phi_i - R_z}, \quad y_* = \frac{\sum y_i \Phi_i - M_x}{\sum \Phi_i - R_z}. \tag{4}$$

It can be shown [1, 14] that the conditions of non-separation (2) are satisfied, if and only if

$$\sum \Phi_i > R_z, \quad K \in \mathrm{co}D, \quad D = \{\mathbf{r}_1, \dots, \mathbf{r}_n\}, \tag{5}$$

where co$D$ is a convex hull of all grippers in the plane Oxy.

The conditions of non-slipping (3) are satisfied [1, 14, 15], if only if

$$\sup_{x,y} Q \leq f, \quad Q = \frac{M_z - xR_y + yR_x}{\sum \rho_i N_i},$$
$$\rho_i = \left[(x - x_i)^2 + (y - y_i)^2\right]^{1/2}, \quad i = 1, \dots, n. \tag{6}$$

Here, sup is taken over all x, y. Thus, the robot will be in equilibrium, if and only if all conditions (5) and (6) for some $N_i > 0$ are satisfied.

Note that the normal reactions N_i cannot be determined uniquely from (2), if $n > 3$. The robot is a statically indeterminate structure, and some additional analysis is necessary. There are two possibilities.

A. The conventional approach takes account of the elastic compliance of the structure. In this case, the stiffness of different parts of the robot must be measured or estimated. The longitudinal flexibility of legs (grippers) and/or the torsional flexibility of joints may be essential. Equations (2) and additional relationships between forces and elastic deformations form a system of equations from which normal reactions N_i can be determined.

B. The alternative (guaranteed) approach [15] does not need additional and often inaccurate information about the elastic properties of the structure. We require instead that conditions (6) should be fulfilled for any admissible N_i satisfying (2). Thus, we come to the condition

$$\sup_{x,y,N_i} Q \leq f, \tag{7}$$

where sup is taken with respect to all x, y and all N_i satisfying (2).

Finally, we come to the following algorithm for verifying the conditions of equilibrium.

1. For a given position of the robot, calculate components of external forces (R_x, R_y, R_z) and of their moment (M_x, M_y, M_z) with respect to the point O.

2. Calculate $\Phi_i = (p - p_i)S$, $i = 1, \dots, n$.

3. Calculate the coordinates (4) of the point K.

4. Check conditions (5). If these conditions are violated, the robot will fall. If they are satisfied, then the next stage should be performed.

5. If we take into account the elastic compliance of the robot, then the normal reactions N_i can be calculated according to the approach A. Then condition (6) should be verified. For the guaranteed approach B, the maximum in (7) should be calculated, and condition (7) must be checked. If conditions (6), (7) are violated for the respective approaches A, B, then the robot may

slide along the plane. If they are satisfied (together with condition (5)), then the robot is in equilibrium.

Note that parameters of the robot such as the stiffness of its structure, coefficient of friction, the pressure p_i under the gripper, etc., are known only approximately and are changeable. Therefore, to be on the safe side, one should satisfy all conditions of equilibrium with a big margin.

Numerical calculations were performed for different versions of climbing robots according to the algorithms described above.

Let us consider an example. The robot attached to a vertical wall has a rectangular platform with a length a and width b. The length of the leg is denoted by h, the radius of the gripper by r, the mass of the robot by m, and the pressure under each of four ($n = 4$) grippers by p_0. The calculations were performed for different positions of the robot on the wall according to both approaches A and B. In approach A, longitudinal and torsional stifnesses were taken into account. According to the approach B, the non-separation and non-slipping conditions are reduced to the respective inequalities:

$$m \leq 2\pi(p - p_0)ar^2(gh)^{-1}, \quad m \leq 4\pi f(p - p_0)r^2g^{-1}.$$

For the following numerical data $a = 0.344m$, $b = 0.151m$, $h = 0.095m$, $r = 0.034m$, $p_0 = 0.1p$, $f = 0.5$, the carrying capacity of the robot, i.e., its admissible total mass, was evaluated. It was found to be between $48kg$ and $67kg$ (for different approaches and values of stiffness). To be on the safe side, the total mass of the robot was chosen equal to $30kg$.

Planning of motions of wall-climbing robots as well as computer simulation of these motions were carried out [1, 2]. The results of these investigations were used for the design of new versions of WCR and also for elaborating control algorithms for the robots.

A quite different type of climbing robots is presented in Figs. 5, 6. This robot was also designed and developed in the IPMech [4]. The robot consists of five links of equal length connected to each other by four identical two-degrees-of-freedom joints, so that the entire system (with a fixed end point) has eight degrees of freedom. All degrees of freedom are independently controlled by identical electric actuators placed at the corresponding joints. The end points of the robot have rigidly attached feet equipped with six vacuum grips, or suction cups (Fig. 5). These feet enable the robot to fix itself to smooth surfaces of arbitrary orientation (e.g., walls or ceilings) and to move over them. Due to its multilink design, the robot has high mobility and can move over surfaces of complex shape. For example, it can begin to move over the floor of a room, then climb up a wall to the ceiling, travel over the ceiling to another wall, and descend. Depending on the control algorithm, the configuration of the robot may vary considerably. The mathematical model of the robot was elaborated and investigated [4]. Using this model, the equilibrium of the robot attached to different surfaces was analyzed, and its kinematics and dynamics in various modes of motion was simulated. Since the robot has a wide range

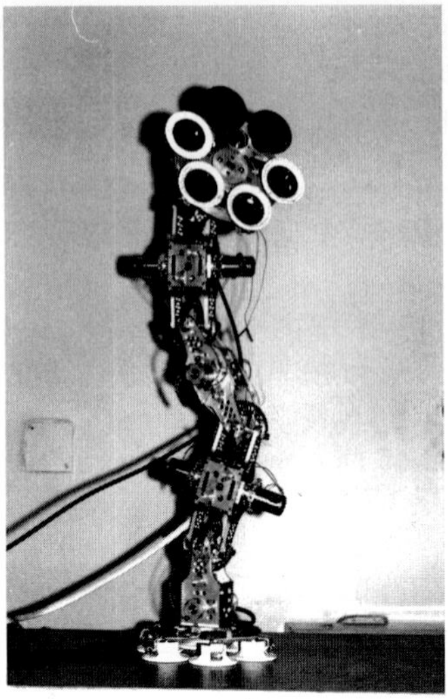

Fig. 5. Multilink climbing robot

Fig. 6. Multilink climbing robot

of possible motions, it is natural to find optimal (in some sense) motion in a given class.

Let us consider briefly optimal periodic two-dimensional motions of the robot along a plane [16]. We assume that at the beginning and at the end of the period the robot is fixed to the plane by means of two feet so that its central link is parallel to the plane (Fig. 6). We assume that the maximum angular velocity ω of relative rotation for each pair of adjacent links is given. We are to choose the initial configuration, i.e., the angle α between the inclined link and the normal to the plane, and the succession of motions in order to maximize the speed of the robot along the plane.

There exist two possible modes of periodic motions. In both modes, each of the feet is first disconnected from the plane, moves, and is again fixed to the plane at a certain distance from its initial position. In the first (simple) mode, the front and rear feet never change places, whereas in the second (overhead) mode they change places at each step (Fig. 7).

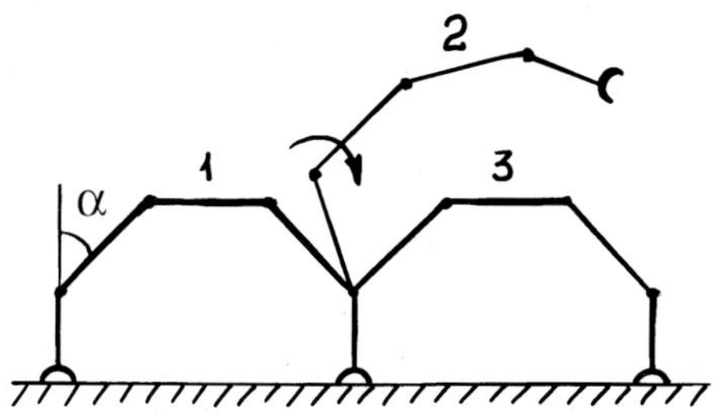

Fig. 7. Optimal motion of the multilink robot

It was shown [16] that the maximum speed for the first mode is $v_1 = L\omega$, where L is the length of the links. This speed is attained, if the robot makes infinitely frequent and infinitely short steps, so that $\alpha \to \pi/2$. The maximum speed for the overhead mode is higher: $v_2 = 2(2^{1/2} + 1)\pi^{-1}L\omega = 1.54L\omega$, it is attained for $\alpha = \pi/4$.

3 Tube-Crawling Robot

The tube-crawling robot considered below (Fig. 8) has been designed and constructed at the Technical University of Munich [5]. The robot can be equipped with measuring instruments, manipulators, and various tools, and can be utilized for nondestructive technical inspection, maintenance and repair work in pipelines.

The machine consists of a body and eight identical legs attached to it. Four legs are arranged at the front part of the body and the other four legs at the rear part. Each leg consists of two links connected by active revolute joints whose axes are parallel to each other. The feet of the legs are balls made of a material that provides high friction between the foot and the tube surface. The robot is driven by 16 electric actuators located at the joints.

All legs lie in two planes perpendicular to each other, each pair of two opposite legs at the front and the corresponding pair of legs at the rear belonging to the same plane. This configuration permits the robot to move inside the tube by pressing the feet of four legs lying in the same plane to the surface of the tube. While the four legs (which are in the support phase) ensure the motion of the robot body along the tube, the other four legs (which are in the transfer phase) do not contact the tube surface and move forward preparing

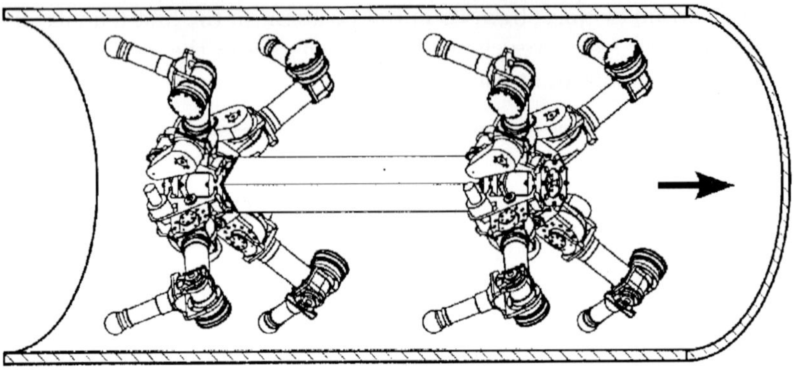

Fig. 8. Tube-crawling robot

for the next step. Then the legs in the support and transfer phases change places. The robot is supplied with a feedback hierarchical control system [5].

In this paper, we restrict ourselves with some aspects of simulation and optimization for the tube-crawling robot.

Note that the driving force of the robot is created by the legs in the support phase and is due to the friction between their feet and the surface of the tube. This force depends significantly on the configuration of the support legs as well as on the torques applied to the joints. Since all four support legs have a similar configuration, we come to the following optimization problems for one leg [17].

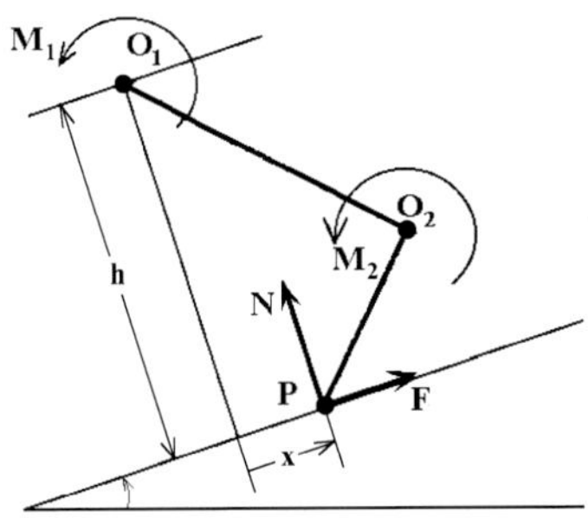

Fig. 9. Robot leg

Consider a planar two-member leg O_1O_2 (Fig. 9). The first link O_1O_2 is connected with a base (the robot body) by a revolute joint O_1. The links are connected to each other by another revolute joint O_2. The torques M_1 and M_2 are applied at the joints O_1 and O_2, respectively, and the end of the second link contacts the surface at the point P which exerts the normal force N and the friction force F upon the linkage. The equations of equilibrium for the linkage can be reduced to the following equations:

$$F = f_1 M_1 + f_2 M_2 + f_0, \quad N = n_1 M_1 + n_2 M_2 + n_0, \tag{8}$$

where f_i and n_i, $i = 0, 1, 2$, are coefficients depending on the geometry of the linkage (they are given in [17]).

Problem 1. Find the torques M_1 and M_2 which satisfy the constraints $|M_1| \le M_1^0$, $|M_2| \le M_2^0$, and such that the friction force F given by (8) satisfies Coulomb's law $|F| \le fN$ and is maximal: $F = \max F = F^*$.

Since Problem 1 is a linear programming problem, its solution $F = F^*$ is attained, when $M_1 = \pm M_1^0$ and $M_2 = \pm M_2^0$, i.e., the both torques are of maximal admissible magnitudes. The detailed solution is presented in [17].

The maximum friction force F^* depends on the maximal admissible torques M_1^0, M_2^0, the lengths of the links $\ell_1 = O_1O_2$ and $\ell_2 = O_2P$, and also on the position x of the foot P with respect to the projection of the point O_1 onto the support surface (Fig. 9). During the step, the robot body moves, and the distance x decreases from its initial value x_0 to $x_0 - s/2$, where s is the length of the step. Note that the full step consists of two half-steps with different support legs. Since the minimal value of F^* over the half-step is important, we come to the following optimization problems. In what follows, we fix the values of the torques M_1^0, M_2^0, the step length s, the clearance h (the distance from the point O_1 to the support surface), and the length ℓ_1 of the first link.

Problem 2. Maximize the minimum

$$\min_{x \in [x_0 - s/2, x_0]} F^* \tag{9}$$

over x_0 (ℓ_2 is fixed).

Problem 3. Maximize the minimum (9) over x_0 and ℓ_2.

These problems were discussed and solved numerically in [17]. Let us consider a numerical example. We take $M_1^0 = 68Nm$, $M_2^0 = 27Nm$, $f = 1$, $s = 0.28m$, $h = 0.24m$, $\ell_1 = 0.15m$. Then the solution of Problem 1 for $\ell_2 = 0.15m$, $x = s/2 = 0.14m$ gives $F^* = 642N$. The solution of Problem 2 (optimization with respect to x_0) yields $x_0 = 0.105m$, $F^* = 786N$. Solving Problem 3 (optimization with respect to x_0 and ℓ_2), we obtain $x_0 = 0.02m$, $\ell_2 = 0.21m$, $F^* = 1047N$. Thus, optimization leads to an essential growth in the driving force. However, the increased length of legs may cause difficulties: the longer legs may touch the tube, especially in the case of curved tubes.

The detailed numerical analysis of the maximum driving force F^* as a function of the torques M_1^0 and M_2^0, the clearance h, the coefficient of friction f, the length ℓ_2 of the second link, and the position of the foot is given in [17].

The obtained results show that the driving force depends on the parameters in a complicated way. For example, as the length ℓ_2 of the second link increases, the driving force F^* can increase for some positions x of the foot and decrease for other positions. A typical example of the obtained results is presented in Fig. 10 for $M_1^0 = 45Nm$, $M_2^0 = 8.6Nm$, $f = 0.2$, $h = 0.24m$. Here, the dependence of the force F^* on x is shown for $\ell_2 = \ell_1 + 0.01(i-1)m$, where $i = 1, \ldots, 8$; each graph is marked with the corresponding number i.

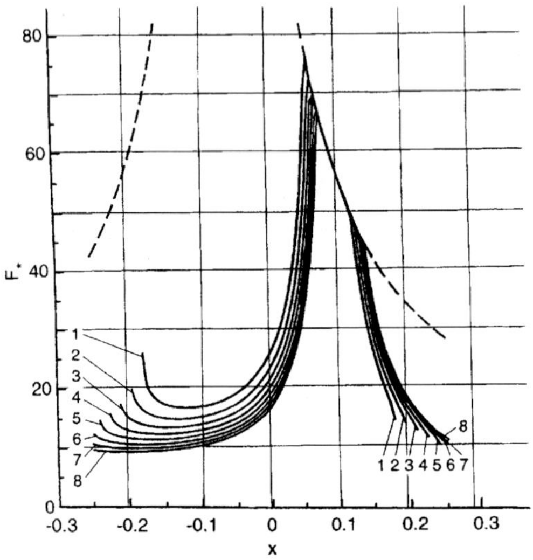

Fig. 10. Driving force as a function of the foot position

Simulation and optimization of the regular (periodic) motions of the robot inside a straight (cylindrical) tube were considered in [18]. In regular motions, the body of the robot travels translationally at a constant velocity v, and the robot axis coincides with the axis of the tube. We assume that all four legs lying in the same plane move synchronously and are either in the support or in the transfer phase. The regular gaits are periodic with the period T, and the duration of the support and transfer phases for each leg are equal to $T/2$.

The equations of motion can be divided into two groups: the equations of the robot as a whole and equations of its legs. The first group consists of the equations for the momentum and angular momentum of the robot. The second group comprises 16 (two for each leg) Lagrange equations. These equations were treated by means of the inverse method.

The motion of the transfer legs was planned in such a way that these legs do not touch the body, and their feet reach the prescribed position on the tube surface in given time $T/2$. Since the motion of the robot body is prescribed, the motion of the support legs is also determined. Then, using a special procedure [18], we obtain the time histories of the control torques applied at the joints.

Thus, the simulation of the regular motions involves the calculation of the positions of the robot feet, the joint angles, and the control torques as grid functions of time with an increment (simulation step) Δt. Since we consider only periodic gaits, it is sufficient to carry out the simulation only for one step. The simulated motion is represented as a computer animation.

As the performance index for the optimization, we take the robot body velocity v to be maximized with respect to design variables. The latter include the position x of the support leg with respect to the joint O_1 (Fig. 9) at the beginning of the support phase, the step length s, and the length of the second link ℓ_2. The length of the first link was fixed: $\ell_1 = 0.15m$. The optimization was performed numerically under rather complicated constraints imposed on the torques and angular velocities of the joints as well as under geometrical constraints.

Table 1. Velocities versus s and x.

$s \backslash x$	0.14	0.11	0.08	0.05	0.02	−0.01	−0.04	−0.07	−0.1	−0.13
0.01	0.003	0.004	0.004	0.005	0.005	0.006	0.006	0.005	0.005	0.004
0.06	0.020	0.026	0.030	0.032	0.034	0.036	0.034	0.032	0.026	
0.11		0.048	0.053	0.055	0.057	0.059	0.057	0.048		
0.16		0.069	0.070	0.070	0.069	0.068	0.065			
0.21		0.081	0.078	0.075	0.071	0.067				
0.26		0.085	0.080	0.074	0.070					
0.31		0.085	0.078	0.073						
0.36		0.082	0.076							
0.41		0.080								
0.46		0.077								

Table 2. Velocities versus l_2, s, and x.

l_2	0.13	0.15	0.17	0.19	0.21	0.23	0.25
s	0.24	0.30	0.30	0.30	0.24	0.24	0.12
x	0.06	0.12	0.15	0.18	0.18	0.18	0.15
v	0.076	0.087	0.103	0.129	0.156	0.196	0.245

Some results for the real tube-crawling robot of the Technical University of Munich are presented in Tables 1 and 2. In Table 1, the parameter ℓ_2 is fixed: $\ell_2 = \ell_1 = 0.15m$. These values correspond to the real robot. Empty boxes in the table correspond to the violation of some constraints. One can see from

Table 1 that the maximum velocity $v = 0.085m/s$ is attained at $s = 0.26m$ and $x = 0.11m$.

In Table 2, all three parameters ℓ_2, s, and x were varied, and the optimal values are presented. It is apparent that taking $\ell_2 = 0.25m$ one can obtain a considerable gain in the speed (by a factor of 3). However, as it was remarked above, longer legs may restrict the manoeuvrability of the robot in curved or branched tubes.

The motion of the tube-crawling robot through a curved (toroidal) tube was investigated in [19]. A class of regular gaits implementing such a motion was defined. In these gaits, the center of mass of the robot body moves by the same distance during each step, and, at the beginning and at the end of each step, the center of mass of the robot body lies on the tube. The parameters of regular gaits are calculated, and some qualitative features of the motion of the robot were analyzed. Computer simulation results are presented in [19]. The analysis of the motion of the robot in a toroidal tube is important for planning and control of the passage of the robot through turns of a pipeline.

4 Snake-like Mechanisms

Crawling motions of snakes and other limbless animals were always of great interest. They were studied in biomechanics [20], and recently the principle of snake-like locomotion was implemented in biologically inspired mobile robots equipped with passive wheels [21].

In papers [6–10], multilink systems without wheels were considered which can move along a horizontal plane in the presence of dry friction between the linkage and the plane. Control torques are created by actuators installed at the joints of the linkage. It was shown that the linkages can perform various motions along the plane and can reach any prescribed position and configuration. For three-link and two-link mechanisms, periodic motions consisting of slow and fast phases were designed [6, 8, 9]. For mechanisms with more than four links, slow wavelike motions were suggested [7]. Here, we restrict ourselves with the simulation and optimization of the three-link mechanism. More results on snake-like mechanisms as well as more references can be found in [22].

Consider a plane three-member linkage $O_1C_1C_2O_2$ moving over a horizontal plane Oxy (Fig. 11). For simplicity, we assume that the links O_1C_1, C_1C_2, and C_2O_2 are rigid massless bars, and the mass of the linkage is concentrated at its joints C_1 and C_2 which are equal point masses m_1, and at the end points O_1 and O_2 which have equal masses m_0. The total mass of the mechanism is $m = 2(m_0 + m_1)$. The length of the central link C_1C_2 is $2a$, and the lengths of the end links are equal to ℓ. Denote by x, y the coordinates of the middle of the central link, by θ the angle between this link and the x-axis, and by α_i the angles between the central link and the end links O_iC_i, $i = 1, 2$ (Fig. 11).

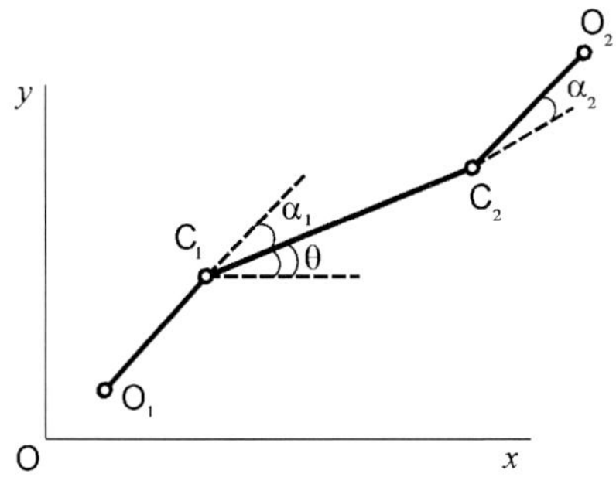

Fig. 11. Three-member linkage

We assume that the dry friction forces obeying Coulomb's law act between the masses O_i, C_i, $i = 1, 2$, and the plane Oxy. The coefficients of friction for the masses m_0 and m_1 are f_0 and f_1, respectively. The control torques M_1 and M_2 are applied at the joints C_1 and C_2. Suppose that these torques can create some prescribed time histories of angles α_1 and α_2.

The desired motions of the linkage will be designed as a combination of simple elementary motions called slow and fast phases. Each phase begins and ends at the state of rest of the mechanism. In slow phases, the end links of the mechanism rotate synchronously either in the same direction or in the opposite directions, so that $\alpha_1(t) = \pm\alpha_2(t)$, whereas the central link stays at rest. In some cases, only one end link rotates in a slow phase. Denote by ω_0 and ε_0 the maximal values of the angular velocities and accelerations during the slow phases:

$$\omega_0 = \max|\dot{\alpha}_i(t)|, \quad \varepsilon_0 = \max|\ddot{\alpha}_i(t)|, \quad i = 1, 2. \tag{10}$$

Here, the maxima are taken along the whole slow phase and do not depend on $i = 1, 2$.

The sufficient condition which ensures that the central link stays at rest during the slow phase can be expressed as follows [8]:

$$m_0\ell\left\{[\omega_0^4 + (\varepsilon_0 + gf_0\ell^{-1})^2]^{1/2} + (\varepsilon_0 + gf_0\ell^{-1})\ell a^{-1}\right\} \leq m_1 g f_1. \tag{11}$$

This condition always holds for very slow motions, i.e., if ω_0 and ε_0 in (11) are sufficiently small, and $m_0 f_0(a + \ell) < m_1 f_1 a$.

In fast phases, the angular velocities and accelerations of the end links are high enough, and the duration τ of this phase is much smaller that the duration T of the slow phase: $\tau \ll T$. The magnitudes of the control torques M_1 and M_2 during the fast phase are high compared to the torques produced by the dry friction:

$$\begin{aligned} |M_1| \gg m^* g f^* \ell^*, \quad i = 1, 2, \quad m^* = \max(m_0, m_1), \\ f^* = \max(f_0, f_1), \quad a^* = \max(a, \ell). \end{aligned} \tag{12}$$

Hence, the friction can be neglected in the fast phases. Therefore, the center of mass of the linkage stays at rest, and its angular momentum is zero during the fast phase. Using these conservation laws, one can easily evaluate the terminal state of the linkage after the fast phase.

Let the linkage be at rest at the initial state, with all its links parallel to the x-axis. Denote slow and fast phases by capital S and F, respectively, indicating the initial and terminal values of angles α_i in each phase as follows: $\alpha_i^0 \to \alpha_i^1$, $i = 1, 2$.

The longitudinal motion of the linkage is implemented by the following succession of phases. First, the auxiliary slow motion is carried out: $S, \alpha_1 : 0 \to \gamma$, $\alpha_2(t) \equiv 0$, where $\gamma \in (-\pi, \pi)$ is a fixed angle. Then, the following four motions are performed (Fig. 12):

$$\begin{aligned} &1.\ F, \alpha_1 : \gamma \to 0, \quad \alpha_2 : 0 \to \gamma;\\ &2.\ S, \alpha_1 : 0 \to -\gamma, \alpha_2 : \gamma \to 0;\\ &3.\ F, \alpha_1 : -\gamma \to 0, \alpha_2 : 0 \to -\gamma;\\ &4.\ S, \alpha_1 : 0 \to \gamma, \quad \alpha_2 : -\gamma \to 0. \end{aligned}$$

After stage 4, the configuration of the linkage coincides with its configuration before stage 1, so that stages 1–4 can be repeated any number of times. To return the linkage to its initial configuration $\alpha_1 = \alpha_2 = 0$, we should perform the motion: $S : \alpha_1 : \gamma \to 0, \alpha_2(t) \equiv 0$.

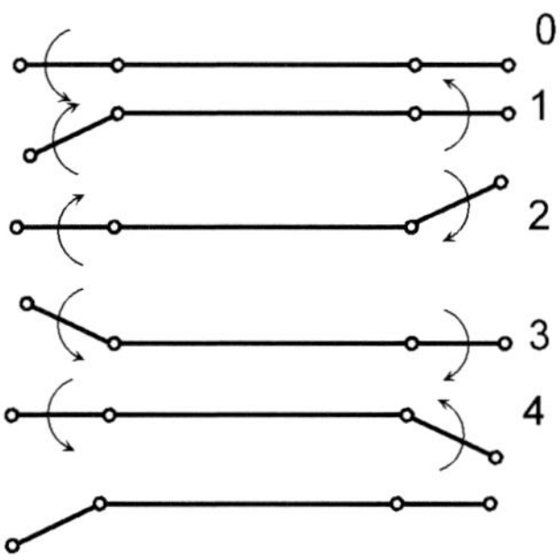

Fig. 12. Longitudinal motion

During the slow phases, central link stays at rest, whereas the the center of mass of the linkage moves forward along the x-axis. During the fast motions, on the contrary, the center of mass stays at rest, while the central link moves. It is shown in [6] that the total displacement of the middle of the central link along the x-axis during the cycle of motions 1–4 is

$$\Delta x = 8 m_0 m^{-1} \ell \sin^2(\gamma/2). \quad m = 2(m_0 + m_1). \tag{13}$$

The total displacement along the y-axis and the total angle of rotation of the central link during the cycle 1–4 are zero: $\Delta y = 0$, $\Delta\theta = 0$. The average speed of the periodic longitudinal motion is

$$v_1 = \Delta x(2T)^{-1}. \tag{14}$$

Similarly, the lateral motion of the linkage and its rotation on the spot are also represented as combinations of slow and fast phases [6]. Using these motions, the linkage can move from any initial state to any prescribed position and configuration in the plane.

Computer simulation of motions of the linkage was based on the complete nonlinear equations of the multibody dynamics. The obtained animation permits to observe the motions with various values of the system parameters.

Since the average speed of the linkage depends significantly on its geometrical and mechanical parameters, it is natural to find the values of the parameters that maximize the speed. Here, we consider one of the optimization problems presented in [10].

Suppose that the following relationships hold for the slow phases of the longitudinal motion:

$$\begin{gathered} \omega(t) = |\dot{\alpha}_i(t)| = \varepsilon_0 t, \quad t \in [0, T/2], \\ \omega(t) = \varepsilon_0 (T - t), \quad t \in [T/2, T], \\ \omega_0 = \varepsilon_0 T/2 = 2\gamma T^{-1}, \quad \varepsilon = 4\gamma T^{-2}. \end{gathered} \tag{15}$$

By substituting (15) into (11), we obtain the inequality

$$m_0 \ell \{[(2\gamma T^{-1})^4 + P^2]^{1/2} + P\ell a^{-1}\} \le m_1 g f_1, \quad P = 4\gamma T^{-2} + g f_0 \ell^{-1}. \tag{16}$$

Let us fix the mass m_1 of the joints, the length $2a$ of the central link, and the angle of rotation γ. The mass m_0, the duration T of the slow phases, the length ℓ of the end links, and the coefficients of friction f_0 and f_1 are to be chosen so as to maximize the speed v_1 given by (13) and (14). The parameters are subjected to the constraints (16) and $f^- \le f_0 \le f^+$, $f^- \le f_1 \le f^+$. It was proved that the optimal values of the coefficients of friction are: $f_0 = f^-$, $f_1 = f^+$.

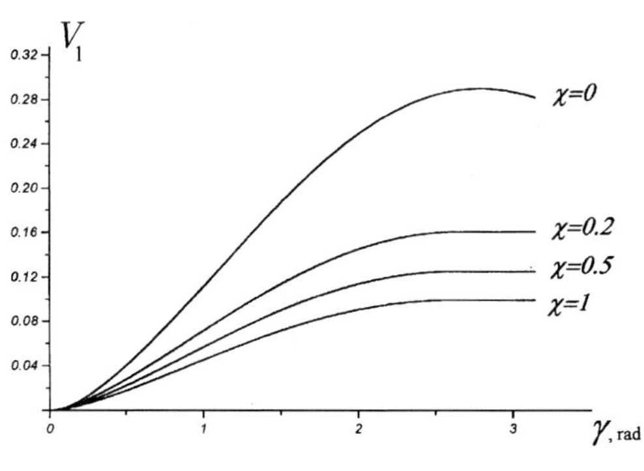

Fig. 13. Maximal speed

The numerical solution of the optimization problem stated above was given in [10]. The non-dimensional optimal speed $V_1 = v_1(gaf^+)^{-1/2}$ as a function of the angle γ and $\chi = f^-/f^+$ is given in Fig. 13. The optimal value of $\lambda = \ell/a$ as a function of the same parameters is presented in Fig. 14. Note that the maximum speed and the ratio $\lambda = \ell/a$ depend significantly on γ and χ.

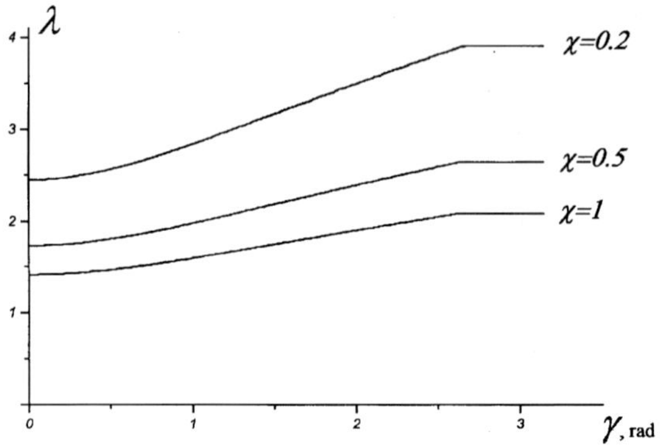

Fig. 14. Optimal λ

The required magnitude of the control torques, according to (12), can be estimated as follows: $|M_i| \sim 10m_1gf^+\ell$.

As a numerical example, consider the linkage having the parameters $a = 0.2m$, $m_1 = 1kg$. Optimal dimensional parameters of the longitudinal motion for some values of the coefficients of friction and angles γ are presented in Table 3. For this example, the required magnitude of control torques is about $40f^+Nm$.

Table 3. Optimal parameters for the longitudinal motion.

f^-/f^+	0.2/0.2	0.1/0.5	0.2/1	0.5/1	0/1
γ, deg	60	60	90	60	90
l, m	0.32	0.57	0.64	0.4	∞
m_0, kg	0.17	0.40	0.39	0.26	1
$m = 2(m_0 + m_1)$, kg	2.34	2.80	2.78	2.52	4
T_1, s	0.77	1.10	1.08	0.49	∞
v_1, m/s	0.030	0.075	0.167	0.084	0.28

5 Conclusion

The paper is based on research carried out in the Institute for Problems in Mechanics of the Russian Academy Sciences. Several types of climbing and crawling mobile robots have been considered. These robots can move in a complex environment: on walls of structures (wall-climbing robots), inside tubes (tube-crawling robot), and perform various technological operations. The principle of snake-like mechanisms can be useful for mini- and micro-robots.

For these types of robots, mechanical and mathematical models are elaborated. A typical feature of these mechanisms is the interaction with the environment (surfaces, walls, tubes) by means of dry friction. Based on the created models, the investigations of the static equilibrium and nonlinear dynamics are performed. An important part of these investigations is the computer simulation of the robot motions.

Since the motion of the robots depends significantly on their parameters, it is natural to optimize the design of the robots and modes of their motions. Several optimization problems for the mobile robots are formulated and solved. It is shown that the optimization gives a substantial gain in the speed and driving force of the robot.

The experimental work carried out in the IPMech in Moscow and in the Technical University of Munich made it possible to check and confirm some of the obtained theoretical and computer results.

The author is grateful to F. Pfeiffer from TU of Munich and to N. N. Bolotnik, V. G. Gradetsky, G. V. Kostin, and S. A. Kumakshev from IPMech for their valuable contributions and useful discussions.

The work was supported by the Russian Foundation for Basic Research, Project 02-01-00201.

References

[1] Abarinov, A. V. *et al*: A robot system for moving over vertical surfaces. Soviet Journal of Computer and Systems Sciences, **27** (3), 130–142 (1989)

[2] Gradetsky, V. G., Rachkov, M. Yu., Sizov, Yu. G., Ulyanov, S. V., Chernousko, F. L.: Mobile systems with vertical displacement robots. Journal of Computer and Systems Sciences International, **31** (1), 126–142 (1993)

[3] Chernousko, F. L., Bolotnik, N. N., Gradetsky, V. G.: Manipulation Robots. Dynamics, Control, and Optimization. CRC Press, Boca Raton (1994)

[4] Chernousko, F. L., Gradetsky, V. G., Bolotnik, N. N., Veshnikov, V. B.: Multilink mobile robot for motion over arbitrarily inclined surfaces. In: Proceedings of the First ECPD International Conference on Advanced Robotics and Intelligent Automation, Athens, Greece (1995)

[5] Rossmann, T., Pfeiffer, F.: Control and design of a pipe crawling robot. In: Proceedings of the 13th World Congress on Automatic Control, San Francisco (1996)

[6] Chernousko, F. L.: The motion of a multilink system along a horizontal plane. Journal of Applied Mathematics and Mechanics, **64** (1), 5–15 (2000)
[7] Chernousko, F. L.: The wave-like motion of a multilink system on a horizontal plane. Journal of Applied Mathematics and Mechanics, **64** (4), 497–508 (2000)
[8] Chernousko, F. L.: On the motion of a three-member linkage along a plane. Journal of Applied Mathematics and Mechanics, **65** (1), 15–20 (2001)
[9] Chernousko, F. L.: Controllable motions of a two-member mechanism along a horizontal plane. Journal of Applied Mathematics and Mechanics, **65** (4), 565–577 (2001)
[10] Chernousko, F. L.: Snake-like locomotions of multilink mechanisms. Journal of Vibration and Control, **9** (1–2), 235–256 (2003)
[11] Fujita, A., Tsuge, M., Mori, K., Sonoda, S., Watahiki, S., Ozaki, N.: Development of inspection robots for spherical gas storage tanks. In: Proceedings of the 16th International Symposium on Industrial Robots, Brussels (1986)
[12] Hirose, S.: Wall climbing vehicle using internally balanced magnetic unit. In: Proceedings of the 6th CISM-IFToMM Symposium ROMANSY-86, Cracow (1986)
[13] Sugiyata, S., Naiton, S., Sato, C., Ozaki, N., Watahiki, S.: Wall surface vehicles with magnetic legs or vacuum legs. In: Proceedings of the 16th International Symposium on Industrial Robots, Brussels (1986)
[14] Chernousko, F. L.: On the mechanics of a climbing robot. Mechatronic Systems Engineering, **1**, 219–224 (1990)
[15] Chernousko, F. L.: Equilibrium conditions for a solid on a rough surface. Mechanics of Solids, **23** (6), 1–12 (1988)
[16] Bolotnik, N. N., Sternin, S. L.: Optimization of motions of a universal multilink walking robot. Journal of Computer and Systems Sciences International, **36** (4), 626–637 (1997).
[17] Bolotnik, N. N., Chernousko, F. L., Kumakshev, S. A., Pfeiffer, F.: Static analysis of a two-member linkage interacting with a given surface. Archive of Applied Mechanics, **69** (7), 429–442 (1999)
[18] Pfeiffer, F., Rossmann, T., Bolotnik, N. N., Chernousko, F. L., Kostin, G. V.: Simulation and optimization of regular motions of a tube-crawling robot. Multibody System Dynamics, **5** (2), 159–184 (2001)
[19] Bolotnik, N. N., Chernousko, F. L., Kostin, G. V., Pfeiffer, F.: Regular motion of a tube-crawling robot in a curved tube. Mechanics of Structures and Machines, **30** (4), 431–462 (2002)
[20] Gray, J.: Animal Locomotion. Weidenfeld & Nicolson, London (1968)
[21] Hirose, S.: Biologically Inspired Robots: Snake-like Locomotors and Manipulators. Oxford University Press, Oxford (1993)
[22] Chernousko, F. L. Modelling of snake-like locomotions. In this volume.

Modelling of Snake-Like Locomotions

Felix L. Chernousko

Institute for Problems in Mechanics of the Russian Academy of Sciences
pr. Vernadskogo 101-1, Moscow 119526, Russia
chern@ipmnet.ru

Summary. Multibody mechanical systems consisting of several links are considered which can perform snake-like locomotions along a horizontal plane in the presence of dry friction between the system and the plane. It is shown that, under certain conditions, these systems can perform longitudinal motions in the plane. Periodic motions are designed and analyzed. For mechanisms with two or three links, these periodic motions consist of slow and fast phases. For multilink mechanisms with more than four links, slow wavelike motions are possible. Displacements, the average speed, and the required control torques are evaluated. Theoretical results are confirmed by the computer simulation of the nonlinear dynamics of the multilink systems as well as by the experimental data.

Key words: Snake-like locomotions, Multibody systems, Dynamics, Computer simulation.

1 Introduction

The motions of snakes and other crawling animals have always been of great interest for specialists in mechanics and biomechanics. By contrast to walking and running creatures who alternate their supporting legs, snakes mostly keep the permanent contact between their bodies and the ground. Though the friction force acting upon each moving segment of the body is directed against the velocity of the segment, the resultant of the friction forces, i.e., the thrust, should be directed along the velocity of the center of mass of the body. To explain this phenomenon, various models of snake-like locomotions have been proposed.

The motion of a snake inside a curved tube has been considered in [1]. It was shown that the required thrust can be created by the normal reactions of the walls of the tube. Locomotions of snakes have been described and classified into three classes in [2]. One of them, a rectilinear one, requires the displacement of the snake's mass along its body, whereas the other two classes (in which the snake twists its body) are possible only in the presence of vertical walls or other vertical or inclined objects. By pressing its body against these objects such as stones, grass, sand slopes, etc., the snake creates horizontal components of reactions along the direction of motion. Snakes always try to use vertical or inclined obstacles and avoid flat surfaces. Biomechanical aspects of snake-like locomotions have been discussed in [3]. Mechanisms of snake-like locomotions in the presence of obstacles have been considered in [4,5].

Nonholonomic multilink snake-like robots have been designed and investigated [6,7]. These mechanisms consist of many elements equipped with passive wheels and connected by joints. By twisting, these robots can perform snake-like locomotions. In these motions, the wheels exert forces directed along their axes and thus produce the desirable thrust in the direction of motion. In fact, the wheels here play the role similar to that of vertical walls for snakes. The kinematics of snake-like locomotions of such nonholonomic mechanisms have been considered in [8-10].

In this paper, based on [11–15], we consider plane multibody systems (linkages) which have no wheels and can move along a horizontal plane in the presence of dry friction between the linkage and the plane. Control torques are created by actuators installed at the joints of the linkage. It is shown that the linkage can perform various motions along the plane. For the two-link mechanism, periodic motions consisting of slow and fast phases are designed. The three-member linkage is analyzed in [16]. Displacements, the average speed of motions, and the required magnitude of control torques are estimated. For the multilink system with more than four links, a slow wavelike motion is proposed and investigated.

2 Mechanical Models

We consider mechanical systems consisting of several rigid bodies connected by revolute joints. The system moves along the horizontal plane Oxy, and the axes of all joints are vertical. Actuators are placed at the joints and create control torques about the joint axes. These torques are applied to two neighbouring links and are internal for the system. The only external forces acting upon the system are the weight and reactions of the plane. We assume that the dry friction force $\mathbf{F}$, acting at each point of contact between the system and the plane, obeys Coulomb's law:

$$\begin{aligned} \mathbf{F} &= -f|\mathbf{N}|v^{-1}\mathbf{v}, \quad \text{if} \quad v \neq 0, \\ |\mathbf{F}| &\leq f|\mathbf{N}|, \quad \text{if} \quad v = 0. \end{aligned} \tag{1}$$

Here, $\mathbf{N}$ is the normal reaction at the contact point, f is the coefficient of friction, $\mathbf{v}$ is the vector of velocity of the point, and $v = |\mathbf{v}|$.

Let us consider first a two-member linkage consisting of two rigid bodies connected by a revolute joint O^* (Fig. 1). The body marked with the subscript 1 will be called the main body, and the body marked with the subscript 2 will

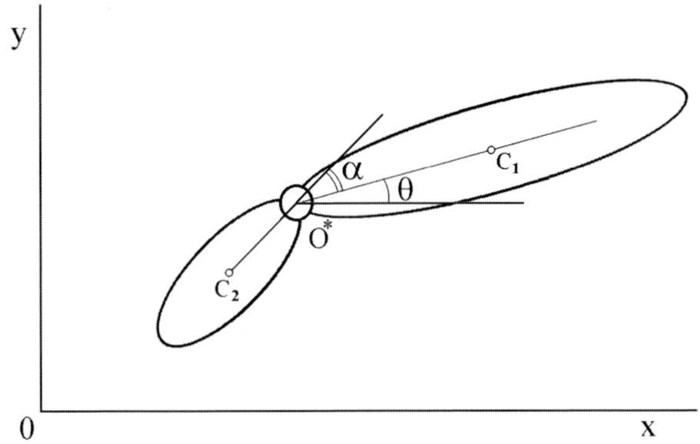

Fig. 1. Two-member linkage

be called the tail. Denote by m_1 and m_2 the masses of the respective bodies, by C_1 and C_2 the projections of their centers of mass upon the plane Oxy, by $a_1 = O^*C_1$ and $a_2 = O^*C_2$ the distances from the axis of the joint to the points C_1 and C_2, and by J_1 and J_2 the moments of inertia of the bodies with respect to the axis of the joint. Suppose that the joint can be treated as a point mass m_0. This assumption reflects the real situation: the electric actuator installed at the joint has a considerable mass. Thus, the total mass of our system is $m = m_0 + m_1 + m_2$.

The coefficients of friction for the points of the main body, tail, and joint are denoted by f_1, f_2, and f_0, respectively. Note that if there are more than three points of contact between the system and the plane, the normal reactions are not determined uniquely. Therefore, the friction forces are also not unique even for moving points; as for the points at rest, they are always not unique by virtue of (1). Thus, we are to deal with the static indeterminacy.

Denote by x_0 and y_0 the coordinates of the point O^*, by θ the angle between the x-axis and the axis O^*C_1 of the main body, and by α the angle between the axis C_2O^* of the tail and the axis O^*C_1 of the main body. The coordinates x_c, y_c of center of mass C of the linkage are

$$\begin{aligned} x_c &= x_0 + m_1 m^{-1} a_1 \cos\theta - m_2 m^{-1} a_2 \cos(\theta + \alpha), \\ y_c &= y_0 + m_1 m^{-1} a_1 \sin\theta - m_2 m^{-1} a_2 \sin(\theta + \alpha). \end{aligned} \tag{2}$$

Denote by ω_i the angular velocities of the bodies, $i = 1, 2$. The angular momentum of the linkage with respect to the point O can be presented as follows:

$$\begin{aligned} K = {}& m(x_0\dot{y}_0 - y_0\dot{x}_0) + m_1 a_1 \dot{\theta}(x_0 \cos\theta + y_0 \sin\theta) \\ & - m_2 a_2 (\dot{\theta} + \dot{\alpha})[x_0 \cos(\theta + \alpha) + y_0 \sin(\theta + \alpha)] \\ & - m_1 a_1 (\dot{x}_0 \sin\theta - \dot{y}_0 \cos\theta) + m_2 a_2 [\dot{x}_0 \sin(\theta + \alpha) \\ & - \dot{y}_0 \cos(\theta + \alpha)] + J_1 \dot{\theta} + J_2 (\dot{\theta} + \dot{\alpha}). \end{aligned} \tag{3}$$

Denote by M the control torque created by the actuator and applied to the tail. The torque applied to the main body is equal to $-M$.

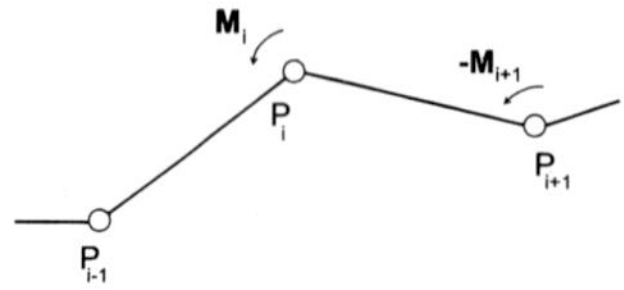

Fig. 2. Multilink system

Consider now a plane multilink system consisting of N equal links which are rigid straight rods of the length a (Fig. 2). For the sake of simplicity, we assume that the masses of the links are negligible compared to the masses of the joints. Each joint is a point mass m, and the end points of the linkage have the same mass. Denote all these point masses by P_i, $i = 0, 1, \ldots, N$. The coefficient of dry friction for each mass P_i is equal to f.

The control torques are applied by actuators at the joints P_i, $i = 1, 2, \ldots, N-1$. Denote by $\mathbf{M}_i$ the torque applied by the actuator at the joint P_i to the link P_iP_{i+1}; then the torque applied to the link $P_{i-1}P_i$ by the same actuator is $-\mathbf{M}_i$.

The motion of the two-member linkage is considered in the framework of dynamics. By contrast, the motion of the multilink system is treated using the quasi-static approach. The angular velocities and accelerations of the links are assumed to be small enough, so that dynamical effects in the motion of the multilink system can be neglected. Denote by ω_0 and ε_0 the maximal values of the angular velocities and accelerations of the links, and by g the gravity acceleration. The quasi-static approach is justified, if $\omega_0^2 a \ll gf$, $\varepsilon_0 a \ll gf$. In the framework of this approach, we treat the motion as a succession of the states of equilibrium.

3 Two-Member Linkage

Let us show how the two-member linkage can perform a longitudinal motion. Suppose that at the initial time instant the linkage is at rest, its both links being parallel to the x-axis. At the initial state 0 in Fig. 3 we have $\theta = 0$, $\alpha = 0$. The periodic longitudinal motion of the linkage will be designed as a succession of alternating *slow* and *fast* phases. Each phase begins and ends at the state of rest of the linkage, and the angle α changes between the prescribed values during each phase.

In a slow phase, the main body does not move. This condition is satisfied, if the maximal angular velocity $\omega_0 = \max_t |\dot{\alpha}(t)|$ and acceleration $\varepsilon_0 = \max_t |\dot{\omega}|$ of the tail during this phase are small enough. More exactly, the following two inequalities comprise a sufficient condition for the slow motion:

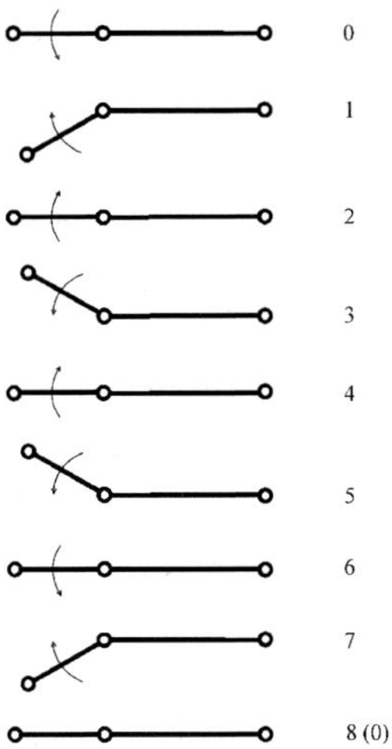

Fig. 3. Motion of the two-member linkage

$$J_2\varepsilon_0 + m_2 g f_2 a_2 \leq m_1 g f_1 a_1,$$
$$J_2\varepsilon_0 + m_2 g f_2 a_2 + m_2 a_1 a_2[\omega_0^4 + (\varepsilon_0 + g f_2 a^{-1})^2]^{1/2} \leq m_0 g f_0 a_1. \tag{4}$$

Inequalities (4) are satisfied, if ω_0 and ε_0 are sufficiently small and $m_2 f_2 a_2 < m_1 f_1 a_1$, $m_2 f_2(a_1 + a_2) < m_0 f_0 a_1$.

In a fast phase, the angular velocity and acceleration are high, and the control torque is much greater than the torques caused by the friction forces:

$$|M| \gg m^* g f^* \ell^*, \quad m^* = \max(m_0, m_1, m_2), \quad \ell^* = \max(a_1, a_2),$$
$$f^* = \max(f_0, f_1, f_2). \tag{5}$$

Under condition (5), the friction forces can be neglected. Therefore, during the fast phase the conservation laws hold: the position of the center of mass C and the angular momentum K of the linkage given by (2) and (3), respectively, are constant:

$$x_c = \text{const}, \quad y_c = \text{const}, \quad K = 0. \tag{6}$$

Denote by S and F the slow and fast phases. Then the longitudinal motion of the linkage is implemented by the following succession of stages (Fig. 3):

$$\begin{array}{ll} 1)\ S,\ \alpha: 0 \to \beta, & 2)\ F,\ \alpha: \beta \to 0, \\ 3)\ S,\ \alpha: 0 \to -\beta, & 4)\ F,\ \alpha: -\beta \to 0, \\ 5)\ S,\ \alpha: 0 \to -\beta, & 6)\ F,\ \alpha: -\beta \to 0, \\ 7)\ S,\ \alpha: 0 \to \beta, & 8)\ F,\ \alpha: \beta \to 0. \end{array} \tag{7}$$

Here, initial and terminal values of the angle α are indicated for each phase, and β is a fixed angle, $\beta \in (-\pi, \pi)$. As a result of each slow phase, the center of mass C of the linkage gets a positive displacement along the x -axis, whereas in fast phases the point C stays at rest (see (6)). Thus, the center of mass C moves in the positive direction along the x-axis. Its total displacement during the cycle of the periodic motion is [14]:

$$\Delta x_c = 8m^{-1} m_2 a_2 \sin(\beta/2) \cos(\gamma/2) \sin[(\beta - \gamma)/2] \tag{8}$$

Here, the following denotations are used

$$\begin{aligned} \gamma &= \beta/2 + A_0(A_+ A_-)^{-1} \arctan[A_+ A_-^{-1} \tan(\beta/2)], \\ A_0 &= m(J_2 - J_1) + m_1^2 a_1^2 - m_2^2 a_2^2, \\ A_\pm &= [m(J_1 + J_2) - (m_1 a_1 \pm m_2 a_2)^2]^{1/2}. \end{aligned} \tag{9}$$

The displacements of the center of mass C along the y-axis and the rotation of the main body also occur during the cycle (7), but it is proved [14] that the total changes of y_c and θ during this cycle are zero: $\Delta y_c = 0$, $\Delta \theta = 0$. Therefore, repeating the cycle (7), we obtain a periodic longitudinal motion, its average speed is $v = \Delta x_c [4(T + \tau)]^{-1}$, where x_c is given by (8), T and τ are the durations of slow and fast phases, respectively, $\tau \ll T$.

More complicated motions of the linkage are also constructed as combinations of slow and fast phases. It is shown [14] that the two-member linkage can move from any initial state to any prescribed terminal state in the horizontal plane. Consider two simple cases of the two-member linkage.

Let the main body and the tail be point masses m_1 and m_2 attached to the joint of the mass m_0 by rigid massless straight rods of the lengths a_1 and a_2, respectively. In this case, we should put $J_1 = m_1 a_1^2$, $J_2 = m_2 a_2^2$ into equations (8) and (9). As a numerical example, consider a mechanism with the following parameters: $m_0 = 0.6kg$, $m_1 = 0.3kg$, $m_2 = 0.3kg$, $m = 1.2kg$, $a_1 = 1m$, $a_2 = 0.2m$, $f_0 = f_1 = f_2 = 0.2$. Suppose the parameters of the motion are: $T = 1s$, $\omega_0 = 2s^{-1}$, $\varepsilon_0 = 4s^{-2}$, $\beta = 1$rad. Checking conditions (4), we find out that they are satisfied. The longitudinal displacement per cycle, evaluated by means of equations (8) and (9), turns out to be $\Delta x_c = 0.085m$. The average velocity of the longitudinal motion is $v = 0.021ms^{-1}$. The required magnitude of the control torque, according to (5), is of the order of $8Nm$.

The second case corresponds to the linkage consisting of two straight rods of the uniform linear density ρ and the lengths ℓ_1 and ℓ_2. The coefficient of friction for all points of the linkage is f. The mass of the joint is taken to be negligible. Thus, this model can be treated as a snake which can bend only at one presrcibed point of its body. In this case, one should substitute the formulas: $m_i = \rho \ell_i$, $a_i = \ell_i/2$, $J_i = \rho \ell_i^3/3$, $m_0 = 0$, $m = m_1 + m_2$, $i = 1, 2$, into equations (9). Since conditions (4) do not hold for $m_0 = 0$, a special analysis of this case was carried out [14]. As a result, it was found that the slow motions can be implemented, if ω_0 and ε_0 are small enough and $\ell_2/\ell_1 < 0.255$.

4 Multilink Mechanism

Consider now the multilink mechanism described in Section 2. Following [12], we will describe two possible types of wavelike quasi-static motions (with three and four moving links) along the horizontal plane Oxy.

At the beginning and at the end of the cycle of motions, the linkage is straight and placed along the x-axis. Its successive configurations for two types of motion are shown in Figs. 4 and 5.

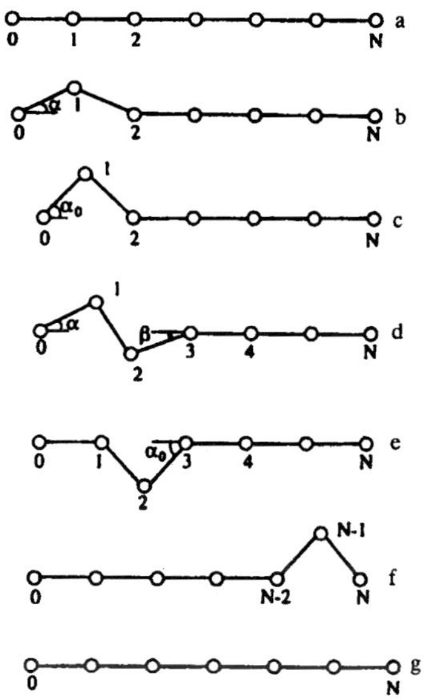

Fig. 4. Wavelike motion with three moving links

At the first stage of the motion with *three moving links* (Fig. 4), the end mass P_0 advances along the x-axis while the points P_i, $i \geq 2$, remain fixed. The angle α between the x-axis and link P_0P_1 grows monotonically from zero to a certain given value α_0, see states a, b, c in Fig. 4. At the next stage, links P_0P_1, P_1P_2, and P_2P_3 are moving. The angle α decreases monotonically from α_0 to zero while the angle β between link P_2P_3 and the x-axis grows from 0 to α_0, see state d in Fig. 4. At the end of this stage, the system will be in state e in which links P_1P_2 and P_2P_3 form an isosceles triangle congruent to the triangle $P_0P_1P_2$ in state c, but with its apex pointing in the opposite direction. Here, all points except P_2 lie on the x-axis. Next, the motion involves links P_1P_2, P_2P_3, and P_3P_4, and is identical to the preceding stage, apart from a displacement along the x-axis and a mirror reflection in the axis. Continuing this process, we see that after each stage all joints except one lie on the x-axis, and that one joint is the apex of the isosceles triangle with angle α_0 at the base. The triangle gradually moves towards the right. Finally, the point P_{N-1}

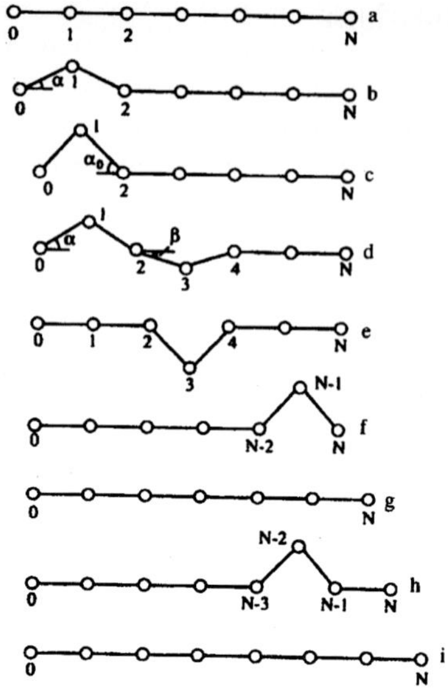

Fig. 5. Wavelike motion with four moving links

becomes the apex of such a triangle, see state f in Fig.4. At the last stage of the motion, the point P_N advances to the right along the x-axis. The angle at the base of the triangle $P_{N-2}P_{N-1}P_N$ decreases from α_0 to zero, and the linkage becomes straight again, see state g in Fig. 4.

As a result of the entire cycle of motions, the linkage advances along the x-axis for a distance

$$L = 2a(1 - \cos\alpha_0). \tag{10}$$

In the motion with *four moving links*, the first stage proceeds exactly as in the previous case (states a–c are the same in Figs. 4 and 5). At the next stage, links P_0P_1, P_1P_2, P_2P_3, and P_3P_4 are involved. The point P_2 moves to the right along the x-axis. As this happens, the angle α at the base of the isosceles triangle $P_0P_1P_2$ decreases from α_0 to zero, while the angle β at the base of the triangle $P_2P_3P_4$ grows from zero to α_0, see state d in Fig. 5. At the end of this stage, all joints of the linkage except P_3 lie on the x-axis, see state e in Fig. 5. Continuing this process, we come to the right end of the linkage. The final stages are somewhat different for the cases of even and odd N, see states f–i in Fig. 5. The resultant displacement of the linkage is again given by formula (10).

Comparing the two types of wavelike motions, we see that the motion with three moving links is somewhat simpler. However, it requires larger angles between links, or more intensive twisting, for the same values of α and L [12].

The wavelike motions whose kinematics is described above were analyzed in a quasi-static formulation [12]. We assume that all velocities and accelerations are extremely small, and hence the external forces applied to the system must balance out. In other words, the friction forces must satisfy three equilibrium conditions (two for the forces and one for the moments). The friction forces applied to the moving points are readily evaluated, whereas for the points at rest they are unknown but bounded by the inequalities. Our equilibrium problem is statically indeterminate and can have a non-unique solution. The simplest distributions of the friction forces are found [12] for which the equilibrium is attained with the participation of the least possible number of points adjacent to the moving ones. It is shown that the wavelike motion with three moving links is feasible, if the linkage has at least five links ($N \geq 5$), whereas the motion with four moving links is possible, if $N \geq 6$. Also, the magnitude M of the required control torques at the joints of the linkage is estimated [12]. It is shown that $M \leq 2mgfa$ where f is the friction coefficient. Comparing this result to (5), we see that the requirements imposed on the control torques for the wavelike quasi-static motions are more moderate than those for the fast phases of the motions of the two-member linkage.

5 Conclusion

A mechanical model of snake-like locomotions is proposed and analyzed. It is shown that a mechanism with two links and one actuator can move in a horizontal plane and reach any prescribed position and configuration in the plane. The periodic motions of the two-member linkage consist of alternating slow and fast phases. Similarly, the three-member linkage can move. Displacements, the average speed, and the required control torques are evaluated. The mechanism consisting of more than four links can perform a wavelike quasi-static motion which requires smaller magnitudes of control torques than the motion of mechanisms with two and three links.

The results of computer simulation confirm the obtained theoretical results. Experiments with models of snake-like mechanisms were performed at the Technical University of Munich and at the Institute for Problems in Mechanics of the Russian Academy of Sciences. The experimental data show that the proposed motions can be implemented. The suggested principle of motion can be of interest for mobile robots, especially for small ones.

References

[1] Lavrentyev, M. A. and Lavrentyev, M.M . On one principle of creating the thrust force in motion. Journal of Applied Mechanics and Technical Physics (4), 6–9 (1962)
[2] Gray, J. Animal Locomotion. Weidenfeld & Nicolson, London (1968)
[3] Dobrolyubov, A. I. Travelling Waves of Deformation. Nauka i Tekhnika, Minsk (1987)

[4] Bayraktaroglu, Z. Y., Butel, F., Pasqui, V., and Blazevic, P. A geometrical approach to the trajectory planning of a snake-like robot. In: G. C. Virk, M. Randall, and D. Howard (eds.), Proceedings of the Second International Conference on Climbing and Walking Robots CLAWAR 1999, Portsmouth, pp. 851–856 (1999)
[5] Bayraktaroglu, Z, Y. and Blazevic, P. Snake-like locomotion with a minimal mechanism. In: M. Armada and P. Gonzalez de Santos (eds.), Proceedings of the Third International Conference on Climbing and Walking Robots CLAWAR 2000, Madrid, pp. 201–207 (2000)
[6] Hirose, S. and Morishima, A. Design and control of a mobile robot with an articulated body. International Journal of Robotics Research **9**(2), 99–114 (1990)
[7] Hirose, S. Biologically Inspired Robots: Snake-like Locomotors and Manipulators. Oxford University Press, Oxford (1993)
[8] Chirikjan, G. S. and Burdick, J. W. Kinematics of hyper-redundant robot locomotion with applications to grasping. In: Proceedings of the IEEE International Conference on Robotics and Automation **1**, Sacramento, pp. 720–725 (1991)
[9] Burdick, J. W., Radford, J., and Chirikjan, G. S. A 'sidewinding' locomotion gait for hyper-redundant robots. In: Proceedings of the IEEE International Conference on Robotics and Automation **3**, Atlanta, pp. 101–106 (1993)
[10] Ostrowski, J. and Burdick, J. Gait kinematics for a serpentine robot. In Proceedings of the IEEE International Conference on Robotics and Automation **2**, Minneapolis, pp. 1294–1300 (1996)
[11] Chernousko, F. L. The motion of a multilink system along a horizontal plane. Journal of Applied Mathematics and Mechanics **64** (1), 5–15 (2000)
[12] Chernousko, F. L. The wavelike motion of a multilink system on a horizontal plane. Journal of Applied Mathematics and Mechanics **64** (4), 497–508 (2000)
[13] Chernousko, F. L. On the motion of a three-member linkage along a plane. Journal of Applied Mathematics and Mechanics **65** (1), 13–18 (2001)
[14] Chernousko, F. L. Controllable motions of a two-link mechanism along a horizontal plane. Journal of Applied Mathematics and Mechanics **65** (4), 565–577 (2001)
[15] Chernousko, F. L. Snake-like locomotions of multilink mechanisms. Journal of Vibration and Control **9**, 235–256 (2003)
[16] Chernousko, F. L. Simulation and optimization of crawling robots. In this volume

Simulation and Visualization of Plant Growth Using Lindenmayer Systems

Somporn Chuai-Aree[1], Willi Jäger[1], Hans Georg Bock[1], and Suchada Siripant[2]

[1] Interdisciplinary Center for Scientific Computing(IWR), University of Heidelberg Im Neuenheimer Feld 368, D-69120 Heidelberg, Germany
Somporn.ChuaiAree@iwr.uni-heidelberg.de,
Wjaeger@iwr.uni-heidelberg.de,
Bock@iwr.uni-heidelberg.de

[2] Advance Virtual and Intelligent Computing Center(AVIC), Chulalongkorn University Phayathai road, Bangkok 10330, Thailand
Suchada.S@chula.ac.th

Summary. Lindenmayer systems (L-systems) were introduced by Aristid Lindenmayer to generate geometrical structures of plants, i.e. shoot, leaf or root. During the last decade of L-systems prototypes, the plant growth has been animated by composing images from all iterations. The problem is that development of a plant model is nonsmooth and discontinuous. In this paper, we solve this problem by adding some mathematical time functions of logistic growth to each component and combine them with an L-system prototype. The stochastic and bracketed L-systems are applied to generate the stochastic structure and branching structure, respectively. Our L-system prototype can generate both plant shoot and root parts. The results of simulation and visualization are presented. These show that the simulation and visualization of the development of the plant growth modeled by using the new proposed method is smoother and more natural.

1 Introduction

The animation of plant growth represents an attractive and challenging problem for computer graphics. Mech and Prusinkiewicz [5] extended L-systems in a manner suitable for simulation to interact between a developing plant and its environment. They developed the plant growth animation by iterating the L-system, but the animation was not smooth. They solved this problem using time L-system and differential L-system [7].

This paper presents a prototype for creating computer models that capture the development of plants using L-systems and a mathematical model incorporating biological data. L-systems are used for a qualitative model in order to represent the plant topology and development. This paper also extends from our previous work in [1], [2], and [3].

The paper is organized into 7 sections. Section 2 summarizes the concept of a general L-system and also stochastic L-system. Section 3 expresses growth function approximation. Section 4 describes plant module design. Section 5 shows the structure of plant simulation. Section 6 describes visualization. The conclusion is given in Section 7.

2 Lindenmayer Systems (L-systems)

Lindenmayer systems (L-systems) were first introduced by Aristid Lindenmayer in 1968 as a mathematical theory of plant development [6]. They have attracted the attention of the computer scientists who investigated them through formal language theory. Specialists in computer graphics, particularly Prusinkiewicz have used the L-systems to produce realistic images of trees, bushes, flowers and some images are well illustrated in [6]. For a three dimensional movement, a component is free to move in any X, Y, or Z direction. Hence, the directional angles in this case are three directions. The initial three directional angles, $\alpha_x, \alpha_y, \alpha_z$ of the first unit movement are set with respect to X, Y, and Z axes, respectively. The directional angles of the other unit movements are computed in a similar fashion to that of the two dimensional case. Three constants, $\delta_x, \delta_y, \delta_z$, are used to adjust the direction of the unit movements. In addition, the physical location of the unit movement must be represented by XYZ coordinates. Therefore, a unit movement described in the Cartesian coordinate system is denoted by a hexaplet (X,Y,Z,$\alpha_x, \alpha_y, \alpha_z$). After adding/subtracting $\delta_x, \delta_y, \delta_z$, the new XYZ coordinates of the movement is computed by multiplying the coordinates of the current movement with the rotation matrices R_x, R_y, R_z shown in the following equation

$$R_x(\theta) = \begin{bmatrix} 1 & 0 & 0 \\ 0 & \cos(\theta) & -\sin(\theta) \\ 0 & \sin(\theta) & \cos(\theta) \end{bmatrix}, \; R_y(\theta) = \begin{bmatrix} \cos(\theta) & 0 & -\sin(\theta) \\ 0 & 1 & 0 \\ \sin(\theta) & 0 & \cos(\theta) \end{bmatrix},$$
$$R_z(\theta) = \begin{bmatrix} \cos(\theta) & \sin(\theta) & 0 \\ -\sin(\theta) & \cos(\theta) & 0 \\ 0 & 0 & 1 \end{bmatrix} \tag{1}$$

The rotation of a unit movement and its direction are captured in a symbolic form similar to the two dimensional case [1] by using these symbols /, \, &, ^, +, -, |. The meaning of each symbol is explained in Table 1.

An example of a simple L-system interpretation in a 3-dimensional space is given below ($\delta_x = \delta_y = \delta_z = 45^o, \alpha_x = \alpha_y = \alpha_z = 0$).

$$\begin{aligned} \text{Shoot} &= I[-IL][+IL]I[+IL]I[-IL]I[+IL]I[-IL][+IL]IF \\ \text{Root} &= I[+I]I[-I]I[+I]I[-I]I[+I]I[-I]I[+I]I \end{aligned}$$

Table 1. Symbols used in L-system prototype

Symbols	Meaning
I	Generate the plant shoot/root internode
i	Generate the plant shoot/root short internode
P	Generate the plant shoot/root petiole
p	Generate the plant shoot/root short petiole
A	Generate the plant shoot apex or root tip
L	Generate the plant shoot leaf
F	Generate the plant shoot flower
$+(\delta_z)$	Roll counterclockwise by angle δ_z, using rotation matrix $R_z(\delta_z)$
$-(\delta_z)$	Roll clockwise by angle δ_z, using rotation matrix $R_z(-\delta_z)$
$\&(\delta_y)$	Roll counterclockwise by angle δ_y, using rotation matrix $R_y(\delta_y)$
$\hat{}(\delta_y)$	Roll clockwise by angle δ_y, using rotation matrix $R_y(-\delta_y)$
$\backslash(\delta_x)$	Roll counterclockwise by angle δ_x, using rotation matrix $R_x(\delta_x)$
$/(\delta_x)$	Roll clockwise by angle δ_x, using rotation matrix $R_x(-\delta_x)$
$\vert$	Roll back, using rotation matrix $R_y(180)$
[	Push the current state of the turtle onto a pushdown stack to create a new branch
]	Pop a state from the stack and make it the current state of the turtle to close the branch

We define the symbol *"I"* for an internode or petiole of shoot and root part, *"L"* for a leaf and *"F"* for a flower of the plant. From the example, there are 13 internodes for the shoot part - six internodes for the main stem, seven internodes for the petiole and their leaves in each direction - and one flower. For the root part, there are 15 internodes - eight internodes for the main root, seven internodes for the lateral roots. The visualized plant is shown in Fig. 1.

2.1 Stochastic L-system

We use the stochastic L-system in our system in order to generate variety of plant structure. The syntax of the production rule is defined as below.

$$\mathbf{Pred} = (\mathbf{Prob_1})\mathbf{Succ_1}, (\mathbf{Prob_2})\mathbf{Succ_2}, \ldots, (\mathbf{Prob_n})\mathbf{Succ_n}$$

where **Pred** is the predecessor of successor $\mathbf{Succ_i}$ with probability value $\mathbf{Prob_i}$ when $i = 1, 2, 3, \ldots, n$. The sum of probabilities of all successors with the same predecessor **Pred** is equal to 1.

For example, $I = (0.40)I[+iL]I, (0.30)I[-iL]I, (0.30)I[-iL][+iL]I$.

The predecessor *"I"* will be replaced by three cases of successors with probability values 0.40, 0.30, 0.30, respectively. The structures of each production rule are shown in Fig. 2.

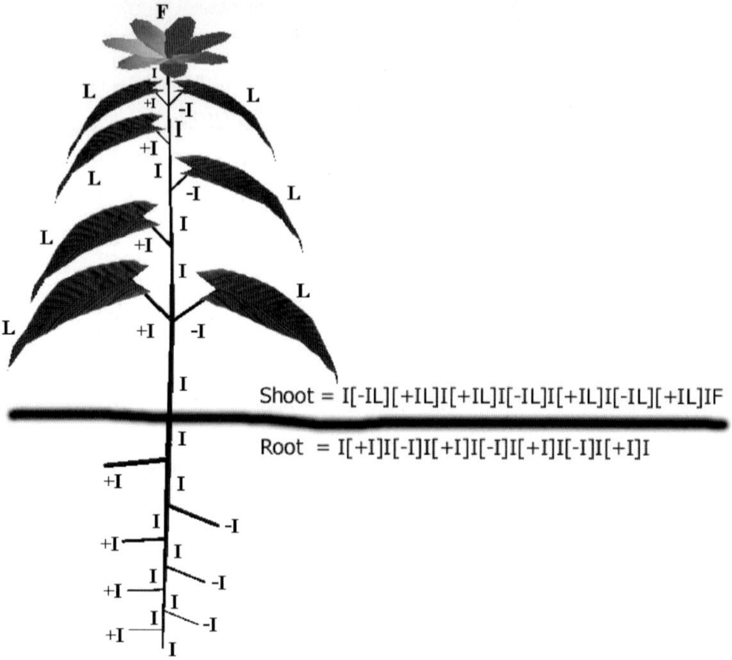

Fig. 1. A simple L-System interpretation. The colour version of this figure can be found in Fig. A.9 on page 581.

Fig. 2. An Example of a stochastic L-system, (a) is *I[+iL]I*, (b) is *I[-iL]I*, and (c) is *I[-iL][+iL]I*, with probabilities 0.4, 0.3, and 0.3, respectively.

3 Growth Function

The data of each component of the plant structure i.e. the internode length, internode diameter, leaf length, leaf width, etc were observed from the actual plants. All data will be approximated by growth function. Each component has their own growth function. A typical growth function is shown in Fig. 3.

The growth function consists of five stages. There are growth stage, stable stage, decreasing stage, stable after decreasing stage and death stage. Each stage is shown in equations (2),(3),(4),(5), and (6).

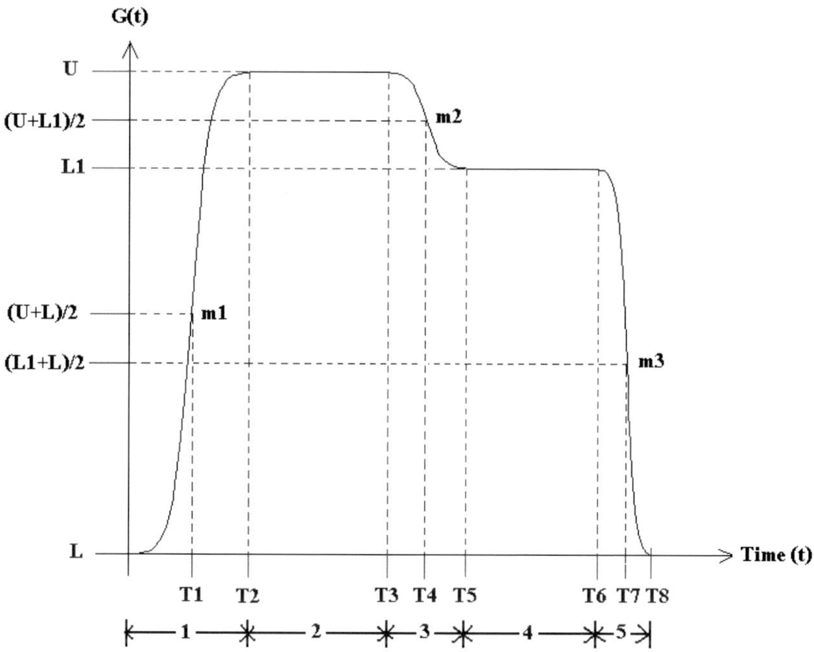

Fig. 3. The Approximated Growth Function.

The starting time of each component is controlled by the behavior of the kind of plant and observed data. The system allows each component to be dead after some period of time. The children of that component will be cut automatically. The decreasing stage, stable stage after decreasing and the death stage are optional for all the components.

The raw data are approximated as a growth function in Fig. 3 using a sigmoidal curve approximation. The raw data are converted to growth function G(t) of length, width or diameter at time t. All equations are given below.

$$G(t) = L + \frac{U - L}{1 + e^{m_1(T_1 - t)}} \quad ; 0 \le t < T_2 \tag{2}$$

$$G(t) = U \quad ; T_2 \le t < T_3 \tag{3}$$

$$G(t) = L_1 + \frac{U - L_1}{1 + e^{-m_2(T_4 - t)}} \quad ; T_3 \le t < T_5 \tag{4}$$

$$G(t) = L_1 \qquad ; T_5 \leq t < T_6 \tag{5}$$

$$G(t) = L + \frac{L_1 - L}{1 + e^{-m_3(T_7 - t)}} \qquad ; T_6 \leq t \leq T_8 \tag{6}$$

where L : the minimum value of length, width or diameter,
L_1 : the minimum value of length, width or diameter in the decreasing stage,
U : the maximum value of length, width or diameter,
m_1 : the approximated slope of raw data in the growth stage,
m_2 : the approximated slope of raw data in the decreasing stage,
m_3 : the approximated slope of raw data in the death stage,
T_1 : the time at $G(t) = (U + L)/2$,
T_2 : the starting time at $G(t) = U$ in the stable stage after growth stage,
T_3 : the ending time at $G(t) = U$ in the stable stage before decreasing,
T_4 : the time at $G(t) = (U + L_1)/2$,
T_5 : the starting time at $G(t) = L_1$ in the stable stage after decreasing,
T_6 : the ending time at $G(t) = L_1$ in the stable stage after decreasing,
T_7 : the time at $G(t) = (L_1 + L)/2$,
T_8 : the ending time at $G(t) = L$ in the death stage, and
t : the independent time variable.

Besides the growth function, there is another function that we use to control all the components of the plant topology, such as the length of each internode from the first to the last internode. The function is

$$Y_i = c(a)^{n_i} \tag{7}$$

where Y_i is the length of internode i, c is a constant, a is a real value greater than zero and n_i is the level of internode i. We also use this equation to control the size of petiole, leaf, flower from the first to the last level. The initial time of each component is specified by the following linear equation.

$$B_i = \beta n_i + b \tag{8}$$

where B_i is the initial time of component i, β is the acceleration rate of B_i, n_i is the level of component i and b is a constant. Every component of the plant is controlled by its growth function with either the same or different slope m, time T, the maximum U, and the minimum L.

4 Plant Module Design

The L-system description of a plant is defined in form of a set of iterations, a set of directions and sizing parameters, an initial string, a set of production rules and a set of terminating productions. This description has the following format.

Table 2. An L-system prototype

Plant_Name {
Shoot {
Iterations=N
Angle=δ
Diameter=D
Axiom=ω
Production 1
Production 2
...
Production n
Endrule
Endproduction 1
Endproduction 2
...
Endproduction m
}
	Root {
	Iterations=N
	Angle=δ
	Diameter=D
	Axiom=ω
	Production 1
	Production 2
	...
	Production n
	Endrule
	Endproduction 1
	Endproduction 2
	...
	Endproduction m
	}
	}

Plant_Name is the name of the plant module. There are two modules of a plant prototype, namely **Shoot** and **Root**. The prototype of the shoot module is similar to the root module and uses the same keyword. The number of iteration is defined by **Iterations=N** using a positive integer **N**. The **Angle=**δ is used to set the angle of the branch corresponding to the rotation symbols. The first internode diameter of shoot and root is described by **Diameter=D**. The string **Axiom=**ω is used to start the status of the plant that will be replaced by all production rules. The **Production 1 ... n** are defined to describe the changing rules of rewriting processes. Each production rule consists of a predecessor and a successor. In case there is only one successor, the probability value is equal to 1. The predecessor is a symbol of character and the successor is a symbol string. **Endrule** and **Endproduction 1 ... m** are optional parts to terminate the rewriting of the production rules and to ignore some symbols that are not defined in Table 1. The characters **{** and **}** are the beginning and ending of L-system's structure, respectively.

5 Structure of Plant Simulation

A plant structure as L-system prototype is obtained after measuring the plant. There are two parts of L-system prototypes. The first part is shoot prototype and the second part is root prototype. These processes are illustrated in Fig. 4. The L-strings of shoot/root part are generated by rewriting of shoot/root prototype (No. 1, No. 3) and these L-strings represent the shoot/root structure (No. 2, No. 4) of the plant, respectively.

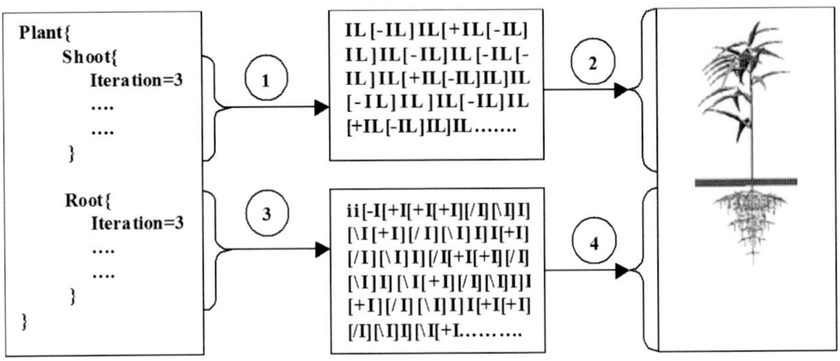

Fig. 4. L-system transformation.

The structure of simulation starts from data collection of the actual plant and also captures the structure of the plant to be the L-system prototype. The measured raw data are approximated by the growth function based on a sigmoidal curve. The L-system prototype is compiled by the rewriting process to be an L-string at the last iteration. The L-string and the growth function of each symbol are combined together and represented the structure of the plant and its development. The system generates the 3D-plant development by visualization. We have to evaluate the plant growth and adjust some parameters, such as the size of each internode, to make the plant model and plant growth to look more realistic.

There are two possibilities for generating the root structure. The first, we can generate from L-system prototype that we know the root structure in advance from the root production rules. The second, we can generate from the nutrient concentration in each small voxel of the soil volume. The roots try to search the maximum value of nutrient concentration and grow in direction of the maximum. In a cubic grid, each voxel has 26 neighbors. If the initial nutrient concentration changes, the algorithm will generate the different root structures. Fig. 5 shows the generated root structure from algorithm. The initial value of nutrient concentration of each voxel is randomly.

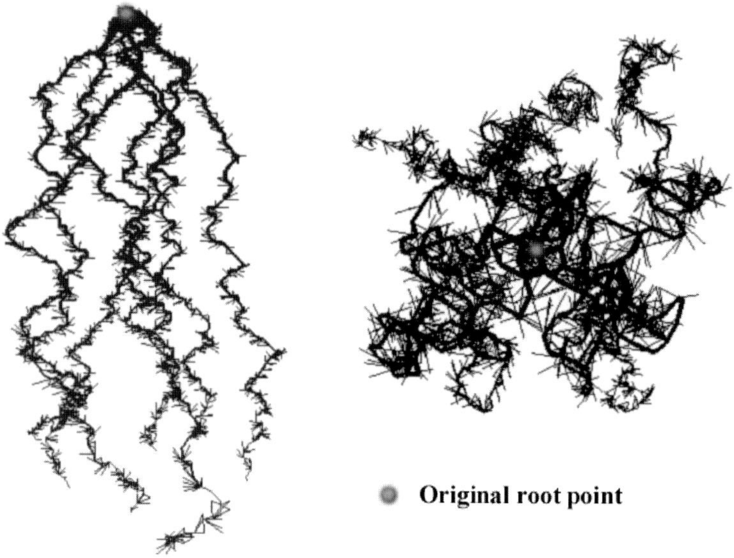

Fig. 5. The example of root structures, front view (left) and top view (right).

6 Visualization

The visualization of an example is shown in vegetative state. Cylinders are used to represent internodes and petioles segments of plant shoots and roots. Spheres are used to represent jointed internodes. Triangular polygons are used to represent leaves and flowers. All the components are allowed a texture mapping technique. Fig. 6 shows many different structures from stochastic L-system. Their root structures are generated by the nutrient concentration with the different initial conditions.

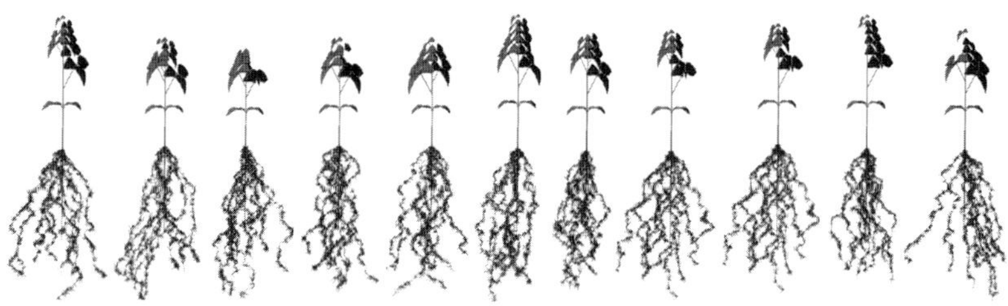

Fig. 6. Some stochastic plant structures. The colour version of this figure can be found in Fig. A.10 on page 582.

From the example prototype, after rewriting, the L-strings of shoot and root become the following shoot and root L-strings.

$$\begin{aligned}\text{Shoot} &= IL[+IL]IL[-IL[+IL]IL]IL[+IL]IL\\ &\quad [-IL[+IL]IL[-IL[+IL]IL]IL[+IL]\\ &\quad IL]IL[+IL]IL[-IL[+IL]IL]IL[+IL]IL\\ \text{Root} &= i\end{aligned}$$

Fig. 7 shows the smooth development of an example from the L-system prototype in Table 3. The roots are generated by the nutrient concentration in the soil voxels under the conditions of having 27 soil voxels in X, Y, Z direction, 25 levels of the root depth, 10 roots and eight hairy roots at each segment randomly. The shoot part and root part are related to each other by linear functions.

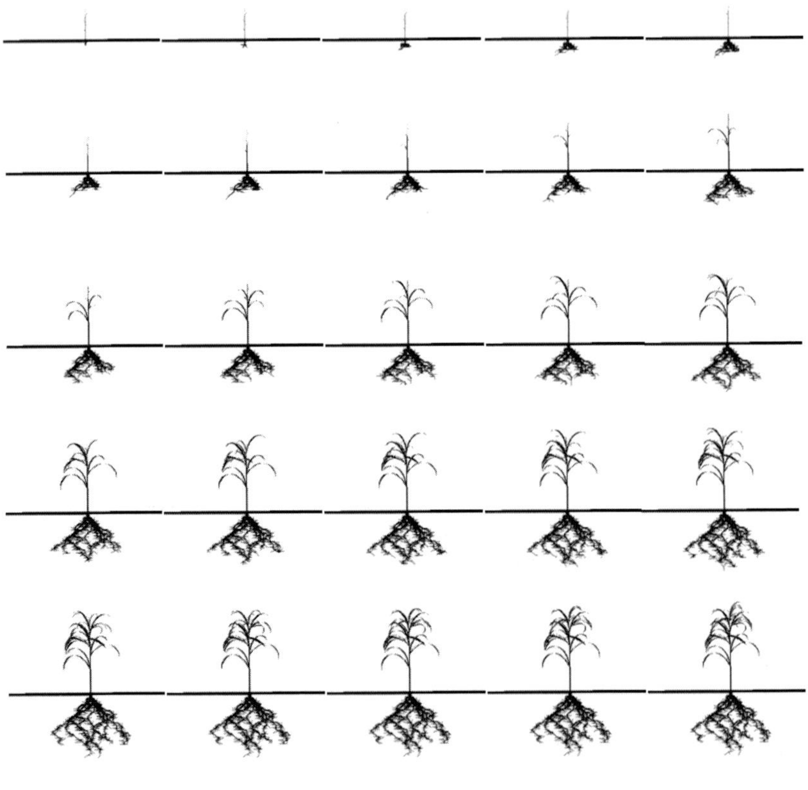

Fig. 7. The development of plant growth.

Table 3. An example of L-system prototype

Plant{	Root {
Shoot {	Iterations=0
Iterations=3	Angle=45
Angle=45	Diameter=0.5
Diameter=0.7	Axiom=i
Axiom=A	}
A=(0.33)A[+A]A[-A]A,(0.33)A[+A]A,(0.34)A[-A]A	}
Endrule	
A=IL	
}	

7 Conclusion

A prototype program called **PlantVR** has been presented to create the continuous development of plant shoot and root models by parametric functional symbols based on the bracketed L-system and stochastic L-system. The optional **Endrule** key word is added to the L-system in order to prevent some symbols that are not defined for plant definition in Table 1. The shoot and root structures are represented by L-system. The root structures can also been generated by nutrient concentration. The nutrient concentration is fixed in this research. The visualization technique makes the plant looking more realistic and every component can be controlled by the mathematical function. The five stages of growth function are allowed the plant components to be reduced their sizes or dead after the period of time. The development of plant growth seems to be smoother and more natural. This prototype can be used to generate a realistic model of the plant. This simulation can expand to the crop growth of a single type of plant without any interaction of environment. Our aims for further work are to investigate how the roots grow during to the changing of nutrient concentration in soil volume. The nutrient uptake will be calculated at each time step.

References

[1] Chuai-Aree, S., Siripant, S., Lursinsap., C.: Animation plant growth in L-system by parametric functional symbols. Proceeding of International Conference on Intelligent Technology 2000, **1**, 135–143 (2000)

[2] Chuai-Aree, S.: An algorithm for simulation and visualization of plant shoots growth. MA Thesis, Chulalongkorn University, Bangkok (2000)

[3] Chuai-Aree, S., Jäger, W., Bock, H.G., Siripant, S.: Smooth Animation for Plant Growth Using Time Embedded Component and Growth Function. East-West Journal of Mathematics, Special Volume, 285–295 (2002)
[4] Hammel, M.S., Prusinkiewicz, P.: Visualization of developmental processes by extrusion in space-time. Proceedings of Graphics Interface '96, 246–258 (1996)
[5] Mech, R., Prusinkiewicz, P.: Visual models of plants interacting with their environment. Proceedings in Computer Graphics (SIGGRAPH'96), 397–410 (1996)
[6] Prusinkiewicz, P., Lindenmayer, A.: The Algorithmic Beauty of Plants. Springer-Verlag, New York (1990)
[7] Prusinkiewicz, P., Hammel, M.S., Kajiya, J.T., Mjolsness, E.: Animation of plant development. Proceedings of Computer graphics SIGGRAPH'93, 351–360 (1993)
[8] Prusinkiewicz, P., James, M., Mech, R.: Synthetic topiary. Proceedings of SIGGRAPH' 94, 351–358 (1994)

Modelling of Time-dependent 3D Weld Pool Due to a Moving Arc

Minh Do-Quang and Gustav Amberg

Mechanics Department, Royal Institute of Technology
SE-100 44 Stockholm, Sweden
minh@mech.kth.se
gustava@mech.kth.se

Summary. It is well recognized that the fluid flow is an important factor in overall heat and mass transfer in molten pools during arc welding, affecting geometry of the pool and temperature distribution in the pool and in the HAZ. These in turn influence solification behavior, which determine the mechanical properties and quality of the weld fusion zone.

Here, a comprehensive numerical model of the time dependent weld pool flow in GTA welding, with a moving heat source has been developed. This model included heat transfer, radiation, evaporation, electromagnetic forces and Marangoni stress in the free surface boundary. With this 3D, fully time dependent model, the true chaotic time dependent melt flow is obtained.

The time dependent properties of flow velocities and temperature of numerical results are examined. It shows that the temperature fields are strongly affected by convection at the weld pool surfaces. The fluid flow in the weld pool is highly complex and it influences the weld pool's depth and width. Moreover, the velocity field at the surface of the specimen determines the streamlines defining the traveling paths of inclusions such as slag particles.

1 Introduction

Traditionally, in modeling welding phenomena there has been a focus either on the complex fluid and thermo-dynamics local to the weld pool, or the global thermo-mechanical behavior of the weld structure. A variety of simplified models is now frequently employed in the simulation of welding processes, but they are totally reliant on the accuracy of model parameters which describe the weld pool's size and shape. For GTA welding, the complex fluid flow in the weld pool is mainly driven by forces due to surface tension gradients, electromagnetic, buoyancy, arc pressure and aerodynamic drag force arising from the shielding gas used in GTA welding.

The main restriction in 3D GTA-welding simulation is the need for a sufficiently resolved grid, combined with the variety of physical mechanism present in the welding process. It also leads to a consideration of computational time limits. Even if the roles of various physical processes are understood in principle in two dimensions, there is some uncertainty in three dimensions, [16] e.g. some works focus on the influence of mass, momentum, and heat transfer on weld pool geometry [5, 11, 12, 13]. In those works, the fluid flow is considered as a laminar flow, while others considered the fluid flow as turbulent [4, 6, 10]. In several cases, the turbulent flow in the weld pool has been taken into account by enhancing both the viscosity and thermal conductivity [8, 9, 10]. More recently, DebRoy and coworkers [15, 14, 18] have also used the $k - \epsilon$ turbulence model in order to model the fluid flow in the weld pool.

There is little direct experimental evidence that suggests whether fluid flow in a weld pool is laminar, turbulent, or transitional. It has been estimated that the Reynolds number was varies from 1400 to 3000 [6] and up to 10000 [12]. On the other hand, neither the characteristic length scale most appropriate for determining the Reynolds number for a weld pool nor the critical Reynolds number for the transition from laminar to turbulent flow in a closed cavity flow such as in a weld pool have yet been established, so arguments concerning whether the flow is laminar or turbulent as based on a criterion for a critical Reynolds number are highly tenuous and indeed not valid yet. The physical dimensions of a weld pool are typically of the order of a few milimetres and the flow velocities in the weld pool can reach $1m/s$ or more. Hence, fluctuating velocities are inevitable when strong recirculating fluid flow occurs in a relatively small weld pool.

The aim of this work is therefore to develop a comprehensive numerical thermo-fluids model of a moving heat source, time dependent, three dimensions GTA welding which included heat transfer, radiation, evaporation, electromagnetic force and viscous stress in free surface boundary conditions. Moreover, fluctuating velocities and temperature are taken into account.

2 Physical phenomena

A schematic diagram of GTA welding is shown in Figure 1. A heat source (arc) is normally incident upon a horizontal flat plate used as the workpiece and is moving with a constant speed U_S along the edges of the two metal plates to be joined and the liquid fusion zone solidifies locally where the heat source has passed. The Cartesian coordinate system (x, y, z) in this study is taken to be fixed with the heat source.

The welding geometry as shown in Figure 1 could be divided into the region between electrode and the surface of workpiece and the workpiece region. Moreover, concerning the time-dependence of the welding process, three separate phases can be distinguished; The initial melting phase, when the electric current is actually melting the metal plate; the quasi-steady operation; and the final phase when the electric current is cut off. This leads to at least three

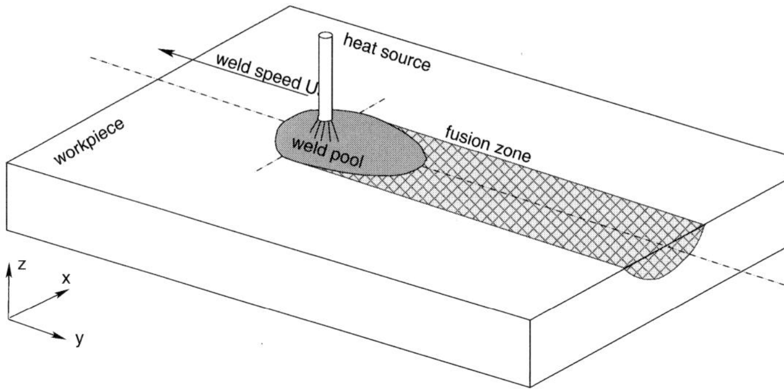

Fig. 1. Schematic drawing of the GTA welding, moving heat source

possibilities to group the physical mechanisms, identified as the initial transient phase, the work phase and the final transient. In this study the work phase is mainly considered. However, the area between electrode and workpiece surface cannot be neglected as there the thermal boundary conditions for the surface of the workpiece are established

3 Modelling approach

3.1 Mathematical Modelling

Governing equations

The governing transport equations are transformed into a coordinate system attached to the arc, which travels along the y direction at a constant speed U_s. The melt is treated as an incompressible Boussinesq liquid.

$$\nabla \cdot \mathbf{u} = 0 \tag{1}$$

$$\frac{\partial \mathbf{u}}{\partial t} + (\mathbf{u} \cdot \nabla)\mathbf{u} = -\frac{1}{\rho}\nabla p + \nu \nabla^2 \mathbf{u} + \mathbf{S} + \frac{\partial}{\partial y}(U_s \mathbf{u}) \tag{2}$$

where, $\mathbf{u}$ is the fluid velocity; p is the pressure, and $\mathbf{S}$ is the source term.

$$\mathbf{S} = \frac{1}{\rho}(\mathbf{J} \times \mathbf{B}) + \frac{\nu}{H(\chi)}\chi \mathbf{u} + \mathbf{g}\beta(T - T_{ref}) \tag{3}$$

The $\mathbf{J} \times \mathbf{B}$ terms are the Lorentz force components in the respective directions, where the solenoidal vectors $\mathbf{J}$ and $\mathbf{B}$ are the current density vector and magnetic flux vector. Those are calculated by solving the Maxwell's equations of the electromagnetic field in the domain of the workpiece. If the effect of fluid motion on the electromagnetic field is neglected, the Lorentz forces are assumed constant and they are computed numerically and included to the system equations as a known body force.

The second terms represent the porous medium model at the solid-liquid interface. χ is fraction of volume occupied by melt, χ has a value of 0 in the solid region, 1 in the completely molten region. $H(\chi)$ is the permeability of the mush [2]. $H(\chi)$ is infinite in the molten region and very rapidly decreases to very small values (say, 10^{-8}) in the solid region, so that the velocity $\mathbf{u}$ is become effectively zero in the solid region.

The additional source term $\mathbf{S_z}$ (equation 3) also includes the natural convection within the weld pool: $g\beta(T - T_{ref})$, where the Boussinesq approximation has been used. β is the volumetric expansion coefficient of welded material, g is the gravity acceleration, whereas T_{ref} is reference temperature ($288K$)

Assuming the material change directly from solid to liquid, we can neglect the presence of a mushy-zone. The fraction of volume occupied by solid can be canceled, according to [2],

$$\frac{\partial \chi}{\partial t} = \frac{1}{\kappa}(T - T_{liq}) \tag{4}$$

This equation allows a simple way of determining the solid fraction χ at the new time level by using the disretized form of the left-hand side: If χ tends to increase beyond 1 or decrease below 0, the new value of χ is taken to be 1 or 0, respectively. This means that T is then allowed to differ from T_{liq}, which, of course, is quite correct in the solid ($\chi = 0$) or the liquid ($\chi = 1$) region. The numerical parameter κ is a small constant, used to amplify any deviation of T from T_{liq} and results in a rapid melting or freezing that restores T to T_{liq}.

The energy conservation is express as follows:

$$\frac{\partial T}{\partial t} + (\mathbf{u} \cdot \nabla)T - \frac{\partial (U_s T)}{\partial y} = \alpha \Delta T + \frac{L^*}{\rho C_p}\left(\frac{\partial \chi}{\partial t} - \frac{\partial (U_s \chi)}{\partial y}\right) \tag{5}$$

where α is the thermal diffusivity, C_p is the specific heat and L^* is the latent heat of fusion. The two terms containing U_s appear as a result of the coordinate transformation.

Boundary conditions

Since the present formulation is based on a single domain for both liquid and solid, the boundary conditions can be specified for entire workpiece domain, there are no explicit boundary condition at the solid-liquid interface. The following are the boundary conditions for the primitive variables (u, v, w, and T).

Surface of the weld pool: At the surface of the weld pool the heat input from the arc has to be simulated and heat losses due to evaporation, convection and radiation has to be considered. The heat flux distribution at the top was supposed to be Gaussian and can be expressed by the following equation

$$q_{gauss} = \frac{3Q}{\pi a^2} exp\left(-\frac{3(x^2+y^2)}{a^2}\right) \tag{6}$$

where $Q = EI\eta$ is the total heat input, with the arc voltage E, the electric current I, the arc efficiency η and a is the effective radius of heat distribution. The evaporation heat flux is computed by follow [7]

$$q_{evap} = W \cdot h_{fg} \tag{7}$$

where W denotes the evaporating heat flux and h_{fg} is the heat of evaporation. The radiative q_{rad} and convective q_{conv} heat losses can be calculated by using

$$q_{rad} = \sigma_b \epsilon (T^4 - T_a^4) \text{ and } q_{conv} = h_c(T - T_a) \tag{8}$$

where σ_b is the Stefan-Boltzmann constant, ϵ the emissivity of steel, h_c the convection heat transfer coefficient, expressing the convective heat exchange between the top surface and the environment, and T_a is the ambient temperature. The complete thermal boundary condition becomes then,

$$-k\frac{\partial T}{\partial z} = -q_{gauss} - q_{evap} - q_{rad} - q_{conv} \tag{9}$$

with the thermal conductivity k.

Because of large temperature gradients on the pool surface, the surface tension, which is temperature dependent, varies greatly. Liquid is pulled in the direction of increasing surface tension. The surface tension-driven flow at the free surface is described by

$$-\mu\frac{\partial u}{\partial z} = \frac{\partial \gamma}{\partial T}\frac{\partial T}{\partial x} \text{ and } -\mu\frac{\partial v}{\partial z} = \frac{\partial \gamma}{\partial T}\frac{\partial T}{\partial y} \tag{10}$$

Here, μ is the dynamic viscosity of the fluid and $\frac{\partial \gamma}{\partial T}$ is the coefficient of surface tension. The vertical velocity w has to be zero at the top surface.

Side wall: At the front and back side walls of the plate, heat transfer is considered. According to the no slip condition, velocity components are zero at the side walls

$$k\Delta T = -\alpha_2(T - T_a) \quad \text{and} \quad u = v = w = 0 \tag{11}$$

where α_2 is a combined heat transfer coefficient for the radiative and conductive boundary condititions.

3.2 Numerical Modelling

The numerical simulations of weld pool heat and fluid flow were carried out using the **femLego** [1] tool to create a set of C and Fortran code from mathematical equations which were written in Maple format. The model in this project can simulate the physical phenomena mentioned above, such as surface tension gradients, electromagnetic, buoyancy, etc., in the weld pool with moving heat source.

The solution of whole set of equations (1)-(5) can be described as follows: With a given temperature at the previous time level, the phase of the material χ (liquid or solid) is computed using equation (4). From the new solid fraction value, the amount of released latent heat of fusion in the energy equation is computed and the temperature can then be obtained at the new time level. Using the new temperature and solid fraction, a pressure and velocity field is obtained from equation (1) and (2). Now all unknowns have been computed at the new time level and the entire procedure restarts.

The equations have been discretised using a finite element approach on an unstructured grid. The type of elements which are used for both velocity, pressure and temperature variables are piecewise linear tetrahedral elements. The restrictions imposed by Babuska-Brezzi condition are avoided by adding a pressure stabilization term. The pressure and velocity solutions are split using a fractional step algorithm. The Poisson equation for the pressure is solved using the well known Conjugate-Gradient method. The convective term in equation (2) and (5) are calculated implicitly using the GMRES method. In this way a reasonably large time step can be used in the computations. In order to add stability without loosing any accuracy we also used a streamline-diffusion method for the convection terms in equations (2) and (5). The spatial derivatives used guarantee a second-order accuracy.

To determine the overall stability of the numerical scheme we introduced a Courant number

$$Co = u_{max} \frac{dt}{dx} \tag{12}$$

where u_{max} is the maximum velocity in the x-direction in the domain and dx is the spatial step size. With the present semi-implicit treatment of convective terms, a practical value of Courant number should not increase above a critical value of 2.5. Here, minimum grid space was 0.02 mm and maximum velocity can reach up to 1 ms^{-1}. Hence, the maximum time step can be applied in this case are 5×10^{-5} s

Parallel computation

A new parallel version of **femLego**, a finite-element tool for partial differential equations, [3] was developed within the scope of this project. The code is implemented using MPI and achieves parallelism by spatially decomposing the problem into an unstructured collection of structured blocks. The algorithm yields excellent scaling on both the shared memory and distributed memory architecture machines, with the latter yielding a performance improvement factor approximately linear with the numbers of processors used. Table 1.

Once a distributed matrix is created, the stiffness matrix A and r.h.s. vector b are computed locally on each processor. Here, the matrix A is already partitioned into blocks of rows, with each block assigned to one processor. The associated components of unknowns and r.h.s. vectors are distributed accordingly. Communication may be needed to transfer some components of x. For example, in $y \leftarrow Ax$, if y_i is update on processor p_1, $A_{ij} \neq 0$, and x_j is owned by processor p_2, then p_2 must send x_j to p_1. In general, a processor may need more than one x component from another processor. The number of components that must be updated is minimized by using the option "Fill-reducing reordering" that provided by Metis.

Table 1. Performance and Speed-up of parallel computing

	SP2 machine			PC-cluster		
CPUs	Time (s)	Speed-up	Efficiency	Time (s)	Speed-up	Efficiency
1	9800.37	1.0000	100.00%	14660.3	1.0000	100.00%
2	5000.88	1.9597	97.99%	7346.3	1.9956	99.78%
4	2816.38	3.4798	86.99%	3848.6	3.8092	95.23%
8	1236.00	7.9291	99.11%	1955.9	7.4952	93.69%
12	937.09	10.4583	87.15%	1472.1	9.9585	82.99%
15	824.50	11.8864	79.24%			

Table 1 shows the calculation times spent on an IBM-SP2 machine and a PC-cluster. Clearly, the speed-up is quite acceptable on expensive machine (IBM-SP2) and also on cheapest one, PC-cluster.

4 Results and Discussion

4.1 3D simulation of the welding pool with a stationary arc

In order to validate the code, a three dimensional simulation of welding pool with stationary heat source is carried out. The material is stainless steel type 304 (AP1) which has an extra-low sulfur (0.0005%). The welding current is 100A, arc voltage 10.6V. In comparison with 2D and experimental results (from [12], $(C), (D)$ and (E)),a quite good qualitative and quantitative agreement in the weld pool shapes can be observed in 3D-solutions after $1s$ (Figure 2 $(A), (B)$). In the 3D simulation, the weld pool radius and depth became 2.3 mm and 0.6 mm, by comparison with the 2D simulation results of 2.07 and 0.58 and experimental results for radius and depth, 2.25 and 0.52, respectively, show a very good agreement.

As being shown in the previous research, there was an inward motion in the 3D weld pool, that drove the fluid flows in a weld pool and created two vortices in opposite directions. They deepened the groove at the periphery even more than in the center. This occurs for a material has low sulfur only, when the sign of temperature coefficient of surface tension ($\frac{\partial \gamma}{\partial T}$) would be changed at low temperature, around $1800K - 2000K$.

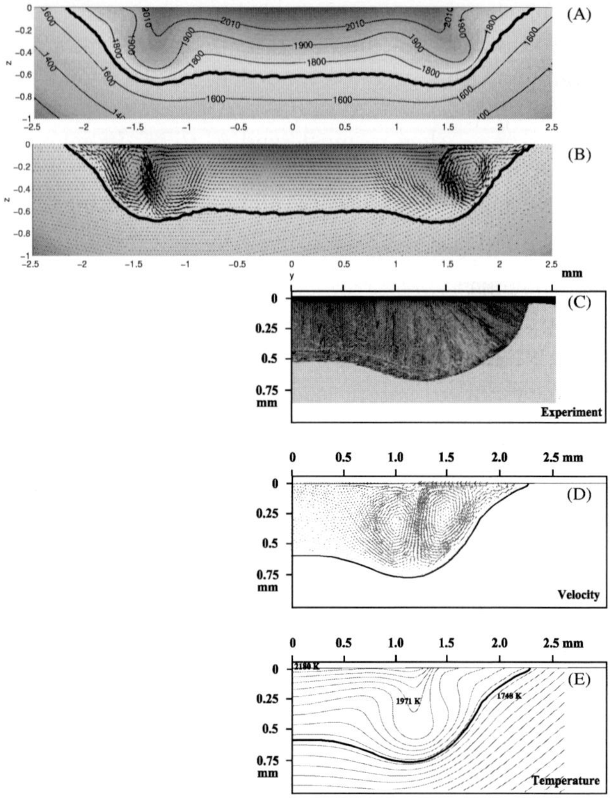

Fig. 2. AP1-100A: Comparison of experimental and 2D, 3D numerical results after 1 second, (C)-(E) from [12]. The colour version of this figure can be found in Fig. A.13 on page 583.

The other numerical result for a high content of the surface active element sulfur in the base material (0.0139%) is shown on figure 3D. In this case, we obtained almost the same conclusions about the depth, width and shape of the weld pool with the previous works [12] and [17]. But the fluid flows in this case became much more complex since it seemed to be unstable at the center of the weld pool. This causes the flow to become asymmetry.

4.2 Effect of moving heat source speed

Figure 3 shows the temperature and velocity distributions for 3D-GTA welding simulations. The electrode (at position $y = 1.25$) is moving from right to left with a constant speed: $U_s = 0mm \cdot s^{-1}$ (A); $3mm \cdot s^{-1}$ (B); $6mm \cdot s^{-1}$ (C) and $9mm \cdot s^{-1}$ (D) respectively. The material is stainless steel type 304 (AP5) with a high sulfur 0.0139%. GTA conditions are: $I = 100A$, $U = 10.6V$. The isotherms are shown on the upper surface of the plate and on the vertical

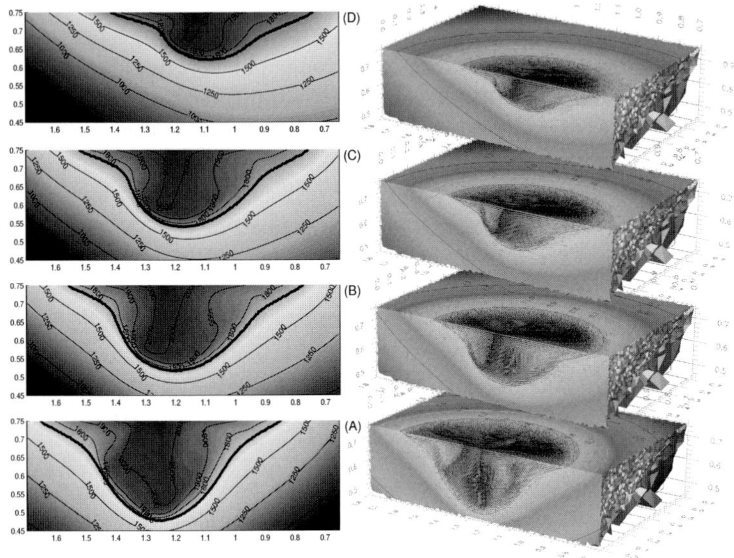

Fig. 3. AP5-100A: Temperature and velocities fields of welding pool. $U_s = 0mm \cdot s^{-1}$ (A); $3mm \cdot s^{-1}$ (B); $6mm \cdot s^{-1}$ (C) and $9mm \cdot s^{-1}$ (D). The colour version of this figure can be found in Fig. A.14 on page 584.

symmetry plane oriented along the traveling direction of the electrode. In this case, there are only one motion toward to the center of the pool, since the temperature that is needed to change the sign of the coefficient of surface tension is $2300K$, greater than the maximum temperature of the weld pool.

It is observed that the temperature fields are strongly affected by convection, at the weld pool surfaces. The fluid flows in the weld pool are highly complex and these influence the weld pool's depth and width. When the heat source is moving, the volume of the weld pool is decreasing and the centroid of the weld pool is moving along behind the arc. Moreover, the velocity field at the surface of the specimen determines the streamlines defining the traveling paths of inclusions such as slag particles.

An overview of the numerical results for the weld pool with moving heat source on stainless steel type 304 can be found in Table 2.

Table 2. Summary of numerical results and effective of moving heat source

Moving speed ($mm\ s^{-1}$)	U_{max} ($m\ s^{-1}$)	T_{max} (K)	pool width (mm)	pool depth (mm)
0	0.9394	2193	3.9808	1.3589
3	0.9214	2170	3.7806	0.9324
6	0.9166	2115	3.6150	0.8354
9	0.9085	2099	2.9778	0.5581

4.3 Time dependence

The simulation of a weld pool with a stationary heat source is used to study the time dependent problem. The material is high sulfur content (0.0139%) stainless steel (AP5-100A). The welding conditions are the same as in the previous sections.

The evolution of temperature of three testing points are shown in Figure 4. The three positions of testing points are defined at the center line in the z direction of the weld pool. It is observed that the weld pool is fully developed after 1 s and the temperature at three testing points are time dependent. The highest amplitude of temperature fluctuations is 50 K at the "top point" and 25 K for the "bottom point". The highest temperature in the weld pool is less than $2300K$, the critical temperature in order to change the sign of temperature coefficient of surface tension ($\frac{\partial\gamma}{\partial T}$). Therefore, there is only one vortex in the pool.

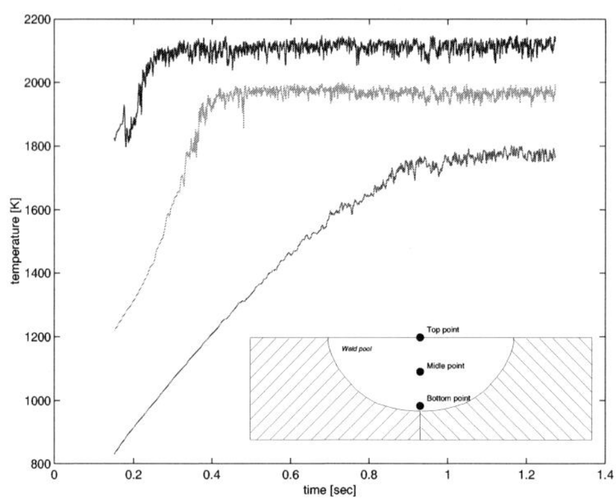

Fig. 4. The temperature evolution of three testing points

The temperature distributation and velocity field at the symmetry plane $x = 0$ at the different times $t = 1.0\ s$, $t = 1.06\ s$, $t = 1,12\ s$ and $t = 1.18\ s$ are shown in Figure 5. The flow fields are in general quite complex and highly time dependent (chaotic), as is visible in all plots. The physical dimension of this weld pool is approximately 4 mm, the flow velocities in the weld pool can reach 1 ms^{-1} and the kinematic viscosity of the steel $\nu = 6.81 \times 10^{-7}\ m^2 s^{-1}$. Hence, the Reynolds number is $Re \approx 5800$. At the center of the weld pool, below the arc position, the flow velocity is maximum ($\approx 1\ ms^{-1}$) and there is an instability flow which occurred by a radially inward flows on the free surface. The area of the liquid that has $T > 2000\ K$ looks like a column at the center of the pool and is obviously unstable. It swings to left then right at the different times. See figures (5A, 5B, 5C, 5D). The depth, width and even the shape of the weld pool remain the same during the study, even through some small oscillation is present here to.

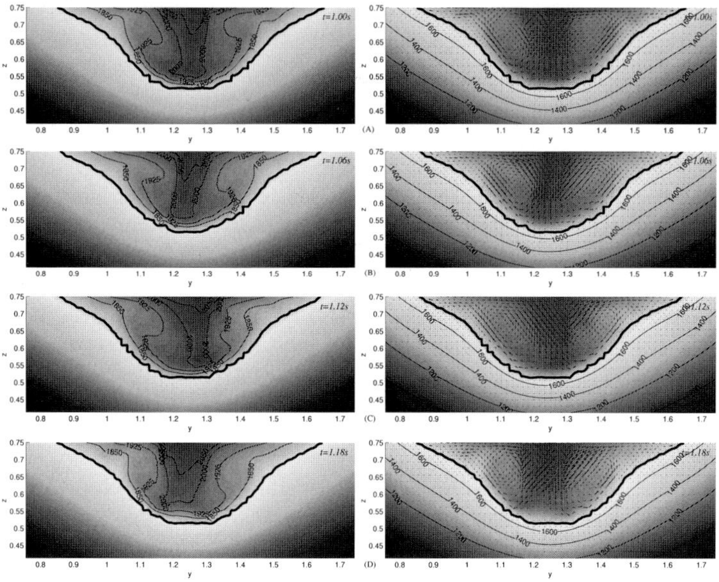

Fig. 5. 3D simulation, temperature distributation and velocity field time dependent. The colour version of this figure can be found in Fig. A.15 on page 584.

There are two big vortex rings in figure 5, that appeared almost in the whole weld pool. They keep the shape of the pool remains unchanged even the fluid flows of the pool at the center is unstable. Otherwise, when a strong recirculating fluid flow occurs in a relatively small weld pool such as this fluctuating velocities are inevitable. This is can be a reason for the appearance of the highly time dependent (chaotic) flow in the weld pool.

5 Conclusions

It has been shown successfully that it is possible to simulate GTA welding in three dimensions by using the **femLego** tool, a numerical simulation tool by symbolic computation.

The present 3D modelling study shows that the welding speed may affect the melt flow and temperature distribution in GTA welding. The melt flow field predicted by the present 3D moddeling is much more complex than that obtained from the previous 2D modelling.

We first show the highly time dependent (chaotic) problem in GTA welding. This appears when a strong recirculating fluid flow occurs in a small weld pool. A systematic research of the properties of this time dependence needs to be further study.

References

[1] Numerical simulation by symbolic computation. *http://www.mech.kth.se/˜gustava/femLego.*

[2] G. Amberg. Computation of macrosergretation in an iron-carbon cast. *Int. J. Heat Mass Transfer*, 34(1):17–227, 1991.

[3] G. Amberg, R. Tönhardt, and C. Winkler. Finite element simulations using symbolic computing. *Mathematics and Computers in Simulation*, 49:149–165, 1999.

[4] R.T.C. Choo and J. Szekely. The posible role of turbulence in gta weld pool behavior. *Weld J.*, 73(2):25s–31s, 1994.

[5] T. DebRoy. Mathematical modeling of fluid flow and heat transfer in fusion welding. *Mathematical Modelling of Weld Phenomena*, 5:1–31, 2001.

[6] K. Hong, D.C. Weckman, A.B. Strong, and W. Zheng. Modelling turbulent thermofluid flow in stationary gas tungsten arc weld pools. *Sci. Technol. Weld. Joi.*, 7(3):125–136, 2002.

[7] C.S Kim. Thermophysical properties of stainless steels. Technical report, Argonne National Laboratory, 1975.

[8] K. Mundra, J.M. Blackburn, and T. DebRoy. Absorption and transport of hydrogen during gma welding of low alloy steel. *Sci. Technol. Weld. Joi.*, 2(4):174–184, 1997.

[9] K. Mundra, T. DebRoy, and K.M. Kelkar. Numerical prediction of fluid flow and heat transfer in welding with a moving heat source. *Numer. Heat Tr. A*, 29(2):115–129, Feb. 1996.

[10] T.A Palmer and T. DebRoy. Numerical modeling of enhanced nitrogen dissolution during gtaw. *Metall. Mater. Trans. B*, 31B:1371–1385, 2000.

[11] M. Ushio and C.S. Wu. Mathematical modeling of three-dimensional heat and fluid flow in a moving gas metal arc weld pool. *Metall. Mater. Trans. B*, 28B:509–516, June 1997.

[12] C. Winkler, G. Amberg, H. Inoue, and T. Koseki. A numerical and experimental investigation of qualitatively different weld pool shapes. *Mathematical Modelling of Weld Phenomena 4*, pages 37–69, 1998.

[13] T. Xacharia, A.H. Eraslan, D.K Aidun, and S.A. David. Three-dimensional transient model for arc welding process. *Metall. Trans. B*, 20B:645–659, 1989.

[14] Z. Yang and T. DebRoy. Modeling macro-and microstructures of gas-metal-arc welded hsla-100 steel. *Metall. Mater. Trans. B*, 30(3):483–493, Jun. 1999.

[15] Z. Yang, J.W. Elmer, J. Wong, and T. DebRoy. Evolution of titanium arc weldment macro and microstructures - modeling and real time mapping of phases. *Weld J.*, 79(4):97S–112S, Apr. 2000.

[16] XH. Ye and X. Chen. Three-dimensional modelling of heat transfer and fluid flow in laser full-penetration welding. *J. Phys. D: Appl. Phys.*, 35:1046–1056, 2002.

[17] T. Zacharia, S.D. David, J.M Vitek, and T: Debroy. Weld pool dvelopment during gta and laser beam welding of type 304 stainless steel, part ii - experimental correlation. *Welding Journal Research Supplement*, 68:510s–519s, 1989.

[18] H. Zhao and T. DebRoy. Weld metal composition change during conduction mode laser welding of aluminum alloy 5182. *Metall. Mater. Trans. B*, 32(1):163–172, Feb. 2001.

Nonlinear Optimization in Gas Networks*

Klaus Ehrhardt and Marc C. Steinbach

Zuse Institute Berlin (ZIB), Takustr. 7, 14195 Berlin, Germany
ehrhardt@zib.de
steinbach@zib.de

Summary. The operative planning problem in natural gas distribution networks is addressed. An optimization model focusing on the governing PDE and other nonlinear aspects is presented together with a suitable discretization for transient optimization in large networks by SQP methods. Computational results for a range of related dynamic test problems demonstrate the viability of the approach.

Key words: Gas networks, operative planning, large-scale optimization

1 Introduction

Natural gas is a primary energy source of increasing relevance, mainly due to its favorable environmental properties. In Germany, gas is primarily used for heating, industries, and power generation. Our industry partner Ruhrgas operates the largest gas network in Germany, a combined transport and distribution network with roughly 11000 km of pipes and 26 compressor stations requiring up to 831 MW. The annual demand in 2002 was 612000 GWh, with daily demands varying between 773 and 3109 GWh.

The current paper is concerned with operative planning, or transient technical optimization (TTO), in gas networks. This planning level addresses the task of controlling the network load distribution over the next 24 to 48 hours to satisfy the actual demand subject to physical, technical, and contractual constraints as well as target values for gas production, storage, purchase, and sale determined by the mid-term planning. The objective is to minimize the variable operating costs, which are dominated by the cost for the gas transport, i.e., the fuel consumption of compressors.

Due to reliable temperature forecasts we can neglect demand uncertainty and hence use a deterministic model, but the operative planning problem involves PDE constraints (gas flow) as well as substantial combinatorial aspects

* This work has been supported by the Federal Ministry of Education and Science (BMBF) under grant 03STM5B4.

(switching compressors on or off, opening or closing valves), leading to a currently intractable mixed-integer PDE boundary value problem. Here we focus on the nonlinear aspects, assuming that combinatorial decisions are externally given—ideally by an enclosing mixed-integer optimization framework.

Typical subjects of the earlier literature are *steady-state* optimization in tree-structured networks by dynamic programming [16], later surveyed in [2], sequential linearization for nonlinear mixed-integer models on general network topologies [10], optimization of single compressor stations by simulated annealing [17], or evaluation criteria for mixed-integer methods and lower bounds on the total energy consumption [1]. For *transient* network optimization with given binary decisions, a gradient method is proposed in [13] based on a highly detailed model [8] that is used in the commercial simulation tool SIMONE; an extended simplex method for a largely simplified model is developed in [15]. The related problem of controlling networks of open water channels is studied in [5]. First approaches for the *mixed-integer TTO* problem, with rather coarse approximations of nonlinearities, are developed in [12] and later in [6]. To address the full TTO problem, our own work aims at a future integration with related mixed-integer approaches that are currently being developed in a partner project [9]. A second partner project studies stochastic aspects [7].

2 Optimization Model

In the following we first describe physical and technical restrictions and the cost function of the TTO problem with prescribed binary decisions. Then we present appropriate space and time discretizations and add contractual restrictions to obtain a large-scale nonlinear optimization problem (NLP).

2.1 Planning Horizon

At Ruhrgas, the end of a *gas day* at 6:00 a.m. is a natural choice for the end of the horizon since the state of the network at this time is well predictable. A reoptimization is to be performed every hour. Therefore we operate with a *breathing horizon* of length 24 to 48 hours that starts at the next full hour and ends after the following gas day. In the current paper we are only concerned with a single optimization problem and denote time by $t \in [0, t_e]$.

2.2 Pipes

Consider a pipe segment with circular cross section. Let D denote the diameter and L the length. Independent variables are the distance $x \in [0, L]$ and time t; relevant dynamic variables are the gas density ρ, velocity v, flow rate $q = \rho v$, pressure p, and temperature T (all depending on x and t). The altitude of the pipe at x is denoted as $h(x)$. The following material is based on [11, 14].

The gas flow in the pipe is governed by thermodynamic conservation laws. Conservation of mass (continuity equation) and of momentum (pressure loss equation) form a hyperbolic PDE system that is coupled with the equation of state for a real gas. Under a given temperature field $T(x,t)$, the latter replaces an (impractical) PDE modeling heat exchange with the ground (cf. [14]),

$$\partial_t \rho + \partial_x q = 0 \,, \tag{1}$$

$$\partial_t q + \partial_x p + \partial_x(\rho v^2) + g\rho\partial_x h = -\frac{\lambda(q)}{2D}\rho v|v| \,, \tag{2}$$

$$p = \gamma(T) z(p,T)\rho \,. \tag{3}$$

Here γ is a constant field given by T, and the compressibility $z(p,T)$ describes the deviation of the real gas from ideal behavior ($z = 1$). For pressures up to 70 bar, the latter is empirically modeled according to the formula of the American Gas Association (AGA),

$$z(p,T) = 1 + 0.257(p/p_{\mathrm{c}}) - 0.533\frac{(p/p_{\mathrm{c}})}{(T/T_{\mathrm{c}})} \,, \tag{4}$$

where p_c and T_c denote the pseudo-critical pressure and temperature of the gas mixture, respectively.

For the usual turbulent flow in gas pipes, the friction coefficient $\lambda(q)$ in (2) is empirically modeled by the implicit formula of Prandtl–Colebrook,

$$\frac{1}{\sqrt{\lambda(q)}} = -2\log_{10}\left(\frac{2.51}{\mathrm{Re}(q)\sqrt{\lambda(q)}} + \frac{k}{3.71D}\right) . \tag{5}$$

Here k is the pipe rugosity (the roughness of its inner surface) and, with η denoting the dynamic viscosity of the gas, $\mathrm{Re}(q)$ is the Reynolds number,

$$\mathrm{Re}(q) = \frac{D}{\eta} q \approx 10^6 \,. \tag{6}$$

In what follows we scale (1)–(3) by the cross-sectional pipe area, $A = \frac{\pi}{4}D^2$, or by $1/A$. Without changing notation, we replace the density with the linear density, $\rho \leftarrow A\rho$ (kg/m), the flow rate with the mass flow, $q \leftarrow Aq$ (kg/s), and $\gamma \leftarrow \gamma/A$, leaving (1) and (3) unchanged. Equation (2) is then simplified by dropping the kinetic energy terms $\partial_t q$ and $\partial_x(\rho v^2)$ (which contribute less than one percent to the sum of all terms under normal operating conditions; see [14]) and substituting $v = q/\rho$ in the hydraulic resistance term,

$$A\partial_x p + g\rho\partial_x h = -\frac{\lambda(q)}{2D}\frac{q|q|}{\rho} \,. \tag{7}$$

2.3 Compressors

The physical process of increasing the pressure of a gas flow q from p_{in} to p_{out} (under adiabatic conditions) requires the theoretical power

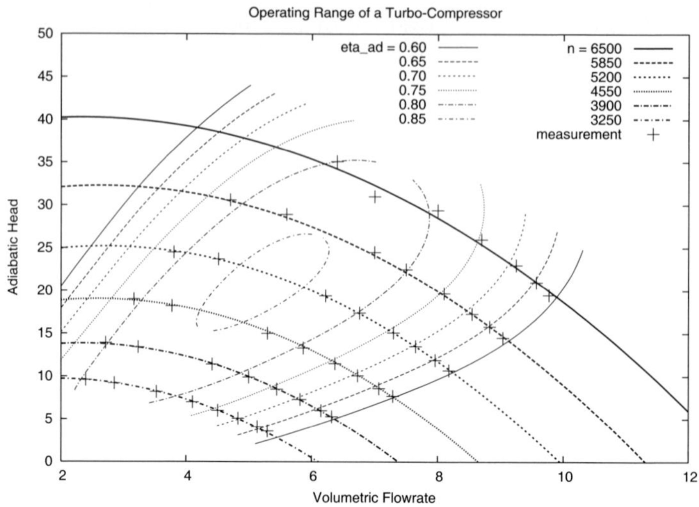

Fig. 1. Characteristic diagram of a turbo-compressor with measurements (+) and isolines for speed (`n`, *thick lines*) and adiabatic efficiency (`eta_ad`, *thin lines*)

$$N_{\text{th}} = c_1 q H_{\text{ad}} = c_1 q\, c_2 z(p_{\text{in}}, T_{\text{in}}) T_{\text{in}} \frac{\kappa}{\kappa - 1} \left[\left(\frac{p_{\text{out}}}{p_{\text{in}}} \right)^{\frac{\kappa-1}{\kappa}} - 1 \right] . \tag{8}$$

Here c_1, c_2 are constants, H_{ad} denotes the adiabatic head, and κ denotes the isentropic exponent of the gas.

In the following we concentrate on turbo-compressors (also called centrifugal compressors) driven by gas turbines. The technical behavior of this most common compressor type is described by a characteristic diagram as shown in Fig. 1. The speed n and the adiabatic efficiency η_{ad} of the compressor can both be modeled as bivariate quadratic polynomials in q and H_{ad} obtained as fits to discrete measurements. The fuel consumption per time unit then reads

$$B = c_3 N b(n, N) \tag{9}$$

where N is the actual power consumption of the compressor, $N = N_{\text{th}}/\eta_{\text{ad}}$, and b is its (empirical) specific fuel consumption. The fuel B is taken directly from the gas flow; it amounts to roughly 0.5% of the compressor throughput.

For our purposes a simplified model is sufficient. We consider entire compressor stations (typically consisting of about half a dozen compressor units) as a single *idealized* compressor satisfying (8) and having constant efficiency and specific fuel consumption, so that $B = \tilde{c}_3 N_{\text{th}}$ with $\tilde{c}_3 = c_3 b/\eta_{\text{ad}}$.

2.4 Further Elements

Real gas networks contain further active and passive elements of which the following are relevant in our context.

Connections are short pipes with a constant relative pressure loss $c \in (0, 1]$ and with identical input and output flows,

$$p_{\text{out}} = c p_{\text{in}} \,, \quad q_{\text{out}} = q_{\text{in}} \,. \tag{10}$$

Valves can be open or closed, with corresponding pressure and flow relations

$$p_{\text{out}} = p_{\text{in}} \,, \quad q_{\text{out}} = q_{\text{in}} \quad (\text{open}) \,, \quad q_{\text{in}} = q_{\text{out}} = 0 \quad (\text{closed}) \,. \tag{11}$$

Regulators (or **control valves**) are valves that reduce the gas pressure by a controlled amount $\Delta p \in [\Delta p_{\min}, \Delta p_{\max}]$ when open,

$$p_{\text{out}} = p_{\text{in}} - \Delta p \,, \quad q_{\text{out}} = q_{\text{in}} \quad (\text{open}) \,, \quad q_{\text{in}} = q_{\text{out}} = 0 \quad (\text{closed}) \,. \tag{12}$$

2.5 Network

The network topology is modeled by a directed graph $G = (\mathcal{N}, \mathcal{A})$ whose vertex set consists of provider nodes $\mathcal{N}_+$ (sources), customer nodes $\mathcal{N}_-$ (sinks), and interior nodes $\mathcal{N}_0$ (junctions),

$$\mathcal{N} = \mathcal{N}_+ \cup \mathcal{N}_- \cup \mathcal{N}_0 \,. \tag{13}$$

The set of arcs consists of the network elements described above: pipes $\mathcal{A}_{\text{pi}}$, connections $\mathcal{A}_{\text{cn}}$, compressors $\mathcal{A}_{\text{cs}}$, valves $\mathcal{A}_{\text{vl}}$, and regulators $\mathcal{A}_{\text{rg}}$,

$$\mathcal{A} = \underbrace{\mathcal{A}_{\text{pi}} \cup \mathcal{A}_{\text{cn}}}_{\text{passive}} \cup \underbrace{\mathcal{A}_{\text{cs}} \cup \mathcal{A}_{\text{vl}} \cup \mathcal{A}_{\text{rg}}}_{\text{active (controlled)}} \,. \tag{14}$$

Individual arcs will be denoted as $a \in \mathcal{A}$ or, using the tail and head $i, j \in \mathcal{N}$, as $ij \in \mathcal{A}$. Below we always require that the flow is directed from i to j.

2.6 Objective

The objective consists in minimizing the variable operating costs, that is, the total fuel costs of all compressors,

$$\text{cost}_{\text{cs}} = \sum_{a \in \mathcal{A}_{\text{cs}}} c_a \int_0^{t_e} B_a(t)\, dt \,. \tag{15}$$

2.7 Terminal constraint

Since the operative planning problem has a finite horizon, some terminal constraint is required to ensure reasonable operating conditions for $t > t_e$. Otherwise the optimization would yield abnormally low values of pressure and network gas content at $t = t_e$ to reduce compressor duties and thus fuel costs.

A straightforward approach to prevent such behavior consists in placing a lower bound on the total gas mass at the end of the horizon (where we neglect the small amount of gas in non-pipe elements),

$$m_{\min} \le \sum_{a \in \mathcal{A}_{\text{pi}}} \int_0^{L_a} \rho_a(x, t_e)\, dx \,. \tag{16}$$

2.8 Discretization

A suitable discretization for transient optimization of large networks should be coarse but reliable (i.e., at least mass-conserving), and stable. We use a full a priori discretization in space and time. Only pipes are relevant for the space discretization; all other arc types are assumed to have zero length. For each pipe we use the "grid" consisting of both end points only, $\Gamma_a = \{0, L_a\}$, $a \in \mathcal{A}_{\text{pi}}$. (A physical pipe may consist of arbitrarily many arcs $a \in \mathcal{A}_{\text{pi}}$.) The planning horizon $I = [0, t_e]$ is divided equidistantly to yield a time grid $\Gamma_I = \{0, 1, 2, \ldots, t_e\}$. (With no loss of generality the length of subintervals is chosen as time unit here, $\Delta t = 1$.) The variable vector is denoted

$$x = (x_0, \ldots, x_{t_e}) \quad \text{with} \quad x_t = (p_t, q_t, s_t, u_t)\,, \quad t = 0, \ldots, t_e\,. \tag{17}$$

Here the states at time t consist of node pressures $p_t = (p_{it})_{i \in \mathcal{N}}$, arc flows $q_t = (q_{it}, q_{jt})_{ij \in \mathcal{A}}$, and further states in pipes and compressors, $s_t = (s_{at})_{a \in \mathcal{A}}$, while the (piecewise constant) controls on subinterval $(t-1, t)$ are the pressure changes $u_t = (u_{at})_{a \in \mathcal{A}}$ in compressors and regulators:

$$\begin{aligned} s_{ijt} &= \rho_{jt}\,,\ ij \in \mathcal{A}_{\text{pi}}\,,\ u_{at} = \Delta p_{at}\,,\ a \in \mathcal{A}_{\text{cs}}\,,\\ s_{at} &= N_{at}\,,\ a \in \mathcal{A}_{\text{cs}}\,,\ \ u_{at} = \Delta p_{at}\,,\ a \in \mathcal{A}_{\text{rg}}\,. \end{aligned} \tag{18}$$

In all other arc types, s_{at} and u_{at} are empty.

The PDE and equation of state are discretized with implicit Euler schemes in space and time, yielding for $a = ij \in \mathcal{A}_{\text{pi}}$, $t \in \{1, \ldots, t_e\}$, and $t_- = t - 1$:

$$\frac{\rho_{jt} - \rho_{jt_-}}{\Delta t} + \frac{q_{jt} - q_{it}}{L_a} = 0\,, \tag{19}$$

$$A_a \frac{p_{jt} - p_{it}}{L_a} + g\rho_{jt} \frac{h_j - h_i}{L_a} + \frac{\lambda(q_{jt})}{2D_a} \frac{q_{jt}^2}{\rho_{jt}} = 0\,, \tag{20}$$

$$p_{jt} - \gamma(T_{jt}) z(p_{jt}, T_{jt}) \rho_{jt} = 0\,. \tag{21}$$

The simplified fuel consumption model for compressor $a = ij \in \mathcal{A}_{\text{cs}}$ reads

$$B_{at} = \tilde{c}_{3a} N_{at} = c_{1a} c_{2a} \tilde{c}_{3a} q_{jt} z(p_{it}, T_{it}) T_{it} \frac{\kappa}{\kappa - 1} \left[\left(\frac{p_{jt}}{p_{it}} \right)^{\frac{\kappa-1}{\kappa}} - 1 \right]. \tag{22}$$

The integrals in the objective and in the terminal constraint are approximated by the trapezoidal rule with respect to the grids Γ_a and Γ_I, respectively,

$$\text{cost}_{\text{cs}} = \sum_{a \in \mathcal{A}_{\text{cs}}} c_a \sum_{t=1}^{t_e} \frac{B_{at_-} + B_{at}}{2} \Delta t\,, \quad m_{\min} \le \sum_{ij \in \mathcal{A}_{\text{pi}}} L_{ij} \frac{\rho_{it_e} + \rho_{jt_e}}{2}\,. \tag{23}$$

This completes the nontrivial arc equations, which are coupled by flow balance equations in every internal node $j \in \mathcal{N}_0$ for $t \in \{1, \ldots, t_e\}$,

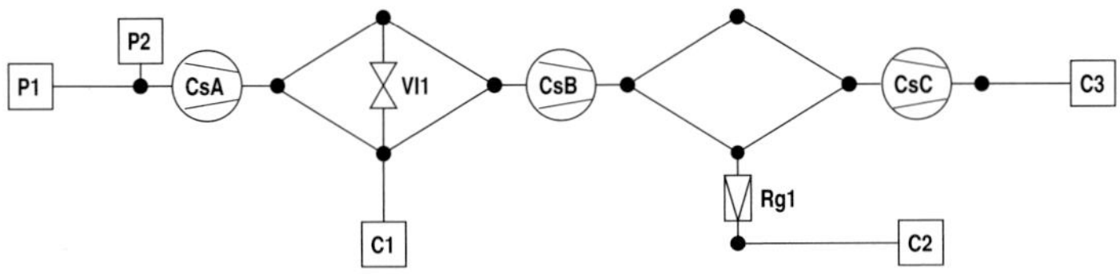

Fig. 2. Flowchart of the test network

$$\sum_{i:\, ij \in \mathcal{A}} q_{ijt} = \sum_{k:\, jk \in \mathcal{A}} q_{jkt} \; . \tag{24}$$

Finally there are initial conditions $x_0 = \hat{x}_0$ and simple bounds on all variables, where contractual constraints are specified as hourly profiles for the pressure bounds in terminal nodes ($i \in \mathcal{N}_+ \cup \mathcal{N}_-$) and for the flow bounds in provider and customer arcs ($ij \in \mathcal{A} : i \in \mathcal{N}_+$ or $j \in \mathcal{N}_-$).

3 Computational Results

All computations are performed on a single 2.4 GHz Pentium IV processor of a dual processor Linux PC workstation with 4 GB RAM. The optimization model is implemented in Fortran 77 and compiled with the GNU g77 compiler.

In the results that follow, the large-scale NLP resulting from the discretization are solved by the general-purpose nonlinear optimization code SNOPT [4] in combination with the automatic differentiation add-on SnadiOpt [3].

3.1 Test Problem

The test network in Fig. 2 has three compressors (Cs A–C), one valve (Vl 1), and one regulator (Rg 1) to supply gas from two providers (P 1–2) to three customers (C 1–3). Pipe segments vary from 0.5–1.1 m in diameter and 50–120 km in length, with a total length of 920 km. Provider P 1 supplies roughly 3–8 times as much gas as P 2, while the main customer C 3 demands roughly 30 times as much gas as each of the other two customers. We assume constant demand profiles for C 1, C 2, and a sinusoidal daily demand profile for C 3. The planning horizon is 48 hours, with subintervals of one hour (a natural choice because of contractual profiles). The length of each pipeline segment $a \in \mathcal{A}_{\mathrm{pi}}$ is $L_a = 10$ km. The NLP has 21168 states, 192 controls, and 21170 constraints. Using stationary solutions or optimal transient solutions of close-by scenarios as initial estimates, SNOPT takes 17–49 major iterations and typically 4–10 minutes (up to 25 minutes in rare cases) for a tolerance of 10^{-6}.

We consider three test cases, each with two scenarios differing in the binary decisions. In scenario one, all three compressors are working all the time and the valve is open all the time. Scenario two differs only in that the second compressor (Cs B) is switched off during 08:00–12:00. In case one, the lower bound

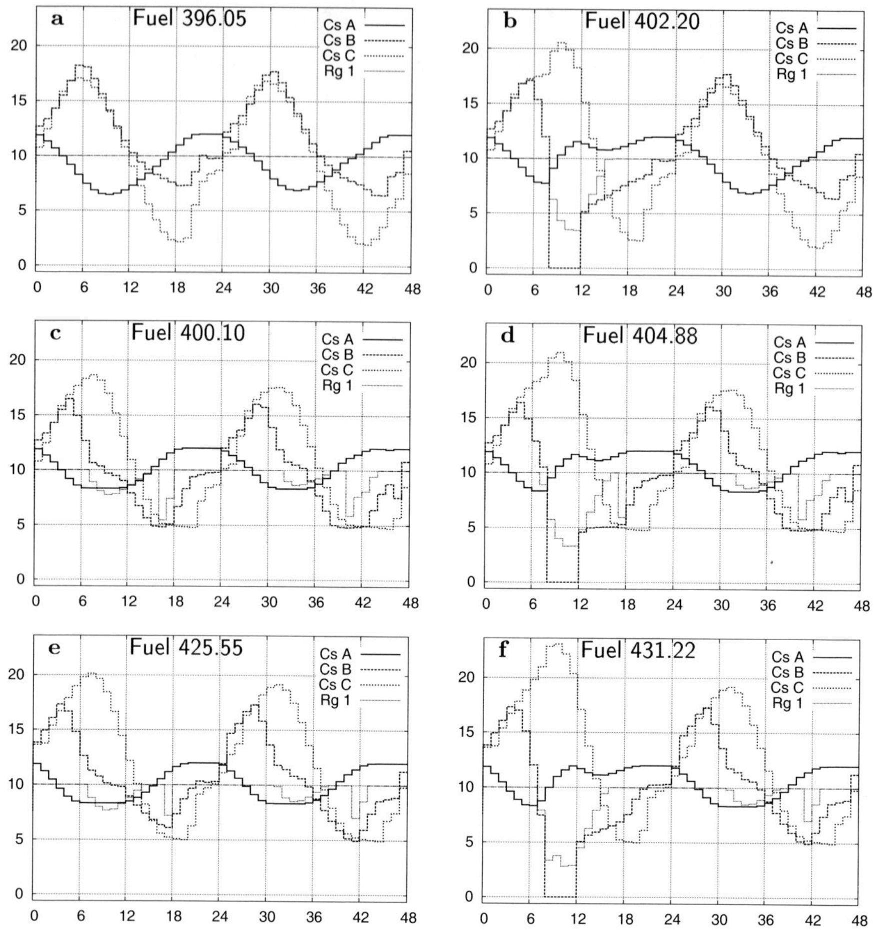

Fig. 3. Optimal pressure changes (bar) versus time (h) and optimal cumulative fuel consumption (1000 kg) for six related test problems

on all compressor powers is zero, which leads to solutions where compressor Cs B occasionally operates below its technical limit, see Fig. 3a/b. This shows that the combinatorial nature of the on/off decision must not be ignored. Case two poses realistic positive lower limits on all compressor powers, yielding the results in Fig. 3c/d. In case three, finally, the altitude increases from sea level to 600 m between P 1 and C 3. Results are shown in Fig. 3e/f.

3.2 Discussion

In case one, all pressure profiles behave approximately periodically, following the biggest customer's demand with different phase shifts. The main burden lies on the second compressor (Cs B) which yields an average pressure increase of 11.6 bar, whereas Cs A and Cs C yield slightly less than 9.6 bar on average. The regulator produces a constant pressure drop of 10 bar (the upper bound). Typical end effects are observable during 42:00–48:00 where the control profiles deviate from the periodical behavior, and at the very beginning where the network leaves its initial stationary state.

When compressor Cs B is switched off during 08:00–12:00 in scenario two, substantial changes in all control profiles are observed roughly during 02:00–18:00 (six hours before and after the interruption), but otherwise the optimal solution is almost identical to scenario one. This shows that perturbations during limited periods of time yield control responses whose magnitude decays with the distance from the perturbation period, an effect which justifies to neglect end effects if the planning horizon is sufficiently long.

Altogether, optimal solutions in case one exhibit precisely the qualitative behavior that one would expect—under the counterfactual assumption, however, that a compressor can produce any nonnegative pressure increase.

Case two shows the effect of placing realistic lower limits on the power consumption of each compressor, resulting in approximate minimal pressure increases of 8 bar for compressor Cs A and 5 bar for Cs B, Cs C. The periodic behavior of the control profiles prevails, but their amplitudes become smaller and the duty is more evenly distributed among the three compressors. Moreover, the end effects become less accentuated. Switching off the second compressor leads again to an optimal solution that is almost identical except during the six hours before and after the interruption.

Increasing the altitude along the network in case three, finally, yields optimal solutions that are very similar to case two except that the peak pressures produced by each compressor are larger than before.

4 Summary

We have presented a nonlinear optimization model and a suitable discretization for minimum-cost operative planning under prescribed binary decisions in natural gas distribution networks. Computational results obtained with an SQP method demonstrate the viability of this approach.

Acknowledgments

We would like to express our sincere thanks to Ruhrgas AG (Essen) and PSI AG (Berlin) for their close cooperation, especially on the practical aspects of the project. Ruhrgas also supplied the measurement data for Fig. 1. We are also indebted to P. E. Gill and E. M. Gertz for free academic licenses of their software packages SNOPT and SnadiOpt, respectively, and finally to an anonymous referee for helpful critique and suggestions.

References

[1] E. A. Boyd, L. R. Scott, and S. Wu, *Evaluating the quality of pipeline optimization algorithms*, in 29th Annual Meeting, Pipeline Simulation Interest Group, 1997. Paper 9709.

[2] R. G. Carter, *Pipeline optimization: Dynamic Programming after 30 years*, in 30th Annual Meeting, Pipeline Simulation Interest Group, 1998. Paper 9803.

[3] E. M. Gertz, P. E. Gill, and J. Muethering, *Users guide for SnadiOpt: a package adding automatic differentiation to SNOPT*, Technical Memorandum ANL/MCS-TM-245, Argonne National Labs, Jan. 2001.

[4] P. E. Gill, W. Murray, and M. S. Saunders, *SNOPT: An SQP algorithm for large-scale constrained optimization*, SIAM J. Optim., 12 (2002), pp. 979–1006.

[5] M. Gugat, G. Leugering, K. Schittkowski, and E. J. P. G. Schmidt, *Modelling, stabilization, and control of flow in networks of open channels*, in Online Optimization of Large Scale Systems, M. Grötschel, S. O. Krumke, and J. Rambau, eds., Springer, Berlin, 2001, pp. 251–270.

[6] P. Hackländer, *Integrierte Betriebsplanung von Gasversorgungssystemen*, Verlag Mainz, Wissenschaftsverlag, Aachen, 2002.

[7] T. Heidenreich, *Linearisierung in transienten Optimierungsmodellen des Gastransports*, Universität Duisburg, 2002.

[8] J. Králik, P. Stiegler, Z. Vostrý, and J. Závorka, *A universal dynamic simulation model of gas pipeline networks*, IEEE Trans. Syst., Man, Cybern., SMC-14 (1984), pp. 597–606.

[9] A. Martin and M. Möller, *Cutting planes for the optimization of gas networks*, Preprint 2253, Fachbereich Mathematik, Technische Universität Darmstadt, 2002.

[10] K. F. Pratt and J. G. Wilson, *Optimization of the operation of gas transmission systems*, Transactions of the Institute of Measurement and Control, 6 (1984), pp. 261–269.

[11] D. Rist, *Dynamik realer Gase*, Springer, Berlin, 1996.

[12] E. Sekirnjak, *Mixed integer optimization for gas transmission and distribution systems.* Vortragsmanuskript, INFORMS-Meeting, Seattle, Oct. 1998.

[13] Z. Vostrý, *Transient optimization of gas transport and distribution*, in Proceedings of 2nd International Workshop SIMONE on Innovative Approaches to Modeling and Optimal Control of Large Scale Pipeline Networks, Prague, 1993, pp. 53–62.

[14] J. F. Wilkinson, D. V. Holliday, E. H. Batey, and K. W. Hannah, *Transient Flow in Natural Gas Transmission Systems*, American Gas Association, 1964.

[15] D. D. Wolf and Y. Smeers, *The gas transmission problem solved by an extension of the simplex algorithm*, Management Sci., 46 (2000), pp. 1454–1465.

[16] P. J. Wong and R. E. Larson, *Optimization of tree-structured natural-gas transmission networks*, J. Math. Anal. Appl., 24 (1968), pp. 613–626.

[17] S. Wright, M. Somani, and C. Ditzel, *Compressor station optimization*, in 30th Annual Meeting, Pipeline Simulation Interest Group, 1998. Paper 9805.

Analysis and Exploitation of Jacobian Scarcity

Andreas Griewank and Olaf Vogel

Technische Universität Dresden, Zellescher Weg 12-14, D-01062 Dresden, Germany
griewank@math.tu-dresden.de
vogel@math.tu-dresden.de

1 Introduction and Motivation

Throughout mathematical modeling and scientific computing a primary objective is to maintain and exploit as much of the inherent system structure as possible. In numerical linear algebra and its application to nonlinear equation solving extensive use has been made of matrix sparsity. We contend here that linearizations of vector functions can have a deeper and more general structure, which we call scarcity. Some of this structure is ingrained in particular algorithms for evaluating the vector functions rather than being a property of the corresponding mathematical map. This algorithmic structure is all the more important in the design of efficient numerical methods for solving equations in various tasks of scientific computing.

We concentrate here on the calculation of Jacobian-vector and vector-Jacobian products in the context of Krylov-based inexact Newton methods. Consequently, the primary aim is to evaluate a long succession of these derivative vectors accurately and with minimal complexity. We will attempt to minimize the number of multiplications while being fully aware that this theoretical measure provides at best an indication of actual computing times. The practical goal of minimizing the multiplications count may be intuitively related to the aesthetic ambition of maintaining structure. This may be motivated and intuitively explained as follows. Disregarding structure of a given problem means embedding it into a larger class of problems, whose representation requires more data and is thus for the more specific class of problems redundant. Hence, reducing the amount of data is also likely to reduce the computational cost of resolving the problems.

2 Evaluation Procedures and Computational Graphs

We shall be concerned with vector functions $y = F(x)$ whose evaluation at a particular argument can be represented by an evaluation procedure of the general form given in Table 1 and explained underneath.

Table 1. General evaluation procedure

v_{i-n}	$= x_i$	for	$i = 1 \ldots n$
v_i	$= \varphi_i(v_j)_{j \prec i}$	for	$i = 1 \ldots l$
y_{m-i}	$= v_{l-i}$	for	$i = m-1 \ldots 0$

Evaluation procedures furnish a mathematical representation of F as a composite function of its elemental constituents φ_i to which the rules of calculus can be unambiguously applied. Without loss of generality we may assume that the intermediate quantities v_i are all real scalars. They themselves and equivalently their indices $i \in V \equiv \{1-n, 2-n, \ldots, l-1, l\}$ can be interpreted as vertices of a *computational graph* $G = (V, E)$. Two vertices $\textcircled{j}$ and $\textcircled{i}$ are connected by an edge (j, i) exactly if the value v_i depends directly on v_j so that

$$(j,i) \in E \quad \Longleftrightarrow \quad \frac{\partial}{\partial v_j}\varphi_i \not\equiv 0 \quad \Longleftrightarrow \quad j \prec i \ .$$

The last precedence notation was already used in Table 1 and will be preferred throughout the paper. The topological ordering condition

$$j \prec i \quad \Longrightarrow \quad j < i$$

ensures that the directed graph G is acyclic and we may assume that the first n and last m indices represent the minimal and maximal vertices, respectively. In Table 1 they are identified as $v_{j-n} = x_j$ for $j = 1 \ldots n$ and $v_{l-i} = y_{m-i}$ for $i = 0 \ldots m-1$ with $x = (x_1, \ldots, x_n)$ and $y = (y_1, \ldots, y_m)$ the vectors of independent and dependent variables. Consequently we have the implication

$$j \prec i \quad \Longrightarrow \quad j \in \{1-n, \ldots, l-m\} \quad \wedge \quad i \in \{1, \ldots, l\}$$

The cardinalities of the predecessor sets

$$n_i \equiv \big|\{j \ : \ j \prec i\}\big|$$

are usually 1 or 2 with φ_i being a univariate intrinsic function or a binary arithmetic operation, respectively. In graph-theoretic terms n_i is the fan-in of vertex $\textcircled{i}$.

For the derivatives of $F(x)$ to be computable by the chainrule it is sufficient (and essentially also necessary) that each elemental function $\varphi_i : \mathbb{R}^{n_i} \to \mathbb{R}$ is continuously differentiable on some open domain $D_i \subset \mathbb{R}^{n_i}$. Therefore we will assume the existence and continuity of the partial derivatives

$$c_{ij} \equiv \frac{\partial}{\partial v_j} \varphi_i(u_i) \quad \text{for} \quad j \prec i$$

at all arguments $u_i \equiv (v_j)_{j \prec i} \in D_i$ of interest. Since we are here only interested in first derivative information about F there is no need to emphasize the *elemental functions* φ_i themselves. Instead we may label the edges (i, j) with the *elemental partials* $c_{ij} = c_{ij}(x)$, which are continuous functions of the global argument vector $x \in \mathbb{R}^n$. We will refer to this edge valued graph as the *linearized computational graph.* To illustrate the relation between an evaluation procedure and its linearized computational graph we consider the so-called *Lion* example due to Naumann in Fig. 1.

Program:

$$\begin{aligned} v_1 &= v_{-1} * v_0 \\ v_2 &= \exp(v_1) \\ v_3 &= \sin(v_2) \\ v_4 &= \mathrm{sqrt}(v_2) \\ v_5 &= \mathrm{atan}(v_2) \\ v_6 &= v_1 - v_2 \end{aligned}$$

(v_{-1}, v_0) independent = minimal
(v_3, v_4, v_5, v_6) dependent = maximal

Linearized Computational Graph:

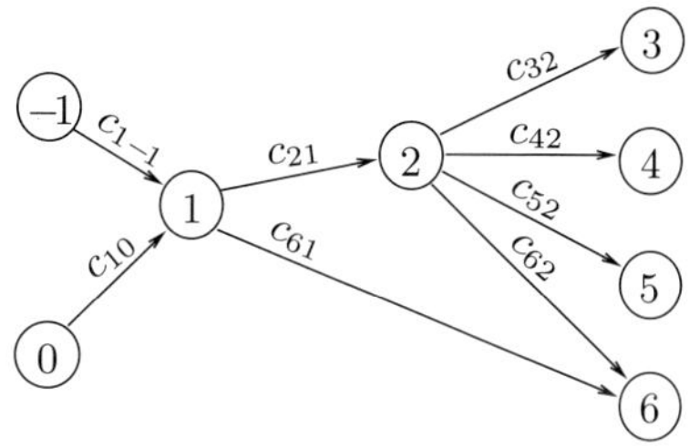

$$\begin{aligned} c_{1-1} &= v_0 \\ c_{10} &= v_{-1} \\ c_{21} &= v_2 \\ c_{32} &= \cos(v_2) \end{aligned} \qquad \begin{aligned} c_{42} &= 0.5/v_4 \\ c_{52} &= 1/(1+v_2^2) \\ c_{62} &= -1 \\ c_{61} &= +1 \end{aligned}$$

Fig. 1. A variant of Naumann's *Lion* example

Notice that in evaluating the elemental partials c_{ij} listed in the lower part of Fig. 1 we have used certain algebraic properties, as for example that the exponential is its own derivative. As a consequence one finds that the evaluation of the c_{ij} on top of the φ_i is usually quite cheap. By setting $c_{ij} \equiv 0$ for $j \not\prec i$ we obtain an $l \times (l - m + n)$ matrix

$$C \equiv (c_{ij})_{j=1-n,\ldots,l-m}^{i=1,\ldots,l} \tag{1}$$

which is usually very sparse. On the above example we have

$$C = \begin{bmatrix} c_{1-1} & c_{10} & 0 & 0 \\ 0 & 0 & c_{21} & 0 \\ 0 & 0 & 0 & c_{32} \\ 0 & 0 & 0 & c_{42} \\ 0 & 0 & 0 & c_{52} \\ 0 & 0 & 1 & -1 \end{bmatrix} .$$

This matrix $C \in \mathbb{R}^{6\times 4}$ has only 8 nonzero elements of which two, namely $c_{62} = -1$ and $c_{61} = 1$ are *unitary*. The other 6 are continuous functions of the two independent variables $(x_1, x_2) \equiv (v_{-1}, v_0)$ so that $C = C(x)$ forms only a two-dimensional manifold in $\mathbb{R}^{6\times 4}$ rather than the affine variety of dimension 6 defined by the sparsity of C. In other words the 6 nonconstant elemental partials c_{ij} are heavily correlated amongst each other. This issue will be discussed more carefully in the following section.

3 Jacobian Set, Jacobian Dimension, and Scarcity

The first derivative of a vector function $F : \mathbb{R}^n \to \mathbb{R}^m$ at a particular point $x \in \mathbb{R}^n$ may be viewed as a linear mapping from $\mathbb{R}^n$ to $\mathbb{R}^m$ or as a matrix in $\mathbb{R}^{m\times n}$. The latter is usually denoted by $F'(x) \in \mathbb{R}^{m\times n}$ and called the Jacobian matrix of F at x. It represents the linear mapping in terms of vector coefficients with respect to the Cartesian bases e_j for $j = 1 \ldots n$ of the domain $\mathbb{R}^n$ and e_i for $i = 1 \ldots m$ of the range $\mathbb{R}^m$. Most numerical analysts would assume that computing the first derivative of F means providing all nonzero entries of the rectangular $F'(x)$. The *global partial derivatives*

$$a_{ij} \equiv e_i^T F'(x) e_j \equiv \frac{\partial}{\partial x_j} e_i^T F(x)$$

are obviously fully determined by the local partial derivatives $c_{\bar{\imath}\bar{\jmath}}$. More specifically we may define for any directed path $P \equiv (i_1, i_2, \ldots i_{\bar{k}})$ with $(i_k, i_{k+1}) \in E$ the path value

$$c_P \equiv \prod_{(j,i)\subset P} c_{ij} = \prod_{k=1}^{\bar{k}-1} c_{i_{k+1}\, i_k} \tag{2}$$

and then obtain by Bauer's formula [1]

$$a_{m-i,j} \equiv \sum_{P\in[j-n \mapsto l-i]} c_P \quad . \tag{3}$$

Here $[j - n \mapsto l - i]$ represents the set of all directed paths connecting the minimal vertex (j-n) to the maximal vertex (l-i) for $0 < j \leq n$ and $0 \leq i < m$. Because many paths have common subsections, evaluating them separately and then summing the ones with common endpoints is about as inefficient as evaluating any other determinant according to its polynomial expansion.

Apart from the computational expense, accumulating the a_{ij} as a rectangular array may be a bad idea, because it can destroy structure. For example on Naumann's Lion example accumulation yields a dense 4×2 matrix with 8 nonzero and nonunitary entries. Hence we have exactly as many global partial derivatives as local nonzero partials including the two unitary ones. However, the accumulated representation of the Jacobian no longer reveals an essential feature ingrained in the linearized computational graph, namely that the rank of $F'(x)$ can nowhere be more than 1. The reason for this is that the generic rank of the Jacobian $F'(x)$ is always equal to the minimal size of a vertex cut of the underlying computational graph. For a proof of this rather elementary fact, see for example [4]. On the other hand, it is quite clear that all rank 1 matrices in $\mathbb{R}^{4\times 2}$ can arise as Jacobian of the Lion example if we let the eight local partials c_{1-1}, c_{10}, c_{21}, c_{32}, c_{42}, c_{52}, c_{62}, and c_{61} roam freely. Moreover, one can see quite easily that this set of Jacobian's is still the same if we impose the artificial conditions $c_{21} = 1$ and $c_{62} = 0$ so that only 5 degrees of freedom are left. More generally, we have the following situation.

Let $X \subset \mathbb{R}^n$ denote some open neighborhood in the domain $D = \mathrm{dom}(F)$, which we may restrict further as required. Then C may be interpreted as matrix valued function

$$C : X \quad \longrightarrow \quad \mathbb{R}^{l\times(l-m+n)} \quad .$$

Assuming that C is real-analytic we find that all its entries are either constant or may be assumed to have open ranges $\{c_{ij}(x) : x \in X\}$ on the possibly restricted open neighborhood $X \subset \mathbb{R}^n$. Once the nonconstant coefficients are evaluated at the current global argument x we have lost all information about their correlation. This will be particularly true when they are subject to roundoff in finite precision arithmetic. Also, the very nature of an evaluation procedure composed of elementary functions suggests that we and nobody else has global information that restricts the values of the φ_i and their partials c_{ij}. Therefore we will suppose that whatever computational procedure uses the values c_{ij} it can only be based on the assumption that

$$C \in \mathcal{C} \equiv \left(\{c_{ij}(x)\}_{x\in X}\right)_{j=1-n\ldots l-m}^{i=1\ldots l} \supset C(X) \quad .$$

In other words $\mathcal{C}$ is the relatively open interval enclosure of the actual range $C(X)$. Consequently $\mathcal{C}$ can be made arbitrary small by restricting X accordingly. The number of nonconstant entries i.e. proper interval components of $\mathcal{C}$ will be denoted by $\dim(\mathcal{C})$ so that certainly

$$\dim(C(X)) \leq \dim(\mathcal{C}) \leq l\,(l-m+n) \quad .$$

Throughout this paper we account for unary functions of the form $\varphi_i(v_j) = \gamma_i * v_i$ with γ_i a floating point constant as follows. Either the constant γ_i is considered as arbitrary number from an interval of small but positive diameter or γ_i is considered as an additional independent variable. While the latter interpretation effects the format of the Jacobian and therefore the cost of its accumulation the complexity of the Jacobian-Vector products discussed in the next section remains the same.

Since the underlying graph G is a DAG with n minimal and m maximal vertices, all $C \in \mathcal{C}$ can be partitioned as

$$C = \begin{bmatrix} B & L \\ R & S \end{bmatrix} \in \mathbb{R}^{((l-m)+m)\times(n+(l-m))} \quad . \tag{4}$$

Where L is strictly lower triangular so that $\det(I - L) \equiv 1$. We will say that C has *lower Hessenberg* $m \times n$ *structure*. This structure occurs naturally in discrete control systems and has first been studied in the context of AD by Coleman and Verma [2].

The application of Bauer's formula (3) with (2) yields the Jacobian matrix $A = \left(a_{ij}\right)_{j=1\ldots n}^{i=1\ldots m}$, which can be expressed as

$$A \equiv A(C) \equiv R + S\left(I - L\right)^{-1} B \quad .$$

This identity has been derived using the implicit function theorem in [3]. Hence we can view A as a matrix valued function

$$A : \mathcal{C} \quad \longrightarrow \quad \mathbb{R}^{m\times n}$$

whose components are polynomial since $\det(I - L) \equiv 1$ due to the strict triangularity of L. The matrices $A(C)$ for $C \in \mathcal{C}$ may have a certain sparsity pattern, which can be represented by the set

$$\mathcal{A} = \left(\left\{a_{ij}(C)\right\}_{C\in\mathcal{C}}\right)_{j=1\ldots n}^{i=1\ldots m} \supset A(\mathcal{C}) \quad .$$

Then $\dim(\mathcal{A})$ gives the number of entries in the Jacobian $A(C)$ for $C \in \mathcal{C}$ that are nonzero and nonunitary. Now we can define the key concepts of this paper as follows.

Definition 1. *For F given by a certain evaluation procedure we call $A(\mathcal{C})$ the* **Jacobian set**, *$\dim(A(\mathcal{C}))$ the* **Jacobian dimension**, *and refer to G and equivalently $\mathcal{C}$ as* **scarce** *if there is a positive codimension*

$$\mathrm{scarce}(G) \equiv \dim(\mathcal{A}) - \dim(A(\mathcal{C})) \quad .$$

Finally we call G and $\mathcal{C}$ **injective** *if $\dim(A(\mathcal{C})) = \dim(\mathcal{C})$.*

In other words scarcity means that the Jacobian set of matrices that can be accumulated from elements $C \in \mathcal{C}$ forms a lower dimensional subset of the sparse matrix set $\mathcal{A} \subset \mathbb{R}^{m\times n}$. Injectivity means that at least locally all elements of the Jacobian set $A(\mathcal{C})$ have a unique inverse w.r.t. C.

In [4] scarcity was defined as a generalization of sparsity, here we have made the two concepts independent. While sparsity is a structure that meets the eye directly, scarcity is much more subtle. Naumann's Lion is scarce but not sparse, and it is also not injective as $\dim(A(\mathcal{C})) = 5 < 6 = \dim(\mathcal{C})$. Hence $\mathcal{C}$ is in some sense a redundant representation of the Jacobian set $A(\mathcal{C})$ on that example.

4 Jacobian-Vector Products and Graph Transformations

Especially in large-scale applications accumulating the Jacobian may be too costly to handle in terms of operation count or storage or both. Then one frequently restricts the evaluation of derivative information to tangents $\dot{y} = F'(x)\,\dot{x} \in \mathbb{R}^m$ or adjoints $\bar{x} = \bar{y}\,F'(x) \in \mathbb{R}^n$ with $\dot{x} \in \mathbb{R}^n$ a given column and $\bar{y} \in \mathbb{R}^m$ a given row vector. Especially in the context of Krylov subspace methods one needs to compute the results $\dot{y}$ and $\bar{x}$ at a fixed argument $x \in \mathbb{R}^n$ for a sequence of directions $\dot{x} \in \mathbb{R}^n$ and weight vectors $\bar{y} \in \mathbb{R}^m$. As shown in Chapter 3 of [3] $\dot{y}$ and $\bar{x}$ can be computed without accumulating the Jacobian by the procedures lated in Table 2.

Since multiplications by 1 or -1 are free the number of costly multiplications needed for the calculation of $\dot{y}$ and $\bar{x}$ is exactly equal to the nonunitary edges in the linearized computational graph. Hence in case of the Lion example where there are 6 nonunitary edges we can save two multiplications compared to the cost incurred by multiplying the Jacobian or transpose by a vector, after they have been accumulated. However, we can do even better by first simplifying the linearized computational graph of the Lion example. Namely we may modify it to the one displayed in Fig. 2.

Table 2. Derivative of the general evaluation procedure of Table 1

tangent procedure

$\dot{v}_{i-n}$	$= \dot{x}_i$	for $i = 1 \ldots n$
$\dot{v}_i$	$= \sum_{j \prec i} c_{ij} \dot{v}_j$	for $i = 1 \ldots l$
$\dot{y}_{m-i}$	$= \dot{v}_{l-i}$	for $i = 0 \ldots m-1$

adjoint procedure

$\bar{v}_{l-i}$	$= \bar{y}_{m-i}$	for $i = 0 \ldots m-1$
$\bar{v}_j$	$= \sum_{i \succ j} \bar{v}_i\, c_{ij}$	for $j = l-m \ldots 1-n$
$\bar{x}_i$	$= \bar{v}_{i-n}$	for $i = 1 \ldots n$

The transformation from the edge valued graph G on the right of Fig. 1 to $\tilde{G}$ in Fig. 2 can be performed in two stages by first eliminating the the vertex ②, and then normalizing the edge $(-1, 1)$ to the value 1. Each of these modification requires a certain adjustment of the edge values and incurs by itself a computational cost. For details see [3], [5] and Section 6 of this paper. Rather than worrying about the cost of these transformation methods we will target here exclusively the quality of the end result, i.e. the number of nonzero and nonunitary edges.

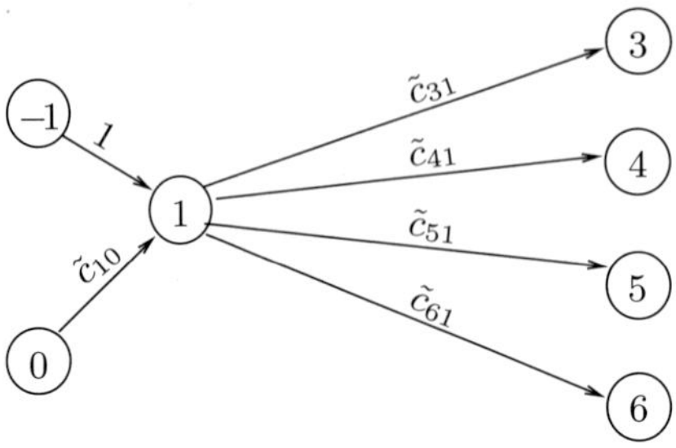

with:

$$\tilde{c}_{10} = c_{10}/c_{1-1}$$

$$\tilde{c}_{31} = c_{32} \cdot c_{21} \cdot c_{1-1}$$

$$\tilde{c}_{41} = c_{42} \cdot c_{21} \cdot c_{1-1}$$

$$\tilde{c}_{51} = c_{52} \cdot c_{21} \cdot c_{1-1}$$

$$\tilde{c}_{61} = (c_{61} + c_{62} \cdot c_{21}) \cdot c_{1-1}$$

Fig. 2. Modified computational graph $\tilde{G}$

In the example above we have computed from $C \in \mathbb{R}^{6\times 4}$ another lower Hessenberg matrix $\tilde{C} = T(C) \in \mathbb{R}^{5\times 3}$, which is *linearly equivalent* in that

$$A(C) = A(\tilde{C}) = A(T(C)) \quad .$$

On the example the transformation $T : C \to \tilde{C}$ is rational and thus not globally defined. Naturally, the big question is how to find a transformation T to a set of Hessenberg matrices $\tilde{\mathcal{C}} = T(\mathcal{C})$ that has a low dimension $\dim(\tilde{\mathcal{C}})$ hopefully as low as $\dim(A(\mathcal{C}))$. It should be noted that any sequence of additions and multiplications, i.e. (homogeneous) linear transformations on scalars, can be written as a lower Hessenberg matrix. This is in particular true for all realizations of the linear transformation from $\dot{x}$ to $\dot{y}$ or equivalently from $\bar{y}$ to $\bar{x}$.

In theory there is an infinite variety of candidate lower Hessenberg structures $\tilde{\mathcal{C}}$, but we can restrict ourselves to a finite number of them by limiting the number $l - m$ of intermediate vertices as follows. We are only interested in Hessenberg structures whose number of free, i.e. nonconstant entries is at most equal to the original $\dim(\mathcal{C})$. Suppose the ith row and the $(n+i)$th column of the structure contains only constants, which corresponds to all edges incident to vertex ⓘ being constant in the computational graph. Then this vertex may be eliminated by multiplying all incoming and outgoing edge values yielding again constant edges. This process can be continued until each intermediate vertex has at least one free edge. Then the number $l-m$ of these vertices equals at most twice the number of free edges $\leq 2 * \dim(\mathcal{C})$. Consequently, for a given computational graph G and the corresponding *natural* matrix representation $\mathcal{C}$ we have to consider only a finite number of related Hessenberg structures $\tilde{\mathcal{C}}$, which may contain linearly equivalent elements $\tilde{C}$ to a given $C \in \mathcal{C}$. We will consider their union as the domain of the accumulation function A.

5 Minimal Jacobian Representation Conjecture

For certain special values of $C \in \mathcal{C}$ there may exist linearly equivalent representations $\tilde{C} = T(C)$ with a very small number of nonzero and nonunitary edges. An extreme case would be the particular linearized graph where all nonunitary edges have the value zero. In other cases we may have fortuitous cancellations during accumulation so that A considered as a special Hessenberg structure with $l = m$ contains very few nonunitary elements. Obviously such extreme cases are of little practical interest. Instead we look for a Hessenberg structure $\tilde{\mathcal{C}}$ and a transformation T defined on a relatively open subset $\mathcal{C}' \subset \mathcal{C}$ so that

$$A(C) = A(T(C)) \quad \text{for all} \quad C \in \mathcal{C}' \quad .$$

For simplicity we identify $\mathcal{C}'$ with $\mathcal{C}$ and assume that on this possibly further restricted domain T is analytic. Hence all entries of $\tilde{C} = T(C)$ may be assumed constant or open w.r.t. C. Then we find immediately that for $\tilde{\mathcal{C}} \equiv T(\mathcal{C})$

$$\dim(A(\mathcal{C})) = \dim(A(T(\mathcal{C})) = \dim(T(\mathcal{C})) \leq \dim(\tilde{\mathcal{C}}) \quad .$$

In other words the number of free entries in $\tilde{\mathcal{C}}$ cannot be lower than the Jacobian dimension which is therefore a lower bound on the number of floating point values we need to represent the Jacobian $F'(x)$ as well as the number of multiplications needed to calculate tangent and adjoint vectors $\dot{y} = F'(x)\dot{x}$ and $\bar{x} = \bar{y}F'(x)$.

In theory it is quite easy to construct a transformation T such that $\dim(T(\mathcal{C})) = \dim(A(\mathcal{C}))$ as follows. Since the mapping $A : \mathcal{C} \to \mathbb{R}^{m \times n}$ is polynomial one can deduce that its Jacobian has a maximal rank $r \leq \dim(\mathcal{A}) \leq m\,n$. Moreover after suitably restricting X and thus $\mathcal{C}$ we have the following generalization of the implicit function theorem [6].

Proposition 1 (Implication of the Rank Theorem).

(i) $A(\mathcal{C}) \subset \mathbb{R}^{m \times n}$ *is a (regular) smooth manifold of dimension* r.
(ii) There exists a transformation $T : \mathcal{C} \to \mathcal{C}$ *with* $T(C) = C$ *for some* $C \in \mathcal{C}$ *such that all but* r *components of* $T(C)$ *are constant and*

$$\dim(A(\mathcal{C})) = \dim(A(T(\mathcal{C})) = \dim(T(\mathcal{C})) \quad .$$

The second assertion means that locally all Jacobians in $A(\mathcal{C})$ can be traced out by varying only r of the edge values and keeping the others constant. Unfortunately, we cannot necessarily choose these constants to be zero or unitary because the proposition above is purely local. The multilinearity of the accumulation function A might possibly help in strengthening the result. Nevertheless, it would seem unlikely that one could always reduce the number of nonzero and nonunitary edges in $\mathcal{C}$ to $\dim(A(\mathcal{C}))$ without changing the structure of the graph, for example by introducing some new edges. Still, we believe the following to be true.

Conjecture 1. *On some neighborhood of every $C \in \mathcal{C}$ there exists a transformation $T : \mathcal{C} \to \tilde{\mathcal{C}}$ such that $\tilde{\mathcal{C}}$ has only* $\dim(A(\mathcal{C}))$ *nonzero and nonunitary edges and $A(\mathcal{C}) = A(T(\mathcal{C}))$ locally.*

If such transformation T can be found the result $T(C)$ would be an optimal representation of the Jacobian $F'(x)$ in terms of both floating point storage and operations.

6 Scarcity Preserving Simplifications

In our quest for the optimal transformation T we consider some local implementations that are guaranteed to preserve the Jacobian set $A(\mathcal{C})$ while reducing the number of free edges. The danger in such local transformations is that structure is lost in that $A(T(\mathcal{C})$ might become a proper superset of $A(\mathcal{C})$. For example this would be the case on the Lion example if we were to eliminate all intermediate edges to arrive at $T(C) \equiv A(C)$, which contains $mn = 8$ nonzero and nonunitary edges compared to $\dim(\mathcal{C}) = 6$ and $\dim(A(\mathcal{C})) = 5$. To avoid these redundancies we must be more cautious in transforming the graph. Below is a list of six modifications of a particular edge $(j, i) \in E$. They are displayed graphically in Fig. 3.

Edge Elimination at front
Delete (j, i) from E after incrementation
$$c_{hj} \mathrel{+}= c_{hi} \cdot c_{ij} \text{ for } h \succ i.$$

Edge Elimination at back
Delete (j, i) from E after incrementation
$$c_{ik} \mathrel{+}= c_{ij} \cdot c_{jk} \text{ for } k \prec j.$$

Edge Normalization forward
With $\gamma = c_{ij} \neq 0$ adjust
$$c_{hi} \mathrel{*}= \gamma \text{ for } h \succ i$$
$$c_{ik} \mathrel{/}= \gamma \text{ for } k \prec i.$$

Edge Normalization backward
With $\gamma = c_{ij} \neq 0$ adjust
$$c_{jk} \mathrel{*}= \gamma \text{ for } k \prec j$$
$$c_{hj} \mathrel{/}= \gamma \text{ for } h \succ i.$$

Edge Prerouting
Delete (j, i) from E after setting for some pivot edge (k, i) with $c_{ik} \neq 0$
$$c_{kj} \mathrel{+}= \gamma \equiv c_{ij}/c_{ik} \text{ and}$$
$$c_{hj} \mathrel{-}= c_{hk} \cdot \gamma \text{ for } h \succ k,\ h \neq i.$$

Edge Postrouting
Delete (j, i) from E after setting for some pivot edge (j, h) with $c_{hi} \neq 0$
$$c_{ih} \mathrel{+}= \gamma \equiv c_{ij}/c_{hj} \text{ and}$$
$$c_{ik} \mathrel{-}= c_{hk} \cdot \gamma \text{ for } k \prec h,\ k \neq j.$$

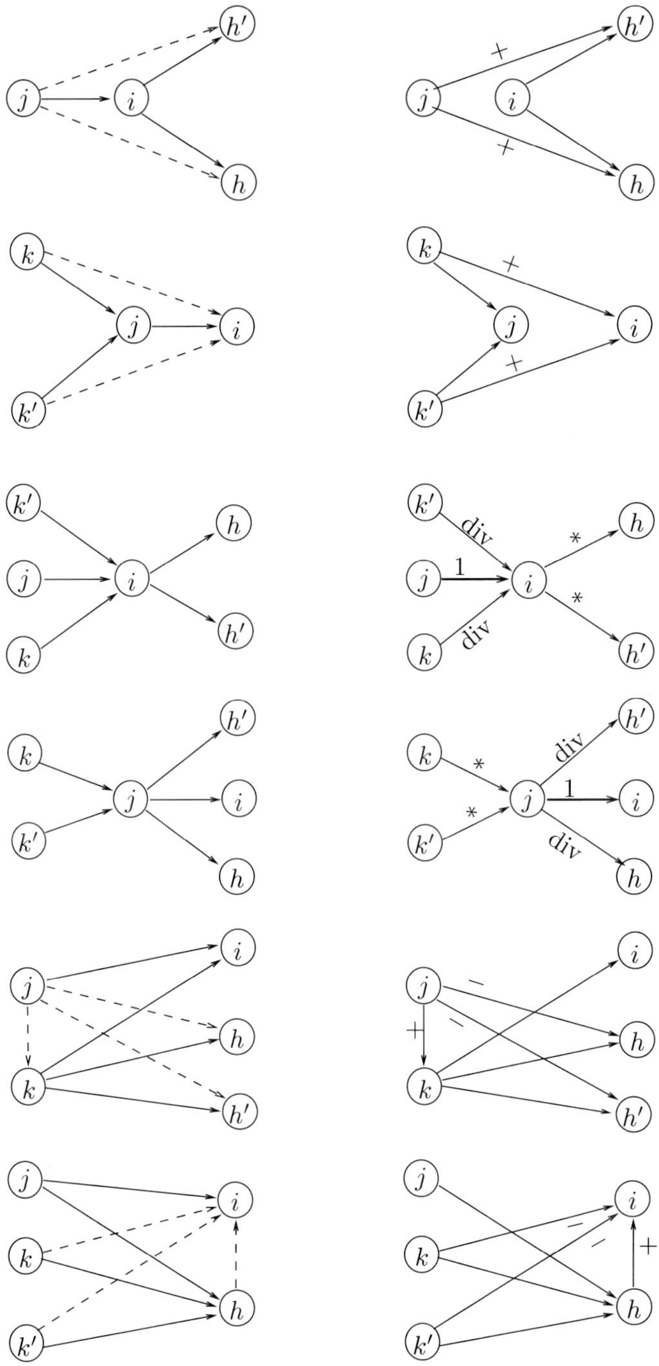

Fig. 3. Local elimination, normalization, and rerouting of edge (j, i)

The local modifications of a computational graph G and the associated Hessenberg matrix C, to a new pair $\tilde{G}$ and $\tilde{C}$ ensure that $A(C) = A(\tilde{C})$. All of them make at least one edge attain a special value in $\{-1, 0, 1\}$, but the modification may result in other edges becoming nonspecial even when they were zero before. Hence the graph $\tilde{G}$ need not necessarily be simpler than the original G in a natural sense. The dashed lines represent edges that may or may not have been present beforehand and are thus newly introduced or merely altered in value, respectively. The resulting edges are labeled with $+$ and $-$ depending they were incremented or decremented. Correspondingly the edges that are multiplied or divided by c_{ij} during the normalization are labeled by $*$ and div , respectively.

Eliminating for fixed ⓘ all edges $(j, i) \in E$ at the front is equivalent to eliminating all edges $(i, k) \in E$ at the back and renders the vertex ⓘ isolated in that it has either no more incoming or no more outgoing edges.

Therefore, it lies no longer on a path from an independent to a dependent vertex and can thus be completely eliminated. Such vertex elimination are the basis of classical methods for Jacobian accumulation [7]. It was demonstrated in [5] that single edge eliminations may yield a more economical elimination process for example on the Lion problem. When an edge or vertex elimination generates no fill-in at all we may speak of absorption. Rerouting all edges $(j, i) \in E$ through the same pivot edge (k, i) is closely related to Gaussian-elimination on linear system of equations. When $m = n$ such modifications can be used to render the linearized graphs *invertible* in that $F'(x)\dot{x} = \dot{y}$ can be solved for the unknown vector $\dot{x} \in \mathbb{R}^n$ given the right hand side $\dot{y} \in \mathbb{R}^n$. In particular one can compute Newton steps in this way without ever accumulating the Jacobian [8].

In the studies mentioned above no attention was payed to the issue of scarcity, i.e. whether or not the Jacobian set was enlarged through these modifications. At first one might think that the Jacobian set is maintained whenever the number of free, i.e. nonspecial, edges does not grow. This need not be the case but we have the following result.

Proposition 2 (Scarcity Preserving Modifications).

(i) *If the back- or front elimination of an edge in a computational graph does not increase the number of free edges the Jacobian dimension remains constant.*

(ii) *If the elimination of a vertex would lead to a reduction in the number of free edges then at least one of the incident free edges can be eliminated via (i) without an increase in the Jacobian dimension.*

(iii) *If the rerouting of an edge via a pivot does not increase the number of nonunitary edges the Jacobian dimension remains constant.*

Obviously normalization does not change the Jacobian set, and it was therefor not mentioned in the proposition.

The application of Proposition (iii) to the 3×3 subgraph depicted in Fig. 4 shows that there are $6 + 5$ edges in the original graph on the left which would be reduced to 9 by the elimination of the central vertex. However, this modification would destroy the property that the leading 2×2 matrix is singular, a scarcity feature that is maintained by the elimination of the two co-incident edges (j, i_2) and (k_2, j).

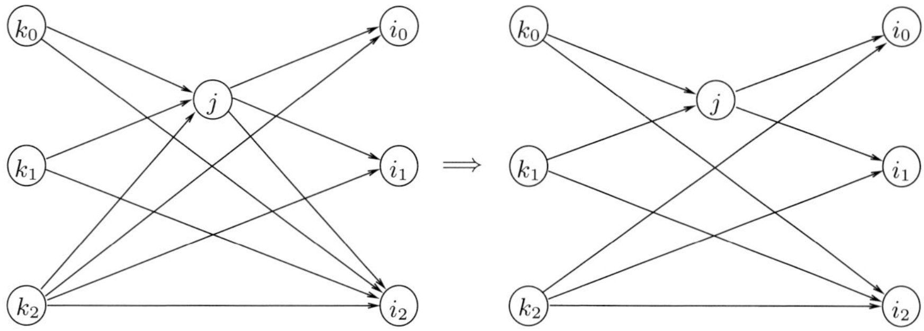

Fig. 4. Scarcity-Preserving Elimination of (j, i_2) and (k_2, j)

On the Lion example the transformation from G to $\tilde{G}$ as displayed in Fig. 1 and Fig. 2 can be interpreted in terms of a two scarcity preserving transformation, namely the front elimination of (1,2) and thus the elimination of vertex ②. The final normalization of $(-1, 1)$ or any other nonzero-valued edge yields the minimal representation $\tilde{G}$ of Fig. 2.

As another example for a successful reduction to a minimal representation we consider the computational graph depicted on the left side of Fig. 5.

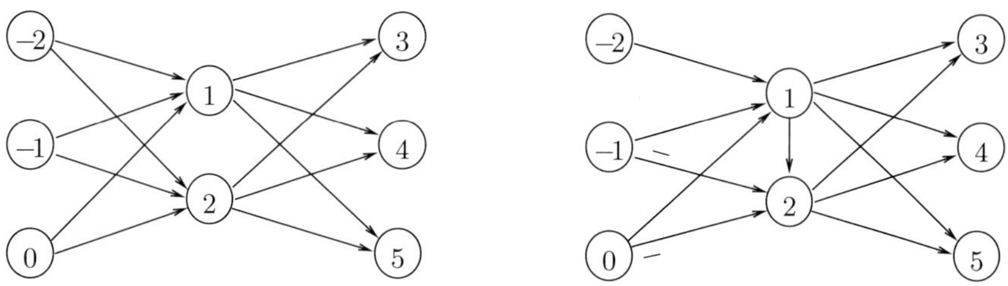

Fig. 5. Postrouting of $(-2, 2)$ via $(-2, 1)$ to be followed by absorption of $(1, 2)$

On the right side of Fig. 5 we see the result of postrouting $(-2, 2)$ via $(-2, 1)$ assuming $c_{1-2} \neq 0$. Whereas the number of free edges stays constant during this modification the subsequent front-elimination of the newly inserted edge $(1, 2)$ yields no fill-in and thus a reduction of the number of free arcs by 1. The result is displayed on the left hand side of Fig. 6. On the right side of Fig. 6 we see the result of prerouting $(2, 3)$ via $(1, 3)$ and its subsequent back-elimination leads to the graph depicted on the left side of Fig. 7.

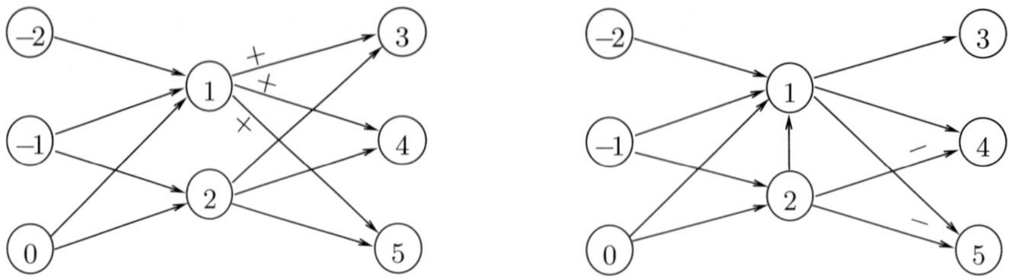

Fig. 6. Prerouting of $(2,3)$ via $(1,4)$ to be followed by absorption of $(2,1)$

There we still have 10 free edges, a number that can be reduced to 8 by normalizing $(-2,1)$ and $(2,4)$ or some other suitable pair of edges. This representation is minimal because the Jacobian set consists of all rank 1 matrices in $\mathbb{R}^{3\times 3}$, whose Jacobian dimension is clearly 8. What we have computed in effect is some kind of LU factorization for a rank deficient matrix. From the above examples one might gain the impression that the structural property of scarcity always manifests itself as singularity of the Jacobian or some of its submatrices. This plausible notion is wrong as can be seen from the following *upwinding* example.

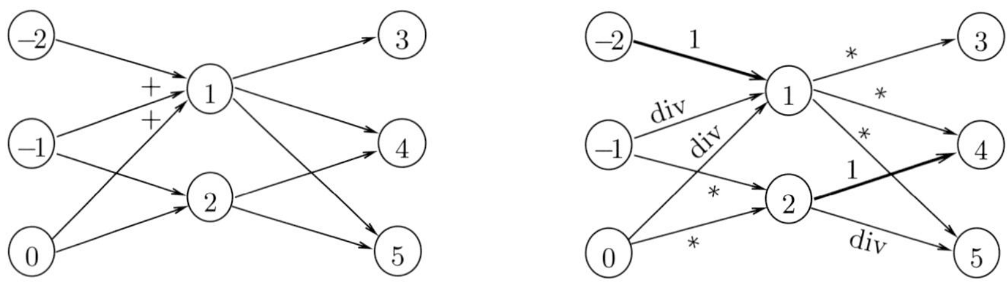

Fig. 7. Normalization of edges $(-2,1)$ and $(2,4)$

Consider the time evolution of a function $v(t,u,w)$ with (u,w) restricted to the unit square and t to the unit interval. Suppose t, u, and w are discretized with a common increment $h = 1/\tilde{n}$ for some $\tilde{n} > 0$ so that

$$t_k = k\,h\;, \quad u_j = j\,h\;, \quad w_i = i\,h \quad \text{for} \quad 0 \le i,j,k \le \tilde{n} \quad .$$

Furthermore we impose periodic boundary conditions in space, i.e.

$$v(t,u_n,w) = v(t,1,w) = v(t,0,w) = v(t,u_0,w)$$

and

$$v(t,u,w_n) = v(t,u,1) = v(t,u,0) = v(t,u,w_0) \quad .$$

Now suppose the underlying evaluation equation can be discretized such that the approximations

$$v_{k,j,i} \approx v(t_k, u_j, w_i)$$

satisfy a difference equation of the form

$$u_{k+1,j,i} = f_h(t_k, u_j, w_i, v_{k,j,i}, v_{k,j-1,i}, v_{k,j,i-1}) \quad .$$

Here $v_{k-1,i} \equiv v_{k,\tilde{n}-1,i}$ and $v_{k,j-1} = v_{k,j,\tilde{n}-1}$. In other words the new value $v_{k+1,j,i}$ depend on the old values of v at the same grid point as well as its immediate neighbors to the West (left), South (underneath), and Southwest. The dependence between new and old values at position $p = (i-1)*3+j$ is depicted in Fig. 8 for $\tilde{n} = 3$.

Considering the initial values $\left(v_{0,j,i}\right)_{j=1\ldots\tilde{n},i=1\ldots\tilde{n}}$ as independents and the final $m = \tilde{n}^2$ values $\left(v_{\tilde{n}-1,j,i}\right)_{j=1\ldots\tilde{n},i=1\ldots\tilde{n}}$ as dependent variables we obtain a directed acyclic graph with $\tilde{n}^3$ vertices and $\dim(\mathcal{C}) = 4\tilde{n}^2(\tilde{n}-1)$ edges. Since each independent vertex is connected by a path to each dependent vertex all $\tilde{n}^4$ entries of the Jacobian are nonzero. Hence there is no sparsity and we have $\dim(\mathcal{A}) = \tilde{n}^4$. However, this number is $\frac{1}{4}\tilde{n}^2/(\tilde{n}-1)$ times larger than $\dim(\mathcal{C}) \geq \dim(A(\mathcal{C}))$, so that we have

$$\text{scarce}(G) \geq \tilde{n}^4 - 4\tilde{n}^2(\tilde{n}-1) = \tilde{n}^2\left[\tilde{n}-2\right]^2 \quad .$$

Hence we see that the Jacobian is scarce for all $\tilde{n} \geq 3$. On closer examination one finds that none of its minors has a vanishing determinant.

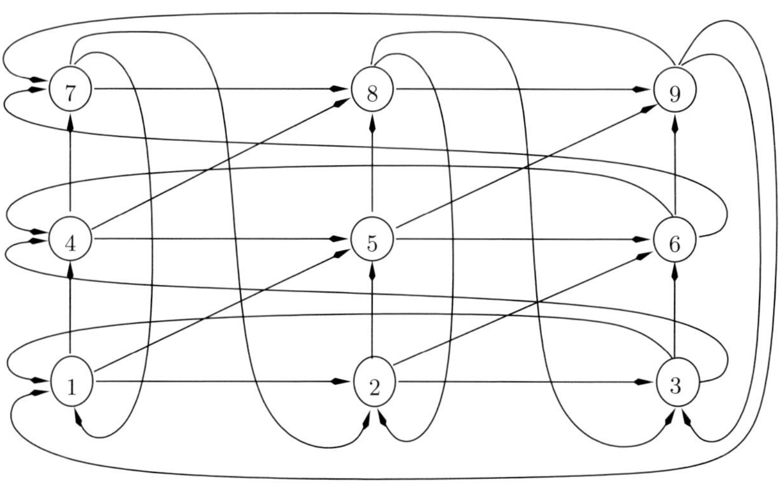

Fig. 8. Update Dependences on the Upwinding Example

Moroever, it is quite easy to see that in general one incoming edge per vertex can be normalized to 1. For example here this can be done quite naturally with the quarter of the edges that connect the new to the old value of v at the same location (u_j, w_i) . Thus a minimal Jacobian representation contains at most $3\tilde{n}^2(\tilde{n}-1)$ floating point numbers compared to the count of $\tilde{n}^2$ for the accumulated matrix representation.

7 Summary and Conclusion

The last upwinding example allows us to assert boldly that certain Jacobians arising in scientific computing should better not be accumulated as rectangular arrays of numbers because this would destroy an internal kind of structure we call here scarcity. As a result Jacobian vector products would be at least $\tilde{n}/3 \approx \sqrt{n}/3$ times as expensive when computed after accumulation as they need be. Possibly further simplifications are possible to reduce the number of free edges further.

In general it is not yet known whether for an arbitrary evaluation procedure for F there exists canonical Hessenberg representations with minimal edge numberes and it seems even less clear how they can be to computed constructively. The resolution of these questions seems to be of theoretical and practical interest.

References

[1] F.L. Bauer, *Computational graphs and rounding error*, SIAM J. Numer. Anal. **11** (1974), 87–96.
[2] T. F. Coleman and A. Verma, *Structure and Efficient Jacobian Calculation* in M. Berz et al *Computational Differentiation, – techniques, applications and tools.* SIAM 1996, pp. 146 – 149.
[3] A. Griewank, *Evaluating Derivatives, Principles and Techniques of Algorithmic Differentiation*, Number 19 in Frontiers in Appl. Math. SIAM, Philadelphia, 2000.
[4] A. Griewank, *A Mathematical View of Automatic Differentiation*, Acta Mathematica, 2003
[5] A. Griewank and U. Naumann, *Accumulating Jacobians by vertex, edge or face elimination*, CARI'02, Proceedings of the 6th African Conference on Research in Computer Science, 2002.
[6] H. Jongen, *Personal Communication.*
[7] M. Iri, *History of automatic differentiation and rounding error estimation*, Automatic Differentiation of Algorithms: Theory, Implementation, and Application (A. Griewank and G. F. Corliss, eds.), SIAM, Philadelphia, Penn., 1991, pp. 1–16.
[8] J. Utke, *Exploiting macro- and micro-structures for the efficient calculation of Newton steps*, Ph.D. thesis, Institut of Scientific Computing, Dresden University of Technology, Germany, 1996.

Exact Numerical Treatment of Finite Quantum Systems Using Leading-Edge Supercomputers

Georg Hager[1], Eric Jeckelmann[2], Holger Fehske[3], and Gerhard Wellein[1]

[1] Regionales Rechenzentrum Erlangen (RRZE)
Martensstraße 1, D-91058 Erlangen, Germany
[2] Johannes-Gutenberg-Universität Mainz, Institut für Physik, KOMET 337
Staudingerweg 7/9, D-55099 Mainz, Germany
[3] Ernst-Moritz-Arndt-Universität Greifswald, Institut für Physik
Domstr. 10a, D-17489 Greifswald, Germany

Summary. Using exact diagonalization and density matrix renormalization group techniques a finite-size scaling study in the context of the Peierls-insulator Mott-insulator transition is presented. Program implementation on modern supercomputers and performance aspects are discussed.

1 Introduction

In the last few years solid-state physics has benefited a lot from scientific computing. The use of numerical techniques is likely to keep on growing quickly in almost all fields of physics as well. Because of the high complexity of many physical problems this will become possible only by the use of modern high-performance supercomputers that are powerful enough to simulate complex systems. Not only do the computational results provide us with important clues on the behaviour of specific materials but they are also widely used as a touchstone to test theoretical approaches, especially in the very difficult regime of strong correlations where the different energy scales in the problem are not well separated. In highly correlated systems, the interaction between the constituents is so strong that they can no longer be considered separately. In other words, the whole is greater than its parts. As a result, the collective behaviour of the microscopic particles, e.g., the electrons in a solid, may scale up to a macroscopic ensemble, exhibiting new and fascinating properties such as high-temperature superconductivity or colossal magnetoresistance [1].

Quasi-onedimensional strongly coupled electron-phonon systems like MX-chain compounds are further examples of electronic systems that are very different from traditional ones [2]. They are particularly rewarding to study for a number of reasons. First they exhibit a remarkably wide range of strengths of competing forces, which gives rise to a rich variety of symmetry-broken ground states. Second these systems share fundamental features with higher-dimensional novel materials, such as high-T_c cuprates or charge-ordered nickelates, i. e. they are complex enough to investigate the interplay of charge, spin, and lattice degrees of freedom which is important for strongly correlated electronic systems in two and three dimensions as well. Nevertheless they are simple enough to allow for a nearly microscopic modeling.

In this context, a frequently used starting point has been the one dimensional Holstein-Hubbard Hamiltonian

$$H_{\mathrm{HHM}} = -t \sum_{\langle i,j \rangle,\sigma} c_{i\sigma}^{\dagger} c_{j\sigma} + U \sum_{i} n_{i\uparrow} n_{i\downarrow} + g\omega_0 \sum_{i,\sigma} (b_i^{\dagger} + b_i) n_{i\sigma} + \omega_0 \sum_{i} b_i^{\dagger} b_i \,. \quad (1)$$

Here the first two terms constitute the conventional Hubbard Hamiltonian with hopping amplitude t and on-site Coulomb repulsion strength U; $c_{i\sigma}^{\dagger}$ creates a spin-σ electron at Wannier site i and $n_{i\sigma} = c_{i\sigma}^{\dagger} c_{i\sigma}$. Since dynamical phonon effects are known to be particularly important in quasi-1D materials, the third and fourth terms take into account the electron-phonon coupling and the elastic energy of a harmonic lattice, respectively, where $g = \sqrt{\varepsilon_p/\omega_0}$ is a dimensionless electron-phonon coupling constant (ε_p gives the polaron binding energy), and ω_0 denotes the frequency of an optical phonon mode. $b_i^{(\dagger)}$ are the usual phonon annihilation (creation) operators. The Hubbard model [3], originally designed to describe the ferromagnetism of transition metals, was subsequently investigated in the context of metal–insulator (Mott) transition, heavy fermions and high-temperature superconductivity as the probably most simple model to account for strong correlation effects. In the single-electron case, the resulting Holstein model [4] has been studied extensively as a paradigmatic model for polaron formation [5]. At half-filling the electron-phonon coupling may lead to a Peierls instability related to the appearance of charge-density-wave (CDW) order (in competition with the spin-density-wave (SDW) instability triggered by U) [6].

As indicated above, in an attempt to close the gap between a microscopic model and its actual ground state, spectral and thermodynamical properties, theorists have turned to the use of large-scale computers (cf. Fig. 1). Nowadays finite-cluster exact diagonalizations (ED), density matrix renormalization group (DMRG) calculations and quantum Monte Carlo simulations have become very powerful and important tools for solving many-body problems with high accuracy. In what follows, we discuss the basic principles, the advantages and weaknesses of the former two techniques by analyzing the 1D half-filled Holstein-Hubbard model exemplarily.

2 Numerical Methods: ED and DMRG

2.1 Exact Diagonalization

In principle, ED is presently the only numerical method which allows an approximation-free study of the Holstein-Hubbard Hamiltonian in the whole

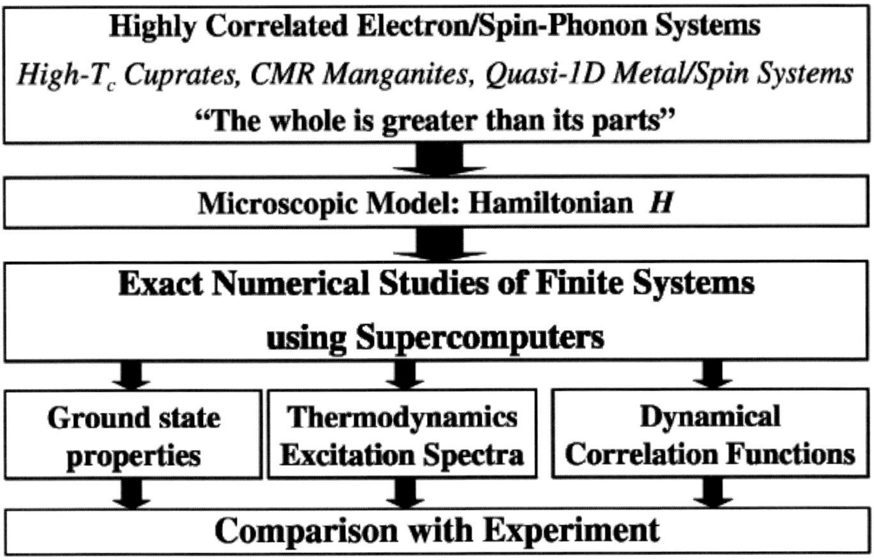

Fig. 1. Numerical approach to strongly correlated systems.

parameter range. A full orthonormal basis set $\{|\Phi_i\rangle\}$ is used to build up the matrix representation of the Holstein-Hubbard Hamiltonian:

$$H_{i,j} = \langle \Phi_i | H_{\mathrm{HHM}} | \Phi_j \rangle \,. \tag{2}$$

Due to the locality of interactions, the matrices $H_{i,j}$ are extremely sparse in a real space basis and standard algorithms such as Lanczos or Davidson can be used to compute exact ground states or low lying excited states while good estimations for excitation spectra are provided by a kernel polynomial method in combination with a maximum entropy method (see [7]). The CPU-time and memory requirements of the three methods are determined by a sparse matrix-vector multiplication (MVM) involving the sparse matrix representation of H_{HHM}. However, the number of degrees of freedom to be considered for a full basis set grows exponentially with increasing cluster size. Even for rather small clusters, e.g. with $N = 8$ to 16 lattice sites, the memory requirements can already exceed the resources of present-day supercomputers. The dimension of the matrix to be diagonalized can be somewhat reduced by exploiting conservation laws. For instance, the conservation of electron number ($\sum_i n_{i\uparrow} + n_{i\downarrow}$) and the z component of the total spin ($\frac{1}{2}\sum_i n_{i\uparrow} - n_{i\downarrow}$) is usually easy to use.

At that point a peculiarity of the Holstein-Hubbard model, where the full basis set can be constructed as a direct product of electronic and phononic basis sets

$$\{|\Phi_{u,v}\rangle = |u\rangle_{\mathrm{el}} \otimes |v\rangle_{\mathrm{ph}};\; u = 1, \ldots, D_{\mathrm{el}};\; v = 1, \ldots, D_{\mathrm{ph}}\}, \tag{3}$$

becomes visible: Because of the phonon degrees of freedom, the total matrix dimension ($D_{\text{tot}} = D_{\text{el}} \times D_{\text{ph}}$) is infinite in principle. However, a well controlled truncation procedure for the phononic basis states

$$|v\rangle_{\text{ph}} = \prod_{i=1}^{N} \frac{1}{\sqrt{m_{i,v}!}} \left(b_i^\dagger\right)^{m_{i,v}} |0\rangle_{\text{ph}}, \; m_{i,v} \in [0, \infty] \tag{4}$$

can be applied by introducing an upper limit for the number of phonons M contained in each basis state, i. e. only states with $\sum_i m_{i,v} \leq M$ are incorporated in the basis set, which thus gets a dimension of $D_{\text{ph}}^M = (M+N)!/M!N!$. Truncating the phonons in such a way is equivalent to setting an upper limit $M\omega_0$ for the elastic energy of the harmonic lattice (fourth term in Eq. (1)). Of course, only the lower part of the spectrum of H_{HHM} with $E \ll M\omega_0$ is described well in this approximation and convergence with respect to M has to be checked carefully for each parameter set $\{t, U, g, \omega_0\}$.

Although a lot of work has been done to reduce the number of basis states in the ED algorithms, even for clusters with only $N = 8$ sites matrix dimensions beyond $D_{\text{tot}}^M = D_{\text{el}} \times D_{\text{ph}}^M \sim 10^{10}$ may be required to achieve sufficient convergence with respect to the truncation of the phonon basis states. Thus the use of powerful supercomputers is indispensable for ED studies of the Holstein-Hubbard model. Our parallel ED package is optimized with respect to memory consumption and does not even store the non-zero elements of the sparse matrix but recomputes them in each MVM step. The parallelization of the MVM step is done by exploiting the natural parallelism of the direct product formulation of the basis set (Eq. 3) (for a detailed discussion cf. [8, 9]).

2.2 Density Matrix Renormalization Group

The DMRG algorithm [10, 11, 12] tries to overcome the limitations of the ED approach by implementing a variational scheme that truncates the Hilbert space used to represent H in an "optimal" way. It is the selection of the basis states that lays the groundwork on which DMRG is built.

DMRG splits the physical system (usually in real space, although a momentum space approach is possible) into two pieces, the so-called *system block* and the *environment block*. Both together form the *superblock* (see Fig. 2).

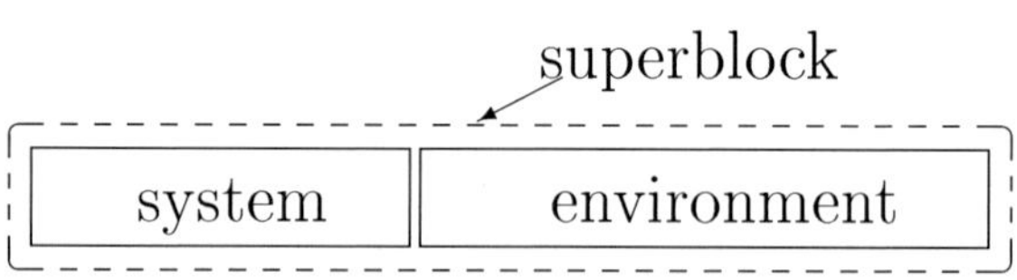

Fig. 2. Division of the complete physical system into "system block" and "environment block". Both blocks together form the "superblock" whose Hamiltonian matrix is diagonalized.

The central entity in the algorithm is the *reduced density matrix*

$$\rho_{ii'} = \sum_j \psi^*_{ij} \psi_{i'j} \quad , \tag{5}$$

where i and j label the states of the system and environment blocks, respectively, so that a superblock state $|\psi\rangle$ can be composed:

$$|\psi\rangle = \sum_{ij} \psi_{ij} \, | i \rangle \, | j \rangle \quad . \tag{6}$$

It can now be shown [12] that the eigenstates of ρ with the largest eigenvalues are those that have the most significant impact on observables, i.e. in order to get a good guess at an optimal basis set for the superblock Hamiltonian one has to

- diagonalize the reduced density matrix for a system block of size l and extract the m eigenvectors with largest eigenvalue,
- construct all relevant operators (system block and environment Hamiltonians, observables) for a system block of size $l + 1$ in the reduced density matrix eigenbasis,
- form a superblock Hamiltonian from the system and environment block (size $l-1$) Hamiltonians plus two single sites (see Fig. 3) and determine its ground state by diagonalization. This is usually the most time-consuming step, performed by a Davidson procedure with sparse MVM as its core.

These steps must be repeated several times, shifting the interface between system block and environment block back and forth until some convergence criterion is fulfilled. This might be e. g. stationarity of the ground-state energy or a sufficiently small *discarded weight*, which is the sum of all density matrix eigenvalues that were not considered when forming the basis. The procedure can be generalized to two dimensions, although it is not quite clear as to how the best "path" for the sweeps through the grid should be chosen [12].

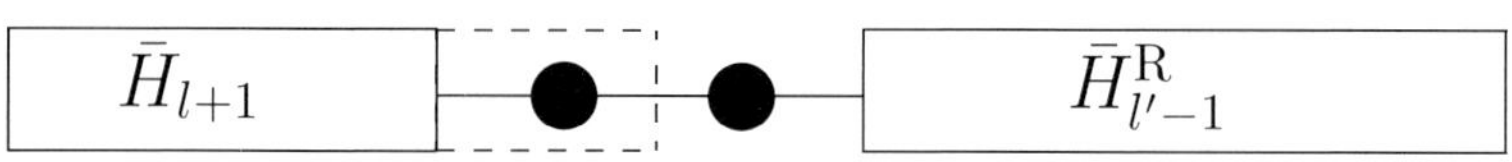

Fig. 3. One step of the finite system DMRG algorithm (left-to-right phase). $\bar{H}_{l+1}$ and $\bar{H}^{\mathrm{R}}_{l'-1}$ are system block and environment block Hamiltonians in the reduced density matrix eigenbasis.

The accuracy of observables like the ground-state energy depends on the number m of density matrix states kept. The discarded weight gives some hint for choosing the right m for a particular problem. Usually one starts with m rather small and increases m every time the ground-state energy has converged. Sensible values for m depend on the physical model under consideration. In the one-dimensional case, where DMRG usually performs best, $m = 500$ to 1000 is often sufficient to get decent data, even for models with electron-phonon interaction like the HHM (1).

It must be stressed that many complications show up in implementing the algorithm for a real-world problem. Fermionic and bosonic commutation rules, reflection and other symmetries, boundary conditions, degeneracies etc. all require special attention [13, 11]. Here we wish to concentrate on the performance aspects alone.

Table 1. One-CPU peak performance in GFlop/s, max. SMP node size and organization of shared memory accesses for all systems used for DMRG calculations. Proprietary, vendor-supplied BLAS and LAPACK implementations were used in all cases. In the case of SGI Origin we ran the calculations on 28 CPU system; the SGI Origin ccNUMA architecture however can be scaled up to 512 processors.

System	**Peak Perf. [GFlop/s]**	**Max. #CPUs per SMP node**	**Memory access**
IBM p690/Power4 (1.3 GHz)	5.2	32	ccNUMA
HP rx5670/Itanium2 (1 GHz)	4.0	4	UMA
Intel Xeon DP (2.4 GHz)	4.8	2	UMA
SunFire 3800 (900 MHz)	1.8	24	ccNUMA
SGI Origin 3400 (500 MHz)	1.0	28 (512)	ccNUMA

Due to the fact that the Hamiltonian is constructed using operators from the system and environment blocks, respectively, the dominant operation in the Davidson diagonalization procedure, i.e. the sparse MVM, is not memory-bound as in the ED case but DGEMM-based and thus cache-bound [14]. This makes DMRG very well suited for contemporary RISC architectures with large caches, because the dense matrix-matrix multiplication can be optimized in a way that yields near-peak speed on virtually all architectures [15]. Table 1 shows some important features of the RISC systems on which calculations were performed. For a standard benchmark case (2-dimensional isotropic half-filled 4x4 Hubbard model with periodic boundary conditions at $U = 4$ and $m = 2000$), Fig. 4 displays absolute performance numbers in MFlop/s for all systems used. Obviously, a significant fraction of peak performance is achievable.

Several parallelization approaches are conceivable for the DMRG algorithm. In the majority of cases, OpenMP parallelization of the MVM step in the Davidson procedure is most suitable, and that is what was used here to obtain the non-dynamical DMRG results.

DMRG can be used to obtain information about the system that requires more than the ground-state properties. This *dynamical* DMRG (DDMRG)

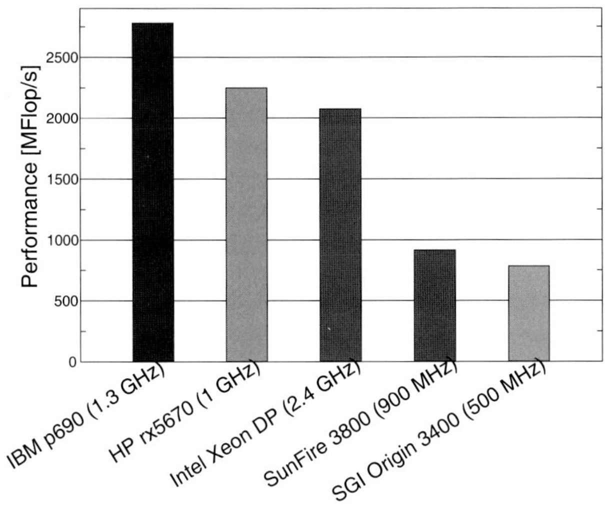

Fig. 4. Sustained single processor performance for DMRG for a standard benchmark case (see text).

approach can calculate important dynamical observables like optical conductivity [16]. Unfortunately, the superblock diagonalization is not any more the dominant step in DDMRG but other parts like the calculation of correction vectors require more computing time. In the present state of our software it does not make sense to use DDMRG in a parallel fashion, i e. all dynamical results have been obtained with 1-CPU runs.

3 Detecting Peierls versus Mott insulating phases in the 1D Holstein Hubbard model

Over the past years the ground-state phase diagram of the Holstein-Hubbard model (1) has been intensely studied using numerical methods. However, despite the apparent simplicity of the model, the quantum lattice and interaction effects are still not completely understood. At half-filling, electron-phonon and electron-electron interactions tend to localize the charge carriers by establishing CDW and SDW ground states, respectively. As a result, Peierls or Mott insulating phases are energetically favoured over the metallic state (Fig. 5). At $U = 0$, the ground state is a Peierls insulator (PI) above a critical electron-phonon coupling $g_c(\omega_0)$, with a vanishing threshold in the adiabatic limit $\omega_0 \to 0$. The PI state has equal spin and charge excitation gaps, and site-parity eigenvalue $P = +1$ [18]. It is mainly characterized by alternating doubly-occupied and empty sites and exhibits long-range order because the CDW phase has broken discrete symmetry. By contrast, at $g = 0$ the ground state is a Mott insulator (MI) with finite charge excitation gap but vanishing spin excitation gap and site-parity $P = -1$ [18]. The corresponding SDW phase has continuous symmetry and hence cannot exhibit true long-range order in 1D.

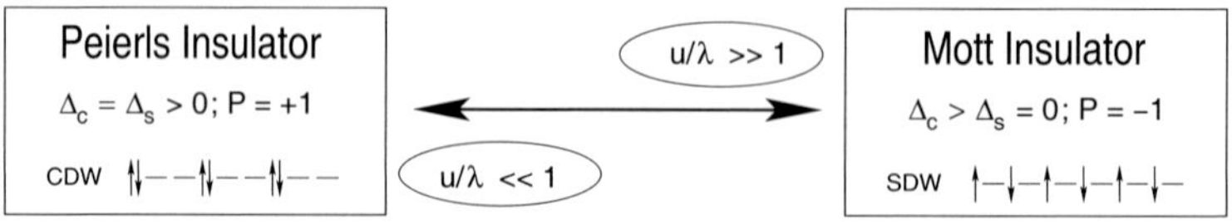

Fig. 5. Schematic phase diagram of the Holstein Hubbard model, where $u = U/4t$ and $\lambda = \varepsilon_p/2t = 2\alpha g^2$ with $\alpha = \omega_0/t$.

An interesting and still controversial question is whether or not only one critical point separates PI and MI phases at T=0 [20, 21]. Furthermore, it is also important to understand how the nature of the PI and MI phases is modified when phonon dynamical effects are taken into account. For a detailed discussion of these problems we refer to recent work [18]. Here we will focus on computational aspects at the determination of ground-state and spectral properties of the HHM.

To spot the transition from the PI to the MI state we first trace the behaviour of the kinetic energy $E_{\mathrm{kin}} = -t\langle\sum_{\langle i,j\rangle\sigma} c_{i\sigma}^{\dagger} c_{j\sigma}\rangle$ when increasing the Coulomb on-site repulsion U (see Fig. 6). Since both CDW and SDW correlations reduce the kinetic energy significantly, E_{kin} reaches a maximum when the system crosses from the PI to the MI regime. At the transition point the optical gap closes. In the non-adiabatic strong electron-phonon coupling regime at small u/λ, the electrons are heavily dressed by phonons, forming bipolarons in real space. As a result the system typifies rather a charge-ordered bipolaronic insulator than a traditional band insulator. This is reflected by the number of phonons needed to reach convergence using the ED approach (see inset). On the other hand, at the transition point and within the MI phase, the ground state is basically a zero-phonon state. We emphasize the extremely weak finite-size dependence of the E_{kin} results.

Further insight into the nature of the PI and MI phases can be obtained by calculating staggered charge- and spin-structure factors:

$$S_{\mathrm{c}}(\pi) = \frac{1}{N^2}\sum_{\substack{i,j\\ \sigma\sigma'}}(-1)^{|i-j|}\langle(n_{i\sigma}-\frac{1}{2})(n_{j\sigma'}-\frac{1}{2})\rangle \;, \tag{7}$$

$$S_{\mathrm{s}}(\pi) = \frac{1}{N^2}\sum_{i,j}(-1)^{|i-j|}\langle S_i^z S_j^z\rangle \;, \quad S_i^z = \frac{1}{2}(n_{i\uparrow}-n_{i\downarrow})\,. \tag{8}$$

Results for $S_{\mathrm{c}}(\pi)$ and $S_{\mathrm{s}}(\pi)$ are given in Fig. 7 for two characteristic Hubbard interactions. Of course, we found pronounced CDW and weak SDW correlations in the Peierls distorted state. Increasing U at fixed g and ω_0, Peierls

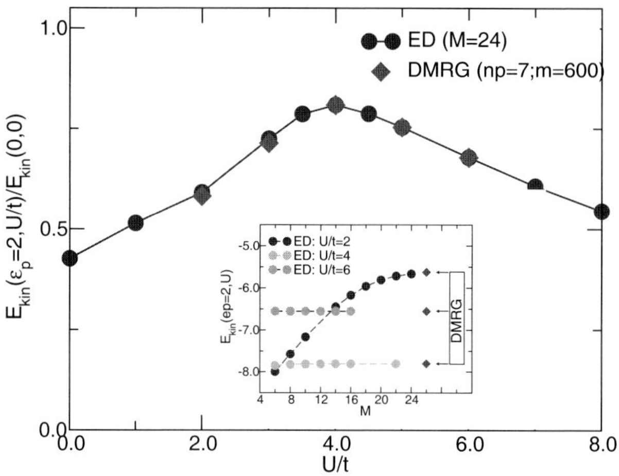

Fig. 6. Kinetic energy as a function of Hubbard interaction U/t with $g^2 = 2$ and $\omega_0/t = 1$. Inset: Comparison of ED and DMRG results for $E_{\rm kin}$ at different U with $g^2 = 2$ and $\omega_0/t = 1$; Convergence of ED results as a function of the cut-off parameter M is demonstrated. The corresponding results from DMRG calculations using $np = 6$ pseudo-sites and $m = 1000$ are represented by stars. The colour version of this figure can be found in Fig. A.16 on page 585.

CDW order is strongly suppressed, whereas the spin structure factor becomes enhanced [18]. Most interesting, however, are the different size dependences of these quantities in the PI and MI phases. For the PI, $S_{\rm c}(\pi)$ shows almost no dependence on the size of the system, which indicates CDW long-range order, whereas $S_{\rm c}(\pi)$ obviously scales to zero as $N \to \infty$. By contrast, in the MI regime our data provides strong evidence for vanishing spin and charge structures in the thermodynamic limit. Clearly the MI is characterized by short-ranged antiferromagnetic spin correlations but nevertheless the staggered spin-spin correlation function shows a slow (algebraic) decay at large distances. Note that such a finite-size scaling (shown here with lattice sizes up to $N = 128$) is definitely out of the range of ED, i.e. in order to obtain reliable results concerning the behaviour of the infinite system we had to apply the DMRG method. The DMRG memory and CPU-time resources required to compute the kinetic energy and structure factor data presented are moderate, at least in relation to the ED algorithm. Each data point in Fig. 6 takes roughly one CPU day on a modern processor like the Intel Itanium 2, at a few GBytes of memory. The DMRG finite-size analysis of structure factors in Fig. 7 is considerably more demanding, with about 5 to 10 GBytes of memory and several CPU days of computer time.

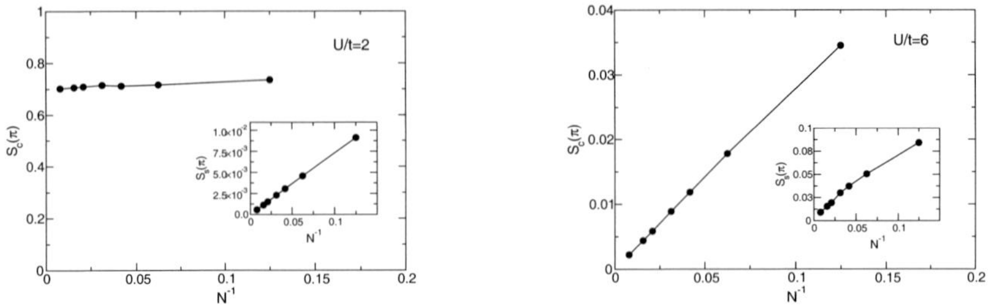

Fig. 7. DMRG study of staggered charge and spin (inset) structure factors as a function of inverse lattice size in the PI (left panel) and MI (right panel) regimes. Generally, $m = 1000$ and 5 boson pseudosites were chosen for the DMRG algorithm, with the exception of the $N = 128$ case where $m = 800$ had to be used for practical reasons with no qualitative loss.

As an example for the calculation of a dynamical (spectral) quantity, we have determined the regular part of the optical conductivity at $T = 0$,

$$\sigma^{\text{reg}}(\omega) = \frac{\pi}{N} \sum_{m \neq 0} \frac{|\langle \psi_0 | \hat{j} | \psi_m \rangle|^2}{E_m - E_0} \, \delta(\omega - E_m + E_0) \, . \tag{9}$$

Here $|\psi_0\rangle$ and $|\psi_m\rangle$ denote the ground state and excited states, respectively, with corresponding energies E_0 and E_m. The current operator $\hat{j} = -iet \sum_{i\sigma} (c^\dagger_{i\sigma} c_{i+1\,\sigma} - c^\dagger_{i+1\,\sigma} c_{i\sigma})$ has finite matrix elements between states of different site-parity only.

Figure 8 displays the optical conductivity of the Holstein-Hubbard model in the MI region, where $\sigma^{\text{reg}}(\omega)$ is dominated by excitations that can be related to those of the pure Hubbard model. In addition phonon side bands appear at multiples of the bare phonon frequency. The optical gap, which represents by its nature a correlation gap, is clearly visible. The optical conductivity can be obtained as well using dynamical DMRG. As our DDMRG code is non-parallel, all calculations were done on one CPU. The DDMRG method yields the optical conductivity (9) convolved with a Lorentzian distribution of width η. The computational effort is smaller for larger broadening η. Of course, the resolution increases as η decreases. As can be seen from the comparison of DDMRG and ED results small values of η are necessary to reproduce the ED peak structure. In fact, DDMRG is better suited for calculating continuous spectral functions in large systems than the discrete spectra of small systems. For the small system investigated here the Lanczos-DMRG approach [22] could be more efficient. We have found that it gets progressively harder (with respect to computing time) to calculate the optical conductivity as the frequency ω increases. This is probably due to the exponentially increasing number of phonon excitations contributing to the optical conductivity at excitation energies larger than ω_0. The DDMRG curves in Fig. 8 were drawn to a point where the required CPU-time is comparable for all values of η (roughly 7 hours on a 2.4 GHz Xeon). Memory requirements for DDMRG are generally quite small. All calculations could be done in less than 200 MBytes of RAM.

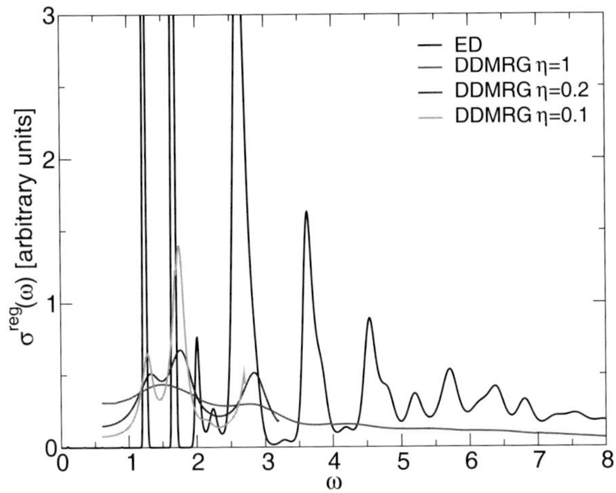

Fig. 8. Optical conductivity for the $N = 8$ site HHM at $U/t = 6$, $g^2 = 2$ and $\omega_0 = 1$. The DDMRG data was calculated at $m = 200$ using three different broadenings $\eta = 1$, $\eta = 0.2$ and $\eta = 0.1$, respectively. Curves for each η were drawn to a point where the CPU-time for the different runs is comparable. The colour version of this figure can be found in Fig. A.17 on page 585.

4 Conclusions

In this work we have presented two methods for the calculation of ground state as well as dynamical properties of low-dimensional physical systems. Where Exact Diagonalization (ED) requires vast computing resources already with very small systems, it is nevertheless free of approximations (apart from a well-controlled phononic cutoff). DMRG methods, on the other hand, are variational by design, but in a way that optimizes the accuracy of measured observables. They allow us to investigate much larger systems with moderate computational effort compared to ED. Non-dynamical DMRG can be used very efficiently on modern, large-memory SMP nodes with high peak performance and outperforms even the best massively parallel ED implementations by far. Dynamical DMRG (DDMRG), on the other hand, while still being able to provide important insights into dynamical properties of much larger systems than ED, has difficulty resolving the discrete peak structure of the small system investigated here and suffers up to now from insufficient parallelization potential. Although results from desktop computer systems look promising, a sensible parallelization approach would be in order to take the step towards supercomputing.

As an example for the reliability and predictive power of such large-scale numerical many-body calculations performed on modern supercomputers, we have analyzed ground state and spectral properties of the one-dimensional half-filled Holstein-Hubbard model with respect to the Peierls-insulator to Mott insulator transition. Significant differences in the finite-size scaling behaviour of the staggered charge and spin structure factors allow to distinguish Peierls and Mott insulating states.

Acknowledgments

We thank the RRZE, the RZG (Computing Center Garching), the URZ (Universitätsrechenzentrum Dresden) and the HLRS (High Performance Computing Center Stuttgart) for providing computational resources. The RRZN (Regionales Rechenzentrum Niedersachsen) and the ZIB (Zuse-Institut Berlin) have granted resources on their HLRN supercomputer complex. This work was partially supported by the Bavarian Competence Network for High Performance Computing (G. H.) and the DAAD (H. F.).

References

[1] For an overview on several important aspects of strongly correlated electron systems see Science Vol. **288** (2000)

[2] A. R. Bishop and B. I. Swanson, Novel Electronic Materials: the MX Family. Los Alamos Science **21**, 133 (1993)

[3] J. Hubbard, Electron Correlations in Narrow Energy Bands. Proc. Roy. Soc. London **A 276**, 238–257 (1963); J. Kanamori, Electron Correlation and Ferromagnetism of Transition Metals, Prog. Theor. Phys. **30**, 275–289 (1963)

[4] T. Holstein, Studies of Polaron Motion. 1. The Molecular Crystal Model. Ann. Phys. (N.Y.) **8**, 325–342 (1959); Studies of Polaron Motion. 2. The Small Polaron. Ann. Phys. (N.Y.) **8**, 343–389 (1959)

[5] G. Wellein and H. Fehske, Self-trapping problem of electrons or excitons in one dimension. Phys. Rev. B **58**, 6208–6218 (1998)

[6] H. Fehske, M. Holicki, and A. Weiße, Lattice dynamical effects on the Peierls transition in one-dimensional metals and spin chains. Advances in Solid State Physics, **40**, 235-249 (2000)

[7] B. Bäuml, G. Wellein, and H. Fehske, Optical absorption and single-particle excitations in the two-dimensional Holstein-tJ model. Phys. Rev. B **58**, 3663-3676 (1998).

[8] G. Wellein, H. Röder, and H. Fehske, Polarons and Bipolarons in Strongly Interacting Electron-Phonon Systems. Phys. Rev. B **33**, 9666-9675 (1996)
[9] G. Wellein and H. Fehske, Towards the limits of present-day supercomputers: Exact diagonalization of strongly correlated electron-phonon systems. In E. Krause and W. Jäger (Eds.): High Performance Computing in Science and Engineering 1999, 112-129, Springer-Verlag Berlin Heidelberg (2000)
[10] S. R. White, Density Matrix Formulation for Quantum Renormalization Groups. Phys. Rev. Lett. **69**, 2863-2866 (1992)
[11] S.R.White, Density Matrix Algorithms for Quantum Renormalization Groups. Phys. Rev. B **48**, 10345-10356 (1993)
[12] R. M. Noack and S. R. White, The Density Matrix Renormalization Group. In I. Peschel, X. Wang, M. Kaulke and K. Hallberg (Eds): Density-Matrix Renormalization: A New Numerical Method in Physics. Lectures of a seminar and workshop, held at the Max-Planck-Institut für Physik Komplexer Systeme, Dresden, Germany, 1998, Lecture Notes in Physics Vol. 528, Springer, Berlin Heidelberg (1999)
[13] E. Jeckelmann and S. R. White, Density-Matrix Renormalization Group Study of the Polaron Problem in the Holstein Model. Phys. Rev. B **57**, 6376-6385 (1998)
[14] G. Hager, E. Jeckelmann, H.Fehske, and G. Wellein, Parallelization Strategies for Density Matrix Renormalization Group Algorithms on Shared-Memory Systems. `arXiv:cond-mat/0305463`
[15] S. Goedecker and A. Hoisie, Performance Optimization of Numerically Intensive Codes. SIAM, Philadelphia (2001)
[16] E. Jeckelmann, Dynamical density-matrix renormalization-group method. Phys. Rev. B **66**, 045114 (2002)
[17] H. Fehske, G. Wellein, A. Weiße, F. Göhmann, H. Büttner, and A. R. Bishop, Peierls insulator Mott-insulator transition in 1D. Physica B 312-313, 562-563 (2002)
[18] H. Fehske, A. P. Kampf, M. Sekania and G. Wellein, Nature of the Peierls- to Mott-insulator transition in 1D. Eur. Phys. J. B **31**, 11-16 (2003)
[19] H. Fehske, G. Wellein, A. P. Kampf, M. Sekania, G. Hager, A. Weiße, H. Büttner, and A. R. Bishop, One-dimensional electron-phonon systems: Mott- versus Peierls-insulators. In S. Wagner et al. (Eds.): High Performance Computing in Science and Engineering Munich 2002, 339-349, Springer-Verlag Berlin Heidelberg (2003)
[20] M. Fabrizio, A. O. Gogolin, and A. A. Nersesyan, From Band Insulator to Mott Insulator in One Dimension. Phys. Rev. Lett. **83**, 2014-2017 (1999)
[21] Ph. Brune, G. I. Japradize, A. P. Kampf, and M. Sekania, Nature of the insulating phases in the half-filled ionic Hubbard model. `arXiv:cond-mat/0304697` (2003)
[22] K. Hallberg, Density-matrix algorithm for the calculation of dynamical properties of low-dimensional systems. Phys. Rev. B **52**, R9827-R9830 (1995)

Numerical Simulation of Solidification Processes in Continuous Casting Processing

Nguyen Hong Hai[1], Nguyen Van Thai[1], and Pham Duc Thang[2]

[1] Hanoi University of Technology
[2] Hanoi Technical College

Summary. The energy equation describing the heat transfer processes (Fourier equation) and the second boundary condition ($-\lambda\overline{n}\cdot$ grad $T(X,t) = q_n(X,t)$), which was determined by simulating experiment, were used to calculate the temperature field of vertical continuous casting process. The Finite Difference Method (FDM) with explicit scheme was applied. The calculating process was much simplified for the cylindrical bar, since in this case one deals with 1D problem. The results are developed for optimization of the technological parameters. The practical applications are made for some high-strength aluminum alloys, such as 6063, 2024, 7075. For 7015 in an army company some modification of the technological process, transferred from abroad, is introduced, giving high-qualified product.

1 Introduction

Continuous casting is now the main process to supply rough materials for the next stage of metals production, such as metal rolling and metal milling. So the problem of optimization of technological parameters of this process is necessary and actual. But to carry out the experiment in real system is almost impossible, that is why the numerical simulation is nearly the only way to approach and to solve the problem. Thus the principal aim of this research work is to verify the main technological parameters of vertical continuous casting process for production of Al-bar of diameter of 150 mm by numerical simulation, and, based on the obtained results, to make the necessary propositions.

2 Mathematical model of continuous casting process

A mathematical model of the continuous casting process differs from the typical foundry technologies, because the energy equation describing the heat transfer processes in the casting domain is the Fourier equation with an additional component called the substantial derivative:

$$C_0\rho_0\left[\frac{\partial T(X,t)}{\partial t}+u\frac{\partial T(X,t)}{\partial z}\right]=\frac{\partial}{\partial x}\left[\lambda_0\frac{\partial T(X,t)}{\partial x}\right]+\frac{\partial}{\partial y}\left[\lambda_0\frac{\partial T(X,t)}{\partial y}\right]+\frac{\partial}{\partial z}\left[\lambda_0\frac{\partial T(X,t)}{\partial z}\right] \quad (1)$$

where $X=\{x,y,z\}$, u is a pulling rate in the z-axis direction (see Figure 1).

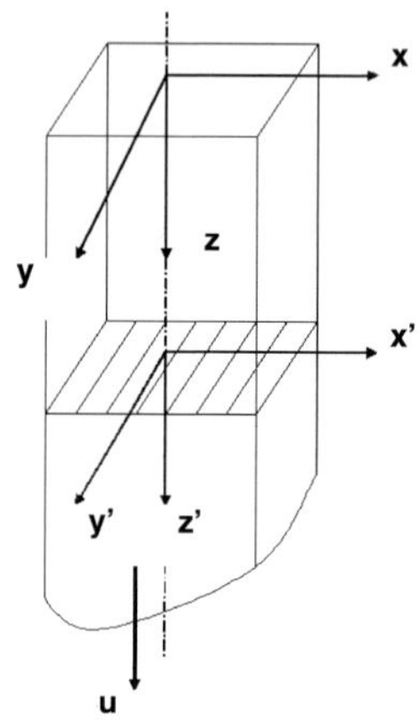

Fig. 1. Rectangular continuous casting

The boundary and initial conditions are as following: on the lateral surface of the cast-slab the 2nd and 3rd boundary conditions can be accepted. On the upper surface of the casting (free surface of molten metal) a boundary condition of the first type (pouring temperature) or the third type can be accepted. On the assumed bottom surface limiting domain Ω_0 (this is the final cooling zone) one can put $q_n=0$, i.e. the adiabatic condition. The initial condition resolves itself into assumption that a layer of molten metal directly over the starter bar has a pouring temperature. It is also possible that the initial temperature in the whole continuous casting domain is equal to the pouring temperature (from a technological point of view this is fiction, of course, but taking into account that solving the problems concerning continuous casting technology we seek the border solutions describing the pseudo-steady temperature fields, the above assumption is acceptable). As it is well known, the border solution does not depend on the initial condition and the final temperature field is determined by physical, geometrical and boundary conditions. In undisturbed conditions of the continuous casting process, i.e. for a constant pulling rate, a constant pouring temperature and fixed boundary conditions on the outer surface of the system, in the domain considered the pseudo-steady temperature field is generated. The temperature at the point $X\in\Omega_0$

is only the function of geometrical coordinates and in spite of the fact that the cast slab shifts through the installation, the isotherms, the position of border surfaces T_L and T_S, the concentration fields etc are fixed for the "observer" from outside. To sum up, it should be pointed out that solution of equation (1) with adequate boundary conditions will asymptotically tend towards to the pseudo-steady solution regardless of the assumed initial condition.

In addition, much experimental research has shown that the conductional component of heat transfer corresponding to the direction of casting displacement is very small (about 5 % of the heat conducted from the axis to lateral surfaces), i.e. that the last equation can be simplified to the form:

$$C_0\rho_0\left[\frac{\partial T(X,t)}{\partial t}+u\frac{\partial T(X,t)}{\partial z}\right]=\frac{\partial}{\partial x}\left[\lambda_0\frac{\partial T(X,t)}{\partial x}\right]+\frac{\partial}{\partial y}\left[\lambda_0\frac{\partial T(X,t)}{\partial y}\right] \quad (2)$$

In this way we obtain a parabolic equation which is typical for the non-steady state heat transfer, but the geometrical coordinate plays a role of time, while the product of the thermal capacity and the mass density is additionally supplemented by the pulling rate u.

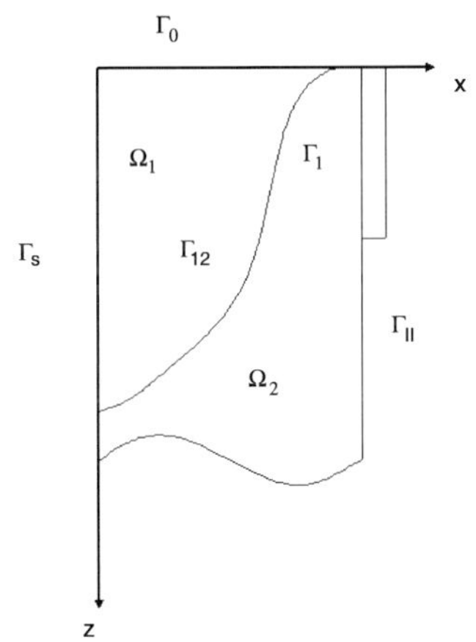

Fig. 2. Division of the boundary and the casting sub-domain

The mathematical model of heat transfer processes in the volume of cylindrical casting is analogous, but the operator div (λgrad T) should be written in the cylindrical coordinate system. We will formulate it for vertical continuous casting made of pure metal. It is assumed that in the domain of the primary cooling zone a heat flux $q = q_I(z)$ is given, while in the domain directly below the crystallizer the third type of the boundary condition is taken into account: $\alpha = \alpha_{II}(z)$, additionally $T_p > T_{cr}$. Thus, we obtain the following system of equations and conditions:

$$
\begin{aligned}
&X \in \Gamma_0 : T(r,z) = T_r \\
&X \in \Gamma_\infty : \frac{\partial T(r,z)}{\partial r} = 0 \\
&X \in \Gamma_1 : -\lambda_2 \frac{\partial T_2(r,z)}{\partial z} = q_1 \\
&X \in \Gamma_2 : -\lambda_2 \frac{\partial T(r,z)}{\partial r} - \alpha_n(z)[T(r,z) - T_{mt}] \\
&X \in \Gamma_s : \frac{\partial T(r,z)}{\partial r} = 0 \\
&X \in \Gamma_{1,2} : \lambda_1 \frac{\partial T_1(r,z)}{\partial r} - \lambda_2 \frac{\partial T_2(r,z)}{\partial r} = U \cdot \rho_2 \cdot L \cdot \frac{F^{m=1,2}}{\partial z} \\
&\Omega_m : C_m \rho_m \cdot U \cdot \frac{\partial T(r,z)}{\partial z} = \frac{1}{z} \frac{\partial}{\partial z} \left[r \cdot \lambda_m \frac{\partial T(r,z)}{\partial z} \right]_{m=1,2}
\end{aligned}
$$

3 Application of FDM for solving the problem

The boundary-initial problems in which the substantial derivative appears (see equation 1) require a separate treatment.

For the vertical, cylindrical continuous casting, because of the fact that the conductional component of heat transfer corresponding to the direction of casting displacement is very small, as mentioned above, and due to the symmetric characteristic of cast-slab the components $\frac{\partial}{\partial y}\left[\lambda_0 \frac{\partial T(Y)}{\partial y}\right]$ and $\frac{\partial}{\partial z}\left[\lambda_0 \frac{\partial T(Z)}{\partial z}\right]$ in equation (1) can be neglected and then one obtains the following mathematical description of the process:

$$
C_0 \rho_0 \left[\frac{\partial T(X,t)}{\partial t} + u \frac{\partial T(X,t)}{\partial z} \right] = \frac{\partial}{\partial x} \left[\lambda_0 \frac{\partial T(X,t)}{\partial x} \right] \tag{3}
$$

with the boundary conditions:

$$
\begin{aligned}
&X \in \quad \Gamma_0 : T(X,t) = T_p \\
&X \in \quad \Gamma_k : -\lambda_0 \frac{\partial T(X,t)}{\partial n} = q_k(X,t) \qquad k = I, II, \dots, \; n = x \text{ or } n = y
\end{aligned} \tag{4}
$$

where Γ_0 is the bottom surface of a molten metal, Γ_k is the part of lateral surface of the cast strand corresponding to the cooling sector k, $k = I, II, \dots$, q_k is the heat flux (boundary condition) for the sector k (the second type boundary conditions are assumed), C_o, ρ_o substitute the thermal capacity and the mass density respectively.

We will use an explicit scheme of the FDM for numerical simulation of heat transfer processes in the continuous casting domain. The left-hand side of the energy equation can be written in the form:

$$\left(\frac{\partial T}{\partial t}+w\frac{\partial T}{\partial z}\right)_0=\frac{T_0^{f+1}-T_0^f}{\Delta t}+w\frac{T_4^{f+1}-T_0^{f+1}}{\Delta z} \tag{5}$$

It can be seen that the temperature derivative with respect to z axis is assumed in an implicit convention, while the index $e = 4$ identifies the direction z^+. The grid step Δz is determined in this way in order to set $\Delta z = u\Delta t$ and then we have

$$\left(\frac{\partial T}{\partial t}+w\frac{\partial T}{\partial z}\right)_0=\frac{T_0^{f+1}-T_0^f}{\Delta t}+\frac{T_4^{f+1}-T_0^{f+1}}{\Delta t}=\frac{T_4^{f+1}-T_0^f}{\Delta t} \tag{6}$$

Thus the FDM equation for the node X_0 is of the form:

$$C_0^f\rho_0^f\frac{T_0^{f+1}-T_0^f}{\Delta t}=\sum_{e=1}^{2}\frac{T_e^f-T_0^f}{R_{oe}^f}\Phi_e\Psi_e+\sum_{e=1}^{2}q_o^f\Phi_e(\Psi_e-1) \tag{7}$$

therefore for $X_0\in\Omega_0$:

$$T_4^{f+1}=T_0^f+\frac{\Delta t}{C_0^f\rho_0^f}\left[\sum_{e=1}^{2}\frac{T_e^f-T_0^f}{R_{0e}^f}\Phi_e\Psi_e+\sum_{e=1}^{2}q^f\Phi_e\left(\Psi_e-1\right)\right] \tag{8}$$

where Φ_e is a sharp function, at the same time $\Psi_e = 1, e = 1, 2$ for the internal nodes, whereas for the boundary nodes $\Psi_e = 1$ for the directions to the interior of domain, and $\Psi_e = 0$ for the direction to the boundary.

For the first layer of nodes ($z = 0$) which corresponds to the bottom surface of a molten metal the initial condition in the form $T(X_0, 0) = T_p$, where T_p is a pouring temperature, can be accepted. On the basic of equation (8) the temperature of the second layer can be found, and this constitutes a pseudo-initial condition for the next layer etc.

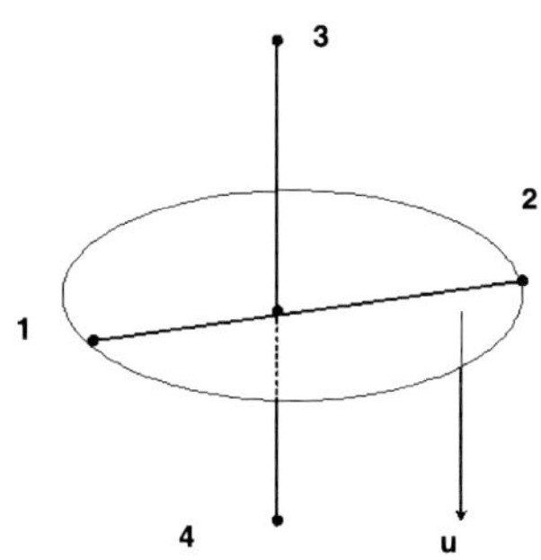

Fig. 3. The mesh of nodes in the continuous casting domain

4 Determination of the heat flux at the boundary slab-crystallizer (Γ_1)

A physical model (see Figure 4) is created to determine the heat flux at the boundary Γ_1 :

1. Thermocouple for measuring temperature on the slab side boundary.
2. Thermocouple for measuring temperature on the crystallizer side boundary.
3. Real crystallizer made of Al-Cu alloy, cooled by water spraying.
4. The steel sleeve simulating continuous casting slab.
5. The heating element for keeping constant temperature on the boundary Γ_1.

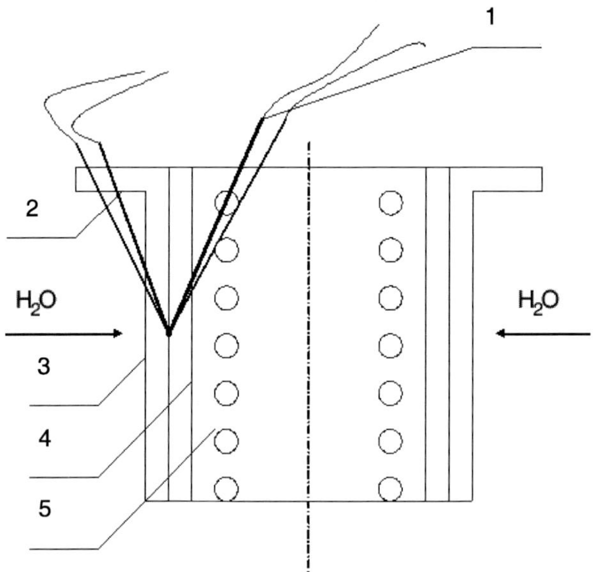

Fig. 4. Physical model for determination of the boundary temperature

The heat flux transferred through the cylindrical 2-layer wall can be determined by the relation, see [2]:

$$q = \frac{t_1 - t_2}{r_1 \left(\frac{1}{\lambda_1} \ln \frac{d_2}{d_1} + \frac{1}{\lambda_2} \ln \frac{d_3}{d_2} \right)} \tag{9}$$

where T_1 is the temperature on the slab side boundary, measured by the thermocouple 5, T_2 is the temperature on the crystallizer side boundary, measured by the thermocouple 1, r_1 and d_1 are the radius and the diameter of the slab at the point for measuring temperature T_1, respectively, d_2 is the diameter of the slab, d_3 is the diameter of the crystallizer at the point for measuring temperature T_2, λ_1 and λ_2 are the conductivity of the slab and the crystallizer, respectively.

The temperature determined by the physical model are presented in the Table 1. Based on these data the heat flux transferred through the boundary slab- crystallizer is calculated and used for calculating program to determine the temperature field of the slab.

Table 1. Temperature measuring results

T_1	T_2	T_1	T_2	T_1	T_2
500	399	580	470	660	540
510	409	590	479	670	549
520	418	600	488	680	558
530	427	610	496	690	566
540	436	620	505	700	575
550	445	630	514	710	584
560	453	640	522		
570	461	650	531		

5 Application for vertical cylindrical continuous casting made of Al-alloy in industry

The principal scheme of vertical continuous casting is shown in Figure 5.

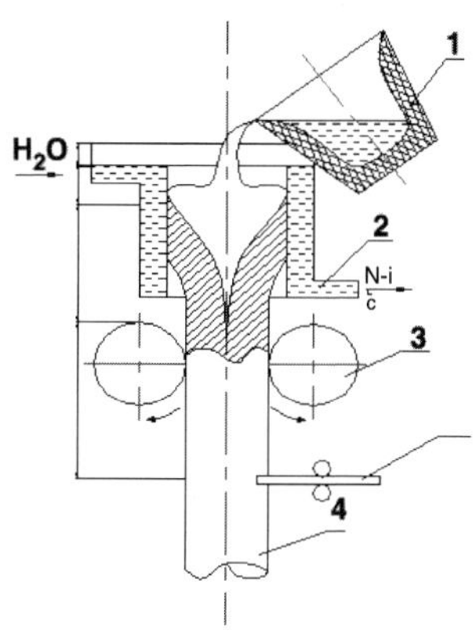

Fig. 5. The principal scheme of vertical continuous casting

The thermo-physical properties of the casting (aluminum alloy 6063) and the crystallizer (made of Al-Cu alloy) are presented in Table 2, see [3].

Table 2. The thermo-physical properties of the slab and crystallizer

Properties	Symbol	Slab AlSi05	Crystallizer AlSi12Cu	Unit
Conductivity coeff.	λ	236	166,6	W/mK
Thermal capacity	C	903,67	890	J/kgK
Latent heat of fusion	L	2500		J/kgK
Density	ρ	2700	2650	kg/m^3
Liquidus temp.	T_L	659	570	^{0}C
Solidus temp.	T_S	638	529	^{0}C

The FDM is applied to calculate the temperature field of the cylindrical slab made of 6063 Al-alloy of the diameter of 150 mm inside the crystallizer made of Al-Cu alloy 90 mm high. The step in the direction from the axis to lateral surface is accepted by 5 mm, meanwhile the step in the pulling direction is of 10 mm. Because of the symmetric characteristic of the cast-slab one can deal with 1D-problem, as mentioned above. The calculating program is written in PASCAL. The results are used to build the temperature field of the slab in the first cooling sector (inside the crystallizer) and to draw the liquidus- and solidus temperature lines. The Figure 6 shows the plots for 2 cases: first one with pulling rate 1,25 mm/s (practiced now in industries) and the last with pulling rate 1,45 mm/s (our proposition).

We should verify if the slab could be broken down with the higher cooling rate.

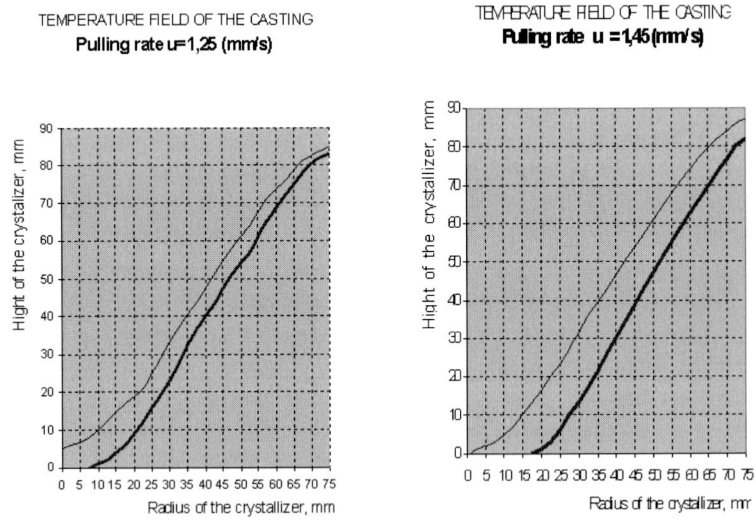

The minimal critical cross-section of the casting while going out of the crystallizer, S_{crit}, should satisfy following condition:

$$S_{cr} = P_p/\sigma_T,$$

where P is a pulling force and σ_T is an ultimate strength of a metal at a high temperature. For Al-alloy 6063 we have $\sigma_T = 1,2$ - $1,5$ MPa at the solidus temperature. The pulling force depends mainly on the friction between the casting and the crystallizer and is equal to 18 kN [4], hence we have $S_{crit} = 163.636 \times 10^{-4}\ m^2$.

The cross-section of the remained liquid area in the center of the casting, S_{liq}, therefore can be determined as: $S_{liq} = S_{cast} - S_{crit}$, where S_{cast} is the casting cross-section area, which is equal to $176.625 \times 10^{-4}\ m^2$, thus $S_{liq} = 12.989 \times 10^{-4}\ m^2$ and the corresponding radius, R_{liq}, is of $2.03.10\times^{-2} m$. Now we can calculate the minimal critical solidified thickness of the casting, L_{crit}, as follows:

$$L_{crit} = R_{cast} - R_{liq} = 75 - 20,3 = 54,7mm.$$

As shown in the Figure 5, the solidified thickness of the cylindrical casting, L_{sol}, in the case of 1,45 mm/s pulling rates is equal to 57 mm and fully satisfies the pulling critical condition.

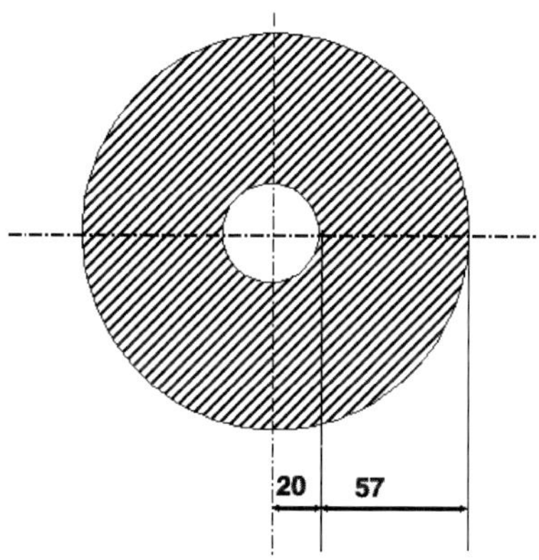

Fig. 6. The solidified thickness of slab while going out of the crystallizer

6 Conclusion

As it is shown, by the numerical simulation the liquidus and solidus lines of an alloy can be plotted, and the critical solidified thickness of a bar or a slab can be calculated. Based on this one can correct the critical pulling rate to obtain higher productivity of an equipment. The model can be applied for other alloy system (steel, copper etc) just by entering the appropriate thermo-physical properties.

The authors would like to thank E.M.T.I (Equipments, Material, Technology for industry) Company for financial support and Plant Z159 for providing the working facilities.

References

[1] Bohdan Mochnacki, Jozef S. Suchy. Numerical methods in computations of foundry processes. Kracow 1995, Polish Foundrymen's Technical Association.
[2] Hoang K.C. Ky thuat nhiet luyen kim. Hanoi University of Technology.
[3] P.R. Saham, P.N. Hansen. Numerical simulation and modeling of casting and solidification processes for foundry and cast-house. CIATF, 1984.
[4] B.Lally, I.Biegler, H.Henein. Finite difference heat transfer modeling for continuous casting. Metallurgical transactions, Volume 21 B, 1990.
[5] All-Russian Institute of Light Alloys, Production of aluminium alloys (in Russian), 1998.

Fast Closed Loop Control of the Navier-Stokes System

Michael Hinze[1] and Daniel Wachsmuth[2]

[1] Technische Universität Dresden, Institut für Numerische Mathematik
Zellescher Weg 12-14, D-01062 Dresden, Germany
hinze@math.tu-dresden.de

[2] Technische Universität Berlin, Institut für Mathematik
Straße des 17. Juni 136, D-10623 Berlin, Germany
wachsmut@math.tu-berlin.de

Summary. We present a construction recipe for closed-loop feedback control of the time-dependent Navier-Stokes system. Its basic idea consists in approximately solving certain instantaneous optimization problems for the discrete-in-time dynamical system. Easy incorporation of control constraints is one key feature of the recipe. We state stabilizing properties of the controllers which we confirm by numerical tests.

1 Introduction

This research is devoted to the construction, numerical validation and stability analysis of nonlinear feedback control policies for the instationary Navier-Stokes system. The governing equations in the primitive setting are given by

$$\begin{cases} y_t - \nu \Delta y + (y \cdot \nabla) y + \nabla p = \mathcal{B}u & \text{in } (0,T) \times \Omega, \\ -\text{div } y = 0 & \text{in } (0,T) \times \Omega, \\ y = 0 & \text{on } (0,T) \times \partial\Omega, \\ y(0) = \varphi & \text{in } \Omega. \end{cases} \tag{P}$$

The control target is to match the given desired state z in the L^2-sense by adjusting the body force $\mathcal{B}u$. In this context $\mathcal{B}$ denotes an abstract control extension operator and $\Omega \subset \mathbb{R}^2$ denotes a bounded domain.

In this work we present recipes for the construction of nonlinear feedback control laws for the time-dependent Navier-Stokes system of the form

$$\mathcal{B}u = K(y)$$

and numerically illustrate their performance.

The construction principle works as follows. The uncontrolled Navier-Stokes system is discretized with respect to time. Then, at selected time instants an appropriate cost functional is approximately minimized with respect to a stationary quasi-(Navier-)Stokes system, whose structure depends on the chosen time discretization method. The obtained control is used to steer the system to the next time instant, where the procedure is repeated. We note that this approach is related to model predictive control techniques, see [6].

Main result: Given a sufficiently smooth desired state z, and a time discretization scheme for the Navier-Stokes system, the above described construction process can be regarded as time discretization of a closed loop feedback policy K, i.e. with A denoting the Stokes operator and $b(y)$ the nonlinearity of the Navier-Stokes equations we get the controlled system

$$y_t + Ay + b(y) = K(y).$$

Under certain assumptions on the initial states the controller K steers the Navier-Stokes system exponentially fast to z. To be more precise, the solution of this system satisfies $|y(t) - z(t)|_{H^1} \leq ce^{-\kappa t}$ with some positive constants c and κ.

It turns out that instantaneous control [9, 11] is a special case of our approach. For applications of instantaneous control we refer to [1, 2, 3, 8, 10, 16, 17, 19], stability analysis of the method is presented in [9, 11]. Further contributions to closed loop control of the Navier-Stokes system can be found in [7], where linear body force feedback control was applied to control the system. The analysis of special case of model predictive control of the Navier-Stokes equations can be found in [14, 15].

The paper is organized as follows. Section 2 contains the analytical preliminaries. In Section 3 we introduce the basic construction recipe which lead to certain discretized closed-loop control laws. These discrete laws are related to continuous closed-loop control laws, whose stability properties are also stated. Finally, in Section 5 numerical examples are presented, which confirm the theoretical results.

Throughout this work c and C denote global generic constants whose dependencies are mentioned when necessary.

2 Analytical preliminaries and time discretization

For given $T > 0$ let $Q = (0,T) \times \Omega$, where $\Omega \subset \mathbb{R}^2$ is a bounded domain. We set $V = \{v \in H_0^1(\Omega)^2, \text{ div } v = 0\}$, $H = \text{clos}_{L^2(\Omega)^2}\{v \in C_0^\infty(\Omega)^2, \text{ div } v = 0\}$ and identify the Hilbert space H with its dual H'. The dual space of V is defined to get a Gelfand-triple $V \hookrightarrow H \hookrightarrow V'$. On H the common inner product is used, and V is endowed with the inner product

$$(\varphi, \psi)_V = (\varphi', \psi')_H \text{ for } \varphi, \psi \in V.$$

Moreover, with Z denoting a Hilbert space, $L^p(Z)$ $(1 \leq p \leq \infty)$ denotes the space of measurable abstract functions $\varphi : (0,T) \to Z$, which are p-integrable $(1 \leq p < \infty)$, or essentially bounded on $(0,T)$ $(p = \infty)$, respectively.

As control space $L^2(\mathcal{U})$ is taken, where $\mathcal{U}$ denotes the Hilbert space of abstract controls. The space $\mathcal{U}$ also is identified with its dual. Furthermore,

$$\mathcal{B} : \mathcal{U} \to V' \tag{1}$$

denotes the control extension operator which is assumed to be bounded. The set of admissible controls is denoted by $\mathcal{U}_{ad} \subseteq \mathcal{U}$ and is required to be closed, convex and bounded. In order to formulate the weak form of the instationary Navier-Stokes equations let

$$W := W(V) = \{\varphi \in L^2(V) : \varphi_t \in L^2(V')\}$$

supplied with the common inner product. Further, we define

$$H^{2,1}(Q) := \{\varphi \in L^2(V \cap H^2(\Omega)), \varphi_t \in L^2(H)\}.$$

For convenience we introduce the tri-linear form

$$b(u, v, w) \;:=\; \int_{\Omega} (u \cdot \nabla) v \, w \, dx.$$

Now, for $y \in L^2(V)$ the function $b(y)$ defined by

$$\langle b(y), v \rangle_{V',V} := -b(y, y, v) \;\text{ for all } v \in V \tag{2}$$

is an element of V' for almost all $t \in (0, T)$ and $b(y) \in L^1(V)$ [18, Lemma 3.1]. If in addition $y \in L^\infty(H)$ holds then $b(y)$ is an element of $L^2(V')$. This statement is true especially for functions $y \in W$, since W is continuously imbedded in $L^\infty(H)$, confer [5].

For controls $u \in L^2(\mathcal{U})$ the solenoidal form of the Navier-Stokes equations reads: Find the state $y \in W$ such that

$$\begin{aligned} &\frac{d}{dt}(y(t), \varphi)_H + \nu (y(t), \varphi)_V \\ &\qquad = \langle b(y) + \mathcal{B}u(t), \varphi \rangle_{V',V} \text{ for all } \varphi \in V \text{ and a.e. } t \in [0, T] \end{aligned} \tag{3a}$$

and

$$(y(0), \chi)_H = (\varphi, \chi)_H \text{ for all } \chi \in H. \tag{3b}$$

With Re denoting the Reynolds number, 1/Re=: $\nu > 0$ is the viscosity parameter. The proof of the following well-known existence theorem can be found in [18].

Theorem 1. *For any $\varphi \in H$ and for every control $u \in L^2(\mathcal{U})$ equations* (3) *admit a unique weak solution $y \in W$.*

2.1 Time discretization

For notational purposes define the Stokes operator $A : V \mapsto V'$ by

$$\langle Ay, v\rangle_{V',V} := \nu(y, v)_V.$$

In this setting the Navier-Stokes equations (3) for $u = 0$ in variational formulation may be rewritten as Burgers equation in the space V',

$$\begin{aligned} y_t + Ay &= b(y), \\ y(0) &= \varphi, \end{aligned}$$

where the nonlinearity $b(y)$ is defined in (2). For $m \in \mathbb{N}$ an equidistant discretization of the time interval $(0, T)$ is defined by $h = \frac{T}{m}$ and $t_k = kh$, $k = 0, 1, \ldots, m$. Now let $z \in W \hookrightarrow C([0, T], H)$ the desired state. We define

$$J^k : V \times \mathcal{U} \to \mathbb{R}, \quad (y, u) \mapsto \frac{1}{2}\,|y - z^k|_H^2 + \frac{\gamma}{2}\,|u|_{\mathcal{U}}^2,$$

where

$$z^k = \frac{1}{h}\int_{t_k - \frac{h}{2}}^{t_k + \frac{h}{2}} z(s, \cdot)\, ds \tag{4}$$

and $z(t, \cdot) = 0$ for $t > T$. Finally, for $k = 1, \ldots, m$ and $i = 1, 2$ introduce the operator $e^k : V \times \mathcal{U} \to V'$ by

$$e^k(y, u) = (I + hA)y \; - \; hb(y^{k-1}) - y^{k-1} - \mathcal{B}u,$$

where y^{k-1} denotes the state at the previous time slice.

The instantaneous optimal control problem for the semi-implicit time integration is given by

$$\text{minimize } J^k(y, u) \text{ subject to } e^k(y, u) = 0 \text{ in } V', \quad u \in \mathcal{U}_{ad}, \tag{$\mathbf{P}^k$}$$

where $y^0 = \varphi$. The initial value φ now is required to be an element of the space V. For given y^{k-1} a pair (y^k, u^k) satisfies the subsidiary condition $e^k(y, u) = 0$ in V' if and only if

$$(y^k, v)_H + \nu h\,(y^k, v)_V = (y^{k-1}, v)_H + \langle \mathcal{B}u^k + hb(y^{k-1}), v\rangle_{V',V} \quad \forall v \in V. \tag{5}$$

Since $\varphi \in V$ holds, the right-hand side in this linear equation defines a bounded linear functional on V. Thus, for every $u^k \in \mathcal{U}$ Eq. (5) admits a unique solution $y^k \in V$ which satisfies the a-priori estimate

$$|y^k|_V \le \frac{C}{\nu h}\left(|y^{k-1}|_H + h|y^{k-1}|^2{}_V + |u^k|_{\mathcal{U}}\right).$$

Since J^k is quadratic, e^k is linear and $\mathcal{U}_{ad}$ is closed and convex every problem $(\mathbf{P}^k)$, $k = 1, \ldots, m$, admits a unique solution $(y_*^k, u_*^k) \in V \times \mathcal{U}$. Furthermore, the unique Lagrange multiplier $\lambda_*^k \in V$ together with the solution (y_*^k, u_*^k) satisfies the first-order necessary optimality conditions (note that A is selfadjoint)

$$(I + hA)y = \mathcal{B}u + y^{k-1} + hb(y^{k-1}), \tag{6a}$$
$$(I + hA)\lambda = -(y - z^k), \tag{6b}$$
$$(\gamma u - \mathcal{B}^\star \lambda, v - u) \geq 0 \text{ for all } v \in \mathcal{U}_{ad}, \tag{6c}$$

where we have set $(y, u, \lambda) = (y_*^k, u_*^k, \lambda_*^k)$. Furthermore, the second-order sufficient optimality condition holds on the whole space $V \times \mathcal{U} \times V$. Hence, the solution (y_*^k, u_*^k) of (6) is the minimum for $(\mathbf{P}^k)$.

The optimal control problem $(\mathbf{P}^k)$ is equivalent with respect to existence to the control-constrained minimization of the functional

$$\hat{J}^k(u) \quad = \quad J^k(y(u), u) \tag{7}$$

over $\mathcal{U}_{ad}$, where for a control $u \in \mathcal{U}$ the state $y(u) \in V$ is given as the unique solution to (5) (indexes dropped). The gradient of $\hat{J}^k$ at u is given by

$$\nabla \hat{J}^k(u) = \gamma u \; - \; \mathcal{B}^\star \lambda,$$

where for given u the function λ is obtained by first solving the linear quasi-Stokes problem (6a) for the state y, and then solving (6b) for λ.

From now onwards let $B := (I+hA)^{-1}$ denote the solution operator of the time-discrete equation (6a), i.e. $e^k(y, u) = 0$ implies $y = B(y^{k-1} + hb(y^{k-1}) + \mathcal{B}u)$.

Observe that the adjoint equation (6b) only depends on the observation $y^k - z^k$. Therefore, gradient information for the functional $\hat{J}^k$ is available utilizing the observations only. In the particular case of boundary observation no information of the state in the whole computational domain is needed at all.

We will now apply the instantaneous control strategy to derive a feedback controller.

3 The closed-loop feedback recipe

The feedback recipe is formulated in terms of a pseudo-algorithm or oracle. The particular form of the feedback strategy depends on the oracle *RECIPE* called in step 2.) of the following algorithm.

Algorithm 1. Feedback recipe.

1.) Set $y^0 = \varphi$, $k = 0$ and $t_0 = 0$.
2.) Given an initial control u_0^k, set

$$u^{k+1} = RECIPE(u_0^k, y^k, z^k, t_k)$$

3.) Solve

$$(I + hA)y^{k+1} = y^k + hb(y^k) + \mathcal{B}u^{k+1}.$$

4.) Set $t_{k+1} = t_k + h$, $k = k + 1$. If $t_k < T$ goto 2.

Next we discuss the *RECIPE*s which are investigated in the present work. We use the instantaneous control problem (**P**) to define the first feedback law. For a given initial control u_0^k one can use a gradient step in direction $-\nabla \hat{J}(u_0^k)$ given by (7). Then one gets the following recipe, see also [9].

RECIPE 1. (Instantaneous control)
The oracle RECIPE in case of the instantaneous control strategy, cf. [9], is defined by $u = RECIPE(v, y^k, z, t)$ iff

1. Solve $(I + hA)y = y^k + hb(y^k) + \mathcal{B}v$,
2. solve $(I + hA)\lambda = -(y - z)$,
3. set $d = \gamma v - \mathcal{B}^\star \lambda$.
4. determine $\rho > 0$,
5. set $RECIPE = v - \rho d$.

To simplify the presentation from now onwards let $\mathcal{U} = L^2(\Omega)^2$ and $\mathcal{B}$ be defined by the injection

$$\langle \mathcal{B}u, v\rangle_{V',V} = (u, v), \quad \text{i.e. } \mathcal{B}u = u. \tag{8}$$

In [9] the following interpretation of RECIPE 1 is given.

Theorem 2. *For $u_0^k = 0$ RECIPE 1 is equivalent to the semi-implicit time discretization with discretization step size h*

$$(I + hA)y^{k+1} = y^k + hb(y^k) - \rho BB(y^k - z^k) - h\rho BB(b(y^k) - Az^k), \quad y^0 = \varphi, \tag{9}$$

of the dynamical system

$$\dot{y} + Ay = b(y) - \frac{\rho}{h} BB(y - z) - \rho BB(b(y) - Az), \quad y(0) = \varphi. \tag{10}$$

Due to Theorem 2 the term

$$K(y) = -\frac{\rho}{h} BB(y - z) - \rho BB(b(y) - Az) \tag{11}$$

in (10) can be interpreted as a non-linear closed-loop control policy for the Navier-Stokes equations. It is important to note that the discretization step-size h and the descent parameter ρ of RECIPE 1 in the continuous case (10) may now be regarded as parameters defining the controller. It is also important to note, that the choice $u_0^k = 0$ guarantees that γ does not enter into (9).

In order to further improve the controller derived in Theorem 2, suppose that K steers y to z, eq. (10) necessarily implies that the desired state z would have to satisfy

$$z_t + Az - b(z) = -\rho BB(b(z) - Az), \quad z(0) = \varphi.$$

This suggests to generalize the control law (11) to

$$K(y) = -\frac{\rho}{h}BB(y - z) - \rho BB(b(y) - b(z)) + z_t + Az - b(z). \tag{12}$$

With this control law the controlled Navier-Stokes equations become the form

$$y_t + Ay - b(y) = K(y) \quad \text{in } L^2(V') \text{ and } \quad y(0) = \varphi. \tag{13}$$

Concerning stability of this system we have from [9, Theorem 6.1,6.2]

$$|w(t)|^2_{H,V} \leq Ce^{-\frac{\rho}{h}t} \quad \forall t \in [0, T],$$

where C is a positive constant and $w := y - z$. It is worth noting that control law (13) can also be obtained by a special choice of u_0^k in Algorithm 1, see [9, 12] and Section 4. Its discrete in time version is given by

$$\begin{aligned}(I + hA)w^{j+1} &= w^j + h\left(b(y^j) - b(z^j)\right) - \rho BBw^j - \rho hBB\left(b(y^j) - b(z^j)\right),\\ w^0 &= \varphi - z(0).\end{aligned} \tag{14}$$

It is easy to extend RECIPE 1 for control constraints where $u \in \mathcal{U}_{ad}$ is required. A simple way to guarantee $u^{k+1} \in \mathcal{U}_{ad}$ is to project the gradient step onto $\mathcal{U}_{ad}$.

RECIPE 2. (Instantaneous control with control constraints)
For the instantaneous control strategy with control constraints the oracle RECIPE is defined as in RECIPE 1 except that (v) is replaced by

(v)′ set $RECIPE = P_{\mathcal{U}_{ad}}(v - \rho d)$.

We now construct a recipe which applied in Algorithm 1 realizes a full optimization step for problem ($\mathbf{P}^k$). We choose the control u^k to be the solution of ($\mathbf{P}^k$) for the unconstrained case, i.e. $\mathcal{U}_{ad} = \mathcal{U}$. Then u^k solves the optimality system, compare (6),

$$\begin{aligned}(I + hA)y^{k+1} &= y^k + hb(y^k) + u^k\\ (I + hA)\lambda^k &= z^k - y^{k+1}\\ \gamma u^k - \lambda^k &= 0.\end{aligned} \tag{15}$$

It is easy to see that

$$u^k = -(BB + \gamma I)^{-1}B(B(y^k + hb(y^k)) - z^k)$$

holds. Now let

$$S = \gamma(BB + \gamma I)^{-1}BB, \tag{16}$$

which defines a continuous linear operator in $L(H, H)$. Further properties of S are investigated in [13]. Exploiting the relation $Bz^k = BB(z^k + hAz^k)$ we get

$$(I + hA)y^{k+1} = y^k + hb(y^k) - \frac{1}{\gamma}S(y^k - z^k + hb(y^k) - hAz^k), \quad y^0 = \varphi, \tag{17}$$

Now it is easy to see that equation (17) in y is the semidiscretization with stepsize h of the dynamical system

$$y_t + Ay - b(y) = -\frac{1}{\gamma h}S(y - z + hb(y) - hAz), \quad y(0) = \varphi. \tag{18}$$

In order to further improve this control law we proceed as in the derivation of (12) and obtain the modified control law

$$y_t + Ay - b(y) = -\frac{1}{\gamma h}S(y - z + hb(y) - hb(z)) + z_t + Az - b(z), \quad y(0) = \varphi, \tag{19}$$

whose semidiscretization is given by

$$\begin{aligned}(I + hA)y^{k+1} = y^k + hb(y^k) - \frac{1}{\gamma}S(y^k - z^k) - \frac{h}{\gamma}S(b(y^k) - b(z^k)) \\ + z^{k+1} - z^k + hAz^k - hb(z^k).\end{aligned} \tag{20}$$

If we define the feedback control operator K by

$$u = K(y) = -\frac{1}{\gamma h}S(y - z + hb(y) - hb(z)) + z_t + Az - b(z) \tag{21}$$

we end up with a closed loop control interpretation. Concerning stability we state from [12]

$$|w(t)|_H^2 \leq C e^{-\frac{\alpha(\gamma)}{h}t}|w(0)|_H^2 \quad \forall t \in [0, T], \text{ where } \alpha(\gamma) = \tfrac{\gamma}{(1+\gamma)^2}.$$

In order to motivate the related recipe we note that we would obtain the feedback operator K in (21) also by investigating the optimal control problem

$$\min J(v^k) = \frac{1}{2}\int_\Omega |w^{k+1}| + \frac{\gamma}{2}|v^k|^2, \tag{$\tilde{\mathbf{P}}^k$}$$

subject to

$$(I + hA)w^{k+1} = w^k + hb(w^k + z^k) - hb(z^k) + v^k.$$

instead of $(\mathbf{P}^k)$. Here $w = y - z$ denotes the difference of the state and the desired state. The related recipe then is given by

RECIPE 3. (Suboptimal control or $(h,1)$ model predictive control)
The oracle RECIPE is defined by $u = RECIPE(v, y^k, z^k, z^{k+1}, t)$ iff

(i) Solve the optimality system for u

$$\begin{aligned}(I+hA)y &= y^k - z^k + hb(y^k) - hb(z^k) + (I+hA)z^{k+1} + u\\ (I+hA)\lambda &= z^{k+1} - y\\ \gamma u - \lambda &= 0.\end{aligned}$$

(ii) set $RECIPE = u$

Here, (h,l) model predictive control for $l \in \mathbb{N}$ means that the lenght of the optimization horizon in each step of the control process is lh. After perfoming the optimization the control horizon is shifted by the factor h. For details we refer to [12].

4 Numerical results

In this section, we numerically investigate the stabilizing properties of Algorithm 1 with RECIPEs 1, 2, and 3. Further numerical results for RECIPE 1 can be found in [9]. In RECIPEs 1 and 2 we choose the initial control as solution of

$$\left(I - \frac{\rho}{1-\rho\gamma} BB\right) u_0^k = \frac{1}{1-\rho\gamma}\left(z^{k+1} - z^k + Az^{k+1} - b(z^k) + \rho BB(b(z^k) - Az^k)\right).$$

In the unconstrained case this choice leads to the continuous controller (12), see [9].

In the control problem considered here we intend to track a desired time dependent state z in the L^2-sense, i.e. our instantaneous cost functional has the form

$$J(y,u) = \frac{1}{2}\int_{\Omega_o} |y(x,t)) - z(x,t)|^2 dx + \frac{\gamma}{2}\int_{\Omega_c} |u(x,t)|^2 dx,$$

where t denotes the time instant, Ω_o denotes the observation domain, and Ω_c the control domain, respectively, which in the present work both are chosen equal to Ω. We choose $\mathcal{U} = L^2(\Omega)$. The controlled state equations are given by the instationary Navier-Stokes system $(\mathbf{P})$ with $\mathcal{B}$ defined in (8). We realize the control $\mathcal{B}u = K(y)$ by the feedback control laws defined through Algorithm 1 with the RECIPEs 1, 2 and 3.

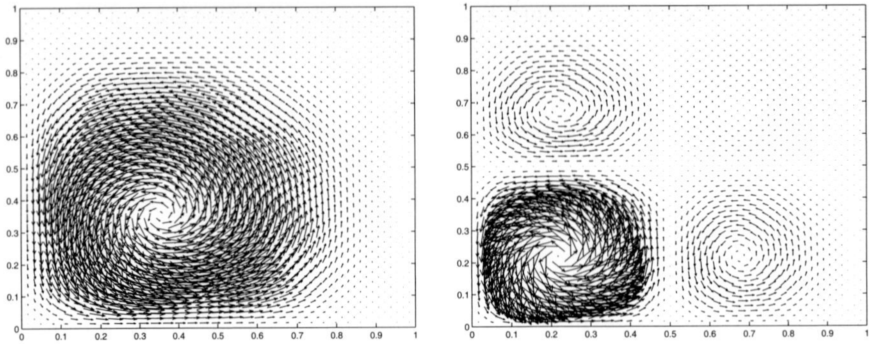

Fig. 1. Desired flow at $T = 1$ and $T = 2$

Let us specify the problem setting. The computation domain is the unit square $\Omega = [0,1]^2$. The initial value is chosen as

$$y(x,0) = \varphi(x) = e \begin{pmatrix} (\cos 2\pi x_1 - 1)\sin 2\pi x_2 \\ -(\cos 2\pi x_2 - 1)\sin 2\pi x_1 \end{pmatrix},$$

where e is the Euler number, and the desired flow is time-dependent and defined by

$$z(t,x) = \begin{pmatrix} \psi_{x_2}(t,x_1,x_2) \\ -\psi_{x_1}(t,x_1,x_2) \end{pmatrix},$$

where ψ is given through the stream function

$$\psi(t,x_1,x_2) = \theta(t,x_1)\theta(t,x_2)$$

with

$$\theta(t,y) = (1-y)^2(1-\cos 2\pi y t).$$

The Reynolds number is set to 10, which results in a viscosity coefficient of $\nu = 1/10$. The final time is chosen as $T = 2$. For the discretization in time a equidistant grid with stepsize $h = 0.01$ is used, whereas for the spatial discretization the Taylor-Hood finite element is applied on a grid containing 1024 triangles with 2113 velocity and 545 pressure nodes.

At first, we compare the results of the RECIPEs 1 and 2. The step-size of the gradient iteration was set to $\rho = 0.1$. The evolution of the costs is shown in Figure 2 for the unconstrained and the box-constrained case, where $|u(x,t)| \leq 10^3$ is required. As one can see, both controllers give the same output if the control constraint is not active. Moreover, we observe exponential decay of $|y(t) - z(t)|_H$ in the unconstrained case, see Fig. 2, left.

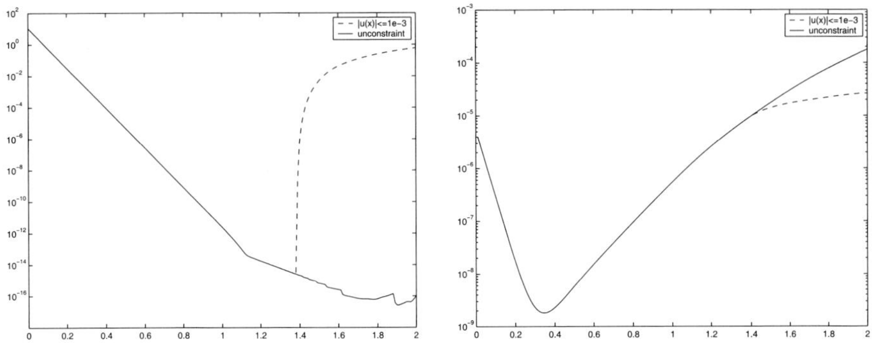

Fig. 2. Evolution of $|y(t) - z(t)|_H^2$ and of $|u(t)|_{L^2(\Omega)}^2$ for unconstrained and constrained control

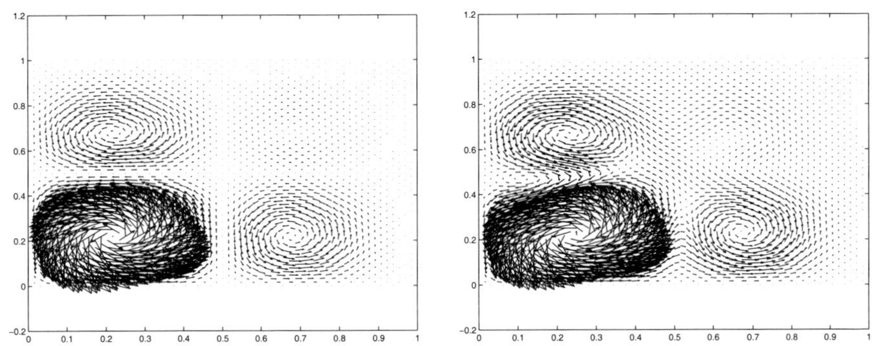

Fig. 3. Flow at $T = 2$ for RECIPEs 1 and 2

Since the desired state becomes more and more dynamic, the constrained controller can not adapt to this situation, and the distance between state and desired state grows for $t \geq 1.4$. On the other hand, the control costs per time step remain constant, see the right-hand diagram in Fig. 2. This fact is also illustrated byFigure 3.

Secondly, we present some results for the feedback controller given by RECIPE 3. In Fig. 4 the evolution of the L^2-Cost $|y(t) - z(t)|_H$ is shown for different values of γ. Exponential decay for both large and small values of the regularization parameter is observed numerically, although the theory is only satisfying for large values. In each application of the control RECIPE 3, an optimization problem has to be solved. It turns out, that with the solution from the previous time-step as initial guess in the cg-method, very few conjugate gradient steps are needed. The optimization process is stopped if either the relative residual norm is less than 10^{4} or the absolute residual norm is less than 10^{8}. At last, we note that the computational time required for the

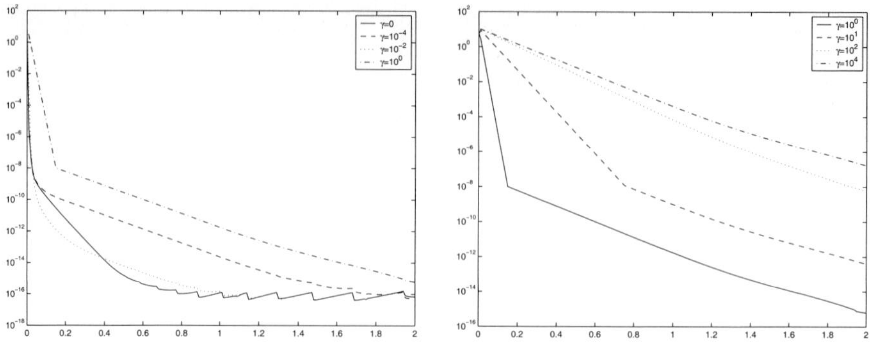

Fig. 4. Evolution of $|y(t) - \bar{y}(t)|_H^2$ for different values of γ

application of the recipe discussed in this research is in the range of 2.5 to 5 times the computational time required for one forward solve of the uncontrolled time dependent Navier-Stokes system, which in fact is *fast* compared to optimal control, say, see [10].

References

[1] Choi, H: Suboptimal Control of Turbulent Flow Using Control Theory. In: Proceedings of the International Symposium. on Mathematical Modelling of Turbulent Flows, Tokyo, Japan. (1995)

[2] Choi, H., Hinze, M., Kunisch, K: Instantaneous control of backward-facing-step flows. Applied Numerical Mathematics, **31**:133–158 (1999)

[3] Choi, H, Temam, R., Moin, P., Kim, J: Feedback control for unsteady flow and its application to the stochastic Burgers equation. J. Fluid Mech., **253**:509–543 (1993)

[4] Constantin, P., Foias, C: Navier-Stokes Equations. The University of Chicago Press (1988)

[5] Dautray, R., Lions, J. L: Evolution problems I, volume 5 of Mathematical analysis and numerical methods for science and technology. Springer, Berlin (1992)

[6] García, C.E., Prett, D.M., Morari, M: Model predictive control: Theory and practice - a survey. Automatica, **25**:335–348 (1989)

[7] Gunzburger, M.D., Manservisi, S: Analysis and approximation for linear feedback control for tracking the velocity in Navier-Stokes flows. Comput. Methods Appl. Mech. Eng., **189**:803–823 (2000)

[8] Hill, D.C: Drag reduction strategies. CTR Annual Research Briefs Center for Turbulence Research, Stanford University/NASA Ames Research Center, 3-14 (1993)

[9] Hinze, M: Instantaneous closed loop control of the Navier-Stokes system. Preprint MATH-NM-09-2002. Institut für Numerische Mathematik, Technische Universität Dresden, Germany (2002)
[10] Hinze, M., Kunisch, K: Control strategies for fluid flows - optimal versus suboptimal control. In: H.G.Bock et al., editor, ENUMATH 97, pages 351–358. World Scientific, Singapore (1997)
[11] Hinze, M., Volkwein, S: Instantaneous control for the Burgers equation: Convergence analysis and numerical implementation. Nonlinear Analysis T.M.A., **50**:1–26 (2002)
[12] Hinze, M., Wachsmuth, D: Controller design for the Navier-Stokes system. Part I: Concept, Numerical Results. Report 11-2003, Institut für Mathematik, Technische Universität Berlin (2003)
[13] Hinze, M., Wachsmuth, D: Controller design for the Navier-Stokes system. Part II: Stability Analysis. in preparation.
[14] Hou, L.S., Yan, Y: Dynamics and approximations of a velocity tracking problem for the Navier-Stokes flows with piecewise distributed controls. SIAM J. Control Optim., **35**:1847–1185 (1997)
[15] Hou, L.S., Yan, Y: Dynamics for controlled Navier-Stokes Systems with distributed controls. SIAM J. Control Optim., **35**:654–677 (1997)
[16] Lee, C., Kim, J., Choi, H: Suboptimal control of turbulent channel flow for drag reduction. J. Fluid Mech., **358**:245–258 (1998)
[17] Min, C., Choi, H: Suboptimal feedback control of vortex shedding at low Reynolds nunmbers. J. Fluid Mech., **401**:123 – 156 (1999)
[18] Temam, R: Navier-Stokes Equations. North-Holland (1979)
[19] Tröltzsch, F., Unger, A: Fast solution of optimal control problems in selective cooling of steel. ZAMM, **81**:447–456 (2001)

Advanced Column Generation Techniques for Crew Pairing Problems

Tran Van Hoai[1], Gerhard Reinelt[2], and Hans Georg Bock[3]

[1] Interdisciplinary Center for Scientific Computing, University of Heidelberg
Im Neuenheimer Feld 368, 69120 Heidelberg, Germany
hoai@iwr.uni-heidelberg.de
[2] Institute for Computer Science, University of Heidelberg
Im Neuenheimer Feld 368, 69120 Heidelberg, Germany
Gerhard.Reinelt@informatik.uni-heidelberg.de
[3] Interdisciplinary Center for Scientific Computing, University of Heidelberg
Im Neuenheimer Feld 368, 69120 Heidelberg, Germany
Bock@iwr.uni-heidelberg.de

Summary. Crew pairing problems are often solved using column generation in a branch-and-price framework. The main idea of this approach consists of solving linear programming relaxations for a subset of variables and employing pricing for introducing additional variables with negative reduced costs. Since the dual variables are not bounded, this leads to instability in the standard implementation of this method. Therefore, a so-called stabilized column generation with respective control parameters is suggested to overcome this problem by reducing the number of redundant variables. In this paper we report about possible realizations of this principle and show that the performance of algorithms can be improved significantly.

1 Introduction

Column generation has been used for solving a variety of scheduling applications because it is able to deal with problems which are modeled with a huge number of variables [2]. The basic idea of this method is to solve linear programming problems first only for a subset of the variables and then, successively, to employ reduced cost pricing for generating further columns and re-solve until all variables not in the current subproblem price out correctly.

In this scheme, the solution of the pricing subproblem, which is NP-hard in this context, is an important and difficult issue. In our experiments, usually the time for pricing amounts to about 90% of the total computing time.

Many algorithms have been suggested to solve the pricing problem exactly as e.g., resource constrained shortest path algorithms [3, 7, 8], k-shortest path algorithms [6], and constraint programming [5, 15]. Furthermore there are also heuristic approaches [13]. One of the big problems for these pricing algorithms (except for constraint programming) is the difficulty to deal with rules and regulations of airline companies. These are very complicated and usually impossible to be presented as linear constraints. Multi-labeling is an idea to tackle that difficulty, but the number of labels in constrained shortest path algorithms could increase exponentially. This is also the case for k-shortest path algorithms because it is quite hard to determine k in order to include at least one feasible pairing.

With the approaches suggested so far, there has still been no breakthrough in the performance of pricing algorithms. Therefore, a different direction for solving the LP relaxation proposed in [11, 4, 12] is to control the dual vector in order to accelerate the solution process. With a modified model, the method aims at reducing the computation time by only generating "good" columns, i.e., columns which will be active in the final solution with high probability.

This paper is organized as follows. In Section 2 we give a brief introduction to the crew pairing problem and describe our test instances. The scheme of the standard column generation method is explained in Section 3 with several remarks on its instability. A *stabilized column generation method* which tries to stabilize the path to the optimal dual point is given in Section 4. We also discuss how to update parameters of the method and give computational results. Some remarks on our approach conclude the paper.

2 The Crew Pairing Problem

Given the time table of an airline and airplanes associated with each flight, the *crew pairing problem* consists of assigning crews to service the flights such that the total service cost is minimal.

The crew pairing problem is usually modeled as *set partitioning problem*

$$\begin{aligned} \min\ & c^T x \\ \text{s.t. } & Ax = 1 \\ & x \in \{0,1\}^n . \end{aligned}$$

Here, every column of A represents a feasible pairing and every row corresponds to a flight that has to be serviced. We have $a_{ij} = 1$, if pairing j covers flight i, and $a_{ij} = 0$, otherwise. The cost of pairing j is given by c_j. The variable x_j states whether its respective pairing is in the optimal solution or not. The cost c_j of a pairing includes the crew costs, the accommodation costs and the penalty costs. Note, that we can have a huge number of columns. In the following we do not distinguish between the terms "column" and "variable".

For testing our algorithmic approaches to the crew pairing problem we have chosen Vietnam Airlines, because it was the only airline for which we could get data on the time table and on crew assignment rules. However, Vietnam Airlines still is a small company with only less than 300 flights for each kind of aircraft per week, and the problems turned out to be too easy.

Therefore, we generated several additional problem instances, but keeping rules and regulations of Vietnam Airlines. Nowadays, the network of airports which Vietnam Airlines operates is small and simple. There are 31 airports which are connected similar to a star graph with 2 main nodes (i.e. there are few connections between subsidiary airports). We added about 20 connections between the subsidiary airports which are likely to be opened in the future. This will create more possibilities of next flights for a crew. In order to create instances with many feasible solutions, the data is generated from flight routes, which are paths that aircrafts will take during a scheduling horizon (for example, day schedule, week schedule). This seems more reasonable than creating flights randomly in the scheduling period without paying attention to the flight routes. Observing the activities of Vietnam Airlines, we classify the flight routes of aircrafts into 2 types: daily route and weekly route. If an aircraft performs a daily route, it departs from a base and returns to that base in the same day. Moreover, that route is repeated almost every day of a week. Carrying out a weekly route, an aircraft could stay far a way from its base several days. This often happens with long haul flights. In order to guarantee the reality of the test instances, two types of routes are involved in which the proportion of the number of daily routes to the number of weekly routes in schedules of Vietnam Airlines is kept unchanged. We chose a flight network which has 31 routes with about 400 flights per week. There are 8 flight sets generated from that flight network. Under rules and regulations of the airlines, there are very many feasible pairings as shown in Table 1.

Table 1. Flight test instances

Instance	# feasible pairings
FS1	267928
FS2	253864
FS3	219119
FS4	283313
FS5	288345
FS6	266370
FS7	259294
FS8	225095

Vietnam Airlines only distinguishes 2 kinds of crew cost: a cost for flying time and another cost for ground time. The company pays a crew a fixed amount of money for a unit of flying time. Therefore, with a given time table, the company must spend a fixed cost to cover all flights. This means there is nothing to be optimized on the cost for flying time. The most important cost which Vietnam Airlines wants to minimize is the cost for ground time which is mainly the cost for the accommodation when crews take a rest far away from their bases. In this paper, the deadheading and some other penalty costs will not be taken into account.

All our computational experiments were executed on an Intel Pentium III 450MHz, with GNU `gcc` compiler using the optimization flag `-O3`. We have used the software framework ABACUS for branch-and-cut-and-price algorithms for embedding our routines. Because we want to assess the performance of the stabilization methods, some options of ABACUS have been deactivated, such as *pool separation*, *branch-and-bound*, *primal heuristics*. Furthermore, the *heuristic pricing* will not be used because it can lead to some phenomena that seem difficult to understand. The linear solver in our experiment is SoPlex [14], but all other linear solvers supported by ABACUS could be used as well. In our computational experiment, we focus on solving the root subproblem only.

3 Standard Column Generation Method

The linear programming relaxation of the set partitioning approach to the crew pairing problem is

$$(P) \quad \begin{aligned} &\min c^T x \\ &\text{s.t. } Ax = 1 \\ &\quad\quad x \geq 0 \end{aligned}$$

with associated dual linear program

$$(D) \quad \begin{aligned} &\max u^T 1 \\ &\text{s.t. } A^T u \leq c. \end{aligned}$$

Since the number of variables is very large, solving the problem directly is not an effective way or even impossible because of memory requirements or insufficient capacity of solution algorithms. Therefore, we have chosen the column generation approach. The idea of this method is to work with restricted models that contain only a subset of columns and to generate additional columns when needed. The standard column generation method can be outlined as follows.

Step 1: Find an initial set A^0 of columns. Add artificial variables to make the problem feasible. Set $k := 0$.

Step 2: Solve the problem (P^k)

$$\begin{aligned} &\min {c^k}^T x^k \\ &\text{s.t. } A^k x^k = 1 \\ &\quad\quad x^k \geq 0, \end{aligned}$$

to obtain solution vectors x^k and u^k where u^k is the solution of the associated dual program (D^k).

Step 3: Solve the pricing problem

$$r^k := \min_{j \in \mathcal{J}} c_j - {u^k}^T A_{.j}$$

where $\mathcal{J}$ is the set of all variables of the master problem (P).

Step 4: If $r^k < 0$, then add the corresponding column to A^k to obtain A^{k+1}. Set $k = k + 1$ and go to Step 2. Otherwise stop, the problem is solved to optimality, since all reduced costs are nonnegative.

The problem (P^k) is called *restricted master problem.* The principle approach shown above only adds the column with minimum reduced cost, but we can add as well further columns with negative reduced costs.

The pricing problem in Step 3 is very difficult. In the context of solving the crew pairing problem, it can be modeled as a shortest path problem with additional constraints originating from the rules and regulations of airlines. For instance, in the case study of Vietnam Airlines, we create a suitable graph as follows. The vertex set V consists of all flights in the intended scheduling period. Note that, because the schedule is the same for following periods and the duration of a pairing is limited within a maximum number of days, we have to add vertices for several next periods in order to cover all possible pairings. Two vertices f_i and f_j are connected if the arrival airport of f_i is the same as the departure airport of f_j and if the take-off time of f_j exceeds the landing time of f_i by at least a given so-called minimum ground time. We clearly obtain an acyclic directed graph. The weight of an arc (f_i, f_j) is $c_{ij} - u_i^k$ as depicted in Figure 1. We denote the cost between two flights by c_{ij}. We create two additional vertices (super source and super sink) and then connect all vertices corresponding to flights departing from a given base to the super source and all vertices of flights arriving at that base to the super sink. We also assign weights to these arcs as shown in Figure 1. So basically the problem amounts to solving a network problem. But, it is difficult due to additional constraints which cannot be embedded into the graph.

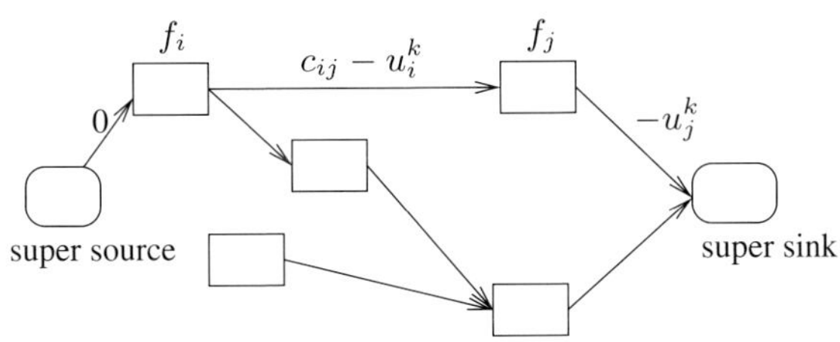

Fig. 1. The network model of the pricing problem

The dual of the column generation method is sometimes considered as Kelley's cutting plane method [9, 12]. Hence we equivalently speak of cutting planes in the dual space and of variables in the primal space. Therefore, in dual space, the pricing step is similar to the process of finding cutting planes that violate the current optimal point. Because of not being bounded, u^{k+1} can be far away from the current point u^k, and also from the optimal dual solution. We call this behavior *unstable.* Figure 2 illustrates the possible poor performance of the standard column generation method with respect to creating redundant cutting planes (in the dual problem). Without any constraint on the dual variables, the cutting plane c^1 is priced out moving the optimal dual point u^1 far away from u^0. The cutting planes c^2 and c^3 will be generated next and c^1 will turn out to be redundant. A crucial point to be observed is that the final dual point u^3 is nearer to the initial point u^0 than u^1.

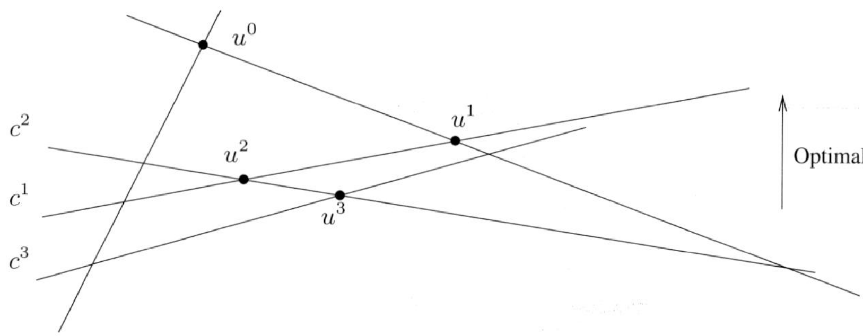

Fig. 2. Unstable behavior of the standard method in dual space

In order to overcome this disadvantageous behavior, we restrict the possible values of the dual variables by defining some box around the current iterate (as in the *boxstep method* of [11]) and ensure that the dual variables do not move too far away. We discuss below the effect of the idea on the number of iterations as well as on the overall CPU time.

4 Stabilized Column Generation

For describing the approach for controlling the dual variables we consider the primal problem $\tilde{P}$ and its dual problem $\tilde{D}$ in a form suggested in [4]:

$$
(\tilde{P}) \quad \begin{array}{lr}
\min c^T\tilde{x} - \delta_-^T y_- + \delta_+^T y_+ & \\
\text{s.t.} \qquad A\tilde{x} - y_- + y_+ & = 1 \\
 y_- & \leq \varepsilon_- \\
 y_+ & \leq \varepsilon_+ \\
 \tilde{x}, y_-, y_+ & \geq 0,
\end{array}
$$

$$(\tilde{D}) \quad \begin{array}{ll} \max \tilde{u} - \varepsilon_-^T v_- - \varepsilon_+^T v_+ & \\ \text{s.t.} & A^T \tilde{u} \leq c \\ & -\tilde{u} - v_- \leq -\delta_- \\ & \tilde{u} - v_+ \leq \delta_+ \\ & v_-, v_+ \geq 0. \end{array}$$

Here the dual problem puts penalty costs on dual variables if they lie outside of the interval $[\delta_-, \delta_+]$.

It is easy to see that the optimum solution $(\tilde{x}^*, \tilde{u}^*)$ of the pair $(\tilde{P}, \tilde{D})$ is the same as the optimum solution (x^*, u^*) of the original (P, D) if

(i) $\varepsilon_- = \varepsilon_+ = 0$, or

(ii) $\delta_- < \tilde{u}^* < \delta_+$ (due to $\varepsilon_- \geq 0, \varepsilon_+ \geq 0$ and strong duality).

This paper focuses on the application of a special variant of the model mentioned above in which ε_- and ε_+ are set to infinity. In that case, $\tilde{P}$ and $\tilde{D}$ change to the boxstep model in which the dual variables are kept in a box. Because of condition (ii), the loop will stop when the current dual point becomes an interior point of the box. The size of the box could be fixed in advance or changed during the algorithm. The primal and dual of the restricted master problem in which the box size can be altered are as follows:

$$(\tilde{P}^k) \quad \begin{array}{l} \min c^k x^k - \delta_-^k y_- + \delta_+^k y_+ \\ \text{s.t. } A^k x^k - y_- + y_+ = 1 \\ \quad x^k, y_-, y_+ \geq 0, \end{array} \qquad (\tilde{D}^k) \quad \begin{array}{l} \max {u^k}^T 1 \\ \text{s.t. } u^k A^k \leq c^k \\ \quad \delta_-^k \leq u^k \leq \delta_+^k. \end{array}$$

At the beginning, the box size is kept small so that only useful columns can be generated. We will increase the box size gradually in order to satisfy the termination condition. The boxstep model will change to the standard model if the size of the box goes to infinity. One could also control the box center, but in this paper, we only consider the so-called *stationary boxstep* (same terms as used in [12]) which keeps the center unchanged. In more detail, the two parameters δ_+ and δ_- are controlled directly and independently of the current dual solution. We will consider an update method different from the one used in [4] in which a sequence of δ_+ and δ_- was chosen in advance. The initial parameter values will be small, but large enough to make sure that the box can intersect with the initial cutting planes. The δ-update method we choose in the stationary boxstep method is

$$\delta_-^{k+1} = \delta_-^k - \Delta\delta_-, \ \delta_+^{k+1} = \delta_+^k + \Delta\delta_+.$$

The values δ_+ and δ_- only change when the column generation step is unable to generate more variables. We believe that the change $\Delta\delta$ will have an impact on the boxstep method. Small values will lead to solving many linear relaxations and pricing subproblems, and on the other hand, large values will quickly turn the boxstep method into the standard column generation method generating many redundant columns.

In the crew pairing problem considered here, the reduced cost of a variable is $c_j - {u^k}^T A_{.j}$. If we keep u^k small, then few negative reduced cost pairings will be priced out. In other words, the pricing step will be faster. The stabilized methods have been discussed with respect to reducing the number of columns in other papers. But in this paper, we will see that they can as well help to improve the speed of pricing algorithms considerably.

4.1 Standard Column Generation Method

Table 2 shows the poor performance of the standard method. For each problem instance, we display the number of linear programs that had to be solved, the number of columns added by pricing, the percentage of the overall CPU time spent in pricing, and the CPU time in seconds. As we will see later when discussing stabilized methods, several generated columns turn out to be redundant.

Table 2. Performance of the standard column generation method

Problem	#LP(s)	#Added Var(s)	% Pricing	Time (secs)
FS1	226	4779	90.74	17746
FS2	196	4193	96.90	11003
FS3	134	3788	96.51	5623
FS4	191	4436	97.42	15168
FS5	181	4746	96.41	9847
FS6	227	4576	96.19	18499
FS7	162	4146	92.98	11586
FS8	101	3214	96.40	3732
avg.	177	4234	95.00	11650

The unstable behavior of dual values may result in a large number of iterations. Figure 3 visualizes the distance $||u - u^*||_2$ between the current dual vector u and the optimal point u^*. The current dual point seems to move far away from the optimal solution in the initial iterations. The tailing-off effect also contributes to the bad performance of the method.

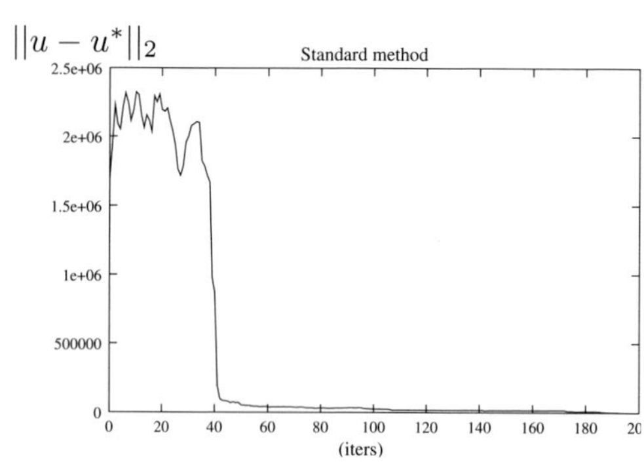

Fig. 3. Behavior of the dual values in standard method

4.2 Stationary Boxstep

Table 3 shows that this technique has reduced the number of pairings added to the restricted master problem. Furthermore, more useful and stronger cutting planes (pairings in the primal problem) have been generated when we pay attention to the number of LPs solved. Apparently, all cutting planes in the dual problem have the form $\sum_{i=1}^{n} u_i \leq c_j$. Therefore, their distance from the origin is $c_j/\sqrt{n}$. If we gradually enlarge the box centered at the origin, more good planes for the restricted problem are generated.

Table 3. Stationary boxstep with $\Delta\delta = 700$

Problem	#LP(s)	#Added Var(s)	% Pricing	Time (secs)
FS1	125	1459	89.47	2414
FS2	135	1490	88.60	3545
FS3	163	1381	89.92	3455
FS4	190	1475	94.60	5964
FS5	166	1471	94.61	4901
FS6	147	1566	94.06	2728
FS7	136	1374	94.69	4294
FS8	153	1348	94.79	4133
avg.	151	1445	92	3929

The rather small number of variables generated proves that the stabilization techniques are very effective. The time needed to solve the relaxation problem is now reduced by more than a factor of 3 in comparison to the standard method. Figure 4 also shows that the dual vector is more stable than that of the standard column generation method. One interesting point is that the average elapsed time of the stationary boxstep in an iteration is about half of that of the standard method. This means that pricing in the stationary approach is faster due to the limitation of the box size and the fixing of the box center. In order to get a better understanding of the effect of the bounded dual values on the pricing algorithms, we also use the multi-labeling method (dynamic programming) [3] in the pricing stage. An interesting result is that the number of labels generated per iteration in the root node is the smallest. With the standard method, about 100,000 labels were generated. The stationary method has only generated approximately 65,000 labels. The number of labels generated in the pricing stage verifies our expectation.

The parameter $\Delta\delta$ plays a very important role in the stabilization technique, as shown in Figure 5. Small values of $\Delta\delta$ will make the column generation step stall many times. Although only few columns are priced out, many linear programs and pricing subproblems have to be solved resulting in high computation time. The poor performance also occurs for large values of $\Delta\delta$, but due to a different reason. When the box size increases very quickly,

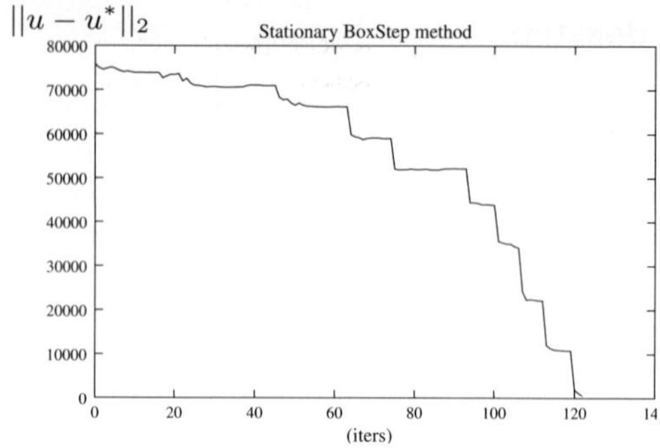

Fig. 4. Progress of the dual values in the stationary boxstep

the stationary boxstep method soon becomes the standard column generation method. Not many iterations are necessary, but too many columns are added and we obtain the same behavior as in the standard method.

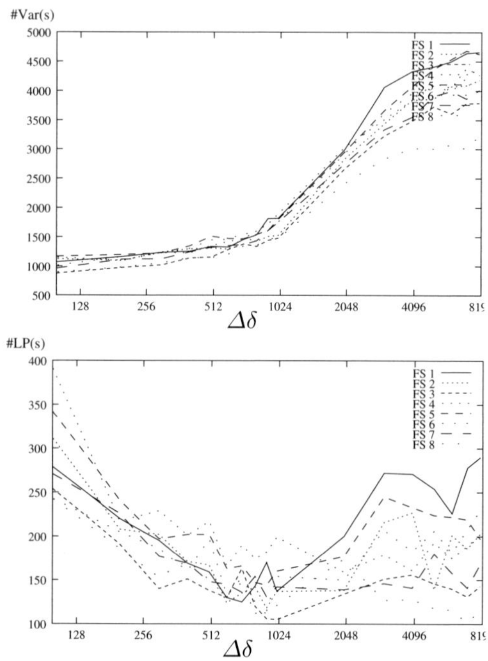

Fig. 5. Number of variables added and LPs solved with respect to different $\Delta\delta$ in stationary boxstep

There is a tradeoff between the number of linear programs to be solved and the number of generated variables leading to the question of which $\Delta\delta$ is the best one. In our application of the crew pairing problem, the cost of a pairing will be less than 10080 (i.e., the number of minutes in a week). Hence the distance between a cutting plane and the origin is less than $10080/\sqrt{2} \simeq 7128$ (a pairing has at least 2 flights). It is unreasonable to choose $\Delta\delta$ larger than this number because the dual restriction mechanism will be disabled in that case. The smallest reasonable value with respect to the airline rules we have here is about 31. Within the interval $[31, 7128]$, the interval $[512, 1024]$ seems to contain the best values as shown by the best computation time in Figure 6. Up to now, we have not found any mathematical relationship between the best $\Delta\delta$ and the properties of our crew pairing problem (cost structure, flight regulations, flight network, etc.). The best parameters can only be chosen based upon empirical tests.

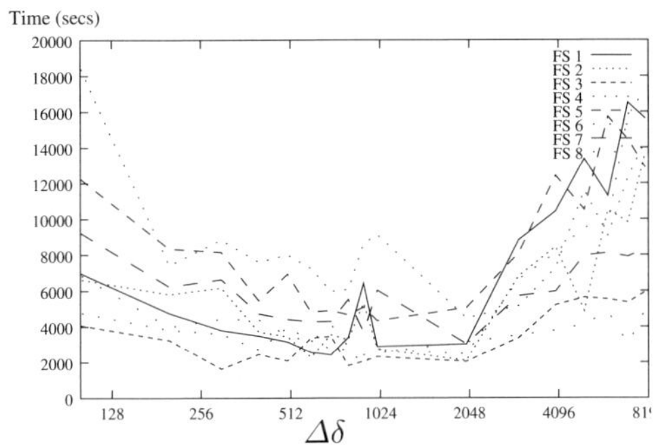

Fig. 6. Time elapsed to solve the root node with respect to different $\Delta\delta$

Dealing with the two steps of the column generation method separately has been a main approach in the research on the crew scheduling problem. This method uses the original mathematical model of the restricted master problem for the linear relaxation step, causing a lot of difficulties to the pricing step. Our approach applies the modified model of the stabilized column generation method with several parameters. With suitable controls of these parameters, the computation time is reduced in our experiments, since stable paths to optimal solutions are generated and the pricing is faster. We believe that the inclusion of stabilization methods is very relevant for practical problem solving.

References

[1] P. Alefragis, C. Goumopoulos, E. Housos, P. Sanders, T. Takkula, and D. Wedelin. Parallel Crew Scheduling in PAROS. In *Proceedings of Euro-Par'98 Parallel Processing: 4th International Euro-Par Conference*, 1998.

[2] C. Barnhart, E. L. Johnson, G. L. Nemhauser, M. W. P. Savelsbergh, and P. H. Vance. Branch-and-price: Column Generation for Solving Huge Integer Programs. *Operations Research*, 46:316–329, 1998.

[3] J. Desrosiers, Y. Dumas, M. M. Solomon, and F. Soumis. *Handbooks in Operations Research and Management Science*, volume 8, chapter Time Constrained Routing and Scheduling, pages 35–139. North-Holland, 1993.

[4] O. du Merle, D. Villeneuve, J. Desrosiers, and P. Hansen. Stabilized Column Generation. *Discrete Mathematics*, 194:229–237, 1999.

[5] N. Guerinik and M. V. Caneghem. Solving Crew Scheduling Problems by Constraint Programming. In *Proceedings of the 1st Conference of Principles and Practice of Constraint Programming*, pages 481–498, 1995.

[6] T. Gustafsson. *A Heuristic Approach to Column Generation for Airline Crew Scheduling.* PhD thesis, Chalmers University of Technology, 1999.

[7] I. Ioachim, S. Gélinas, F. Soumis, and J. Desrosiers. A Dynamic Programming for the Shortest Path Problem with Time Windows and Linear Node Costs. *Networks*, 31, 1997.

[8] B. Jaumard, F. Semet, and T. Vovor. A Two-Phase Resource Constrained Shortest Path Algorithm for Acyclic Graphs. Technical report, Les Cahiers du GERAD, 1999.

[9] J. E. Kelley. The Cutting Plane Method for Solving Convex Program. *Journal of SIAM*, 8:703–712, 1960.

[10] D. Klabjan and K. Schwan. Airline Crew Pairing Generation in Parallel. Technical report, The Logistics Institute, Georgia institute of Technology, 1999.

[11] R. E. Marsten, W. W. Hogan, and J. W. Blankenship. The Boxstep Method for Large-Scale Optimization. *Operations Research*, 23(3):389–405, 1975.

[12] P. J. Neame. *Nonsmooth Dual Methods in Integer Programming.* PhD thesis, The University of Melbourne, 1999.

[13] P. H. Vance, A. Atamtürk, C. Barnhart, E. Gelman, E. L. Johnson, A. Krishna, D. Mahidhara, G. L. Nemhauser, and R. Rebello. A Heuristic Branch-and-Price Approach for the Airline Crew Pairing Problem, 1997.

[14] R. Wunderling. *SoPlex, The Sequential Object-Oriented Simplex Class Library.* ZIB, 1997.

[15] T. H. Yunes, A. V. Moura, and C. C. de Souza. *Practical Aspects of Declarative Languages*, volume 1753 of *Lecture notes in computer science*, chapter A Hybrid Approach for Solving Large Scale Crew Scheduling Problems, pages 293–307. Springer, 2000.

The Study of Pores and Free Volume in Amorphous Models

Pham Khac Hung[1], Vo Van Hoang[2], Hoang Van Hue[1], Le Van Vinh[1], and Ngyuen Van Hong[1]

[1] Hanoi University of Technology, Institute of Engineering Physics
No 1 Dai Co Viet Road, Hanoi, Vietnam
vinhlv@mail.hut.edu.vn

[2] Department of Physics, National University of HoChiMinh City

Summary. We presented the results of computing simulation of microspores and free volumes surrounding an atom in the amorphous model which constructed by statistic relaxation method on parallel computers. On purpose to accurately determine the amount of vacancy-like pores, several models containing from 10^4 to 4.10^5 atoms with boundary periodic condition have been constructed corresponding to the density of 85.56 atoms/nm^3 or 83.90 atoms/nm^3. The calculations showed that amorphous models have about 0.0009-0.0075 vacancy-like pores per atom depending on the atomic density and local metstable states.

Key words: amorphous metals, multimillion computer simulation

1 Introduction

The microscopic nature of atomic transport in amorphous materials has been the subject of numerous studies [1-9]. The great interest is possible vacancy mechanism or interstitial one that concerns the existence of different kinds of pores in amorphous structure. Based on the comparison of the activation energy obtained with experiment, the study of the properties of crystal-like: defects, formation and migration energies, and entropies also lead to conclusion that the vacancy mechanism can be considered as a more possible one. The analysis of microspores is usually done by calculating the radii of hard spheres that can be embedded in amorphous model without intersecting any atomic spheres. As indicated in [1-7], the amorphous models constructed by standard molecular dynamics have some large pores, which can exchange their positions with the neighboring atoms, and therefore they behave like vacancy in diffusion. The existence of large vacancy-like pores in the amorphous Co-B, Co-P and Ni-P alloys has been predicted in [4-6]. As shown in [4], the number of large pores in the amorphous Co-B alloy became larger when the boron concentration is bigger than 18.5 at %. Meanwhile, the number of large pores in the amorphous Ni-P [6] and Co-P [7] alloys monotonically increases with the increasing phosphorus content. On the other hand, the microspores may generate the microcrack and some point defect in amorphous structure. The problem of point defects and its relation with local inhomogeneities has been summarized detail in [1]. Nevertheless, all simulations in [1-9] have been performed on too small models containing just several thousand atoms, and those are unclear about the accurate amount of large cavities. The parallel computing technique gives us a new ability to treat large collections of atoms [10-14]. Multimillion atomic simulations show interesting results about the consolidation, fracture and oxidation in the amorphous materials [15-18]. However, the problem of pores and free volume around an atom in amorphous structure has not been investigated for the multimillion atomic models yet. Especially, the influence of local metastable states, atomic density and size of models on microstructure has not been carried out. In this work, the simulation of microstructure included the spherical pores, free volume and coordination number in amorphous Fe models containing 10^4 , 2.10^4, 4.10^4 and 4.10^5 atoms has been done and presented.

2 Calculation

Four models of the amorphous iron containing 2.10^4 and 4.10^5 atoms in a simple cube with periodic boundary conditions were constructed by the statistical relaxation method. The size of cube were adopted corresponding to the real density of the amorphous Fe (85.56 atoms/nm^3). The calculations were carried out in LAN connecting 24 personal computers supported by parallel computing PVM. The Paka-Doyama pair potential was used to obtain a good agreement with the experimental radial distribution function (RDF) as was successfully done for small models in [8]. This potential has the form as follows:

$$\varphi(r) = -0.18892(r - 1.82709)^4 + 1.70192(r - 2.50849)^2 - 0.79829 \quad (1)$$

Where r and $\varphi(r)$ are in A^0 and eV, respectively. The potential cut-off radius is $3.44A^0$. We have treated three models A, B and D with the same density 85.56 at. atoms/nm^3, but different number of atoms. The model A and D have been constructed as follows: Initially we randomly arranged all atoms in a simple cube, then moved each atom of the system on the force direction with displacement length of $0.01A^0$. This force is caused by all remain atoms. The movement of atoms repeated until the energy of the system reached minimum. The model B was obtained from reproducted model A by relaxation with two displacement length of $0.05A^0$ and $0.01A^0$. Large displacement ($0.05A^0$) has been used to move the system out the local equilibrium. Then, the system step by step reached the equilibrium by relaxation with the small displacement of $0.01A^0$. The model C contained 2.10^4 atoms with the density 83.9 at. atoms/nm^3. The characteristic of the models is given in table 1.

To calculate the pores we take three neighboring atoms surrounding each ith atom of the system. Then, we insert the sphere pore in contact with i^{th} atom and its neighbors. The neighboring atom has considered as located at

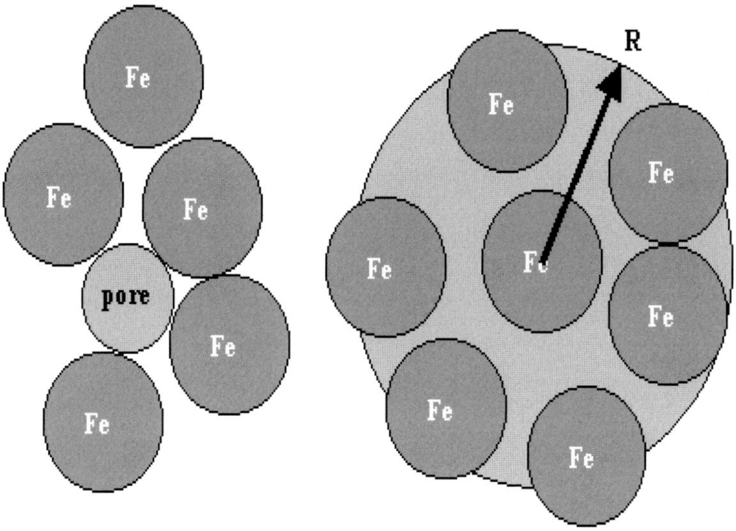

Fig. 1. The scheme of calculation of pores (left) and free volume (right) around an atom.

distance less than $3.4A^0$ from i^{th} atom. If a inserted pore overlaps any atomic spheres or any bigger pores, we remove this one. This process has taken over all group of ith atom and its three neighbors. The radius of atomic sphere has accepted to equal to $1.26A^0$. The schematic calculation of pore has shown in fig.1.

The free volume around an atom has been investigated by inserting a sphere with radius R and center located at the chosen i^{th} atom. After that, we randomly distributed 2000 points in this sphere. The part of these points have been belong atomic spheres. It means that they located at distance less than $1.26A^0$ from some atomic centers. Let us define the number of points belong atomic sphere by N_v. Therefore, the free volume surrounding i^{th} atom can be calculated as

$$V_f = \frac{(2000 - N_v)4\pi R^3}{3\times 2000} \tag{2}$$

In order to convenient we used a non-unit parameter V_f equal to (2000-N_v)/2000.

3 Results and discussions

RDF of the models A, B, C and D is presented in fig.2. The position and height of first peak in RDF have shown in table 1. Those are almost close to the RDF of the amorphous iron [7] and in good agreement with experiment [19]. The model A and B have different energy of 0.6%, but their RDF are similar to each other. The slight difference of the height of 1st peak can be observed in table 1.

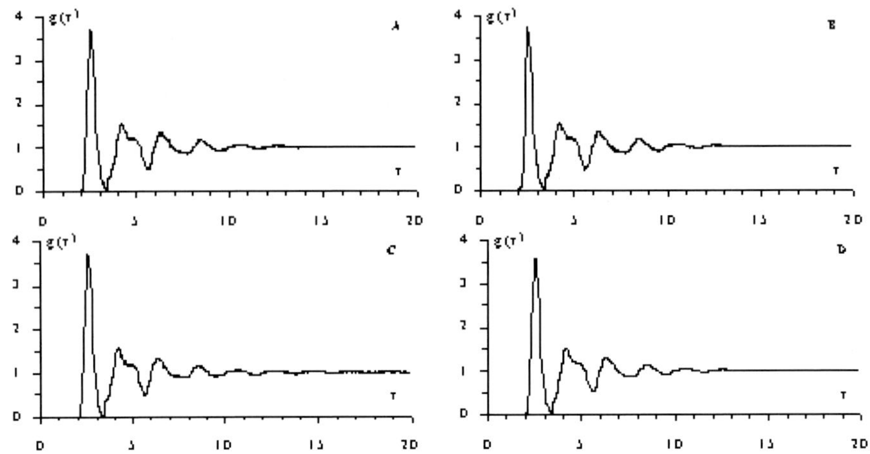

Fig. 2. The radial distribution functions of models A, B, C and D. r is in A^0

Table 1. The position and height of first peak in RDF

Model	Model A	Model B	Model C	Model D	Exp.[19]
Position of 1st peak of RDF,A^0	2.52	2.50	2.52	2.50	2.52
Height of 1st peak of RDF	3.682	3.734	3.724	3.584	3.6
Position of 2st peak of RDF,A^0	4.16	4.18	4.18	4.18	4.16
Height of 2st peak of RDF	1.548	1.535	1.572	1.504	1.55

The local microstructural characteristics surrounding an atom including: the radii (left pictures), the free volume, the coordination and number of cavities (right pictures) have depicted in fig.3. Analogously the RDF, the significant variety of local microstructure of all models has not been observed in fig.3. The radii distribution has two peaks: the first peak located at 0.32 A^0 and the second at 0.52 A^0. As indicated in [1] for amorphous iron, if the radii of pores bigger than 0.98 A^0 then these pores can play role of vacancy in the diffusion. We define these pose as vacancy-like ones. The amount of vacancy-like pores is listed in table 1. Here we can notice that for the models A, B and D having the same density, the more less the potential energy, the more bigly the amount of vacancy-like pores. In the case of model C, we found about 0.0049 pores per atom. It is rather bigger than one of models A and B despite its energy is the smallest ones. This effect can be understood because of the atomic density of model C is the smallest. Its value is less than other models on $\sim$ 2%. The statistic relaxation model can be considered as atomic state at 0 K. Therefore, the vacancy-like pores in this model are the "native" vacancy and not normal thermal vacancy which is created by thermal activated process. Hence, the vacancy diffusion mechanism in amorphous iron is characterized by the participation of both kinds of vacancy.

In contrast with vacancy-like pore, the total number of pores in the model is bigger if its energy is less. The same behavior has been observed to the height of 1st peak of f(r).

Table 2. The local microstructure of amorphous models

Parameters	Model A	Model B	Model C	Model D
Energy per atom, eV	-1.3742	-1.3811	-1.3838	-1.3615
Length of model 's cube,A^0	61.16	61.16	62.0	167.197
Number of atoms in the model	2.10^4	2.10^4	2.10^4	4.10^5
Density, atom/nm^3	85.6	85.6	83.9	85.6
Position of 1st peak of f(r) ,A^0	0.32	0.32	0.34	0.32
Height of 1st peak of f(r)	3.360	3.435	3.340	3.210
Position of 2st peak of f(r) ,A^0	0.52	0.52	0.52	0.52
Height of 2st peak of f(r)	0.368	0.380	0.445	0.376
Number of cavities with r $\geq$ 1.0 A^0	0.0027	0.0009	0.0049	0.0075
Position of a peak of fc(z)	20	20	20	20
Height of the peak of fc(z)	0.1999	0.1967	0.2040	0.1925
The averaged number of cavities	20.5118	20.6249	20.2219	20.384
Position of a peak of fa(z)	14	14	14	14
Height of the peak of fa(z)	0.4133	0.4274	0.4237	0.4083
Coordination number	14.2040	14.2285	14.1003	14.1920
Position of a peak of f(V_c)	0.5	0.5	0.5	0.5
Height of the peak of f(V_c)	0.3112	0.3125	0.3019	0.2962
Position of a peak of f(V_f)	0.27	0.27	0.28	0.27
Height of the peak of f(V_f)	0.1838	0.1916	0.1837	0.1734

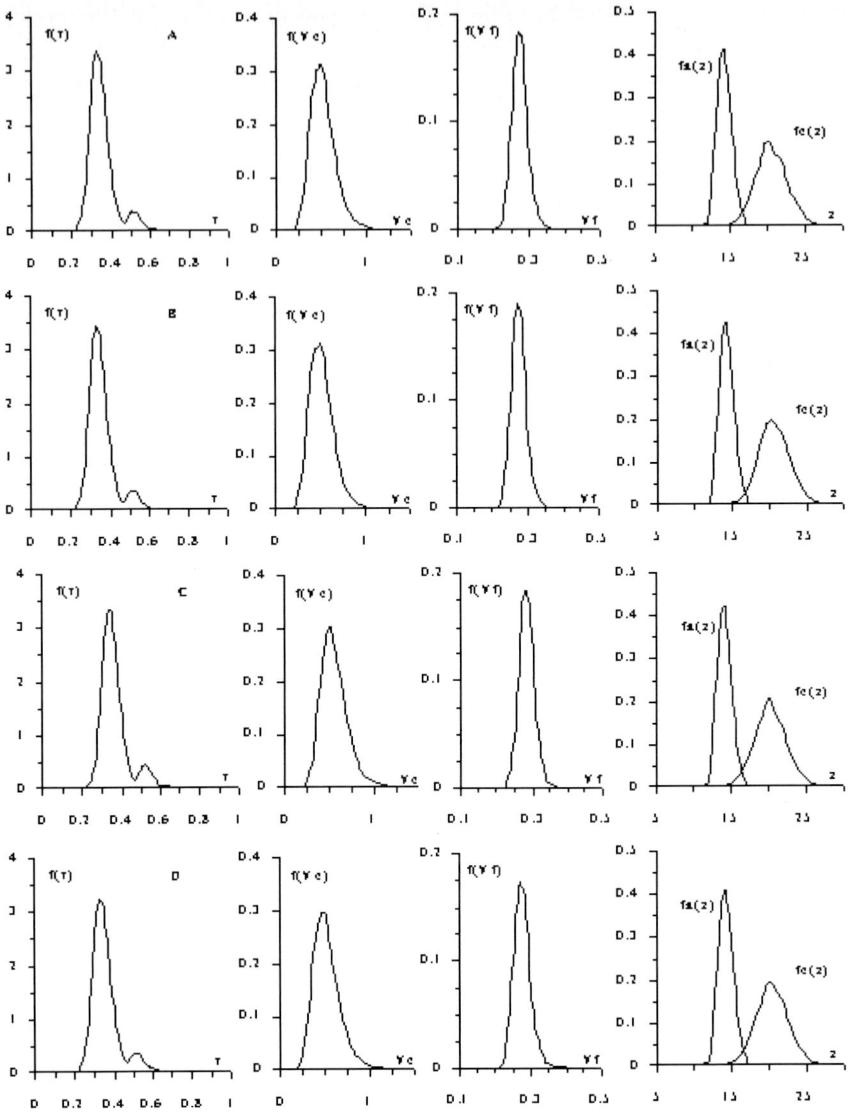

Fig. 3. The radii, the cavity volume, the free volume, the cavities number and coordination number distribution of the amorphous models A, B, C and D. r is in A^0

To calculate the distribution f(V_f) we have chosen the radii R= 3.4 A^0. The average free volume surrounding an atom of models A, B, C and D is equal to 0.2749, 0.2746, 0.2859 and 0.2763, respectively. It means that the free volume is about 27-29 percent of the space of sphere with radii 3.4 A^0. The percent of atomic spheres in this space is a remainder: 73-71. It can be seen in fig.3 that the distribution f(V_f) is spread in the range from 0.2 to 0.45. The fraction of atoms with the value of V_f less than 0.22 is 0.0062, 0.0047, 0.0007 and 0.0015 for the models A, B, C and D, respectively. If the value of V_f is bigger 0.38 then the fraction is 0.0009, 0.0001, 0.0022 and 0.0018 to the models A, B, C and D, respectively. Obviously, the large value of V_f corresponds to "compressed positions", where the atomic density is high. Oppositely, the small value of V_f relates to "expanded positions" that corresponds to the low atomic density. The "expanded positions" and the "compressed positions" can be considered as point defect in amorphous structure.

The parameter V_c in fig. 3 has been calculated as

$$V_c = \frac{V_t}{V_a} \tag{3}$$

Where V_t is the total volume of all cavities surrounding an atom, and V_a is atomic volume $\sim 4\pi\times1.26^3\ A^0$.

Another characteristics of local microstructure are the coordination number and the cavity number surrounding atom. The coordination number is determined as an amount of atoms separated from chosen atom with the distance less than 3.4 A^0. The average values of cavity and coordination number are 14.1-14.3 and 20.2-20.7, respectively. The distribution of cavity number $f_c(z)$ is wider than the distribution of coordination number $f_a(z)$, and $f_c(z)$ spreads in range from 11 to 28.

In order to investigate the effect of model's size on calculated microstructure characteristics we also constructed the models E and F containing 10^4 and 4.10^4 atoms, respectively. The majority pores characteristics of these models have shown in table 3 in compare with model A and D. The results showed that despite the large different in the number atoms between the models A and D, their characteristics of pores were the same. Especially, the amount of vacancy-like pores in these models was in the range of 0.0025-0.0075.

Table 3. The local microstructure of amorphous models with the same atomic density and different number of atoms

Parameters	Model E	Model A	Model F	Model D
Energy per atom, eV	-1.3788	-1.3742	-1.3768	-1.3615
Number of atoms in the model	10^4	2.10^4	4.10^4	4.10^5
Density, atom/nm^3	85.6	85.6	85.6	85.6
Position of 1st peak of f(r) ,A^0	0.32	0.32	0.32	0.32
Height of 1st peak of f(r)	3.4617	3.360	3.391	3.210
Number of cavities with r $\geq$ 1.0 A^0	0.0033	0.0027	0.0025	0.0075
The averaged number of cavities	20.5484	20.5118	20.5633	20.384
Coordination number	14.2108	14.2040	14.2142	14.1920

We calculated the radial distribution of average free volume surrounding an atom depending on radii R for the model D. The result of this calculation has been shown in fig.4. Here we can see the free volume surrounding an atom has as the short order as the RDF. The 1st peak located at the radii R=1.76 A^0 and has a height of 0.381. At the radii R bigger 4.0 A^0, the free volume surrounding an atom became closed to the value of 0.286. We also calculated the radial distribution of average free volume surrounding an atom depending on radii R for the crystal model with bcc structure and atomic density of 85.6 at.$/nm^3$. This distribution (f0(R)) is almost similar to the distribution of the amorphous model (f1(R)). However, they have some small different at the first and second peaks.

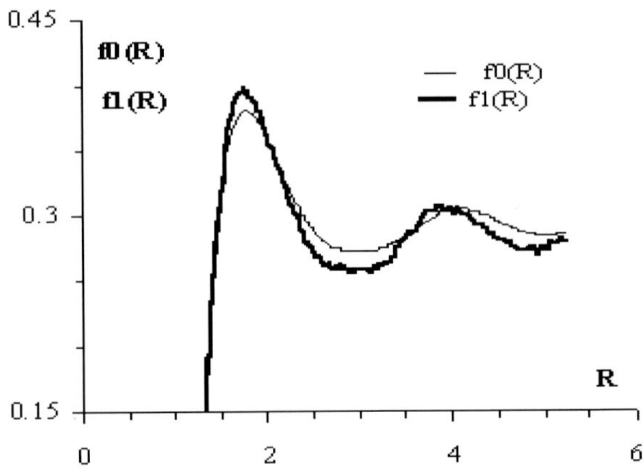

Fig. 4. The radial distribution of free volume surrounding an atom in the crystal model with bcc structure f0(R) and amorphous model f1(R).

4 Conclusions

The local microstructure of several amorphous models iron containing from 10^4 to 4.10^5 atoms have been investigated as follows: cavities, free volume surrounding an atom and coordination number. The results of calcualations showed that one of the characteristic amorphous iron models is the amount of vacancy-like pores which is in the range from 0.0009-0.0075 depending on the atomic density and metastable states. These vacancy-like pores have indicated the participation of 'native' and 'thermal' vacancy for difusion in amorphous iron.

References

[1] Belashchenko D. K, Physics-Uspekhi 42 (4) 297-319 (1999)
[2] N.Mousseau. Defect and Diffusion forum,Vol.194-199(2000) 775-778.
[3] Delaye J. M, Limoge I. J. Phys.I. France 3 (1993) 2079
[4] Belashchenko D. K, Hoang V. V, Hung P. K. J. Non-Cryst. Solids, 276 (2000) 169
[5] Hoang V. V, Van T. B, Hung P. K. J. Metastable and Nanocrystalline Materials, vol. 2-6 (1999) 551
[6] Hoang V. V, Van T. B. J. Met. and Nanocrystalline Materials (2001), e-vol. 9, 5
[7] Hung P. K, Belashchenko D. K, Nguyen P. N. Izv. Akd. Nauk Russii, Metally 4 (1996) 155
[8] Belashchenko D. K. Physica metalov and metalovedenie, Vol.60, 6(1985) 1076
[9] Hoang V.V.,Nguyen H.H,Nguyen H.T.A. J. Metastable and Nanocrystalline Materials,e-Vol. 18 (2003) 43
[10] Kalia R. K, Campell T. J et al. Computer Physics Communications 128 (2000) 245
[11] Nakano A, Bachlechner M. E, et al. IEEE Computational Science and Engineering 5 (4) (1998) 68
[12] Vashishta P, Kalia R. P et al. Computational Materials Science 2 (1994) 180
[13] Nakano A, Kalia R. P, Vashishta P. Computer Physics Communications 77 (1993) 303
[14] Kalia R. P, Nakano A, Greenwell D. L, Vashishta P, de Leeuw S. W. Supercomputer 54 (X-2) (1993) 11
[15] Vashishta P, Bachlechner M. E et al. Progress of Theoretical Physics Supplement 138 (2000) 175
[16] Vashishta P, Kalia R. K, Nakano A. Computing in Science and Engineering B37 (1996) 56
[17] Vashishta P, Nakano A, Kalia R. P, EbbsjO I. Mater. Sciences and Engineering B 37 (1996) 56
[18] Vashishta P, Kalia R. K et al. Current Opinion in Solid State and Mater. Sciences 1 (1996) 853
[19] Y.Loirat et al. Defect and Diffusion forum,Vol.194-199(2000) 850-860.
[20] Ichikawa T. Phys.Stat.Sol A.19 (1973) 707

A Two-Stage, High-Accuracy, Finite Element Technique of the Two Dimensional Horizontal Flow Model

Nguyen The Hung

Department of Water Resources Engineering, University of Danang
Lien Chieu, Danang, Vietnam
ngthehung@dng.vnn.vn

Summary. An algorithm and essential subroutines are presented which implement a two stage finite element Galerkin method for integrating the complete two dimensional horizontal flow model. In the method a high accuracy is obtained by combining the Galerkin product with a high-order difference approximation to the derivatives in the nonlinear advection operator. The program includes the use of a weighted selective lumping scheme in the finite element method and the use of the Gauss - Seidel iterative method for solving the resulting systems of linear equations. Small scale noise is eliminated by using a shuman filter.

1 Introduction

The two dimensional horizontal flow model or shallow water equation system are very popular for treatment of flow of lakes, rivers, oceans etc.

One of the difficult issues for solving the two dimensional horizontal flow model is how to treat the nonlinear advective terms (see Cullen and Morton, 1980).

The finite element method when applied to hydrodynamic problems gives an accurate solution. This method is conservative and therefore avoids aliasing errors associated with nonlinear terms. Moreover, it has the advantage over the finite difference method of being flexible in the treatment of irregular domains and allows a variable resolution, thus permitting to focus on regions of interest.

Navon (1982, 1983) introduced the Numerov-Galerkin finite element method for the shallow water equations with an Augmented Lagrangian constrained optimization method to enforce integral invariants conservation. Similar work was done by Zienkiewicz and Heinrich (1979) with a finite element penalty method, and by Zienkiewicz and others (1984). Navon (1987) applied a two-stage finite element Galerkin method combined with a high-accuracy compact approximation to the first derivative for solving the shallow water equations.

In this paper, this method is applied for solving the full shallow water equations.

2 The Finite Element Galerkin Solution of the Two Dimensional Flow Model

2.1 Governing Equation

Under the assumptions of a homogeneous 2D incompressible flow with a hydrostatic pressure distribution over a cross section, the governing unsteady flow equations can be expressed as follows:

$$\begin{aligned} u_t + u.u_x + v.u_y + \varphi_x - f.v + T^x &= 0 \\ v_t + u.v_x + v.v_y + \varphi_y + f.u + T^y &= 0 \\ \varphi_t + (\varphi u)_x + (\varphi u)_y - r &= 0 \end{aligned} \tag{1}$$

Here u and v are the velocity components in the x and y directions, respectively, f is the Coriolis parameter, r is the precipitation intensity, $\varphi = gh$ is the geopotential, h is the depth of the fluid, g is the acceleration of gravity,

$$T^x = (\tau_b^x - \tau_s^x)/\rho h \ , \ T^y = (\tau_b^y - \tau_s^y)/\rho h \tag{2}$$

with

$$c = \tfrac{1}{n} R^{1/6} \ , \ \tau_b^y = \rho g v \sqrt{u^2 + v^2}/c^2 \ , \ \tau_b^x = \rho g u \sqrt{u^2 + v^2}/c^2$$

$$\tau_s^x = \rho_a C_z W_z^2 cos\varphi_x \ , \ \tau_s^y = \rho_a C_z W_z^2 cos\varphi_y,$$

τ_b^x, τ_b^y is the bottom friction force in the x and y directions, respectively; n is the Manning hydraulic roughness, R is the hydraulic radius, τ_s^x , τ_s^y are the components of the wind stress at the water surface in the x and y directions, respectively, ρ is the mass density of water ρ_a is the mass density of air, C_z is the drag coefficient, W_z is the wind speed φ_x , φ_y are the angles between the wind direction and x, y directions, repectively.

2.2 Finite Element Algorithm

The formulation of the finite element Galerkin method for the equation systems (1) is as follows:

$$\langle f(x,y), V_i \rangle = \sum_{elements}^{m} \iint f(x,y) \cdot V_i \; dxdy$$
$$= \iint_{global} f(x,y) \cdot V_i \; dxdy \ , \ i = i,j,k \tag{3}$$

This notation defines the inner product when a function is multiplied by a trial function.

Linear piecewise polynomials on triangular elements are used where for a given triangular element each variable is represented as a linear sum of interpolating functions, for example

$$u^e = \sum_{j=i,j,k} u_j(t) \cdot V_j(x,y) \tag{4}$$

where u_j(t) represents the scalar nodal value of the variable u at the node of three vertices of the triangular element, V_j is a basis function also called a trial function. The basis function can be expressed in natural coordinate as follows:

$$V_i = \frac{1}{2A}(a_i y + b_i x + c_i) \tag{5}$$

where A is the area of the triangle,

$$b_i = y_j - y_k, a_i = x_k - x_j, c_i = x_j \cdot y_k - x_k \cdot y_j (6)$$

i, j, k are cyclically permuted (i, j, k = 1, 2 , 3).
The derivatives of the shape functions, V_i, are given by

$$\frac{\partial V_i}{\partial x} = \frac{b_i}{2A} \ ; \ \frac{\partial V_i}{\partial y} = \frac{a_i}{2A} \ , \ (i = 1,2,3) \tag{7}$$

A time extrapolated Crank-Nicolson time differencing scheme is applied for integrating in time the system of the ordinary differential equations resulting from the application of the Galerkin finite element method to the continuity equation

$$[M] \left\{\dot{\varphi}\right\} + [K_1] \{\varphi\} = [M] \{r\} \tag{8}$$

where

$$[M] = \sum_1^m [M^e] \ ; \ [K_1] = \sum_1^m [K_1^e] \ ; \ \left\{\dot{\varphi}\right\} = \sum_1^m \left\{\dot{\varphi^e}\right\} ; \ \{\varphi\} = \sum_1^m \{\varphi^e\} \tag{9}$$

M^e = element mass matrix

$$[M]^e = \begin{bmatrix} \iint\limits_A V_iV_i dxdy & \iint\limits_A V_iV_j dxdy & \iint\limits_A V_iV_k dxdy \\ \iint\limits_A V_jV_i dxdy & \iint\limits_A V_jV_j dxdy & \iint\limits_A V_jV_k dxdy \\ \iint\limits_A V_kV_i dxdy & \iint\limits_A V_kV_j dxdy & \iint\limits_A V_kV_k dxdy \end{bmatrix} = \frac{A}{12}\begin{bmatrix} 2 & 1 & 1 \\ 1 & 2 & 1 \\ 1 & 1 & 2 \end{bmatrix} \tag{10}$$

$[K_1^e]$ = element matrix

$$[K_1^e] = \iint\limits_A \left(V_i \cdot \sum_{j=i,j,k} (V_j \cdot u_j^*) \cdot \sum_{j=i,j,k} \left(\frac{\partial V_j}{\partial x}\varphi_j \right) \right) dxdy$$

$$+ \iint\limits_A \left(V_i \cdot \sum_{j=i,j,k} (V_j \cdot v_j^*) \cdot \sum_{j=i,j,k} \left(\frac{\partial V_j}{\partial y}\varphi_j \right) \right) dxdy$$

$$[K_1^e] = [K_{11}^e] + [K_{12}^e] \tag{11}$$

$$[K_{11}^e] = \frac{1}{24}\begin{bmatrix} (2u_i^* + u_j^* + u_k^*)b_i & (2u_i^* + u_j^* + u_k^*)b_j & (2u_i^* + u_j^* + u_k^*)b_k \\ (u_i^* + 2u_j^* + u_k^*)b_i & (u_i^* + 2u_j^* + u_k^*)b_j & (u_i^* + 2u_j^* + u_k^*)b_k \\ (u_i^* + u_j^* + 2u_k^*)b_i & (u_i^* + u_j^* + 2u_k^*)b_j & (u_i^* + u_j^* + 2u_k^*)b_k \end{bmatrix}$$

$$[K_{12}^e] = \frac{1}{24}\begin{bmatrix} (2v_i^* + v_j^* + v_k^*)a_i & (2v_i^* + v_j^* + v_k^*)a_j & (2v_i^* + v_j^* + v_k^*)a_k \\ (v_i^* + 2v_j^* + v_k^*)a_i & (v_i^* + 2v_j^* + v_k^*)a_j & (v_i^* + 2v_j^* + v_k^*)a_k \\ (v_i^* + v_j^* + 2v_k^*)a_i & (v_i^* + v_j^* + 2v_k^*)a_j & (v_i^* + v_j^* + 2v_k^*)a_k \end{bmatrix}$$

In the notation , u^* and v^* are given by

$$\begin{aligned} u^* &= u^{n+1/2} = 3u^n/2 - u^{n-1}/2 \\ v^* &= v^{n+1/2} = 3v^n/2 - v^{n-1}/2 \end{aligned} \tag{12}$$

After introducing the time discretization in the continuity equation, which is the first to be solved at a given time step, one obtains:

$$[M]\frac{\{\varphi\}^{n+1} - \{\varphi\}^n}{\Delta t} + [K_1]\frac{\{\varphi\}^{n+1} + \{\varphi\}^n}{2} = [M]\{r\}^{n+1/2} \tag{13}$$

Here n, $n+1$ is the time level $t^{n+1} = t^n + \Delta t$, $t^n = n\Delta t$

$$\left(\frac{2[M]}{\Delta t} + [K_1]\right)\left(\{\varphi\}^{n+1} - \{\varphi\}^n\right) = 2\left([M]\{r\}^{n+1/2} - [K_1]\{\varphi\}^n\right) \tag{14}$$

3 The Two Stage Numerov-Galerkin Scheme

The two stage Galerkin method (see Cullen and Morton, 1980) is applied to the nonlinear advective terms of form $u\frac{\partial v}{\partial x}$. If we conside the advective operator

$$L(u, v) = u\partial v/\partial x \tag{15}$$

then, as shown by Cullen and Morton (1980), we may consider two methods: a direct Galerkin approximation and a two stage Galerkin approximation.

3.1 A Direct Galerkin Approximation

Using two function

$$u = exp^{(ikx)} \quad , \quad v = exp^{(ilx)} \tag{16}$$

with $\xi = kh$, $\eta = lh$ where h is a positive mesh length, the asymptotic truncation error of $u(\partial v/\partial x)$ can be found as follows (by assuming Fourier modes)

$$|T.E.| \quad \sim \quad \frac{[4\eta^4+8\eta^2+7\eta^2\xi^2-2\eta\xi^3]}{720} \tag{17}$$

and if $\xi = \eta$ then $|\text{T.E.}| \sim \frac{17\eta^4}{720}$.

3.2 A Two Stage Galerkin Approximation

In this approach we calculate the Galerkin approximation for the term $Z := \partial v/\partial x$) as follows

$$\frac{h}{6}\begin{bmatrix} 2 & 1 & & & & \\ 1 & 4 & 1 & & & \\ & \cdots & \cdots & & & \\ & 1 & 4 & 1 & & \\ & & \cdots & \cdots & & \\ & & & 1 & 4 & 2 \\ & & & & 2 & 1 \end{bmatrix} \begin{Bmatrix} Z_1 \\ Z_2 \\ \cdots \\ Z_j \\ \cdots \\ Z_m \\ Z_{m+1} \end{Bmatrix} = \frac{1}{2}\begin{Bmatrix} v_2 - v_1 \\ v_3 - v_1 \\ \cdots \\ v_{j+1} - v_{j-1} \\ \cdots \\ v_{m+1} - v_{m-1} \\ v_{m+1} - v_m \end{Bmatrix} \tag{18}$$

Then at the second stage of the finite element Numerov-Galerkin, we calculate the product $w = u\partial v/\partial x$ by solving the tridiagonal system of the form

$$\begin{bmatrix} 2 & 1 & & & & \\ 1 & 4 & 1 & & & \\ & \cdots & \cdots & & & \\ & 1 & 4 & 1 & & \\ & & \cdots & \cdots & & \\ & & & 1 & 4 & 1 \\ & & & & 1 & 2 \end{bmatrix} \begin{Bmatrix} w_1 \\ w_2 \\ \vdots \\ w_j \\ \vdots \\ w_m \\ w_{m+1} \end{Bmatrix} = \frac{1}{12}\begin{Bmatrix} 3u_1Z_1 + u_1Z_2 + u_2Z_2 + u_2Z_2 \\ u_1Z_1 + u_1Z_2 + u_2Z_1 + u_2Z_3 \\ +u_3Z_2 + u_3Z_3 + 6u_2Z_2 \\ M \\ u_{j-1}Z_{j-1} + u_jZ_{j-1} + u_{j-1}Z_j \\ +u_{j+1}Z_j + u_jZ_{j+1} \\ +u_{j+1}Z_{j+1} + 6u_jZ_j \\ M \\ u_{m-1}Z_{m-1} + u_mZ_{m-1} \\ +u_{m-1}Z_m + u_{m+1}Z_m \\ +u_mZ_{m+1} + u_{m+1}Z_{m+1} \\ +6u_mZ_m \\ u_mZ_m + u_mZ_{m+1} \\ +u_{m+1}Z_m + 3u_{m+1}Z_{m+1} \end{Bmatrix} \tag{19}$$

It can be shown that this algorithm has an asymptotic truncation error of

$$|T.E.| \quad \sim \quad \frac{[2\xi^3\eta+3\eta^2\xi^2+2\eta^3\xi-4\eta^4]}{720} \tag{20}$$

and if $\xi = \eta$ then $|\text{T.E.}| \sim \frac{3}{720}\eta^4$ that is, an error is at least six times smaller than the error (17).

The u and v momentum equation of system (1) undergo changes due to the use of the Numerov-Galerkin finite element method. Denoting

$$\partial u/\partial x = Z_{xu} \ , \ \partial v/\partial x = Z_{xv} \tag{21}$$

the intermediate Numerov approximation representing the first stage derivetives $\partial u/\partial x$ and $\partial v/\partial x$ respectively and the similar notation Z_{yu} , Z_{yv} for the y derivatives corresponding to the intermediate stage of the Numerov-Galerkin we get the following modified matrix u-momentum equation

$$[M]\,.\left[\frac{\{u\}^{n+1} - \{u\}^n}{\Delta t} + \{(u.Z_{xu})\} + \{(v.Z_{yu})\} - \{f.v\} + \{T^x\}\right] + [K_2]\,.\,\{\varphi\} = 0 \tag{22}$$

and in a similar manner we obtain the modified v-momentum equation

$$[M]\,.\left[\frac{\{v\}^{n+1} - \{v\}^n}{\Delta t} + \{(u.Z_{xv})\} + \{(v.Z_{yv})\} + \{f.u\} + \{T^y\}\right] + [K_3]\,.\,\{\varphi\} = 0 \tag{23}$$

with

$$[K_2] = \sum_1^m [K_2^e] \ ; \ [K_3] = \sum_1^m [K_3^e] \tag{24}$$

where

$$[K_2^e] = \frac{1}{2A}\begin{bmatrix} b_i \iint\limits_A V_i dxdy & b_j \iint\limits_A V_i dxdy & b_k \iint\limits_A V_i dxdy \\ b_i \iint\limits_A V_j dxdy & b_j \iint\limits_A V_j dxdy & b_k \iint\limits_A V_j dxdy \\ b_i \iint\limits_A V_k dxdy & b_j \iint\limits_A V_k dxdy & b_k \iint\limits_A V_k dxdy \end{bmatrix} = \frac{1}{6}\begin{bmatrix} b_i & b_j & b_k \\ b_i & b_j & b_k \\ b_i & b_j & b_k \end{bmatrix} \tag{25}$$

$$[K_3^e] = \frac{1}{2A}\begin{bmatrix} a_i \iint\limits_A V_i dxdy & a_j \iint\limits_A V_i dxdy & a_k \iint\limits_A V_i dxdy \\ a_i \iint\limits_A V_j dxdy & a_j \iint\limits_A V_j dxdy & a_k \iint\limits_A V_j dxdy \\ a_i \iint\limits_A V_k dxdy & a_j \iint\limits_A V_k dxdy & a_k \iint\limits_A V_k dxdy \end{bmatrix} = \frac{1}{6}\begin{bmatrix} a_i & a_j & a_k \\ a_i & a_j & a_k \\ a_i & a_j & a_k \end{bmatrix} \tag{26}$$

4 Test problem

The test problem used in this paper is a truncation of a rectangular channel with regular grid domain (5 x 5). The time and the space increments are:

$$\Delta x = 100\ m\ ,\ \Delta y = 50\ m\ ,\ \Delta t = 60\ sec\ ,\ \ \text{coriolis force}\ f = 2w \sin\varphi$$

where $w = 7.29x10^{-5}\ sec^{-1}$,$\varphi = 15^0$, $\tau_s^x = \tau_s^y = 0$, Manning coefficient $n = 0.035$, acceleration of gravity $g = 10\ ms^{-2}$, precipitation intensity $r = 0$.

Boundary conditions

The boundary conditions are chosen as follows: upstream $u = u(0, y, t) = 0.500 - 1.000\ m/s$, $v = v(0, y, t) = 0.0\ m/s$, downstream $h = 4.00 + 0.5\sin(\pi t/21600)\ m$, and rigid boundary conditions $v = v(x, 100, t) = 0.0\ m/s$, $v = v(x, 300, t) = 0.0\ m/s$.

Figures 1 and 2 show the calculated water surface elevation and the velocity distribution at the nodes of the grid domain using single stage Galerkin and two stage Galerkin FEM method. The estimated errors of the solution of two methods obtained are:

Water surface elevation: $\varepsilon_h = \max(|h_i^{(k+1)} - h_i^{(k)}|) = 0.001\ m$

Velocity distribution : $\varepsilon_v = \max(|v_i^{(k+1)} - v_i^{(k)}|) = 0.003\ m$

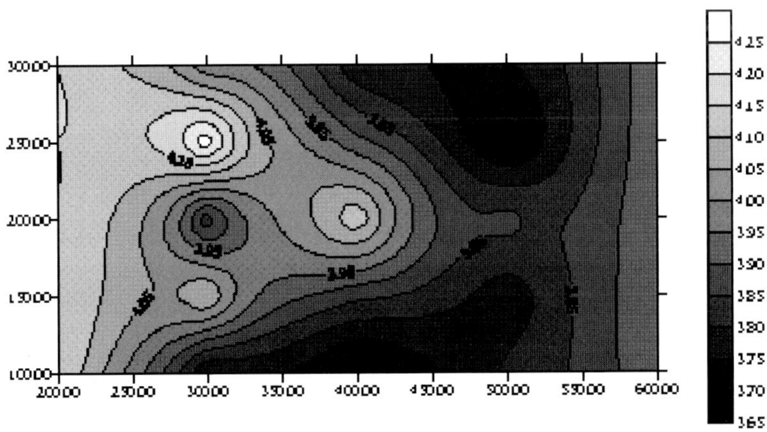

Fig. 1. Calculus time step at $n\Delta t = 4320$ of five day forecast, Contour of water surface elevation (h)

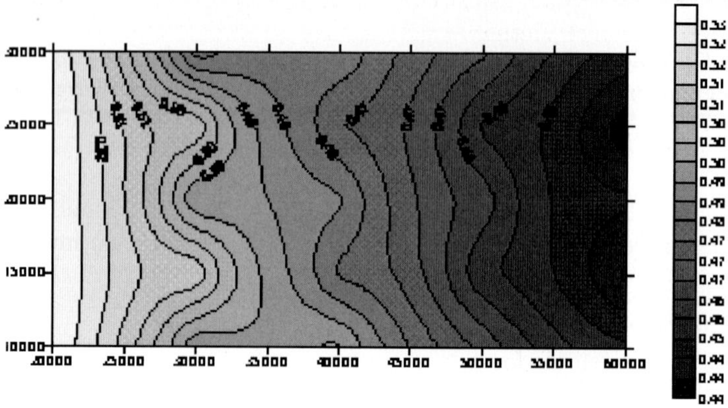

Fig. 2. Calculus time step at $n\Delta t = 4320$ of five day forecast, Contour of velocity distribution (u)

5 Conclusion

In comparison with the single stage Galerkin finite element method (Navon and Muller, 1979 and Navon 1987), it is observed that equation (22) and (23) result in reducing computational efforts since the mass matrix M is time independent and is calculated only once.

Moreover, the two stage Galerkin approximation gives more accurate results compared to the direct Galerkin approximation.

References

[1] Cullen, M. J. P., and Morton, K., K., 1980, Analysis of evolutionary error in finite element and other methods: Jour. Comput. Phys., V.34, p. 245-267.

[2] Navon, I. M., 1982, A Numerov-Galerkin technique applied to a finite element shallow water equations model with exact conservation of integral invariants, in Kaway, T., ed. Finite element flow analysis: Univ. Tokyo Press, Tokyo, p. 75-86.

[3] Navon, I. M., 1983, A Numerov-Galerkin technique applied to a finite element shallow water equations model with enforced conservation of integral invariants and selective lumping: Jour. Comput. Phys., V.52, No.2, p. 313-339.

[4] Navon, I. M., 1987, A two stage, high accuracy, finite element Fortran program for solving shallow water equations: Computers & Geosciences, V.13, No. 3, p. 255-285.

[5] Nguyen The Hung, Mathematical model of the two dimensional vertical flow, Journal of Vietnam National Science & Technology, No. 7+8, Hanoi 1990.

[6] Nguyen The Hung, Mathematical modeling of sediment transport two dimensional horizontal, Proceedings of International Conference on Engineering Mechanics Today, Vol. 1, p. 541-548, Hanoi 1995.

[7] Zienkiewicz, O. C., and Heinrich, J. C., 1979, A unified treatment of steady stage shallow water equations and two dimensional Navier-Stocks equations – a finite element penalty function approach : Computer Math. Appl. Mech. and Eng., V. 17/18, p. 673-688.

[8] Zienkiewicz, O. C., Vilotte, J. P., Nakazawa, S., and Toyoshima, S., 1984, Iterative methods for constrained and mixed approximation : An inexpensive improvement of F.E.M. performance : Inst. Numerical methods in Engineering Report C/R/489/84, Swansea, United Kingdom, 20 p.

Solenoidal Discrete Initialization for Magnetohydrodynamics

Rolf Jeltsch and Manuel Torrilhon

ETH Zurich, Seminar for Applied Mathematics, CH-8092 Zurich, Switzerland
jeltsch@math.ethz.ch
manuel@math.ethz.ch

Summary. We propose a procedure to initialize the magnetic flux, possibly discontinuous, on a finite volume grid in an exactly solenoidal way. That is, a certain discrete divergence operator will vanish on each cell. Combined with a new locally divergence preserving numerical scheme we are able to conduct MHD simulations which have an *exactly vanishing* discrete divergence. In this paper we describe the new scheme and the initialization procedure and present the results of a simulation of a shock interaction with a magnetized cloud.

1 Introduction

The field of the magnetic flux $\mathbf{B}$ in magnetohydrodynamic simulations is subject to the constraint $\operatorname{div}\mathbf{B} = 0$. To be precise, the divergence of $\mathbf{B}$ has to be zero initially and, there after, it is the character of the evolution equation for the magnetic flux which analytically preserves the divergence during the evolution. In MHD simulations this is no longer the case due to the discretization. One major task in the design of numerical methods for magnetohydrodynamics is the control of the divergence errors. Certainly, in the limit of finer grids any consistent scheme would reproduce the preservation properties of the exact evolution since it is an analytical property. However, MHD calculations are threatened by severe problems due to possible discontinuities which may generate divergence errors, see e.g. [2]. These errors usually accumulate and lead to a breaking down of classical numerical schemes making it impossible to calculate MHD solutions with those methods.

Recently, in [13] and [14] the construction of locally divergence preserving finite volume methods was presented. The first part of this paper will summarize the construction of these schemes for magnetohydrodynamics on rectangular grids. The divergence preservation will be incorporated directly into the fluxes of the scheme reproducing the analytical preservation properties of the evolution. As a result the values of the discrete divergence operator given in Equation (16) will stay exactly the same during the entire calculation. Since the scheme is based only on special distributions of intercell fluxes, any finite volume scheme can be modified into a locally divergence preserving scheme. For the construction of the schemes we rely on the flux distribution framework given in [13] to formulate constraint preserving numerical methods. Flux distributions are piecewise constant basis shape functions in the grid for which a given discrete constraint vanishes. The update in a numerical scheme has to be built out of linear combinations of such flux distributions in order to provide constraint preservation. We demonstrate the flux distributions for divergence preservation and give the necessary modifications for finite volume fluxes.

The locally divergence preserving methods exactly preserve the discrete value of the divergence during the entire calculation. This implies special demands on the discrete initial conditions since its discrete divergence must vanish. Especially in the case of discontinuous data this is in general not easily obtained. In the second part of this paper we propose a new initialization procedure which provides directly solenoidal discrete fields. Using this initialization and the locally divergence-free scheme we present the simulation of a shock interaction with a magnetized cloud. The solution at different times is discussed and provides insight into the special behavior of MHD flows.

The paper is organized as follows: After we briefly introduce the equations of magnetohydrodynamics in Sec. 2, we discuss the major existing approaches of MHD divergence cleaning in Sec. 3. In Sec. 4 the deduction of locally divergence preserving schemes is summarized. Afterwards we discuss the problem of solenoidal initialization in Sec. 5 and present the simulation of the shock-cloud-interaction in the last section, Sec. 6.

2 Magnetohydrodynamic Equations

The equations of ideal magnetohydrodynamics consider the conservative variables density ρ, momentum density $\rho\mathbf{v}$, energy density E and magnetic flux $\mathbf{B}$ to describe the flow of a plasma. As system of field equations we have

$$\begin{aligned}
&\partial_t \rho && + \operatorname{div} \rho\mathbf{v} && = 0 \\
&\partial_t \rho\mathbf{v} && + \operatorname{div}\left(\rho\mathbf{v}\mathbf{v} + \left(p + \tfrac{1}{2}\mathbf{B}^2\right)\mathbf{I} - \mathbf{B}\mathbf{B}\right) && = 0 \\
&\partial_t E && + \operatorname{div}\left(\left(E + p + \tfrac{1}{2}\mathbf{B}^2\right)\mathbf{v} - \mathbf{B}\mathbf{B}\cdot\mathbf{v}\right) && = 0 \\
&\partial_t \mathbf{B} && + \operatorname{div}\left(\mathbf{B}\,\mathbf{v} - \mathbf{v}\,\mathbf{B}\right) && = 0
\end{aligned} \tag{1}$$

that is, the balance laws of mass, momentum and energy and the induction equation. The system is closed by the equation of state of an ideal plasma

$$E = \frac{1}{\gamma - 1} p + \frac{1}{2}\rho\mathbf{v}^2 + \tfrac{1}{2}\mathbf{B}^2 \tag{2}$$

where γ is the adiabatic coefficient of the plasma. The system (1) forms a hyperbolic system of conservation laws (see e.g. [8] for hyperbolic properties) a fact which suggests to use a finite volume scheme in instationary MHD flow simulations.

The difficulty of such simulations is to handle the intrinsic constraint which follows by rewriting the induction equation $(1)_4$ with a curl so that we have

$$\partial_t \mathbf{B} + \operatorname{curl}(\mathbf{B} \times \mathbf{v}) = 0 \qquad \Rightarrow \qquad \operatorname{div} \mathbf{B} = \mathit{const} \text{ in time} \tag{3}$$

This means that the divergence of the magnetic flux remains untouched during the evolution. Since the magnetic flux has to be solenoidal in the initial conditions, it follows that it will be divergence-free for all times.

3 Divergence Cleaning Approaches

The construction of divergence-free methods for magnetohydrodynamics is vastly discussed in the literature. We may distinguish between three major approaches. See also the paper of Toth [15] for an overview and comparison of the major approaches.

The first one, originally described by Brackhill and Barnes in [2], uses a classical numerical method but cleans the field of the magnetic flux after every time step or after a certain number of time steps. The cleaning procedure solves the elliptic equation

$$\begin{aligned} \operatorname{div} \operatorname{grad} \psi &= \operatorname{div} \tilde{\mathbf{B}} \quad \text{in } \Omega \\ \psi &= 0 \qquad \text{on } \partial\Omega \end{aligned} \tag{4}$$

for the auxiliary discrete field ψ. The spoiled discrete field $\tilde{\mathbf{B}}$ is corrected by $\mathbf{B}_{i,j} = \tilde{\mathbf{B}}_{i,j} - \operatorname{grad} \psi|_{i,j}$. This method leads to solenoidal fields during the calculation and avoids divergence errors. However, the method is expensive due to the solution of a global elliptic equation. Moreover, inspection of the analytical equation shows that the preservation of the divergence *is not connected to an elliptic problem.* Analytically the divergence is locally preserved and it should be possible to construct a locally divergence preserving numerical method as well.

The second approach constructs divergence free methods by special discretization of the evolution equation of $\mathbf{B}$. Originally described by Evans and Hawley in [6], these ideas have been used and further developed by Balsara and Spicer [1], as well as by Dai and Woodward [3]. In those methods again a correction step follows each time step of a classical numerical method. This correction considers the magnetic field components $b^{(x)}_{i+\frac{1}{2},j}$ and $b^{(y)}_{i,j+\frac{1}{2}}$ stored at edges (in two dimensions), the so-called staggered grid. The relevant discrete divergence operator is given by

$$\operatorname{div}^{(0)} \mathbf{b}\Big|_{i,j} := \frac{b^{(x)}_{i+\frac{1}{2},j} - b^{(x)}_{i-\frac{1}{2},j}}{\Delta x} + \frac{b^{(y)}_{i,j+\frac{1}{2}} - b^{(y)}_{i,j-\frac{1}{2}}}{\Delta y} \tag{5}$$

formulated with the staggered variables rather than the cell mean values. Each time step provides a divergence preserving evolution for the magnetic flux on the staggered grid. The operator (5) is exactly preserved. The correction character and the staggered grid appear as disadvantages since they spoil the cell average approach in the finite volume method. In his article [15] Toth showed that the staggered grid may be eliminated by explicit extrapolation and interpolation. The apparent restriction of the staggered approach to structured meshes is somewhat relaxed by DeSterck in [5].

The third approach is due to Powell [11] who constructed a modified analytical MHD system based on the assumption that $\operatorname{div}\mathbf{B} \neq 0$. This new system contains additional terms which advect the divergence errors out of the computational domain. In [4] Dedner et al. and in [10] Munz et al. elaborate Powells ideas for MHD as well as for electrodynamics on unstructured grids. They introduce a new variable ψ which is coupled to the system of MHD by

$$\begin{aligned} \partial_t \mathbf{B} + \operatorname{curl}(\mathbf{B} \times \mathbf{v}) + \nabla\psi &= 0 \\ \mathcal{D}(\psi) + \nabla \cdot \mathbf{B} &= 0 \end{aligned} \tag{6}$$

where $\mathcal{D}$ is a linear operator. Different choices of $\mathcal{D}$ lead to either the advection method of Powell, an elliptic cleaning as in the first approach or a diffusive cleaning. However, Toth demonstrated in [15] that the artificial source terms arising in Powells system lead to wrong shock speeds in certain MHD Riemann problems.

All of the approaches are mainly concerned with eliminating an arising divergence error during the time steps such that the magnetic flux becomes solenoidal. However the solenoidal magnetic flux should not be seen as main issue. Inspired by the analytical properties, it is the *update or residual* in the numerical method that has to be divergence-free. Thus, the main problem is to construct numerical methods that exactly preserve the divergence during the evolution irrespective of the actual divergence of the field that is evolved. This, of course, will only be possible on a discrete level for a certain discretization of the divergence. For such an divergence preserving method, the solenoidality of the magnetic flux is a problem only for the initial data.

In [13] Torrilhon and Fey and in [14] Torrilhon worked out locally divergence preserving numerical schemes. These schemes use only one single finite volume grid to represent the fields, the updates and the divergence. In the following section we will shortly summarize the derivation of locally divergence preserving methods for MHD.

We restrict ourselves to the two-dimensional case. The extension of the presented algorithms to three dimensions is possible and mostly straightforward, see [14]. In two dimensions the divergence of the magnetic flux is only influenced by the components $B^{(x)}$ and $B^{(y)}$. Hence, the evolution of $B^{(z)}$ is not of interest for the divergence preservation.

4 Local Divergence Preservation

In [13] a general frame work is given of numerical methods for general evolution equations with inherent constraints. In that frame work a vector field $\mathbf{u} \in \Omega \subseteq \mathrm{I\!R}^D$ (D: space-dimension) and a generic evolution

$$\partial_t \mathbf{u} + \mathcal{F}(\mathbf{u}) = 0 \quad \text{in } \Omega \tag{7}$$

with transport operator $\mathcal{F}$ is considered. The generic constraint $\mathcal{C}$ is assumed to be inherent to (7), that is the relation

$$\mathcal{C}(\mathcal{F}(\mathbf{u})) \equiv 0 \tag{8}$$

holds, which directly implies

$$\mathcal{C}(\mathbf{u}) = const \text{ in time} \tag{9}$$

for any solution of (7). In the frame work it is also assumed that the constraint is linear, which is fortunately the case in most applications, e.g. in MHD.

Here, we proceed with constructing divergence preserving schemes for magnetohydrodynamics on rectangular grids. Hence, the vector field is given by $\mathbf{u} \hat{=} \mathbf{B}$, and the constraint is $\mathcal{C}(\mathbf{u}) \hat{=} \operatorname{div} \mathbf{B}$. The evolution equation is given by the induction equation $(1)_4$. The computational domain Ω is covered by a grid $\mathcal{T}$ with cells are denoted by $K = (i, j)$ at positions (x_i, y_j) and size $\Delta x \times \Delta y$. In cases of accuracy considerations we refer to $h = \max(\Delta x, \Delta y)$. A time discretization by Δt leads to a cell-wise constant grid function $\tilde{\mathbf{B}}^n : \mathcal{T} \to \mathrm{I\!R}^D$ which approximates $\mathbf{B}$ after n time steps by cell mean values. Cell averages of $\mathbf{B}$ at time level n are denoted by $\mathbf{B}^n_{i,j}$. Locally divergence preserving methods can also be constructed on triangular grids, see [14].

4.1 Flux Distribution Schemes

The central quantity of constraint preserving schemes is the so called "flux distribution". It is the structure of the flux distribution that determines wether a certain scheme is constraint preserving or not.

Definition 1 (flux distribution). *Given the space of vector-valued grid functions denoted by* $V = \{g : \mathcal{T} \to \mathrm{I\!R}^D\}$, *we define a "flux distribution"* $\mathbf{\Phi}_K : V \to V$ *which is attached to a grid cell* (i, j) *and maps the grid function* $\tilde{\mathbf{B}}$ *into another grid function, that is*

$$\mathbf{\Phi}_{i,j}(\tilde{\mathbf{B}}) : \mathcal{T} \to \mathrm{I\!R}^D. \tag{10}$$

The evaluation $\mathbf{\Phi}_{i,j}\left(\tilde{\mathbf{B}}\right)\Big|_{k,l}$ *gives the change of* $\tilde{\mathbf{B}}$ *at cell* (k, l) *caused by cell* (i, j), *that is the flux.*

A flux distribution is assigned to each cell of the grid and depends on the solution $\tilde{\mathbf{B}}$ in a local manner. The definition is more general than that of usual intercell fluxes, since it admits fluxes to any neighbouring cell, especially across corners. This incorporates multidimensionality from the very beginning. Conservation of $\tilde{\mathbf{B}}$ may be expressed by the statement that the integral over $\mathbf{\Phi}_{i,j}(\tilde{\mathbf{B}})$ vanishes.

A certain form of the flux distribution and its dependency on $\tilde{\mathbf{B}}$ is usually constructed from consistency with the evolution equation, that is the induction equation in the present case. Once the flux distribution is defined an explicit evolution scheme follows by simply collecting contributions of all flux distributions, viz.

$$\mathbf{B}_{i,j}^{n+1} = \mathbf{B}_{i,j}^{n} + \sum_{\substack{\text{cells } (k,l) \\ \text{surrounding } (i,j)}} \mathbf{\Phi}_{k,l}\left(\tilde{\mathbf{B}}^n\right)\Big|_{i,j}. \tag{11}$$

The value of $\tilde{\mathbf{B}}$ in a cell (i,j) is updated by contributions of all neighbouring cells. The contributions are given by evaluations of flux distributions. Note, that virtually any finite volume scheme can be written in the form (11), and the flux distribution may then be identified.

Since the divergence is linear we expect a discretization which may be written as matrix operation

$$\operatorname{div}\mathbf{B}|_{i,j} = \widetilde{\operatorname{div}}_{i,j} \cdot \tilde{\mathbf{B}} + O\left(h^m\right) \tag{12}$$

on the grid function $\tilde{\mathbf{B}}$. Here, $\widetilde{\operatorname{div}}_{i,j}$ represents the stencil of the discrete divergence operation in the grid. If preservation of the divergence should be achieved for the scheme (11) the following lemma gives sufficient conditions, see [13].

Lemma 1. *If the flux distribution of the scheme is built by linear combinations of shape functions $\hat{\mathbf{\Phi}}_{i,j}^{(g)}$ which satisfy the condition*

$$\widetilde{\operatorname{div}}_{k,l} \cdot \hat{\mathbf{\Phi}}_{i,j}^{(g)} = 0 \quad \forall \text{ cells } (i,j),(k,l) \tag{13}$$

then the resulting scheme (11) is exactly locally divergence preserving.

As the system (13) is homogeneous we generally hope for a solution space from which we only consider an appropriate basis set of shape functions $\{\hat{\mathbf{\Phi}}_{i,j}^{(g)}\}$ with $g = 1, 2, ...$ which all are constraint preserving. The final flux distribution has to be assembled from these solutions via

$$\mathbf{\Phi}_{i,j}\left(\tilde{\mathbf{u}}\right) = \sum\nolimits_g \varphi_{i,j}^{(g)}\left(\tilde{\mathbf{B}}\right)\hat{\mathbf{\Phi}}_{i,j}^{(g)} \tag{14}$$

with unknown coefficients $\varphi_{i,j}^{(g)}$, which give the amplitudes of the flux distributions. Note, that the choice of $\varphi_{i,j}^{(g)}$ does not affect the preservation of the constraint. The expression for $\mathbf{\Phi}_{i,j}$ enters the scheme (11) and the remaining coefficients $\varphi_{i,j}^{(g)}$ have to follow from consistency.

It follows from the lemma given above that the divergence preserving property of a scheme depends on the choice of the discrete divergence operator $\widetilde{\operatorname{div}}_{k,l}$ used in the conditions (13). Since the divergence can be evaluated discretely in many ways, certain numerical schemes will be preserve one discretization of the divergence but not another one.

4.2 Flux Distributions for Divergence Operators

We proceed to present solutions of the conditions (13), that is divergence preserving flux distributions for specific discretizations of the divergence.

In [13] it was shown that the classical divergence operator on a rectangular grid

$$\operatorname{div}_{i,j}^{(0)} \tilde{\mathbf{B}} := \frac{B_{i+1,j}^{(x)} - B_{i-1,j}^{(x)}}{2\Delta x} + \frac{B_{i,j+1}^{(y)} - B_{i,j-1}^{(y)}}{2\Delta y}. \tag{15}$$

admit only a single flux distribution which does not give rise to practical schemes. However, the so-called extended operator $\operatorname{div}^{(\star)}$ given by

$$\operatorname{div}_{i,j}^{(\star)} \tilde{\mathbf{B}} := \frac{\{B_{i+1,j}^{(x)}\}_y - \{B_{i-1,j}^{(x)}\}_y}{2\Delta x} + \frac{\{B_{i,j+1}^{(y)}\}_x - \{B_{i,j-1}^{(y)}\}_x}{2\Delta y} \tag{16}$$

turned out to be more fruitful. Here, curled brackets stand for

$$\begin{aligned} \{\psi_{i,j}\}_y &= \tfrac{1}{4}\left(\psi_{i,j+1} + 2\psi_{i,j} + \psi_{i,j-1}\right) \\ \{\psi_{i,j}\}_x &= \tfrac{1}{4}\left(\psi_{i+1,j} + 2\psi_{i,j} + \psi_{i-1,j}\right) \end{aligned} \tag{17}$$

i.e. averaging between vertical or horizontal cells. Like the classical operator, $\operatorname{div}_{i,j}^{(\star)} \tilde{\mathbf{B}}$ gives a second order approximation to the divergence on cell (i,j) using a 3×3 stencil. The difference lies only in the second order residual terms. It follows that both operators are equivalent up to second order for smooth solutions. Hence, if one operator is exactly preserved by a numerical scheme the other one will give a result of $\mathcal{O}(h^2)$ in smooth regions of the flow. The exact control of a single operator is enough also in the case of non-smooth solutions in order to keep the solution free from divergence error.

At discontinuities the equivalence of different discrete operators is no longer true. Indeed, the difference between discrete operators is of order $\mathcal{O}(1)$ due to the blow-up of the residual terms in the vicinity of discontinuities. However, if the scheme is divergence preserving for a specific divergence operator the value of the divergence given by this operator will be *exactly zero also across discontinuities.* Furthermore, any $\mathcal{O}(1)$-result of the other operators will stick to the discontinuities, since behind and in front of discontinuities the solution is smooth and the operators give equivalent results. Thus, no divergence error will spoil the solution. To some extend, the $\mathcal{O}(1)$-result of

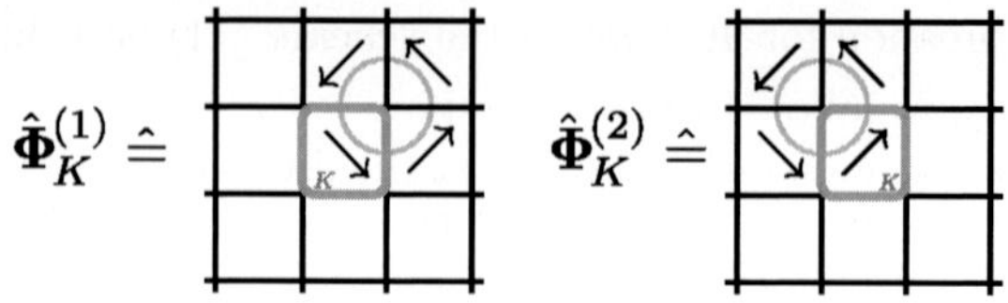

Fig. 1. Two out of four possible flux distributions of the cell $K = (i, j)$ which exactly preserve the extended divergence operator $\mathrm{div}^{(\star)}$. Note that these flux distributions approximate closed curves, which is intuitive, since they represent a solenoidal update for $\mathbf{B}$.

non-preserved operators in divergence preserving schemes in the vicinity of shocks simply indicate the presence of blowing-up derivatives of $\mathbf{B}$ and not the presence of a divergence error. Note, that for non-smooth solutions the operators (16) and (15) still give a result consistent to the divergence in the sense of distributions.

The possible flux distributions of $\mathrm{div}^{(\star)}$ follow from the solution of the homogeneous system given in (13). The system arises by fixing the cell (i, j) and successive evaluation of the divergence operator in the neighbourhood of (i, j) (see [13] for details). We obtain four possible shape functions for flux distributions, whose support consists of four cells. The non-vanishing values of the first flux distribution are given by

$$\begin{aligned} \hat{\mathbf{\Phi}}_{i,j}^{(1)}\Big|_{i+1,j+1} &= (-\Delta x, \Delta y), & \hat{\mathbf{\Phi}}_{i,j}^{(1)}\Big|_{i,j+1} &= (-\Delta x, -\Delta y), \\ \hat{\mathbf{\Phi}}_{i,j}^{(1)}\Big|_{i,j} &= (\Delta x, -\Delta y), & \hat{\mathbf{\Phi}}_{i,j}^{(1)}\Big|_{i+1,j} &= (\Delta x, \Delta y) \end{aligned} \tag{18}$$

and the others follow by translation as indicated in Fig. 1. From the flux distributions follow that the fluxes from one cell into another are not any more independent if we want to control the divergence. The sketch in Fig. 1 demonstrates how the fluxes are coupled. A flux from K into its right neighbour, i.e. a change of the magnetic flux in that cell, immediately implies a flux into e.g. the upper right corner. If this coupling is not respected, the divergence will be spoiled.

The update in a numerical method which is constraint preserving must be built out of linear combinations of flux distributions $\hat{\mathbf{\Phi}}_{i,j}^{(g)}$ as given in (18). Once such a scheme is constructed the local value of the discrete divergence operator $\mathrm{div}^{(\star)}$ in (16) will remain completely unchanged during the time steps. The values of the operator given by the discrete initial conditions will be exactly preserved.

4.3 Modification of Flux Distributions

Equipped with the information of the last section we will now modify a generic finite volume scheme (see e.g. [7]) for MHD such that it is locally divergence preserving. It is sufficient to consider only the part of the scheme updating the magnetic flux given by

$$\mathbf{B}_{i,j}^{n+1} = \mathbf{B}_{i,j}^{n} + \frac{\Delta t}{\Delta x}(\mathbf{F}_{i-\frac{1}{2},j} - \mathbf{F}_{i+\frac{1}{2},j}) + \frac{\Delta t}{\Delta y}(\mathbf{G}_{i,j-\frac{1}{2}} - \mathbf{G}_{i,j+\frac{1}{2}}) \tag{19}$$

where $\mathbf{F}$ and $\mathbf{G}$ are magnetic flux components of intercell fluxes which are obtained using the full set of conservative variables. These intercell fluxes are assumed to be given by any Riemann solver (e.g. HLLE, Roe, ...). Clearly, these fluxes depend on *all* MHD variables of the adjacent cells. However, this dependency is suppressed in this section. Since the type of Riemann solver remains unspecified, the following modifications may be applied to virtually any finite volume scheme.

As first step the flux distributions $\mathbf{\Phi}^{(\text{class})}$ of the classical scheme given in (19) are identified. Due to the curl-structure of the induction equation (3) the flux $\mathbf{F}$ in x-direction changes only the y-component of the magnetic flux and vice versa for the flux $\mathbf{G}$. Furthermore the amplitude of both intercell fluxes is given by a single scalar function f, since the flux in x-direction and the flux in y-direction is governed by the same function in the induction equation. Hence, we write

$$\mathbf{F}_{i+\frac{1}{2},j} = -f_{i+\frac{1}{2},j}\begin{pmatrix}1\\0\end{pmatrix} \qquad \text{and} \qquad \mathbf{G}_{i,j+\frac{1}{2}} = f_{i,j+\frac{1}{2}}\begin{pmatrix}0\\1\end{pmatrix}. \tag{20}$$

The flux distributions are most easily defined on the cell interfaces. The definitions

$$\mathbf{\Phi}^{(\text{class})}_{i+\frac{1}{2},j}\Big|_{i,j} = f_{i+\frac{1}{2},j}\begin{pmatrix}0\\ \Delta y\end{pmatrix} \qquad \mathbf{\Phi}^{(\text{class})}_{i+\frac{1}{2},j}\Big|_{i+1,j} = f_{i+\frac{1}{2},j}\begin{pmatrix}0\\ -\Delta y\end{pmatrix} \tag{21}$$

$$\mathbf{\Phi}^{(\text{class})}_{i,j+\frac{1}{2}}\Big|_{i,j} = f_{i,j+\frac{1}{2}}\begin{pmatrix}-\Delta x\\ 0\end{pmatrix} \qquad \mathbf{\Phi}^{(\text{class})}_{i,j+\frac{1}{2}}\Big|_{i,j+1} = f_{i,j+\frac{1}{2}}\begin{pmatrix}\Delta x\\ 0\end{pmatrix} \tag{22}$$

lead to the equivalent flux distribution formulation

$$\mathbf{B}_{i,j}^{n+1} = \mathbf{B}_{i,j}^{n} + \frac{\Delta t}{\Delta x \Delta y}\left(\mathbf{\Phi}^{(\text{class})}_{i+\frac{1}{2},j} + \mathbf{\Phi}^{(\text{class})}_{i-\frac{1}{2},j} + \mathbf{\Phi}^{(\text{class})}_{i,j+\frac{1}{2}} + \mathbf{\Phi}^{(\text{class})}_{i,j-\frac{1}{2}}\right)\Big|_{i,j} \tag{23}$$

of the scheme (19). So far nothing has happened except a reformulation of the finite volume scheme. A flux distribution of the classical scheme is depicted at the left hand side of Fig. 2. Note that the evaluation of the divergence on neighbouring cells in the picture will vanish neither using $\text{div}^{(0)}$ nor $\text{div}^{(\star)}$, hence the classical scheme (19) does not preserve the divergence, as expected.

A divergence preserving scheme may be established by modifying the flux distributions such that they form linear combinations of the shape functions $\mathbf{\Phi}^{(g)}_{i,j}$ given in the previous section. The difficulty is to obtain a consistent method. We suggest to use

$$\mathbf{\Phi}^{(\text{div})}_{i,j+\frac{1}{2}} = -\tfrac{1}{8} f_{i,j+\frac{1}{2}}\left(\mathbf{\Phi}^{(1)}_{i,j} + \mathbf{\Phi}^{(2)}_{i,j}\right) \tag{24}$$

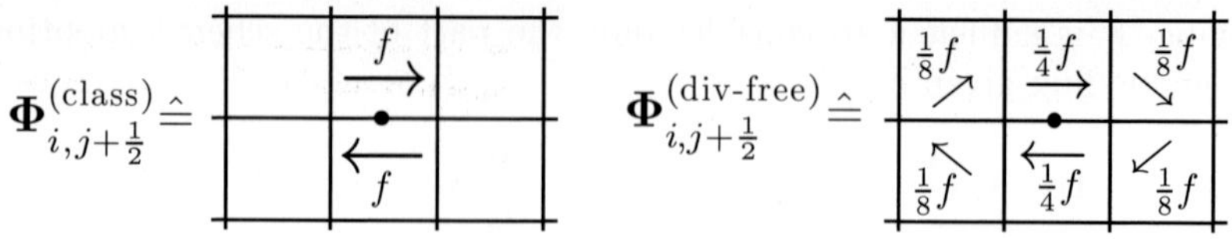

Fig. 2. Left: the flux distribution of a classical finite volume scheme applied to the induction equation. This flux distribution is *not* divergence preserving. Right: a modified flux distribution based on divergence preserving shape functions as shown in Fig. 1.

as divergence preserving flux distribution. Its sketch may be found at the right hand side of Fig. 2. The flux distribution $\mathbf{\Phi}^{(\text{div})}_{i+\frac{1}{2},j}$ is built analogously. These flux distributions use the same amplitude of the intercell fluxes as the classical distribution except they distribute this flux on more cells. The enlarged support results in a more lengthy formulation of the scheme, since the value $\mathbf{B}^{n+1}_{i,j}$ is influenced by intercell fluxes of neighbouring cells. The scheme with divergence preserving flux distributions read

$$\begin{aligned}
\mathbf{B}^{n+1}_{i,j} = \mathbf{B}^{n}_{i,j} &+ \frac{\Delta t}{\Delta x \Delta y}\left(\mathbf{\Phi}^{(\text{div})}_{i+\frac{1}{2},j} + \mathbf{\Phi}^{(\text{div})}_{i-\frac{1}{2},j} + \mathbf{\Phi}^{(\text{div})}_{i,j+\frac{1}{2}} + \mathbf{\Phi}^{(\text{div})}_{i,j-\frac{1}{2}}\right)\Big|_{i,j} \\
&+ \frac{\Delta t}{\Delta x \Delta y}\left(\mathbf{\Phi}^{(\text{div})}_{i+\frac{1}{2},j+1} + \mathbf{\Phi}^{(\text{div})}_{i-\frac{1}{2},j+1} + \mathbf{\Phi}^{(\text{div})}_{i+1,j+\frac{1}{2}} + \mathbf{\Phi}^{(\text{div})}_{i+1,j-\frac{1}{2}}\right)\Big|_{i,j} \\
&+ \frac{\Delta t}{\Delta x \Delta y}\left(\mathbf{\Phi}^{(\text{div})}_{i+\frac{1}{2},j-1} + \mathbf{\Phi}^{(\text{div})}_{i-\frac{1}{2},j-1} + \mathbf{\Phi}^{(\text{div})}_{i-1,j+\frac{1}{2}} + \mathbf{\Phi}^{(\text{div})}_{i-1,j-\frac{1}{2}}\right)\Big|_{i,j} .
\end{aligned} \tag{25}$$

Obviously the new scheme has a larger stencil and is expected to introduce slightly more diffusion into the numerical solution. However a significant decrease of resolution has not been observed in the numerical experiments. It is also important to note that the preservation of the divergence requires the coupling of the changes in the neighbouring cells and leads necessarily to a larger stencil. For the same reason the preserving scheme appears with a multidimensional flavour. It becomes evident that multidimensionality is a key issue to control the divergence constraint.

To show consistency of the divergence preserving scheme we make the flux distributions in (25) explicit and rearrange the resulting terms. Finally we obtain the equivalent formulation

$$\mathbf{B}^{n+1}_{i,j} = \mathbf{B}^{n}_{i,j} + \Delta t \begin{pmatrix} \frac{1}{\Delta y}\left(\langle f_{i,j-\frac{1}{2}}\rangle - \langle f_{i,j+\frac{1}{2}}\rangle\right) \\ \frac{1}{\Delta x}\left(\langle f_{i+\frac{1}{2},j}\rangle - \langle f_{i-\frac{1}{2},j}\rangle\right) \end{pmatrix} \tag{26}$$

where the angular brackets stand for the averaging of certain neighbouring intercell fluxes, see [14] for details. Assuming that the intercell flux amplitudes $f_{i+\frac{1}{2},j}$ are at least second order approximations to the exact values, e.g. by linear reconstruction, we proceed with a Taylor expansion of the above given scheme. Finally this leads to the statement

$$\begin{pmatrix} \frac{1}{\Delta y}(\langle f_{i,j-\frac{1}{2}}\rangle - \langle f_{i,j+\frac{1}{2}}\rangle) \\ \frac{1}{\Delta x}(\langle f_{i+\frac{1}{2},j}\rangle - \langle f_{i-\frac{1}{2},j}\rangle) \end{pmatrix} = \begin{pmatrix} -\partial_y f \\ \partial_x f \end{pmatrix}_{i,j} + \mathcal{O}(h^2) \tag{27}$$

which shows second order consistency in space with the induction equation (3) if f is substituted by $(\mathbf{v} \times \mathbf{B})^{(z)}$. Second order in time may now be obtained by Runge-Kutta integration of the residual.

5 Discrete Initial Conditions

The new locally divergence-free scheme preserves the value of the divergence in each time step *exactly*. Any field of divergence from the initial conditions will be frozen during the calculation. Hence, the requirement $\operatorname{div}\mathbf{B} = 0$ is a problem of the initial conditions: Like in the analytical case a (discrete) solenoidal field remains solenoidal. Usually, the analytical prescription of the initial values of the magnetic flux is divergence-free. However, the discretization of these initial conditions, especially discontinuous ones, may introduce divergence errors which spoil the entire simulation. We will discuss this problem in the case of a Riemann problem and propose a solution.

5.1 Example: Riemann Problem

For simplicity let us consider only the induction equation for $\mathbf{B}$. The velocity field shall be assumed to be constant and given by

$$\mathbf{v} = v_0 \begin{pmatrix} \sin\varphi \\ \cos\varphi \end{pmatrix} \tag{28}$$

pointing in the direction of an angle φ. Then, the induction equation becomes an equation for the $\mathbf{B}$-field alone. The initial magnetic flux is prescribed in the form of a Riemann problem

$$\mathbf{B}(\mathbf{x}, t=0) = \begin{cases} \mathbf{B}_0 & x < -y\tan\varphi \\ \mathbf{B}_1 & x > -y\tan\varphi \end{cases} \tag{29}$$

where the line of discontinuity is perpendicular to the velocity vector. The vectors $\mathbf{B}_{0,1}$ are constant. As a result the problem is essentially one-dimensional in the direction $(\cos\varphi, \sin\varphi)^T$. The solution of this Riemann problem is simply the advection of the discontinuity by the velocity field.

As example we consider the values $\mathbf{B}_0 = \mathbf{R}(\varphi)(1,1)^T$ and $\mathbf{B}_1 = \mathbf{R}(\varphi)(1,2)^T$ where the rotation matrix

$$\mathbf{R}(\varphi) = \begin{pmatrix} \cos\varphi & -\sin\varphi \\ \sin\varphi & \cos\varphi \end{pmatrix} \tag{30}$$

rotates the vectors such that the normal component across the discontinuity is constant. Hence, the initial conditions have zero divergence in the weak sense.

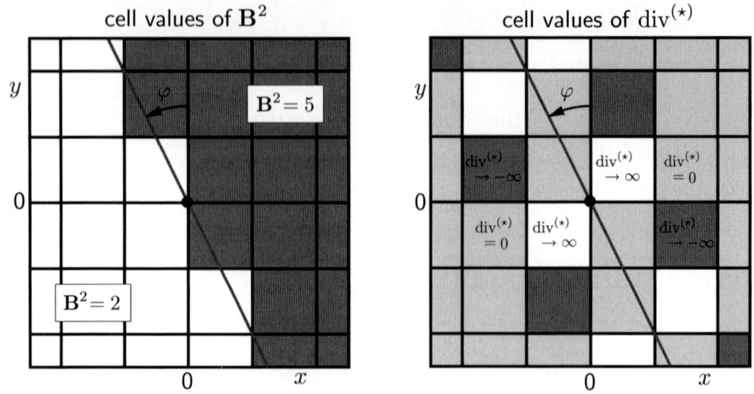

Fig. 3. Sketch of the distribution of the cell values of $\mathbf{B}^2$ and of $\mathrm{div}^{(\star)}\,\mathbf{B}$ in the vicinity of the discontinuity given by the initial conditions (29) in the case of $\tan\varphi = \frac{1}{2}$. The values of $\mathrm{div}^{(\star)}\,\mathbf{B}$ in the cells along the discontinuity tend to $\pm\infty$ if the grid is refined.

If we calculate the solution to this Riemann problem with the locally divergence-free scheme for the angle $\varphi = 0$ or $\varphi = \frac{\pi}{4}$ the simulation runs through smoothly and the result gives an approximation to the propagating discontinuity. The value of the divergence operator $\mathrm{div}^{(\star)}$ gives exactly zero for all time steps.

However, if we turn to intermediate values of φ, e.g. $\tan\varphi = \frac{1}{2}$, the situation is different. Already the *initialization* of the grid cells leads to a significant divergence error. At the left hand side of Fig. 3 the initial distribution of the cell values for $\mathbf{B}^2$ is sketched for the case $\tan\varphi = \frac{1}{2}$. Each cell is constantly initialized by the value of $\mathbf{B}$ of the initial conditions *at the cell center*. We remark, that an integration of the initial conditions on each cell will not essentially change the situation. The corresponding values of the divergence evaluated by the extended operator $\mathrm{div}^{(\star)}$ on each cell is shown on the right hand side of Fig. 3. Along the discontinuity the divergence gives a pattern of alternating large positive and negative values. Due to the divergence preservation property of our numerical scheme this divergence field will be *exactly frozen* and stay in place during the calculation. The result for such a calculation is shown at the left hand side of Fig. 4. The grid consists of 40x40 cells in the domain $[-1, 1]^2$ and the result is shown after 10 time steps. Note, that the pattern of the non-zero divergence spoils clearly the field of $\mathbf{B}$ and of $\mathbf{B}^2$ during the simulation. A similar calculation based on the complete MHD system would immediately fail.

5.2 Solenoidal Initialization

In order to get rid of the divergence in the initial conditions, the discrete initial field has to be corrected.

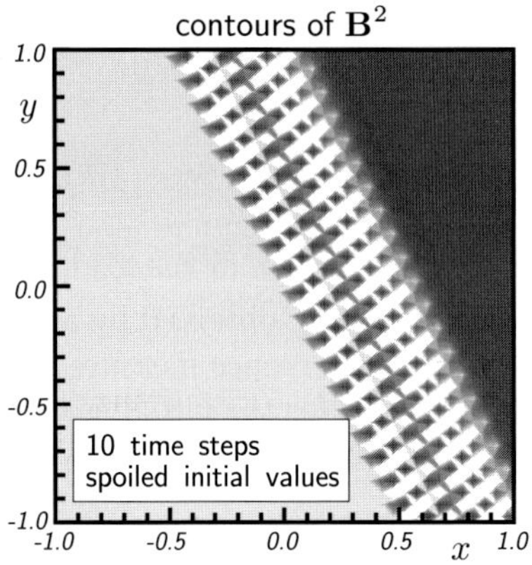

Fig. 4. Results for the Riemann problem given in (29) for the case $\tan\varphi = \frac{1}{2}$ after 10 time steps. Left: direct initialization of the magnetic flux leads to a spurious solution spoiled due to non-zero divergence. Right: the true solution is recovered by use of solenoidal discrete initialization.

A first idea is to use the global cleaning approach as proposed e.g. in [2]. We stress that this cleaning procedure would *only* be needed for the *initial* conditions with non-vanishing divergence due to discontinuities. Hence, the procedure is only applied once in the beginning of the calculation. The cleaning procedure solves the elliptic equations (4) for the auxiliary discrete field ψ. The discrete *initial* field $\tilde{\mathbf{B}}$ is afterwards corrected by $\mathbf{B}_{i,j} = \tilde{\mathbf{B}}_{i,j} - \operatorname{grad}\psi|_{i,j}$ which gives a solenoidal field.

The differential operator div grad has to be built from the extended divergence operator (16) since the result should give a divergence-free field according the extended operator. The use of the traditional discretization of the Laplace operator will not result in this property. The construction of the Laplace operator by applying an appropriate discrete gradient and afterwards $\operatorname{div}^{(\star)}$ to the field ψ results in a special discrete Laplace operator which assures that the evaluation of $\operatorname{div}^{(\star)}$ on the corrected solution will be zero. The discretized form of (4) may be solved by using iterative linear solvers.

This elliptic cleaning of the discrete initial conditions has been implemented and lead to satisfactory results, see [13]. However, it is quite expensive, though it is only applied once at the beginning of the calculation. Furthermore, we note that the analytic initial conditions need no cleaning. We would like to use the solenoidal character of the initial conditions in order to obtain a directly solenoidal initialization. The following lemma turns out to be useful in this situation. It relates the extended divergence operator to the operator (5) formulated with the staggered grid, that is magnetic field components $b^{(x)}_{i+\frac{1}{2},j}$ and $b^{(y)}_{i,j+\frac{1}{2}}$ which are stored at edges.

Lemma 2. *If* $\operatorname{div}^{(0)} \mathbf{b}\Big|_{i,j} = 0 \quad \forall\, i,j$, *we have the statement*

$$\mathbf{B}_{i,j} = \frac{1}{2}\begin{pmatrix} b^{(x)}_{i+\frac{1}{2},j} + b^{(x)}_{i-\frac{1}{2},j} \\ b^{(y)}_{i,j+\frac{1}{2}} + b^{(y)}_{i,j-\frac{1}{2}} \end{pmatrix} \quad \forall i,j \quad \Rightarrow \quad \operatorname{div}^{(\star)} \mathbf{B} = 0 \tag{31}$$

for the extended divergence operator.

Hence, if we have staggered variables $b^{(x)}_{i+\frac{1}{2},j}$ and $b^{(y)}_{i,j+\frac{1}{2}}$ in the grid which have a zero discrete divergence, the cell mean values $\mathbf{B}_{i,j}$ obtained by averaging out of these staggered variables will have zero divergence measured with the extended operator. This is exactly what is needed for the initialization. It remains to construct a solenoidal distribution of staggered variables. We use the analytic representation

$$\mathbf{B}\,(x,y) = \begin{pmatrix} \partial_y \varphi\,(x,y) \\ -\partial_x \varphi\,(x,y) \end{pmatrix} \tag{32}$$

with a potential $\varphi(x,y)$, which is possible thanks to $\operatorname{div}\mathbf{B} = 0$. With this representation the staggered variables are given by

$$b^{(x)}_{i+\frac{1}{2},j} = \frac{\varphi_{i+\frac{1}{2},j+\frac{1}{2}} - \varphi_{i+\frac{1}{2},j-\frac{1}{2}}}{\Delta y}, \quad b^{(y)}_{i,j+\frac{1}{2}} = \frac{\varphi_{i+\frac{1}{2},j+\frac{1}{2}} - \varphi_{i-\frac{1}{2},j+\frac{1}{2}}}{\Delta x} \tag{33}$$

which may be inserted into the formula for $\mathbf{B}_{i,j}$ given in (31). Finally, the initial cell mean values are obtained depending on the values of φ in the four corners of the cell, which are easily evaluated, since $\varphi(x,y)$ is a function explicitly known from the initial conditions.

For the piecewise constant Riemann data (29) we use

$$\varphi\,(x,y) = B^{(x)}y - B^{(y)}x \tag{34}$$

in each domain of the initial condition. Since the normal component of $\mathbf{B}$ is constant, $\varphi(x,y)$ is continuous across the discontinuity of $\mathbf{B}$. The evaluation of the extended divergence operator on the initial distribution gives now exactly zero in every cell. The calculation of 10 time steps with the divergence-free initial conditions is shown at the right hand side of Fig. 4. It is free from divergence errors.

6 Simulation of a Shock-Cloud-Interaction

To conclude this paper we present the simulation of a shock interaction with a plasma cloud. This simulation will demonstrate the abilities of the divergence-preserving scheme as well as provide some insight into the physical behavior of plasma flows. The results of the last section are applied in order to obtain a directly divergence-free initialization.

Consider the configuration sketched in Fig. 5. At position $x = -0.6$ a shock is located which travels to the right with a Mach number $M_s = 8$ calculated for

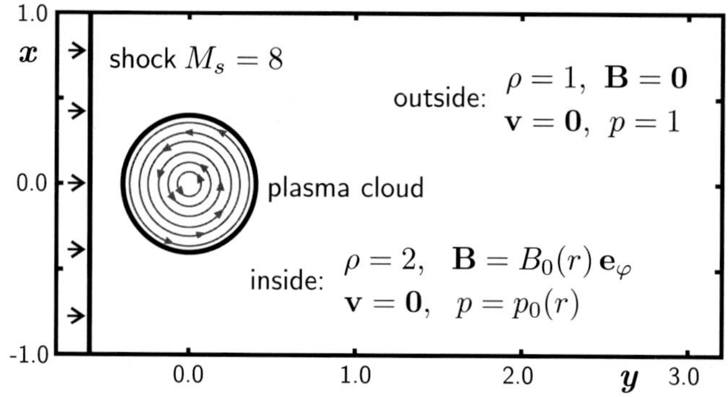

Fig. 5. Initial configuration for the simulation of a shock interaction with a magnetized dense plasma cloud. A normal shock with Mach number 8 is located at $x = -0.6$. The cloud has an radius of $r_0 = 0.4$ and is in equilibrium with the surrounding.

the right hand state. The domain right to the shock is at rest with pressure and density equal to unity. At both sides of the shock the magnetic flux vanishes, thus the state to the left of the shock is given by the Rankine-Hugoniot-conditions of the Euler equations. Inside the circle with radius r_0 around the origin we construct a magnetized cloud of dense plasma. This cloud is at rest as well, but its density is twice as large as in the surrounding. The magnetic field lines form circles. For the distribution of the magnetic flux **B** we assume

$$\mathbf{B}_{\text{inside}} = B_{\max} \frac{r}{r_0} \mathbf{e}_\varphi \tag{35}$$

where $\mathbf{e}_\varphi = (-y/r, x/r)^T$ is the unit vector in φ-direction. The boundary of the cloud represents a tangential discontinuity in which the tangential component of **B** drops to zero. The plasma cloud shall be in equilibrium with the surrounding, hence the complete plasma pressure $p^\star$ must be constant. It follows the relation

$$p_{\text{inside}} = p_{\text{outside}} - \frac{1}{2} \mathbf{B}^2_{\text{inside}} \tag{36}$$

for the pressure inside the cloud. Note that there is a pressure decrease across the boundary of the cloud due to the discontinuous magnetic flux. Hence, the cloud may be interpreted as a bubble with surface tension. This will also become visible in the simulation.

Since the initial conditions of **B** are discontinuous we use the initialization procedure presented in the last section. The potential of the magnetic flux is given by

$$\varphi(x, y) = \begin{cases} B_{\max} \frac{r_0^2 - (x^2+y^2)}{2r_0} & x^2 + y^2 < r_0^2 \\ 0 & \text{else} \end{cases} \tag{37}$$

which represents the distribution (35). In the simulation presented we used $B_{\max} = 1.3$ and $r_0 = 0.4$. The simulation was conducted up to the time

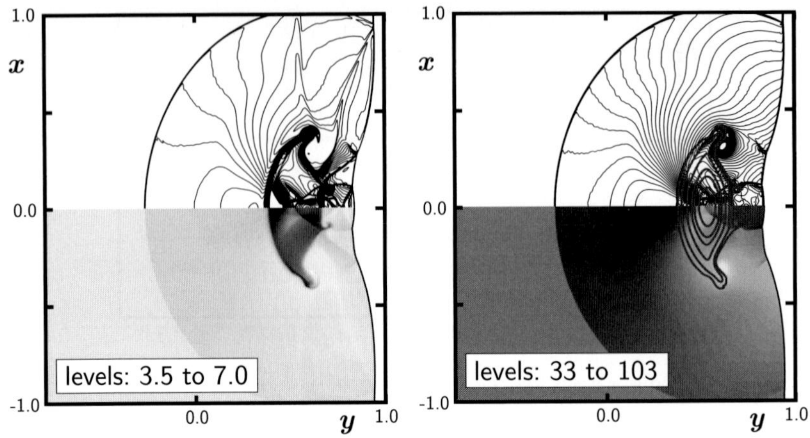

Fig. 6. Shock interaction with a magnetized cloud at time $t = 0.3$. Left: density contours and contour lines. Right: pressure contours and contour lines. The field lines of the magnetic flux are superimposed in the pressure plot. The colour version of this figure can be found in Fig. A.18 on page 586.

$t = 0.5$. Due to symmetry only the upper half of Fig. 5 needs to be calculated. For the boundaries to the left and right and at the top constant extrapolation is used. To reduce the influence of spurious reflections the computational domain has been expanded in y-direction. The complete domain simulated was $[-0.8, 4.2] \times [0, 2]$. At $t = 0.5$ the shock is approximately at $x = 4.4$, that is it has shortly left the domain.

The computational domain was discretized with 1000×2500 cells. The maximal allowed Courant number was chosen to be 0.9 and the time step was adjusted adaptively. The intercell flux was computed by the HLLE Riemann solver as described in [12] or [16], while the linear reconstruction was limited using the WENO limiter of [9]. Second order time integration is done by Heun's method. The implementation of the divergence-preserving scheme has been parallelized via MPI and the simulation was conducted on 6 PC's with 2.4GHz CPU's. The entire simulation used 4500 time steps and took approximately 35 hours.

In Fig. 6 the pressure and density fields are displayed at $t = 0.3$. Superimposed to the contours of the pressure at the right hand side the field lines of $\mathbf{B}$ are drawn. The shock has just passed the cloud.

The flow field around the cloud is dominated by the reflected shock created by the incident shock hitting the cloud and spreading out to all sides. The cloud is strongly compressed and accelerated. It will continue drifting in the flow behind the shock. Along the incident shock six Mach reflection points are visible each consisting of three shocks and a slip line. The two most inner reflection points travel along the incident shock line towards each other while the other four reflection points travel outwards. Inside the cloud a complicated shock system has formed consisting of several magnetohydrodynamic shocks in which the magnetic field lines are bent. Note that the field lines essentially follow the deformation of the cloud. The magnetic field lines of **B** *stick to the matter.*

However, for physical reasons the magnetic flux goes against the deformation and tries to restore a circular shape. This is similar to a bubble whose surface tension leads to a spherical shape. Hence, the stickiness of the field lines leads to a restoring force to the matter of the cloud. At the left hand side of Fig. 7 we see the contours of the density field at time $t = 0.5$ with the field lines of **B** superimposed.

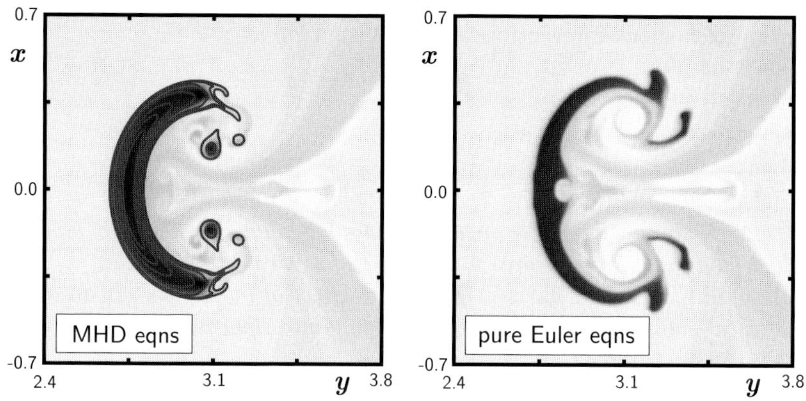

Fig. 7. Simulation results for the shock interaction with a cloud at time $t = 0.5$. Left: magnetized cloud. Density contours with magnetic field lines superimposed. Right: non-magnetized cloud. Density contours. In both plots the contour levels have the range 3.5 - 6.5. The colour version of this figure can be found in Fig. A.19 on page 586.

At the right hand side the result for the case of initially vanishing magnetic flux inside the cloud is shown. It is obtained from a second simulation using the Euler equations alone. In the MHD case on the left hand side the cloud is re-contracting in order to restore a circle. It has lost some magnetized matter due to the strong vorticity and the upper and lower tips. This vorticity leads to a ongoing deformation in the case of the pure Euler equations. Also, the two strong kinks in the back of the cloud visible in Fig. 6 for the MHD case are still present. In the MHD case the magnetic forces let these kinks disappear.

During the entire calculation the divergence of **B** evaluated with the extended divergence operator $\mathrm{div}^{(\star)}$ gives locally *exactly* the same value as in the initial conditions due to the divergence preservation property of our flux distribution scheme. Thanks to the initialization of **B** which was done completely divergence-free by use of (31) and (33) we have $\mathrm{div}^{(\star)}\,\mathbf{B} = 0$ for all times up to machine precision.

7 Conclusions

We presented new flux modifications that turn an arbitrary MHD finite volume method into a locally divergence preserving scheme. That scheme preserves the value of a certain discrete divergence operator and keeps the calculation free of divergence errors. We deduced the necessary modifications for rectangular in two dimensions using the flux distribution framework. Details and generalization of this approach to triangles may be found in [13] and [14].

The new scheme preserves exactly the value of the divergence as given by discrete initial conditions. We proposed an easy method how to initialize the magnetic flux in the grid such that the discrete divergence vanishes exactly. The abilities of the new scheme and initialization are demonstrated by the simulation of a shock interaction with a plasma cloud.

References

[1] Balsara, D. S. and Spicer, D. S., *A staggered mesh algorithm using high order Godunov fluxes to ensure solenoidal magnetic fields in magnetohydrodynamic simulations*, J. Comp. Phys. **149**, (1999) p.270

[2] Brackhill, J. U. and Barnes, D. C., *The effect of nonzero $\nabla \cdot B$ on the numerical solution of the magnetohydrodynamic equations*, J. Comp. Phys. **35**, (1980) p.426

[3] Dai, W. and Woodward, P. R., *On the divergence-free condition and conservation laws in numerical simulations for supersonic magnetohydrodynamic flows*, Astrophys. J. **494**, (1998) p.317

[4] Dedner, A., Kemm, F., Kröner, D., Munz, C.-D., Schnitzer, T., and Wesenberg, M., *Hyperbolic Divergence Cleaning for the MHD Equations*, J. Comp. Phys. **175**(2), (2002) p.645

[5] DeSterck, H., *Multi-Dimensional Upwind Constrained Transport on Unstructured Grids for Shallow Water Magnetohydrodynamics*, AIAA Paper 2001-2623, (2001)

[6] Evans, C. R. and Hawley, J. F., *Simulation of Magnetohydrodynamic Flows: A Constrained Transport Method*, Astrophys. J. **332**, (1988) p.659

[7] Godlewski, E. and Raviart, P.-A., *Numerical Approximation of Hyperbolic Systems of Conservation Laws*, Springer, New York (1996)

[8] Jeffrey, A. and Taniuti, T., *Non-linear Wave Propagation*, Academy Press, New York (1964)

[9] Jiang, G.-S. and Shu, C.-W., *Efficient Implementation of weighted ENO schemes*, J. Comp. Phys. **126**, (1996) p.202

[10] Munz, C.-D., Omnes, P., Schneider, R., Sonnendrücker, E., and Voss, U., *Divergence Correction Techniques for Maxwell Solvers Based on a Hyperbolic Model*, J. Comp. Phys. **161**(2), (2000), p.484

[11] Powell, K. G., *An approximate Riemann solver for magnetohydrodynamics (that works in more than one dimension)*, ICASE Report No. 94-24, (1994)

[12] K. G. Powell, P. L. Roe, T. J. Linde, T. I. Gombosi, and D. L. DeZeeuw, *A solution-adaptive upwind scheme for ideal magnetohydrodynamics*, J. Comp. Phys. **154**/2 (1999) p.284

[13] Torrilhon, M. and Fey, M., *Multidimensional Upwind Methods for Advection Equations with Constraints*, submitted to SIAM J. Num. Anal., (2003)

[14] Torrilhon, M., *Locally Divergence-preserving Upwind Schemes for Magnetohydrodynamic Equations*, submitted to SIAM J. Sci. Comp., (2003)

[15] Toth, G., *The $\nabla \cdot B$ Constraint in Shock-Capturing Magnetohydrodynamics Codes*, J. Comp. Phys. **161**, (2000)

[16] M. Wesenberg, *Efficient MHD Riemann Solvers for Simulations on Unstructured Triangular Grids*, East West J. Numer. Math. in press (2002)

Numerical Methods for Nonlinear Experimental Design

Stefan Körkel and Ekaterina Kostina

Interdisciplinary Center for Scientific Computing, University of Heidelberg
Im Neuenheimer Feld 368, D-69120 Heidelberg, Germany
stefan@koerkel.de*
ekaterina.kostina@iwr.uni-heidelberg.de

Summary. Nonlinear experimental design leads to a challenging class of optimization problems which occur in the procedure of the validation of process models. This paper discusses the formulation of such problems for a general class of underlying process models, presents numerical methods for the solution and shows their successful application to industrial processes.

Key words: experimental design, parameter estimation, variance-covariance matrix, multiple experiments, nonlinear constrained optimization

1 Introduction

To obtain reliable simulations of dynamic processes, e.g. in physics, chemistry, biology, or engineering, good models of the processes are required. A promising approach is to use models based on laws of nature. Usually the model equations contain quantities whose values are known only very roughly, we call them *parameters*. They may appear in a highly nonlinear way. To determine their values, experiments are carried out and the resulting data is analyzed by the method of nonlinear parameter estimation. The statistical reliability of the estimates can be described by confidence regions and the variance-covariance matrix. Data from numerous and costly experiments are required to obtain reliable parameter estimates.

Our aim is to design experiments whose data yield parameter estimates with maximal statistical reliability. For this purpose, we minimize functionals on the variance-covariance matrix subject to given constraints on operability and costs. The dynamic process model appears as constraint of this nonlinear optimization problem.

* This work was supported by the German Research Foundation (DFG) and the German Federal Ministry of Education and Research (BMBF).

We formulate the class of nonlinear optimum experimental design problems for a general class of process models. We present suited numerical methods for the solution of these intricate problems and apply them to practical processes from industry. Computational results show that enormous savings of experimental costs are possible by the application of our methods.

1.1 Structure of this Paper

This paper opens with a short overview on previous work. Then we present the general problem statement for the class of parameter estimation and experimental design problems we consider. We discuss the numerical difficulties these problems contain and give numerical strategies to solve them. We give a short description of our software package VPLAN where these strategies are implemented. And finally we present numerical results from industry application problems treated by our methods.

1.2 Overview on Previous Work

Experimental design for linear models is well established and discussed e.g. in the text books of Fedorov [11], Atkinson and Donev [1] or Pukelsheim [22]. But for many processes a description which is valid over a higher range requires nonlinear models. Nonlinear experimental design for examples with small nonlinear equation systems has been investigated e.g. by Haines [14], Rudolph and Herrendörfer [25] and Doví, Reverberi and Acevedo-Duarte [10]. Reilly, Bajramovic, Blau, Branson and Sauerhoff [24], Qureshi, Ng and Goodwin [23] and Oinas, Turunen and Haario [21] treat very small special dynamic processes which allow an analytic solution of the differential equations. Lohmann, Bock and Schlöder [19] present a numerical method for experimental design for ordinary differential equations which leads to a simplified formulation, because only the sampling design is optimized. Baltes, Schneider, Sturm and Reuss [9] use a simple search method for the optimization of a nonlinear experimental design problem for a special process. Until our work, a method which allows the efficient numerical solution of a general class of optimum experimental design problems for nonlinear dynamic process models is missing.

2 Problem Statement: Parameter Estimation and Experimental Design for Nonlinear Dynamic Process Models

2.1 Process Model

The processes we want to treat in this paper are dynamic or stationary, homogeneous or inhomogeneous deterministic systems. We describe their states by variables x, defined on the domain Ω of the independent variables $\tau \in \Omega$, which can be time, space or (time, space): $\tau = t$, z or (t, z). Thus the state variables are mappings $x : \Omega \to \mathbb{R}^{n_x} : \tau \mapsto x(\tau)$.

The states are solutions of the system model equations

$$F[x, p, q] = 0 \tag{1}$$

can e.g. be systems of nonlinear equations, ordinary differential equations (ODE), differential algebraic equations (DAE), or partial differential equations (PDE).

The model contains additional variables:

- the parameters $p \in \mathbb{R}^{n_p}$ which are determined by laws of nature, and the
- controls $q : \Omega \to \mathbb{R}^{n_q}$ determined by experimental setup and processing.

In general, both states as well as parameters as well as controls enter the model equations (1) in a nonlinear way.

Usually, constraints on the states are given, including e.g. boundary values, initial conditions, interior point constraints, or path constraints. We denote them as

$$D[x, p, q] = 0 \tag{2}$$

with $D[x, p, q] \in \mathbb{R}^{n_D}$. We assume that for given p and q the solution x of (1)-(2) exists and is unique.

2.2 Experiments and Data

An experiment is a processing of the system with given settings for the controls. We assume that multiple experiments are performed, consisting of series of N_1 experiments with control settings q^j, $j = 1, \ldots, N_1$. Usually, experiments provide measurement data. Let in experiment $j \in \{1, \ldots, N_1\}$ experimental data η_i^j be acquired, measured at points $\tau_i^j \in \Omega$ with standard deviations σ_i^j, $i = 1, \ldots, M^j$.

Remark 1 *Different experiments can be described by different models*

$$F^j[x^j, p, q^j] = 0, \quad D^j[x^j, p, q^j] = 0 \tag{3}$$

with $x^j : \Omega^j \to \mathbb{R}^{n_x^j}$, $q^j : \Omega^j \to \mathbb{R}^{n_q^j}$, $j = 1, \ldots, N_1$. *We assume that the parameters* p *are common to all these models.*

2.3 Simulation

For given parameter values and the control settings of an experiment, the process can be simulated by solving the model equations (3). For experiment $j \in \{1, \ldots, N_1\}$ we obtain the solution x^j. Let $x_i^j := x(\tau_i^j)$ be the state variable at point τ_i^j, $i = 1, \ldots, M^j$.

These values x_i^j allow the computation of model responses $h_i^j(x_i^j, p, q^j)$ for the measurement values. We assume a nonlinear data regression model

$$\eta_i^j = h_i^j(x_i^j, p, q^j) + \epsilon_i^j, \quad i = 1, \ldots, M^j$$

with additive measurement errors ϵ_i^j. We assume that the measurement errors are independent and normally distributed

$$\epsilon_i^j \sim \mathcal{N}(0, {\sigma_i^j}^2), \quad i = 1, \ldots, M^j.$$

The quality of a simulation can be described by the residuals, the weighted differences between data and model responses $(\eta_i^j - h_i^j(x_i^j, p, q^j))/{\sigma_i^j}^2$.

2.4 Parameter Estimation

To fit the model to the data, we want to find parameter values which minimize the residuals. We obtain a maximum likelihood estimator minimizing the sum of least squares of the residuals. This yields a constrained least squares optimization problem:

$$\min_{x,p} \sum_{j=1}^{N_1} \sum_{i=1}^{M^j} \frac{(\eta_i^j - h_i^j(x_i^j, p, q^j))^2}{{\sigma_i^j}^2} \tag{4}$$

$$F^j[x^j, p, q^j] = 0, \quad j = 1, \ldots, N_1 \tag{5}$$

$$D^j[x^j, p, q^j] = 0, \quad j = 1, \ldots, N_1 \tag{6}$$

Here the vector x summarizes the states of the N_1 experiments:

$$x = \left(x^j, \ j = 1, \ldots, N_1\right).$$

The model equations appear as constraints implicitly describing the variables.

2.5 Parameterization and Solution of the Parameter Estimation Problem

To solve problem (4)-(6), we have to discretize it to obtain a finite dimensional optimization problem. We assume that for the process in experiment j, the solution of system $F^j[x^j, p, q^j] = 0$ has a finite number of degrees of freedom — maybe after some approximation.

We parameterize these degrees of freedom by variables $s^j \in \mathbb{R}^{n_{sj}}$. Then the implicit function theorem gives a representation ξ^j of the solution x^j of $F^j[x^j, p, q^j] = 0$: $x^j = \xi^j(p, q^j, s^j)$, which we denote as $x^j = x^j(p, q^j, s^j)$ resp. $x^j(\tau) = x^j(\tau, p, q^j, s^j)$.

If we insert this representation into problem (4)-(6), we obtain the weighted least squares sum

$$\sum_{i=1}^{M^j} \frac{(\eta_i^j - h_i^j(x^j(\tau_i^j, p, q^j, s^j), p, q^j))^2}{{\sigma_i^j}^2}$$

and the constraints $D^j[x^j(p, q^j, s^j), p, q^j] = 0$, which are maybe augmented by additional constraints $\hat{D}^j[x^j(p, q^j, s^j), p, q^j] = 0$ given by the discretization and parameterization methods.

We summarize the variables as $v = \left(s^1, \ldots, s^{N_1}, p\right) \in \mathbb{R}^n$ and write the weighted least squares terms as $\|r_1^j(v)\|_2^2$ and the constraints as $r_2^j(v) = 0$. ($r_1 : \mathbb{R}^n \to \mathbb{R}^{n_1}$, $r_2 : \mathbb{R}^n \to \mathbb{R}^{n_2}$, $n_2 < n \le n_1 + n_2$).

Thus we arrive at a finite dimensional least squares problem:

$$\min_v \sum_{j=1}^{N_1} \|r_1^j(v)\|_2^2$$
$$r_2^j(v) = 0, \quad j = 1, \ldots, N_1$$

Let $r_1 = \left(r_1^1, \ldots, r_1^{N_1}\right)$ and $r_2 = \left(r_2^1, \ldots, r_2^{N_1}\right)$. For solution and further discussion we need the Fréchet derivative

$$J = \begin{pmatrix} J_1 \\ J_2 \end{pmatrix} = \frac{\mathrm{d}}{\mathrm{d}v} \begin{pmatrix} r_1 \\ r_2 \end{pmatrix}$$

where J_1 is the derivative of the least squares terms and J_2 the derivative of the constraints with respect to the parameter estimation variables v.

In the case of multiple experiments, J_1 and J_2 have the following structures:

$$J_1 = \begin{pmatrix} \frac{\mathrm{d}r_1^1}{\mathrm{d}s^1} & 0 & \cdots & 0 & \frac{\mathrm{d}r_1^1}{\mathrm{d}p} \\ & & \ddots & & \vdots \\ 0 & \cdots & 0 & \frac{\mathrm{d}r_1^{N_1}}{\mathrm{d}s^{N_1}} & \frac{\mathrm{d}r_1^{N_1}}{\mathrm{d}p} \end{pmatrix}, \quad J_2 = \begin{pmatrix} \frac{\mathrm{d}r_2^1}{\mathrm{d}s^1} & 0 & \cdots & 0 & \frac{\mathrm{d}r_2^1}{\mathrm{d}p} \\ & & \ddots & & \vdots \\ 0 & \cdots & 0 & \frac{\mathrm{d}r_2^{N_1}}{\mathrm{d}s^{N_1}} & \frac{\mathrm{d}r_2^{N_1}}{\mathrm{d}p} \end{pmatrix}$$

We assume regularity, i.e. J_2 has full rank n_2 (CQ) and J has full rank n (PD).

Like Bock [8], we employ Gauss-Newton methods solving a sequence of constrained linear weighted least squares problems. The solution operator of the constrained linearized problem, the generalized inverse, can be written as

$$J^+ = \begin{pmatrix} I & 0 \end{pmatrix} \begin{pmatrix} J_1^T J_1 & J_2^T \\ J_2 & 0 \end{pmatrix}^{-1} \begin{pmatrix} J_1^T & 0 \\ 0 & I \end{pmatrix}.$$

2.6 Sensitivity Analysis of the Solution of a Constrained Parameter Estimation Problem

Now we are interested in the statistical reliability of the solution $\hat{v}$, (e.g. given by the Gauss-Newton method). Because the input of the parameter estimation problem, the experimental data, is random, also the output, the parameter estimate, is a random variable.

In first order $\hat{v} \sim \mathcal{N}(v^*, C)$ with the true values v^* as expected value and a variance-covariance matrix C given by

$$C = \begin{pmatrix} I & 0 \end{pmatrix} \begin{pmatrix} {J_1}^T J_1 & {J_2}^T \\ J_2 & 0 \end{pmatrix}^{-1} \begin{pmatrix} {J_1}^T J_1 & 0 \\ 0 & 0 \end{pmatrix} \begin{pmatrix} {J_1}^T J_1 & {J_2}^T \\ J_2 & 0 \end{pmatrix}^{-T} \begin{pmatrix} I \\ 0 \end{pmatrix}, \tag{7}$$

evaluated in $\hat{v}$.

Remark 2 *[8] With a probability of $\alpha \in [0; 1]$, the true variables v^* lie in the $(100 \cdot \alpha)\%$ confidence region*

$$G(\alpha, v^*) := \{v \in \mathbb{R}^n : \; F_2(v) = 0, \; \|F_1(v)\|_2^2 - \|F_1(v^*)\|_2^2 \leq \gamma^2(\alpha)\}$$

where $\gamma^2(\alpha) := \chi^2_{n-n_2}(1-\alpha)$ is the quantile of the χ^2 distribution with value α and $n - n_2$ degrees of freedom. Because the true values v^* are unknown, the estimate $\hat{v}$ is used instead as an approximation. Linearizing $G(\alpha, \hat{v})$, we obtain

$$\begin{aligned} G_L(\alpha, \hat{v}) := \{v \in \mathbb{R}^n : \; & F_2(\hat{v}) + J_2(\hat{v})(v - \hat{v}) = 0, \\ & \|F_1(\hat{v}) + J_1(\hat{v})(v - \hat{v})\|_2^2 - \|F_1(\hat{v})\|_2^2 \leq \gamma^2(\alpha)\}. \end{aligned}$$

It holds that

$$G_L(\alpha, \hat{v}) = \{v \in \mathbb{R}^n : \; v - \hat{v} = -J^+(\hat{v}) \begin{pmatrix} \delta\omega \\ 0 \end{pmatrix}, \; \delta\omega \in \mathbb{R}^{n_1}, \; \|\delta\omega\|_2 \leq \gamma(\alpha)\}.$$

2.7 Design of New Experiments

In order to improve the reliability of the parameter estimation, we want to evaluate data from additional experiments. We want to design these experiments in a way that the variance-covariance matrix of the parameter estimate from all data is minimal w.r.t. to a suited criterion. Optimization variables are the experimental settings and the measurement placement of the new experiments.

We consider N_2 additional experiments: $j = N_1 + 1, \ldots N_1 + N_2$. The variables for experimental design then model

- how to perform the experiments, i. e. controls q^j and
- where to measure. We specify some points where measurements are possible τ_i^j, $i = 1, \ldots, M^j$ and assign weights for measurement selection $w_i^j \in \{0; 1\}$, $i = 1, \ldots, M^j$ where 1 means the measurement is carried out and 0 means the measurement is omitted.

Thereby we modify the distribution model of the measurement errors:

$$\epsilon_i^j \sim \mathcal{N}(0, {\sigma_i^j}^2 / w_i^j), \quad i = 1, \ldots, M^j$$

where weight zero leads to an infinite variance, i.e. the measurement is not performed.

As constraints the number of measurements

$$a \leq \sum_{i \in I_k^j} w_i^j \leq b$$

or the costs of the measurements

$$\sum_{i \in J_k^j} c_i^j w_i^j \leq c_{max}$$

are restricted for some index sets $I_k^j, J_k^j \subseteq \{1, \ldots, M^j\}$, $k = 1, \ldots, K^j$. The 0-1-conditions $w_i^j \in \{0; 1\}$ can be relaxed by replacing the discrete by continuous weights $w_i^j \in [0; 1]$, $i = 1, \ldots, M^j$. This relaxation allows a practical interpretation: For $0 < w_i^j < 1$ a w_i^j-fold measurement is carried out, i. e. a measurement with variance ${\sigma_i^j}^2 / w_i^j$ and costs $c_i^j w_i^j$.

2.8 Parameter Estimation Problem for Experimental Design

Thus altogether we now consider $N_1 + N_2$ experiments:

- N_1 old experiments with available data η^j, $j = 1, \ldots, N_1$, and
- N_2 new experiments. Their realization would lead to assumed data η^j $j = N_1 + 1, \ldots, N_1 + N_2$.

Estimating the parameters from all — available and assumed — data leads to the following least squares problem:

$$\min_{x,p} \left(\sum_{j=1}^{N_1} \sum_{i=1}^{M^j} \frac{(\eta_i^j - h_i^j(x_i^j, p, q^j))^2}{{\sigma_i^j}^2} + \sum_{j=N_1+1}^{N_1+N_2} \sum_{i=1}^{M^j} w_i^j \cdot \frac{(\eta_i^j - h_i^j(x_i^j, p, q^j))^2}{{\sigma_i^j}^2} \right)$$

$$\begin{aligned} F^j[x^j, p, q^j] &= 0, \quad j = 1, \ldots, N_1 \\ D^j[x^j, p, q^j] &= 0, \quad j = 1, \ldots, N_1 \\ F^j[x^j, p, q^j] &= 0, \quad j = N_1 + 1, \ldots, N_1 + N_2 \\ D^j[x^j, p, q^j] &= 0, \quad j = N_1 + 1, \ldots, N_1 + N_2 \end{aligned}$$

with state variables $x = \left(x^j,\ j = 1, \ldots, N_1 + N_2\right)$.

Again a parameterization yields the formulation

$$\min_{v} \sum_{j=1}^{N_1} \|r_1^j(v)\|_2^2 + \sum_{j=N_1+1}^{N_1+N_2} \|r_1^j(v)\|_2^2 \tag{8}$$

$$r_2^j(v) = 0, \quad j = 1, \ldots, N_1 \tag{9}$$

$$r_2^j(v) = 0, \quad j = N_1 + 1, \ldots, N_1 + N_2 \tag{10}$$

with the variables $v = (s^1, \dots, s^{N_1+N_2}, p)$.

Note that the Jacobians J_1 resp. J_2 of the least squares terms resp. the constraints of this problem consist of blocks belonging to the old and of blocks belonging to the new experiments.

Analogously to the considerations above, the statistical reliability of the parameter estimate is described by the variance-covariance matrix which can be computed from J_1 and J_2 via formula (7).

Remark 3 *The Jacobians J_1 and J_2 and therewith the variance-covariance matrix do not depend on the experimental data, especially not on the unknown, only assumed data for the new experiments.*

2.9 Constraints on Experimental Design

Choosing experimental setups is usually restricted by several constraints of different type, describing limitations on operability, safety, costs and model validity:

- state constraints
$$\psi_L^j \le \psi^j(x^j(t), p, q^j(t)) \le \psi_U^j,$$
- control constraints
$$\vartheta_L^j \le \vartheta^j(q^j) \le \vartheta_U^j,$$
- measurement and cost constraints
$$a \le \sum_{i \in I_k^j} w_i^j \le b, \quad \sum_{i \in J_k^j} c_i^j w_i^j \le c_{max},$$
$$I_k^j, J_k^j \subseteq \{1, \dots, M^j\}, \quad k = 1, \dots, K^j,$$
- and integrality constraints
$$w_i^j \in \{0; 1\}, \quad i = 1, \dots, M^j,$$
maybe in a relaxed form
$$w_i^j \in [0; 1], \quad i = 1, \dots, M^j,$$

for the new experiments $j = N_1 + 1, \dots, N_1 + N_2$.

2.10 Nonlinear Experimental Design Optimization Problem

We want to compute experiments $j = N_1 + 1, \dots, N_1 + N_2$, determined by the controls q^j and the measurement weights w^j, which minimize, under the given constraints, the variance-covariance matrix of the parameter estimation problem (8)-(10).

As objective function, we employ uncertainty criteria defined on the variance-covariance matrix C, e.g.

$$\varphi(C) = \begin{cases} \frac{1}{n} \cdot \text{trace}(C) & \text{A-criterion} \\ \left(\det(K^T C K)\right)^{\frac{1}{n}} & \text{D-criterion} \\ \max\{\lambda : \lambda \text{ eigenvalue of } C\} & \text{E-criterion} \\ \max\{\sqrt{C_{ii}},\ i = 1, \dots, n\} & \text{confidence interval criterion} \end{cases}$$

where K is a full ranked projection matix on a subspace such that $K^T C K$ is regular.

We summarize all experimental design variables, the controls and weights of the new experiments, in the vector $\xi = (q^j, w^j,\ j = N_1 + 1, \dots, N_1 + N_2)$. Let $x = \left(x^j,\ j = 1, \dots, N_1 + N_2\right)$ be the state variables of the process model.

Then the experimental design optimization problem reads as follows

$$\min_{\xi, x} \varphi(C) \tag{11}$$

where

$$C = \begin{pmatrix} I & 0 \end{pmatrix} \begin{pmatrix} {J_1}^T J_1 & {J_2}^T \\ J_2 & 0 \end{pmatrix}^{-1} \begin{pmatrix} {J_1}^T J_1 & 0 \\ 0 & 0 \end{pmatrix} \begin{pmatrix} {J_1}^T J_1 & {J_2}^T \\ J_2 & 0 \end{pmatrix}^{-T} \begin{pmatrix} I \\ 0 \end{pmatrix} \tag{12}$$

is the variance-covariance matrix in the solution point

of the constrained parameter estimation problem (8)-(10)

subject to the process model for $j = 1, \dots, N_1 + N_2$

$$F^j[x^j, p, q^j] = 0 \tag{13}$$

$$D^j[x^j, p, q^j] = 0 \tag{14}$$

and subject to constraints for $j = N_1 + 1, \dots, N_1 + N_2$

$$\psi_L^j \leq \psi^j(x^j(t), p, q^j(t)) \leq \psi_U^j \tag{15}$$

$$\vartheta_L^j \leq \vartheta^j(q^j) \leq \vartheta_U^j \tag{16}$$

$$a \leq \sum_{i \in I_k^j} w_i^j \leq b, \quad \sum_{i \in J_k^j} c_i^j w_i^j \leq c_{max},$$

$$I_k^j, J_k^j \subseteq \{1, \dots, M^j\}, \quad k = 1, \dots, K^j \tag{17}$$

$$w_i^j \in \{0; 1\}, \quad i = 1, \dots, M^j \tag{18}$$

Remark 4 *Scaling of the parameters influences the variance-covariance matrix and the uncertainty criteria, and consequently the result of the experimental design optimization in the way that parameters with large absolute values are taken into account stronger. Hence scaling can be used to aim at particular high reliability of selected parameters. For details, see [18].*

The new designed experiments can now be realized in laboratory. From the resulting experimental data, together with the old data, a new parameter estimate can be computed. This sequence of alternated processing of experiments, evaluation of the available data by parameter estimation and design of additional new experiments under the consideration of the previous old experiments can be repeated until the reliability of parameter estimation is good enough or a given cost limit is reached. We call this "sequential approach of parameter estimation and experimental design", see [16].

2.11 Robust Experimental Design

In the case, where the parameters enter the model nonlinearly, the variance-covariance matrix depends not only on the experimental design variables ξ, but also on values of the variables of the parameter estimation problem v. To stress this, we write $C = C(\xi, v)$ in this section.

We now want to take into account that the true parameter values are not known. We assume that we only know that the parameters lie within a confidence region $\left\{v : \|v - v_0\|^2_{2,\Sigma^{-1}} := (v - v_0)^T \Sigma^{-1}(v - v_0) \leq \gamma^2\right\}$.

Minimizing the statistical uncertainty under this parameter distribution in a worst case sense leads to a min-max problem with objective function

$$\min_{\xi,x} \max_{\|v-v_0\|_{2,\Sigma^{-1}} \leq \gamma} \varphi(C(\xi, v))$$

The solution of this problem requires methods of semi-infinite programming, see e.g. [15] which, in general, are very costly. Hence our approach is to simplify the formulation by an expansion of the objective function with respect to the parameters

$$\min_{\xi,x} \max_{\|v-v_0\|_{2,\Sigma^{-1}} \leq \gamma} \varphi(C(\xi, v_0)) + \frac{\mathrm{d}}{\mathrm{d}v}\varphi(C(\xi, v_0))(v - v_0),$$

where we can solve the inner problem explicitly and obtain a modified robust experimental design problem:

$$\min_{\xi,x} \varphi(C(\xi, v_0)) + \gamma \left\| \frac{\mathrm{d}}{\mathrm{d}v}\varphi(C(\xi, v_0)) \right\|_{2,\Sigma}. \tag{19}$$

Remark 5 *The probability factor γ gives a weighting between the "reliability part" $\varphi(C(\xi, v_0))$ and the "robustness part" $\left\|\frac{d}{dv}\varphi(C(\xi, v_0))\right\|_{2,\Sigma}$ of the objective function.*

For further details, especially the treatment of constraints, we want to refer to [17]

3 Properties, Difficulties and Solution Approaches

3.1 For Parameter Estimation

To solve the nonlinear constrained parameter estimation problems we apply the methods introduced by Bock [8] and Schlöder [26]. The model equations are parameterized, in the case of Ordinary Differential Equations or Differential Algebraic Equations by a multiple shooting approach. The resulting high dimensional constrained nonlinear least squares problem is solved by a generalized Gauss-Newton method with structure exploiting linear algebra and efficient globalization strategies [7]. These methods are implemented in the software package PARFIT.

3.2 For Experimental Design

The experimental design optimization problem (11)-(18) (resp. with the modified objective function of (19)) belongs to the class of nonlinear constrained optimal control problems. Due to the w-variables it is a mixed integer optimization problem.

The objective function is implicitly defined on the derivatives of a constrained parameter estimation problem and hence depends on derivatives of the solution of the model equations. This non-standard objective function makes the application of the indirect approach of optimal control problematic.

We apply the relaxation of the integrality constraints as discussed above. After parameterization of the model equations, parameterization of the control functions by the direct approach and discretization of the state constraints we obtain a finite dimensional constrained nonlinear optimization problem.

We use the SQP method of Gill [12]. This gradient based method requires among other things derivatives of the objective function with respect to the experimental design variables ξ. To compute these gradients, we use formulas for derivatives of φ with respect to C, see [20] and of C with respect to J, see [18]. In this paper, we want to put emphasis on the derivative of J with respect to ξ. E.g., the entries in J_1 corresponding to measurement $i \in \{1, \ldots, M^j\}$ in experiment $j \in \{N_1+1, \ldots, N_1+N_2\}$ and derived w.r.t. to the parameters p are

$$\begin{aligned}\frac{\mathrm{d}}{\mathrm{d}p} r^j_{1i} &= \frac{\mathrm{d}}{\mathrm{d}p} \frac{\sqrt{w^j_i}}{\sigma^j_i} \cdot \left(\eta^j_i - h^j_i(x^j(\tau^j_i, p, q^j, s^j), p, q^j)\right) \\ &= -\frac{\sqrt{w^j_i}}{\sigma^j_i} \left(\frac{\partial h^j_i}{\partial x} \frac{\partial x^j_i}{\partial p} + \frac{\partial h^j_i}{\partial p} \right)\end{aligned}$$

where $h^j_i := h^j_i(x^j(\tau^j_i, p, q^j, s^j), p, q^j)$ and $x^j_i := x^j\ (\tau^j_i,\ p,\ q^j,\ s^j)$. For the derivative of that with respect to q^j, we obtain

$$\frac{\mathrm{d}}{\mathrm{d}q^j} \frac{\mathrm{d}}{\mathrm{d}p} r^j_{1i} = \frac{\mathrm{d}}{\mathrm{d}q^j} \frac{\mathrm{d}}{\mathrm{d}p} \frac{\sqrt{w^j_i}}{\sigma^j_i} \cdot \left(\eta^j_i - h^j_i(x^j(\tau^j_i, p, q^j, s^j), p, q^j)\right)$$

$$= -\frac{\sqrt{w_i^j}}{\sigma_i^j}\left(\frac{\partial^2 h_i^j}{\partial x \partial x}\frac{\partial x_i^j}{\partial p}\frac{\partial x_i^j}{\partial q^j} + \frac{\partial^2 h_i^j}{\partial q^j \partial x}\frac{\partial x_i^j}{\partial p} + \frac{\partial h_i^j}{\partial x}\frac{\partial^2 x_i^j}{\partial q^j \partial p}\right.$$
$$\left. + \frac{\partial^2 h_i^j}{\partial x \partial p}\frac{\partial x_i^j}{\partial q^j} + \frac{\partial^2 h_i^j}{\partial q^j \partial p}\right).$$

From this follows that the derivatives of the state variables $\frac{\partial x_i^j}{\partial p}$, $\frac{\partial x_i^j}{\partial q^j}$ and especially the second derivatives $\frac{\partial^2 x_i^j}{\partial q^j \partial p}$ are needed and have to be provided by the solver of the model equations.

For DAE models, we use a tailored combination of Internal Numerical Differentiation [8] and Automatic Differentiation [13] to compute these derivatives by the BDF integrator DAESOL [5]. For details, see [3] and [4].

We exploit the multiple experiment structures by structured computation, decomposition and derivation of the blocks of the Jacobian. Here we can take into account that some parts are fixed because they belong to the available old experiments, see [16] and [4].

Due to the relaxation of the integrality constraints we cannot expect in general that the solutions have w-variables with values in $\{0; 1\}$. Numerical results show that the violation of the integrality is only very modest. Heuristics, see [18], can be applied to generate integer feasible solutions.

Our methods are implemented in the software package VPLAN [18].

Remark 6 *A special case of experimental design is the so called sampling design where the optimization only aims at selecting measurements, i.e. the experimental design variables only consist of the weights w. In this case the model states do not depend on the experimental design, what makes the solution of the problem numerically easier because the model states remain unchanged during the SQP iterations and derivatives of the model states with respect to controls are not required. This kind of problems is e.g. discussed by Lohmann [19].*

4 Practical Applications

The methods discussed in this paper and the software package VPLAN have been applied to several industrial processes. Detailed descriptions and results are presented in [18]. In this section, we want to show two examples which give an idea of the possibilities and the performance of our approach.

4.1 The Reaction of Urethane

The reaction of urethane was investigated by the author together with BASF AG, Ludwigshafen [6]. This reaction is an important prototype for the understanding of the processes in the formation of polyurethane plastics. Having a validated model means being able to influence product quality and properties. Figure 1 shows reaction scheme, species and reactor of the reaction of urethane.

$$\begin{aligned} A + B &\rightarrow C \\ A + C &\rightleftharpoons D \\ 3A &\rightarrow E \end{aligned}$$

Educts: Phenylisocyanate A, Butanole B,
Products: Urethane C, Allophanate D,
Isocyanurate E,
Solvent: Dimethylsulfoxide L

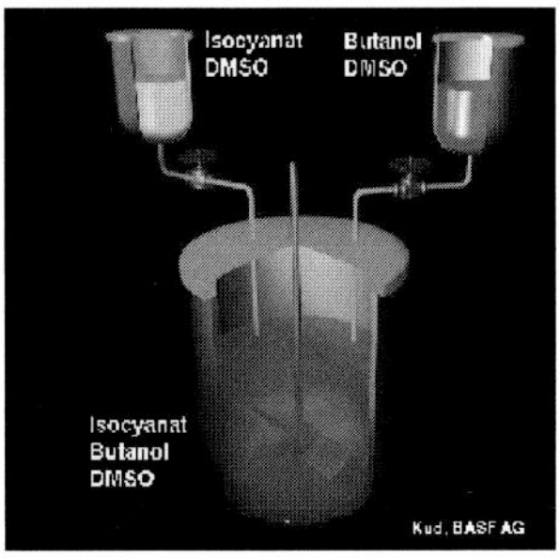

Fig. 1. Reaction scheme, species and reactor of the reaction of urethane.

We model this process as a DAE initial value problem:

$$\begin{aligned}
\dot{n}_C(t) &= V(t) \cdot (r_1(t) - r_2(t) + r_3(t)) \\
\dot{n}_D(t) &= V(t) \cdot (r_2(t) - r_3(t)) \\
\dot{n}_E(t) &= V(t) \cdot r_4(t) \\
n_A(t) &= n_{A,0} + n_{A,e}(t) - n_C(t) - 2 \cdot n_D(t) - 3 \cdot n_E(t) \\
n_B(t) &= n_{B,0} + n_{B,e}(t) - n_C(t) - n_D(t) \\
n_L(t) &= n_{L,0} + n_{L,e}(t)
\end{aligned}$$

$$n_C(t_0) = n_D(t_0) = n_E(t_0) = 0$$

The reaction velocities follow Arrhenius law:

$$\begin{aligned}
r_1 &= k_1 \cdot \frac{n_1}{V} \cdot \frac{n_2}{V} \\
r_2 &= k_2 \cdot \frac{n_1}{V} \cdot \frac{n_3}{V} \\
r_3 &= k_3 \cdot \frac{n_4}{V} \\
r_4 &= k_4 \cdot \left(\frac{n_1}{V}\right)^2
\end{aligned}
\qquad
\begin{aligned}
k_1 &= k_{ref1} \cdot \exp\left(-\frac{E_{a,1}}{R} \cdot \left(\frac{1}{T} - \frac{1}{T_{ref1}}\right)\right) \\
k_2 &= k_{ref2} \cdot \exp\left(-\frac{E_{a,2}}{R} \cdot \left(\frac{1}{T} - \frac{1}{T_{ref2}}\right)\right) \\
k_4 &= k_{ref4} \cdot \exp\left(-\frac{E_{a,4}}{R} \cdot \left(\frac{1}{T} - \frac{1}{T_{ref4}}\right)\right) \\
K_C &= K_{C2} \cdot \exp\left(-\frac{\Delta H_2}{R} \cdot \left(\frac{1}{T} - \frac{1}{T_{C2}}\right)\right).
\end{aligned}$$

Further quantities in the model are given by $k_3 = \frac{k_2}{K_C}$,

$$\begin{aligned}
n_{A,e} &= n_{A,e1,0} \cdot feed_1 \quad n_{B,e} = n_{B,e2,0} \cdot feed_2 \\
n_{L,e} &= n_{L,e1,0} \cdot feed_1 + n_{L,e2,0} \cdot feed_2,
\end{aligned}$$

$$V = \frac{n_A \cdot M_A}{\rho_A} + \frac{n_B \cdot M_B}{\rho_B} + \frac{n_C \cdot M_C}{\rho_C} + \frac{n_D \cdot M_D}{\rho_D} + \frac{n_E \cdot M_E}{\rho_E} + \frac{n_L \cdot M_L}{\rho_L}$$

Independent variable is the time $t \in [t_0; t_{end}] = [0\,h, 80\,h]$. The state variables are the molar numbers of the species n_A, n_B, n_C, n_D, n_E and n_L. The parameters describe the reaction velocities, namely activation energies $E_{a,1}$, $E_{a,2}$, $E_{a,4}$, steric factors k_{ref1}, k_{ref2}, k_{ref4} , equilibrium constant K_{C2} and reaction enthalpy ΔH_2. The controls consist of initial molar numbers in reactor: $n_{A,0}$, $n_{B,0}$, $n_{L,0}$, feed 1: $n_{A,e1,0}$, $n_{L,e1,0}$ and feed 2: $n_{B,e2,0}$, $n_{L,e2,0}$ and the feed profiles $feed_1(t)$, $feed_2(t)$ and the temperature $T(t)$.

Three measurement methods are available to determine • the mass percent of Phenylisocyanate with variance 0.5 (titration), • the mass percent of Urethane with variance 0.5 and the mass percent of Allophanate with variance 0.005 (HPLC I), and • the mass percent of Isocyanurate with variance 0.0005 (HPLC II). There is a limitation of 16 measurements per experiment.

Several constraints are given to model restrictions on the controls to ensure safety requirements during a 4 day-and-night shift.

An optimized sequential experimental design was compared with an heuristic design planned intuitively by an experienced experimenter from BASF. He suggested altogether 15 experiments with 90 measurements. The optimum experimental design was terminated after the sequential optimization of 2 experiments with 32 measurements. A detailed description of these experiments can be found in [18]. Table 1 compares the results and statistical reliabilities of the parameter estimation from the data of the two approaches. It can be observed that the intuitive approach is not able to give reliable estimators for the parameters describing the chemical equilibrium, whereas the optimum experimental design allows the estimation of all parameters with variances smaller than 1%.

Parameter	from expert design	from optimized design
k_{ref1}	2.50 $\pm$0.02	2.504 $\pm$0.006
$E_{a,1}$	0.835 $\pm$0.007	0.836 $\pm$0.001
k_{ref2}	91.3 $\pm$0.7	91.25 $\pm$0.02
$E_{a,2}$	0.834 $\pm$0.002	0.83537 $\pm$0.00008
k_{ref4}	57.991 $\pm$0.009	58.004 $\pm$0.006
$E_{a,4}$	0.65725 $\pm$0.00007	0.6574 $\pm$0.0002
ΔH_2	0.9 $\pm$0.3	1.082 $\pm$0.006
K_{C2}	1.1 $\pm$0.3	1.29 $\pm$0.01

Table 1. Comparison of the parameter estimation from experimental data from expert design with the parameter estimation from experimental data from optimized design.

As another study, we have computed a robust experimental design for the reaction of urethane. Detailed results are published in [17]. Figure 2 shows that the sensitivity of the statistical reliability with respect to parameter perturbations can be reduced drastically.

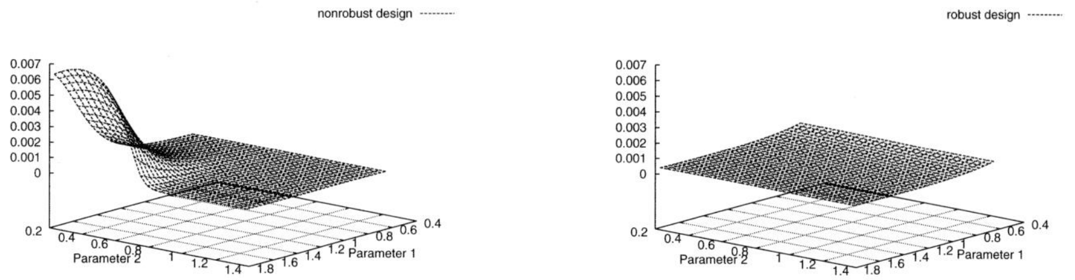

Fig. 2. Left: Parameter dependence of the A criterion for perturbations of the (scaled) parameters $E_{a,1}$ and $E_{a,2}$. Right: Improvement of the parameter sensitivity by robust experimental design.

4.2 Transport and Degradation Processes of Xenobiotics in Soil

In this project together with BASF AG, Ludwigshafen, Altmann-Dieses et al. [2] investigated the correlation between pesticides and water quality. Transport and degradation in soil is usually simulated with soil columns and lysimeters. Due to cost reasons, it is desirable to substitute these experiments more and more by simulations on computers. In order to use these simulations in the so-called registration procedure for admission of xenobiotics, EU law requires validated models.

For given experimental data η_{kij}, $(k = \psi, \theta, c)$, $(i = 1, \ldots, m_1)$, $(j = 1, \ldots, m_2)$ with measurement functions of ψ, θ, c and the nonlinear regression

$$\eta_{kij} = h_k(t_i, k(t_i, z_j; p), p) + \epsilon_{kij}, \quad \epsilon_{kij} \sim \mathcal{N}(0, \sigma_{kij}^2)$$

the estimation problem for the unknown model parameters p can be written as

$$\min \sum_{k=\psi,\theta,c} \sum_{i=1}^{m_1} \sum_{j=1}^{m_2} \sigma_{kij}^{-2} \left(\eta_{kij} - h_k(t_i, k(t_i, z_j, p), p)\right)^2$$

$$\begin{aligned} C(\psi;p)\frac{\partial\psi}{\partial t} &= \frac{\partial}{\partial z}\left(K(\psi;p)\frac{\partial}{\partial z}(\psi - z)\right) + S(\psi;p) \\ \frac{\partial\theta}{\partial t} &= \frac{\partial}{\partial z}\left(\bar{D}(\theta;p)\frac{\partial\theta}{\partial z} - \bar{K}(\theta;p)\right) + \bar{S}(\theta;p) \\ \frac{\partial(\theta c)}{\partial t} - &= \frac{\partial}{\partial z}\left(\theta D_h(\theta;p)\frac{\partial c}{\partial z} - qc\right) - \frac{\partial(\rho s)}{\partial t} + Q(c;p) \\ &+ \text{initial and boundary conditions} \end{aligned}$$

The process model consists of two parts, the water transport which is described by the Fokker-Planck equation for the water potential ψ and the Richards equation for the water content θ, and an equation for transport of substance with concentration c. The model contains retention curves with van-Genuchten-Mualem parameterization $K(\psi;p)$, $C(\psi;p)$, $\bar{D}(\theta;p)$, $\bar{K}(\theta;p)$, $D_h(\theta;p)$, the degradation $Q(c;p)$ and source terms $S(\psi;p)$ and $\bar{S}(\theta;p)$. Using semi-discretization in space we transform the in-stationary PDE to a large structured ODE system. In order to design experiments, water flux and substance input concentration can be used as controls. The duration of each experiment is 12 days. The sampling scheme consists of measurements for outflow data and an investigation of the sliced column at the end of the experiment.

One experiment is planned by the method of nonlinear optimum experimental design. Table 2 shows that a reliable estimation of all parameters is possible with data from only one experiment. This result cannot be achieved by intuitively designed experiments.

	nominal	scaled	optimal design	design A	design B
A Criterion			0.0472	3.8859	0.750
p_1	1.2	1.0	± 0.0191	± 0.1604	± 0.0800
p_2	0.0102	1.0	± 0.2438	± 1.5968	± 0.8600
p_3	10.0	1.0	± 0.4348	± 3.3751	± 1.6735
p_4	0.05	1.0	± 0.0086	± 0.0124	± 0.0083
p_5	10.0	1.0	± 0.1852	± 3.0460	± 0.9712
p_6	20.0	1.0	± 0.0108	± 0.2644	± 0.1032

Table 2. Parameter variances for the optimal soil column experiment compared to two intuitive designs.

5 Conclusion

In this paper we have discussed optimization problems and numerical methods for nonlinear experimental design for a general class of process models. We have seen that these problems are quite challenging and demand state-of-the-art numerical methods for their solution. The author has developed a software package which is successfully applied to relevant industrial processes.

References

[1] A. C. Atkinson and A. N. Donev. *Optimum Experimental Designs.* Oxford University Press, 1992.

[2] A. E. Altmann-Dieses, J. P. Schlöder, H. G. Bock, and O. Richter. Optimal experimental design for parameter estimation in column outflow experiments. *Water Resources Research*, 38(10), 2002.

[3] I. Bauer, H. G. Bock, S. Körkel, and J. P. Schlöder. Numerical methods for initial value problems and derivative generation for DAE models with application to optimum experimental design of chemical processes.

In F. Keil, W. Mackens, H. Voss, and J. Werther, editors, *Scientific Computing in Chemical Engineering II*, volume 2, pages 282–289, Berlin, Heidelberg, 1999. Springer-Verlag.

[4] I. Bauer, H. G. Bock, S. Körkel, and J. P. Schlöder. Numerical methods for optimum experimental design in DAE systems. *Journal of Computational and Applied Mathematics*, 120:1–25, 2000.

[5] I. Bauer, H. G. Bock, and J. P. Schlöder. DAESOL — a BDF code for the numerical solution of differential-algebraic equations. Preprint, IWR der Universität Heidelberg, SFB 359, November 1999.

[6] I. Bauer, M. Heilig, S. Körkel, A. Kud, A. Mayer, and O. Wörz. Versuchsplanung am Beispiel einer Phosphin- und Urethanreaktion. 14. 10. 1998.

[7] H. G. Bock, E. Kostina, and J. P. Schlöder. On the role of natural level functions to achieve global convergence for damped newton methods. In M. J. D. Powell and S. Scholtes, editors, *System Modelling and Optimization: Methods, Theory and Applications, 19th IFIP TC7 Conference on System Modelling and Optimization, July 12-16, 1999, Cambridge, UK*, volume 174 of *IFIP Conference Proceedings*, pages 51–74. Kluwer, 2000.

[8] H. G. Bock. Randwertproblemmethoden zur Parameteridentifizierung in Systemen nichtlinearer Differentialgleichungen. *Bonner Mathematische Schriften 183*, 1987.

[9] M. Baltes, R. Schneider, C. Sturm, and M. Reuss. Optimal Experimental Design for Parameter Estimation in Unstructured Growth Models. *Biotechnol. Prog.*, 10:480–488, 1994.

[10] V. G. Doví, A. P. Reverberi, and L. Acevedo-Duarte. New Procedure for Optimal Design of Sequential Experiments in Kinetic Models. *Ind. Eng. Chem. Res.*, 33:62–68, 1994.

[11] V. V. Fedorov. *Theory of Optimal Experiments.* Probability And Mathematical Statistics. Academic Press, London, 1972.

[12] P. E. Gill, W. Murray, and M. A. Saunders. SNOPT: An SQP algorithm for large-scale constrained optimization. *SIAM J. Opt.*, 12:979–1006, 2002.

[13] A. Griewank. *Evaluating Derivatives. Principles and Techniques of Algorithmic Differentiation.* Frontiers in Applied Mathematics. SIAM, 2000.

[14] L. M. Haines. Optimal Design For Nonlinear Regression Models. *Commun. Statist.-Theory Meth.*, 22(6):1613–1627, 1993.

[15] R. Hettich and K. O. Kortanek. Semi-infinite programming: Theory, methods, and applications. *SIAM Review*, 35(3):380–429, 1993.

[16] S. Körkel, I. Bauer, H. G. Bock, and J. P. Schlöder. A sequential approach for nonlinear optimum experimental design in DAE systems. In F. Keil, W. Mackens, H. Voss, and J. Werther, editors, *Scientific Computing in Chemical Engineering II*, volume 2, pages 338–345, Berlin, Heidelberg, 1999. Springer-Verlag.

[17] S. Körkel, E. Kostina, H. G. Bock, and J. P. Schlöder. Numerical methods for optimal control problems in design of robust optimal experiments

for nonlinear dynamic processes. *Optimization Methods and Software*, to appear in 2004.

[18] S. Körkel. *Numerische Methoden für Optimale Versuchsplanungsprobleme bei nichtlinearen DAE-Modellen.* PhD thesis, Universität Heidelberg, available at http://www.ub.uni-heidelberg.de/archiv/2980, 2002.

[19] T. W. Lohmann, H. G. Bock, and J. P. Schlöder. Numerical Methods for Parameter Estimation and Optimal Experiment Design in Chemical Reaction Systems. *Ind. Eng. Chem. Res.*, 31:54–57, 1992.

[20] T. W. Lohmann. *Ein numerisches Verfahren zur Berechnung optimaler Versuchspläne für beschränkte Parameteridentifizierungsprobleme.* Reihe Informatik. Verlag Shaker, Aachen, 1993.

[21] P. Oinas, I. Turunen, and H. Haario. Experimental Design With Steady-State And Dynamic Models Of Multiphase Reactors. *Chemical Engineering Science*, 47(13/14):3689–3696, 1992.

[22] F. Pukelsheim. *Optimal Design of Experiments.* John Wiley & Sons, Inc., New York, 1993.

[23] Z. H. Qureshi, T. S. Ng, and G. C. Goodwin. Optimum experimental design for identification of distributed parameter systems. *Int. J. Control*, 31(1):21–29, 1980.

[24] P. M. Reilly, R. Bajramovic, G. E. Blau, D. R. Branson, and M. W. Sauerhoff. Guidelines for the Optimal Design of Experiments to Estimate Parameters in First Order Kinetic Models. *The Canadian Journal of Chemical Engineering*, 55:614, 1977.

[25] P. E. Rudolph and G. Herrendörfer. Optimal Experimental Design and Accuracy of Parameter Estimation for Nonlinear Regression Models Used in Long-term Selection. *Biom. J.*, 37(2):183–190, 1995.

[26] J. P. Schlöder. *Numerische Methoden zur Behandlung hochdimensionaler Aufgaben der Parameteridentifizierung.* Dissertation, Hohe Mathematisch-Naturwissenschaftliche Fakultät der Rheinischen Friedrich-Wilhelms-Universität zu Bonn, 1987.

Controlling the Continuos Positive Airway Pressure-Device Using Partial Observable Markov Decision Processes

Clemens Kreutz[1] and Josef Honerkamp[2]

[1] Freiburg Center for Data Analysis and Modeling FDM
Eckerstr. 1, D-79104 Freiburg, Germany
Faculty of Physics, Hermann-Herder-Str. 3, D-79104 Freiburg, Germany
ckreutz@fdm.uni-freiburg.de

[2] Freiburg Center for Data Analysis und Modeling FDM
Eckerstr. 1, D-79104 Freiburg, Germany
Freiburg Materials Research Center FMF
Stefan-Maier-Str. 21, D-79104 Freiburg, Germany
Faculty of Physics, Hermann-Herder-Str. 3, D-79104 Freiburg, Germany
hon@physik.uni-freiburg.de

Summary. Partial Observable Markov Decision Processes (POMDP's) allow finding an optimal control of a hidden Markov model. We develop within this framework a control of a CPAP-device (Continuous Positive Airway Pressure). People who suffer from an obstructive sleep apnoea syndrome may use such a device by which a small additive air pressure is applied to the patient through a mask so that the occlusion of the respiratory tract is prevented. Different values of the pressure will correspond to different actions in the POMDP-approach.

We study the performance of a control determined by a POMDP-approach in comparison with other, more traditional control strategies. In commercial devices the applied pressure is held on a constant level, carefully determined in a sleep laboratory. More recently, the pressure is controlled by a classification of the airflow during breathing and then choosing the action in dependence on the classification features. Within the POMDP approach, we introduce the unobservable state of the airway as the hidden state and regard the features of the airflow during breathing as the observations. The parameters of the hidden models are chosen with help of experience from laboratories, where such devices are developed.

We will compare the POMDP control with direct control strategies in dependence of parameters of a POMDP model. We will show that the advantage of POMDP control is substantial in general, however, that there are also cases, where the performance of the belief control does not differ too much from some direct control. We discuss, when this does happen and thus find a criterion for the implementation of the POMDP-control instead of a direct control.

1 Introduction

The CPAP (Continuous Positive Airway Pressure) is a respirator for people, who suffer from obstructive sleep apnoea syndrome (OSAS). The treatment is of high relevancy because more than approximately two percent of people between 30 and 60 years suffer from this sleep disorder. The word 'apnoea' is used to describe a collapse of the upper airways in the respiratory tract which blocks the air supply to the lungs. The resulting lack of oxygen provokes an arousal. Though the person does not become aware of it, the sleeping quality deteriorates dramatically. Affected patients have up to ten apnoea events per hour. To help such people one may apply a positive air pressure to the respiratory tract to prevent blockades or, eventually, to open them after a occlusion. The technical basis for such an artificial ventilation was developed since 1980 [15].

However, applying an additional air pressure during the sleep affects the condition of the patient negatively. Another negative aspect is the noise of the device. Furthermore, because of the high pressure air can flow trough the middle ear which causes further discomfort. Therefore the goal is to find an optimal respiration policy which avoids too much apnoea events with a mean pressure which is as small as possible.

By using such a CPAP-device the air flow during breathing can be measured and thus, by extracting suitable feature components, classified. The development of these sophisticated methods took researchers a long time [12], [14], [13]. A control of the applied pressure according to this classification is a first step to obtain a lower mean value for the applied pressure during a night. But from the form of the air flow one cannot infer unambiguously the true state of the respiratory tract so that false actions are to be expected.

A more realistic setting would be to consider the true states of the respiratory tract as hidden states. The transitions between theses states may be modelled by a Markov process dependent on the action (the applied pressure). Thus a Markov decision process would be the appropriate model for the situation in which the states of the respiratory tract would be completely observable. However, because this observability is not given, we have to introduce an observation equation which relates the true states to the observations, i.e. to the feature vector which characterizes the airflow through the respiratory tract. Hence we have to consider a hidden Markov process depending on the action and we thus arrive at the task to control such a hidden Markov process. This setting is usually known as Partially Observable Markov Decision Process (POMDP).

Within a POMDP framework we are able to improve the control of the CPAP device in comparison with existing control strategies. In order to demonstrate this we compare the performance of different policies.

2 Background

2.1 Hidden Markov Models (HMM) and Partial Observable Markov Decision Processes (POMDP)

We consider a hidden Markov process with discrete states $s, s = 1, \ldots, n$, which are not directly observable, and with discrete observations $z, z = 1, \ldots, m$. The transition probabilities can be arranged into a matrix T with

$$T(s, s') := \rho(s_{t+1} = s' | s_t = s) \tag{1}$$

and the probabilities $\rho(z_t = z | s_t = s)$ for observing z given state s constitute diagonal observation matrices

$$O^z(s, s') := \rho(z_t = z | s_t = s)\delta_{s,s'} \ , \quad z = 1, ..., m \tag{2}$$

with

$$\delta_{s,s'} = \begin{cases} 1 & \text{for } s = s' \\ 0 & \text{else} . \end{cases} \tag{3}$$

The belief $b_t(s) = \rho(s_t = s)$ about the hidden state is the probability that at time t state s is realized. It holds $\sum_s b(s) = 1$ and $0 \leq b(s) \leq 1$.

A controllable hidden Markov process is called a Partial Observable Markov Decision Process (POMDP). In this case an agent has the possibility to choose actions a at each time t which affects transition and observation probabilities. Thus, we have to introduce transition probabilities $\rho(s'|s, a)$ depending on action a respectively transition matrices $T^a(s, s')$. With Bayes' Theorem, after an action a and an observation z' the belief can be updated:

$$b_{t+1}(s) = \frac{1}{N} \sum_{s',s''} b_t(s')\, T^a(s', s'')\, O^{z'}(s'', s) \quad \text{with} \tag{4}$$

$$N := \sum_{s,s',s''} b_t(s')\, T^a(s', s'')\, O^{z'}(s'', s) \ .$$

The policy determines the action at each time. Within a POMDP model the policy depends on the belief $a_t = \pi(b_t)$. We call this strategy 'belief control'. In the case where a POMDP model is not used, the action can be a function of the immediate observation $a_t = \pi(z_t)$. This strategy is called 'direct control'. Direct control is frequently used in applications.

To assess the benefit of a chosen action a one has to introduce a reward $r(s, a)$, which may also depend on the actual state and on the state reached after action. The value function

$$V^\pi(b_t) := E\left(\sum_{i=0}^{T} \gamma^i r(b_{t+i}, a_{t+i})\right) \quad \text{with } r(b_t, a_t) = \sum_s r(s, a_t) b_t(s) \tag{5}$$

is the expectation value of the rewards, following b_t while acting according to π. $\gamma \in [0,1]$ is a discount factor which takes into account that future rewards may be less worth than actual ones. In the POMDP approach one further introduces the action value function for a policy π

$$Q^{\pi}(b_t, a_t) := r(b_t, a_t) + \gamma \sum_{z_{t+1}} \rho(z_{t+1}|b_t, a_t)\, V^{\pi}(b_{t+1}) \;, \tag{6}$$

which measures the cumulative reward for a process controlled by the policy π, starting the process in belief b and taking a as the first action. Here

$$\rho(z_{t+1}|b_t, a_t) = \sum_{s,s',s''} b_t(s) T^{a_t}(s, s') O^{z_{t+1}}(s', s'') \tag{7}$$

is the probability of observing z_{t+1} given b_t and a_t. b_{t+1} is given by (4). For an optimal policy π^* one can find an implicit equation, the Bellman equation [2], which reads

$$Q^{\pi^*}(b_t, a_t) := r(b_t, a_t) + \gamma \sum_{z_{t+1}} \rho(z_{t+1}|b_t, a_t)\, V^{\pi^*}(b_{t+1}) \tag{8}$$

with

$$V^{\pi^*}(b_t) = \arg\max_{a_t} Q^{\pi^*}(b_t, a_t). \tag{9}$$

Several methods for solving this equation are investigated. The best known exact methods are value iteration [4] and policy iteration [11]. Most algorithms of these methods make use of dynamic programming updates. Some fast algorithms are 'one pass' [10], 'linear support' [5] and 'witness' [8]. For this paper we applied 'incremental pruning' introduced by Cassandra et al in 1997 [1].

Choosing an appropriate hidden Markov model and solving the Bellman equation requires a great effort. One may ask whether there is a benefit in the sense that the control strategy achieved is significantly better than policies which can be elaborated much simpler, as e.g. constant control, any MDP based approximation [3] or direct control. In the case of direct control, there are only a limited number of features characterizing the measurements. Therefore, the possible mappings from the feature space to the action space are manageable and one may study one after the other by Monte Carlo simulations to find out the best one given the hidden Markov model and the rewards.

Considering direct control within a POMDP approach will lead to a lower value than the policy given by the solution of the Bellman equation, because this is just by construction the policy with maximum value [10]. However, nothing is known about the difference in efficiency and further properties of strategies outside the POMDP framework.

3 Comparison of control variables

The quantity which the policy function π depends on is called control variable. In the POMDP framework the belief is the natural control variable ('belief control'). Strategies outside the POMDP framework use other control variables. In this section we compare belief control with direct control, that uses the immediate observation as control variable.

To exemplify, which different features the various control strategies can show, we use two models:

Tiger Model

First we use a slight modification of the well-known Tiger problem [4]. There are two states (tiger is behind a left door, tiger is behind a right door) with two corresponding observations (one hears the tiger behind the left/right door) and three actions (open the left door, open the right door, listen). The aim of the agent is to open the door, where the tiger is not located. If one door is opened, the game restarts. The action 'listen' is done to get more information about the tiger's location. Thus the model is formulated as:

$$T^{a=1} := \begin{pmatrix} c_t & 1-c_t \\ 1-c_t & c_t \end{pmatrix},\ T^{a=2} := \begin{pmatrix} 0.5\,c_t & 1-0.5\,c_t \\ 1-0.5\,c_t & 0.5\,c_t \end{pmatrix},\ T^{a=3} := \begin{pmatrix} 1 & 0 \\ 0 & 1 \end{pmatrix}$$

$$O^{z=1} := \begin{pmatrix} c_o & 0 \\ 0 & 1-c_o \end{pmatrix},\ O^{z=2} := \begin{pmatrix} 1-c_o & 0 \\ 0 & c_o \end{pmatrix},$$

$$r(s,a) := \begin{pmatrix} -100\,c_r & 10 & -1 \\ 10 & -100\,c_r & -1 \end{pmatrix}.$$

In literature [4] $c_o = 0.85$, $c_r = 1$, $c_t = 0.5$ and a discount factor of $\gamma = 0.75$ are chosen. We compare belief control with two interesting direct policies:

$$\pi_1(z) = \begin{cases} 2 & \text{for } z = 1 \\ 1 & \text{for } z = 2 \end{cases} \qquad \text{and} \qquad \pi_2(z) = 3\ \forall z\ .$$

Smoothing Model

We introduce a further simple model with two states, two observations and two actions, to show a smoothing effect of belief control. The tiger problem can not show this effect because opening a door resets the system. We define:

$$T^{a=1} := \begin{pmatrix} 0.995 & 0.005 \\ 0.02 & 0.98 \end{pmatrix},\ T^{a=2} := \begin{pmatrix} 0.97 & 0.03 \\ 0.005 & 0.995 \end{pmatrix}$$

$$O^{z=1} := \begin{pmatrix} 0.9 & 0 \\ 0 & 0.1 \end{pmatrix},\ O^{z=2} := \begin{pmatrix} 0.1 & 0 \\ 0 & 0.9 \end{pmatrix},$$

$$r(s,a) := 20 \begin{pmatrix} -1 & -1 \\ 0 & 0 \end{pmatrix} + \begin{pmatrix} -1 & -2 \\ -1 & -2 \end{pmatrix}.$$

The reward is similar to the afterwards used CPAP model, because state $s = 1$ and actions that favor the state $s = 2$ are penalized.

3.1 Smoothing effect and discreteness of control variables

The properties of a control variable affect the corresponding optimal policy. By applying Bayes' theorem, the belief is the probability of the hidden state given a set of observations. Estimating the hidden state using the maximum of the belief leads to the Viterbi algorithm [16], [9] in the Hidden Markov case and to the Kalman filter [7], [6] in the case of a steady state space model. Since POMDP control is based on the belief, it shows the same optimal smoothing properties as the Kalman filter and the Viterbi algorithm. Therefore, by means of the POMDP framework, we are able to transfer optimality properties of parameter estimation procedures to the control of a dynamical system. This we call 'smoothing effect' of the belief.

Figure 1 shows that this optimal smoothing property of the belief leads to a smooth course of the action in contrast to direct control, in which the dependence on the observations leads to frequent changes in the applied action. Thus, if a lot of alterations in the action are undesirable, the belief control has to be preferred.

There is a drawback of discrete control variables: In the discrete case there is only a finite number of possible strategies. In contrast to continuous control variables, small alterations in the model cannot be adjusted. This becomes serious, for example, in the case where there are fewer observations than hidden states. In addition, a continuous control variable permits to control with cheap actions in the case of uncertainty. Figure 2 shows in which way, in contrast to direct control, belief control can be adjusted to changes in a parameter of the model. Similar plots can be achieved by changing other parameters of the model.

Belief control: Direct control:

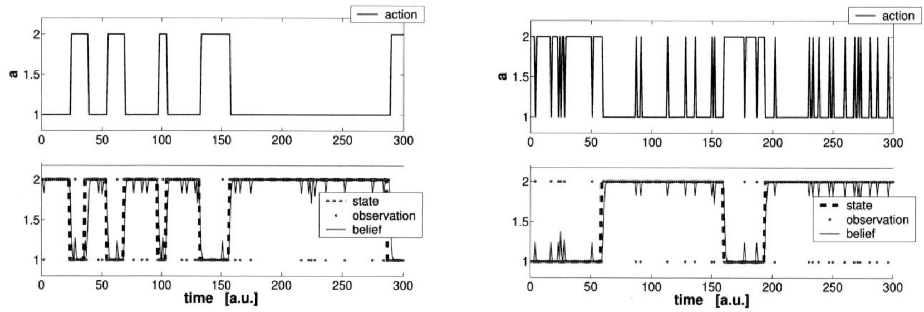

Fig. 1. The figures show typical realizations of belief control and direct control. Belief control (upper left plot) shows a clearly smoother policy than direct control (upper right plot), because of optimal smoothing properties of the belief. The colour version of this figure can be found in Fig. A.20 on page 587.

Belief control: Direct control:

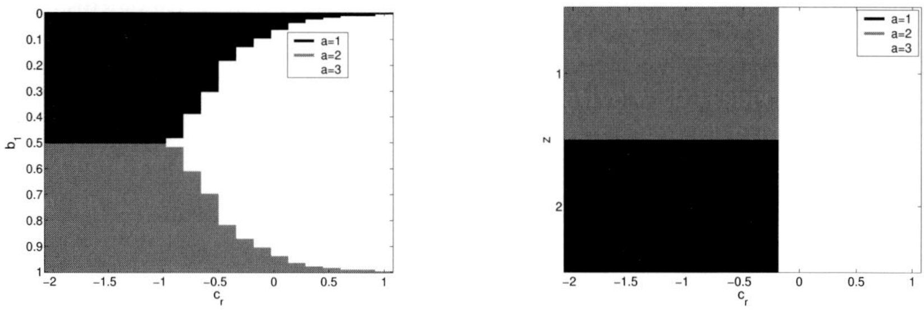

Fig. 2. This figures show the policy of belief control (left) and direct control in dependence on a model parameter, in this case c_r. The parameter c_r is a measure for the penalty of choosing the wrong door. The disadvantage of a discrete control variable is, that small changes of the model cannot be adjusted. If c_r increases above 0.2, the optimal direct control changes abruptly, whereas belief control changes continuously.

3.2 Performance of control variables

To compare the performance of different control variables it is necessary to assess policies independent of initial conditions. All possible values $V^\pi(s)$ are therefore weighted with $\rho(s|\pi)$. We assess the performance of a policy π by means of the expected reward

$$\langle R^\pi \rangle := \frac{\langle V^\pi \rangle}{\sum_t \gamma^t} \text{ with } \langle V^\pi \rangle = \sum_t \gamma^t r_{t_0+t} \ , \tag{10}$$

where t_0 is the length of a transient which ensures that initial states are distributed after that time like $\rho(s|\pi)$. We estimate this expected reward by a Monte-Carlo simulation.

The expectation value of the reward certainly depends on the observability of the process. Noise leads to lower values. Therefore, the accessible value is a monotonically increasinging function of this observation probability. In the case of the tiger problem $\langle R \rangle$ is a monoton increasing function of c_o. Figure 3 shows this dependency and the advantage of belief control versus the direct control $\pi_1(z)$ (dotted line) and the direct control $\pi_2(z) = 3$ (dashed line).

The advantage of belief control depends in a similar way on any other parameter of the model. Figure 4 shows the dependency of the expected reward on the penalty of choosing the wrong door in the case of the tiger problem and Figure 5 shows the performance of the strategies for various transition probabilities.

The belief constitutes an optimal control variable due to its continuity and the optimal smoothing properties. But in the case in which no further information can be achieved from the past of the process and discontinuity of a control variable is no disadvantage, the belief control becomes equivalent to the direct control. However, if one receives further information from past observations with respect to the current, the belief control becomes superior.

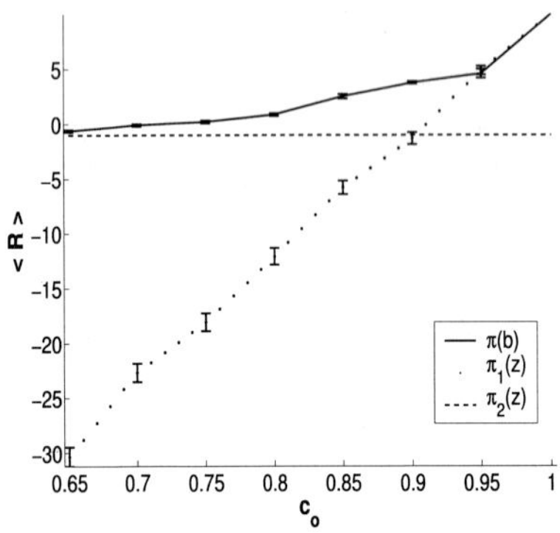

Fig. 3. Effect of observation probabilities on the performance of direct and belief control in the case of the tiger problem. In the absence of noise ($c_O = 1$, compare dotted and solid lines) and in case of no information ($c_O = 0.5$, cp. dashed and solid lines) about the hidden state, belief control and direct control become equivalent.

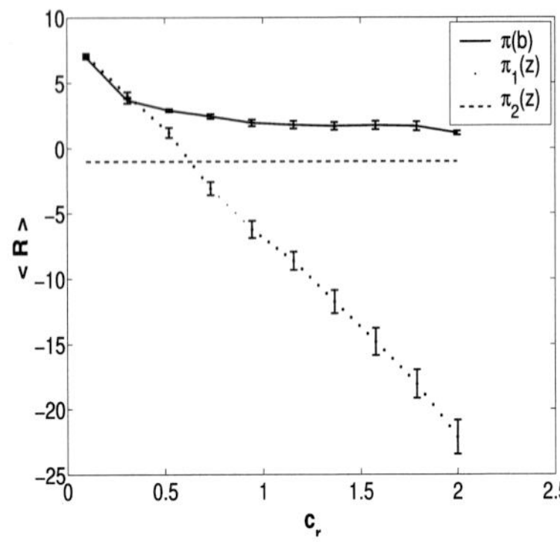

Fig. 4. Connection between the reward and performance of direct and belief control. Especially in the case of heavy penalties for choosing the wrong door (increasing c_r) direct control becomes worse.

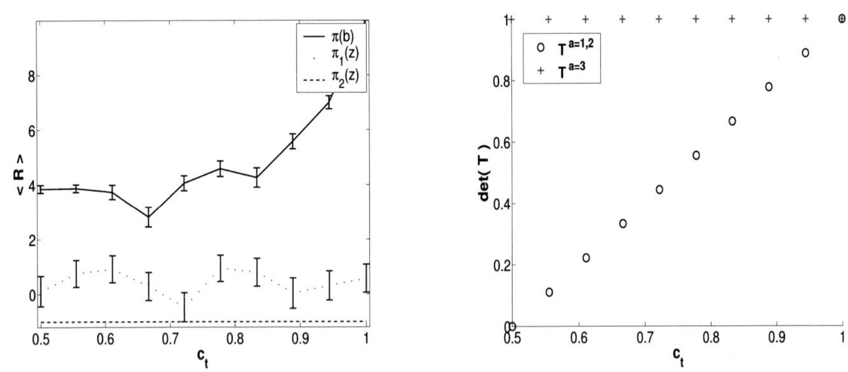

Fig. 5. Connection between transition probabilities and performance of direct and belief control in the case of the tiger problem. Increasing c_t leads to lower transition probabilities. This enables using information of the history of the process and therefore the advantage of belief control increases. Simultaneously, the absolute values of the determinants of $T^{a=1,2}$ become closer to 1.

Quantities which can serve as a measure for the information about the hidden states are the absolute values of the determinants of the transition matrices, $|\det(T^a)|$. It holds $0 \leq |\det(T^a)| \leq 1$. Note that in the case where the transition probability to both states is equal, i.e. if

$$T = \begin{pmatrix} 0.5 & 0.5 \\ 0.5 & 0.5 \end{pmatrix} , \qquad (11)$$

the determinant $\det(T)$ is zero. Thus, if $|\det(T^a)|$ is close to zero, the history of the process provides only little information about the present hidden state, if action a is performed. On the other hand, if $|\det(T)|$ is close to 1, the current state is determined by the past state. Therefore observations and actions of the history of the process can be used to predict the current hidden state. Thus, if the model contains transition matrices with small determinants the advantage of the belief control is presumably not large enough to justify the effort involved in solving the Bellman equation.

In the following section the CPAP model is discussed. We will see again that belief control is superior, but this is an example, where the determinant of one transition matrix is zero and therefore the performance of direct control is not much inferior to POMDP control.

4 Controlling the CPAP: An Application of POMDP

4.1 The model

Having discussed various strategies in order to get an insight into their characteristic capability we will discuss in this section realistic models for the CPAP-device. The CPAP (Continuous Positive Airway Pressure) is a respirator, that is used by people which suffer from an obstructive sleep apnoea syndrome. By such a device a small additive air pressure is applied to the patient through a mask. Different actions consist on choosing different possible pressures. We suppose that there are three pressures available (6, 8 and 10 mbar) which are denoted by action $a = 1$, $a = 2$ and $a = 3$.

There are two undesirable states with their corresponding observations: Apnoea ($s = 1$) and hypopnoea ($s = 2$).Taking into account the fact that there are special sleeping periods (for example in REM phase) where transitions to undesirable states are much lower than in other periods, we additionally suppose two healthy states, but they are not distinguishable by observations ($s = 3, 4$). Thus the hidden Markov model is an aggregated one. The observations $z = 1, 2, 3$ correspond to classes obtained by a classification of the measured air flow and applied pressure [12].

In addition we introduce two extreme cases of affection, S (sligth) and H (heavy), which determines the transition probabilities to undesired states.

With help of experts we suppose the following transition model:

$$T_S^{a=1} := \begin{pmatrix} 1.0000 & 0 & 0 & 0 \\ 0.0500 & 0.9492 & 0.0008 & 0 \\ 0.0333 & 0.0333 & 0.9308 & 0.0025 \\ 0.0033 & 0.0033 & 0.0025 & 0.9908 \end{pmatrix}$$

$$T_H^{a=1} := \begin{pmatrix} 1.0000 & 0 & 0 & 0 \\ 0.0500 & 0.9492 & 0.0008 & 0 \\ 1.0000 & 0 & 0 & 0 \\ 0.1000 & 0 & 0 & 0.9000 \end{pmatrix}$$

$$T^{a=2} := \begin{pmatrix} 1.0000 & 0 & 0 & 0 \\ 0.0250 & 0.9733 & 0.0017 & 0 \\ 0.0017 & 0.0017 & 0.9942 & 0.0025 \\ 0.0002 & 0.0002 & 0.0025 & 0.9972 \end{pmatrix}$$

$$T^{a=3} := \begin{pmatrix} 0 & 0 & 1.0000 & 0 \\ 0 & 0 & 1.0000 & 0 \\ 0.0008 & 0.0008 & 0.9958 & 0.0025 \\ 0.0033 & 0.0001 & 0.0025 & 0.9941 \end{pmatrix}$$

A general degree of affection d can be introduced by interpolation leading to the transition matrix

$$T_d^{a=1} := d\, T_H^{a=1} + (1-d)\, T_S^{a=1}\,, d \in [0,1]\ .$$

This distinction is only necessary when small pressures are applied. Furthermore, we assume the following observation model:

$$O^{z=1} := \begin{pmatrix} 0.9 & 0 & 0 & 0 \\ 0 & 0.1 & 0 & 0 \\ 0 & 0 & 0.1 & 0 \\ 0 & 0 & 0 & 0.1 \end{pmatrix} O^{z=2} := \begin{pmatrix} 0.08 & 0 & 0 & 0 \\ 0 & 0.6 & 0 & 0 \\ 0 & 0 & 0.2 & 0 \\ 0 & 0 & 0 & 0.2 \end{pmatrix}$$

$$O^{z=3} := \begin{pmatrix} 0.02 & 0 & 0 & 0 \\ 0 & 0.3 & 0 & 0 \\ 0 & 0 & 0.7 & 0 \\ 0 & 0 & 0 & 0.7 \end{pmatrix} .$$

There are two targets. We aim at a low mean pressure and at as few as possible apnoea states. In this situation $r(s,a)$ is a sum of $r(s)$ and $r(a)$:

$$r(s,a) = r(s) + r(a)\ . \tag{12}$$

The penalty of action a should be proportional to the mean pressure and the penalty of an apnoea should be proportional to the sum of all apnoea. One apnoea state should be penalized in the same way as an increase of the pressure of 10 mbar. Therefore we are weighting the part of $r(s)$ with a factor of 5 and we have $r(s,a) = r(a) + 5\,r(s)$.

$$r(a) := \begin{pmatrix} -6 & -8 & -10 \\ -6 & -8 & -10 \\ -6 & -8 & -10 \\ -6 & -8 & -10 \end{pmatrix}, r(s) := \begin{pmatrix} -1 & -1 & -1 \\ 0 & 0 & 0 \\ 0 & 0 & 0 \\ 0 & 0 & 0 \end{pmatrix}, r(s,a) = \begin{pmatrix} -11 & -13 & -15 \\ -6 & -8 & -10 \\ -6 & -8 & -10 \\ -6 & -8 & -10 \end{pmatrix}$$

We use an infinite time horizon $T \to \infty$ and a discount factor of $\gamma = 0.9$. A discount factor $\gamma = 1$ would be desirable but with respect to the convergence of (5) we have choosen $\gamma < 1$.

4.2 Comparison of the solutions

Now we solve the control problem for the proposed models for the CPAP device. Similar to section 3.2 we calculate and compare the performance of direct control and belief control. As mentioned, the optimal POMDP strategy is obtained by using the 'Incremental Pruning'. The results are shown in table 1.

Table 1. Optimal direct control (upper row) and optimal belief control (figures below) for CPAP in the case of slight affection (left) and heavy affection (right). The colour version of the figures can be found in Fig. A.21 on page 587.

Slight affection	Heavy affection
$\pi(z_t) = \begin{cases} 3 & \text{for } z_t = 1 \\ 1 & \text{for } z_t = 2 \\ 1 & \text{for } z_t = 3 \end{cases}$	$\pi(z_t) = \begin{cases} 3 & \text{for } z_t = 1 \\ 2 & \text{for } z_t = 2 \\ 2 & \text{for } z_t = 3 \end{cases}$
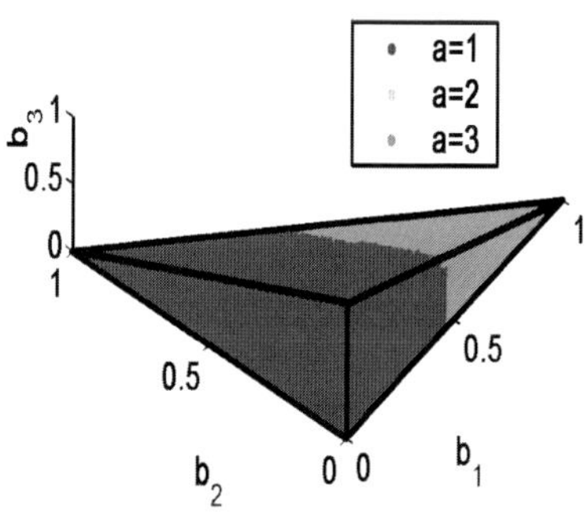	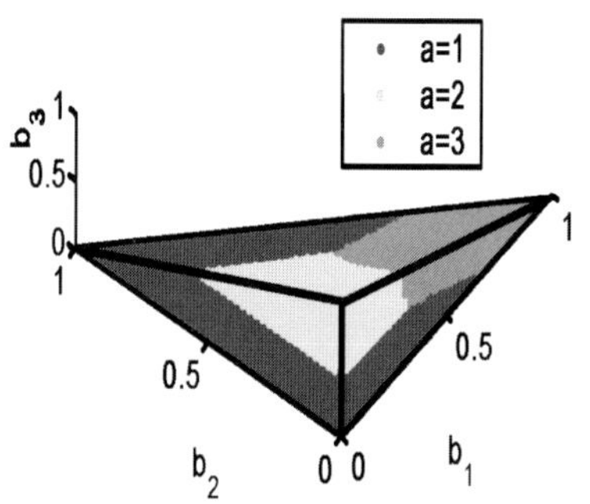

In the case of slight affection, direct control and belief control are quite similar. Only actions $a = 1$ and $a = 3$ are used to treat a slightly affected patient. In the case of heavy affection, direct control clearly differs from belief control. If direct control is used to treat a heavily affected patient the low pressure $a = 1$ will never be applied.

These solutions are compared by using the expected reward (10) of a strategy. To measure the smoothing effect, we look at the alterations of the pressure $\sum_t |a_t - a_{t-1}|$. Too many changes affect the behavior of the patient during sleep in a negative way. A comparison of direct control and POMDP control

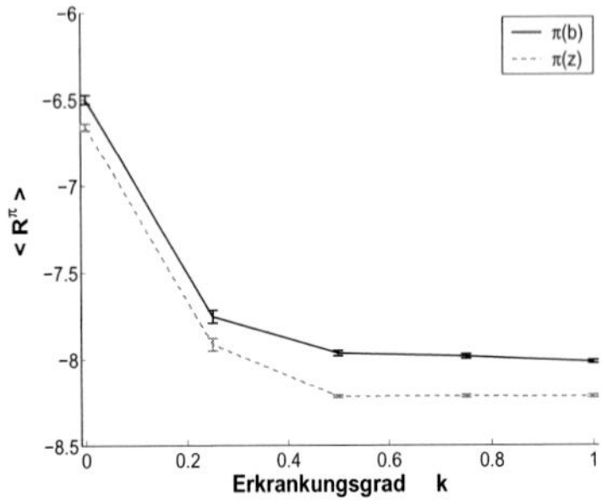

Fig. 6. Comparison of direct control and belief control for different degrees of affection d. Belief control shows a significant better performance $\langle R \rangle$.

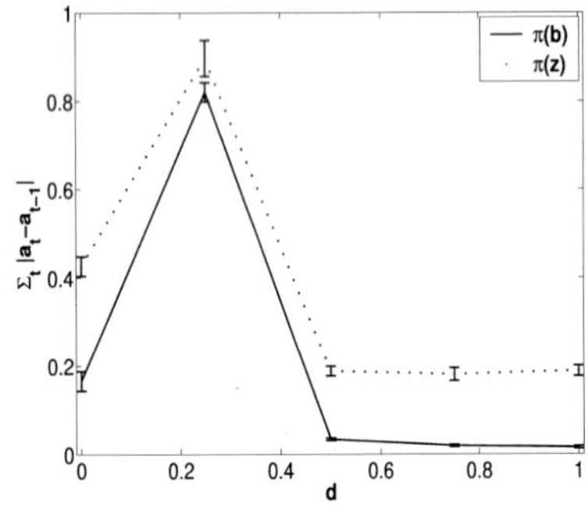

Fig. 7. Smoothing effect in the case of the CPAP model. Belief control $\pi(b)$ decreases the difference of following actions $\sum_t |a_t - a_{t-1}|$ in comparison to direct control $\pi(z)$.

shows, that both strategies lead to the same number of apnoea states. But in the case of POMDP control, the applied mean pressure is about 0.2 mbar lower than the mean pressure in the case of direct control. Therefore POMDP control has a superior expected reward (see figure 6). Furthermore we can see that, as expected, belief control produces fewer alterations in pressure than direct control.

But, especially in the case of slight affection, the difference to direct control becomes small. This is one example for a case where POMDP control becomes similar to a strategy based directly on observations.

5 Conclusion

We compared POMDP control with direct control strategies in dependence of parameters of a POMDP model. The advantage of POMDP control is substantial in general. Especially the smoothness of the action variable within the belief control is notable, it has the same source as the Kalman-filter or the Viterbi algorithm. However, there are cases, where the performance of the belief control does not differ too much from some direct control. Such cases can be detected by the property, that the determinant of some transition matrices is close to zero. That means that the information from the past is very low and then the belief as control variable is not very helpful.

We used this experience for the discussion of the POMDP-control for the CPAP-device. We showed, that the belief control is also superior to the common used direct control. By using the new strategy, in most cases the mean pressure, necessary for avoiding the apnoea events, can be reduced significantly. When there is only a slight affection, however, this difference between the two strategies may become negligibly small.

References

[1] M.L. Littman A. Cassandra and N.L. Zhang. Incremental pruning: A simple, fast, exact method for partial observable markov decision processes. *Proceedings of the Thirteenth National Conference on Uncertainity in Artificial Intelligence*, pp 54–61, 1997.

[2] R.E. Bellman. Dynamic programming. *Princeton University Press, Princeton, N.Y.*, pp 1–70, 1957.

[3] A. Cassandra. Exact and approximate algorithms for partially observable markov decision processes. *PhD thesis, Brown University*, 1998.

[4] A. Cassandra and M.L. Littman. Acting optimally in partially observable stochastic domains. *Proceedings of the Twelfth National Conference on Uncertainity in Artificial Intelligence*, pp 1023–1028, 1994.

[5] H.-T. Cheng. Algorithms for partially observable markov decision processes. *PhD thesis, University of British Columbia*, 1988.

[6] G.Welch and G. Bishop. An introduction to the kalman filter, March 2002.

[7] R.E. Kalman. Planning and acting in partial observable stochastic domains. *Transactions of ASME-Journal of Basic Engineering*, 82:35–45, 1960.

[8] M. L. Littman. The witness algorithm: Solving partially observable markov decision processes. Technical Report CS-94-40, 1994.

[9] L.R. Rabiner. A tutorial on hidden markov models and selected applications in speech recognition. *IEEE*, 77:257–285, 1989.

[10] E. Sondik. The optimal control of partially observable markov decision processes. *Operations Research*, 1971.

[11] E. Sondik. The optimal control of partially observable markov processes over the inifinte horizon: Discounted costs. *Operations Research*, 26(2):282–304, 1978.

[12] R. Staats, H. Steltner, M. Vogel, J. Guttmann, H. Matthys, J. Timmer, and J.C. Virchow. Classification of sleep apnoeas using nasal pressure respiratory input impedance and cardiogenic oscillations. *Respiratory Journal*, 18:205s, 2002.

[13] H. Steltner. *Data Analysis in Respiratory Medicine.* PhD thesis, University of Freiburg, 1/2001.

[14] H. Steltner. Diagnosis of sleep apnea by automatic analysis of nasal pressure and forced oscillation impedance. *American Journal Respiratory Crit Care Med*, 165:940–944, 2002.

[15] C. Sullivan. Reversal of obstructive sleep apnea by continuous positive airway pressure. *Lancet*, 1:862–865, 1981.

[16] A. J. Viterbi. Error bounds for convolutiogal codes and an asymptotically optimal decoding algorithm. *IEEE Trans. Informat. Theory*, 13:260–269, 1967.

Implementing Hydrodynamic N-Body Codes on Reconfigurable Computing Platforms

Gerhard Lienhart

Department of Computer Science V, Chair Prof. Dr. Reinhard Männer
University of Mannheim, B6-23-29A, 68131 Mannheim, Germany
lienhart@ti.uni-mannheim.de

Summary. N-Body codes for scientific applications like astrophysics are usually highly demanding in computing power, and even the application of cutting edge computer technologies still doesn't satisfy the need of calculation power for lots of most interesting computational problems. This work investigates the utilization of new computing methods based on reconfigurable logic devices in order to overcome the limitations of current computer technology in the case of hydrodynamic N-Body simulation. The implementation of the central part of a state-of-the-art hydrodynamic simulation code with a resulting performance of 3.9 GFlops on a single reconfigurable chip is presented which demonstrates the prospects of this approach.

1 Introduction

With the enormous development of computer technology in the last years, the computational investigation of physical models became a research field of major importance. In this paper we focus on the treatment of physical problems where hydrodynamic effects in three-dimensional physical systems with very high contrasts in density are to be simulated. Our current work deals with N-Body simulations for astrophysics, where the dynamical behavior of galaxy collisions shall be simulated. For such systems the so called smoothed particle hydrodynamics method (SPH) has shown to be a very powerful algorithm with a simple computing structure and nice dynamical behavior of the simulated system. It fits well together with the simulation of gravitation. In our case highly efficient special-purpose computers called GRAPE (GRAvity PipE) are applied to calculate the gravitational forces (see [1]). These machines, developed at the university of Tokyo, are based on ASICs (Application Specific Integrated Circuits) which consist of highly parallelized calculation units specialized for the gravitational interaction. Because of the specialization on one fixed application these chips perform extremely fast and cost efficient – a single modern GRAPE-board has a peak performance of 1 TFlops.

Using standard computing technology for treating SPH in astrophysical simulations including long range gravitational interaction has shown severe inefficiencies. Single node computers are not fast enough but it is hard to gain a speedup by parallelizing such simulation systems due to a communication bottleneck. The scaling of speedup of mainstream computing hardware with time, as it is expected for the next years, will not allow to treat the most interesting simulation problems we are facing. The computing power for SPH simulations could be increased to the highest possible level by applying special-purpose computing platforms with optimized data paths and parallelized calculation units like in the case of GRAPE, but the diversity of SPH-Algorithms foils this approach. Recent progress in Field Programmable Gate Array (FPGA) technology opened the door for building reconfigurable machines based on these chips which can be used like special-purpose processors while being applicable for several computing problems. These chips offer an alternative design strategy besides constructing CPU- or ASIC-based machines, as we can build optimally suited high performance arithmetic units with these devices while being flexible to reconfigure the hardware at any time. We applied this new technique for building an accelerator for SPH.

In this paper the algorithm which is to be accelerated will be briefly presented and the FPGA-based platform on which we performed our implementations and tests will be introduced. The focus will be on the description of the requirements and methods of the implementation on the prototype platform. The achieved performance will be shown and we will conclude with some remarks of further work and our expectations for the future.

2 Algorithm

The algorithm which became implemented on the reconfigurable computing platform is part of astrophysical N-Body simulations where gas-dynamical effects are treated by the smoothed particle hydrodynamics method (SPH). Here this method will only be described very briefly. A thorough review of this method has been published by W. Benz [5]. In SPH the gaseous matter is represented by particles, which have a position $\mathbf{r}_i$, velocity $\mathbf{v}_i$ and mass m_i. To form a continuous distribution of gas from these particles they become smoothed by convoluting their discrete mass distribution with a smoothing kernel W which is a strongly peaked function around zero and is non-zero only in a limited area. Then the discretized formula for the density ρ_i is given by

$$\rho_i = \sum_{j=1}^{N} m_j W(\mathbf{r_i} - \mathbf{r_j}, h_{ij}). \tag{1}$$

The smoothing kernel has a parameter h which controls the width of the smoothing function. This allows for adapting the smoothing to the density of particles.

According to the SPH method any gas-dynamical variables and even their derivatives can also be calculated by a simple summation over the particle data multiplied by the smoothing kernel or its derivative. This enables a simple calculation scheme for the hydrodynamic equations which are subject to simulation. Equation 2 shows the physical formula for the time derivative of the velocity ($\mathbf{v}$) given by the force from the gradient of the pressure (P) and the so-called artificial viscosity $\mathbf{a_i}^{visc}$. This artificial viscosity is needed to be able to simulate the behavior of shock waves.

$$\frac{d\mathbf{v_i}}{dt} = -\frac{1}{\rho_i}\nabla P_i + \mathbf{a_i}^{visc} \tag{2}$$

The SPH method transforms this equation to the following formulation, which is only one of several possibilities

$$\frac{d\mathbf{v_i}}{dt} = -\sum_j m_j \left(\frac{P_i}{\rho_i^2} + \frac{P_j}{\rho_j^2} + \Pi_{ij} \right) \nabla_i W\left(|\mathbf{r_{ij}}|, h_{ij} \right). \tag{3}$$

The term Π_{ij} is due to the artificial viscosity and is a complicated function of the particle data, highly dependent on the used SPH formulation. Formula 1 and 3 exhibit the same structure of computation and lead to the hardware scheme shown in Fig. 1, where the particle data input – here generally called x_k – is stored locally in RAM.

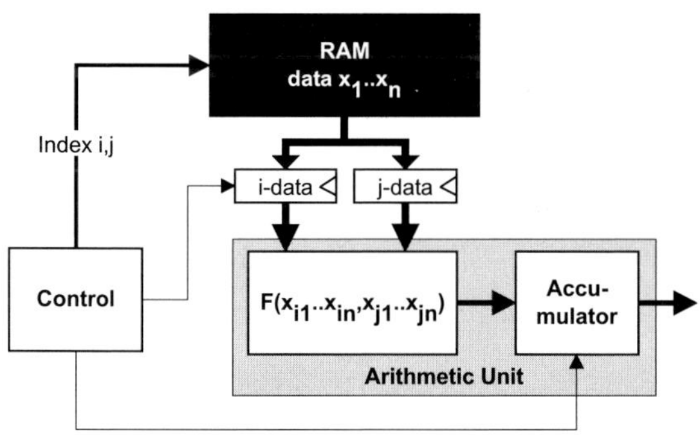

Fig. 1. Hardware scheme for abstract computation pattern

For the arithmetic unit floating-point calculation is needed, because the astrophysical variables span an enormous range. As the resources of an FPGA are limited, the impact of floating-point precision on hardware implementation is a critical aspect, and will therefore be discussed in section 4. But before our current hardware platform will be introduced.

3 Prototype Hardware Architecture

As mentioned in Sect. 1 our decision of using FPGAs for tackling the challenge of SPH-based simulation is motivated by the requirement of introducing new special-purpose hardware in order to outperform the mainstream technology, but which is also applicable for different SPH formulations.

FPGAs consist of an array of configurable logic elements which can perform combinatorial as well as sequential logics and are usually enhanced by additional arithmetic resources like carry chains for building fast adders. The logic elements become interconnected by a configurable routing network. With these building blocks arbitrary digital circuits can be implemented, only limited by the size of the FPGA. With their programmable logic resources, the design of hardware architectures is similar to ASIC design. But these chips can be reconfigured at any time in a fraction of a second, which makes them almost as flexible as CPUs (you can even embed a CPU with the FPGA resources).

The prototype platform which is being used for SPH implementation is a FPGA processor PCI-card, which has been developed in our institute (see Fig. 2, [8]). The card is based on the 66 MHz, 64 bit PCI bus specification. It contains a modern Virtex2 FPGA (XC2V3000), four 36 bit wide 133 MHz SRAM banks and a connector to a standard 133 MHz SDRAM bank. Moreover it has two connectors to daughter cards and board-to-board interfaces for multiple-board designs. For Xilinx FPGAs of the Virtex Series, the basic configurable logic elements are called 'Slices' and the present FPGA has 14.336 of them.

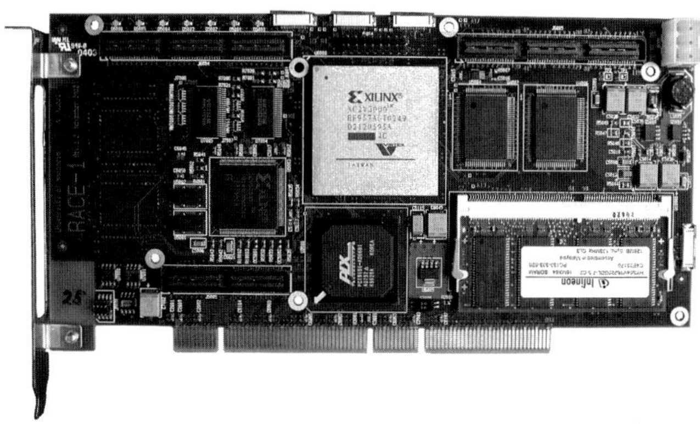

Fig. 2. Target platform – PCI plug in card with Virtex2 FPGA

We supplemented this platform with an interface board for connecting directly to the special-purpose hardware GRAPE which treats the gravitational interaction of the simulated particles (see Fig. 3). This approach of routing the data flow through our FPGA-board saves communication bandwidth on the PCI-bus, as the particle data is needed both for SPH and gravitation. Moreover some preprocessing of data can be performed in the FPGA which results in less overhead for communication with the special-purpose hardware.

Fig. 3. Structure of hybrid computing system

4 SPH Implementation

4.1 Floating-Point Arithmetic on FPGAs

As already mentioned we need floating-point arithmetic for calculating the SPH formulas. This requirement is due to the enormous density contrasts in astrophysical problems. The arithmetic precision we need for the calculation units is critical for the implementation as the amount of FPGA resources for the operators is directly related with the precision. To illuminate the problem of resource requirements we have to discuss the hardware aspects of floating-point arithmetics.

Figure 4 shows the general composition of a floating-point number (the IEEE 754 standard for single precision numbers gives: ExpBits=8, SignifBits=24). Accordingly such a number consists of sign, biased exponent (Exp) and a fractional part (Fract). The latter is the mantissa of the floating-point number without the most significant bit, which always equals one for normalized numbers. The formula for the real number which is represented by this form is given by

$$f = \left\{ \begin{array}{ll} +1 & \text{for Sign}=0 \\ -1 & \text{for Sign}=1 \end{array} \right\} \times \left(1 + Fract \times 2^{-(SignifBits-1)}\right) \times 2^{Exp-bias} \tag{4}$$

with $bias = 2^{(ExpBits-1)} - 1$.

In Fig. 5 the abstract layout of a floating-point operator is shown. Generally we can divide the processing into the treatment of sign, exponents and mantissas, which may interact. While the operations on the sign and exponents are comparatively cheap, the operations on the mantissa are utilizing most of the resources. The contrast between floating-point operations and

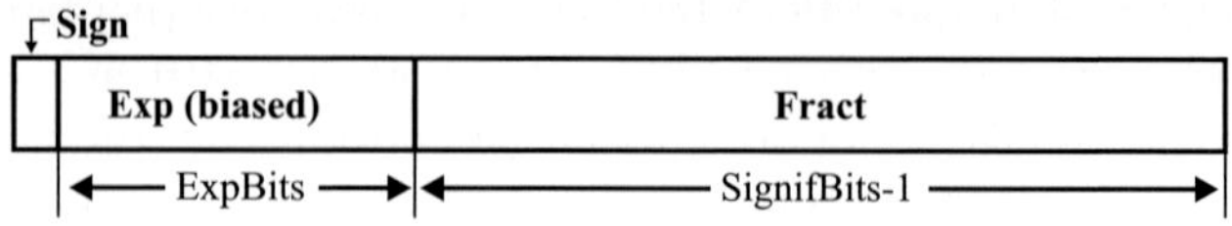

Fig. 4. General layout of a floating-point number

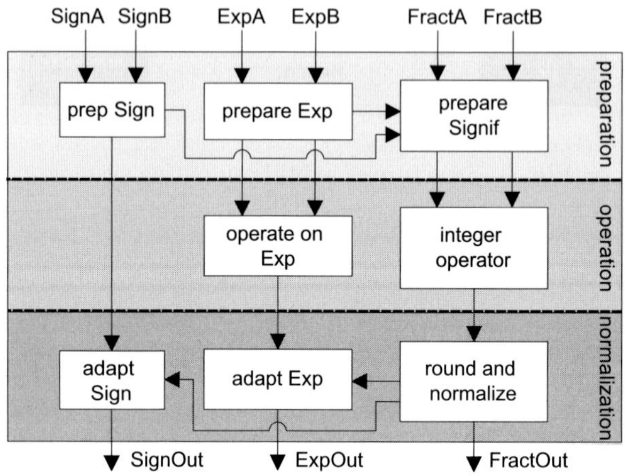

Fig. 5. General structure of floating-point operator

fixed-point operations can be best illustrated in the case of the floating-point adder. There the mantissa of one of the input numbers has to become shifted when the operands have different exponents. Also normalization after integer summation or subtraction (depending on signs) of the preprocessed mantissas may be required if a MSB carry or digit-cancellation occurs. This leads to the fact that floating-point adders utilize several times the resources of an integer adder. The resource scaling is proportional $N \times log\ N$ with the number of significand bits. Floating-point operators like multipliers, dividers, square-root and log-units are clearly dominated by the resources for the corresponding integer operators. For these elements parallel implementations for one result per clock cycle exist. Here, the scaling of the required hardware resources on the FPGA is roughly quadratic.

In order to be be able to easily adapt the internal precision for different applications, a parameterized library of floating-point operators for FPGA-synthesis was built. Tables 1 and 2 show the results in respect of resource utilization and applicable clock frequencies for the most important floating-point operators with different levels of accuracy (width of significand varies from 16 to 24 bits) and pipelining depth. In the left Table 1 the difference of logic requirements between unsigned and signed addition of almost a factor of two is a result of the additional pre- and postprocessing for normalized input and output numbers in a signed adder, where additionally two's complement and shifter operations have to be performed. The right side of the right Table 1 demonstrates the dramatical saving of basic logic elements when hardwired block multipliers are present, as they are in the Virtex2 FPGAs. The numbers in Table 2 show the impact of increasing the pipeline depth of the divide and square root unit on the utilized resources and the achieved clock frequency.

Table 1. Resource utilization and speed of floating-point adder (left) and multiplier (right) for XC2V3000-4 FPGA

Unsigned add.			Signed add.		
Signif bits	Slices	Design freq. (MHz)	Signif bits	Slices	Design freq. (MHz)
16	111	112	16	195	76
18	119	109	18	222	93
20	126	101	20	232	68
22	137	100	22	260	74
24	152	106	24	290	85

No block mult.			With block mult.		
Signif bits	Slices	Design freq. (MHz)	Signif bits	Slices	Design freq. (MHz)
16	193	97	16	30	76
18	246	69	18	61	61
20	287	73	20	66	66
			22	72	66
			24	78	63

Table 2. Resource utilization and speed of floating-point divider (left) and square root (right) for XC2V3000-4 FPGA

4 quotient bits per pipe stage			2 quotient bits per pipe stage		
Signif bits	Slices	Design freq. (MHz)	Signif bits	Slices	Design freq. (MHz)
16	224	50	16	262	73
18	269	50	18	320	70
20	328	50	20	384	61
22	382	44	22	454	70
24	452	44	24	530	61

4 sqrt bits per pipe stage			2 sqrt bits per pipe stage		
Signif bits	Slices	Design freq. (MHz)	Signif bits	Slices	Design freq. (MHz)
16	129	54	16	144	93
18	152	51	18	175	86
20	188	53	20	209	86
22	216	50	22	246	76
24	255	50	24	286	78

4.2 Design Strategy

In section 2 the general computing scheme of our application has been put out. The core of the scheme shown in Fig. 1 is the arithmetic unit. It will now be discussed, how the central part – the computing unit for $F(x_{i1}..x_{in}, x_{j1}..x_{jn})$ – becomes implemented.

The most efficient and also straightforward strategy of implementing a special-purpose arithmetic unit for computing one specific formula is to build a completely parallel architecture where each operation of the formula is mapped to a separate operator in hardware. Here we have an algorithm where there is no feedback of intermediate results to the processing of further results. Therefore the latency of the arithmetic unit doesn't matter for the performance of the system. These are optimal preconditions for a deeply pipelined and therefore fast design. We can layout the data-flow of input and intermediate data with the operators at the nodes and we get a fully parallel calculation unit with sustained performance, as long as the flow of input data continues.

4.3 Design Sample of SPH Application

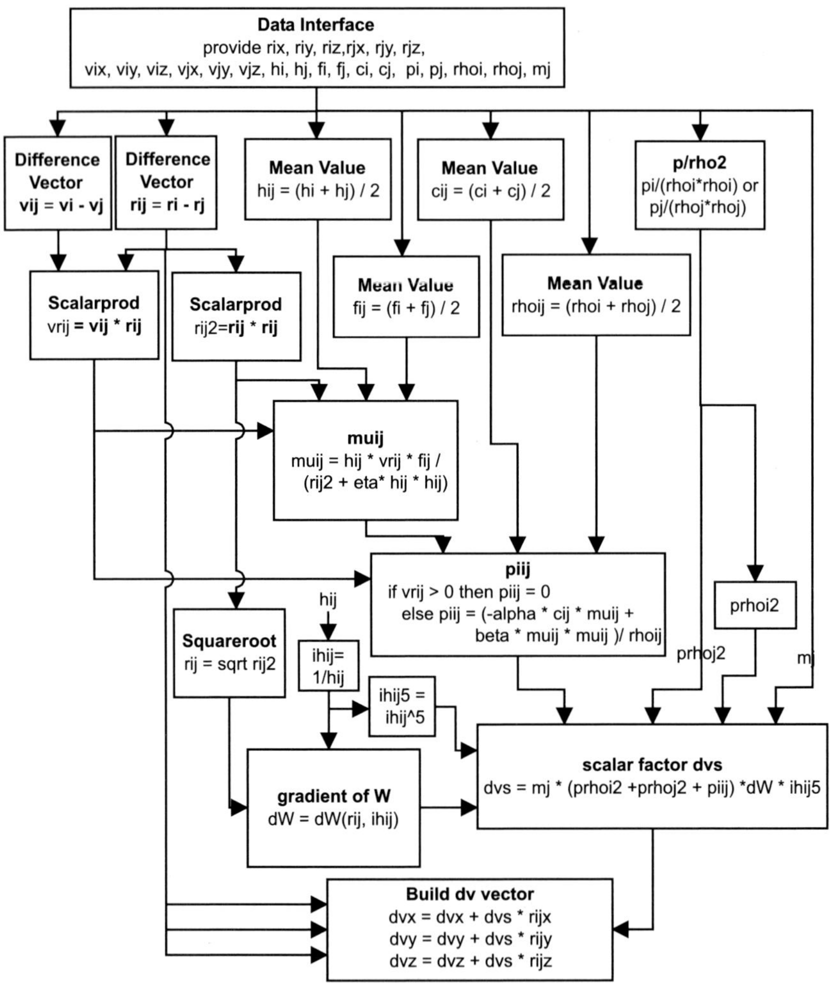

Fig. 6. Structure of pipeline for calculating the acceleration formula of SPH

The proposed design will be illustrated by its application on the most complex formula of our current simulation application – the calculation of the particle accelerations. In section 2 the formula was given in (3). Completely broken down to elementary operators we find that a fully parallel computation unit requires 60 floating-point operators, including expensive operations like dividers (4) and square-root (1). The flow-graph for the design is shown in Fig. 6. For calculating the terms for the artificial viscosity, parameters like sound speed (c), density (ρ = rho), pressure (p), smoothing length (h) and the Balsara parameter f are required individually for each particle in addition to the variables for mass, position and velocity.

5 Performance

The total design for the circuit outlined in Fig. 6 with a mantissa width of 16 bits fits in 49 % of the logic resources of a Xilinx Virtex2 XC2V3000-4 FPGA and is able to process one input data set per clock cycle at 65 MHz design frequency. With 60 parallel floating-point units we get a performance of 3.9 GFlops. On the other hand the floating-point performance for our kind of application using general-purpose processors is around 150–300 MFlops, so this result means a substantial speedup.

6 Conclusion

The paper outlined the application of reconfigurable computing technology for accelerating the hydrodynamic part (SPH) of astrophysical N-body simulation. The hardware and technique of using FPGAs as reconfigurable processing elements for treating complex formulas of a structurally simple algorithm has been introduced. The capabilities and limitations of floating-point arithmetic on FPGAs were presented and the important aspect of resource requirements depending on operator precision has been worked out. The key data of the parameterized floating-point library were given for providing an estimation base for future applications. The performance of the approach was demonstrated with the implementation of the formula for the time derivative of the velocity. The Design fit in an off-the-shelf FPGA and achieves a performance of 3.9 Gflops.

This work is the base of a variety of future developments and potentialities. The current hardware implementations are prototypes for a big scale future system consisting of multiple reconfigurable computing boards with several GRAPE-boards attached. This system shall provide superior computing performance for astrophysical N-body simulations as well as similar structured simulation tasks e.g. molecular dynamics.

7 Acknowledgments

This research work has been performed at the Dept. of Computer Science V of the University of Mannheim, Chair Prof. Dr. R. Männer in cooperation with Dr. R. Spurzem, Astronomisches Recheninstitut, Heidelberg, Dr. A. Burkert and M. Wetzstein, Max-Planck Institute for Astronomy, Heidelberg.

References

[1] Toshikazu Ebisuzaki et al.: GRAPE Project: An Overview. Publications of the Astronomical Society of Japan, 1993, **45**, 269–278 (1993)
[2] Spurzem, R., Kugel, A.: Towards the Million-Body Problem on the Computer - no news since the three-body-problem? Molecular Dynamics on Parallel Computers, Proc. Workshop NIC Juelich, World Scientific Press, Singapore, (1999)
[3] Cook, T.A., Kim, H.-R., Louca, L.: Hardware Acceleration of N-Body Simulations for Galactic Dynamics. Proc. SPIE2607 on FPGAs for Fast Board Development and Reconfigurable Computing, 42–53 (1995)
[4] Hamada, T., Fukushige, T., Kawai, A., Makino, J.: PROGRAPE-1: A Programmable, Multi-Purpose Computer for Many-Body Simulations. Publ. of the Astronomical Society of Japan, **52**, 943–954 (2000)
[5] Benz, W.: Smooth Particle Hydrodynamics: A Review. J.R.Buchler(ed), The Numerical Modelling of Nonlinear Stellar Pulsations. Kluwer Academic Publishers, 269–288 (1990)
[6] Bate, M.R., Burkert, A.: Resolution Requirements for Smoothed Particle Hydrodynamics Calculations with Self-gravity. Mon. Not. R. Astron. Soc., **288**, 1060–1072 (1997)
[7] Kuberka, T., Kugel, A., Männer, R., Singpiel, H., Spurzem,R., Klessen, R.: AHA-GRAPE: Adaptive Hydrodynamic Architecture - GRAvity PipE. Proc. International Conf. on Parallel and Distributed Processing Techniques and Applications, **3**, 1189–1195 (1999)
[8] Kugel, A.: MPRACE – A PCI-64 based High Performance FPGA Co-Processor. http://www-li5.ti.uni-mannheim.de/fpga/?race/ (2003)
[9] Lienhart, G., Kugel, A., Männer, R.: Using Floating-Point Arithmetic on FPGAs to Accelerate Scientific N-Body Simulations. Proc. FCCM'02, 182–191 (2002)
[10] Parhami, Behrooz: Computer Arithmetic, Algorithms and Hardware Designs. Oxford University Press (2000)

Design of a Noncausal FIR Model Inverse as a Compensator in Repetitive Control

Richard W. Longman and Benjamas Panomruttanarug

Columbia University, New York, 10027 NY, USA
RWL4@columbia.edu
bp270@columbia.edu

Summary. A new method of designing compensators for repetitive controllers is presented. The ideal compensator is a filter that is the inverse of the plant, but this is usually unstable, and therefore cannot be used in practice. The approach used here works on a restricted class of transfer functions, and bypasses this difficulty by making a noncausal FIR model of the plant inverse. This model has poles only at the origin, and is therefore stable. Methods are presented to adjust the three parameters of the design for stability and good learning speed, i.e. the repetitive control gain, and the number of causal, and the number of noncausal gains chosen to compose the finite impulse response model. A third order system is studied, which models the closed loop behavior of one link of a commercial robot. One can produce a stable design with a number of gains ranging from 11 to 15 (this gives the number of real time computations for control update), and these numbers are not particularly sensitive to sample rate. Using 18, 20, or 30 gains can produce a quite reasonable plant inverse model that gives fast learning in repetitions at all frequencies.

1 Introduction

Repetitive control is a relatively new field that aims to create control systems that can produce zero tracking error of periodic commands, or that eliminate the effects of periodic disturbances of known period [1-6]. The original application in [1] was to particle physics accelerators to eliminate the effects of 60 Hz and harmonics in the rectified voltage driving the magnets. Perhaps the majority of applications are to motion control systems and mechatronics, including the use for improved positioning control in computer disk drives for higher density storage, for improved mechanical machining, for timing belt drive systems in copy machines, and for many applications of robots doing repetitive motions. Normally a repetitive controller is an extra loop placed around a feedback control system. The command or disturbance is periodic, and the repetitive controller looks back one period, and based on the error observed, adjusts the command for the current time step, aiming to converge to zero error. The major impediment to applications of repetitive control is the difficulty in obtaining a stable learning process. It is the purpose of this paper to develop methods of designing a finite impulse response (FIR) filter that approximates the plant inverse, as a method of designing compensators for repetitive controllers.

2 Mathematical Formulation and Stability of Repetitive Control Systems

Repetitive control makes use of stored information from the previous period, and hence is necessarily discrete time. Suppose that the period of the command, or the period of the disturbance we wish to cancel is p time steps. Let $G(z)$ be the z-transfer function of the feedback control system, with command input $u(k)$ or $U(z)$ in the transform domain, and response $y_0(k)$ or $Y_0(z)$. The periodic disturbance if present might appear, for example, before the plant in this feedback control system, but wherever it appears there is an equivalent periodic disturbance $d(k)$ or $D(z)$ that one can add to the output of $G(z)$ to represent its influence. Then the actual output is $y(k) = y_0(k) + d(k)$. Now we place a unity feedback repetitive control loop around this. The command is the desired output $y_d(k)$, and the error signal is $e(k) = y_d(k) - y(k)$. The repetitive control law forms the command to the feedback controller $u(k)$ based on this error information. The feedback control system is the plant in the repetitive control loop. The simplest form of repetitive control can be stated in words as follows, for a robot application: if the robot link was two degrees too low at this time in the previous period, add two degrees, or a repetitive control gain φ times two degrees, to the command in this repetition. In mathematics this is written as

$$u(k) = u(k-p) + \varphi e(k-p+1) \tag{1}$$

The one time step shift in the argument of the error is based on the assumption that there is a one time step delay from the time of a change in the command given $G(z)$ to the first time a change in the output is observed. The z-transform of the control law can be written as

$$U(z) = z^{-p}\left[U(z) + \varphi F(z)E(z)\right] \tag{2}$$

It will become clear below that repetitive control law (1) is usually unstable for all gains. The main situations when it can work are first order plants and plants (feedback control systems) which in continuous time have relative degree one. Hence, in (2) the simple time shift z has been replaced by $F(z)$ which is a compensator designed to produce stability. This paper develops a method of generating such a compensator. In practice one often needs in addition a zero phase low pass filter on the entire control signal to make the repetitive controller robust to model inaccuracies or parasitic poles at high frequencies [7]. In this paper we concentrate on the design of $F(z)$. The better this design is, the higher the frequency cutoff can be, and with a good enough model one might be able to do without a cutoff filter.

Combining the transfer functions around the repetitive control loop produces the following difference equation for the error

$$\left\{1 - z^{-p}\left[1 - \varphi G(z)F(z)\right]\right\} E(z) = \left(1 - z^{-p}\right)\left[y_d(z) - D(z)\right] \tag{3}$$

The right hand side is the difference of the desired output and the disturbance at the current time and one period back, and since both are periodic with p time steps, the forcing function becomes zero. Then the error will converge to zero provided all roots of the characteristic polynomial are inside the unit circle. We can develop a stability condition by paralleling the development of Nyquist criterion. Select a contour that goes counterclockwise around the unit circle, then out to infinity along a branch cut on the negative real axis, clockwise around at infinity, and then comes in along the branch cut. Apply this to $P(z) = 1 - z^{-p}\left[1 - \varphi G(z)F(z)\right] = 0$. The numerator of this expression, after clearing fractions, is the closed loop characteristic equation. If there is any root of the numerator inside the contour then the system is unstable. The phase angle of $P(z)$ will decrease by 2π going once around the contour for every root inside, and will increase by 2π for any pole inside. Assume that the feedback control system is stable so there are no poles of $G(z)$ inside, and agree to make the compensator $F(z)$ have the same property. Then the number of decreases by 2π is the number of unstable roots of the repetitive control system. It is convenient to modify the procedure looking at $P^*(z) = z^{-p}\left[1 - \varphi G(z)F(z)\right]$ instead, and counting how many times its plot for z going around the contour encircles the point $+1$. If we agree to make $F(z)$ have the property that $z^{-p}G(z)F(z)$ has more poles than zeros (since p is normally a large number, this is essentially automatic), then the part of the contour at infinity goes to zero, and the in and out portions along the branch cut cancel each other. Hence, we can restrict out contour to just the unit circle. The stability condition becomes, the plot of $P^*(z)$ for z going counterclockwise around the unit circle should not encircle the point $+1$. Certainly this condition will be satisfied if we require that

$$|1 - \varphi G(e^{i\omega T})F(e^{i\omega T})| < 1 \tag{4}$$

for all ω up to Nyquist, where T is the sample time interval. In [8] it is shown that the difference between this sufficient condition and the stability boundary is negligible in most applications, so that one can consider (4) as the stability condition that must be satisfied in applications. For design one can plot the complex number $\varphi G(e^{i\omega T})F(e^{i\omega T})$ for changing ω, and if the plot stays inside the unit circle centered at $+1$, then the repetitive control system is stable. We will make such plots in designing $F(z)$.

Note that (3) can be rewritten as $z^p E(z) = \left[1 - \varphi G(z)F(z)\right] E(z)$, so that $\left[1 - \varphi G(z)F(z)\right]$ appears as a transfer function of the error from one period to the next. Then satisfying (4) suggests that there will be monotonic decay of every frequency component of the error, provided that the response within each period can be thought of as represented by steady state frequency response. This is not a rigorous development, but it is a useful concept in obtaining good transients. For any ω, the distance from $+1$ to the plot of $\varphi G(e^{i\omega T})F(e^{i\omega T})$ is the factor by which the amplitude of the error at that frequency is decreased each period.

3 The Desirability and the Difficulty of using $G^{-1}(z)$ as a Compensator

If one can make $F(z) = G^{-1}(z)$ and use $\varphi = 1$, then the distance from $+1$ is zero for all frequencies, suggesting convergence to zero error in one repetition by this frequency response thinking. Looking at the characteristic polynomial after the pole-zero cancellations one has $z^p = 0$, meaning all roots are at the origin, and the response is deadbeat. When a less aggressive learning rate is used with $0 < \varphi < 1$, then the roots are on a circle of radius $1-\varphi$, and again the system is stable. The root locus plot is obtained by using $\varphi F(z)G(z)/(z^p-1) = -1$. There are p poles on the unit circle, and if $F(z)G(z) = 1$, as the gain φ is increased all p roots move radially inward, reaching the origin when the gain reaches one. From all of these considerations, one can say that $F(z) = G^{-1}(z)$ is the ideal compensator.

Unfortunately, pole-zero cancellation only makes sense when the inverse of the system being cancelled is asymptotically stable, and this is rarely the case in discrete time systems. Suppose that the feedback control system $G(z)$ is a continuous time system fed by a zero order hold. The majority of continuous time systems have no zeros, and when they do have zeros, the pole excess is still normally larger than one. The equivalent difference equation will have a pole excess of only one, since a change in the input at one time step will normally first influence the output one time step later. The extra zeros introduced by the discretization are images of zeros at infinity, and [9] shows the asymptotic locations as the sample time tends to zero. For a pole excess of 2 in continuous time, one extra zero is introduced asymptotically located at -1. For a pole excess of 3, one has zeros asymptotically at -3.732 and -0.268, for pole excess 4, at -9.899, -0.101, and -1.000, and for pole excess five, at -23.204, -2.322, -0.431, and -0.043. Furthermore these asymptotic zero locations are reached relatively fast as the sample time decreases. Clearly the inverse of most continuous time systems after discretizing is unstable, and can be very unstable – a root at -23.204 at a sample rate of 1000 Hz creates a solution to the inverse system difference equation of the form of a constant times $(-23.204)^{1000}$ at the end of one second of operation.

In this paper we examine a third order difference equation model of the closed loop response of each link of a commercial robot, and we develop a method of finding a noncausal FIR filter that is a good approximation of $G^{-1}(z)$. The approximation becomes arbitrarily good as one takes more terms. And this compensator has no poles outside the unit circle. For purposes of comparison as we develop this method, we comment on the main alternative to the approach developed here, which is the zero phase algorithm of [5]. In that algorithm all poles and zeros of $G(z)$ that are inside the unit circle are cancelled. For a zero factor $(z - z_1)$ with z_1 outside the unit circle and on the negative real axis as above, the compensator multiplies by a constant times $(z^{-1} - z_1)$. For z on the unit circle, this product is real and positive. Hence the plot of $\varphi G(e^{i\omega T})F(e^{i\omega}T)$ will be on the positive real axis, and within the unit circle centered at $+1$ for small enough gains. Note however that it does not cancel the system – which would put the entire plot at the single point $+\varphi$, and when φ is set to unity, convergence to zero error is completed in one repetition. The robot model is $G(s) = a\omega_0^2/\left[(s+a)(s^2 + 2\zeta\omega_0 s + \omega_0^2)\right]$ fed by a zero order hold running at 200 Hz sample rate. This is a reasonable model of the closed loop control for each link of a Robotics Research Corporation robot [6], with $a = 8.8$ and $\omega_0 = 37$. The pole excess is two, and one zero is introduced outside the unit circle. Figure 1 shows the Bode magnitude plot of $\varphi G(z)F(z) = \varphi^*(z - z_1)(z^{-1} - z_1)$ with φ^*, the gain product of the filter gain and φ, picked to produce unity DC gain. This figure indicates that at low frequencies the image is essentially at the center of the unit circle at $+1$, but at high frequencies the learning becomes slow. A true inverse filter would have this plot be unity for all frequencies. As described at the beginning of this section, a true inverse filter creates a root locus plot where all p poles on the unit circle go radially inward to the origin. Figure 2 gives the root locus plot for the approach of [5] for the case of $p = 10$, and one sees that the roots depart radially inward but soon bend and go out of the unit circle, making the system go unstable at a smaller gain. In the next section we develop a noncausal FIR filter approximation of the plant inverse, and see how close we can get to the ideal frequency response plot and ideal root locus plot.

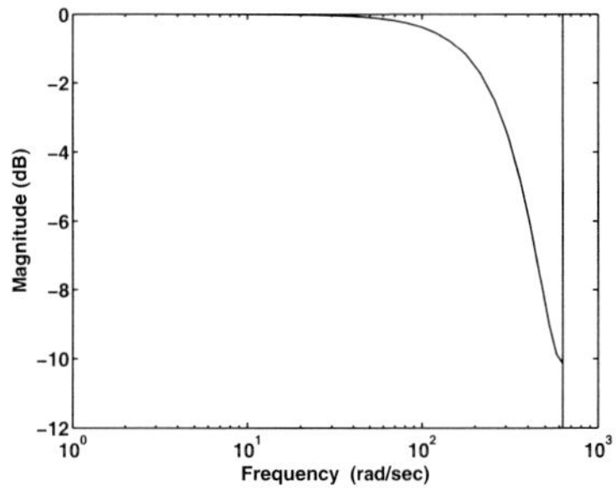

Fig. 1. Bode magnitude plot for zero phase algorithm applied to 3rd order system.

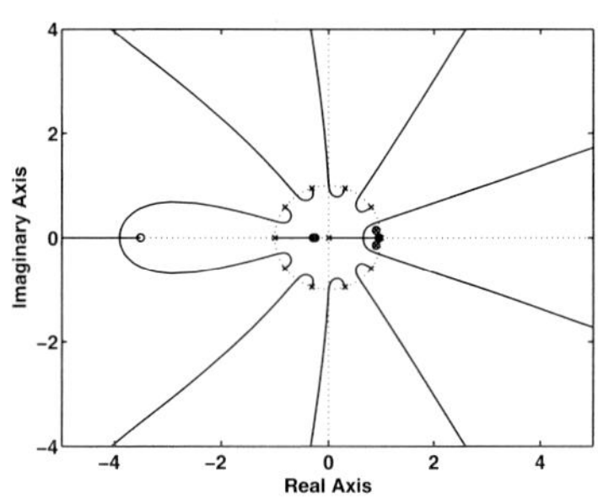

Fig. 2. Root locus plot for zero phase algorithm applied to 3rd order system.

4 Designing a Noncausal FIR Approximation of the Plant Inverse

The third order system, written in difference equation form is

$$y_0(k+3)+a_2y_0(k+2)+a_1y_0(k+1)+a_0y_0(k)=b_2u(k+2)+b_1u(k+1)+b_0u(k) \tag{5}$$

Given a desired output sequence that would have zero tracking error, one knows the left hand side of this equation for all time steps, which forms a forcing function. To solve for the input sequence $u(k)$ one needs to solve the resulting nonhomogeneous difference equation whose characteristic equation is the numerator of $G(z)$, and is unstable. There are an infinite number of particular solutions to a difference equation, one for each set of initial conditions. Generally there is one solution that does not have any term that satisfies the homogeneous equation, and this solution will not grow with time. The FIR filter developed here will approximate such a solution.

Whatever the disturbance $d(k)$ is, in order to cancel it we need to find input $u(k)$ that makes $y_0(k) = -d(k)$. We will generate this by superposition of the solution obtained by setting $d(100) = 1$ and the disturbance for all other time steps is zero. Write equation (5) for time steps $k = 1, 2, 3, \ldots, 198$, extending the solution far enough in both directions that one might have small values and be able to truncate both ends. The equations are

$$\begin{bmatrix} b_0 & b_1 & b_2 & 0 & \ldots & 0 \\ 0 & b_0 & b_1 & b_2 & \ldots & 0 \\ \vdots & \vdots & \vdots & \vdots & \vdots & \vdots \\ 0 & \ldots & 0 & b_0 & b_1 & b_2 \end{bmatrix} \begin{bmatrix} s(1) \\ s(2) \\ \vdots \\ s(200) \end{bmatrix} = \begin{bmatrix} a_0 & a_1 & a_2 & 1 & 0 & \ldots & 0 \\ 0 & a_0 & a_1 & a_2 & 1 & \ldots & 0 \\ \vdots & \vdots & \vdots & \vdots & \vdots & \vdots & \vdots \\ 0 & \ldots & 0 & a_0 & a_1 & a_2 & 1 \end{bmatrix} \begin{bmatrix} \vdots \\ 0 \\ 1 \\ 0 \\ \vdots \end{bmatrix} \tag{6}$$

The solution for the $u(k)$ for this particular disturbance cancellation is relabeled $s(k)$ for k from 1 to 200. The right most matrix has entry 1 in the 100th row, and all other entries zero. There are 198 equations. The first matrix on the left is of dimension 198×200 while the second is 200×1. The first matrix on the right is 198×201 and the second is 201×1. There are fewer equations than unknowns. We find the solution by Guassian elimination using the MATLAB algorithm which sets $s(1) = s(200) = 0$ and solves the remaining equations. Essentially the same results are produced by taking the Moore-Penrose pseudoinverse. Figure 3 gives the solution which is large near the time step of the nonzero disturbance and gets small in each direction. It does not however become zero in either direction except at the first and last entries. Figure 4 shows a detail of time steps from 70 to 80. The corresponding plot for 80 to 90 looks essentially identical but the scale now goes from -0.12 to +0.04. A plot from 110 to 120 looks similar in shape but reflected about the vertical and going from -10 to +4 times 10^{-3}. For use when

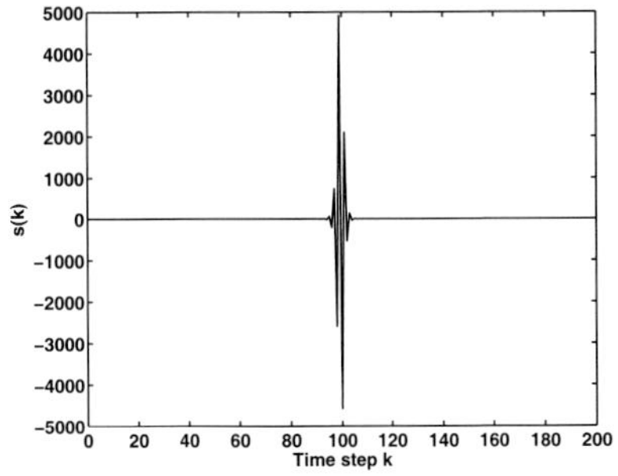

Fig. 3. Solution to equation(6) for the $s(k)$ sequence.

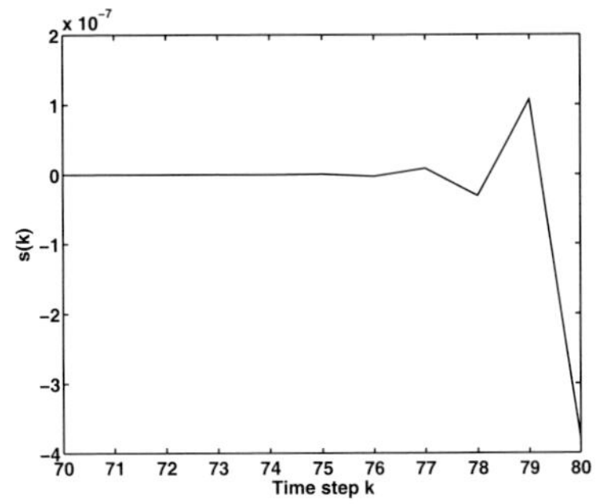

Fig. 4. A detail of Fig.3 for time steps 70 to 80.

applying this result in the repetitive control law, we rename the solution as $a_{-99} = s(1), \ldots, a_0 = s(100), \ldots, a_{100} = s(200)$.

We will need to truncate this solution on both sides of the 100th time step. And then we need to do superposition to handle disturbances at all time steps (starting after the start-up phase, and provided the number of noncausal gains needed is less than one period). For ease of analysis, suppose that we truncate to just three terms, a_{-1}, a_0, a_{+1}. And suppose the disturbance/error for time steps 99, 100, and 101 is to be cancelled one period later, where the period is 200 steps. Then the control action at time step 300 is the control at time step 100 plus the learning gain φ, times the sum of three terms: $a_0e(100)$ to address the error one period back, $a_{+1}e(99)$ needed to complete addressing the error for time step 99, and $a_{-1}e(101)$ to start addressing the error for time step 101. Generalizing this thinking to retain the following terms $a_{-n2}, \ldots, a_0, \ldots, a_{+n1}$ and truncate the rest, gives the repetitive control law and the filter $F(z)$ approximation of $G^{-1}(z)$:

$$u(k+p) = u(k) + \varphi\left[a_{+n1}e(k-n_1) + \ldots + a_0e(k) + \ldots + a_{-n2}e(k+n_2)\right] \quad (7)$$

$$\begin{aligned} F(z) &= a_{n1}z^{-n_1} + \ldots + a_0z^0 + \ldots + a_{-n2}z^{n_2} \\ &= \left[a_{n1}z^0 + \ldots + a_0z^{n_1} + \ldots + a_{-n2}z^{n_1+n_2}\right]/z^{n_1} \end{aligned} \quad (8)$$

Note that all of the poles introduced by this filter are at the origin, so they do not cause instability. The number of zeros introduced is $n_1 + n_2$ which is one less than the number of terms taken in the solution of (6), and n_1 is the number of poles (for the stability analysis to apply, one limits the choice to ensure the number of poles of $z^{-p}\left[1 - \varphi G(z)F(z)\right]$ exceeds the number of zeros, but this is not normally an active constraint).

The problem of designing the compensator reduces to picking the repetitive control gain φ and picking the two numbers n_1 and n_2. These are to be picked first to establish stability by ensuring that the plot of $\varphi G(z)F(z)$ for z going around the unit circle, stays inside the unit circle centered at $+1$. Then to optimize the learning speed, one can aim to have the plot stay as close as possible to +1 for all frequencies around the unit circle. In the example computation of the next section it is seen that a relatively small number of gains can produce stability, and adding several more can get a plot that deviates little from +1.

5 Example Repetitive Controller Designs

Initially adjust the repetitive control gain φ so that $\varphi G(z)F(z)$ is at +1 for $z = 1$, i.e. unity DC gain, and then examine the range of values of n_1 and n_2 having a plot that stays inside the unit circle centered at +1. If the plot goes into the left half plane, then no adjustment of the positive gain φ can stabilize the system. If it goes out of the unit circle but does not go into the left half plane, then decreasing the gain will stabilize the system. Table 1 gives a set of results for use of 11 gains up through 18 gains. The shaded region corresponds to stable choices for the repetitive control law (the DC gains used are unity except that the gain had to be reduced for the entries 93-107 through 93-110). The entries in the table are the time step arguments on the $s(k)$ solution to equation (6) for the selected gains. The minimum number of gains needed to produce stability is 12, and as the number of gains allowed is increased one has a choice of various stabilizing controllers. Since the number of gains picked is the number of gains that must be handled in real time, it is of interest to know how sensitive this minimum number is to the system being considered and to the sample rate selected. The sample rate for Table 1 is 200 Hz. Decreasing to 100 Hz changes the minimum number of gains needed from 12 to 11. This suggests that the minimum number is not highly sensitive to sample rate. Another set of runs was made that decrease the damping ratio from 0.5 to 0.1 in the complex conjugate roots of $G(s)$. For this lightly damped system sampling at 200 Hz, the minimum number of gains is 15. With such small numbers of gains, this design approach applied to the third order robot model appears not to be constrained by real time computation limits.

Table 1. The set of possible stabilizing choices for the compensator gains

	Disturbance at step 100							
# of gains	Gain time step							
11						93-103	94-104	95-105
12						93-104	94-105	95-106
13						93-105	94-106	95-107
14					92-105	93-106	94-107	95-108
15				91-105	92-106	93-107	94-108	95-109
16			90-105	91-106	92-107	93-108	94-109	95-110
17		89-105	90-106	91-107	92-108	93-109	94-110	95-111
18	88-105	89-106	90-107	91-108	92-109	93-110	94-111	95-112

Now examine how good the inverse is for different choices for the number of gains. To do this we make the disturbance equal $+1$ at $k = 100, 300, \ldots$ periodic with $p = 200$ time steps and set the command to zero. For the first 200 steps the feedback control system is operating without repetitive control so the error is just the $+1$ at time step 100. And then the repetitive control starts. If the compensator were a perfect inverse of $G(z)$ there would be no error resulting from the disturbance coming in at 300. Figure 5 shows the result using 20 gains (90 to 109) and a DC gain of 1.3. The error is reduced from a maximum of 1 to a maximum of 8×10^{-4}, a very substantial improvement for one repetition. If we reduce the number of gains to 16 and use 92-107 with a DC gain of 1.45, then the plot looks very similar but the scale is increased by about a factor of 10. So, not many gains are needed in order to eliminate several orders of magnitude of the error in one repetition. Figure 6 gives the root locus plot associated with 91-108 when $p = 10$. By comparison to Fig. 2 this plot is much closer to the ideal root locus where all roots go radially inward to the origin. The process of designing the compensator is to pick a number of gains, i.e. a row on the Table, and look at the Nyquist plot of $\varphi G(z)F(z)$ for differenct values of n_1 and n_2, looking for one that keeps the plot close to $+1$ for all frequencies (one might later lower the gain to be less aggressive in the design). Add more gains until satisfied with the result. For 18 gains the best result is 90-107 which is given in Fig. 7. Figure 8 gives an example unstable plot for 18 gains, 95-112. The design process is simple and straighforward and works well on the third order robot model. In future work, this approach will be studied in detail to assess which classes of systems can be handled. Also, practical considerations will be evaluated relating to the conditioning of the repetitive control updates.

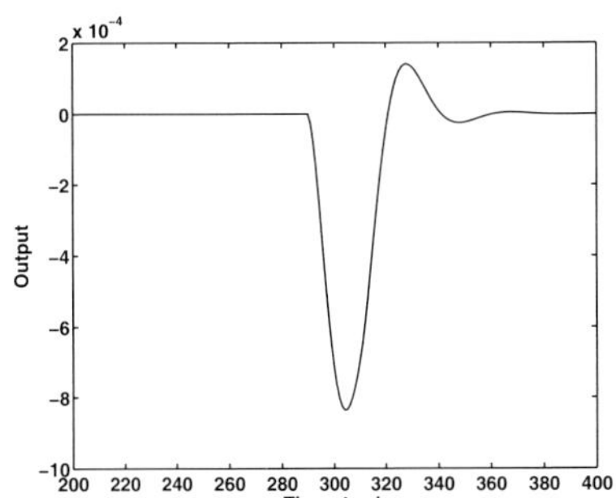

Fig. 5. Output vs. time step for a 20 gain compensator.

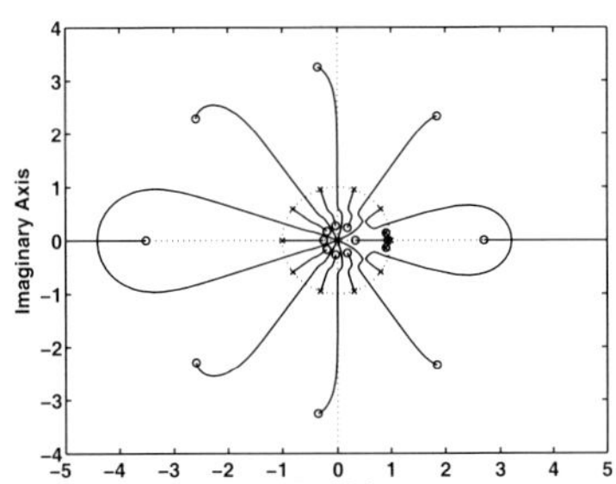

Fig. 6. Root locus plot of the non-causal compensator using 18 gains.

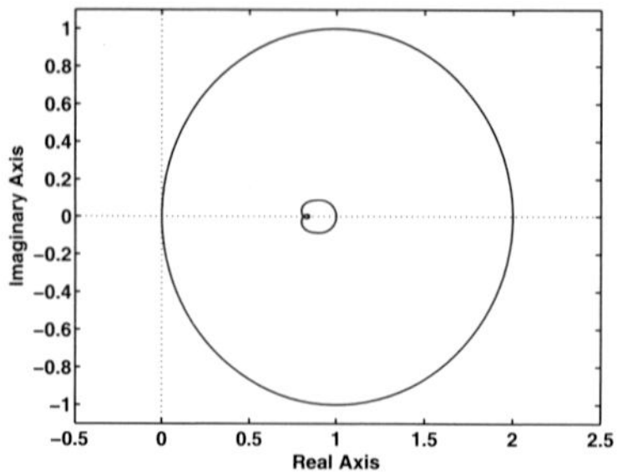

Fig. 7. Nyquist plot of $\varphi F(z)G(z)$ using steps 90 through 107.

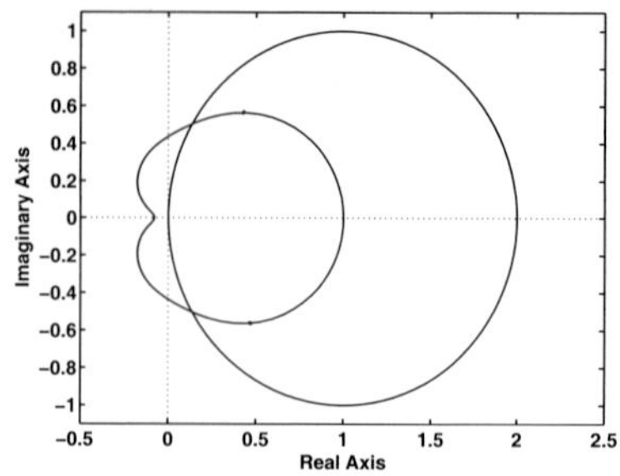

Fig. 8. Nyquist plot of $\varphi F(z)G(z)$ using steps 95 through 112.

References

[1] T. Inoue, M. Nakano, and S. Iwai, "High Accuracy Control of a Proton Synchrotron Magnet Power Supply, " Proceedings of the 8th World Congress of IFAC, Vol. 20, 1981, pp. 216-221.

[2] R. H. Middleton, G. C. Goodwin, and R. W. Longman, "A Method for Improving the Dynamic Accuracy of a Robot Performing a Repetitive Task," International Journal of Robotics Research,, Vol. 8, No. 5, 1989, pp. 67-74. Also, University of Newcastle, Newcastle, Australia, Department of Electrical Engineering Technical Report EE8546, 1985.

[3] S. Hara, and Y. Yamamoto, "Synthesis of Repetitive Control Systems and its Applications," Proceedings of the 24th IEEE Conference on Decision and Control, 1985, pp. 326-327.

[4] S. Hara, T. Omata, and M. Nakano, "Stability of Repetitive Control Systems," Proceedings of the 24th IEEE Conference on Decision and Control, 1985, pp. 1387-1392.

[5] Tomizuka M., Tsao T.-C., and Chew K.-K., "Analysis and Synthesis of Discrete-time Repetitive Controllers, " Journal of Dynamic Systems, Measurement, and Control, Vol. 111, 1989, pp. 353-358.

[6] Longman, R. W., "Iterative Learning Control and Repetitive Control for Engineering Practice," International Journal of Control, Special Issue on Iterative Learning Control, Vol. 73, No.10, July 2000, pp. 930-954.

[7] S. J. Oh and R. W. Longman, "Methods of Real-Time Zero-Phase Low-Pass Filtering for Robust Repetitive Control, " Proceedings of the AIAA/AAS Astrodynamics Specialist Conference, Monterey, CA, August 2002.

[8] S. Songschon and R. W. Longman, "Comparison of the Stability Boundary and the Frequency Response Condition in Learning and Repetitive Control," International Journal of Applied Mathematics and Computer Science, to appear.

[9] K. Astrom, P. Hagander, and J. Strenby, "Zeros of Sampled Systems," Proceedings of the Nineteenth IEEE Conference on Decision and Control, 1980, pp. 1077-1081.

Cutting Planes for the Optimisation of Gas Networks

Alexander Martin and Markus Möller

Research Group of Discrete Optimization, Technical University Darmstadt
Schlossgartenstrasse 7, 64289 Darmstadt, Germany

Summary. This paper presents cutting planes which are useful or potentially useful for solving mixed integer programs that arise in the optimisation of gas networks. We consider polyhedra that are defining essential parts of the model and give a polynomial algorithm for the calculation of the set of vertices of such polyhedra implying a polynomial separation algorithm for the convex hull of the polyhedra can be developed.

Key words: Mixed Integer Programming, Cutting Planes, Gas Optimisation, Piecewise Linear Functions

1 Introduction

The following studies of a special polyhedron have arisen from our researches on the problem of the Transient Technical Optimisation (TTO) in gas networks. A gas network basically consists of a set of compressors and valves that are connected by pipes. Since the gas pressure in the pipes decreases due to the friction in the pipes the compressors are used in order to increase the gas pressure again since the consumers want to get gas of a certain pressure value and quality. The task of the Transient Technical Optimisation is to optimise the drives of the gas and to set in the compressors cost-efficiently such that the required demands are satisfied. Modelling this problem will lead to a complex mixed integer nonlinear optimisation problem. We have approached it by approximating the nonlinearities (the most important nonlinear functions in this model describe the fuel gas consumption of the compressors and the pressure loss in the pipes) by piece-wise linear functions leading to a huge mixed integer program. We want to solve the mixed integer program via a branch-and-cut algorithm. Therefore we have studied the polyhedral consequences of this model.

In this paper we present some new cutting planes for polyhedra that describe important substructures of the gas network model. We also point out how this knowledge can be generalised to more complex structures. Finally our preliminary computational results show the benefits when incorporating these cuts into a general mixed integer programming solver.

2 The Polyhedron

The above described problem of the Transient Technical Optimisation is evidently modelled in a graph $G = (V, E)$. The set E consists of the set of compressors, the set of valves and control valves and the set of pipes. The set V of nodes consists of the set of intersection points of the segments, the set of sources (the gas delivering points) and the set of sinks (which are the gas demanding points) of the gas network. We point out the most important kind of variables which are necessary to understand the succeeding ventilations. At first we introduce flow variables $q_e, e \in E$. These variables describe the mass flow of gas in each segment. Second we consider pressure variables $p_v, v \in V$. The pressure variables describe the pressure of the gas in each node. Clearly, p_{in} describes the gas pressure in the node at the beginning of a segment and p_{out} describes the gas pressure in the node at the end of a segment. A very important principle is that the pressure at the end of all ingoing segments of a node must be equal the pressure at the beginning of all outgoing segments of the same node. This principle will be very important for our further discussions. There are a lot of other kinds of variables in the whole model, but for the here presented analysis of a polyhedron that describes only a special part of the whole model these two kinds of variables suffice.

Now let us shortly describe how the polyhedron under investigation comes upon in the global model. The physics of the gas is basically described by three partial differential equations. The momentum equation, the continuity equation and the energy equation. We focus on the momentum equation that describes the pressure loss in pipes. Under some mild assumptions, which we do not want to discuss here, the momentum equation can be simplified to a nonlinear function of the following shape:

$$p_{out}^2 = p_{in}^2 - \mathrm{ff}\, q\, |q|,$$

where

$$\mathrm{ff} = \mathrm{ff}(p_{out}, p_{in})$$

is the friction factor. After simplifying the friction factor to a constant we get $p_{out} = p_{out}(p_{in}, q)$, where p_{out} means the pressure at the end of the pipe, p_{in} means the pressure at the beginning of the pipe and q means the gas flow through the pipe.
The other important nonlinearity is the fuel consumption of the compressors (which has to be minimised as mentioned in the introduction): The fuel consumption is described by a nonlinear function f of the form: $f = f(p_{in}, p_{out}, q)$. Here f describes the fuel consumption of the compressor, p_{in} the pressure of

the gas at the beginning of the compressor, p_{out} the gas pressure which the compressor has to constitute at the endpoint of the compressor and q stands for the gas flow through the compressor. In order to come up with a mixed integer linear program these two nonlinear functions are approximated by suitable triangulations as pointed out in the following demonstrations.

The first substructure of the model we have studied are sequences of pipes. The situation is shown in Figure 1. We have already mentioned one important

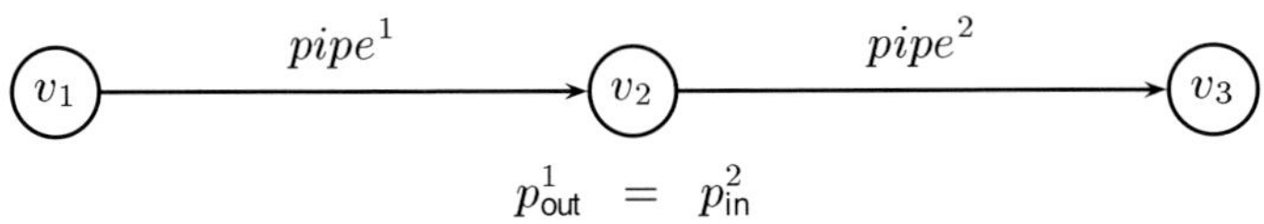

Fig. 1. Sequence of pipes

aspect of the model that the pressure p^1_{out} at the end of the ingoing pipe ($pipe^1$) must be equal the pressure p^2_{in} at the beginning of the outgoing pipe ($pipe^2$). We already know that p^1_{out} is a nonlinear function depending on the flow through the pipe and the pressure at the beginning of the pipe. We approximate the pressure loss in pipes by determing a **triangulation** of the 2-dimensional manifold describing the pressure loss in the pipes. We denote by Λ^{pipe} the set of grid points and by Y^{pipe} the set of triangles. We approximate the 2−dimensional function $p_{out}(p_{in}, q)$ by linearising it within each triangle. Modelling this piecewise linear approximation results in the following non convex polyhedron:

$$
\begin{aligned}
P_\Delta = \Big\{ \tbinom{\lambda^1}{\lambda^2} \in \mathbb{R}^{|\Lambda^1|+|\Lambda^2|} \mid \sum_{j\in\Lambda^1} \lambda^1_j \qquad\qquad\qquad &= 1 \\
\sum_{j\in\Lambda^2} \lambda^2_j &= 1 \\
\sum_{j\in\Lambda^1} p^1_{\text{out},j}\,\lambda^1_j - \sum_{j\in\Lambda^2} p^2_{\text{in},j}\,\lambda^2_j &= 0 \\
\lambda^1_j, \qquad \lambda^2_j &\geq 0 \\
\lambda^1, \lambda^2 \text{ satisfy the triangle condition} \Big\}&,
\end{aligned}
$$

where the triangle condition states that the set of λ−variables which are strictly positive must belong to grid points of a distinct triangle.

Figure 2 describes the situation of the polyhedron P_Δ: The numbers in the left triangulation (for the ingoing pipe) stand for the pressure values $p^1_{out,j}$ at the grid points $j \in \Lambda^1$ and the numbers in the right triangulation (for the outgoing pipe) stand for the pressure values $p^2_{in,i}$ at the grid points $i \in \Lambda^2$. Let us consider a simple example (see Figure 3) for a little calculation. Here is $p^1_{out,1} = 10$, $p^1_{out,2} = 8$, $p^1_{out,3} = 4$ and so on, analogously we have for $p^2_{in,1} = p^2_{in,2} = p^2_{in,3} = 10, \ldots$, etc. Consider

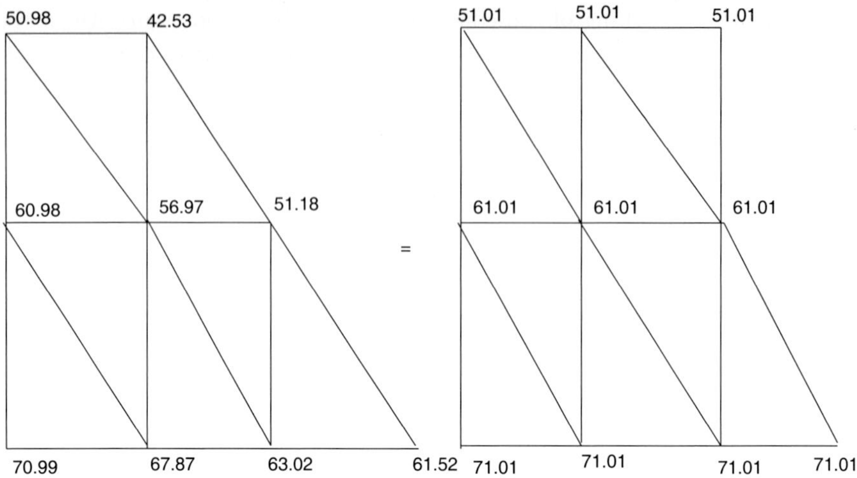

Fig. 2. Typical triangulation of the pressure loss in a pipe

$$\lambda_1^1 = \frac{1}{4}, \lambda_2^1 = 0, \lambda_3^1 = 0, \lambda_4^1 = \frac{1}{2}, \lambda_5^1 = \frac{1}{4}, \lambda_6^1 = 0$$

and

$$\lambda_1^2 = \frac{7}{20}, \lambda_2^2 = 0, \lambda_3^2 = 0, \lambda_4^2 = 0, \lambda_5^2 = \frac{13}{20}, \lambda_6^2 = 0.$$

This setting for the $\lambda-$variables fulfils all conditions, especially the triangle condition.

But if we take

$$\lambda_1^1 = \frac{1}{4}, \lambda_2^1 = 0, \lambda_3^1 = 0, \lambda_4^1 = \frac{1}{2}, \lambda_5^1 = \frac{1}{4}, \lambda_6^1 = 0$$

and

$$\lambda_1^2 = \frac{7}{20}, \lambda_2^2 = 0, \lambda_3^2 = 0, \lambda_4^2 = 0, \lambda_5^2 = 0, \lambda_6^2 = \frac{13}{20},$$

we see that the triangle condition is not satisfied since the nonzero variables λ_1^2 and λ_6^2 belong to two **different** triangles of the triangulation. In the following we want to generalise this approach. Clearly the sequence of two pipes is of course only the simplest case we are faced with. We want to examine the problem more general, where we consider the case that we have an arbitrary number *in* of ingoing segments and an arbitrary number *out* of outgoing segments. A segment can now be either a pipe or a compressor (but we can as a matter of principle take valves or control valves as segments as we will see later). For every in- and outgoing segment we determine a certain triangulation. In the general case these triangulations do not need to consist only of such regular triangles as in Figure 2. The structure can be much more complicated. Perhaps we can not only consider triangles but also squares, pentagons, sexangles, heptagons and so on. Even arbitrary mixtures in the triangulations

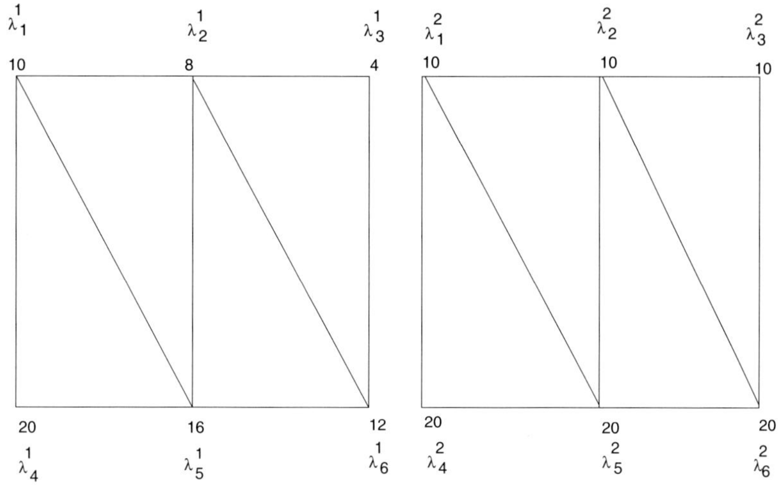

Fig. 3. An easy example for a triangulation

are possible although this is not interesting for a concrete gas network. And we do not only describe the pressure in the segments but also the gas flow in the segments. Very important for the general formulation is the first law of Kirchhoff which means that the sum of the ingoing gas flows must be equal to the sum of the outgoing gas flows. So in principle (see e.g. [1], [2]) we get the situation which is shown in Figure 4.

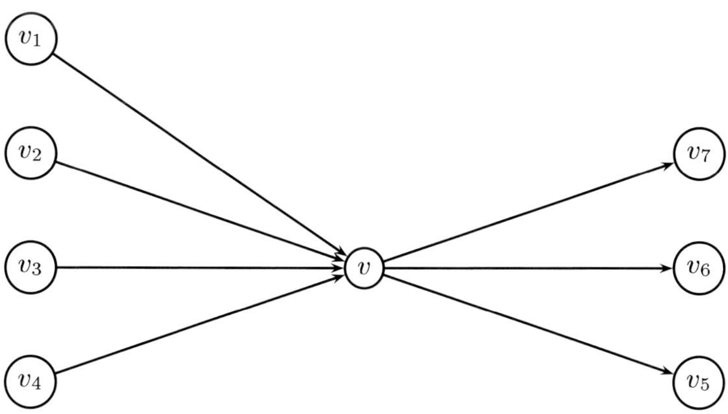

Fig. 4. Ingoing and outgoing segments in a node

The requirements of the triangle conditions of $P_\triangle$ are now generalised in the following way:

The triangle conditions mean that for every segment only special combinations of $\lambda-$variables are allowed. For $P_\triangle$ this means that only $\lambda-$variables may be positive that belong to exactly one certain triangle. In the general case only the elements of special sets of $\lambda-$variables may not vanish (Indeed: the reader can recognise that our conditions are a generalised form of Special Ordered Sets (SOS) of type 2, see [3]). Before going into the details we need to fix some notation.

Notation

In this section we give some mathematical notation which is necessary in order to formalise and generalise the above approach.

Let $in \in \mathbb{N}$ be the number of ingoing segments and $out \in \mathbb{N}$ be the number of outgoing segments. A segment may be a pipe or a compressor but also the other types of segments, i.e., valves, control valves and connections (short pipes without pressure loss) can be included in this model. In the mathematical formulation of the model we are no longer bounded to the physical background of the model.

We define a set N^i of grid points for every segment $i \in \{1, 2, \dots, in+out\}$. W.l.o.g. we assume the ingoing segments to be $1, 2, \dots, in$ and the outgoing segments to be $in+1, in+2, \dots, in+out$. Furthermore we assume:

$$N^i \cap N^j = \emptyset \qquad \forall i \neq j.$$

We denote by

$$\mathcal{N} = \{N^i \mid i = 1, 2, \dots, in + out\}$$

the **list** of sets of grid points.
$\mathbb{R}^{N^i}$ denotes the $|N^i|-$dimensional vector space where the components are indexed by N^i and $\mathbb{R}^{\mathcal{N}}$ is defined as:

$$\mathbb{R}^{\mathcal{N}} = \bigotimes_{i=1}^{in+out} \mathbb{R}^{N^i}.$$

We remark that for $\lambda \in \mathbb{R}^{\mathcal{N}}$ we write

$$\lambda = \begin{pmatrix} \lambda^1 \\ \lambda^2 \\ \vdots \\ \lambda^{in+out} \end{pmatrix}$$

with $\lambda^i \in \mathbb{R}^{N^i}$ for all $i \in \{1, 2, \dots, in + out\}$.

For a list $\mathcal{S}$ of sets of the form

$$\mathcal{S} = \{S^1, S^2, \dots, S^{in+out}\}$$

we say for some index $j \in \bigcup_{i=1}^{in+out} N^i$:

$$j \in \mathcal{S} \quad \text{iff} \quad \exists i \in \{1, 2, \ldots, in + out\} \quad \text{with} \quad j \in S^i.$$

We define

$$\mathcal{S} \subseteq \mathcal{N} \quad \Leftrightarrow \quad \emptyset \neq S^i \subseteq N^i \quad \forall i \in \{1, 2, \ldots, in + out\}.$$

The cardinality of $\mathcal{S}$ is set to

$$|\mathcal{S}| = \sum_{i=1}^{in+out} |S^i|.$$

The characteristic vector of $\mathcal{S}$, which we denote by $\mathcal{X}^{\mathcal{S}} \in \mathbb{R}^{\mathcal{N}}$, is obtained by setting

$$\mathcal{X}_j^{\mathcal{S}} = \begin{cases} 1 & \text{, if } j \in \mathcal{S} \\ 0 & \text{else.} \end{cases}$$

For each $N^i, i \in \{1, 2, \ldots, in+out\}$ we define n_i subsets $N_k^i, k \in \{1, 2, \ldots, n_i\}$ with

$$N^i = \bigcup_{k=1}^{n_i} N_k^i \quad \text{and} \quad |N_k^i| \geq 2.$$

As an example: In Figure 2 holds $n_1 = 8$, $n_2 = 9$ and $|N_k^i| = 3$ for all i, k. We say that a vector $\lambda \in \mathbb{R}^{\mathcal{N}}, \lambda \geq 0$ satisfies the **set condition** if for all $i = 1, 2, \ldots, in + out$ there exists one $k_i \in \{1, 2, \ldots, n_i\}$ such that

$$\{j \in N^i | \lambda_j^i > 0\} \subseteq N_{k_i}^i.$$

In other words, the set condition holds if for all in- and outgoing segments the non vanishing $\lambda-$variables belong to exactly one of the subsets N_k^i. We say that $\mathcal{S}$ fullfills the set condition if $\mathcal{X}^{\mathcal{S}}$ fullfills the set condition.

Now we define a polyhedron P by

$$P = \{\lambda \in \mathbb{R}^{\mathcal{N}} | A\lambda = b, \lambda \geq 0\},$$

where $A \in \mathbb{R}^{M \times \mathcal{N}}, b \in \mathbb{R}^M$ for some finite set M. We will say something about the cardinality of the set M in the next subsection when we discuss the special structure of the matrix A.

Let us continue with the following definition:
For $A \in \mathbb{R}^{M \times \mathcal{N}}$ with

$$A = (a_{ij})_{\substack{\mathrm{i}=1,\ldots,\mathrm{m} \\ \mathrm{j}=1,\ldots,\mathrm{n}}}$$

and a subset $J \subseteq \{1, 2, \ldots, n\}$ we define

$$A_J = (a_{ij})_{\substack{i \in M \\ j \in J}}$$

Here $m = |M|$ and $n = |\mathcal{N}|$. Analogously we use for $\lambda \in \mathbb{R}^{\mathcal{N}}$ and a subset $J \subseteq \{1, 2, \ldots, n\}$

$$\lambda_J = (\lambda_j)_{j \in J}.$$

For $x \in \mathbb{R}^{\mathcal{S}}$ with $\mathcal{S} \subseteq \mathcal{N}$ we define the **zero-extension** $x^0(\mathcal{S}) \in \mathbb{R}^{\mathcal{N}}$ of x by

$$x^0(\mathcal{S}) = \begin{cases} x_i & , \text{ if } i \in \mathcal{S}, \\ 0 & , \text{ if } i \in \mathcal{N} \setminus \mathcal{S}. \end{cases}$$

We remark that in the definition of the zero-extension we clearly define $\mathcal{N} \setminus \mathcal{S}$ by

$$\mathcal{N} \setminus \mathcal{S} :\Leftrightarrow N^i \setminus S^i \qquad \forall i \in \{1, 2, \ldots, in + out\}$$

At the end of this section we define a list $\mathcal{S}$ to be a subset of another list $\bar{\mathcal{S}}$:

Let

$$\mathcal{S} = \{S^1, S^2, \ldots, S^{in+out}\}$$

and

$$\bar{\mathcal{S}} = \{\bar{S}^1, \bar{S}^2, \ldots, \bar{S}^{in+out}\}$$

two sets (in the sense of this section). We define:

$$\begin{aligned} \mathcal{S} \subseteq \bar{\mathcal{S}} :\Leftrightarrow S^1 &\subseteq \bar{S}^1 \\ S^2 &\subseteq \bar{S}^2 \\ &\vdots \\ S^{in+out} &\subseteq \bar{S}^{in+out}. \end{aligned}$$

3 The Problem

Using the above notation we are now ready to introduce the polyhedron we are going to investigate in this paper. Remember that we want to model the situation that there are *in* ingoing and *out* outgoing segments at some node in the gas network.

So we consider a polyhedron P with the following structure:

$$P = \{\lambda \in \mathbb{R}^{\mathcal{N}} | A\lambda = b, \lambda \geq 0, \lambda \text{ satisfies the set conditions}\}.$$

We remark that from our introductory examples it is easy to see that this polyhedron in general is **not** convex.

The special form of the matrix A and the vector b for the general case becomes

$$A = \begin{pmatrix}
(\mathbb{1}^1)^T & & & & & & & \\
& (\mathbb{1}^2)^T & & & & & & \\
& & \ddots & & & & & \\
& & & (\mathbb{1}^{\mathsf{in}})^T & & & & \\
& & & & (\mathbb{1}^{\mathsf{in+1}})^T & & & \\
& & & & & (\mathbb{1}^{\mathsf{in+2}})^T & & \\
& & & & & & \ddots & \\
& & & & & & & (\mathbb{1}^{\mathsf{in+out}})^T \\
(p^1)^T & & & & -(p^{in+1})^T & & & \\
(p^1)^T & & & & & -(p^{in+2})^T & & \\
\vdots & & & & & & \ddots & \\
(p^1)^T & & & & & & & -(p^{in+out})^T \\
& (p^2)^T & & & -(p^{in+1})^T & & & \\
& (p^2)^T & & & & -(p^{in+2})^T & & \\
& \vdots & & & & & \ddots & \\
& (p^2)^T & & & & & & -(p^{in+out})^T \\
& & \ddots & & \vdots & \vdots & \vdots & \vdots \\
& & & (p^{in})^T & -(p^{in+1})^T & & & \\
& & & (p^{in})^T & & -(p^{in+2})^T & & \\
& & & \vdots & & & \ddots & \\
& & & (p^{in})^T & & & & -(p^{in+out})^T \\
(q^1)^T & (q^2)^T & \ldots & (q^i n)^T & -(q^{in+1})^T & -(q^{in+2})^T & \ldots & -(q^{in+out})^T
\end{pmatrix}$$

The entries $p^i \in \mathbb{R}_+^{N^i}, i \in \{1, 2, \ldots, in + out\}$ are vectors describing the pressure at the end of the ingoing and at the beginning of the outgoing segments. Analogously the vectors $q^i \in \mathbb{R}_+^{N^i}, i \in \{1, 2, \ldots, in + out\}$ describe the gas flow in the in- and outgoing segments. These vectors are used to formulate the mentioned first law of Kirchhoff.

The form of the vector b is

$$b = \begin{pmatrix} \mathbb{1}_{\mathsf{in+out}} \\ 0_{in \cdot out+1} \end{pmatrix}$$

$\mathbb{1}^{\mathsf{i}} \in \mathbb{R}^{\mathsf{N^i}}$ denotes the vector of all ones. In addition $\mathbb{1}_{\mathsf{m}}$ denotes the vector of all ones in $\mathbb{R}^m$ and 0_m the zero vector in $\mathbb{R}^m$. Let us shortly describe the structure of the matrix A more detailed:

The first in rows describe the sum of the $\lambda-$variables of each ingoing segment. Analogously the next out rows describe the sum of the $\lambda-$variables of each outgoing segment. All sums must be one. In each node there must be a certain pressure. So the rows $in + out + 1$ up to $in + 2out$ describe that the pressure at the end of the first segment must be equal the pressure at the beginning of the outgoing segments. The rows $in + 2out + 1$ up to $(in+1)(out+1) - 1$ describe the same situation for the other ingoing segments combined with the outgoing segments. The last row describes the gas flow in the distinct segments. The gas flow in the outgoing segments is multiplied by -1 because the sum of the gas flows of the ingoing segments must be equal the sum of the gas flows of the outgoing segments. It is easy to see that the matrix A and the vectors λ and b are generalisations of the first discussed situation of one ingoing and one outgoing segment.

As a side remark we want to mention that there are some additional types of segments in a gas network, for example valves, control valves and connections without pressure loss or fuel gas consumption (i.e., there no nonlinear function has to be linearised). In the situation that such an additional segment is an essential part of a subsystem of the gas network also this types of segments can be modelled. Here the vectors for the pressure p or the gas flow q reduce to vectors that are elements from $\mathbb{R}^1$ (In this case the set of grid points for such a segment consists only of one element. Here it is very important to know that it is our aim that we want to cut off LP-solutions, so we can set for this types of segments the pressure and flow values that are calculated in the last iteration. This solution then can be cut off.) because such a segment can in every LP-iteration be interpreted with constant pressure and constant flow and so can be modelled via one single $\lambda-$variable which then has to be one. So the generality of the model is ensured.

When we do not want to include the first law of Kirchhoff, i.e., the gas flow preservation equation in this model, we forget about the last line in $A\lambda = b$. The rang of the Matrix A reduces by one in this case. We also remark that holds $|M| = in + (in + 1)out + 1 = (in + 1)(out + 1)$.

For the following considerations it is important to mention that

$$P \subseteq [0,1]^{\mathcal{N}}$$

holds, which is easy to see since $\lambda \geq 0$ and because of the definition of the first $in + out$ rows of matrix A and vector b. So the polyhedron P is bounded.

3.1 The vertices of the polyhedron

Let us introduce the idea of calculating the vertices of the polyhedron before we describe the general case formally in the case of the polyhedron P_Δ: If we want to find a vertex we take one triangle from the triangulation of the ingoing pipe ($pipe^1$) and one triangle from the triangulation of the outgoing pipe ($pipe^2$). Hereto we choose some $\lambda-$variables from the selected triangles. Due to the triangle condition the non vanishing $\lambda-$variables at a vertex of P_Δ must belong to exactly one triangle of $pipe^1$ and one triangle of $pipe^2$. Concentrating on two triangles we investigate the extreme points for the selected $\lambda-$variables that fulfil the remaining properties of P_Δ, i.e., if the sum of the selected $\lambda-$variables of *pipe*[1] and the sum of the selected $\lambda-$variables of *pipe*[2] are equal 1, if the pressure equation is fulfilled and of course all $\lambda-$variables we have selected must be nonnegative. We will show that this results in a vertex. By repeating this procedure for all possible selections of $\lambda-$variables we will see that we obtain all vertices of P_Δ.

Now we give the formal algorithm how the vertices of the polyhedron P can be calculated: Let us begin with the following definition ($rg(A)$ denotes the rang of matrix A):

Definition 1. *We say a subset $\mathcal{S} \subseteq \mathcal{N}$ is feasible if*

- $|\mathcal{S}| \leq rg(A)$.
- *$\mathcal{S}$ satisfies the set condition.*

Algorithm 1.

1. Let $L = \emptyset$ be the list of all vertices of P.
2. For all feasible subsets $\mathcal{S} \subseteq \mathcal{N}$ do
   a) Solve $A_\mathcal{S}\,\lambda_\mathcal{S} = b$.
   b) If the system has a unique solution $\bar{\lambda}_\mathcal{S}$ with $\bar{\lambda}_\mathcal{S} \geq 0$, add the zero-extension of $\bar{\lambda}_\mathcal{S}$ to L.

In the following we want to prove that this algorithm runs in polynomial time and computes all vertices of P. As a consequence we obtain that P has only polynomially many vertices. But at first let us make the following remark. The matrix A on page 315 can be simplified to the form as depicted in Figure 5 and the vector b to

$$b = \begin{pmatrix} \mathbb{1}_{\text{in+out}} \\ 0_{in+out} \end{pmatrix}$$

From this we conclude that

$$in + out \leq rg(A) \leq 2(in + out)$$

Before we prove that the algorithm is correct we discuss the following

Lemma 1. *The described algorithm reduced by the postulation of the set condition can principally also be used in order to calculate the vertices of the polyhedron P without the set condition.*

This lemma is a direct consequence of well known results of linear programming, namely that the support of a vertex of a polyhedron $P = \{x \mid Ax = b, x \geq 0\}$ is at most the number of rows of A. When we consider the polyhedron P without the set conditions this polyhedron is completely described by (in-) equalities and thus the above argument applies. The problem in the case of P (with set conditions) is that we do not know the complete description of the polyhedron in form of equalities or inequalities and thus this simple argument cannot be used.

In our case we formulate the following

$$
\begin{pmatrix}
(\mathbb{1}^1)^T & & & & & & & & & \\
& (\mathbb{1}^2)^T & & & & & & & & \\
& & \ddots & & & & & & & \\
& & & \ddots & & & & & & \\
& & & & (\mathbb{1}^{\mathsf{in}})^T & & & & & \\
& & & & & (\mathbb{1}^{\mathsf{in}+1})^T & & & & \\
& & & & & & (\mathbb{1}^{\mathsf{in}+2})^T & & & \\
& & & & & & & \ddots & & \\
& & & & & & & & \ddots & \\
& & & & & & & & & (\mathbb{1}^{\mathsf{in}+\mathsf{out}})^T \\
(p^1)^T & & & & & -(p^{in+1})^T & & & & \\
(p^1)^T & & & & & & -(p^{in+2})^T & & & \\
\vdots & & & & & & & \ddots & & \\
\vdots & & & & & & & & \ddots & \\
(p^1)^T & & & & & & & & & -(p^{in+out})^T \\
(p^1)^T & -(p^2)^T & & & & & & & & \\
& (p^2)^T & -(p^3)^T & & & & & & & \\
& \ddots & \ddots & & & & & & & \\
& & \ddots & \ddots & & & & & & \\
& & & (p^{in-1})^T & -(p^{in})^T & & & & & \\
(q^1)^T & (q^2)^T & \cdots & \cdots & (q^{in})^T & -(q^{in+1})^T & -(q^{in+2})^T & \cdots & \cdots & -(q^{in+out})^T
\end{pmatrix}
$$

Fig. 5. Simplified Matrix A

Theorem 1. *The above algorithm is correct, i.e., it calculates all vertices of the polyhedron P.*

For the proof of Theorem 1 we formulate

Lemma 2. *Let $\mathcal{S}$ be a feasible set in the sense of Definition 1. If $A_{\mathcal{S}}\lambda_{\mathcal{S}} = 0_{|\mathcal{N}|}$ has a nontrivial solution then the zero-extension of a* <u>*positive*</u> *solution of $A_{\mathcal{S}}\lambda_{\mathcal{S}} = b$ is not a vertex.*

Proof. Let $\bar{\lambda}_{\mathcal{S}}$ be a positive solution of $A_{\mathcal{S}}\lambda_{\mathcal{S}} = b$, i.e., it holds $A_{\mathcal{S}}\bar{\lambda}_{\mathcal{S}} = b$ with $\bar{\lambda}_{\mathcal{S}} > 0_{|\mathcal{S}|}$. Let $\bar{\lambda}$ be the zero extension of $\bar{\lambda}_{\mathcal{S}}$. We will show that $\bar{\lambda}$ is a nontrivial convex combination of two other points in P (which are elements of an $\epsilon-$environment ($\epsilon > 0$) of $\bar{\lambda}$). This shows that $\bar{\lambda}$ cannot be a vertex. We define for $\mathcal{S}$ a vector $\epsilon \in \mathbb{R}^{\mathcal{S}}$ with $\mathcal{S} \subseteq \mathcal{N}$ as follows:

From $A_{\mathcal{S}}\bar{\lambda}_{\mathcal{S}} = b$ and the condition $A_{\mathcal{S}}(\bar{\lambda}_{\mathcal{S}} + \epsilon) = b$ we get:

$$A_{\mathcal{S}}\epsilon = 0_{|\mathcal{N}|}$$

Obviously $\bar{\epsilon} = 0_{|\mathcal{S}|}$ is a solution. We know from the assumptions of Lemma 2 that $A_{\mathcal{S}}\epsilon = 0_{|\mathcal{N}|}$ has a nontrivial solution. Because of this we also know that the set of solutions of $A_{\mathcal{S}}\epsilon = 0_{|\mathcal{N}|}$ is a vector space (with nontrivial solutions). Therefore, there exists $\bar{\epsilon} \neq 0_{|\mathcal{S}|}$ such that $A_{\mathcal{S}}(\bar{\lambda}_{\mathcal{S}} + \bar{\epsilon}) = b$, with $\bar{\lambda}_{\mathcal{S}} + \bar{\epsilon} > 0_{|\mathcal{S}|}$ and $\bar{\lambda}_{\mathcal{S}} - \bar{\epsilon} > 0_{|\mathcal{S}|}$.

Now we built the zero-extension $(\bar{\lambda}_{\mathcal{S}} + \bar{\epsilon})^0(\mathcal{S})$ of $\bar{\lambda}_{\mathcal{S}} + \bar{\epsilon}$ and we get $(\bar{\lambda}_{\mathcal{S}} + \bar{\epsilon})^0(\mathcal{S}) \in P$. Observe that for $\mathcal{S}$ all $\lambda-$variables must fulfil the set condition by construction. Similarly, $A_{\mathcal{S}}(\bar{\lambda}_{\mathcal{S}} - \bar{\epsilon}) = A_{\mathcal{S}}\bar{\lambda}_{\mathcal{S}} - A_{\mathcal{S}}\bar{\epsilon} = A_{\mathcal{S}}\bar{\lambda}_{\mathcal{S}} - 0_{|\mathcal{N}|} = b$ and hence also $A(\bar{\lambda}_{\mathcal{S}} - \bar{\epsilon})^0(\mathcal{S}) = b$. We conclude $(\bar{\lambda}_{\mathcal{S}} - \bar{\epsilon})^0(\mathcal{S}) \in P$.
Finally,

$$\frac{1}{2}(\bar{\lambda}_{\mathcal{S}} + \bar{\epsilon}) + \frac{1}{2}(\bar{\lambda}_{\mathcal{S}} - \bar{\epsilon}) = \bar{\lambda}_{\mathcal{S}}.$$

and

$$\frac{1}{2}(\bar{\lambda}_{\mathcal{S}} + \bar{\epsilon})^0(\mathcal{S}) + \frac{1}{2}(\bar{\lambda}_{\mathcal{S}} - \bar{\epsilon})^0(\mathcal{S}) = (\bar{\lambda}_{\mathcal{S}})^0(\mathcal{S}) = \bar{\lambda}.$$

Since $\bar{\lambda}$ can be written as a convex sum of two other points of P it cannot be a vertex. □

We use the lemma in the following

Proof. We show Theorem 1 in two steps. At first we show that all calculated points are vertices of P and then we show that there cannot exist other vertices of P.

1) The calculated points are vertices of P.

From the first $in + out$ rows of A it is clear that for every segment at least one variable must be greater than zero. We define for a feasible subset $\mathcal{S} \subseteq \mathcal{N}$ and its characteristic vector $\mathcal{X}^{\mathcal{S}}$ the following inequality:

$$(\mathcal{X}^{\mathcal{S}})^T\lambda \leq in + out.$$

From the definition of P we see that this inequality is valid for P since the sum of all $\lambda-$variables of a point in P is always equal to $in + out$.

Let $\bar{\lambda} = \lambda^0(\mathcal{S}) \in P$ be the zero extension of $\lambda_{\mathcal{S}}$ calculated according the algorithm corresponding to $\mathcal{S}$. We show the following:

$$\{\bar{\lambda}\} = P \cap \{(\mathcal{X}^{\mathcal{S}})^T\lambda = in + out\}.$$

Since $\bar{\lambda} \in P$ by construction and $(\mathcal{X}^{\mathcal{S}})^T\bar{\lambda} = in + out$ by definition of $\mathcal{X}^{\mathcal{S}}$, the first inclusion $\{\bar{\lambda}\} \subseteq P \cap \{(\mathcal{X}^{\mathcal{S}})^T\lambda = in + out\}$ is trivial.

Now we show $\{\bar{\lambda}\} \supseteq P \cap \{(\mathcal{X}^{\mathcal{S}})^T\lambda = in + out\}$. Suppose there exists another point

$$\tilde{\lambda} \in (P \cap \{(\mathcal{X}^{\mathcal{S}})^T\lambda = in + out\}) \setminus \{\bar{\lambda}\}$$

Observe that $\tilde{\lambda}_i = 0$ for all $i \notin \mathcal{S}$. This implies that $\tilde{\lambda}$ is another solution to $A_{\mathcal{S}}\lambda_{\mathcal{S}} = b$, a contradiction to the construction of $\bar{\lambda}$.

2) There are no other vertices of P.

We have seen in the first part of this proof that the constructed points are indeed vertices of P. From Lemma 2 it is now easy to see that there are no other vertices of P. W.l.o.g. we can restrict ourselves to feasible sets $\mathcal{S}$ that produce a positive solution $\bar{\lambda}_{\mathcal{S}}$ of $A_{\mathcal{S}}\lambda_{\mathcal{S}} = b$ which is not unique. In this case we apply Lemma 2, because in this case $A_{\mathcal{S}}\lambda_{\mathcal{S}} = 0_{|\mathcal{N}|}$ must have a non-trivial solution. Let $\bar{\lambda} \in P$ be the zero-extension of $\bar{\lambda}_{\mathcal{S}}$. Thus $\bar{\lambda} \in P$ cannot be a vertex. □

From the theorem and its proof above we conclude that the non-convex polyhedron P can be written as a union of convex polytopes. This can be understood in the following way: In the case of the polyhedron P_Δ every selection of a triangle of the ingoing pipe combined with a selection of a triangle of the outgoing pipe defines a small polyhedron. By the zero-extension we get a polyhedron in the space of all λ-variables. The non-convex polyhedron P_Δ can evidently be understood as the union of all polyhedra in the space of all λ-variables that arise from all possible combinations of a triangle of the ingoing pipe and a triangle of the outgoing pipe. It is obvious that this idea can be extended to the general case of polyhedron P.

Example 1. We consider a simple example in order to demonstrate the essential parts of the notation (not all elements because the notation is much more complex than the idea behind it ...). Let us consider the following case of polyhedron P_Δ (a picture is shown in Figure 6. According to the picture holds: $n_1 = n_2 = 1$ with $|N_1^1| = |N_1^2| = 3$. Let the matrix A be:

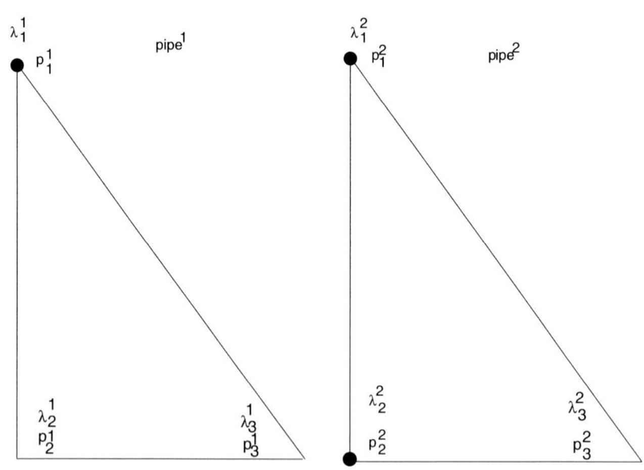

Fig. 6. Building vertices of the polyhedron P_Δ

$$A = \begin{pmatrix} 1 & 1 & 1 & 0 & 0 & 0 \\ 0 & 0 & 0 & 1 & 1 & 1 \\ 15 & 10 & 10 & -10 & -10 & -20 \\ 0 & 0 & 0 & 0 & 0 & 0 \end{pmatrix}$$

with

$$p^1 = \begin{pmatrix} 15 \\ 10 \\ 10 \end{pmatrix},$$

$$p^2 = \begin{pmatrix} 10 \\ 10 \\ 20 \end{pmatrix}.$$

Also

$$q^1 = q^2 = \begin{pmatrix} 0 \\ 0 \\ 0 \end{pmatrix}.$$

The vector b becomes

$$b = \begin{pmatrix} 1 \\ 1 \\ 0 \\ 0 \end{pmatrix}$$

We have already mentioned that if we do not want to model the gas flow preservation the last row of A can be omitted. We will do this from now on and in all upcoming examples. Because $rg(A) = 3$ we take as a first selection $\mathcal{S}_1 = \{S^1, S^2\}$ with $S^1 = \{1\}$ and $S^2 = \{4, 6\}$. Here $A_{\mathcal{S}_1}$ becomes

$$A_{\mathcal{S}_1} = \begin{pmatrix} 1 & 0 & 0 \\ 0 & 1 & 1 \\ 15 & -10 & -20 \end{pmatrix}$$

and according to our algorithm we have to solve:

$$A_{\mathcal{S}_1}\lambda_{\mathcal{S}_1} = \begin{pmatrix} 1 & 0 & 0 \\ 0 & 1 & 1 \\ 15 & -10 & -20 \end{pmatrix} \begin{pmatrix} \lambda_1^1 \\ \lambda_1^2 \\ \lambda_3^2 \end{pmatrix} = \begin{pmatrix} 1 \\ 1 \\ 0 \end{pmatrix}$$

We get as a unique (and also nonnegative) solution:

$$\lambda_{\mathcal{S}_1} = \begin{pmatrix} 1 \\ \frac{1}{2} \\ \frac{1}{2} \end{pmatrix}$$

The zero-extension of $\lambda_{\mathcal{S}_1}$

$$\begin{pmatrix} 1 \\ 0 \\ 0 \\ \frac{1}{2} \\ 0 \\ \frac{1}{2} \end{pmatrix}$$

is a vertex of P_Δ. If we take as a second selection $\mathcal{S}_2 = \{S^1, S^2\}$ with $S^1 = \{2\}$ and $S^2 = \{4, 5\}$ we have to solve:

$$A_{\mathcal{S}_2} \lambda_{\mathcal{S}_2} = \begin{pmatrix} 1 & 0 & 0 \\ 0 & 1 & 1 \\ 10 & -10 & -10 \end{pmatrix} \begin{pmatrix} \lambda_2^1 \\ \lambda_1^2 \\ \lambda_2^2 \end{pmatrix} = \begin{pmatrix} 1 \\ 1 \\ 0 \end{pmatrix}.$$

Here $rg(A_{\mathcal{S}_2}) = 2$ and so we know from Theorem 1 that $\mathcal{S}_2$ does not lead to a vertex, since $|\mathcal{S}_2| > 2$. But if we reduce $\mathcal{S}_2$ to $\mathcal{S}_3 = \{S^1, S^2\}$ with $S^1 = \{2\}$ and $S^2 = \{4\}$ we have to solve

$$A_{\mathcal{S}_3} \lambda_{\mathcal{S}_3} = \begin{pmatrix} 1 & 0 \\ 0 & 1 \\ 10 & -10 \end{pmatrix} \begin{pmatrix} \lambda_2^1 \\ \lambda_1^2 \end{pmatrix} = \begin{pmatrix} 1 \\ 1 \\ 0 \end{pmatrix},$$

and we get a unique (and nonnegative) solution

$$\begin{pmatrix} 1 \\ 1 \end{pmatrix},$$

which fullfills all demanded properties that we have pointed out in Algorithm 1. Thus the zero-extension of this vector

$$\begin{pmatrix} 0 \\ 1 \\ 0 \\ 1 \\ 0 \\ 0 \end{pmatrix}$$

yields a vertex of P_Δ. We will discuss the general case of P_Δ in the next example in a more detailed way.

3.2 The construction of cuts and the separation algorithm

The algorithm we have described above can now be used to construct cutting planes for our MIP model. Unfortunately, we do not know the facets that are describing P, since they are relatively difficult to describe even in quite easy situations like the sequence of two pipes described at the beginning of the paper.

Here we give a simple example for this situation. Clearly in more complicated or in realistic situations for the Transient Technical Optimisation the problem of describing the facets normally becomes bigger and bigger.

The following table gives an impression how the complexity of the vertices and the facets of P_Δ increases with the increasing number of grid points.

The first column describes the number of triangles (the sum in the in- and outgoing pipe), the second column the number of $\lambda-$variables, the third column describes the number of vertices, the fourth the number facet-defining inequalities and the last column the maximal coefficient within the facet-defining inequalities. The facets have been calculated with the program **Porta Version 1.3**, see [4].

Δ	λ	vertices	facets	max. coeff.
8	12	16	18	25
16	18	49	47	42
24	24	73	90	670
32	32	142	10492	50640

We have tested several other examples and usually we get the situation that if we add to P_Δ the first law of Kirchhoff the number of vertices and facets is lower than in the case above (this is clear because we have to fullfill more conditions) but the coefficients of the facets are getting worse (but this cannot be proved in general).

So we cannot yet calculate the facets until now but –blessing in disguise– we have seen that we can calculate the vertices of the polyhedron P. Now it is on time to show what we can do with them.

In order to use the vertices it is first very important to see that in all interesting cases there are only polynomially many of vertices which we can calculate algorithmically in addition.

Lemma 3. *For the polyhedron P (with the usual definitions and notations as used before) exist numbers l, c such that the maximal number of vertices of P is less than or equal to cl^{in+out}.*

Proof. Define a number l^* as

$$l^* := \prod_{j=1}^{in+out} n_j \tag{1}$$

where the values $n_j, j = 1, 2, \ldots, in + out$ were defined as the number of subsets in which the set of $\lambda-$variables of the in- and outgoing segments are divided. It is clear that l^* is the number of possible combinations of subsets N_k^i from all in- and outgoing segments were from every segment exactly one subset according to Algorithm 1 is taken. We remark that in the special case P_Δ the values n_j are the number of triangles of the triangulation for the in- and outgoing pipe.

It is necessary for a vertex that the non vanishing $\lambda-$variables belong to exactly one such subset for each segment. Let $m \leq rg(A)$ be the maximal number of non vanishing $\lambda-$variables as it was pointed out in Algorithm 1 (remark that this number already has been calculated). Only in order to blow up the notation not too much we define for $j \in \{1, 2, \ldots, in + out\}$ numbers N^j_{max}:

$$N^j_{max} := max\{|N^j_1|, |N^j_2|, \ldots, |N^j_{n_j}|\}$$

Then take for $j \in \{1, 2, \ldots, in + out\}$ variables x_j which can be positive (natural) numbers and after that define a number c as:

$$c := \sum_{\sum_{j=1}^{in+out} x_j \leq m} \prod_{j=1}^{in+out} \binom{N^j_{max}}{x_j}$$

We remark that by construction $\sum_{j=1}^{in+out} x_j \geq in + out$. The interpretation of c is as follows:
c is an upper bound for the maximal number of possible vertices for the selection of subsets in (4). This is clear because we sum over all selections of $\lambda-$variables (resp. the chosen subsets in $\mathcal{S}$) for which the number of selected $\lambda-$variables is not greater than m. Additionally the product of the binomial coefficients calculates the maximal number of possibilities how we can choose the $\sum_{j=1}^{in+out} x_j$ $\lambda-$variables out of the sets of $\lambda-$variables belonging to the selected feasible subsets. We conclude that the number of vertices cannot be greater than cl^*.
Define

$$l := max\{n_1, n_2, \ldots, n_{in+out}\}$$

Summarising our argumentation we finally conclude that the number of vertices cannot be greater than cl^{in+out}. □

Note that a trivial upper bound for c is

$$c = 2^{\sum_{j=1}^{in+out} N^j_{max}}.$$

But this value for c is a good deal worse than the (even not quite good) value we have given in Proof 3.2.

We see that in the case of the polyhedron P_Δ the polynomiality of Algorithm 1 follows since $m = 3$. Also the polynomiality of Algorithm 1 in the general case of polyhedron P follows since N^j_{max} and m are bounded from above. The estimation in the above lemma will be much bigger than the real number of vertices in the polyhedron. For the example in Figure 7 we obtain:

$$c = \binom{3}{1}\binom{3}{1} + \binom{3}{1}\binom{3}{2} + \binom{3}{2}\binom{3}{1} = 27$$

and $l = 4$, and thus the maximal number of vertices is $27 * 4^{1+1} = 432$.

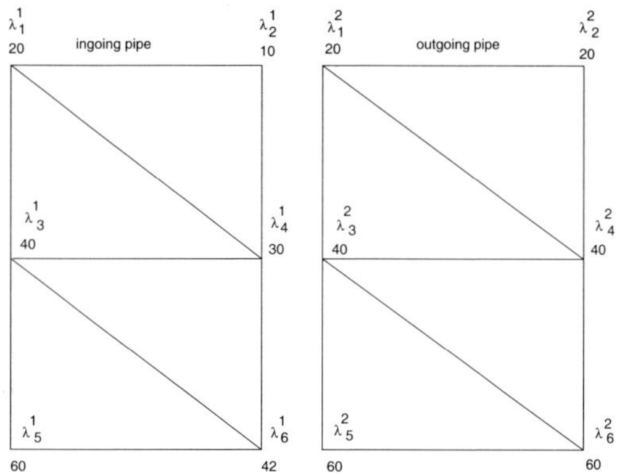

Fig. 7. Example for comparing vertices and facets

Indeed there are only 16 vertices. Although our estimation is bad it suffices to show that the vertices can be calculated in polynomial time.
The number of vertices is usually noticeable lower than the maximal number of vertices. To give a reason for this consider the following

Lemma 4. *Let $\mathcal{S}, \bar{\mathcal{S}}$ be two feasible sets (of $\lambda-$variables) in Algorithm* 1 *with $\mathcal{S} \subseteq \bar{\mathcal{S}}$. If both sets lead to a vertex of P according to Algorithm* 1 *they are identical.*

Proof. A vertex of P regarding to $\mathcal{S}$ is the zero-extension of a unique, non-negative and not vanishing solution of $A_{\mathcal{S}}\lambda_{\mathcal{S}} = b$ (see the description of the algorithm). The same holds for $\bar{\mathcal{S}}$. Adding to the solution of $A_{\mathcal{S}}\lambda_{\mathcal{S}} = b$ the $\lambda-$variables of $\bar{S}^i \setminus S^i$ for all $i \in \{1, 2, \ldots, in + out\}$ which we set to zero. We get a solution of $A_{\bar{\mathcal{S}}}\lambda_{\bar{\mathcal{S}}} = b$. This solution must be the unique solution of $A_{\bar{\mathcal{S}}}\lambda_{\bar{\mathcal{S}}} = b$ by assumption.

Analogously we argue when we start from a vertex calculated from $\bar{\mathcal{S}}$. If there would be a vertex belonging to the selection $\mathcal{S}$ we can conclude in the same way as above that the vertices must be equal.

Lemma 4 has an interesting consequence: If we have found a vertex for a feasible set $\mathcal{S}$ (of a selection of $\lambda-$variables) it is not necessary to search for vertices in a superset of $\mathcal{S}$. Therefore we can start with the feasible sets in which we take exactly one $\lambda-$variable for each segment, i.e., $|S^i| = 1 \, \forall i \in \{1, 2, \ldots, in + out\}$, and then look for "bigger" (with respect to set inclusion) feasible sets of selected $\lambda-$variables. In this way we can find all needed vertices in a systematic way.

Another possibility is to start with feasible sets $\mathcal{S}$ for which $|\mathcal{S}| = rg(A)$ holds. If for such a set a vertex is found you do not need to search for a vertex in any subset of this set. This procedure starts from the "biggest" selections whereas the first one starts from the "smallest". In realistic cases (of course you can always construct some pathological cases) this strategy will find the vertices much faster as we have studied in the case of a sequence of two pipes where we modelled the gas flow equation. It turns out that in data sets from gas networks mostly $rg(A) = rg(A_{\mathcal{S}})$ holds for a feasible set $\mathcal{S}$.

For the polyhedron P_Δ Lemma 4 has a nice consequence for the maximal number of vertices:

Corollary 1. *An upper bound for the number of vertices of P_Δ is*

$$9\, n_1\, n_2.$$

Proof. We have described the possibilities for constructing vertices of the polyhedron P_Δ. We have seen that for each choice of two triangles there are 9 possibilities for the selection of one λ–variable from the ingoing pipe and one λ–variable from the outgoing pipe, i.e., $|S^1| = |S^2| = 1$. Now either this or one of the extensions where we add one λ–variable either in the chosen triangle of the ingoing pipe or the chosen triangle of the outgoing pipe may result in a vertex, cf Lemma 4. From this argument directly follows Lemma 1.

Note from Lemma 3 we just obtain a bound of $27\, n_1\, n_2$, since

$$c = \binom{3}{1}\binom{3}{1} + \binom{3}{1}\binom{3}{2} + \binom{3}{2}\binom{3}{1} = 27.$$

Let us come back now to our primal aim. All the previous ventilations give us the possibility to develop the following **separation algorithm** for P:

Let $v_1, \ldots, v_k$ be the constructed vertices for P (in realistic situations they can be calculated very fast as we have described above).

Let $\bar{\lambda}$ be an optimal LP-solution to be cut off. We look for a cut of the form $a^T x \leq \alpha$ by solving

$$\begin{aligned} z^* &= max\ a^T \bar{\lambda} - \alpha \\ &\quad s.t.\ a^T v_i \leq \alpha \quad \text{for } i = 1, \ldots, k \end{aligned}$$

We remark that w.l.o.g we can assume $\alpha \in \{0, 1, -1\}$.

Let $(\bar{a}, \bar{\alpha})$ be such that $\bar{a}^T \bar{\lambda} - \bar{\alpha} = z^*$.

(a) $\bar{a}^T \lambda \leq \bar{\alpha}$ is valid for P.

Proof. We know from the theory of linear optimisation that every feasible point of the polytope P can be combined as a convex combination of its vertices $v_1, v_2, \ldots, v_k$ (this is correct although P is not convex). That is for $\lambda^* \in P$ there exist nonnegative real numbers $\beta_1, \beta_2, \ldots, \beta_k$ with $\sum_{i=1}^k \beta_i = 1$ such that:

$$\lambda^* = \sum_{i=1}^{k} \beta_i v_i$$

We then calculate:

$$\bar{a}^T \lambda^* = \bar{a}^T \sum_{i=1}^{k} \beta_i v_i = \sum_{i=1}^{k} \beta_i (\bar{a}^T v_i) \leq \sum_{i=1}^{k} \beta_i \bar{\alpha} = \bar{\alpha} \sum_{i=1}^{k} \beta_i = \bar{\alpha}.$$

Therefore $\bar{a}^T \lambda \leq \bar{\alpha}$ is valid for P.

(b) There exists a violated cut if and only if $z^* > 0$.

Proof. If $z^* > 0$ then due to (a) $\bar{a}^T \lambda \leq \alpha$ is such a cut. On the other hand, suppose $\tilde{a}^T \lambda \leq \tilde{\alpha}$ is a valid inequality violated by $\bar{\lambda}$ then $z^* \geq \tilde{a}^T \bar{\lambda} - \bar{\alpha} > 0$.

4 Computational Results

We have tested our implementation of the algorithm for the polyhedron $P_\triangle$ for a gas network which consists of three compressors and ten pipes. This gas network is shown in Figure 8. In our first formulation of the model we used the

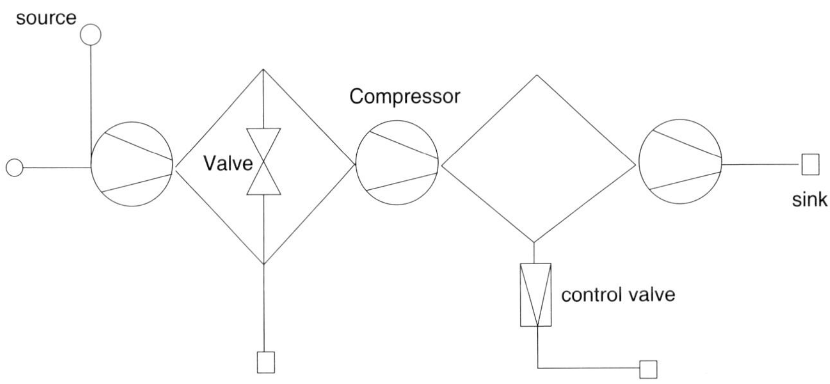

Fig. 8. A simple gas network

traditional way of the introduction of binary variables for modelling piecewise linear functions. That is we introduce for each triangle $i \in \Lambda$ a binary variable y_i and model the fact that all positive $\lambda-$variables must belong to the same triangle. The computational results for this model are indicated by y in the following table. The table shows our experiences of the computational progress when incorporating the polynomial separation algorithm instead of binary variables in a branch-and-cut algorithm (here the compressors are formulated with binary variables but the pipes are using already the cuts obtained from the separation algorithm). $p_{in,C}$ is the number of grid points used for the pressure at the beginning of a compressor. $p_{out,C}$ analogously describes the number of grid points for the pressure at the end of a compressor. q_C is the number of grid points for the gas flow of the compressor. $p_{in,P}$ is the number of grid points used for pressure at the beginning of a pipe and q_P means the number of grid points for the gas flow in the pipe. In the rows in which the number of user cuts (constructed by the separation algorithm) is zero the problem was calculated by the formulation with binary variables. We see that the use of cuts constructed by the separation algorithm reduces the calculation time about factor 10. Let us give a short comment about our implementation: We have used CPLEX as the MIP solver for our test instances and used the cutcallback capabilities of CPLEX to add our own cutting planes.

compressors ($\square, y$)			pipes ($\triangle$)		Solution			
$p_{in,C}$	$p_{out,C}$	q_C	$p_{in,P}$	q_P	CPLEX cuts	User cuts	Opt val	sec
3	3	7	4	10	29	0	9.39	3.07
3	3	7	4*	10*	6	10	9.36	0.79
3	3	7	8	20	28	0	9.16	295.9
3	3	7	8*	20*	7	204	9.15	23.09

Figure 9 shows the situation before using the constructed cuts. The solid lined pipes do not fulfill the triangle (set) conditions whereas the dotted pipes do. In Figure 10 we can see the situation after the use of the separation algorithm. We see that in Figure 10 still one pipe does not fulfill the triangle condition. The reason for this is that the polyhedron P_Δ (in general the polyhedron P) is not convex. So in some cases it can be possible that the solution to be cut off lies in the interior of the convex hull of the polyhedron P but not in P itself. Such points cannot be cut off by a valid inequality constructed by the separation algorithm.

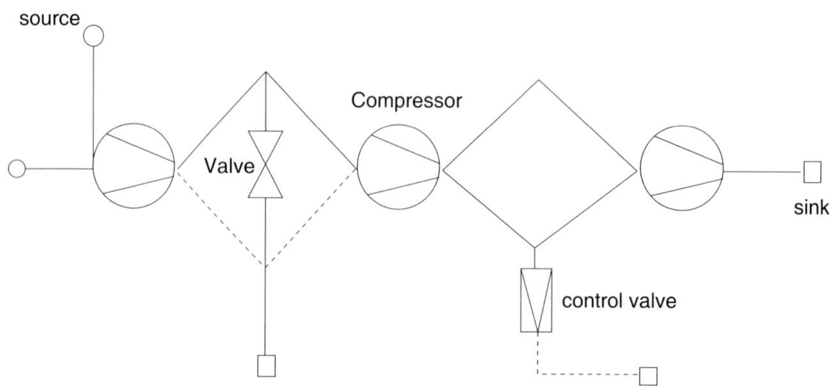

Fig. 9. The test modell before the separation algorithm

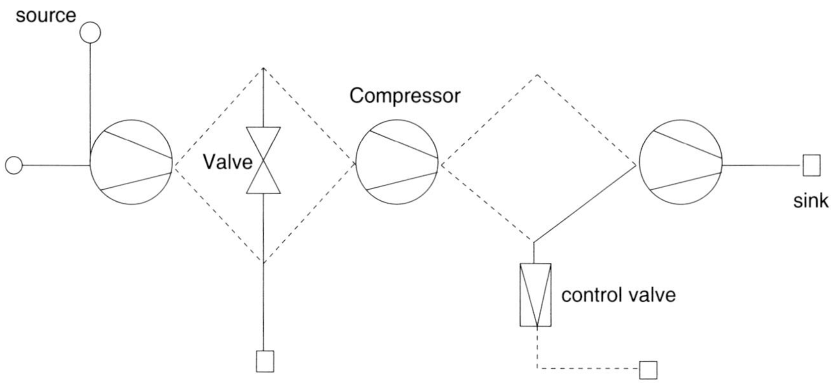

Fig. 10. The test modell after the separation algorithm

5 Conclusions

Although we have not yet implemented the separation algorithm for compressors or for more complex subsystems than the sequence of two pipes the theoretical knowledge of the vertices and the separation algorithm gives us the possibility to extend our branch-and-cut algorithm to complexer gas networks and it is very supposable that in this way the solution time can be reduced significantly.

References

[1] SEKIRNJAK, E. (1998): *Mixed Integer Optimization for Gas Transmission and Distribution Systems.* INFORMS Meeting, Seattle, October 1998

[2] SEKIRNJAK, E. (2000): *Transiente Technische Optimierung (TTO) Prototyp* Technical Report for use in the PSI AG, Ruhrgas AG

[3] BEALE, E.L.M, TOMLIN, J.A. (1970): *Special Facilities in a General Mathematical Programming System for Nonconvex Problems Using Ordered Sets of Variables* Proceedings of the Fith International Conference on Operations Research, Tavistock Publications, pp. 447-454

[4] CHRISTOF, T., LÖBEL, A. (2000): *PORTA: Polyhedron Representation Transformation Algorithm, Version 1.3* Konrad-Zuse-Zentrum für Informationstechnik, Berlin

Clustering Algorithms for Parallel Car-Crash Simulation Analysis

Liquan Mei[1] and Clemens A. Thole[2]

School of Science, Xi'an Jiaotong University, 710049, Xi'an, China
lqmei@mail.xjtu.edu.cn
Fraunhofer Institute for Algorithms and Scientific Computing
Schloss Birlinghoven, 53754, St. Augustin, Germany
clemens-august.thole@scai.fhg.de

Summary. Buckling and certain contact situations cause scattering results of numerical crash simulation: For a BMW model differences between the position of a node in two simulation runs of up to 10 cm were observed, just as a result of round-off differences in the case of parallel computing. An engineer has to measure this scatter, to check whether important parts of the car show such indeterministic behavior and to find the origins. The tool DIFF-CRASH compares simulation results and uses data mining technology to cluster those nodes of the car model, which show similar scatter among the simulation runs. For the BMW model the indeterministic behavior could be traced back to a certain part of the motor carrier and was removed by a redesign. DIFF-CRASH is the only activity using data mining technology for crash simulation stability analysis. In this paper we present the clustering algorithm and illustrate its usage in car crash simulation analysis.

Key words: Crash Simulation, Data Mining, Cluster Analysis

1 Introduction

Nowadays the car manufacturing industry relies heavily on simulation results. By simulation the number of real prototypes is reduced, the insight into the features of the actual design is increased and the turn-around time between model changes is much shorter than in the case of real tests. Numerical crash simulation is the most computer-time consuming simulation task in car design. Therefore it is obvious that crash simulation codes were among the first industrial simulation codes, which were ported onto parallel distributed memory architectures during the EUROPORT project[1] [10] (1994-96).

[1] The EUROPORT Project was funded by the European Commission as part of the Esprit programme.

Using the mpp-versions of industrial crash simulation codes the engineers made a surprising discovery for certain models: The result of numerical simulation changed from one parallel execution to the next by more than 10 cm for the node positions, although the input decks and the simulation parameters were identical. Figure 1 shows a model provided by BMW consisting of about 60.000 shell elements and maximal and average differences between several simulation runs. Actually this observation has stopped car manufacturing companies from using mpp-system for crash simulation for more than 5 years.

As part of the PROMENVIR project[2] [8] (1996-97) a stochastic analysis tool was developed (now named STORM), which automatically changes certain parameters of input deck, performs simulations, extracts a set of parameters of the results and analyses the dependency between input parameters and result parameters. Using PROMENVIR it was possible to show that small changes in the input deck may result in substantial changes of the simulation results and no correlation between changes and results may be available.

Geometric scatter analysis is performed in weather forecast, and stability analysis is usually performed by stochastic variation of some design parameters and correlation analysis for some key measures like the intrusion. As part of the AUTOBENCH Project[3] [14] (1998-01) and the AUTO-OPT Project[4](2002-05), the reasons for the scatter of the results were investigated in detail. It turned out that numerical properties of the simulation codes as well as certain features of the car design may be responsible for the "butterfly effects". Typical sources of instabilities are buckling and contact of different parts under an angle of 90°. During the investigation an analysis tool named DIFF-CRASH^{TM} was developed.

2 DIFF-CRASH^{TM} Overview

DIFF-CRASH^{TM} is a tool for the detailed analysis of the scatter of simulation results. Currently it supports PAM-CRASH, one of four leading commercial simulation codes. The preprocessing module of DIFF-CRASH modifies node positions in an input deck in order to generate a set of simulation results. The postprocessing module takes the output files of several PAM-CRASH runs, performs a detailed analysis, computes values for each node and reported time step and adds these functions to one result file (see Figure 2). Standard postprocessing tools can be used for the visualisation of DIFF-CRASH results. Typical result functions of DIFF-CRASH are:

[2] The PROMENVIR Project was funded by the European Commission as part of the Esprit programme.

[3] The AUTOBENCH Project was funded by the German Minister for Education and Research (BMB+F).

[4] The AUTO-OPT Project was funded by the German Minister for Education and Research (BMB+F).

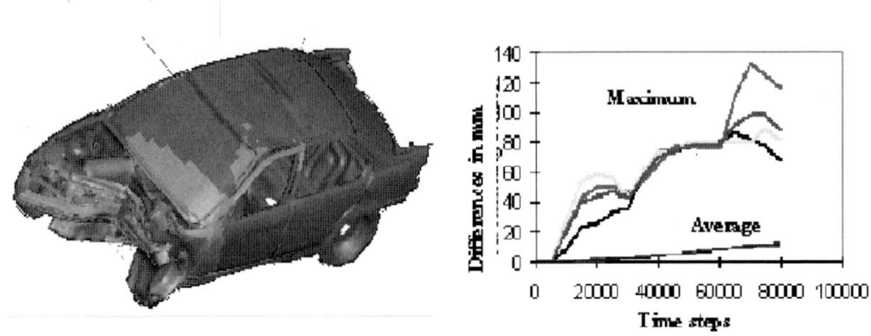

Fig. 1. BMW model after a 40% offset crash using PAM-CRASH and the differences between simulation runs on a 32 node IBM SP2. The colour version of this figure can be found in Fig. A.22 on page 587.

- Maximal difference of the position of a node at a specific time step over all simulation runs.
- The sequence number of the extreme simulation runs for each node and time step.
- The dependence of the scatter of the results of each node and time step to a selected reference point.

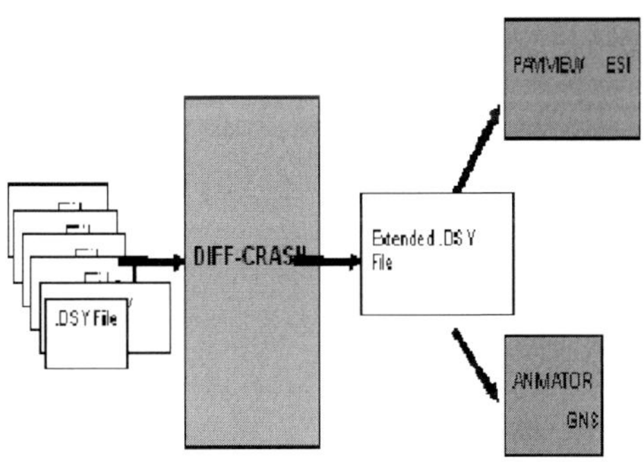

Fig. 2. DIFF-CRASH postprocessing principle of operation.

A detailed discussion of the different functions is contained in [3]. Figure 1 shows a typical DIFF-CRASH result for the testcase at 81 ms. The color indicates the scatter of the results at each point. Red areas are those with the largest values. The large scatter of the wheel in Figure 1 is not important for design engineers. More important is the scatter of parts of the fire wall close to the driver´s feet shown in Figure 3. In order to improve the model, it is

Fig. 3. Scatter of simulation results on the motor carrier and in the interior. The colour version of this figure can be found in Fig. A.23 on page 588.

important to distinguish between local effects and impacts of scatter sources at other places, like the part of the motor carrier also shown in Figure 3.

3 Similarity function

In order to distinguish source and impact of indeterministic behavior a set of functions was developed, each of which computes the relation of the scatter of the results at two nodes and (potentially different) time steps with each other.

Let $X(p,t,s)$ be the position of node p at time t during the crash simulation process of run s out of a total number of S simulation runs. $X(p,t) = \{X(p,t,s)|s \in (1 \cdots S)\}$ is the ordered set of node positions and

$$M_{ij}(p,t) = \frac{\|X(p,t,i) - X(p,t,j)\|}{\max\limits_{k,l}(\|X(p,t,l) - X(p,t,k)\|)} \tag{1}$$

is the related scaled distance matrix. The similarity function $sim(X1, X2)$ then defines a relation between two sets of node positions. A computed value of 1 indicates that the scatter of the nodes is the same and a value over a certain threshold b_{sim} indicates a significant relation between the scatter of these two node position sets. If the scatter of the node positions of two nodes at the two time steps $(p1, t1)$ and $(p2, t2)$ is similar, the scatter at these two points is itself likely to have the same origin [12].

For a given reference (p_{ref}, t_{ref}) at any node and time step (p, t), DIFF-CRASH implements a function, which returns $sim(X(p,t), X(p_{ref}, t_{ref}))$.

A whole set of similarity functions is discussed in [12]. For the concrete results of the clustering algorithm in the following, the function sim with a threshold b_{sim} of 0.666 is used. This function is based on scaled distance matrixes.

4 Clustering function

The similarity function turned out to be quite useful. However, it is necessary to select a reference node and time step. The analysis process would be faster, if a function could identify sets of nodes and time steps (p, t), which are related to each other.

Clustering is to Propose to develop algorithms that automate the process of finding the number of group(clusters), and find high-quality solutions to group memberships. Data Mining for scientific applications in particular cluster analysis is used in many different application areas, especially for simulation data here. One challenge for this process is the problem size. A typical crash simulation code consists of 500,000 nodes and for a dependence analysis up to 150 time steps may be used. The data bases of objects to be compared therefore contain 70 million objects. Fortunately, the geometric position of the nodes provides some structure, which can be exploited.

Another difficulty known from the literature [6] is the fact that the assignment to clusters is not a deterministic process. One node might be related to two others, which are not related with each other. Therefore this node might be assigned to any of the two reference clusters. In order to reduce these effects, the clustering process is performed twice.

The algorithm used for crash simulation analysis is performed in the following steps:

1. Select a reference time step, a minimal cluster size and a threshold.
2. Perform a preclustering for this time step:
   · Perform clustering $(Cluster(N))$ for this time step
   · Find the center node for each cluster
   · Perform the second $Cluster(N)$ to check for all related nodes to these center nodes
   · Sort center nodes by the size of clusters to get a new node sequence.
3. For each time step
   · Perform $Cluster(N)$ by the new node sequence.
4. Output.

In the following, each of these steps is detailed.

Selection of reference time step

In order to reduce the complexity of the analysis, a reference time step t_{ref} is selected. The algorithm starts clustering the nodes for this time step and then uses the result as a basis for all other time steps. In practice, this has not been a major problem. The last time step usually contains the most significant and largest amount of scatter and will be selected as reference.

In addition, a minimal size of a cluster $S_{cluster}$ needs to be specified, in order to limit the total number of clusters.

Preclustering

This step builds a first set of clusters $C_i(t_{ref})$ for the selected time step. Each cluster consists of a number of nodes. Nodes are tested to be included in a cluster, if one of its neighboring nodes has successfully been added to this cluster. In this context two nodes are defined to be neighbors, if they belong to the same finite element. As representation for a cluster, a distance matrix is used. This distance matrix is the average of the scaled distance matrixes of all nodes. This means, that

$$M(C) = \frac{1}{size(C)} \sum_{p \in C} M(p, t_{ref}), \tag{2}$$

where $size(C)$ returns the number of nodes in C.

For the preclustering the following algorithm has been chosen:

Cluster(N): N is an ordered list of nodes
- clear list A of active nodes
- loop until list N is empty
  - ∘ get first node P from list N and form a new cluster C containing P
  - ∘ add all neighboring nodes of P, which are not assigned to a cluster, to list A of active nodes.
  - ∘ loop until A is empty
    - ⋄ get first node Q from list A
    - ⋄ if $sim(M(C), X(Q, t_{ref})) \geq b_{sim}$ then
      - ⋆ add node Q to cluster C
      - ⋆ take node Q from list N
      - ⋆Add all neighboring nodes of P, which are not assigned to a cluster and not already on list A of active nodes, to this list.
    - ⋄ end if
  - ∘ end loop
- end loop

Cluster(N) is executed with N representing a list of all nodes of the car model. As a result all nodes are assigned to clusters.

Selection of a representative node for each cluster and performing a second clustering

The clustering algorithm is very indeterministic and might result in too many clusters. Therefore the clustering process is started again. The clustering is based on the representative node

$$P(C) = \{p \in C | \begin{matrix} sim(M(C), X(p, t_{ref})) \geq sim(M(C), X(q, t_{ref})) \ \forall q \in C, \\ \text{and } p \text{ is the first such node in } C \end{matrix} \} \tag{3}$$

The new ordered list of nodes N consists of the list of representative nodes in the sequence of the size of the cluster (largest first), followed by all other nodes.

The algorithm **cluster(N)** is executed again with this new list.

Assign all other nodes of all other time steps to these clusters

The resulting set of clusters from the last step is ordered by size and all clusters smaller than S_{cluster} are eliminated.

For each cluster a list $T(C)$ of tuples (p, t) is set up. Each tuple references a node and a time step. $T(C)$ is initialised with tuples of all nodes of cluster C and time step t_{ref}.

For each time step t and for all nodes p, $T(C)$ is extended by (p, t), if $sim(M(C), X(p, t)) \geq b_{sim}$ and if C is the first cluster in the list, for which this condition holds.

The *simcluster* function assigns an integer value to each tuple (p, t) as follows:

$$simcluster(p, t) = \{ \begin{matrix} \text{index of } C \text{ in the list of clusters,} & \text{if} (p, t) \in T(C) \\ 0, & \text{else} \end{matrix} \} \tag{4}$$

5 Results and experiences

Figure 4 shows the result of the cluster algorithm as colors on the test case. The time at the reference time step was 81 ms. Each color represents a new cluster. Dark blue nodes were not assigned to any cluster because of the threshold for the minimal cluster size (300 nodes). In total the algorithm identifies 13 different clusters. The figure shows several clusters in the frontal part of the car and one dominating cluster covering most of the rest of the body.

One cluster dominates the fire wall in the area of the driver's legs for those nodes with the largest scatter (Figure 4(right)).

According to Figure 5, the "green" cluster shows up between 40 and 50 ms for the first time on the fire wall and gets larger. Figure 6 contains a picture of the motor carrier at 38 ms, part of which is covered by the green cluster. This cluster starts when the shock absorber pipe hits the rest of the motor carrier near 28 ms. At this point in time the shock absorber pipe puts pressure on the motor carrier parallel to the direction of this part. This causes a buckling effect. The results for the motor carrier are shown in Figure 7. The form of the motor carrier at the end of the simulation is completely different. The simcluster function indicates, that this causes the substantial scatter at the fire wall.

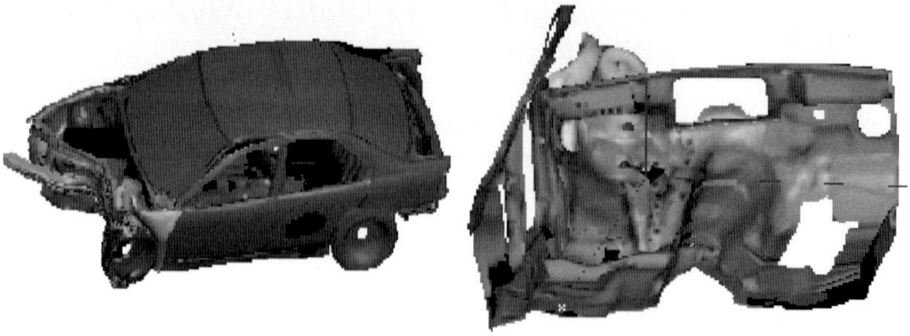

Fig. 4. Simcluster as color for BMW testcase of whole car and the interior at time 80ms. The colour version of this figure can be found in Fig. A.24 on page 588.

Fig. 5. Development of the clusters with time (70ms, 60ms, 50ms, 40ms). The colour version of this figure can be found in Fig. A.25 on page 588.

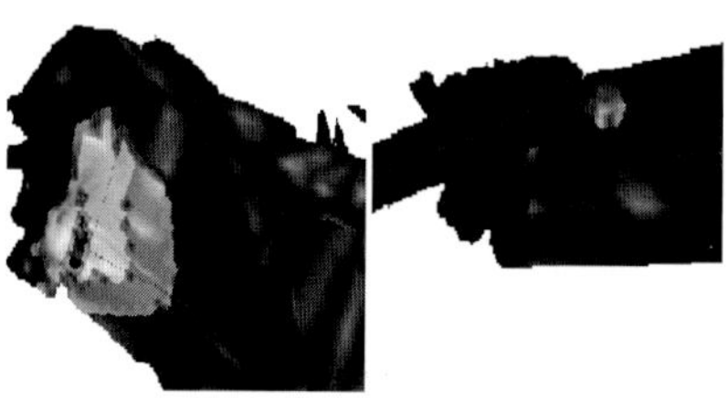

Fig. 6. Simcluster for the motor carrier at time 35ms (left) and at 28ms (right) for its inner part. The colour version of this figure can be found in Fig. A.26 on page 589.

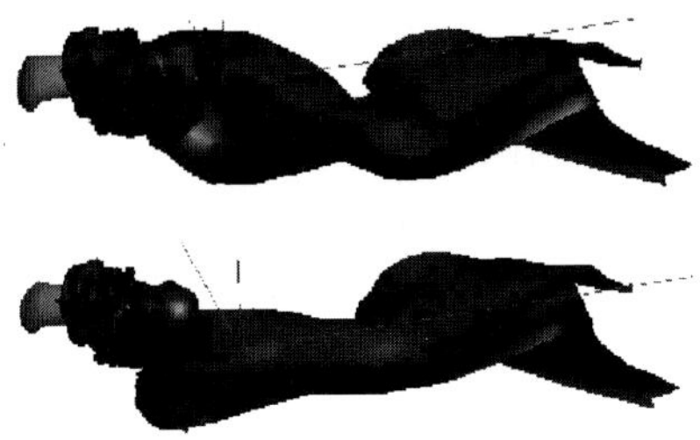

Fig. 7. Top view on the motor carrier at 80ms for two extremely different simulation runs. The colour version of this figure can be found in Fig. A.27 on page 589.

Conclusions and perspectives

DIFF-CRASH is the only activity using data mining technology for crash simulation stability analysis. Stability analysis is usually performed by stochastic variation of some design parameters and correlation analysis for some key measures like the intrusion. The paper has shown that clustering algorithms can be useful in identifying the origin of the scatter of simulation results in crash simulation. As a result of the analysis using DIFF-CRASH the motor carrier was modified in the areas of the origin of the instability. As a consequence the scatter of the result due to parallel computing on the fire wall was reduced substantially [12].

This application is only a first step using data mining technology in the context of industrial crash simulation. Car manufacturing companies store the complete outputs of their crash simulation runs. This provides an excellent basis for data mining applications, like design of experiments, optimisation, parameter fitting for coarser models used in concept studies.

Acknowledgement

The authors thank Dr. Luebbing of BMW and C. Thibaud from ESI Germany for their fruitful collaboration and support during the AUTOBENCH project. Thanks also to BMW for the permission to publish the results. The authors point out that the model does not have any relation to an existing production car of BMW.

References

[1] Berry, M.W.; Drmac, R.; Jessup E.R.: Matrices, vector spaces and information retrieval. SIAM review, 41(2):335:362 (1999)

[2] Berry, M.W.; Dumais, S.; O'Brien, G.: Using linear algebra for intelligent information retrieval, SIAM Review, 37(4):573-595 (1995)

[3] Bendisch, J.; von Trotha, H.: Investigation on car stability in crash simulations. In ATTCE Proceedings Volume 1 Safety, SAE International (2001)
[4] Hamerly, G.: Learning structure and concepts in data through data clustering, doctoral thesis of Uni. of California, San Diego (2002)
[5] Han, J.; Kamber, M.: Data Mining Concepts and Techniques, Morgan Kaufmann (2000)
[6] Jain, A.K.; Dubes, R.C.: Algorithms for Clustering Data, Prentice Hall (1988)
[7] Karypis, G.; Han,J.; Kumar, V.: CHAMELEON: A hierarchical clustering algorithm using dynamic modeling, Computer,32:68-75 (1999)
[8] Marczyk, J.: Principles of simulation-based computer-aided engineering. FIM Publications, Barcelona (1999)
[9] Marczyk, J.: Computational stochastic mechanics in meta-computing perspective, International Center for Numerical Methods in Engineering, Barcelona (1997)
[10] Mierendorff, H.; Stueben, K.; Thomas, O.: EUROPORT-1 Porting industrial codes to parallel architectures. In: Hertzberger, B.; Serazzi, G. (eds.); High Performance Computing and Networking, Lecture Notes in Computer Science, Number 919, Springer, Berlin, Heidelberg (1995)
[11] Pizzuti, C., Talia, D.: Using SVD for data mining of high dimensional data sets, 2nd workshop on mining scientific datasets, AHPCRC (2000)
[12] Thole, C.A., Mei, L.: Comparison of several similarity functions for stability analysis of crash simulation results. SCAI Report, FhG-SCAI Sankt Augustin (2002)
[13] Thomas, R.S., Nolan, N.W.: Once is not enough - A few more thoughts, Sound and Vibration (1994)
[14] Thole, C., Kolibal, S.; Wolf, K.: AUTOBENCH - Virtual prototypes for automotive industry. In Deville, M.; Owen, R. (eds): 16th IMACS World Congress 2000 Proceedings, IMACS, Rutgers University, New Brunswick (2000)

A General-Purpose Finite Element Method for 3D Line Transfer Problems with Application on Galaxies in the Early Universe

Erik Meinköhn[1,2]

[1] Institute f. Theoret. Astrophysics, University of Heidelberg
Tiergartenstr. 15, D–69121 Heidelberg, Germany
[2] Institute of Applied Mathematics, University of Heidelberg
Im Neuenheimer Feld 294/293, D–69120 Heidelberg, Germany
erik.meinkoehn@iwr.uni-heidelberg.de

1 Introduction

Hydrogen Lyman-α (Lyα) as a prominent emission line of high-redshift galaxies is important for the understanding of galaxy formation and evolution in the early universe. Aside from being a good redshift indicator, Lyα emission also bears information on the distribution and kinematics of the interstellar gas as well as the nature of the energy source. Lyα observations of high redshift radio galaxies (cf. [14], [15]) reveal extended Lyα halos with sizes $> 3 \times 10^{15}$ km. These halos usually show an extremely clumpy density structure, with typical length scales which are several order of magnitude smaller than the halo size. The observed line profiles are often double-peaked and the line spectra point to complex kinematics involving velocities $> 1000\,\mathrm{kms}^{-1}$.

The interpretation of Lyα observations is difficult, because high-redshift radio galaxies tend to be in the center of proto clusters, where the radio jet interacts with a clumpy environment influenced by merging processes. Actually, the three-dimensional structure of the objects and the fact that Lyα is a resonance line require detailed radiative transfer modeling.

The transfer of resonance line photons is profoundly determined by scattering in space and frequency. Analytical [12] as well as early numerical methods were restricted to one-dimensional, static media (cf. [1],[6]). Only recently, codes based on the Monte Carlo method were developed which are capable to investigate the more general case of a multi-dimensional, scattering medium (see [2],[3]).

In this paper a finite element method for solving the resonance line transfer problem in moving media is presented. The algorithm works in three spatial dimensions on unstructured grids which are adaptively refined by means of an a posteriori error indicator. Additionally, an ordinate parallelization is performed to deal with the resulting extremely large linear system of equations. The solution is obtained using an iterative procedure, where several monochromatic radiative transfer problems are solved successively. Thus, a fast and accurate solution strategy for the monochromatic transfer problem is crucial to simulate the extended frequency-dependent model efficiently (for more details see [13]). Section 2 provides a detailed description of the finite element discretzation of the monochromatic 3D radiative transfer problem. In Sect. 3 an algorithm for the solution of the frequency-dependent line transfer problem is presented. The code accounts for arbitrary velocity fields up to 10% of the speed of light and complete redistribution. The latter is critical for the correct modeling of Lyα line profiles in optically thick media. A sample of line profiles for a spherically symmetric 3D density distribution is presented in Sect. 5 illustrating the underlying physics. More realistic and complex 3D configurations are published in [13, 8, 9].

2 Monochromatic 3D Radiative Transfer

2.1 The Radiative Transfer Problem

Our aim is the calculation of the radiation field in diffuse matter in space. Assuming the matter is surrounded by a vacuum, we will only consider a convex domain containing the area of interest. Radiation leaving this domain will not enter it again. Inside this 3D domain $\Omega \subset R^3$, the specific intensity I satisfies the monochromatic radiative transfer equation

$$\begin{aligned} &\mathbf{n} \cdot \nabla_x I(\mathbf{x}, \mathbf{n}) + \kappa(\mathbf{x}) I(\mathbf{x}, \mathbf{n}) \\ &+\mathbf{s}(\mathbf{x})\Big(I(\mathbf{x}, \mathbf{n}) - \frac{1}{4\pi}\int_{S^2} P(\mathbf{n}', \mathbf{n}) I(\mathbf{x}, \mathbf{n}')\, d\mathbf{t}'\Big) = f(\mathbf{x}), \end{aligned} \tag{1}$$

where $\mathbf{x} \in \Omega$ is the space variable and $\mathbf{n}$ the direction pointing to the solid angle $d\mathbf{t}$ of the unit sphere S^2. The optical properties of the matter are given by the absorption coefficient $\kappa(\mathbf{x})$ and the scattering coefficient $\mathbf{s}(\mathbf{x})$. The angular phase function P occurring in the scattering integral is normalized, such that $\frac{1}{4\pi}\int P(\mathbf{n}', \mathbf{n})\, d\mathbf{t}' = 1$. The source term

$$f(\mathbf{x}) = \kappa(\mathbf{x}) B(T(\mathbf{x})) + \epsilon(\mathbf{x}) \tag{2}$$

consists of thermal emission depending on a temperature distribution $T(\mathbf{x})$ and an additional emissivity $\epsilon(\mathbf{x})$ of a point source or an extended object. B is the Planck-Function. To be able to solve (1), boundary conditions of the form $I(\mathbf{x}, \mathbf{n}) = g(\mathbf{x}, \mathbf{n})$ must be imposed on the "inflow boundary" $\Gamma_- = \{(\mathbf{x}, \mathbf{n}) \in \Gamma | \mathbf{n}_\Gamma \cdot \mathbf{n} < 0\}$, where $\mathbf{n}_\Gamma$ is the unit vector perpendicular to the boundary surface Γ. The sign of the product $\mathbf{n}_\Gamma \cdot \mathbf{n}$ describes the "flow direction" of the photons across the boundary. If there are no light sources outside the modeled domain, the function g will be zero everywhere.

The left hand side of Eq. (1) will be abbreviated as an operator **A** applied to the intensity function I, yielding the very compact operator form of the radiative transfer equation

$$\mathcal{A}I(\mathbf{x}, \mathbf{n}) = f(\mathbf{x}). \tag{3}$$

This equation is five-dimensional, three variables characterizing the spatial domain and two variables describing the photon propagation. In addition, the numerical solution is complicated by rapidly changing values of I in very small parts of the domain and smooth transport in larger regions. Using a reasonable resolution of $h = 1/1000$ in space and 1000 directions or *ordinates* already results in 10^{12} unknowns in the 3D case. Since this amount of data cannot be handled even on most advanced supercomputers, the application of efficient error estimation and grid adaption techniques is necessary both to reduce memory requirements and CPU time and to obtain reliable quantitative results. Finite element methods, in particular so-called Galerkin methods, are most suitable for these techniques.

2.2 Finite Element Discretization

Equation (1) is analyzed in [5] and the natural space for finding solutions is

$$W = \{I \in L^2(\Omega \times S^2) \mid \mathbf{n} \cdot \nabla_x I \in L^2(\Omega \times S^2)\}, \tag{4}$$

where L^2 is the Lebesgue space of the square integrable functions. If we consider homogeneous vacuum boundary condition, i.e. $g(\mathbf{x}, \mathbf{n}) = 0$, the solution space is

$$W_0 = \{I \in W \mid I = 0 \text{ on } \Gamma_-\}. \tag{5}$$

In order to apply a finite element method, we have to use a weak formulation of (3). Therefore, we multiply both sides of (1) by a trial function $\varphi(\mathbf{x}, \mathbf{n})$ and integrate over the whole domain $\Omega \times S^2$. By extending the definition of the L^2-scalar product we introduce the abbreviation

$$(I, \varphi) = (I, \varphi)_{\Omega \times S^2} = \int_\Omega \int_{S^2} I\varphi \, d\mathbf{t} \, d^3x. \tag{6}$$

Thus, the weak formulation reads: Find $I \in W_0$, such that $\forall \varphi \in W_0$

$$(\mathbf{A}I, \varphi) = (f, \varphi) \qquad \forall \varphi \in W_0. \tag{7}$$

If there is no scattering, i.e. $\mathbf{s}(\mathbf{x}) = 0$ on Ω, the problem decouples to a system of convection equations on Ω. These equations are hyperbolic. If the solutions are not smooth, standard finite element techniques applied to this type of equations are known to produce spurious oscillations. We can achieve stability by applying the streamline diffusion modification:

$$(\mathbf{A}I, \varphi + \delta\mathbf{n}\cdot\nabla_x\varphi) = (f, \varphi + \delta\,\mathbf{n}\cdot\nabla_x\varphi) \quad \forall\,\varphi \in W_0. \tag{8}$$

The cell-wise constant parameter function δ depends on the local mesh width and the coefficients κ and $\mathbf{s}$. Note that the solution of (7) solves (8), too. No additional consistency error is induced by the stabilization. In the following, the streamline diffusion discretization term will be omitted but implicitly assumed.

Applying standard Galerkin finite elements to solve (7), we choose a finite dimensional subspace W_h of W consisting of functions that are piecewise polynomials with respect to a subdivision or *triangulation* T_h of $\Omega \times S^2$. The mesh size h is the piecewise constant function defined on each triangulation cell K by $h|_K = h_K = \mathrm{diam}K$. The discrete analogue of (7) is finding $I_h \in W_h$, such that

$$(\mathbf{A}I_h, \varphi_h) = (f, \varphi_h) \quad \forall\,\varphi_h \in W_h. \tag{9}$$

The construction of the subspace W_h needs some further consideration (see [7]). The discretized domain is a tensor product of two sets of completely different length scales: While Ω represents a domain in physical space, S^2 is the unit sphere in the Euclidean space R^3. Therefore, we use a tensor product splitting of the five-dimensional domain $\Omega \times S^2$, such that a grid cell of the five-dimensional grid will be the tensor product of a two-dimensional cell $K_\mathbf{t}$ and a three dimensional cell K_x. Accordingly, the mesh sizes with respect to $\mathbf{x}$ and $\mathbf{t}$ will be different.

For the space domain Ω we use locally refined hexahedral meshes. The mesh size with respect to the space variable will be denoted by h in the course of this publication. Since the boundaries are arbitrary for our astrophysical application, we can choose a unit cube for Ω and do not have to worry about boundary approximation. We use continuous piecewise trilinear trial functions in space.

3 Polychromatic 3D Line Transfer

3.1 Line Transfer in Moving Media

We calculate the frequency-dependent radiation field in moving media by solving the non-relativistic radiative transfer equation in the comoving frame for a three-dimensional domain Ω which in operator form simply reads

$$(\mathcal{T} + \mathcal{D} + \mathcal{S} + \chi(\mathbf{x},\nu))\mathcal{I}(\mathbf{x},\mathbf{n},\nu) = f(\mathbf{x},\nu). \tag{10}$$

The transfer operator $\mathcal{T}$, the "Doppler" operator $\mathcal{D}$, and the scattering operator $\mathcal{S}$ are defined by

$$\begin{aligned}
\mathcal{T}\mathcal{I}(\mathbf{x},\mathbf{n},\nu) &= \mathbf{n}\cdot\nabla_x\mathcal{I}(\mathbf{x},\mathbf{n},\nu),\\
\mathcal{D}\mathcal{I}(\mathbf{x},\mathbf{n},\nu) &= w(\mathbf{x},\mathbf{n})\,\nu\frac{\partial}{\partial\nu}\mathcal{I}(\mathbf{x},\mathbf{n},\nu),\\
\mathcal{S}\mathcal{I}(\mathbf{x},\mathbf{n},\nu) &= -\frac{\sigma(\mathbf{x})}{4\pi}\int_0^\infty\int_{S^2} R(\hat{\mathbf{n}},\hat{\nu};\mathbf{n},\nu)\mathcal{I}(\mathbf{x},\hat{\mathbf{n}},\hat{\nu})\,d\hat{\omega}\,d\hat{\nu}.
\end{aligned}$$

Considering three dimensions in space, the relativistic invariant specific intensity $\mathcal{I}$ is six-dimensional and depends on the space variable $\mathbf{x}$, the direction $\mathbf{n}$ (pointing to the solid angle $d\omega$ of the unit sphere S^2), and the frequency ν.

The extinction coefficient $\chi(\mathbf{x},\nu) = \kappa(\mathbf{x},\nu) + \sigma(\mathbf{x},\nu)$ is the sum of the absorption coefficient $\kappa(\mathbf{x},\nu) = \kappa(\mathbf{x})\varphi(\nu)$ and the scattering coefficient $\sigma(\mathbf{x},\nu) = \sigma(\mathbf{x})\varphi(\nu)$. The frequency-dependence is given by a normalized profile function $\varphi \in L^1(R^+)$. Usually, we adopt a Doppler profile

$$\varphi(\nu) = \frac{1}{\sqrt{\pi}\,\Delta\nu_D} \exp\left[-\left(\frac{\nu-\nu_0}{\Delta\nu_D}\right)^2\right], \tag{11}$$

where ν_0 is the frequency of the line center. The Doppler width $\Delta\nu_D$ and the Doppler velocity v_D are determined by a thermal velocity v_{therm} of the hydrogen atoms as well as a macroscopic turbulent velocity v_{turb}

$$\Delta\nu_D = \frac{\nu_0}{c} v_D = \frac{\nu_0}{c}\sqrt{v_{\text{therm}}^2 + v_{\text{turb}}^2}. \tag{12}$$

c is the speed of light. For situations, where $v_{\text{turb}} \gg v_{\text{therm}}$, the Doppler core is very broad and would dominate the Lorentzian wings of a Voigt profile. Then, the Doppler profile is a reasonable description of a line profile.

For the source term

$$f(\mathbf{x},\nu) = \kappa(\mathbf{x},\nu)B(T(\mathbf{x}),\nu) + \epsilon(\mathbf{x},\nu), \tag{13}$$

we can consider thermal radiation and non-thermal radiation. In the case of thermal radiation, f is calculated from a temperature distribution $T(\mathbf{x})$, where $B(T,\nu)$ is the Planck function.

The "Doppler" operator $\mathcal{D}$ is responsible for the Doppler shift of the photons. A derivation of the operator for non-relativistic velocities ($v/c < 0.1$) can be found in [16]. In contrast to the full relativistic transfer equation (cf. [11]), we neglect any terms responsible for aberration or advection effects. The function

$$w(\mathbf{x},\mathbf{n}) = -\mathbf{n}\cdot\nabla_x\left(\mathbf{n}\cdot\frac{\mathbf{v}(\mathbf{x})}{c}\right) \tag{14}$$

is the gradient of the velocity field $\mathbf{v}(\mathbf{x})$ in direction $\mathbf{n}$. Note that the sign of w may change depending on the complexity of the velocity field $\mathbf{v}$.

The scattering operator $\mathcal{S}$ depends on the general redistribution function $R(\hat{\mathbf{n}},\hat{\nu};\mathbf{n},\nu)$ giving the probability that a photon $(\hat{\mathbf{n}},\hat{\nu})$ is absorbed and a photon $(\mathbf{n},\nu)$ is emitted. In the following, we assume isotropic scattering

$$\mathcal{S}\mathcal{I} = -\frac{\mathbf{s}(\mathbf{x})}{4\pi}\int_0^\infty R(\hat{\nu},\nu)\int_{S^2}\mathcal{I}(\mathbf{x},\hat{\mathbf{n}},\hat{\nu})\,d\hat{\omega}\,d\hat{\nu}, \tag{15}$$

where $R(\hat{\nu},\nu)$ is the angle-averaged redistribution function

$$R(\hat{\nu},\nu) = \frac{1}{(4\pi)^2} \int_{S^2} \int_{S^2} R(\mathbf{x},\hat{\mathbf{n}},\hat{\nu};\mathbf{n},\nu)\, d\hat{\omega}\, d\omega. \tag{16}$$

The function defined by (16) is normalized such that

$$\int_0^\infty \int_0^\infty R(\hat{\nu},\nu)\, d\hat{\nu}\, d\nu = 1. \tag{17}$$

In [8] two limiting cases are considered: strict coherence and complete redistribution in the comoving frame. In the former case, we have

$$R(\hat{\nu},\nu) = \varphi(\hat{\nu})\delta(\nu - \hat{\nu}) \tag{18}$$

and in the latter

$$R(\hat{\nu},\nu) = \varphi(\hat{\nu})\varphi(\nu). \tag{19}$$

Thus, for coherent isotropic scattering, the scattering term simplifies to

$$\mathcal{S}^{\mathrm{coh}}\mathcal{I}(\mathbf{x},\mathbf{n},\nu) = -\frac{\sigma(\mathbf{x},\nu)}{4\pi} \int_{S^2} \mathcal{I}(\mathbf{x},\hat{\mathbf{n}},\nu)\, d\hat{\omega} \tag{20}$$

In the case of complete redistribution, the photons are scattered isotropically in angle, but are randomly redistributed over the line profile. Then, the scattering term reads

$$\mathcal{S}^{\mathrm{crd}}\mathcal{I}(\mathbf{x},\mathbf{n},\nu) = -\frac{\sigma(\mathbf{x},\nu)}{4\pi} \int_0^\infty \varphi(\hat{\nu}) \int_{S^2} \mathcal{I}(\mathbf{x},\hat{\mathbf{n}},\hat{\nu})\, d\hat{\omega}\, d\hat{\nu}. \tag{21}$$

In the course of this paper, we would like to discuss complete redistribution only, since this is the appropriate approximation for modeling resonance lines like Lyα (see [10]).

3.2 Boundary Conditions

For the modeling of prominent resonance lines, in particular Lyα, we restrict the frequency discretization to a finite interval $\Lambda := [\nu_0, \nu_{N+1}]$, where ν_0 and ν_{N+1} are located far out of the line center in the continuum. The function $w(\mathbf{x},\mathbf{n})$ from Eq. (14) defines the Doppler shift of the spectrum towards smaller ($w < 0$) or larger ($w > 0$) frequencies in the comoving frame. Therefore, boundary conditions of the form $\mathcal{I}(\mathbf{x},\mathbf{n},\nu) = \mathcal{I}_{\mathrm{cont}}(\mathbf{x},\mathbf{n},\nu)$ are necessary on the upper and lower frequency interval boundary of the whole computational domain $\Sigma = \Omega \times \mathcal{S}^2 \times \Lambda$. Furthermore, to be able to solve Eq. (10), boundary conditions of the form $\mathcal{I}(\mathbf{x},\mathbf{n},\nu) = \mathcal{I}_{\mathrm{in}}(\mathbf{x},\mathbf{n},\nu)$ must be imposed on the "inflow boundary" $\Gamma^- \times \Lambda = \{(\mathbf{x},\mathbf{n},\nu) \in \Gamma | \, \mathbf{n}_\Gamma \cdot \mathbf{n} < 0\}$,where $\mathbf{n}_\Gamma$ is the unit vector perpendicular to the boundary surface Γ of the spatial domain Ω. The sign of the product $\mathbf{n}_\Gamma \cdot \mathbf{n}$ describes the "flow direction" of the photons across the boundary. If we neglect any continuum emission ($\mathcal{I}_{\mathrm{cont}} = 0$) and assume that there are no light sources outside the modeled domain as in the case of a non-illuminated atmosphere ($\mathcal{I}_{\text{in}} = 0$), the two boundary conditions for the solution of the transfer equation (10) in moving media are

$$\mathcal{I}(\mathbf{x}, \mathbf{n}, \nu) = 0 \qquad \text{on } \Sigma, \tag{22}$$

$$\mathcal{I}(\mathbf{x}, \mathbf{n}, \nu) = 0 \qquad \text{on } \Gamma^{-} \times \Lambda. \tag{23}$$

3.3 Finite Element Discretization

A full Galerkin discretization of the frequency-dependent radiative transfer problem is presented. Considering memory and CPU requirements, the solution of the complete system is by far too "expensive". Thus, N quasi-monochromatic radiative transfer problems are solved successively for each discrete frequency point $\nu_i \in \{\nu_1, \nu_2, ..., \nu_N\} \subset \Lambda$. The additional frequency derivative, introduced by the velocity field, is discretized via a discontinous Galerkin (DG) method of grad zero, which can be identified with an implicit Euler method for N equidistantly distributed frequency points (see [4]).

Considering complete redistribution, we use a simple quadrature method for the frequency integral in the scattering operator $\mathcal{S}^{\text{crd}}$ in Eq. (21). Starting from N equidistantly distributed frequency points $\nu_i \in \{\nu_1, \nu_2, ..., \nu_N\} \subset \Lambda$ and N weights $q_1, q_2, ..., q_N$, we define a quadrature method

$$Q(\nu_i) := \sum_{j=1}^{N} q_j \xi(\nu_j) \tag{24}$$

for integrals $\int_\Lambda \xi(\nu')d\nu'$. In the case of complete redistribution the kernel is

$$\xi(\nu_j) = \frac{\varphi(\nu_j)}{4\pi} \int_{S^2} \mathcal{I}(\mathbf{x}, \hat{\mathbf{n}}, \nu_j) d\hat{\omega}. \tag{25}$$

Separating the terms with the unknown intensities $\mathcal{I}_i$ from the known quantities $\mathcal{I}_j$ the discretized scattering integral (21) reads

$$\frac{\sigma_i}{4\pi} \varphi_i q_i \int_{S^2} \mathcal{I}_i d\hat{\omega} + \frac{\sigma_i}{4\pi} \sum_{j \neq i}^{N} \varphi_j q_j \int_{S^2} \mathcal{I}(\mathbf{x}, \hat{\mathbf{n}}, \nu_j) d\hat{\omega}. \tag{26}$$

Using this quadrature scheme and employing the Euler method for the frequency derivatives we get a semi-discrete formulation of the transfer problem for each frequency point including complete redistribution

$$\begin{aligned} &\left(\mathbf{A}_i^{\text{crd}} + \frac{|w|\nu_i}{\Delta\nu}\right)\mathcal{I}_i \\ &= \tilde{f}_i + \frac{\sigma_i}{4\pi} \sum_{j \neq i}^{N} \varphi_j q_j \int_{S^2} \mathcal{I}(\mathbf{x}, \hat{\mathbf{n}}, \nu_j) d\hat{\omega}, \end{aligned} \tag{27}$$

where

$$\mathbf{A}_i^{\mathrm{crd}} = \mathcal{T} + \chi_i + \varphi_i q_i \mathcal{S}^{\mathrm{coh}}. \tag{28}$$

The additional terms on the right hand side of Eq. (27) must be interpreted as artificial source terms. Eq. (27) can also be written in a compact operator form

$$\tilde{\mathbf{A}}_i^{\mathrm{crd}} \mathcal{I}_i = \hat{f}_i. \tag{29}$$

The total discrete system has the matrix form

$$\mathbf{A}^{\mathrm{crd}} \mathbf{u} = \mathbf{f}. \tag{30}$$

Unfortunately, the global frequency coupling via the scattering integral (21) results in a full block matrix

$$\mathbf{A}^{\mathrm{crd}} = \begin{pmatrix} \tilde{\mathbf{A}}_1^{\mathrm{crd}} & \mathbf{R}_1 + \mathbf{Q}_2 & \mathbf{Q}_3 & \dots \mathbf{Q}_N \\ \mathbf{B}_2 + \mathbf{Q}_1 & \tilde{\mathbf{A}}_2^{\mathrm{crd}} & \mathbf{R}_2 + \mathbf{Q}_3 & \ddots \;\; \vdots \\ \mathbf{Q}_1 & \ddots & \ddots & \ddots \;\; \vdots \\ \vdots & \ddots & \ddots & \ddots \;\; \vdots \\ \mathbf{Q}_1 & \dots & \dots & \dots \tilde{\mathbf{A}}_N^{\mathrm{crd}} \end{pmatrix}. \tag{31}$$

According to the sign of $w(\mathbf{x}, \mathbf{n})$ the block matrices $\mathbf{R}_i$ and $\mathbf{B}_i$ hold entries of $w(\mathbf{x}, \mathbf{n})\nu_i/\Delta\nu$ causing a redshift and blueshift of the photons in the medium, respectively. The block matrices $\mathbf{Q}_j$ hold the terms from the quadrature scheme.

Requiring a reasonable resolution, the resulting linear system of equations of the total system is too large to be solved directly. Hence, we are treating N "monochromatic" radiative transfer problems

$$\tilde{\mathbf{A}}_i^{\mathrm{crd}} \mathbf{u}_i = \hat{\mathbf{f}}_i, \tag{32}$$

with the right hand side

$$\hat{\mathbf{f}}_i = \mathbf{f}_i + \mathbf{R}_i \mathbf{u}_{i+1} + \mathbf{B}_i \mathbf{u}_{i-1} + \sum_{j \neq i} \mathbf{Q}_j \mathbf{u}_j. \tag{33}$$

3.4 Full Solution Algorithm

Equation (29) is of the same form as the monochromatic radiative transfer equation, cf. Eq. (3), for which we proposed a finite element discretization in Sect. 2. The finite element method employs unstructured grids which are adaptively refined by means of an a posteriori error estimation strategy (for details see [7, 8, 9]). Now, we apply this method to Eq. (29). In brief, the full solution algorithm reads:

1. Start with $\mathcal{I} = 0$ for all frequencies.

2. Solve Eq. (29) for $i = 1, .., N$.
3. Repeat step 2 until convergence is reached.
4. Refine grid and repeat step 2 and 3.

We start with a relatively coarse grid, where only the most important structures are pre-refined, and assure that the frequency interval $[\nu_1, \nu_N]$ is wide enough to cover the total line profile. Then, we solve Eq. (29) for each frequency several times depending on the choice of the redistribution function and the velocity field. During this fix point iteration, we monitor the changes of the resulting line profile in a particular direction $\mathbf{n}_{\text{out}}$. Not until the line profile remains unchanged, we go over to step 4 and refine the spatial grid. We apply the so-called fixed fraction grid refinement strategy: The cells are ordered according to the size of the local refinement indicator $\eta_K = max(\eta_K(\nu_i))|_{\nu_i}$ and a fixed portion of the cells with largest η_K is refined. $\eta_K(\nu_i)$ is an indicator for the error of the solution in cell K at frequency ν_i (for further details see [7, 8, 9]).

4 Test calculations

The test calculations are performed using a spherically symmetric 3D distribution of the extinction coefficient $\chi(\mathbf{x}) = \chi(x, y, z)$ of the form

$$\chi(\mathbf{x}) = \begin{cases} \chi_0/(1+\alpha r_c^2) & \text{for } r \le r_c \\ \chi_0/(1+\alpha r^2) & \text{for } r_c < r \le r_h \\ \chi_0/(1+\alpha r_h^2)/10^3 & \text{for } r > r_h \end{cases}, \tag{34}$$

where $r^2 = x^2 + y^2 + z^2$. Note, that the value of the extinction coefficient is constant in the center within the core radius r_c and outside the halo radius r_h. χ_0 is determined from the line center optical depth

$$\tau = \int_{r_c}^{r_h} \chi(\mathbf{x})\varphi(\nu_0)\,\mathbf{n}\,d\mathbf{x} \tag{35}$$

between r_c and r_h along the photon direction $\mathbf{n}$. In total, the spatial distribution of χ is determined by three parameters: the radii r_c and r_h, the dimensionless parameter α, and the optical depth τ. For r_c, r_h and α we use the values given in Tab. 1.

Since we are predominantly interested in the transfer of radiation in resonance lines like Lyα, we assume $\sigma(\mathbf{x}) = \chi(\mathbf{x})$ and $\kappa(\mathbf{x}) = 0$ for all calculations presented here. This restricts us to the use of purely non-thermal source functions. In particular, we consider one spatially confined source region with radius r_s centered at the position $\mathbf{x}_i = 0$:

$$f(\mathbf{x}, \nu) = \begin{cases} \varphi(\nu) & \text{for } \quad |\mathbf{x} - \mathbf{x}_i| \le r_s \\ 0 & \text{for } \quad |\mathbf{x} - \mathbf{x}_i| > r_s \end{cases}. \tag{36}$$

Table 1. Parameters used for all calculations. Distances are given in units of the computational unit cube.

r_h	r_c	α	r_s	v_D	v_0	r_0	R_0
1.0	0.2	10^3	0.2	$10^{-3}c$	$-10^{-3}c$	0.2	1.0

The function $\varphi(\nu)$ is the Doppler profile defined in Eq. (11).

In general, the finite element code is able to consider arbitrary velocity fields. For velocity fields defined on a discrete grid, e.g. resulting from hydrodynamical simulations, the velocity gradient in direction **n** must be obtained numerically. Here, we use two simple velocity fields and calculate the function w analytically.

The first velocity field describes a spherically symmetric inflow ($v_0 < 0$) or outflow ($v_0 > 0$) and is of the form

$$\mathbf{v}_{\mathrm{io}} = v_0 \left(\frac{r_0}{r}\right)^l \frac{\mathbf{x}}{r}, \tag{37}$$

where $r = |\mathbf{x}|$ and v_0 the scalar velocity at radius r_0. The corresponding w function is

$$w(\mathbf{x}, \mathbf{n}) = v_0 \left(\frac{r_0}{r}\right)^l \left(\frac{1}{r} - (l+1)\frac{|\mathbf{nx}|}{r^3}\right). \tag{38}$$

For the second velocity field, we assume rotation around the z-axis

$$\mathbf{v}_{\mathrm{rot}} = v_0 \left(\frac{R_0}{R}\right)^l R^{-1} \begin{pmatrix} y \\ -x \\ 0 \end{pmatrix}, \tag{39}$$

where $R^2 = x^2 + y^2$ is the distance from the rotational axis and v_0 the scalar velocity at distance R_0. If $\mathbf{n} = (n_x, n_y, n_z)$, the w function reads

$$w = v_0 \left(\frac{R_0}{R}\right)^l (l+1) \left(\frac{xy(n_y^2 - n_x^2) + n_x n_y (x^2 - y^2)}{R^3}\right). \tag{40}$$

Fig. 1 shows the results of the finite element code for different optical depths, velocity fields and redistribution functions. We used 41 frequencies equally spaced in the interval $(\nu - \nu_0)/\Delta\nu_D = [-4, 6]$ and 80 directions. Starting with a grid of 4^3 cells and a pre-refined source region, we needed 3–5 spatial refinement steps.

The simplest case is a static model with coherent isotropic scattering. Fig. 1a displays the emergent line profiles for different τ. As expected, the Doppler profile is preserved and the flux F_ν is independent on τ. The deviation of the numerical results from the analytical solution indicated with crosses is very small. The line profiles for $\tau = 0.1$ and $\tau = 1$ are identical even in the little window which shows the peak of the line in more detail. For $\tau = 100$,

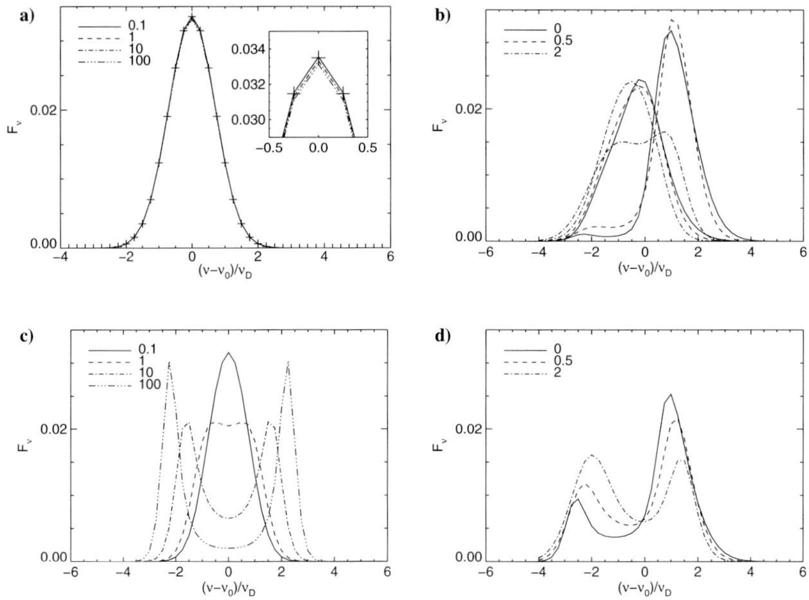

Fig. 1. Lyα line profiles calculated with the finite element code for a spherically symmetric model configuration: a) a static halo with coherent scattering, b) an inflowing halo with coherent scattering, c) a static halo with complete redistribution, and d) an inflowing halo with complete redistribution. For the static cases a) and c) the line styles refer to calculations with different optical depth τ as indicated. The small window in a) enlarges the peak of the line. The crosses mark the results of the analytical solution. For the moving halos we show in b) the results for $\tau = 1$ (thin lines) and $\tau = 10$ (thick lines) and in d) only for $\tau = 10$. Here, the line styles refer to the exponent l used for the velocity fields.

the total flux is still conserved better than 99%. This result demonstrates that the frequency-dependent version of our finite element code works correctly.

Next, we consider an inflowing halo with coherent scattering and show the effects of frequency coupling due to the Doppler term. The emergent line profiles in Fig. 1b are plotted for different exponents l of the velocity field $\mathbf{v}_{\rm io}$ defined in Eq. (37). The line profiles are redshifted for $\tau = 1$ (thin lines). Most of the photons directly travel through the halo moving away from the observer. Since the Doppler term is proportional to the gradient of the velocity field, the redshift is larger for a greater exponent l. For $\tau = 10$ (thick lines), the line profiles are blueshifted. Before photons escape from the optically thick halo in front of the source, they are back-scattered and blueshifted in the approaching halo behind the source. The blueshift is less pronounced for the accelerated inflow with $l = 2$, because the strong gradient of the velocity field leads to a slight redshift in the very inner parts of the halo. In this region, the total optical depth is still small. Further out, where the total optical depth increases, the line profile becomes blueshifted.

Complete redistribution is another method of frequency coupling which leads to a stronger coupling than the Doppler effect (see Sect. 3). The line profiles obtained for a static model with complete redistribution are displayed in Fig. 1c for different τ. With increasing optical depth the photons more and more escape through the line wings. For $\tau \geq 1$, we get a double-peaked line profile with an absorption trough in the line center. The greater τ the larger the distance between the peaks and the depth of the absorption trough. Since our frequency resolution is too poor for the pointed wings, the flux conservation is only 96% for $\tau = 100$.

Fig. 1d shows the results for an inflowing halo with complete redistribution for $\tau = 10$ and different exponents l. For $l = 0$ and $l = 0.5$ the inflowing motion of the halo enhances the blue wing of the double-peaked line profile. Equally, an outflowing halo would enhance the red peak. But for $l = 2$, the red peak is slightly enhanced for an inflowing halo due to the strong velocity gradient, as explained above. This example affords an insight into the mechanisms of resonance line formation and shows the necessity of a detailed treatment of multi-dimensional radiative transfer problems.

5 Summary

A finite element code for solving the resonance line transfer problem in moving media is presented. Simple velocity fields and complete redistribution are considered. The code is applicable to any three-dimensional model configuration with optical depths up to 10^{3-4}.

The application to the hydrogen Lyα line of slightly optically thick model configurations ($\tau \leq 10^2$) are shown and the resulting line profiles are discussed in detail. The systematic approach from very simple to more complex models gave the following results:

- An optical depth of $\tau \geq 1$ leads to the characteristic double peaked line profile with a central absorption trough as expected from analytical studies (e.g. [12]). This form of the profile is the result of scattering in space and frequency. Photons escape via the line wings where the optical depth is much lower.
- Global velocity fields destroy the symmetry of the line profile. Generally, the blue peak of the profile is enhanced for models with inflow motion and the red peak for models with outflow motion. But there are certain velocity fields (e.g. with steep gradients) and spatial distributions of the extinction coefficient, where the formation of a prominent peak is suppressed.

The applications demonstrate the capacity of the finite element code and show that the complex angle-frequency coupling and the kinematics of the model configurations are very important. Additionally, more complex three-dimensional configurations strongly affect the shape of the line profiles. For the sake of simplicity, the test calculations were preformed only for spherically symmetric density distributions. More realistic three-dimensional configurations and modeled line profiles are published in [13, 8, 9].

Acknowledgements. Erik Meinköhn would like to thank Sabine Richling, Rainer Wehrse and Guido Kanschat for their collaboration. This work is supported by the Deutsche Forschungsgemeinschaft (DFG) within the SFB 359 "Reactive Flows, Diffusion and Transport".

References

[1] Adams, T.F.: The escape of resonance-line radiation from extremely opaque media. ApJ 174, 439 (1972)

[2] Ahn, S.-H., Lee, H.-W., Lee, H. M.: Lyα line formation in starburst galaxies. I. Moderately thick, dustless and static H I media. ApJ 554, 604-614 (2001)

[3] Ahn, S.-H., Lee, H.-W., Lee, H. M.: Lyα line formation in starburst galaxies. II. Extremely thick, dustless and static H I media. ApJ 567, 922-930 (2002)

[4] Böttcher K.: Adaptive Schrittweitenkontrolle beim unstetigen Galerkin-Verfahren für gewöhnliche Differentialgleichungen. Diplomarbeit, Universität Heidelberg (1996)

[5] Dautray R., Lions J.-L.: Mathematical Analysis and Numerical Methods for Science and Technology. Vol. 6. Springer, Berlin Heidelberg New York (2000)

[6] Hummer, D.G., Kunasz, P.B.: Energy loss by resonance line photons in an absorbing medium. ApJ 236, 609-618 (1980)

[7] Kanschat G.: Parallel and adaptive Galerkin methods for radiative transfer problems. Ph.D. Thesis, University of Heidelberg (1996), http://www.iwr.uni-heidelberg.de/sfb359/Preprints1996.html

[8] Meinköhn E., Richling S.: Radiative transfer with finite elements. II. Lyα line transfer in moving media. A&A 392, 827-839 (2002)

[9] Meinköhn E.: Modeling three-dimensional radiation fields in the early Universe. Ph.D. Thesis, University of Heidelberg (2002), http://www.iwr.uni-heidelberg.de/sfb359/Preprints2002.html

[10] Mihalas, D.:Stellar Atmospheres. Freeman, San Francisco (1978)

[11] Mihalas, D., Weibel-Mihalas, B.: Foundation of Radiation Hydrodynamics. Oxford University Press, New York (1984)

[12] Neufeld, D.A.: The transfer of resonance-line radiation in static astrophysical media. ApJ 350, 216-241 (1990)

[13] Richling, S., Meinköhn, E., Kryzhevoi, N., Kanschat, G.: Radiative transfer with finite elements. I. Basic method and tests. A&A 380, 776-788 (2001)

[14] van Ojik, R., Röttgering, H.J.A., Carilli, C.L., Miley, G.K., Bremer, M.N., Macchetto, F.: A powerful radio galaxy at z= 3.6 in a giant rotating Lyman-α halo. A&A 313, 25-44 (1996)

[15] van Ojik, R., Röttgering, H.J.A., Miley, G.K., Hunstead, R.W.: The nature of the extreme kinematics in the extended gas of high redshift radio galaxies. A&A 317, 358-384 (1997)

[16] Wehrse R., Baschek B., von Waldenfels, W.: The diffusion of radiation in moving media: I. Basic assumptions and formulae. A&A 359, 780-787 (2000)

Design and control of MEMS for microfluidic applications

Bijan Mohammadi

Montpellier univeristy and Institut Universitaire de France
Bijan.Mohammadi@math.univ-montp2.fr

Summary. We present a new global semi-deterministic recursive minimization algorithm based on the solution of over-determined boundary value problems. The paper also deals with low-complexity evaluation of gradient through incomplete sensitivity concept. The minimization algorithm is first validated on several multi-minima configurations. Then, its applications to shape optimization for microfluidic devices is presented.

1 Introduction

Global solution of minimization problems is of great practical importance and this is one of the reason why evolutionary algorithms received a large deal of interest in recent years [5, 3]. The main difficulty with these algorithms is their complexity in term of the number of functional evaluations which makes their use difficult for large optimization problems.

A fundamental remark on classical gradient based minimization algorithms, having a continuous representation as a cauchy problem for a first order dynamic system [7, 1], is that they can find the global minimum if the initial condition belongs to the attraction bassin of the infimum and that otherwise the minimizing sequence they build is in principle captured by a local minimum. In that sense, the problem of global minimization with a gradient based algorithm becomes the prescription of an initial condition for the mentioned Cauchy problem in the suitable attraction bassin.

This paper presents a formulation of global minimization problems in term of over-determined boundary value problems. The problem is then solved using a semi-deterministic shooting method. It is shown how successive functionals can be built from the solution of multi-level shooting for nested boundary value problems. We notice that a first order dynamic system is not a suitable continuous representation for a global minimization algorithm and that using a method coming from the discretization of a second order dynamic system, the complexity of the algorithm can be reduced by reducing the non-deterministic aspects of the algorithm.

2 Global minimization and dynamic systems

Consider the minimization of a functional $J(x), x \in \mathcal{O}_{ad}$, x is the optimization parameter and belongs to a compact admissible space $\mathcal{O}_{ad}$.

Engineers like GAs because these algorithms do not require sensitivity computation, perform global and multi-objective optimization and are easy to parallelize. Their drawbacks remain their weak mathematical background, their computational complexity due to a size of the population proportional to the size of the control space and that, as we will see for our shape optimization problem below, for complex state equations, often it is not possible to get an answer from the solver for the set of parameters proposed by GAs. The semi-deterministic algorithm (SDA) presented below here aims to address these issues.

2.1 Semi-deterministic multi-level optimization

Most deterministic minimization algorithms can be seen as discretizations of the following dynamical system:

$$\begin{cases} M(\zeta)x_\zeta = -d(x(\zeta)) \\ x(\zeta = 0) = x_0 \end{cases} \quad (1)$$

For example:

- If $d = \nabla J$ and $M = Id$, we recover the classical steepest descent method.
- If $d = \nabla J$ and $M = \nabla^2 J$ (resp. its approximation), we recover the Newton (resp. Quasi-Newton) method.

In addition, we make the following assumptions:

- H1: $J \in C^1(\Omega_{ad}, \mathbb{R})$.
- H2: the infimum J_m is known. This is often the case in industrial applications.
- H3: the problem is admissible: the infimum is reached inside the admissible domain:
  $\exists x_m \in \Omega_{ad}, s.t. J(x_m) = J_m$.
- H4: J is coercive (i.e. $J(x) \to \infty$ when $|x| \to \infty$).

We consider that system (1) has a solution if for a given $x_0 \in \Omega_{ad}$, we can find a finite Z_{x_0} such that $J(x(Z_{x_0})) = J_m$:

$$\begin{cases} M(\zeta)x_\zeta = -d(x(\zeta)) \\ x(0) = x_0 \\ J(x(Z_{x_0})) = J_m \end{cases} \quad (2)$$

This is an over-determined boundary value problem which can be solved using classical techniques for BVPs (e.g. shooting, finite differences,...). Because we are interested by constrained global optimization we prefer to express the condition at Z_{x_0} on the functional instead of its gradient. Indeed, in our context first order optimality condition is usually not satisfied at infimum.

This over-determination is an explanation of why we should not solve global optimization problems with methods which are particular discretizations of first order differential systems. We could use variants of classical methods after adding second order derivatives:

$$\begin{cases} \eta x_{\zeta\zeta} + M(\zeta)x_\zeta = -d(x(\zeta)), \\ x(0) = x_0, \quad \dot{x}(0) = \dot{x}_0, \\ J(x(Z_{x_0})) = J_m \end{cases} \tag{3}$$

In practice, we consider $|\eta| << 1$ and decreasing with ζ in order not to introduce too much perturbation in the method.

The over determination can be removed, for instance, by considering $x_0 = v$ for (1) (resp. $\dot{x}(0) = v$ for (3)) as a new variable to be found by the minimization of $h(v) = J(x_v(Z_v)) - J_m$, where $x_v(Z_v)$ is the solution of (1) (resp. (3)) found at $\zeta = Z_v$ starting from v.

The algorithm $A_1(v_1, v_2)$ reads:

- *(v_1, v_2) given,*
- *Find $v \in argmin_{w\in\mathcal{O}(v_2)} h(w)$ where $h(w) = J(x_w(Z_w)) - J_m$, with $x_w(Z_w)$ solution of system (1) found at $\zeta = Z_w$ starting from w, and $\mathcal{O}(v_2) = \{t\overrightarrow{v_1v_2}, t \in \mathbb{R}\} \cap \Omega_{ad}$.*
- *return v*

The line search minimization might fail. For instance, a secant method degenerates on plateau and critical points. In that case, we add an external level to the algorithm A_1, keeping v_1 unchanged, and looking for v_2 by minimizing a new functional h^2 defined by $h^2(v_2^2) = \min_{v_2^2} h(v_2^2)$ by algorithm $A_1(v_1, v_2^2)$.

This leads to the following two-level algorithm $A_2(v_1, v_2^2)$:

- *(v_1, v_2^2) given,*
- *Find $v_2 \in argmin_{w\in\mathcal{O}(v_2^2)} h^2(w)$ where $h^2(w) = h(A_1(v_1, w))$ and $\mathcal{O}(v_2^2) = \{t\overrightarrow{v_1v_2^2}, t \in \mathbb{R}\} \cap \Omega_{ad}$.*
- *return v_2*

The choice of initial conditions in this algorithm contains the non-deterministic feature of the algorithm. The construction can be pursued building recursively $h^i(v_2^i) = \min_{v_2^i} h^{i-1}(v_2^i)$ using $A_{i-1}(v_1, v_2^i)$, with $h^1(v) = h(v)$ where i denotes the external level.

In practice, the algorithm gives satisfaction if the trajectory passes close enough to the infimum (i.e. in $B_\varepsilon(x_m)$ where ε defines the accuracy in the capture of the infimum). This means that we should consider for h a functional of the form

$$h(v) = \int_{T_1}^{T} (J(x_v(\tau)) - J_m)^2 d\tau, \quad \text{for } 0 < T_1 < \tau < T$$

where $x_v(\tau)$ is the trajectory generated by (1) and $T_1 = T/2$ for instance. Also, in the algorithm above $x_w(Z_w)$ is replaced by the best solution found over $[0, Z_w]$.

In cases J_m is unknown, we set $J_m = -\infty$ and look for the best solution for a given complexity and computational effort. This is the approach adopted here where we predefine the effort we would like to make in each level of the algorithm.

2.2 One dimensional geometric interpretation

We give a simple geometric interpretation of the approach above where shooting is used to solve the external minimization problems. For the shooting method to converge we need the functional we minimize (here $h(v) = J(x_v) - J_m$) not too sensitive to small perturbation of v (i.e. h continuous) and strictly monotonic. In one dimension, if a first order discrete dynamic system is used with $d^n = \nabla J^n$, we find the same local minimum starting from all the points of an attraction bassin (see Fig. (1)).

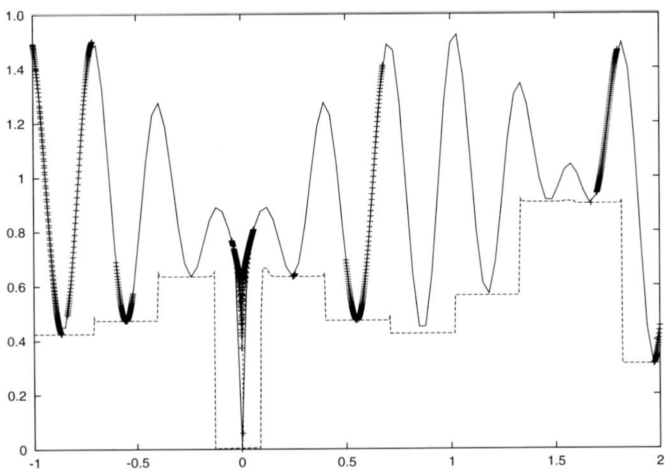

Fig. 1. Plots of $J(x) = x\sin(20x)\cos(x) + |x|^{0.1}$, non-differentiable at origin (here $J_m = 0$), (continuous curve) and of $h(v) = J(x_v)$ (dashed curve). Here we consider v from an uniform sampling of the parameter space ($[-1, 2]$). Points reached by the current minimization method with a constant step size for different initial guess provided by the shooting method (cross along the continuous curve) are also reported. Here a two-level construction was necessary to reach the global minimum.

In other words, $h(v)$ is piece-wise constant with values corresponding to the local minima of $J(x_v) - J_m$. But, $h(v)$ is discontinuous where the functional reaches a local maximum, or has a plateau. For these points, the shooting degenerates. In picture 1, we show the graph of $J(x) = x\sin(20x)\cos(x)+|x|^{0.1}$ (here $J_m = 0$) which is non-differentiable at origin and $h(v)$ constructed for illustration using an uniform sampling of the parameter space $[-1, 2]$. This construction leads to a convexification of the initial functional.

Fig. 2 illustrates the effect of a second order system on the the recursive minimization.

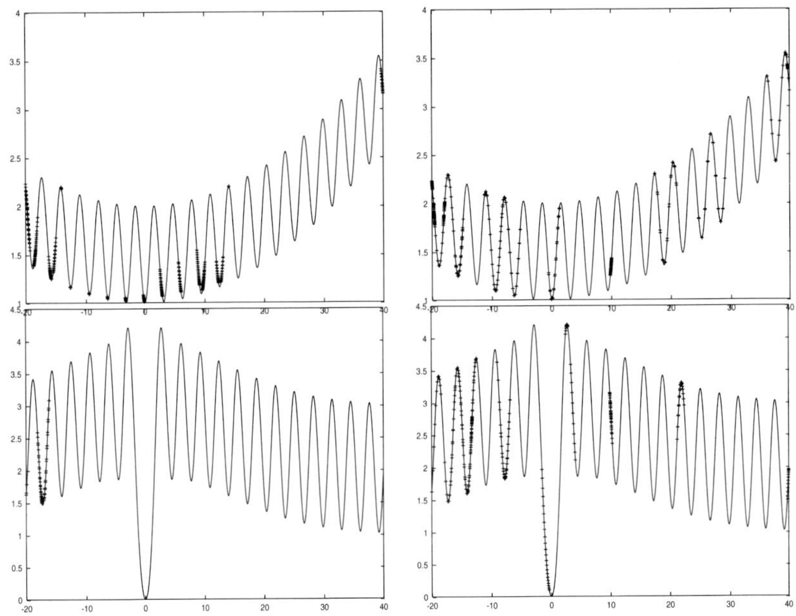

Fig. 2. Behavior of the recursive algorithm for two functionals having a convex (upper) and a non-convex hull (lower). The random search has been turned off. The trajectories of the solution are also reported (crosses). In the convex hull case, minimization algorithms based on the discretization of either first or second order dynamic systems are efficient. While in the non-convex hull case, only minimization algorithms based on the discretization of second order dynamic systems (right pictures) succeed. This shows that these algorithms reduce needs for non-deterministic aspects while minimization methods based on a first order dynamic system need an efficient random search to succeed. This is because in the non-convex hull case the shooting method alone always points to the wrong direction.

We consider the application of a two-level construction to the minimization of two functionals having a convex and non-convex hulls. We use central differences for the discretization of the systems and constant step size. To see the efficiency of the deterministic part to find the global minimum, we turn off the non-deterministic part. In the convex hull case, minimization algorithms based on the discretization of either first or second order dynamic systems are efficient. While in the non-convex hull case, only minimization algorithms based on the discretization of second order dynamic systems detect the infimum.

3 Implementation issues and academic examples

In practice, any user-defined, black-box or commercial minimization package minimizing from an initial condition can be used to build the lowest level minimization sequences. In that sense, the algorithm permits to use what is known by the user on his optimization problem. In the same way, preconditioning can be introduced at any level, and in particular in the lowest one.

We report here a few minimization problems where the functional is known explicitly. Then the application of the algorithm to a shape optimization problem is shown with incomplete evaluation of gradients. For all these examples, the gradients have been evaluated either by finite differences or by automatic differentiation in reverse mode.

3.1 Modified Rosenbrock function

The Rosenbrock function has been modified from its original form for the global minimum to be outside the banana shape attraction bassin of the initial Rosenbrock function (see Fig. 3):

$$J(x) = 74 + 100(x_2 - x_1^2)^2 + (1 - x_1)^2 - 400\exp(-10((x_1 + 1)^2 + (x_2 + 1)^2)).$$

Descent algorithms converge to the local minimum (global minimum of the original Rosenbrock function at $(1, 1)$ where $J(1, 1) = 74$) while the global minimum is reached around $(-0.90, -0.95)$ where $J = 34$ (see Fig. 4). In this example, we used $J_m = 0$ a global minorant for $J(x)$.

3.2 Generalized Griewank function

To see the behavior of the algorithm on an academic example, we consider the minimization of $J(x) = 1 - \Pi_{i=1}^{I} \cos(x_i - 100) + 10^{-6} \sum_{i=1}^{I} (x_i - 100)^2, x \in [-600, 600]^I$ for $I = 5, 10$ and 20. The graph of the functional along the first and second variables is shown in 5.

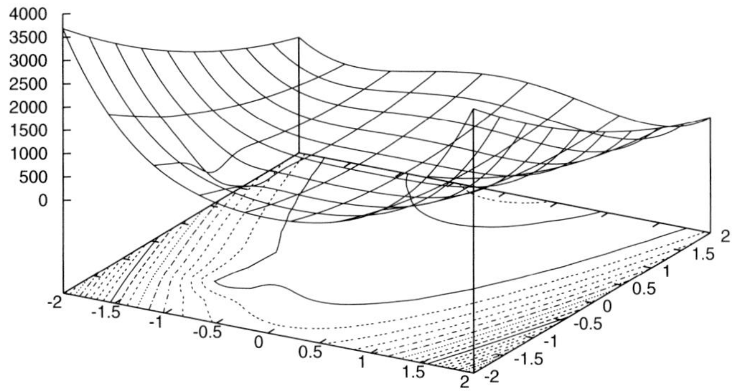

Fig. 3. Modified Rosenbrock function: graph of $J(x) = 74 + 100(x_2 - x_1^2)^2 + (1 - x_1)^2 - 400\exp(-10((x_1+1)^2 + (x_2+1)^2))$. The global minimum is outside the large banana shape attraction bassin of the local minimum.

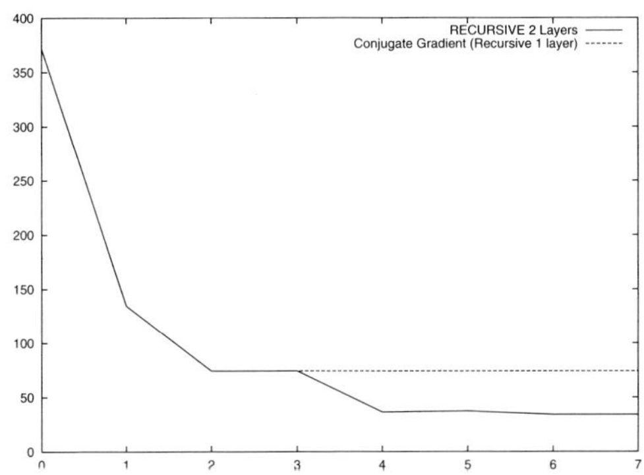

Fig. 4. Modified Rosenbrock function: Convergence history vs. the accumulation of the optimization iterations for the initializations provided by a two-level shooting algorithm.

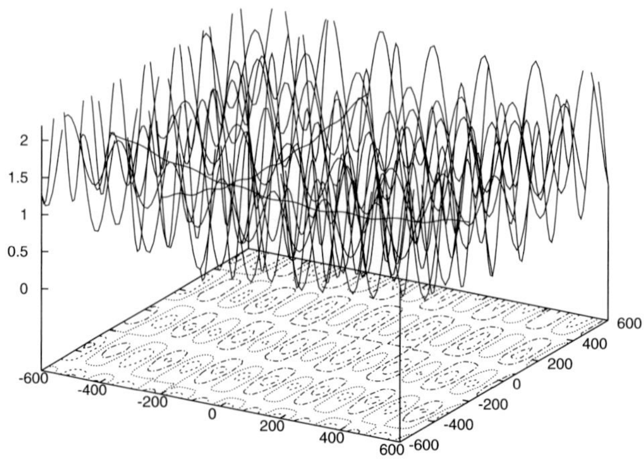

Fig. 5. Modified Griewank function: graph of $J(x) = 1 - \Pi_{i=1}^{I} \cos(x_i - 100) + \rho \sum_{i=1}^{I}(x_i - 100)^2, x \in [-600, 600]^I$ along the first and second variables. We consider the cases of $I = 5, 10$ and 20.

Figure 6 shows the convergence history vs. the accumulation of the optimization iterations for the initializations provided by a three-level shooting algorithm. We see that several local minima have been visited and that the total number of optimization iterations grows sub-linearly with the parameter space dimension.

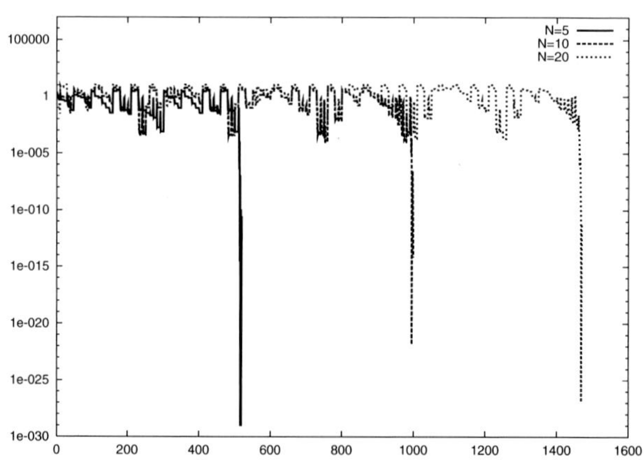

Fig. 6. Modified Griewank function: convergence history vs. the accumulation of the optimization iterations for the initializations provided by a three-level shooting algorithm.

4 Design of microfluidic systems

Microfluidic channel systems used in bio-analytical applications are fabricated using technologies derived from micro-electronics industry including lithography, wet etching and bonding of substrates. One important class of these channel system uses capillary zone electrophoresis to separate and detect chemical species. This technique separates chemical species suspended in a liquid buffer based on their electrophoretic mobility. The electric field in these systems is applied in the axis of the channel using electrode immersed at reservoirs at the end of the micro-channels. The ability to discriminate between sample species of nearly equal mobility is enhanced by increasing the channel length [2, 9]. In order to achieve channel lengths of order 1 m and yet confine the micro-channel system to a compact configuration with dimensions less than about 10 cm, curved channel geometries are required. Unfortunately, curved channel geometries introduce skews which imply a dispersion of the electrophoretic sample bands in the flow . This curved-channel dispersion has been identified as an important factor in the decrease of separation efficiency of electrophoretic micro-channel systems.

The optimization formulation for such devices has to include the following points:

- minimize the skew due to turns,
- minimize the residual dispersion associated with band advection,
- avoid too much variations in walls curvature,
- maximize the length of the channel,
- minimize the occupied surface.

We show design of a 90 and 180 degrees corner minimizing the dispersion of chemical species in motion in the electric field. These turns are important as their combination permits to maximize the length to surface ratio for a channel. Typical cross-section sizes for these channels are 100 μm in cross-section width and 10 μm in depth.

This problem is multi-model in the sense that several PDE are involved in the definition of the state variables. In particular, we solve the Poisson-Boltzmann equation for the electric field, the Stokes equation with an adequate boundary condition on the wall [9], to avoid computing the electric double-layer, for flow variables and finally advection and diffusion equations for each species.

The cost function we consider to qualify the skew uses the difference between migration times for the internal Γ_i and external Γ_o walls of the channel:

$$J(x) = (\int_{\Gamma_i} \frac{ds}{V.\tau} - \int_{\Gamma_o} \frac{ds}{V.\tau})^2. \quad (4)$$

$V.\tau$ denotes the tangential velocity component along channel walls. The shape of the turns is parameterized in CAD-Free [8]. This choice is motivated by the fact that we would like to use incomplete definition of the gradient as described below.

Consider a general simulation loop, leading from shape parametrization to the cost functional:

$$J(x) : x \rightarrow q(x) \rightarrow U(q(x)) \rightarrow J(x, q(x), U(q(x))). \tag{5}$$

The Jacobian of J is given by:

$$\frac{dJ}{dx} = \frac{\partial J}{\partial x} + \frac{\partial J}{\partial q}\frac{\partial q}{\partial x} + \frac{\partial J}{\partial U}\frac{\partial U}{\partial q}\frac{\partial q}{\partial x}. \tag{6}$$

An incomplete definition of the sensitivity [8] can be used, neglecting state variations, if the cost function is, or can be reformulated, to have the following characteristics:

- The cost function J and the parameterization x are defined on the shape (or a same part of it),
- J is of the form

$$J(x) = \int_{\text{shape or part of the shape}} f(x, q) g(u) d\gamma,$$

  which means that it involves a product of geometrical and state based functions.
- The shape curvature is not too high (this has to be quantified).

This leads to neglecting the last term in (6).

We can illustrate this idea on the following simple example. Consider as cost function $J = a^n u_x(a)$ and as state equation the following diffusion equation:

$$-u_{xx} = 1, \quad \text{on } \ ]\epsilon, 1[, \quad u(\epsilon) = 0, \quad u(1) = 0,$$

which has as solution $u(x) = -x^2/2 + (\epsilon + 1)/2 - \epsilon/2$. We are in the domain of application of the incomplete sensitivities:

- the cost function is product of state and geometrical quantities (larger is n, better is the approximation),
- it is defined at the boundary,
- the curvature of the boundary is small (here no curvature at all).

The gradient of J with respect to ϵ is given by:

$$J_\epsilon(\epsilon) = \epsilon^{n-1}(nu_x(\epsilon) + \epsilon u_{x\epsilon}(\epsilon)) = \frac{\epsilon^{n-1}}{2}(-n(\epsilon + 1) - \epsilon).$$

The second term between parenthesis is the state linearization contribution which is neglected in incomplete sensitivities. We can see that the sign of the gradient is always correct and the approximation is better for large n.

One way to improve incomplete evaluation of sensitivities is to use reduced complexity models which provide an inexpensive approximation of the missing term in (6) (i.e. the last term). For instance, consider the following reduced model for the definition of $\tilde{U}(x) \sim U(q(x))$. Suppose $\tilde{U}$ is a wall function to be used instead of the full flow equation on the wall and giving wall values knowing local internal flow description. The incomplete gradient of J with respect to x can be improved evaluating the former term in (6) linearizing the simple model. Note that $\tilde{U}$ is never used in the definition of the state U, but only in an approximation of $\partial\tilde{U}/\partial x$. It is also important to notice that the reduced model needs to be valid only over the support of the control parameters. More precisely, we linearize the following approximate simulation loop

$$x \rightarrow q(x) \rightarrow \tilde{U}(x)(\frac{U(q(x))}{\tilde{U}(x)}). \tag{7}$$

freezing $U(q(x))/\tilde{U}(x)$ which gives

$$\frac{dJ}{dx} \approx \frac{\partial J(U)}{\partial x} + \frac{\partial J(U)}{\partial q}\frac{\partial q}{\partial x} + \frac{\partial J(U)}{\partial U}\frac{\partial \tilde{U}}{\partial x}\frac{U(q(x))}{\tilde{U}(x)}. \tag{8}$$

A simple example shows the importance of the scaling introduced in (7). Consider $U = \log(1+x)$ scalar for simplicity and $J = U^2$ with $dJ/dx = 2UU' = 2\log(1+x)/(1+x) \sim 2\log(1+x)(1-x+x^2...)$ and consider $\tilde{U} = x$ as the reduced complexity model, valid around $x = 0$. To see the impact of the scaling factor we compare $J' \sim 2U\tilde{U}' = 2\log(1+x)$ with $J' \sim 2U\tilde{U}'(U/\tilde{U}) = 2\log(1+x)(\log(1+x)/x) \sim 2\log(1+x)(1-x/2+x^2/3...)$.

A direct application of this in the present context is to introduce in (4) the fact that the velocity is always parallel to the walls (slip condition) and therefore parallel to the electric field E. The cost function we consider for derivation becomes

$$J(x) = (\int_{\Gamma_i} \frac{ds}{\tau\mu_{ek}|E|} - \int_{\Gamma_o} \frac{ds}{\tau\mu_{ek}|E|})^2. \tag{9}$$

To monitor mesh quality, we use a Delaunay mesh adaptation technique by local metric control for unsteady phenomena [4, 6]. The impact of the adaptation has been shown on the advection of a passive scalar by the field (figure 8). The remeshing is also important and absolutely necessary as the large deformations introduced for the shape makes the mesh too distorted to be effective for finite element simulations.

We show the skews produced by 90 and 180 degrees turns in pictures (9-10). The optimized shapes for the 90 and 180 degree turns are shown in (Fig. 11-12).

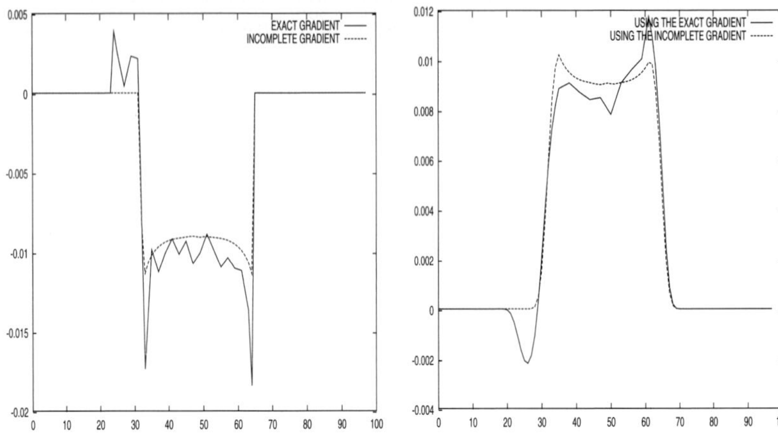

Fig. 7. Sensitivity evaluation around the initial 90 degree turn for control points along the inner channel wall. Comparison between the complete (by finite differences) and incomplete gradient evaluations (left) around a given state. The deformations obtained using these gradients (right).

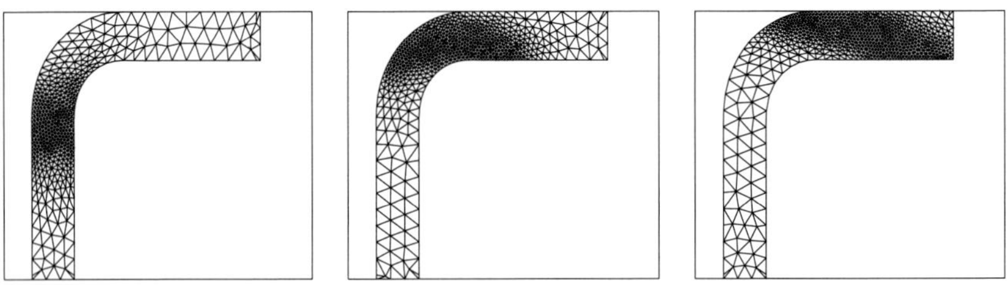

Fig. 8. Adaptive simulation for an accurate capture of the skew.

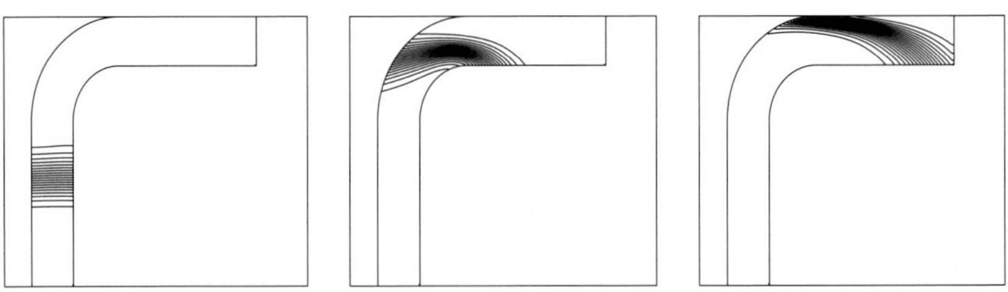

Fig. 9. Initial shape for the 90 degrees turn and a skewed band.

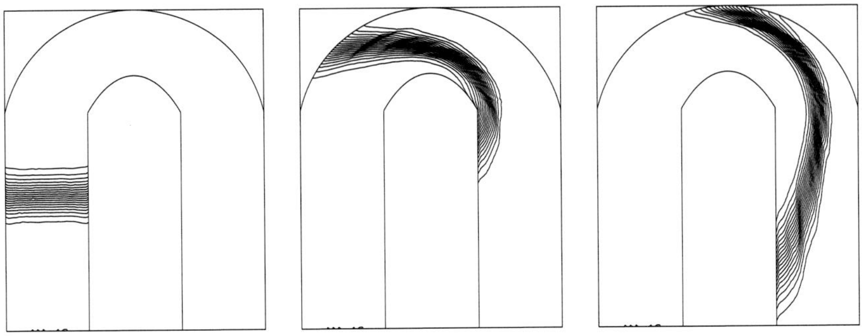

Fig. 10. Initial shape for the 180 degrees turn and a skewed band.

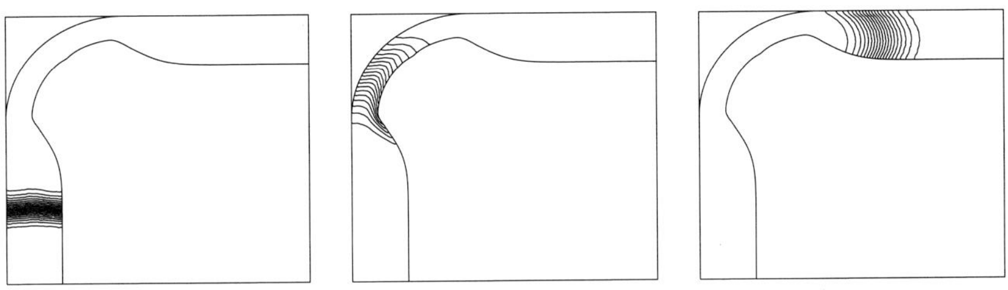

Fig. 11. Optimized shapes for the 90 degree turn. The magnitude of the skew has been reduced by one order.

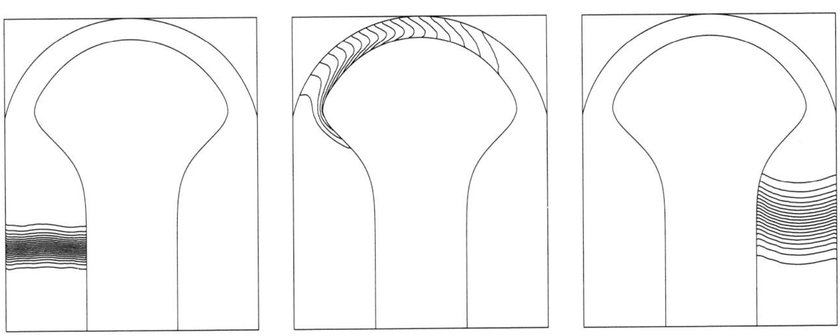

Fig. 12. Optimized shapes for the 180 degree turn. The magnitude of the skew has been reduced by more than one order.

5 Concluding Remarks

A new low-complexity semi-deterministic global minimization algorithm has been presented. The algorithm is recursive and builds successive functionals by solution of nested boundary value problems. Each level provides initial condition for the level below. The aim for the lowest level is to give an initial condition in the attraction bassin of the infimum to a user-chosen minimization method. We suggest this user-defined algorithm to be issued from the discretization of a second order ODE. To reduce the complexity of shape optimization problems, it has been shown how to combine incomplete sensitivity and the present algorithm for the design of optimal channels for microfluidic transport.

6 Acknowledgments

This work has been realized through collaborations with Stanford microfluidic laboratory and professor J. Santiago. The author would like to thank the organizers of the international conference on high performance computing in Hanoi for their interest in this work.

References

[1] Attouch, H., Cominetti, R.: A dynamical approach to convex minimization coupling approximation with the steepest descent method, J. Differential Equations, **128-2**, 519–540 (1996).
[2] Culbeston, C.T., Jacobson, S. C., Ramsey, J. M.: Dispersion Sources for Compact Geometries on Microchips, Analytical Chemistry, **70**, 3781–3789 (1998).
[3] Fonseca, C.M., Fleming, J.: An overview of evolutionary algorithms in multi-objective optimization. Evolutionary Computation, **3-1**, 1–16 (1995).
[4] Frey P., George P.L.: Maillages, Hermes, Paris (1999).
[5] Goldberg, D.: Genetic algorithms in search, optimization and machine learning. Addison Wesley, New York (1989).
[6] Hecht, F., Mohammadi, B.: Mesh Adaptation for Time Dependent Simulation and Optimization Revue Europèenne des Elements Finis, **10**, 5/2001, 575–595 (2000).
[7] Mohammadi, B, Saiac, J.H.: Pratique de la simulation numérique. Dunod, Paris (2002).
[8] Mohammadi, B, Pironneau, O.: Applied shape optimization for fluids. Oxford University Press, Oxford (2000).
[9] Probstein, R.F.: Physicochemical Hydrodynamics, Wiley (1995).

Open-loop Stable Control of Periodic Multibody Systems

Katja D. Mombaur[1], Hans Georg Bock[1], Johannes P. Schlöder[1], and Richard W. Longman[2]

[1] IWR, Universität Heidelberg
Im Neuenheimer Feld 368, 69117 Heidelberg, Germany
katja.mombaur@iwr.uni-heidelberg.de

[2] Dept. of Mechanical Engineering, Columbia University
New York, 10027 NY, USA
rwl4@columbia.edu

Summary. We investigate the open-loop stability of periodic mechanical systems with state discontinuities and multiple phases of motion. Examples for such systems are walking and running motions in robotics and biomechanics. In this paper, we focus an an ostrich-like running robot that consists of a trunk and two legs and has very small feet. It is capable of open-loop stable periodic running motions without any feedback even though it has no statically stable standing position. Running as opposed to walking involves flight phases which makes stability a particularly difficult issue. The concept of open-loop stability implies that the actuators receive purely periodic torque or force inputs that are never altered by any feedback in order to prevent the robot from falling. The choice of model parameter values and actuator inputs leading to stable motions is a difficult task that could only be accomplished using newly developed stability optimization methods.

1 Introduction

Walking and running motions in robotics and biomechanics are complex examples of periodic mechanical systems that are interesting both from a practical and a mathematical point of view. The resulting mathematical models involve distinct model phases with possibly different degrees of freedom and each described by a different set of nonlinear differential and (possibly) algebraic equations and state variable or right hand side discontinuities between phases.

In this paper we investigate a specific robot model the design of which has been inspired by the biological example of the ostrich: It has a large and quite heavy trunk and long and thin legs with very small feet. In contrast to the ostrich, it has telescopic instead of kneed legs and neither a head nor a neck. However, it can perform ostrich-like periodic motions involving flight

Fig. 1. Running ostriches in New Zealand (http://www.nzsouth.co.nz/ostrich)

phases and single-leg contact phases. During a running motion, the system is never in a statically stable state such that some form of dynamic stabilization is required. Stability control of the real ostrich is based on a very sophisticated feedback control system involving a number of senses, the brain and the nervous system of the ostrich. Since technical systems still lag behind their biological role models in speed, the exact ostrich control system is hard to mimic.

Therefore, we use a completely different approach to stability control instead. We determine, in a first step, what can be achieved without any active feedback, and to search in fact for purely open-loop controlled, self-stabilizing system configurations and running motions. An open-loop stable system does neither require sensors nor does it use active reaction to respond to perturbations but entirely relies on the mechanical system's natural kinematics and dynamics to stabilize the trajectory. Actuator histories are a priori determined, prescribed and not changed by any feedback interference. Its outstanding advantages are low cost and speed of control. The motivation behind this approach is that even for irregular motions on rough terrain where closed-loop control is a necessity, a better exploitation of the systems natural stability properties leads to a significant reduction of feedback control effort.

Bipedal running robots have been investigated by a few authors. Running bipeds in 2D and 3D with designs similar to the robot presented in this paper, but using feedback control have been built and operated at the MIT Leg Lab by Hodgins and Playter, respectively [3]. In the field of passive dynamic robots, i.e. robots that have neither actuators nor active feedback, but are purely mechanical devices moving down inclined slopes, McGeer [5] investigated bipedal running and found stable solutions for some sets of parameters.

Not much research has been done in the field of actuated open-loop stable running bipeds. Ringrose [11] was the first to discover that one-legged hopping is possible without active feedback and also extended this concept to two-

and four-legged models. However, these robots rely on very large circular feet placing the center of the foot radius above or at least close to the center of mass.

To our knowledge, we present here for the first time an actuated two-legged running robot that is capable of self-stabilizing motions without any feedback while having only point feet. The self-stabilizing effects exploited by this robot are too complex to be explained in an intuitive manner. This also implies that finding these stable configurations is a difficult task that cannot be solved intuitively. Determining model parameter values that allow a stable periodic hopping motion was only possible by means of numerical stability optimization methods that we developed specifically for this purpose. Stability is defined in terms of the spectral radius of the monodromy matrix. This criterion is non-differentiable, it may be non-Lipschitz at points of multiple maximum eigenvalue, and it involves the derivatives of the Jacobian mapping, thus representing a difficult non-standard optimization criterion. The same methods have previously been used to determine open-loop stable e. g. human-like walking (Mombaur et al. [7]) and one-legged hopping with small circular and point feet(Mombaur et al. [9]). The methods presented in this paper are taken from the recent thesis of Mombaur [6].

The rest of the paper is organized as follows: The model of the running biped robot is described in detail in section 2. The stability optimization methods are outlined in section 3. In section 4, we finally present the most stable solution for this robot.

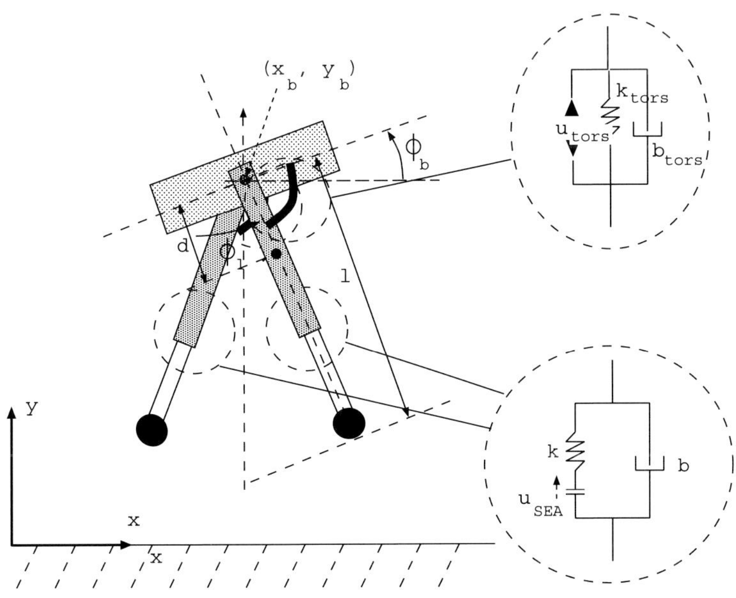

Fig. 2. Parameters, controls, and coordinates of bipedal runner

2 Mathematical Models of Periodic Running Motions

The two-legged hopping robot consists of a trunk and two telescopic legs which are coupled to the trunk by actuated hinges. The two parts of each leg are connected by an actuated spring-damper element. The foot is fixed to the lower leg without articulation. This two-legged robot is a straightforward extension of the open-loop stable hopping monopod (by just adding an identical second leg) that has been investigated earlier by the same authors [9]. The robot can perform stable running motions including non-sliding contact phases and flight phases without any feedback controllers. In this paper we concentrate on the study of two-dimensional motions and their stability. The system is holonomic, but non-conservative due to damper forces and inelastic impacts. The latter property may promote stability of the system.

A sketch of the model and its parameters is given in figure 2. Parameters are trunk mass and inertia m_b and Θ_b, leg mass and inertia m_l and Θ_l, distance between centers of mass of trunk and leg d, leg rest length l_0, torsional spring and damper constants k_{tors} and b_{tors}, rest location of torsional spring $\Delta\varphi$, and translational spring and damper constants k and b. The feet are assumed to be massless.

We model one step - and not a full physical cycle consisting of two steps - because we are only interested to find symmetric motions. The cycle modeled starts right after touchdown with leg number one, goes through a full contact phase and a flight phase and ends right after touchdown with leg number two.

During the flight phase, the robot has got five degrees of freedom. As state variables we choose the uniform set of coordinates $q = (x_b, y_b, \varphi_b, \varphi_{l_1}, \varphi_{l_2})^T$, and the corresponding velocities $\dot{q}$, where x_b and y_b are two-dimensional position coordinates of the trunk center of mass, and φ_b and φ_{l_1}, φ_{l_2} are the orientations of trunk and legs. The coordinates of the leg centers of mass x_{l_i} and y_{l_i} can be eliminated using the distance parameter d by

$$x_{l_i} = x_b + d \sin \varphi_{l_i} \tag{1}$$

$$y_{l_i} = y_b - d \cos \varphi_{l_i}. \tag{2}$$

The leg length l is fixed to $l_0 + u_0$ during the major part of the flight phase (since the foot is massless) and depends on the other coordinates during the contact phase (with leg i) as follows:

$$l_i = \frac{y_b}{\cos \varphi_{l_i}} \Rightarrow \tag{3}$$

$$\dot{l}_i = \frac{\dot{y}_b}{\cos \varphi_{l_i}} + y_b \frac{\sin \varphi_{l_i}}{\cos^2 \varphi_{l_i}} \dot{\varphi}_{l_i}. \tag{4}$$

The robot has four actuators of two different types:

1. $u_{SEA,1}$, $u_{SEA,2}$ – series elastic actuator (SEA) in the prismatic joint: as described by Pratt et al. [10], this is an actuated spring-damper element with spring constant k and damping constant b (see figure 2). The control

$u_{SEA,i} \geq 0$ actively changes the spring's length which has the same effect as changing the spring's rest length in the opposite direction:

$$\Delta l_i = \frac{y_b - r}{\cos \varphi_{l_i}} + r - u_{SEA,i} - l_0 \qquad (5)$$

The control $u_{SEA,i}$ is only effective during the contact phase of leg i - due to the massless foot it can be brought back to zero position during flight without any effect. $u_{SEA,i}$ is equal to zero at touchdown of leg i and must be positive at liftoff to compensate for the energy loss in the damper. Instantaneous compressions and general control histories can be modeled.

2. $u_{tors,1}$, $u_{tors,2}$– torque control between trunk and leg (in parallel with a spring-damper-element k_{tors}, b_{tors}, see figure 2).

The equations of motion during the flight phase are described by the following set of ODEs:

$$\begin{pmatrix} m & 0 & 0 & m_l d \cos \varphi_{l_1} & m_l d \cos \varphi_{l_2} \\ 0 & m & 0 & m_l d \sin \varphi_{l_1} & m_l d \sin \varphi_{l_2} \\ 0 & 0 & \theta_b & 0 & 0 \\ m_l d \cos \varphi_{l_1} & m_l d \sin \varphi_{l_1} & 0 & \theta_l + m_l d^2 & 0 \\ m_l d \cos \varphi_{l_2} & m_l d \sin \varphi_{l_2} & 0 & 0 & \theta_l + m_l d^2 \end{pmatrix} \begin{pmatrix} \ddot{x}_b \\ \ddot{y}_b \\ \ddot{\varphi}_b \\ \ddot{\varphi}_{l_1} \\ \ddot{\varphi}_{l_2} \end{pmatrix} =$$

$$\begin{pmatrix} m_l d(\sin \varphi_{l_1} \dot{\varphi}_{l_1}^2 + \sin \varphi_{l_2} \dot{\varphi}_{l_2}^2) \\ -m_l d(\cos \varphi_{l_1} \dot{\varphi}_{l_1}^2 + \cos \varphi_{l_2} \dot{\varphi}_{l_2}^2) - mg \\ \sum_{i=1}^{2} (u_{tors,i} - k_{tors}(\varphi_b - \varphi_{l_i} - \Delta\varphi) - b_{tors}(\dot{\varphi}_b - \dot{\varphi}_{l_i})) \\ -u_{tors,1} - m_l g d \sin \varphi_{l_1} + k_{tors}(\varphi_b - \varphi_{l_1} - \Delta\varphi) + b_{tors}(\dot{\varphi}_b - \dot{\varphi}_{l_1}) \\ -u_{tors,2} - m_l g d \sin \varphi_{l_2} + k_{tors}(\varphi_b - \varphi_{l_2} - \Delta\varphi) + b_{tors}(\dot{\varphi}_b - \dot{\varphi}_{l_2}) \end{pmatrix} \qquad (6)$$

where m is the total mass $m = m_b + 2m_l$ and $u_{tors,1}$ and $u_{tors,2}$ are the torques between trunk and leg 1 and 2, respectively. During the contact phase the contact point is assumed to be fixed due to friction, but the previously fixed leg length now varies influenced by the SEA spring-damper forces. This leads to a reduction from four to three DOFs during contact phase which is described by the additional kinematic constraint in velocity space

$$\dot{x}_b + (y_b + y_b \tan^2 \varphi_{l_1}) \, \dot{\varphi}_{l_1} + \tan \varphi_{l_1} \, \dot{y}_b = 0. \qquad (7)$$

A corresponding equation for the differences in position space can be formulated. The equations of motion for the contact phase are described by the following DAE:

$$\begin{pmatrix} m & 0 & 0 & m_l d \cos \varphi_{l_1} & m_l d \cos \varphi_{l_2} & 1 \\ 0 & m & 0 & m_l d \sin \varphi_{l_1} & m_l d \sin \varphi_{l_2} & tan \varphi_{l_1} \\ 0 & 0 & \theta_b & 0 & 0 & 0 \\ m_l d \cos \varphi_{l_1} & m_l d \sin \varphi_{l_1} & 0 & \theta_l + m_l d^2 & 0 & y_b(1 + \tan^2 \varphi_{l_1}) \\ m_l d \cos \varphi_{l_2} & m_l d \sin \varphi_{l_2} & 0 & 0 & \theta_l + m_l d^2 & 0 \\ 1 & \tan \varphi_{l_1} & 0 & y_b(1 + \tan^2 \varphi_{l_1}) & 0 & 0 \end{pmatrix} \begin{pmatrix} \ddot{x}_b \\ \ddot{y}_b \\ \ddot{\varphi}_b \\ \ddot{\varphi}_{l_1} \\ \ddot{\varphi}_{l_2} \\ \lambda \end{pmatrix}$$

$$= \begin{pmatrix} m_l d(\sin\varphi_{l_1}\dot\varphi_{l_1}^2 + \sin\varphi_{l_2}\dot\varphi_{l_2}^2) + (F_k + F_d)\sin\varphi_{l_1} \\ -m_l d(\cos\varphi_{l_1}\dot\varphi_{l_1}^2 + \cos\varphi_{l_2}\dot\varphi_{l_2}^2) - mg - (F_k + F_d)\cos\varphi_{l_1} \\ \sum_{i=1}^2 (u_{tors,i} - k_{tors}(\varphi_b - \varphi_{l_i} - \Delta\varphi) - b_{tors}(\dot\varphi_b - \dot\varphi_{l_i})) \\ -u_{tors,1} - m_l g d\sin\varphi_{l_1} + k_{tors}(\varphi_b - \varphi_{l_1} - \Delta\varphi) + b_{tors}(\dot\varphi_b - \dot\varphi_{l_1}) \\ -u_{tors,2} - m_l g d\sin\varphi_{l_2} + k_{tors}(\varphi_b - \varphi_{l_2} - \Delta\varphi) + b_{tors}(\dot\varphi_b - \dot\varphi_{l_2}) \\ -2\cdot\cos^{-2}\varphi_{l_1}\dot\varphi_{l_1}(\dot y_b + y_b\tan\varphi_{l_1}\dot\varphi_{l_1}) \end{pmatrix} \tag{8}$$

with spring and damper forces F_k and F_d

$$F_k = k\,(\frac{y_b}{\cos\varphi_{l_1}} - l_0 - u_{SEA,1}) \tag{9}$$

$$F_d = b\,(\frac{\dot y_b}{\cos\varphi_{l_1}} + y_b\frac{\tan\varphi_{l_1}}{\cos\varphi_{l_1}}\dot\varphi_{l_1}). \tag{10}$$

Phase change from contact phase (with leg 1) to flight pase, i.e. liftoff takes place, when the spring length is equal to the (modified) rest length

$$s_{liftoff} = l_0 + u_{SEA,1} - \frac{y_b}{\cos\varphi_{l_1}} = 0 \tag{11}$$

and, at the same time, the trunk has a positive vertical speed

$$c_{liftoff} = \dot y_b > 0. \tag{12}$$

Touchdown, i.e. phase change from flight phase to contact phase with leg 2, occurs when the height of the prospective contact point is equal to zero

$$s_{touchdown} = y_b - l_0\cos\varphi_{l_2} = 0. \tag{13}$$

The vertical speed of the contact point at touchdown must be negative:

$$c_{touchdown} = \dot y_b + l_0\sin\varphi_{l_2}\dot\varphi_{l_2} < 0. \tag{14}$$

There may be a discontinuity in the velocities at touchdown because friction is assumed to be large enough to instantaneously set the velocity of the contact point to zero. There are no discontinuities in the position variables if the full cycle is considered, however, the one-step model includes position discontinuities in φ_{li} due to the leg swap ($\varphi_{l_1} \leftrightarrow \varphi_{l_2}$). The five velocities after the touchdown-discontinuity are determined by the following five conditions:

- non-sliding ground contact combined with spring-damper action:

$$\dot x_{contact} = \dot x_b + l_0\cos\varphi_{l,2}\dot\varphi_{l_2} + \dot y_b\tan\varphi_{l_2} + y_b\dot\varphi_{l_2}\tan^2\varphi_{l_2} = 0 \tag{15}$$

- conservation of angular momentum of trunk about hip:

$$H_{trunk,hip} = \Theta_b\dot\varphi_b = const. \tag{16}$$

- conservation of angular momentum of prospective swing leg (leg 1) about hip

$$H_{swingleg,hip} = (\Theta_l + m_l d^2)\dot\varphi_{l_1} = const.$$

- conservation of angular momentum of full robot about prospective contact point

$$H_{robot,contact} = \Theta_b \dot{\varphi}_b - m_b(y_b - y_c)\dot{x}_b + m_b(x_b - x_c)\dot{y}_b$$
$$+\Theta_{l,1}\dot{\varphi}_{l,1} - m_l(y_{l,1} - y_c)\dot{x}_{l,1} + m_l(x_{l,1} - x_c)\dot{y}_{l,1} \quad (17)$$
$$+\Theta_l \dot{\varphi}_{l,2} - m_l(y_{l,2} - y_c)\dot{x}_{l,2} + m_l(x_{l,2} - x_c)\dot{y}_{l,2} = const \quad (18)$$

where x_{l_i} and y_{l_i} are determined by equations (1) and (2) and

$$x_c = x_b + l_0 \sin \varphi_{l_2} \quad (19)$$
$$y_c = y_b - l_0 \cos \varphi_{l_2}. \quad (20)$$

- conservation of translational momentum in direction of prospective stance leg (considering spring-damper-force)

$$m(\dot{x}_b \sin \varphi_l - \dot{y}_b \cos \varphi_l) - F_{kd} = const. \quad (21)$$

There is no velocity discontinuity at liftoff.

The variable x_b describes the forward motion of the robot and is non-periodic. All other state variables have to satisfy periodicity constraints ($q_{red}(T) = q_{red}(0)$ and $\dot{q}(T) = \dot{q}(0)$) where the period T is to be determined by the optimization.

3 Optimization of Open-Loop Stability

We have developed and implemented a numerical method for the optimization of open-loop stability of a periodic system. A detailed description of the method can be found in Mombaur [6] and Mombaur et al. [8]. This is the first time stability optimization is combined with the simultaneous solution of a periodic optimal control problem. As shown in figure 3, we introduce a two-level approach splitting the problem of periodic gait generation and stabilization of the periodic system.

3.1 Outer Loop: Stabilization of a periodic gait

In the outer loop of the optimization procedure, model parameters are determined according to stability aspects. Stability is defined in terms of the spectral radius of the Jacobian C of the Poincaré map associated with the periodic solution. We have proven that this criterion based on linear theory and typically applied to simple smooth systems can also be used to demonstrate the stability of solutions of a nonlinear multiphase system with discontinuities (Mombaur et al. [8]). If the spectral radius is smaller than one, the solution is asymptotically stable, and if it is larger than one, the solution is unstable. If the models include non-periodic variables, a projection of the monodromy

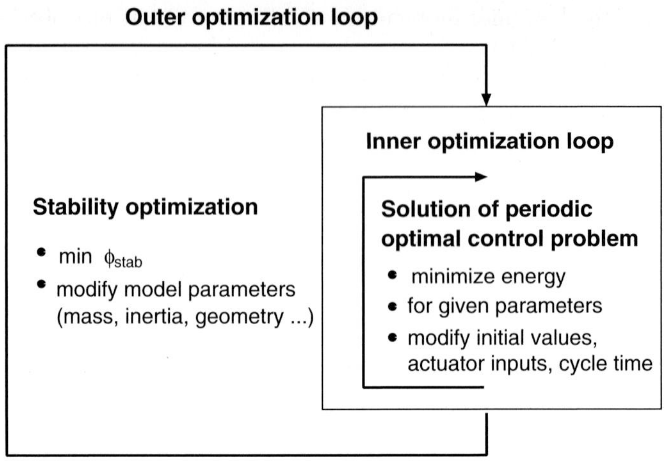

Fig. 3. Sketch of stability optimization procedure

matrix to the subspace of the periodic variables has to be performed to compute the correct Jacobian. We use the spectral radius as objective function of our optimization

$$\min_p |\lambda_{max}(C(p))|, \tag{22}$$

in the intention to decrease it below one.

This is a difficult optimization criterion for two different reasons:

- The maximum eigenvalue function of the non-symmetric matrix C is non-differentiable and possibly even non-Lipschitz at points where multiple eigenvalues coalesce.
- The determination of the matrix C involves the computation of first order sensitivities of the discontinuous trajectories.

Any gradient-based optimization method would thus require second order derivatives of the trajectory which are extremely hard to compute, especially due to the discontinuities in the dynamics. For all reasons mentioned, a direct search method has proven to be a very good choice for the solution of this outer loop optimization problem. Direct search methods are optimization methods that solely use function information and do neither compute nor explicitly approximate derivatives. We have implemented a modification of the Nelder-Mead algorithm which is based on a polytope with $n + 1$ vertices for optimization in n-dimensional space. According to the function information collected at its vertices the polytope expands in directions promising descent and contracts in bad directions. In contrast to the original method, we allow for multiple expansions in a promising direction, we use a different direction of contraction, and we only apply full polytope shrinking after multiple one-dimensional contractions. In addition, we consider the different nature of optimization variables by appropriate scaling of the initial polytope, we use a modified termination criterion, and we rely on a restart procedure as globalization strategy. Other than the original Nelder-Mead method, our algorithm can directly handle box constraints on the optimization variables not requiring a penalty function.

3.2 Inner Loop: Generation of periodic gaits

The task of the inner loop is to find – for the set of parameters prescribed by the outer loop – actuator patterns, initial values and cycle time leading to a periodic trajectory. The choice of those variables is governed by energy consumption considerations (in terms of actuator inputs u). We also have imposed a lower bound on the trunk forward speed at all points, and bounds on the leg inclination angle at touchdown and liftoff instants. Together with the equations of motion, the periodicity constraints and phase switching conditions, box constraints on all variables etc. this leads to a multi-phase optimal control problem of the following form:

$$\min_{x,u,T} \quad \int_0^T ||u||_2^2 dt \tag{23}$$

$$\text{s. t.} \qquad \dot{x}(t) = f_j(t, x(t), u(t), p) \quad \text{or DAE} \tag{24}$$

$$x(\tau_j^+) = h(x(\tau_j^-)) \tag{25}$$

$$g_j(t, x(t), u(t), p) \geq 0 \tag{26}$$

$$\text{for} \quad t \in [\tau_{j-1}, \tau_j], \quad j = 1, ..., n_{ph}, \; \tau_0 = 0, \tau_{n_{ph}} = T$$

$$r_{eq}(x(0), .., x(T), p) = 0 \tag{27}$$

$$r_{ineq}(x(0), .., x(T), p) \geq 0. \tag{28}$$

We solve this problem using a variant of the optimal control code MUSCOD (Bock & Plitt [2], Leineweber [4]) suited for periodic gait problems. It is based on

- a direct method for the optimal control problem discretization using in this case a piecewise constant control discretization
- and a multiple shooting state parameterization which transforms the original boundary value problem into a set of initial value problems with corresponding continuity and boundary conditions.

When choosing identical grids for both discretization steps, one obtains a large but very structured non-linear programming problem. The solution of the discretized problem finally rests on two pillars:

- an efficient tailored SQP algorithm exploiting the structure of the problem (also compare Leineweber [4]).
- fast and reliable integration of the trajectories on the multiple shooting intervals including a computation of sensitivity information (Bock [1]).

4 Most stable solution of ostrich-like running robot

In this paper, we only present results for the stability optimization problem with an energy optimization in the inner loop as described in the previous section. In the context of ostrich running several other optimization criteria come into mind, when ignoring the stability issue, like

- maximum speed
- minimum vertical motion range of trunk
- maximum step length etc.

These problems can all be addressed using our models and methods, but they will be treated at another time.

In the outer stability optimization loop, ten out of twelve model parameters are varied whereas leg length l_0 and trunk mass m_b are fixed for scaling reasons. Physically reasonable bounds are imposed on all parameters.

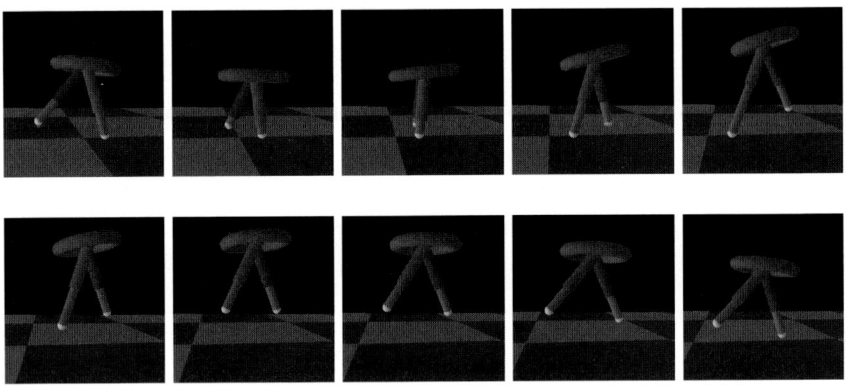

Fig. 4. Most stable open-loop controlled running motion

The most stable solution found for the two-legged running robot has a spectral radius of 0.8168. Figure 4 shows one model cycle (i.e. one step) of this solution. The model parameter values of the solution are $m_b = 2.0$, $\Theta_b = 0.3465$, $m_l = 0.2622$, $\Theta_l = 0.182$, $d = 0.11$, $l_0 = 0.5$, $k_{tors} = 11.08$, $\Delta\varphi_l = 0.5$, $b_{tors} = 9.989$, $k = 606.8$, and $b = 42.48$ (in ISO units). The corresponding cycle time is $T = 0.5476s$ for one step with $T_{contact} = 0.2533s$ and $T_{flight} = 0.2943s$ and the initial values are

$$x_0^T = (0.0, 0.4777, -0.1, 0.3, -0.7, 1.240, -2.490, -0.3941, -0.8908, 2.032)$$

Since $x_b(0)$ is fixed to zero, $x_b(T)$ is the step length of a running step, which is for this solution $0.4637m$. The full trajectories for all position and velocity variables are given in figure 6 while the corresponding actuator inputs are

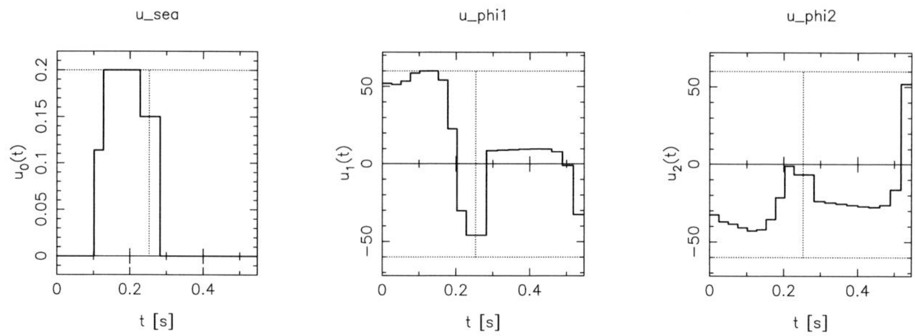

Fig. 5. Actuator inputs for most stable running solution

given in figure 5. Note the u_{SEA2} is identically zero during the step in which leg No. 2 has no ground contact and is therefore not visualized.

Due to non-periodicity of x_b, only nine out of ten eigenvalues are relevant for stability. These eigenvalues are, by magnitude:

$$\begin{aligned} |\lambda_1| &= 0.6228 \\ |\lambda_{2,3}| &= 0.8168 \\ |\lambda_4| &= 0.8168 \\ |\lambda_5| &= 0.5373 \end{aligned} \qquad \begin{aligned} |\lambda_6| &= 0.0515 \\ |\lambda_7| &= 0.0001 \\ |\lambda_{8,9}| &= 0.0 \end{aligned}$$

In the optimum, one real eigenvalue and a conjugate complex couple have the same maximum value.

The size of the spectral radius does not say anything about the size of perturbations from which the system can recover, but they are particularly interesting for the practical use of the computational solution. We determine these stability margins numerically by applying one-dimensional perturbation to the initial values of the trajectory and simulating the resulting behavior of the system checking if it stumbles or if it returns to the periodic motion. If these perturbations are not consistent with the initial phase-separating manifold, it is often customary to apply coupled consistent perturbations instead, like in the case of the hopper to y_b and φ_{l_1} (compare eqn. (13), and note the brace in the table below). The robot can recover from the following maximum perturbations of its initial values under the invariant influence of its periodic actuations

$$\begin{array}{lll} \varphi_b & +400\% & -3\% \\ \left.\begin{array}{l} y_b \\ \varphi_{l_1} \end{array}\right\} & \begin{array}{l} -0.046\% \\ +0.184\% \end{array} & \begin{array}{l} +0.05\% \\ -2\% \end{array} \\ \varphi_{l_2} & +3\% & -0.1\% \end{array} \qquad \begin{array}{lll} \dot{x}_b & +1\% & -0.1\% \\ \dot{y}_b & +3\% & -0.5\% \\ \dot{\varphi}_b & +12\% & -1\% \\ \dot{\varphi}_{l_1} & +7\% & -0.1\%. \\ \dot{\varphi}_{l_2} & +0.2\% & -5\%. \end{array}$$

Arbitrary start values are obviously possible for the non-periodic variable x_b.

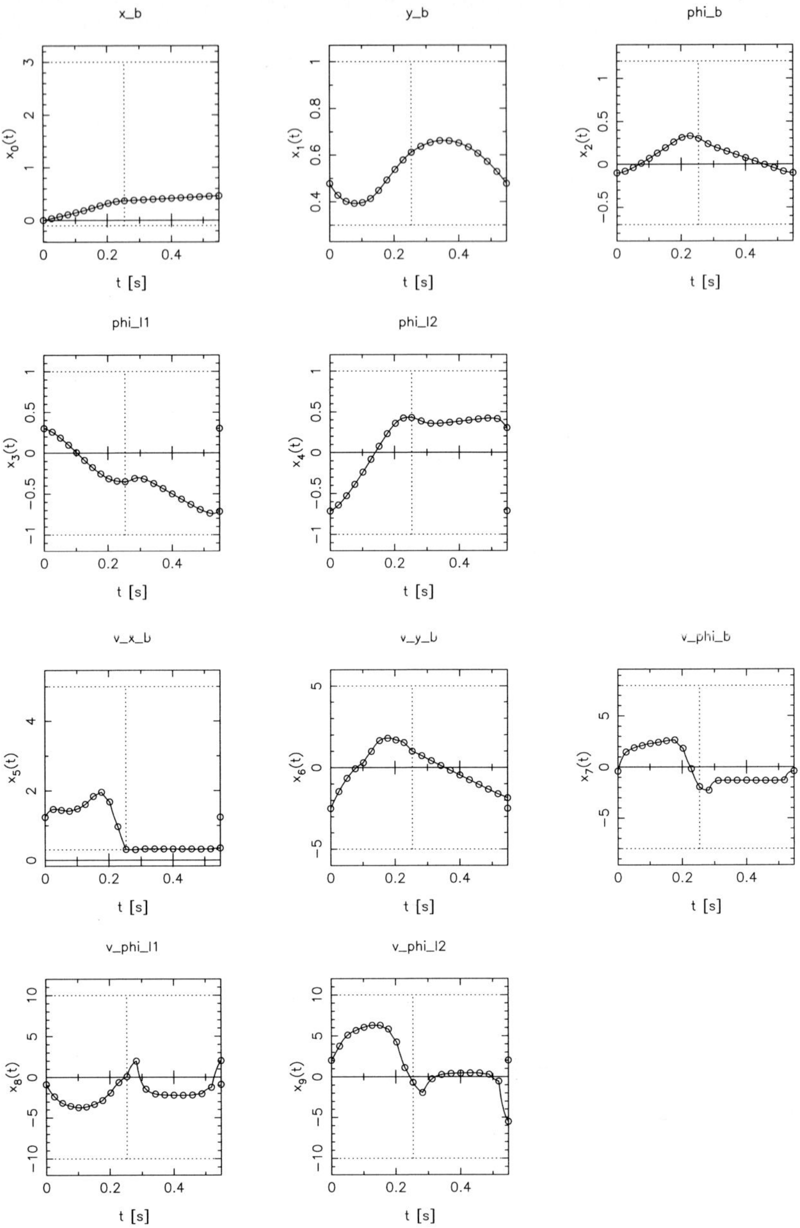

Fig. 6. Position and velocity histories of most stable biped solution

In figure 7 we compare the trajectory starting from a perturbed value of φ_b (+400%) with the corresponding unperturbed solution.

The following table finally lists the maximum possible perturbations of model parameters:

Θ_b	+0.5%	-5%	$\Delta\varphi$	+100%	-1%
m_l	+1%	-0.1%	b_{tors}	+0.2%	-0.05%
Θ_l	+0.1%	-0.5%	k	+0.05%	-2%
d	+1%	-5%	b	+0.02%	-0.3%
k_{tors}	+1%	-0.1%			

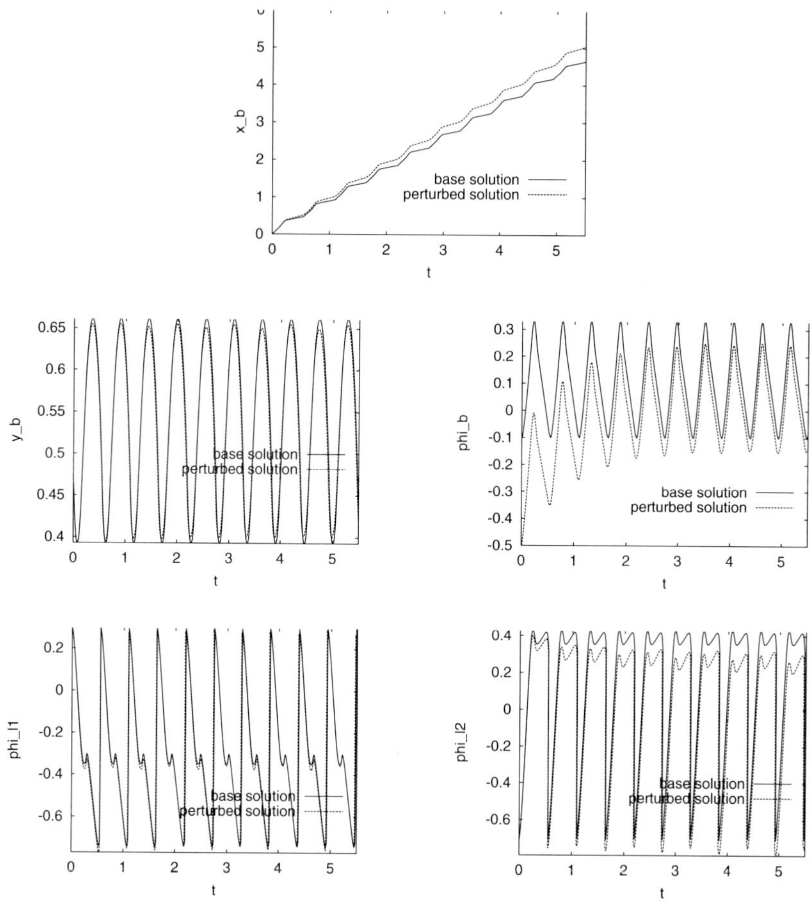

Fig. 7. Most stable periodic solution for running biped with and without perturbation (in φ_b) in all position variables

5 Conclusions

We have presented an ostrich-like robot that can perform stable running motions involving flight phases while not relying on any feedback. Self-stabilizing motions are possible even though there are no statically stable standing positions. This robot - as well as other walking and running models - is a complex example of a periodic mechanical system with state discontinuities and multiple phases of motion. Furthermore we have presented in this paper recently developed numerical optimization methods for the computation of such open-loop stable solutions of periodic mechanical systems.

References

[1] H. G. Bock. *Randwertproblemmethoden zur Parameteridentifizierung in Systemen nichtlinearer Differentialgleichungen.* PhD thesis, Universität Bonn, 1985.

[2] H. G. Bock and K.-J. Plitt. A multiple shooting algorithm for direct solution of optimal control problems. In *Proceedings of the 9th IFAC World Congress, Budapest*, pages 242–247. International Federation of Automatic Control, 1984.

[3] MIT Leg Lab. Leg lab robots. *http://www.ai.mit.edu/projects/leglab/robots/robots-main.html*, 2003.

[4] D. B. Leineweber. *Efficient Reduced SQP Methods for the Optimization of Chemical Processes Described by Large Sparse DAE Models.* PhD thesis, University of Heidelberg, 1999. VDI-Fortschrittbericht, Reihe 3, No. 613.

[5] T. McGeer. Passive bipedal running. *Proceedings of the Royal Society of London*, B 240:107 – 134, 1990.

[6] K. D. Mombaur. *Stability Optimization of Open-loop Controlled Walking Robots.* PhD thesis, University of Heidelberg, 2001. www.ub.uni-heidelberg.de/archiv/1796. VDI-Fortschrittbericht, Reihe 8, No. 922, ISBN 3-18-392208-8.

[7] K. D. Mombaur, H. G. Bock, J. P. Schlöder, and R. W. Longman. Human-like actuated walking that is asymptotically stable without feedback. In *Proceedings of IEEE International Conference on Robotics and Automation*, Seoul, Korea, May 2001.

[8] K. D. Mombaur, H. G. Bock, J. P. Schloeder, and R. W. Longman. Open-loop stable solution of periodic optimal control problems in robotics. *submitted to Zeitschrift für Angewandte Mathematik und Mechanik (ZAMM)*, 2003.

[9] K. D. Mombaur, R. W. Longman, H. G. Bock, and J. P. Schlöder. Stable one-legged hopping without feedback and with a point foot. In *IEEE International Conference on Robotics and Automation*, Washington, USA, May 2002.

[10] G. A. Pratt and M. M. Williamson. Series Elastic Actuators. In *Proceedings of IROS*, Pittsburgh, 1995.

[11] R. P. Ringrose. Self-stabilizing running. Technical report, Massachusetts Institute of Technology.

Stability of Higher Order Repetitive Control

Sang June Oh and Richard W. Longman

Columbia University, New York, NY 10027, USA
sjo0@columbia.edu
RWL4@columbia.edu

Summary. Repetitive control (RC) and iterative learning control (ILC) apply to repeating situations, and iteratively adjust the command to a feedback control system aiming to converge on zero tracking error. Thus, these forms of control solve an inverse problem, but instead of doing so numerically with a mathematical model, it is done iteratively with the real world hardware. ILC restarts the system before each run, while repetitive control applies to executing a periodic command or eliminating the effects of a periodic disturbance. The ILC literature has many contributions developing what are called higher order ILC laws, where the control action in the current repetition is a function of the errors observed in multiple previous runs. It is the purpose of this paper to develop the higher order repetitive control, and in particular to develop the relevant theory of stability. It is proved that the simple frequency response based sufficient condition for convergence in first order RC, is also a sufficient condition for convergence in higher order RC, and it is independent of the order chosen. Furthermore, this condition is very close to being a necessary condition for stability. The result is that the typical design process in the frequency domain for first order RC can now be directly extended to the design of higher order RC as well.

1 Introduction

Repetitive control (RC) and iterative learning control (ILC) develop controllers that learn from previous experience performing a specific command. In most applications these controllers adjust the command to a feedback control system, aiming to converge on that command that gives zero tracking error. In ILC the control system performs the same task repeatedly, and the system is returned to the same initial condition before each run. There can be disturbances that repeat every time the task is executed. Repetitive control executes a periodic command, possibly with a disturbance of the same period. The majority of RC systems deal with the special case of a constant command (periodic with any period) together with a periodic disturbance, and the RC learns from previous periods to cancel the influence of the disturbance. There are many applications of these fields, e.g. to robots performing repetitive operations, to improved machining, to improved positioning in electron beam accelerators, to increased density in computer disk storage devices, etc.

The problem of finding the command input to a feedback control system that will make the output match the desired output, eliminating deterministic tracking errors and cancelling the influence of repeating disturbances, is an inverse problem. One could pose the problem as the solution of a set of linear algebraic equations, one equation for each time step. Provided that one knows a good mathematical model of the feedback control system, then it might appear as a simple matrix problem to create a numerical solution for the needed input. Unfortunately, the equations involved are extremely ill-conditioned [1]. Iterative methods such as Tikhonov regularization are of some assistance in obtaining the solution, but the sensitivity to inaccuracy in the mathematical model used remains. What RC and ILC do is the equivalent of a numerical iterative solution of these ill-conditioned equations – but the iteration is done with the real world hardware rather than with a numerical model. Thus, the sensitivity to model inaccuracy is eliminated provided one has a convergent iteration. Creating iterative solutions of inverse problems using the hardware rather than a computer, is perhaps a rather unorthodox form of scientific computing.

The usual RC and ILC laws look at the error in the previous repetition (ILC) or the previous period (RC) to update the current control action. However, in the ILC literature there are many contributions developing what are called higher order ILC laws, where the control action in the current repetition is a function of the errors observed in multiple previous repetitions [2] - [12]. Reference [13] gives a general formulation for higher order ILC. Recently a special conference session was organized to investigate the potential advantages of higher order ILC [14] - [19].

It is the purpose of this paper to parallel the development of a stability theory for higher order ILC, and present a corresponding stability theory of higher order repetitive control. First order RC has a simple frequency response based stability condition [20]. It is a sufficient condition for stability, and in typical applications it is very close to being a necessary condition as well. It serves as the main criterion in making frequency response based RC designs. This paper proves that the same stability condition applies to higher order repetitive control laws, and that this condition is independent of the order of the control law. Furthermore, for the vast majority of applications the difference between this sufficient condition for convergence of the iterative process and the actual stability boundary is negligible. This makes the process of design of stable higher order repetitive controllers in the frequency domain become essentially the same as that of first order RC. In a separate work, the tradeoffs in performance in making higher order RC will be studied.

2 Mathematical Formulation of First Order RC

Figure 1 shows a basic block diagram of a typical repetitive control system.

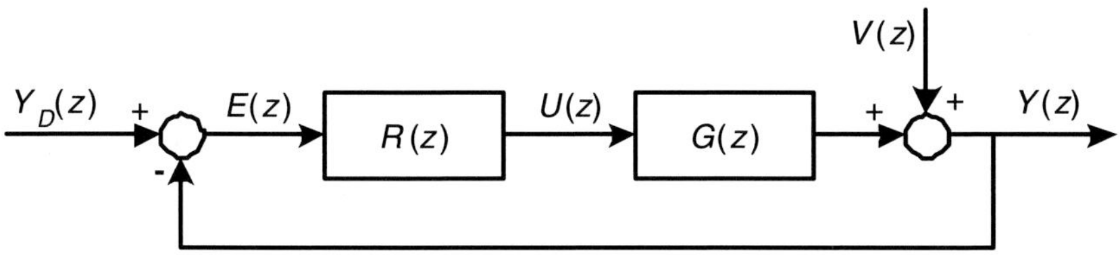

Fig. 1. Typical Repetitive Control Block Diagram

The $Y_D(z), R(z), U(z), G(z), V(z)$ and $Y(z)$ represent: the desired periodic output, the repetitive controller transfer function, the command to a feedback control system that is adjusted by the repetitive controller, the closed loop transfer function of the feedback control system, a deterministic periodic disturbance, and the output, respectively. A periodic disturbance might enter somewhere within the feedback control loop, but there is always an equivalent periodic output disturbance which is used here. Let p be the number of time steps in a period of the periodic disturbance and the desired trajectory. The simplest form of repetitive control can be written as

$$u(k) = u(k-p) + \varphi e(k-p+\gamma) \tag{1}$$

where k is the current time step and φ is the repetitive control gain (learning gain). The γ is normally equal to the time step delay from input to output in the feedback control system, but can be set to a larger value as a linear phase lead in the repetitive controller design. Stated in words, this law says that if the error was 2 units too low in the previous period at this time (adjusted for delay), add a gain times 2 units to the command in this period. In practice, it is convenient to design the repetitive control law in the frequency domain, using the form

$$z^p U(z) = F(z)[U(z) + z^\gamma \Phi(z) E(z)] \tag{2}$$

This generalizes (1) by replacing φ by $\Phi(z)$ that can include not only a gain, but also a compensator. Also, the $F(z)$ has been introduced which can be a zero phase low pass filter used to make a frequency cutoff of the learning process. Such a cutoff makes it much easier to obtain a convergent process. Then the repetitive control law transfer function becomes

$$R(z) = \frac{U(z)}{E(z)} = \frac{F(z) z^\gamma \Phi(z)}{z^p - F(z)} \tag{3}$$

and the relationship of the command and the periodic disturbance to the resulting error can be written as

$$[1+G(z)R(z)]E(z) = Y_D(z) - V(z)$$
$$\{z^p - F(z)[1 - z^\gamma \Phi(z)G(z)]\}E(z) = [z^p - F(z)][Y_D(z) - V(z)] \quad (4)$$

Suppose for the moment that there is no filter $F(z)$ so that it is replaced by the number 1. Then the right hand side of equation (4) becomes zero because it represents the difference between periodic functions at the present time and time shifted by one period. Then the equation can be thought of as a homogeneous difference equation. The tracking error will converge to zero as the time step k progresses, for all initial conditions, if and only if all roots of the characteristic polynomial

$$z^p - F(z)[1 - z^\gamma \Phi(z)G(z)] = 0 \quad (5)$$

are inside the unit circle. If $F(z)$ is a perfect low pass filter, then below the cutoff it is one, and above the cutoff it is zero, and there is no phase change. In this case, all frequency components of the command and the disturbance are eliminated on the right hand side of (4), but components above the cutoff remain. Convergence is still determined by the roots of (5) being inside the unit circle, but the system will no longer converge to zero error because the equation is no longer homogeneous. There will be obvious modifications when the $F(z)$ is not a perfect low pass filter.

3 Mathematical Formulation of Higher Order RC

3.1 Second Order RC Case

We consider a rather general class of higher order RC that can be thought of as applying the repetitive control law (1) or (2) to multiple previous periods and taking a weighted average to compute the command update. Paralleling the previous section, the second order analog of (1) is

$$u(k) = \alpha_1 u(k-p) + \alpha_2 u(k-2p) + \varphi[\alpha_1 e(k-p+\gamma) + \alpha_2 e(k-2p+\gamma)] \quad (6)$$

To correspond to the concept of a weighted average, the α_i coefficients should be non-negative, and should sum to unity. We will see that summing to unity is necessary in order to converge to zero error when there is no frequency cutoff of the learning. Generalizing this law to include a compensator and a filter in the z-transform domain produces the generalized version of (3)

$$R(z) = \frac{U(z)}{E(z)} = \frac{F(z)z^\gamma \Phi(z)(\alpha_1 z^p + \alpha_2)}{z^{2p} - F(z)(\alpha_1 z^p + \alpha_2)} \quad (7)$$

The difference equation of (4) becomes

$$\{z^{2p} - F(z)[1 - z^\gamma \Phi(z)G(z)](\alpha_1 z^p + \alpha_2)\}E(z)$$
$$= [z^{2p} - F(z)(\alpha_1 z^p + \alpha_2)][Y_D(z) - V(z)] \quad (8)$$

When one does not use a filter so that $F(z)$ is replaced by unity, then the fact that the sum of the weights is one makes the right hand side become zero. And the comments about how this is changed when there is an ideal low pass filter again apply here. Convergence of the repetitive control process as time step k goes to infinity for all initial conditions requires that all roots of the characteristic polynomial

$$z^{2p} - F(z)[1 - z^{\gamma}\Phi(z)G(z)](\alpha_1 z^p + \alpha_2) = 0 \tag{9}$$

lie inside the unit circle in the z-plane.

3.2 General Nth Order RC Case

Generalizing to Nth order RC using N previous repetitions with non-negative weights that sum to one, produces the following equations equivalent to (1), (3), (4) and (5) :

$$\begin{aligned} u(k) &= \alpha_1 u(k-p) + \alpha_2 u(k-2p) + \cdots + \alpha_N u(k-Np) \\ &\quad + \varphi[\alpha_1 e(k-p+\gamma) + \alpha_2 e(k-2p+\gamma) + \cdots + \alpha_N e(k-Np+\gamma)] \end{aligned} \tag{10}$$

$$R(z) = \frac{U(z)}{E(z)} = \frac{F(z)z^{\gamma}\Phi(z)[\alpha_1 z^{(N-1)p} + \alpha_2 z^{(N-2)p} + \cdots + \alpha_N]}{z^{Np} - F(z)[\alpha_1 z^{(N-1)p} + \alpha_2 z^{(N-2)p} + \cdots + \alpha_N]} \tag{11}$$

$$\begin{aligned} &\{z^{Np} - F(z)[1 - z^{\gamma}\Phi(z)G(z)][\alpha_1 z^{(N-1)p} + \alpha_2 z^{(N-2)p} + \cdots + \alpha_N]\}E(z) \\ &= \{z^{Np} - F(z)[\alpha_1 z^{(N-1)p} + \alpha_2 z^{(N-2)p} + \cdots + \alpha_N]\}[Y_D(z) - V(z)] \end{aligned} \tag{12}$$

$$z^{Np} - F(z)[1 - z^{\gamma}\Phi(z)G(z)][\alpha_1 z^{(N-1)p} + \alpha_2 z^{(N-2)p} + \cdots + \alpha_N] = 0 \tag{13}$$

4 Extending the Heuristic Monotonic Decay Condition to Higher Order RC

For first order RC one can write an approximate condition for monotonic decay of the error (see e.g. [20]). Examine (4). If there is no filter and $F(z)$ is equal to one, then the error satisfies the homogeneous equation, and one can write $z^p E(z) = [1 - z^{\gamma}\Phi(z)G(z)]E(z)$. If there is a perfect low pass filter, then there is a periodic forcing function containing frequencies above the cutoff, and associated with this is a steady state particular solution. The decay of the error is then associated with the solution of the homogeneous equation. Call this solution $E_h(z)$and one can write

$$z^p E_h(z) = F(z)[1 - z^\gamma \Phi(z)G(z)]E_h(z) \tag{14}$$

This formulation suggests that $M(z) \equiv F(z)[1 - z^\gamma \Phi(z)G(z)]$ can be thought of as a transfer function from one repetition to the next, since z^p is a time shift forward by one period. Then if

$$\left|M(e^{i\omega T})\right| = \left|F(e^{i\omega T})[1 - (e^{i\omega T})^\gamma \Phi(e^{i\omega T})G(e^{i\omega T})]\right| < 1 \tag{15}$$

for all frequencies ω up to Nyquist (T is the sample time interval), all steady state frequency components of the error $E_h(z)$ decay monotonically from period to period. This argument is not rigorous, since it considers $M(e^{i\omega T})$ as a frequency transfer function from one period to another, and that the response in each period can be characterized by steady state frequency response. Actually, $E_h(z)$ is the transform of the entire history of the homogeneous equation solution, and using a frequency transfer function will only apply once this has reached steady state, i.e. once the repetitive control process has converged. Nevertheless, if the learning process is sufficiently slow, and the dominant time constant or settling time of the system is sufficiently short compared to the period, then it can be valid to make this quasi-static assumption. And then (15) can be used as a condition to satisfy to obtain stability and good learning transients, i.e. monotonic decay of the transients from one period to the next.

Now consider how this thinking can be generalized to higher order RC. The equivalent of (14) for Nth order RC comes from (12)

$$z^{Np}E_h(z) = M(z)[\alpha_1 z^{(N-1)p} + \alpha_2 z^{(N-2)p} + \cdots + \alpha_N]E_h(z) \tag{16}$$

Then $M(z)[\alpha_1 z^{(N-1)p} + \alpha_2 z^{(N-2)p} + \cdots + \alpha_N]$ appears like a transfer function from the error in one period to the error N periods ahead, starting with any period. Hence, keeping the frequency transfer function version of this less than one in magnitude suggests monotonic decay from one set of N periods to the next. Note that if (15) is satisfied, then the desired inequality on this transfer function is also satisfied, because the new term is bounded by one for z on the unit circle:

$$\begin{aligned}&\left|\alpha_1 z^{(N-1)p} + \alpha_2 z^{(N-2)p} + \cdots + \alpha_N\right| \\ &\leq \alpha_1 |z|^{(N-1)p} + \alpha_2 |z|^{(N-2)p} + \cdots + \alpha_N\end{aligned}$$

$$\begin{aligned}&\left|\alpha_1 (e^{i\omega T})^{(N-1)p} + \alpha_2 (e^{i\omega T})^{(N-2)p} + \cdots + \alpha_N\right| \\ &\leq \alpha_1 + \alpha_2 + \cdots + \alpha_N = 1\end{aligned} \tag{17}$$

We conclude that, under the same kind of quasi-static assumption as above, one can expect that satisfying (15) will produce stability and monotonic decay of each frequency component of the error $E_h(z)$ from one set of N periods to the next, for an Nth order repetitive control system.

5 The Heuristic Monotonic Decay Condition as a Sufficient Condition for Stability for Higher Order RC

A logical approach to determining an if and only if condition for asymptotic stability of a repetitive control system, is to directly apply the Nyquist stability criterion for digital control to the feedback loop of Fig. 1, i.e. apply the criterion to $R(z)G(z) = -1$. Because of the p poles on the unit circle, and the need to go around each of these with the Nyquist contour, this approach is inconvenient. Reference [20] presents a different approach that avoids this difficulty, and presents a proof that satisfying the heuristic monotonic decay condition does in fact ensure asymptotic stability, even though it does not necessarily establish monotonic decay unless the quasi-static assumption is satisfied. However, in that treatment the filter $F(z)$ was not included, so that here we revisit the argument to include the filter, and then generalize the result to higher order RC. Dividing (5) by z^p produces

$$P(z) \equiv 1 - z^{-p}F(z)[1 - z^{\gamma}\Phi(z)G(z)] = 0 \tag{18}$$

Once this is put over a common denominator, the numerator is the characteristic polynomial of the repetitive control system. The contour needed to enclose all parts of the z-plane outside the unit circle, goes from $+1$ around the upper half of the unit circle, out to infinity on a branch cut along the negative real axis, then clockwise around at infinity, back in along the branch cut, and around the lower half of the unit circle. We assume that the roots of the denominators of $F(z), \Phi(z)$ and $G(z)$ are all inside the unit circle, and furthermore, assume that the degree of the denominator of $P(z)$ is greater than that of the numerator. If $G(z)$ is a feedback control system is should have all poles inside the unit circle, and one would normally pick the compensator and the filter to be stable as well. Also, p is normally a large number. Therefore, these assumptions are normally satisfied. Then the contour at infinity is mapped onto the origin, and the branch cut goes back and forth along the real axis from $P(z) = P(-1)$ (corresponding to Nyquist frequency) to the origin. By the principle of the argument, the change in the phase angle of $P(z)$ as z goes around the contour will then equal the number of roots of the numerator inside the contour, i.e. the number of unstable roots of the characteristic polynomial of the repetitive control system. If there is no encirclement of the origin, then the repetitive control system is asymptotically stable. Now rewrite the $P(z)$ of (18) as

$$\begin{aligned} P(z) &= 1 - P^*(z) \\ P^*(z) &= z^{-p}M(z) \end{aligned} \tag{19}$$

Note that

$$\left|P^*(e^{i\omega T})\right| = \left|e^{i\omega T}\right|^{-p}\left|M(e^{i\omega T})\right| = \left|M(e^{i\omega T})\right| \tag{20}$$

From this we conclude that if (15) is satisfied, then the plot of $P(z)$ for z going around the contour, cannot possibly encircle the origin. Hence, the monotonic decay condition is a sufficient condition for asymptotic stability of a first order repetitive control system.

Now consider how to generalize the above argument to apply to higher order repetitive control. The analog of (18) for an Nth order RC system is obtained by dividing (13) by z^{Np} to obtain

$$P(z) = 1 - M(z)(\alpha_1 z^{-p} + \alpha_2 z^{-2p} + \cdots + \alpha_N z^{-Np}) = 0 \tag{21}$$

Then $P^*(z)$ becomes

$$P^*(z) = M(z)(\alpha_1 z^{-p} + \alpha_2 z^{-2p} + \cdots + \alpha_N z^{-Np}) \tag{22}$$

For z on the unit circle

$$\begin{aligned} |P^*(z)| &= |M(z)| \left|\alpha_1 z^{-p} + \alpha_2 z^{-2p} + \cdots + \alpha_N z^{-Np}\right| \\ &\leq |M(z)| \left(\alpha_1 |z|^{-p} + \alpha_2 |z|^{-2p} + \cdots + \alpha_N |z|^{-Np}\right) = |M(z)| \end{aligned} \tag{23}$$

Therefore, if condition (15) is satisfied, then the magnitude of $P^*(e^{i\omega T})$ is less than one for all frequencies up to Nyquist, and the plot of $P(z)$ cannot encircle the origin. We conclude, that under the same assumptions as in the first order case, (15) is a sufficient condition for asymptotic stability of any Nth order repetitive control system of the weighted average type considered here, no matter what value of N is chosen.

In [21] the relationship between condition (15) and the stability boundary for first order RC was investigated for first, second, and third order plants. It was shown that only in the unusual case of very small values of p and a first order system, there is any significant difference between (15) and the actual stability boundary. This is even more true in the case of higher order RC. This can be easily illustrated. Note that p is usually a large number, for example, if the period is one second at a sample rate of 1000 Hz, then p is 1000. Suppose the plot of $M(z)$ violates (15) for a frequency interval $\Delta\omega$ such that $\Delta\omega T$ is 10 degrees, i.e. the plot violates (15) for 1/18th of the frequency range from zero to Nyquist. Then the $P^*(z)$ of (19) for the first order RC case will be outside magnitude one for an interval $\Delta\omega Tp$. A p of 36 or more guarantees that the origin is encircled no matter where the frequency interval appears. For the case of higher order RC, the $P^*(z)$ may encircle much faster, since the term z^{-Np} will have a phase change of $\Delta\omega TNp$, going N times as fast as in the first order case. We conclude that in nearly all practical problems, the distinction between the sufficient condition for stability (15) and the condition for the stability boundary is negligible. Furthermore, the fact that (15) is independent of the period under consideration is an important simplifying property.

6 Conclusion

The majority of practical repetitive control designs are done in the frequency domain. A simple sufficient condition for stability is used that suggests monotonic decay of the tracking error with periods, under appropriate conditions. This condition is simple, and independent of the number of time steps in a period of the command or disturbance. And it is clear what the design objective is in terms of designing a compensator and a low pass filter. In this paper it is proved that this same stability condition, also establishes stability of higher order repetitive control laws (of the weighted sum type), and this is independent of the order of the controller. This allows the design methods for first order repetitive control to be directly applied to higher order repetitive control. A separate work will study the way in which the extra design variables, the order of the RC and the weights chosen for each of the previous periods, influence performance.

References

[1] Longman, R.W., Peng, Y.-T., Kwon, T., Lus, H., Betti, R. and Juang, J.-N.: Adaptive Inverse Iterative Learning Control, *Advances in the Astronautical Sciences*, Vol. 114, to appear (2003)

[2] Bien, Z. and Huh, K.M.: Higher-Order Iterative Control Algorithm, *IEE proceedings Part D, Control Theory and Applications*, Vol. 136, pp. 105-112 (1989)

[3] Xu, J.-X., Heng, L.T. and Nair, H.: A Revised Iterative Learning Control Strategy for Robotic Manipulator, *Proceedings of 1993 Asia-Pacific Workshop on Advances in Motion Control*, pp. 88-93 (1993)

[4] Xu, J.-X, Wang, X.-W. and Lee, T.-H.: Analysis of Continuous Iterative Learning Control Systems using Current Cycle Feedback, *Proceedings of the 1995 American Control Conference*, Vol. 6, pp. 4221-4225 (1995)

[5] Chen, Y., Xu, J.-X. and Lee, T.-H.: Current Iteration Tracking Error Assisted High-Order Iterative Learning Control of Discrete-Time Uncertain Nonlinear Systems, *Proceedings of 2nd Asian Control Conference* (1997)

[6] Chien, C.-J.: A Discrete Iterative Learning Control of Nonlinear Time-Varying Systems, *Proceedings of the 35th IEEE Conference on Decision and Control*, Vol. 3, pp. 3056-3061 (1996)

[7] Chien, C.-J.: A Discrete ILC for a Class of Nonlinear Time-Varying Systems, *IEEE Transactions on Automatic Control*, Vol. 43, No. 5 (1998)

[8] Kurek, J.E. and Zaremba, M.B.: Iterative Learning Control Synthesis Based on 2-D System Theory, *IEEE Transactions on Automatic Control*, Vol. 38 (1993)

[9] Liang, Y.-J. and Looze, D.P.: Performance and Robustness Issues in Iterative Learning Control, *Proceedings of the 32nd IEEE Conference on Decision and Control*, Vol. 3, pp. 1990-1995 (1993)

[10] Owens, D.H., Amann, N. and Rogers, E.: Iterative Learning Control – an Overview of Recent Algorithms, *Applied Mathematics and Computer Science*, Vol. 5, No. 3, pp. 425-438 (1995)

[11] Owens, D.H.,Amann, N. and Rogers, E.: Systems Structure in Iterative Learning Control, *Proceedings of International Conference on System Structure and Control*, pp. 500-505 (1995)
[12] LeVoci, P., Longman, R.W.: Frequency Domain Prediction of Final Error Due to Noise in Learning and Repetitive Control, *Advances in the Astronautical Sciences*, Vol. 112, pp. 1341-1360 (2002)
[13] Phan, M.Q., Longman, R.W., and Moore, K.L.: A Unified Formulation of Linear Iterative Learning Control, *Advances in Astronautical Sciences*, Vol. 105, pp. 93-111 (2000)
[14] Moore, K.L. and Chen, Y.: On Monotonic Convergence of a High-Order Iterative Learning Update Law, *Proceedings of 15th Triennial World Congress of IFAC '02*, Barcelona, Spain (2002)
[15] Norrlof, M. and Gunnarsson, S.: Disturbance Aspects of High Order Iterative Learning Control, *Proceedings of 15th Triennial World Congress of IFAC '02*, Barcelona, Spain (2002)
[16] Al-Towaim, T., Lewin P., and Rogers, E.: Higher Order ILC versus Alternatives Applied to Chain Conveyer Systems, *Proceedings of 15th Triennial World Congress of IFAC '02*, Barcelona, Spain (2002)
[17] Rzewuski, M., Rogers, E. and Owens, D.H.: Prediction in Iterative Learning Control Schemes versus Learning along the Trials, *Proceedings of 15th Triennial World Congress of IFAC '02*, Barcelona, Spain (2002)
[18] Xu, J.-X. and Tan, Y.: On the Robust Optimal Design and Convergence Speed Analysis of Iterative Learning Control Approaches, *Proceedings of 15th Triennial World Congress of IFAC '02*, Barcelona, Spain (2002)
[19] Phan, M.Q. and Longman, R.W.: Higher-Order Iterative Learning Control by Pole Placement and Noise Filtering, *Proceedings of 15th Triennial World Congress of IFAC '02*, Barcelona, Spain (2002)
[20] Longman, R.W.: Iterative Learning Control and Repetitive Control for Engineering Practice, *International Journal of Control*, Vol. 73, No. 10, pp. 930-954 (2000)
[21] Songchon, T. and Longman, R.W., Comparison of the Stability Boundary and the Frequency Response Condition in Learning and Repetitive Control, *International Journal of Applied Mathematics and Computer Science*, to appear

An Approach to Parameter Estimation and Model Selection in Differential Equations

Michael R. Osborne

Mathematical Sciences Institute, Australian National University, ACT 0200, Australia

Summary. Two distinct and essentially independent sources of error occur in the parameter estimation problem – the error due to noisy observations, and the error due to approximation or discretization effects in the computational procedure. These give contributions of different orders of magnitude so the problem is essentially a two grid problem and there is scope for balancing these to minimize computational effort. Here the underlying computing procedures that determine these errors are reviewed. There is considerable structure in the integration of the differential system, and the role of cyclic reduction in unlocking this is discribed. The role of the stochastic effects in the optimization component of the computation is critically important in understanding the success of the Gauss-Newton algorithm, and the importance both of adequate data and of a true model is stressed. If the true model must be sought among a range of competitors then a stochastic embedding technique is suggested that converts under-specified models into "non-physical" consistent models for which the Gauss-Newton algorithm can be used.

1 Introduction

The estimation problem for parameterized systems of differential equations starts with data acquired through observations of system trajectories made in the presence of noise, and it seeks to estimate the parameter values by solving an optimization problem which matches computed solutions of the differential equation to this observed data. Note that the explicit assumption that the data is noisy means that there is an explicit stochastic component to the problem. This has the immediate consequence that two distinct grids are relevant in this problem formulation. These are:

- the grid defined by the observation process; and
- the grid defined by the discretization of the differential system.

The need to distinguish between these is a consequence of the different rates of convergence of the estimates implied by the two different sources of error – the stochastic component deriving from the measurement process with its characteristic $O\left(n^{-1/2}\right)$ rate, and the rate deriving from the truncation error in the numerical procedure where an $O\left(n^{-2}\right)$ estimate is readily achieved. It follows that there could well be scope for economizing on the work done in integrating the differential system because of the distinctly lower accuracy achievable in the estimates based on the observation grid. Our aim here is to examine the key computational components of the estimation problem (the discretization of the differential equations and the parameter estimation optimization problem) to expose structure that might be used in the economization process.

To define the problem let the differential equation be

$$\frac{d\mathbf{x}}{dt} = \mathbf{w}(t, \mathbf{x}, \boldsymbol{\beta}) \tag{1}$$

where $\mathbf{x}, \mathbf{w} \in R^m$, $\boldsymbol{\beta} \in R^p$. An important case is the equation linear in the state variable $\mathbf{x}$

$$\mathbf{w} = M(t, \boldsymbol{\beta})\mathbf{x} + \mathbf{f}(t). \tag{2}$$

This is not only significant in its own right, but through a process of successive linearization it provides the "enabling technology" needed for the fully nonlinear problem. For this reason most of the discussion here is in terms of the linear equation system. The second component of the problem formulation is the observed data:

$$y_i = \boldsymbol{\varphi}^T \mathbf{x}(t_i) + \varepsilon_i, \; i = 1, 2, \cdots, n, \tag{3}$$

where $\boldsymbol{\varphi}$ defines the "observation functional" and the observational error is given by ε_i. These errors are assumed to be independent and identically distributed and typically normal.

Typically, two classes of method are considered:

embedding method : Here explicitly computed solution trajectories are compared directly with the observations in an unconstrained optimization procedure. Typically this has two main components:

1. Given trial $\boldsymbol{\beta}$, plus *auxiliary information* $\mathbf{b}$, generate trial solution $\mathbf{x}(t, \boldsymbol{\beta}, \mathbf{b})$.
2. Using trial solution make adjustments to $\boldsymbol{\beta}$ and auxiliary information $\mathbf{b}$ to improve estimate of $\boldsymbol{\beta}$ and $\mathbf{x}(t, \boldsymbol{\beta}, \mathbf{b})$. Measure goodness of fit by

$$F(\boldsymbol{\beta}, \mathbf{b}) = \sum_{i=1}^{n} r_i(t_i, \boldsymbol{\beta}, \mathbf{b})^2, \tag{4}$$

$$r_i = y_i - \boldsymbol{\varphi}^T \mathbf{x}(t_i, \boldsymbol{\beta}, \mathbf{b}). \tag{5}$$

To set up the auxiliary information first select boundary matrices B_1, B_2 and guess the variable part of the auxiliary information $\mathbf{b}$. This data is required to permit the solution of the boundary (or initial) value problem

$$B_1\mathbf{x}(0) + B_2\mathbf{x}(1) = \mathbf{b},$$
$$\frac{d\mathbf{x}}{dt} = M(t,\beta)\mathbf{x} + \mathbf{f}(t).$$

It is important to choose B_1, B_2 so that the Green's matrix is nicely bounded. This is always possible if the system possesses a known, well-defined dichotomy [1]. However, this requires significant additional structural information about the differential system. Typically it is required of the imposed conditions that the fast solutions of the differential equation be pinned down at $t = 1$, and slow solutions be similarly pinned down at $t = 0$. Simple shooting corresponds to $B_1 = I,\ B_2 = 0$. It requires the initial value problem to be stable.

This formulation leads to the nonlinear least squares problem:

$$\min_{\mathbf{b},\boldsymbol{\beta}} \sum_{i=1}^{n} (y_i - \boldsymbol{\varphi}^T \mathbf{x}(t_i, \boldsymbol{\beta}, \mathbf{b}))^2. \tag{6}$$

It is recommended that the correction to $\boldsymbol{\beta}$, $\mathbf{b}$ be computed using the Gauss-Newton or Scoring method [2]. This requires the integration of the variational equations:

$$\begin{cases} \frac{d\boldsymbol{\Delta}_\beta}{dt} = M\boldsymbol{\Delta}_\beta + \nabla_\beta M\mathbf{x}, \\ B_1\boldsymbol{\Delta}_\beta(0) + B_2\boldsymbol{\Delta}_\beta(1) = 0, \end{cases}$$
$$\begin{cases} \frac{d\boldsymbol{\Delta}_b}{dt} = M\boldsymbol{\Delta}_b, \\ B_1\boldsymbol{\Delta}_b(0) + B_2\boldsymbol{\Delta}_b(1) = I, \end{cases}$$

where

$$\boldsymbol{\Delta}_\beta = \frac{\partial\mathbf{x}}{\partial\boldsymbol{\beta}},\ \boldsymbol{\Delta}_b = \frac{\partial\mathbf{x}}{\partial\mathbf{b}}.$$

simultaneous method : In this approach the system of differential equations is imposed as explicit constraints on the optimization problem (6) [3]. The resulting equality constrained mathematical program typically is solved by a variant of sequential quadratic programming. For the linear differential equation the mathematical program can be formulated as

$$\min_{\boldsymbol{\beta}} \sum_{i=1}^{n} (y_i - \boldsymbol{\varphi}^T \mathbf{x}(t_i, \boldsymbol{\beta}))^2, \tag{7}$$

subject to the equality constraints

$$\mathbf{x}_{i+1} - X_i(t_{i+1}, t_i)\mathbf{x}_i = \mathbf{v}_i,\ i = 1, 2, \cdots, n-1, \tag{8}$$

where $X_i(t, t_i)$ is the fundamental matrix, and $\mathbf{v}_i(t)$ is the particular integral for equation (2) given by

$$\frac{dX_i}{dt} = MX_i, \ X_i(t_i, t_i) = I, \tag{9}$$

$$\mathbf{v}_i = \int_{t_i}^{t_{i+1}} X_i(t_{i+1}, u)\mathbf{f}(u)du. \tag{10}$$

In practice, the differential equation constraints would be replaced by an appropriate discretization. In this formulation the additional information comes from the Lagrange multipliers, but now the dimension of the constraint system grows with n. This can be avoided to some extent at least by the use of a coarser solution grid to generate data which can be interpolated linearly to obtain values to compare with the observed data. However, the differential system has only m degrees of freedom. This suggests a more compact specification could be available.

This brief sketch of the algorithmic possibilities suggests a number of significant problems including the determination of suitable boundary matrices $B_1, \ B_2$, finding problem formulations that might aid the selection of appropriate integration grids, and reducing the degrees of freedom information in the mathematical programming formulation. A possible approach to these problems is considered in the next section which is based around the possible forms of cyclic reduction in this context. It is shown that, in certain circumstances at least, an optimal reduction in the degrees of freedom in the constraint system is possible. The applicability of the Gauss-Newton algorithm has been indicated above. A general treatment is sketched in the following section (including for constrained problems). Part of the context required here is that the model for the system trajectory be correctly specified. To rescue the Gauss-Newton method when this is not the case introduces the interesting class of problems involving model selection. One possible approach, involving a stochastic embedding of the tentative model equations, is sketched in the final section.

2 Cyclic reduction

Cyclic reduction [4] is an elimination scheme applied to the block bidiagonal recurrence

$$A_i^0\mathbf{x}_{i+1} + B_i^0\mathbf{x}_i = \mathbf{c}_i^0 \tag{11}$$

which combines adjacent rows using techniques such as partial pivoting or orthogonal reduction as follows:

$$\begin{bmatrix} B_{i-1}^0 & A_{i-1}^0 & 0 & \mathbf{c}_{i-1}^0 \\ 0 & B_i^0 & A_i^0 & \mathbf{c}_i^0 \end{bmatrix} \rightarrow \begin{bmatrix} B_{i/2}^1 & 0 & A_{i/2}^1 & \mathbf{c}_{i/2}^1 \\ V_i^1 & -I & W_i^1 & \mathbf{w}_i^1 \end{bmatrix}. \tag{12}$$

The procedure can be applied recursively to give

Interpolation equations

$$\mathbf{x}_t = V_t\mathbf{x}(0) + W_t\mathbf{x}(1) + \mathbf{w}_t, \tag{13}$$

Constraint equation

$$G_1^k \mathbf{x}(0) + G_2^k \mathbf{x}(1) = \mathbf{c}_1^k. \tag{14}$$

The process is simplest if $n = 2^k$, but this restriction is not necessary in the bidiagonal case considered here [5]. The resulting equations (13), (14) are intrinsic properties of the differential equation system in the sense that they do not depend on the boundary conditions.

The constraint equation (14) gives immediate information concerning the choice of the boundary matrices in the embedding approach. It is required to choose B_1, B_2 to ensure that

$$\begin{bmatrix} B_1 & B_2 \\ G_1^k & G_2^k \end{bmatrix}$$

has a 'nicely' bounded inverse for then $\mathbf{x}(0)$, $\mathbf{x}(1)$ can be found stably and the remaining values filled in using the interpolation equations. Thus G_1^k, G_2^k must reflect the dichotomy properties of the ODE system.

The interpolation equations (13) produced by the recursive cyclic reduction transformations allow the reformulation of the equality constrained estimation problem with minimum degrees of freedom (minimum numbers of equality constraints):

$$\min_{\beta} \sum_{t=t_i, i=1}^{n} \left(y_t - \varphi^T \left(V_t \mathbf{x}(0) + W_t \mathbf{x}(1) + \mathbf{w}_t \right) \right)^2 \tag{15}$$

subject to the constraints

$$G_1^k \mathbf{x}(0) + G_2^k \mathbf{x}(1) = \mathbf{c}_1^k. \tag{16}$$

This reduces the Lagrangian form of the problem to solving an optimization problem involving a fixed number m of equality constraints.

Certain properties are an immediate consequence of the basic process:

- Boundary conditions on the interpolation equations (13) follow directly:

$$\begin{array}{l} V(0) = I, \ V(1) = 0, \\ W(0) = 0, \ W(1) = I, \\ \mathbf{w}(0) = 0, \ \mathbf{w}(1) = 0. \end{array} \tag{17}$$

- $V_t, W_t, \mathbf{w}_t, G_1, G_2, c$ are not uniquely defined by the cyclic reduction process. Let C_t be the transformation that combines adjacent block rows. Then there is an equivalence class of transformations:

$$C_t \leftarrow \begin{bmatrix} R_1(t) & 0 \\ R_{21}(t) & R_2(t) \end{bmatrix} C_t \tag{18}$$

  that preserve the basic structure in the elimination tableau (12). Freedom in the interpolation is in $R_2^{-1} R_{21}$. Freedom in the constraint is in R_1.

- Relationships in the constraint equation can also be identified

$$\left(G_2^k\right)^{-1} G_1^k = X(1,0),\ \left(G_2^k\right)^{-1} \mathbf{c}_1^k = \mathbf{v}_1.$$

The simplest transformation introducing a zero in the cyclic reduction step is given by

$$C = \begin{bmatrix} I & X(t_{i+1}, t_i) \\ I & -X(t_{i+1}, t_i) \end{bmatrix}. \tag{19}$$

Assume $\delta = t_{i+1} - t_i$ is small. Substitute for the state variable using the interpolation equations, apply the transformation, and expand in powers of δ. Equating leading terms gives second order differential systems for the interpolation equation quantities (note this means the boundary condition (17) can be satisfied):

$$\frac{d^2}{dt^2}\left(X^{-1}\begin{Bmatrix} V \\ W \end{Bmatrix}\right) = 0,\ \Rightarrow\ V = X(t,0)(1-t),\ W = X(t,1)t.$$

Other possibilities can be found by fixing

$$R_2^{-1} R_{21} = S_1 = \delta S + O(\delta^2). \tag{20}$$

Substituting this into $C \leftarrow RC$ and repeating the calculation gives for V (W is similar)

$$\frac{d^2 V}{dt^2} + 2(S - M)\frac{dV}{dt} + \left(M^2 - 2SM - \frac{dM}{dt}\right) V = 0. \tag{21}$$

The interesting reduction is based on the use of orthogonal transformations as we know that bidiagonal systems with coupled boundary conditions provides a practical example of the potential instability of partial pivoting. Orthogonal transformation requires

$$C^T R^T R C = I$$

Substituting and expanding in powers of δ gives

$$S = \frac{M + M^T}{2} \tag{22}$$

Substituting in the general equation (21) gives (here order is important)

$$\left(\frac{d}{dt} + M^T\right)\left(\frac{d}{dt} - M\right)\begin{Bmatrix} V \\ W \end{Bmatrix} = 0$$

The general equation (21) corresponds to the first order system (write Y for either V, W)

$$\frac{d}{dt}\begin{bmatrix} Y \\ Z \end{bmatrix} = N \begin{bmatrix} Y \\ Z \end{bmatrix},\ N = \begin{bmatrix} M & I \\ & -(2S - M) \end{bmatrix}. \tag{23}$$

To see where the constraint equation (14) fits in write the particular integral equation in form

$$\frac{d}{dt}\begin{bmatrix}\mathbf{w}\\ \mathbf{z}\end{bmatrix} = N\begin{bmatrix}\mathbf{w}\\ \mathbf{z}\end{bmatrix} + \begin{bmatrix}\mathbf{f}\\ 0\end{bmatrix}.$$

The interpolation equation (13) gives $\mathbf{x}$ as a combination of solutions of the higher order ($2m \times 2m$) first order systems for V, W, $\mathbf{w}$. The function of the constraint equation is to remove unwanted components of the expanded fundamental matrix. It is required that

$$\begin{aligned} 0 &= \frac{d\mathbf{x}}{dt} - M\mathbf{x} - \mathbf{f}, \\ &= \left(\frac{dV}{dt} - MV\right)\mathbf{x}(t_1) + \left(\frac{dW}{dt} - MW\right)\mathbf{x}(t_n) + \frac{d\mathbf{w}}{dt} - M\mathbf{w} - \mathbf{f}, \\ &= Z_V(t)\mathbf{x}(t_1) + Z_W(t)\mathbf{x}(t_n) + \mathbf{z}(t). \end{aligned} \tag{24}$$

Although the constraint must hold for every t there is really only one condition here because the equations for Z_V, Z_W, $\mathbf{z}$ are homogeneous so the constraint equation determined for $t = t_2$ is obtained from that for $t = t_1$ by multiplying by $Z(t_2, t_1)$ where Z is a fundamental matrix for the system

$$\frac{dZ}{dt} = -\left(2S - M\right)Z.$$

3 Optimization methodology

When the model is known then the Gauss-Newton or scoring method appears the method of choice for the embedding methods, and good reasons for this are presented which are a consequence of the stochastic setting. Similar approximations appear to work well in sequential quadratic programming applied to the simultaneous class of methods. Results from Zengfeng Li's thesis [6] are summarized.

Scoring [2] is a generalisation of the Gauss-Newton algorithm. It is based on two main ideas:

Maximum likelihood for parameter estimation This starts with:

- *events*: $\mathbf{y}_t \in R^m$, $t \in T$,
- *probability density*: $f(\mathbf{y}_t, \boldsymbol{\eta}(\mathbf{x}, t))$,
- *exact model*: $\boldsymbol{\eta}(\mathbf{x}, t) : R^p \times T \to R^q$ (the parameter and covariate information).

The parameter estimates $\mathbf{x}$ are computed by minimizing the negative of the likelihood

$$\mathbf{x}_T = \arg\min \mathcal{K}_T(\mathbf{x}), \tag{25}$$

$$\mathcal{K}_T(\mathbf{x}) = -\sum_{t\in T} \mathcal{L}_t, \tag{26}$$

$$\mathcal{L}_t = \log f(\mathbf{y}_t, \boldsymbol{\eta}(\mathbf{x}, t)). \tag{27}$$

The context which generalises that typically assumed for least squares involves:

- independent events and an appropriately structured sampling regime,
- $n = |T| \gg m = \dim \mathbf{x}$, and
- a model with suitable analytic properties.

Newton's method for function minimization This computes:

$$\mathcal{J} = \nabla^2 \mathcal{K}(\mathbf{x}), \tag{28}$$

$$\mathbf{h} = -\mathcal{J}^{-1}(\mathbf{x}) \nabla \mathcal{K}(\mathbf{x})^T, \tag{29}$$

$$\mathbf{x} \to \mathbf{x} + \mathbf{h}. \tag{30}$$

Advantages:

1 It has a fast rate of ultimate convergence to $\hat{\mathbf{x}} \ni \nabla \mathcal{K}(\hat{\mathbf{x}}) = 0$ provided $\mathcal{J}(\hat{\mathbf{x}})$ is nonsingular.
2 It has good transformation invariance properties.

Disadvantages:

1 Convergence is local and could be, at least in theory, just to a stationary point.
2 The method requires $\nabla^2 \mathcal{K}(\mathbf{x})$. In the past it has often been regarded as uneconomical or inconvenient to compute this.

Scoring aims to maintain or even improve on the advantages of Newton's method while avoiding the disadvantages. The key step is the replacement of $\nabla^2 \mathcal{K}(\mathbf{x})$ by its expectation. This gives the modified iteration the basic (full step) form:

$$\mathcal{I} = \frac{1}{n} \mathcal{E} \left\{ \nabla^2 \mathcal{K}(\mathbf{x}) \right\}, \tag{31}$$

$$\mathbf{h} = -\mathcal{I}(\mathbf{x})^{-1} \frac{1}{n} \nabla \mathcal{K}(\mathbf{x})^T, \tag{32}$$

$$\mathbf{x} \to \mathbf{x} + \mathbf{h}. \tag{33}$$

In [2] it is shown that the scoring iteration gives consistent estimates $\hat{\mathbf{x}}_n$ which tend to the true parameter vector $\mathbf{x}^*$ in an appropriate stochastic sense as the number of observations increases without limit provided

- the sampling procedure is structured appropriately, and
- the assumed model is correct.

The important identity

$$\mathcal{E}\{\nabla^2 \mathcal{K}(\mathbf{x})\} = \mathcal{E}\{\nabla \mathcal{K}(\mathbf{x})^T \nabla \mathcal{K}(\mathbf{x})\} \tag{34}$$

shows that $\mathcal{I}$ is generically positive definite under reasonable modelling assumptions. In [2] it is shown that $\lim_{n \to \infty} \mathcal{I}_n$ is essentially a bounded Gram matrix. A second consequence of the disappearance of second derivatives in $\mathcal{I}_n$ is that the modified algorithm has even better transformation invariance properties. If $\mathcal{I}$ has to be estimated because integration is difficult then the law of large numbers can help.

$$\begin{aligned}\frac{1}{n}\mathcal{E}\{\nabla^2\mathcal{K}_n\} &= \frac{1}{n}\sum_i \mathcal{E}\{\nabla\mathcal{L}_i^T\nabla\mathcal{L}_i\}\\ &= -\frac{1}{n}\sum_i\left(\nabla\mathcal{L}_i^T\nabla\mathcal{L}_i - \mathcal{E}\{\nabla\mathcal{L}_i^T\nabla\mathcal{L}_i\}\right) + \frac{1}{n}\sum_i \nabla\mathcal{L}_i^T\nabla\mathcal{L}_i\\ &\to \frac{1}{n}\sum_i \nabla\mathcal{L}_i^T\nabla\mathcal{L}_i,\ n\to\infty.\end{aligned}$$

As $\mathcal{I}_n$ is positive (semi) definite so $\nabla\mathcal{K}_n\mathbf{h} < (=)0$. This last property ensures that the scoring step is necessarily downhill for minimizing $\mathcal{K}_n$ when $\mathcal{I}_n$ is nonsingular, and that $\mathcal{K}_n$ is a suitable function to use in a linesearch step to stabilize the iteration. This has the consequences:

1 $\frac{\nabla\mathcal{K}\mathbf{h}}{\|\nabla\mathcal{K}\|\|\mathbf{h}\|} < -\frac{1}{\mathrm{cond}\mathcal{I}}$,
2 Limit points of the iteration are stationary points of $\mathcal{K}$.
3 A full step will be acceptable in the line search eventually provided n is large enough.

The final point in favour of scoring is that the rate of convergence to a consistent estimate is very satisfactory. Once a full step is acceptable in the linesearch then the iteration can be written as the fixed point iteration:

$$\mathbf{x}_{i+1} = \mathbf{F}(\mathbf{x}_i);\quad \mathbf{F}(\mathbf{x}) = \mathbf{x} - \mathcal{I}_n(\mathbf{x})^{-1}\frac{1}{n}\nabla\mathcal{K}_n(\mathbf{x})^T$$

Here $\hat{\mathbf{x}}_n$ is a point of attraction provided the spectral radius

$$\varpi(\nabla\mathbf{F}(\hat{\mathbf{x}}_n)) < 1$$

As $\nabla\mathcal{K}_n(\hat{\mathbf{x}}_n) = 0$ it follows that

$$\begin{aligned}\nabla\mathbf{F}(\hat{\mathbf{x}}_n) &= I - \mathcal{I}_n(\hat{\mathbf{x}}_n)^{-1}\frac{1}{n}\nabla^2\mathcal{K}_n(\hat{\mathbf{x}}_n)\\ &= (\mathcal{I}_n(\hat{\mathbf{x}}_n))^{-1}\left((\mathcal{I}_n(\hat{\mathbf{x}}_n) - \frac{1}{n}\nabla^2\mathcal{K}_n(\hat{\mathbf{x}}_n))\right)\\ &= \nabla\mathbf{F}(\mathbf{x}^*) + \mathcal{O}(\|\hat{\mathbf{x}}_n - \mathbf{x}^*\|),\ \text{a.s.},\ n\to\infty\end{aligned}$$

But $\nabla\mathbf{F}(\mathbf{x}^*) = o(1),\ n\to\infty$ where $\mathbf{x}^*$ is the true vector of parameters using the strong law of large numbers

$$\Rightarrow \varpi(\nabla\mathbf{F}(\hat{\mathbf{x}}_n)) \to 0,\ n\to\infty$$

showing an arbitrary fast rate of (first order) convergence provided the effective sample size is large enough. Note that consistency of the estimate is used explicitly here.

4 Extension to constrained problems

For simplicity consider the linearly constrained problem [7]:

$$\min_{\mathbf{x}} \mathcal{K}_n;\ C\mathbf{x} = \mathbf{d},\ C : R^p \to R^m,\ \operatorname{rank}(C) = m. \tag{35}$$

The necessary conditions for a minimum of (35) give

$$\nabla \mathcal{K}_n = \boldsymbol{\lambda}^T C \tag{36}$$

where $\boldsymbol{\lambda}$ is the vector of Lagrange multipliers. The limiting form as $n \to \infty$ follows by an application of the law of large numbers to

$$\frac{1}{n}\{\nabla \mathcal{K}_n - \mathcal{E}\{\nabla \mathcal{K}_n\}\} + \frac{1}{n}\mathcal{E}\{\nabla \mathcal{K}_n\} = (\boldsymbol{\lambda}/n)^T C$$

The left hand side has the limiting form

$$-\int_0^1 \mathcal{E}\{\nabla \mathcal{L}(\mathbf{y}, \mathbf{x}, t)\} d\omega(t),$$

where ω is a limiting weight function characterizing the design of the observation process. Thus the limiting system is

$$-\int_0^1 \mathcal{E}\{\nabla \mathcal{L}(\mathbf{y}, \mathbf{x}, t)\} d\omega(t) = \boldsymbol{\lambda}^{*T} C \tag{37}$$

$$C\mathbf{x} = \mathbf{d} \tag{38}$$

where $\boldsymbol{\lambda}^* = \lim_{n\to\infty} \boldsymbol{\lambda}/n$. This has the solution

$$\mathbf{x} = \mathbf{x}^*,\ \boldsymbol{\lambda}^* = 0.$$

Thus if there is a fixed finite number of constraints then the associated Lagrange multipliers asymptote to zero. In particular, the limiting (correctly scaled) multipliers associated with (15), (16) are zero.

An equality constrained sequential quadratic programming approach to constrained problems is now sketched. The target problem here is the simultaneous method. Needed results are straightforward if cyclic reduction can be applied directly (that is rather than used to simplify a current linearization), but this is not assumed. Let the problem have the form:

$$\min_{\mathbf{x}} \mathcal{K}(\mathbf{x});\ \mathbf{c}(\mathbf{x}) = 0.$$

We introduce the Lagrangian

$$l(\mathbf{x}, \boldsymbol{\lambda}) = \mathcal{K}(\mathbf{x}) + \boldsymbol{\lambda}^T \mathbf{c}(\mathbf{x}).$$

Let B_k be an approximation to $\nabla^2_{\mathbf{x}} l(\mathbf{x}_k, \boldsymbol{\lambda}_k)$ and solve linear subproblem

$$\min_{\mathbf{d}\in S} \nabla\mathcal{K}(\mathbf{x}_k)\mathbf{d} + \frac{1}{2}\mathbf{d}^T B_k \mathbf{d}, \; S = \{\mathbf{d};\; \mathbf{c}(\mathbf{x}_k) + A(\mathbf{x}_k)\mathbf{d} = 0\}$$

(typically cyclic reduction would have an application in the constraint reduction step). To make progress take the guarded step

$$\mathbf{x}_{k+1} := \mathbf{x}_k + \gamma \mathbf{d}_k.$$

Here Li implemented the Byrd and Omojokum trust region strategy following [8]. The current iteration is completed by updating the Lagrange multiplier vector $\boldsymbol{\lambda}$ using

$$\boldsymbol{\lambda}_{k+1} = -A_k^+(\nabla\mathcal{K}_k^T + B_k\mathbf{d}_k))$$

The Gauss-Newton approximation to B_k (possibly guarded to avoid too large changes) involves:

i ignoring the term $\sum_{i=1}^{n} r_i \frac{\partial^2 r_i}{\partial x_j \partial x_k}$, and this is justified as in the unconstrained case;

ii ignoring also the term $\sum_{i=1}^{n-1} \boldsymbol{\lambda}_i^T \frac{\partial^2 \mathbf{c}_i}{\partial x_j \partial x_k}$.

Numerical experience is that this Gauss-Newton style approximation works. It is not too difficult to show that a suitably scaled $\boldsymbol{\lambda}_i \to 0$, but this is not quite enough to justify omitting the constraint contributions for a potentially unbounded number of constraints. However, there is more structure which includes a stochastic differential equation for a limiting multiplier vector:

$$d\boldsymbol{\lambda} = -\nabla_{\mathbf{x}}\mathbf{w}(t, \mathbf{x}, \boldsymbol{\beta})^T \boldsymbol{\lambda} dt + \sigma\varphi d\omega.$$

It is worth observing that the modified iteration just gives the unconstrained minimization step. This iteration can be shown to give consistent but not feasible estimates in the case of a fixed finite number of constraints. Here the trust region step serves to force feasibility.

The following example was given in [6] to illustrate the Gauss-Newton strategy. The distinctly unstable equation is due to Matheij [1]. The comparison is between a Newton strategy where the Hessian is computed exactly and the Gauss-Newton strategy summarized above.

$$M(t, \boldsymbol{\beta}) = \begin{bmatrix} 1 - \beta_1\cos(\beta_2 t) & 0 & 1 + \beta_1\sin(\beta_2 t) \\ 0 & \beta_1 & 0 \\ 1 + \beta_1\sin(\beta_2 t) & 0 & 1 + \beta_1\cos(\beta_2 t) \end{bmatrix}$$

$$\mathbf{f}(t) = e^t \begin{bmatrix} -1 + 19(\cos(2t) - \sin(2t)) \\ -18 \\ 1 - 19(\cos(2t) + \sin(2t)) \end{bmatrix}$$

$$\mathbf{x}(t) = e^t\mathbf{e}$$

The data is chosen in the form $\mathbf{x}(t) + \sigma\mathbf{rnd}$ where $\mathbf{rnd}$ is a vector of standardized normal variates and $\sigma = 5.,\ 1.,\ .01$. The initial parameter vector has values 20% larger than true - $[19, 2]$. The results are displayed in the following table, and show clearly the reduction in number of iterations as the number of observations is increased for each of the values of σ.

Table 1. Newton and scoring compared

n	Ne	GN	Ne	GN	Ne	GN
2^5+1	15	55	6	11	4	4
2^7+1	16	20	6	10	3	4
$2^{10}+1$	7	13	4	5	3	3

5 Model selection

It all becomes harder if the only information available is that the model is known to lie within a parameterized class of systems. Presumably one should start the searching with the simpler members of this class (the potentially under-specified systems) as an aid to numerical stability. However, the scoring method requires a consistency result and thus looses its justification in this case. A stochastic embedding procedure which produces spline-like fits to the data, and which offers the possibility of overcoming this difficulty, is being studied for systems linear in the state variables.

The smoothing spline $\eta(t)$ is defined by:

$$\min_{\eta} \sum_{i=1}^{n} (y_i - \eta(t_i))^2 + \tau \int_0^1 \left(\frac{d^k \eta}{dt^k} \right)^2 dt. \tag{39}$$

Here the value of τ can be chosen to provide a compromise between data fit and smoothness. An alternative stochastic formulation is given by Wahba [9]:

$$\begin{aligned} \eta(t) &= \mathcal{E}\left\{ y(t) | y_1, y_2, \cdots, y_n, \lambda \right\}, \\ \frac{d^k \eta}{dt^k} &= \sigma \sqrt{\lambda} \frac{d\omega}{dt}. \end{aligned} \tag{40}$$

Here $\lambda = 1/\tau$. The consistency result available is $\eta(t) \rightarrow \mathcal{E}\{y(t)\}$, $n \rightarrow \infty$ provided λ is chosen appropriately. For our purposes a key step is the generalisation to more general differential operators (g-splines)

$$\min_{\eta} \sum_{i=1}^{n} (y_i - \eta(t_i))^2 + \tau \int_0^1 (\mathcal{M}_k \eta)^2 \, dt. \tag{41}$$

As τ gets large the minimizing η is forced to the null space of $\mathcal{M}_k$. This suggests choosing $\mathcal{M}_k$ to provide the possibility of identifying a linear model for the underlying signal.

An alternative approach has been considered by Wecker, Ansley, and Kohn ([10], [11] for example). They write $\mathcal{M}_k$ in first order system form

$$\frac{d\mathbf{x}}{dt} = M_k \mathbf{x}$$

giving the stochastic form corresponding to (40) (here $\mathbf{b} = \mathbf{e}_k$)

$$\eta(t) = \mathcal{E}\left\{x_1(t)|y_1, y_2, \cdots, y_n, \lambda\right\},$$
$$d\mathbf{x} = M_k \mathbf{x} dt + \sigma\sqrt{\lambda}\mathbf{b} d\omega. \tag{42}$$

This invites generalisations to the more general observation data: $y_i = \varphi^T \mathbf{x}(t_i)$, to general linear systems characterized by the matrix $M(t, \boldsymbol{\beta})$, and to more general smoothness controls characterized by the steering vector $\mathbf{b}$ [12], [13]. Let $X(t, \xi)$ be a fundamental matrix of the deterministic equation. Then variation of parameters gives the relations

$$\mathbf{x}_{i+1} = X(t_{i+1}, t_i)\mathbf{x}_i + \sigma\sqrt{\lambda}\mathbf{u}_i, \tag{43}$$
$$\mathbf{u}_i = \int_{t_i}^{t_{i+1}} X(t_{i+1}, s)\mathbf{b} d\omega(s), \tag{44}$$
$$\sim N(0, \sigma^2 R(t_{i+1}, t_i)), \tag{45}$$
$$R(t_{i+1}, t_i) = \lambda \int_{t_i}^{t_{i+1}} X(t_{i+1}, s)\mathbf{b}\mathbf{b}^T X(t_{i+1}, s)^T ds. \tag{46}$$

The system is now in a form suitable for computing the conditional expectation $\mathbf{x}(t|n)$ using the Kalman filter and interpolation smoother. The filter is a forward recursion for $\mathbf{x}_{i|i} = \mathcal{E}\left\{\mathbf{x}(t_i)|y_1, y_2, \cdots, y_i, \lambda\right\}$, and $\sigma^2 S_{i|i}$, the corresponding covariance. The interpolation smoother gives the dependence on all the data. If $t_i \leq t \leq t_{i+1}$ then:

$$\mathbf{x}(t|n) = X(t, t_i)\mathbf{x}_{i|i} + A(t, t_i)\left(\mathbf{x}_{i+1|n} - \mathbf{x}_{i+1|i}\right), \tag{47}$$
$$A(t, t_i) = \left\{X(t, t_i)S_{i|i}X_i + \Gamma(t, t_i)\right\} S_{i+1|i}^{-1}, \tag{48}$$
$$\Gamma(t, t_i) = R(t, t_i)X(t_{i+1}, t)^T. \tag{49}$$

Two choices of the initial condition $\mathbf{x}_{1|0}$ are possible – either to choose it as constant or to assume the diffuse prior ($\mathbf{x}_{1|0} = 0,\ S_{1|0} \uparrow \infty$). Both possibilities have been considered for computing smoothing splines. The filter is an initial value process involving a possibly unstable deterministic component. However, it does involve a correction step to take account of the new data. Still stability is a legitimate question. In this connection note that there is a "multiple shooting" form which has proved convenient in the case of the diffuse prior [14]:

$$\min_{\mathbf{x}} \left\{\mathbf{r}_1^T V^{-1}\mathbf{r}_1 + \mathbf{r}_2^T R^{-1}\mathbf{r}_2\right\}, \tag{50}$$

$$\begin{bmatrix} \mathbf{r}_1 \\ \mathbf{r}_2 \end{bmatrix} = \begin{bmatrix} \varphi_1^T & & & & \\ & \varphi_2^T & & & \\ & \cdots & \cdots & & \\ & & & & \varphi_n^T \\ -X_1 & I & & & \\ & -X_2 & I & & \\ & & \cdots & \cdots & \\ & & & X_{n-1} & I \end{bmatrix} \mathbf{x} - \begin{bmatrix} \mathbf{y} \\ 0 \end{bmatrix}, \tag{51}$$

where $V = \sigma^2 I$, $R = \sigma^2 \text{diag}\{R_1, R_2, \cdots, R_{n-1}\}$, $R_i = R(t_{i+1}, t_i)$. Here the conditioning of R is a potential source of problems.

Interaction between the choices of φ, $\mathbf{b}$ turns out to determine the smoothness properties of the estimated state variable $\mathbf{x}(t|n)$ and thus provides useful information on how to choose them. Differentiating the interpolation smoother gives

$$\frac{d\mathbf{x}(t|n)}{dt} = M\mathbf{x}(t|n) + \mathbf{b}\mathbf{b}^T X(t_{i+1}, t)^T \mathbf{v}, \tag{52}$$
$$\mathbf{v} = S_{i+1|i}^{-1}(\mathbf{x}_{i+1|n} - \mathbf{x}_{i+1|i}).$$

It follows that if the deterministic equation has smooth solutions then the state can fail to be smooth only at the points t_i. The interesting term in (52) is that involving $\mathbf{b}\mathbf{b}^T X(t_{i+1}, t)^T$. The calculations need the derivatives of $X(t_i, t)$:

$$\frac{d^j X(t_i, t)}{dt^j} = X(t_i, t) P_j(M), \tag{53}$$
$$P_0 = I, \; P_j = \frac{dP_{j-1}}{dt} - MP_{j-1}, \; j = 1, 2, \cdots .$$

Smoothness of $\frac{d^j \mathbf{x}(t|n)}{dt^j}$ at t_i requires [13]

$$\varphi^T P_{j-1}(M)\mathbf{b} = 0, \; j = 1, 2, \cdots . \tag{54}$$

Limits to smoothness follow necessarily if the $P_{j-1}(M)^T \varphi$ are linearly independent for $j < k$. For example, the standard smoothing spline results follow by setting

$$\varphi = \mathbf{e}_1, \; \mathbf{b} = \mathbf{e}_k,$$
$$M = \begin{bmatrix} 0 & 1 & & & \\ & 0 & 1 & & \\ & & \cdots\cdots & & \\ & & & 0 & 1 \\ -m_k & -m_{k-1} & \cdots\cdots & & -m_1 \end{bmatrix}$$

The smoothness results have implications for the conditioning of $R(t_{i+1}, t_i)$ and hence for possible difficulties in methods in which $R(t_{i+1}, t_i)$ appears explicitly. If δ is small then Taylor expansion gives

$$R(t+\delta, t) = \int_t^{t+\delta} \sum_{i,j} \frac{(s-(t+\delta))^{i+j}}{i!j!} P_i(M)\mathbf{b}\mathbf{b}^T P_j(M)^T ds \tag{55}$$

Here the Rayleigh quotient gives eigenvalue information as $\delta \to 0$.

1 Largest eigenvalue of $R(t_{i+1}, t_i)$:

$$\pi_k = \lambda \delta \mathbf{b}^T \mathbf{b} + O(\delta^2).$$

The corresponding eigenvector $\to \mathbf{b}$.

2 Smallest eigenvalue of $R(t_{i+1}, t_i)$: If the orthogonality conditions

$$\mathbf{b}^T P_{j-1}(M)^T \varphi = 0,\ j = 1, 2, \cdots, k-1,$$

are satisfied then the eigenvector associated with the smallest eigenvalue $\rightarrow \varphi$. The corresponding Rayleigh quotient is

$$\pi_1 = \frac{\lambda}{((k-1)!)^2} \frac{(\mathbf{b}^T P_{k-1}(M)^T \varphi)^2}{\varphi^T \varphi} \frac{\delta^{2k-1}}{2k-1} + O(\delta^{2k}).$$

This is an upper bound for the smallest eigenvalue.

Two main approaches have been used for computing parameter estimates – generalised cross validation (GCV) [15],and generalised maximum likelihood (GML) [10],[11].The latter involves a "likelihood" approach. It takes its starting point from the observation that the innovations $\zeta_i = y_i - \varphi^T \mathbf{x}_{i|i-1}$ are independent, normally distributed with variance $\sigma^2 \mathcal{V}_i$ where $\mathcal{V}_i = (1 + \varphi^T S_{i|i-1} \varphi)$. The idea is to minimize

$$\sum_i{}' \left\{ \log \sigma^2 + \log \mathcal{V}_i + \frac{\zeta_i^2}{\sigma^2 \mathcal{V}_i} \right\}.$$

Minimizing with respect to σ^2 gives (here $N \leq n$ and the summation limits depend on the form of the initial conditions):

$$\hat{\sigma}^2 = \frac{1}{N} \sum_i{}' \frac{\zeta_i^2}{\mathcal{V}_i}.$$

Substituting back gives the concentrated likelihood

$$GML = \sum_i{}' \log \mathcal{V}_i + N \log \left(\sum_i{}' \frac{\zeta_i^2}{\mathcal{V}_i} \right).$$

The alternative uses generalised cross validation. The ojective function is

$$GCV = \frac{\sum_{i=1}^{n} \left(y_i - \varphi^T \mathbf{x}_{i|n} \right)^2 / n}{\left\{ \mathrm{trace}\{I - T\}/n \right\}^2},$$

were T is the influence matrix mapping observations y_i into the estimated signal $\varphi^T \mathbf{x}_{i|n}$. Advantages are claimed for its use in estimating λ. Problem is in finding an implementation that can be used for parameter estimation which requires less than $O(n^2)$ cost. In contrast, GML is relatively easy to calculate with $O(n)$ cost.

References

[1] Ascher, U., Mattheij, R., Russell, R.: Numerical Solution of Boundary Value Problems for Ordinary Differential Equations. SIAM, Philadelphia (1995)

[2] Osborne, M.: Fisher's method of scoring. Int. Stat. Rev. **86** (1992) 271–286

[3] Tjoa, I., Biegler, L.: Simultaneous solution and optimization strategies for parameter estimation of differential-algebraic systems. Ind. Eng. Chem. Res. **30** (1991) 376–385

[4] Osborne, M.: Cyclic reduction, dichotomy, and the estimation of differential equations. J. Comp. and Appl. Math. **86** (1997) 271–286

[5] Hegland, M., Osborne, M.R.: Wrap-around partitioning for block bidiagonal linear systems. IMA J. Numer. Anal. **18** (1998) 373 – 383

[6] Li, Z.: Parameter Estimation of Ordinary Differential Equations. PhD thesis, School of Mathematical Sciences, Australian National University (2000)

[7] Osborne, M.: Scoring with constraints. ANZIAM J. **42** (2000) 9–25

[8] Lalee, M., Nocedal, J., Plantenga, T.: On the implementation of an algorithm for large-scale equality constrained optimization. SIAM J. Optim. **8** (1998) 682 – 706

[9] Craven, P., Wahba, G.: Smoothing noisy data with spline functions. Numer. Math. **31** (1979) 377–403

[10] Wecker, W., Ansley, C.F.: The signal extraction approach to nonlinear regression and spline smoothing. J. Amer. Statist. Assoc. **78** (1983) 81–89

[11] Ansley, C.F., Kohn, R.: Estimation, filtering, and smoothing in state space models with incompletely specified initial conditions. The Annals of Statistics **13** (1985) 1286–1316

[12] Osborne, M.R., Prvan, T.: On algorithms for generalised smoothing splines. J. Austral. Math. Soc. B **29** (1988) 322–341

[13] Osborne, M.R., Prvan, T.: Smoothness and conditioning in generalised smoothing spline calculations. J. Austral. Math. Soc. B **30** (1988) 43–56

[14] Paige, C.C., Saunders, M.A.: Least squares estimation of discrete linear dynamic systems using orthogonal transformations. SIAM J. Numer. Anal. **14** (1977) 180–193

[15] Wahba, G.: A comparison of GCV and GML for choosing the smoothing parameter in the generalised spline smoothing problem. Ann. Statist. **13** (1985) 1378–1402

Comparison of Parallel Programming Models on Clusters of SMP Nodes

Rolf Rabenseifner[1] and Gerhard Wellein[2]

[1] High-Performance Computing-Center (HLRS), University of Stuttgart
Allmandring 30, D-70550 Stuttgart, Germany
rabenseifner@hlrs.de, www.hlrs.de/people/rabenseifner/

[2] Regionales Rechenzentrum Erlangen, Martensstraße 1, D-91058 Erlangen, Germany
gerhard.wellein@rrze.uni-erlangen.de

Summary. Most HPC systems are clusters of shared memory nodes. Parallel programming must combine the distributed memory parallelization on the node interconnect with the shared memory parallelization inside of each node. Various hybrid MPI+OpenMP programming models are compared with pure MPI. Benchmark results of several platforms are presented. This paper analyzes the strength and weakness of several parallel programming models on clusters of SMP nodes. There are several mismatch problems between the (hybrid) programming schemes and the hybrid hardware architectures. Benchmark results on a Myrinet cluster and on recent Cray, NEC, IBM, Hitachi, SUN and SGI platforms show, that the hybrid-masteronly programming model can be used more efficiently on some vector-type systems, but also on clusters of dual-CPUs. On other systems, one CPU is not able to saturate the inter-node network and the commonly used masteronly programming model suffers from insufficient inter-node bandwidth. This paper analyses strategies to overcome typical drawbacks of this easily usable programming scheme on systems with weaker inter-connects. Best performance can be achieved with overlapping communication and computation, but this scheme is lacking in ease of use.

Key words: OpenMP, MPI, Hybrid Parallel Programming, Threads and MPI, HPC, Performance

1 Introduction

Most systems in High Performance Computing are clusters of shared memory nodes. Such hybrid systems range from small clusters of dual-CPU PCs up to largest systems like the Earth Simulator consisting of 640 SMP nodes connected by a single-stage cross-bar and with SMP nodes combining 8 vector CPUs on a shared memory [3, 5]. Optimal parallel programming schemes enable the application programmer to use the hybrid hardware in a most efficient way, i.e., without any overhead induced by the programming scheme. On distributed memory systems, message passing, especially with MPI [4, 12, 13], has shown to be the mainly used programming paradigm. One reason of the success of MPI was the clear separation of the optimization: communication could be improved by the MPI library, while the numerics had to be optimized by the compiler. On shared memory systems, directive-based parallelization was standardized with OpenMP [15], but there is also a long history of proprietary compiler-directives for parallelization. The directives handle mainly the work sharing; there is no data distribution.

On hybrid systems, i.e., on clusters of SMP nodes, parallel programming can be done in several ways: one can use pure MPI, or some schemes combining MPI and OpenMP, e.g., calling MPI routines only outside of parallel regions (which is herein named the *masteronly* style), or using OpenMP on top of a (virtual) distributed shared memory (DSM) system. A classification on MPI and OpenMP based parallel programming schemes on hybrid architectures is given in Section 2. Unfortunately, there are several mismatch problems between the (hybrid) programming schemes and the hybrid hardware architectures. Often, one can see in publications, that applications may or may not benefit from hybrid programming depending on some application parameters, e.g., in [7, 10, 22].

Section 3 gives a list of major problems often causing a degradation of the speed-up, i.e., causing that the parallel hardware is utilized only partially. Section 4 shows, that there isn't a silver bullet to achieve an optimal speed-up. Measurements show that different hardware platforms are more or less prepared for the hybrid programming models. Section 5 discusses optimization strategies to overcome typical drawbacks of the hybrid masteronly style. With these modifications, efficiency can be achieved together with the ease of parallel programming on clusters of SMPs. Conclusions are provided in Section 6.

2 Parallel programming on hybrid systems, a classification

Often, *hybrid MPI+OpenMP programming* denotes a programming style with OpenMP shared memory parallelization inside the MPI processes (i.e., each MPI process itself has several OpenMP threads) and communicating with MPI between the MPI processes, but *only outside of parallel regions.* For example, if the MPI parallelization is based on a domain decomposition, the MPI communication mainly exchanges the halo information after each iteration of the outer numerical loop. The numerical iterations itself are parallelized with OpenMP, i.e., (inner) loops inside of the MPI processes are parallelized with OpenMP work-sharing directives. However, this scheme is only one style in a set of different hybrid programming methods. This hybrid programming scheme will be named *masteronly* in the following classification, which is based on the question, when and by which thread(s) the messages are sent between the MPI processes:

1. **Pure MPI**: each CPU of the cluster of SMP nodes is used for one MPI process. The hybrid system is treated as a flat massively parallel processing (MPP) system. The MPI library has to optimize the communication by using shared memory based methods between MPI processes on the same SMP node, and the cluster interconnect for MPI processes on different nodes.
2. Hybrid MPI+OpenMP without overlapping calls to MPI routines with other numerical application code in other threads:

2a. **Hybrid masteronly**: MPI is called only outside parallel regions, i.e., by the master thread.

2b. **Hybrid multiple/masteronly**: MPI is called outside the parallel regions of the application code, but the MPI communication is done itself by several CPUs: The thread parallelization of the MPI communication can be done

- automatically by the MPI library routines, or
- explicitly by the application, using a full thread-safe MPI library.

In this category, the non-communicating threads are sleeping (or executing some other applications, if non-dedicated nodes are used). This problem of idling CPUs is solved in the next category:

3. Overlapping communication and computation: While the communication is done by the master thread (or a few threads), all other non-communicating threads are executing application code. This category requires, that the application code is separated into two parts: the code that can be overlapped with the communication of the halo data, and the code that must be deferred until the halo data is received. Inside of this category, we can distinguish two types of sub-categories:

   • How many threads communicate:

   (A) Hybrid funneled: Only the master thread calls MPI routines, i.e., all communication is funneled to the master thread.

   (B) Hybrid multiple: Each thread handles its own communication needs (B1), or the communication is funneled to more than one thread (B2).

   • Except in case B1, the communication load of the threads is inherently unbalanced. To balance the load between threads that communicate and threads that do not communicate, the following load balancing strategies can be used:

   (I) Fixed reservation: reserving a fixed amount of threads for communication and using a fixed load balance for the application between the communicating and non-communicating threads; or **(II) Adaptive**.

4. **Pure OpenMP**: based on virtual distributed shared memory systems (DSM), the total application is parallelized only with shared memory directives.

Each of these categories of hybrid programming has different reasons, why it is not appropriate for some classes of applications or classes of hybrid hardware architectures. The paper focuses on pure MPI and hybrid masteronly programming style. Overlapping communication and computation is studied in more detail in [16, 17]. Regarding pure OpenMP approaches, the reader is referred to [1, 6, 8, 11, 18, 19, 20]. Different SMP parallelization strategies in the hybrid model are studied in [21] and in [2] for the NAS parallel benchmarks. The following section shows major problems of mismatches between programming and hardware architecture.

3 Mismatch problems

All these programming styles on clusters of SMP nodes have advantages, but also serious disadvantages based on mismatch problems between the (hybrid) programming scheme and the hybrid architecture:

- With pure MPI, minimizing of the inter-node communication requires that the application-domain's neighborhood-topology matches with the hardware topology.
- Pure MPI also introduces intra-node communication on the SMP nodes that can be omitted with hybrid programming.
- On the other hand, such MPI+OpenMP programming is not able to achieve full inter-node bandwidth on all platforms for any subset of inter-communicating threads.
- With masteronly style, all non-communicating threads are idling.
- CPU time is also wasted, if all CPUs of an SMP node communicate, although a few CPUs are already able to saturate the inter-node bandwidth.
- With hybrid masteronly programming, additional overhead is induced by all OpenMP synchronization, but also by additional cache flushing between the generation of data in parallel regions and the consumption in subsequent message passing routines and calculations in subsequent parallel sections.

Overlapping of communication and computation is a chance for an optimal usage of the hardware, but

- causes serious programming effort in the application itself to separate numerical code that needs halo data and that cannot be overlapped with the communication therefore,
- causes overhead due to the additional parallelization level (OpenMP), and
- communicating and non-communicating threads must be load balanced.

A few of these problems will be discussed in more detail and based on benchmark results in the following sections.

3.1 The inter-node bandwidth problem

With hybrid masteronly or funneled style, all communication must be done by the master thread. The benchmark measurements in Fig. 3 and the inter-node results in Tab. 1 show, that on several platforms, the available aggregated inter-node bandwidth can be achieved only, if more than one thread is used for the communication with other nodes.

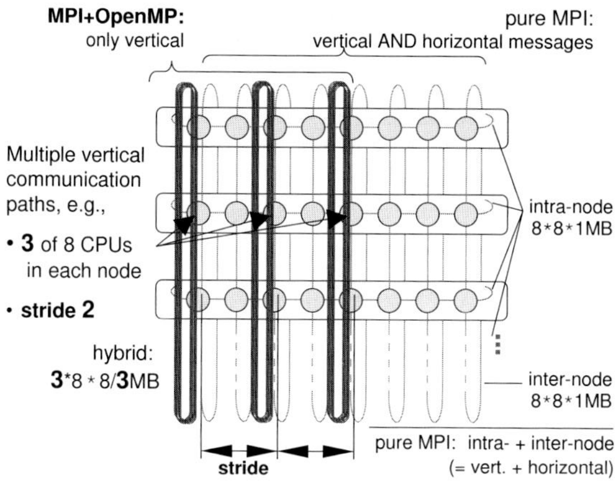

Fig. 1. Communication pattern with *hybrid MPI+OpenMP* style and with *pure MPI* style. The colour version of this figure can be found in Fig. A.28 on page 589.

In this benchmark, all SMP nodes are located in a logical ring. Each CPU sends messages to the corresponding CPU in the next node and receives from the previous node in the ring. The benchmark is done with pure MPI, i.e., one MPI process on each CPU, except for Cray X1, where we used as smallest entity an MSP (which itself has 4 SSPs [=CPUs]). Fig. 1 shows the communication patterns. The aggregated bandwidth per node is defined as the number of all bytes of all messages on the inter-node network divided by the time needed for the communication and divided by the number of nodes. Note that in this definition, each message is counted only once, and not twice.[3]

Fig. 2 shows the absolute bandwidth over the number of CPUs (or MSPs at Cray X1), Fig. 3 shows relative values, i.e., the percentage of the achieved peak bandwidth in each system over the percentage of CPUs of a node. One can see, that only on the NEC SX-6, Cray X1 systems, and on the Myrinet based cluster of dual-CPU PCs, one can achieve more than 75 % of the peak

[3] The hardware specification typically presents the duplex inter-node bandwidth by counting each message twice, i.e. as incoming and outgoing message at a node, e.g., on a Cray X1, 25.6 GB/s = 2*12.8 GB/s; the measured 12 GB/s (shmen_put) must be compared with the 12.8 GB/s of the hardware specification.

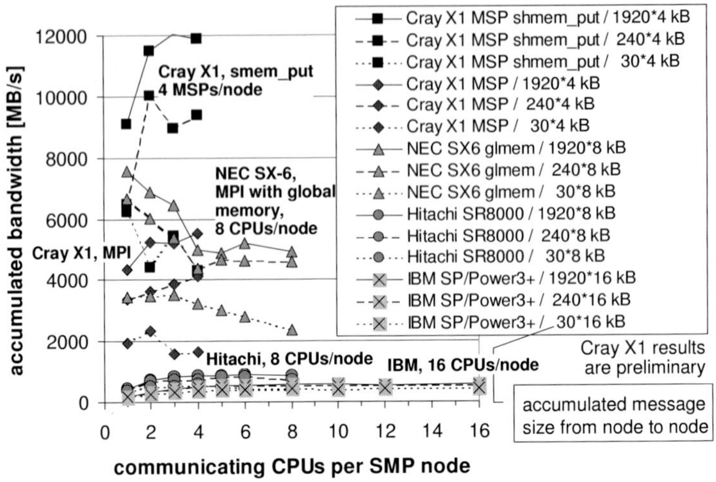

Fig. 2. Aggregated bandwidth per SMP node. The colour version of this figure can be found in Fig. A.29 on page 590.

bandwidth already with **one** CPU (or MSP on Cray X1) per node (see highlighted values in Tab. 1, Col. 3).

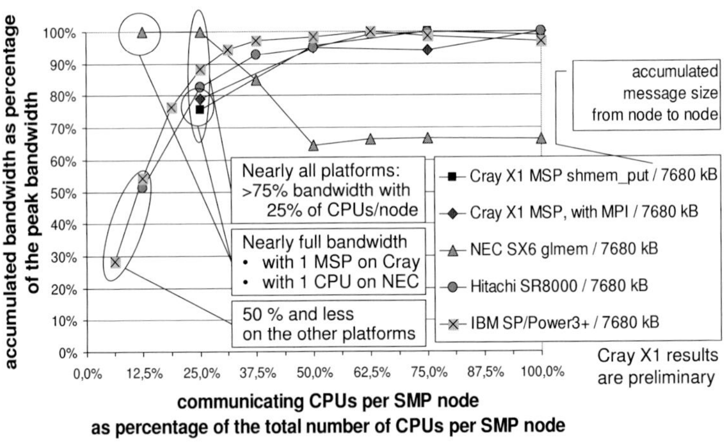

Fig. 3. Aggregated bandwidth per SMP node. The colour version of this figure can be found in Fig. A.30 on page 590.

On the other systems, the hybrid masteronly or funneled programming scheme can achieve only a small percentage of the peak inter-node bandwidth. [16] has compared the pure MPI with the masteronly scheme. For this compar-

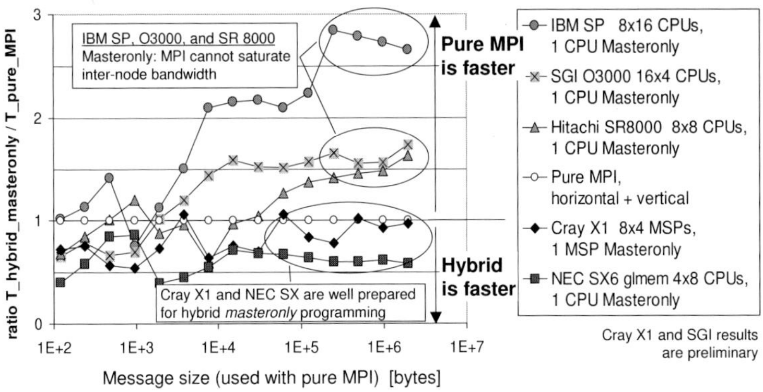

Fig. 4. Ratio of hybrid communication time to pure MPI communication time. The colour version of this figure can be found in Fig. A.31 on page 590.

ison, each MPI process in the pure MPI scheme has also to exchange messages between the processes in the same node. These intra-node messages have the same size as the inter-node messages, (c.f. Fig. 1). Fig. 4 shows the ratio of the inter-node communication time of hybrid MPI+OpenMP masteronly style divided by the time needed for the inter- and intra-node communication with pure MPI. In case of hybrid masteronly style, the messages must transfer the accumulated data of the parallel inter-node messages in pure MPI style, i.e., the message size is multiplied with the number of CPUs of an SMP node. In Fig. 4, one can see a significantly better communication time with pure MPI, on those platforms, on which the inter-node network cannot be saturated by the master thread. In this benchmark, the master thread in the masteronly scheme was emulated by an MPI process (and the other threads by MPI process waiting in a barrier). On most platforms, the measurements were verified with a benchmark using hybrid compilation and hybrid application start-up. The diagram compares experiments with the same aggregated message-size in the inter-node communication; on the x-axis, the corresponding number of bytes in the pure-MPI experiment is shown. This means, e.g., that the message size in the hybrid-masteronly experiment on a 16-CPU-per-node system is 16 times larger than in the experiments with pure MPI.

Benchmark platforms were: a Cray X1 with 16 nodes at Cray; the NEC SX-6 with 24 nodes and IXS interconnect at the DKRZ, Hamburg, Germany; the Hitachi SR8000 with 16 nodes at HLRS, Stuttgart, Germany; the IBM SP-Power3 at NERSC, USA; the SGI Origin 3000 (400 MHz) *Lomax* with 512 CPUs at NASA/Ames Research Center, NAS, USA; an SGI Origin 3800 (600 MHz) at SGI; the SUN Fire 6800 cluster with Sun Fire Link at the RWTH Aachen, Germany; and HELICS, a Myrinet 2 GBit/s full bisection network based cluster of 256 dual AMD Athlon 1.4 GHz PCs at IWR, University of Heidelberg.

3.2 The sleeping-threads problem and the saturation problem

The two most simple programming models on hybrid systems have both the same problem although they look quite different: With hybrid masteronly style the non-master threads are sleeping while the master communicates, and with pure MPI, all threads try to communicate while only a few (or one) threads already can saturate the inter-node network bandwidth (expecting that the application is organized in communicating and computing epochs). If one thread is able to achieve the full inter-node bandwidth (e.g., NEC SX-6, see Fig. 2), then both problems are equivalent. If one thread can only achieve a small percentage (e.g., 28 % on IBM SP), then the problem with masteronly style is significantly higher. As example on the IBM system, if an application communicates 1 sec in the pure MPI style (i.e. $1 \times 16 = 16$ CPUsec), then this program would need about $16/0.28 = 57$ CPUsec in masteronly style, and if one would use 4 CPUs for the inter-node communication (4 CPUs achieve 88.3 %) and the other 12 threads for overlapping computation, then only $4/0.883 = 4.5$ CPUsec would be necessary.

If the inter-node bandwidth cannot be achieved by one thread, then it may be a better choice to split each SMP node into several MPI processes that are itself multithreaded. Then, the inter-node bandwidth in the pure MPI and hybrid masteronly model are similar and mainly the topology, intra-node communication, and OpenMP-overhead problems determine which of both programming styles are more effective. With overlapping communication and computation, this splitting can also solve the inter-node bandwidth problem described in the previous section.

3.3 OpenMP overhead and cache effects

OpenMP parallelization introduces additional overhead both in sequential and in MPI parallel programs. Using a fine-grained OpenMP parallelization approach, the frequent creation of parallel regions and synchronization at the end of parallel regions as well as at the end of parallel worksharing constructs may sum up to a substantial part of the total runtime. Moreover, many of the OpenMP constructs imply automatic OMP FLUSH operations to provide a consistent view of the memory at the cost of additional memory operations. These effects may impact, in particular, the *Hybrid masteronly* style where OpenMP and/or automatic parallelization of inner loops is often done. Using the sparse-matrix-vector application described in Ref. [16], Fig. 5 clearly demonstrates the drawback of these effects when scaling processor count at fixed problem size:

Contrary to the *pure MPI* approach where a superlinear performance increase occurs if the aggregate L2 cache size becomes larger than the data size of the application, no cache effect is seen at all for the *Hybrid masteronly* style.

The overhead associated with the OpenMP parallelization can be reduced by a coarse-grained approach: The parallel regions are started only once at the beginning of the application, and OMP MASTER and OMP BARRIER directives are used for synchronizing before and after the MPI communication. Of course, this approach is more appropriate for *Hybrid funneled* and *Hybrid multiple* styles but will still suffer from the OMP FLUSH and OMP BARRIER operations which are necessary to establish a consistent memory view among the threads.

4 Bite the bullet

Each parallel programming scheme on hybrid architectures has one or more significant drawbacks. Depending on the needed resources of an application, the drawbacks may be major or only minor.

Programming without overlap of communication and computation

One of the two problems, *sleeping-threads* and *saturation problem* is indispensable. The major design criterion may be the topology problem:

- If it cannot be solved, pure MPI may cause too much inter-node traffic, but the masteronly scheme implies on some platforms a slow inter-node communication due to the inter-node bandwidth problem described above.

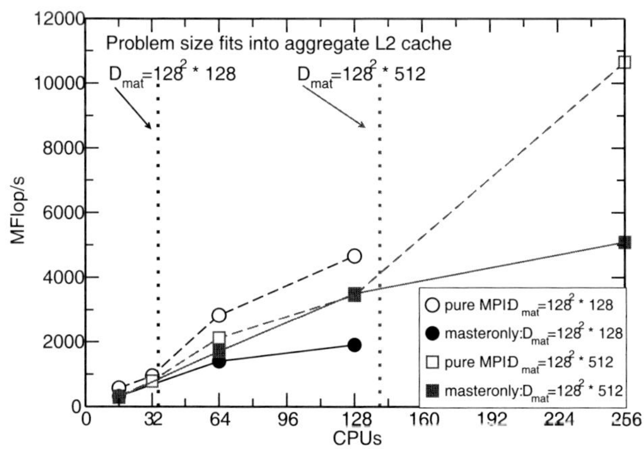

Fig. 5. Sparse matrix-vector-multiplication: Scalability of *pure MPI* style and *Hybrid masteronly* style for two different problem sizes on IBM SP-Power3/NERSC. The vertical dotted lines denote the CPU counts where the aggregate L2 cache sizes are large enough to hold all the data.

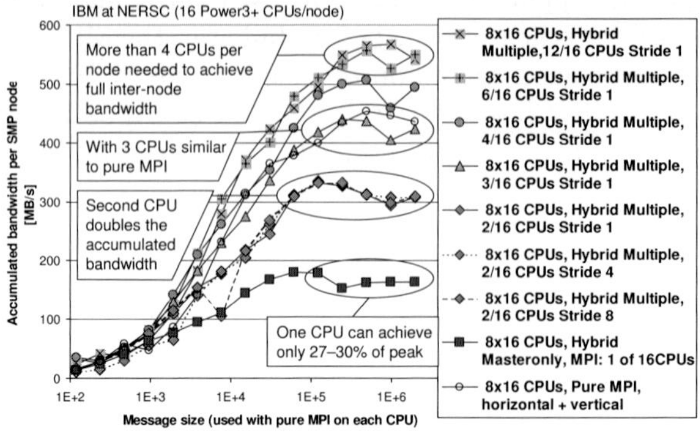

Fig. 6. Aggregated bandwidth per SMP node on IBM SP with 16 Power3+ CPUs per node. The colour version of this figure can be found in Fig. A.32 on page 591.

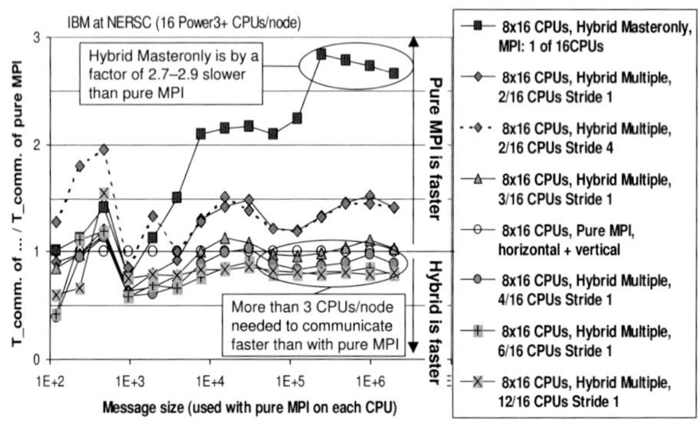

Fig. 7. Ratio of communication time in hybrid models to pure MPI programming on IBM SP. The colour version of this figure can be found in Fig. A.33 on page 591.

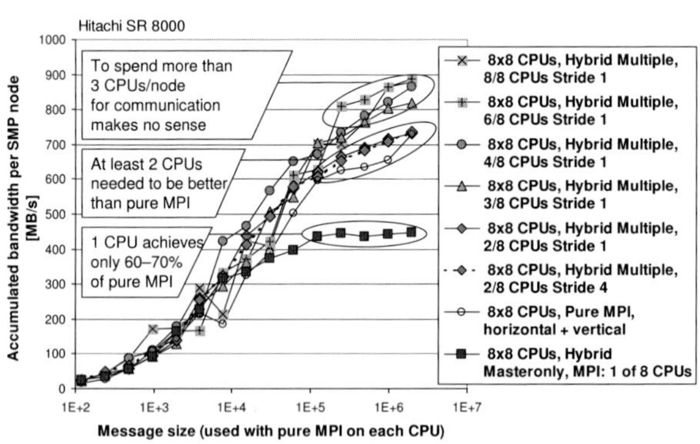

Fig. 8. Aggregated bandwidth per SMP node on Hitachi SR 8000. The colour version of this figure can be found in Fig. A.34 on page 591.

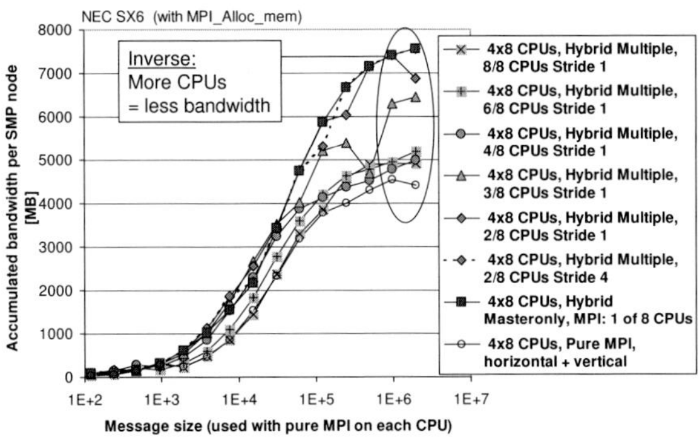

Fig. 9. Aggregated bandwidth per SMP node on NEC SX-6. The colour version of this figure can be found in Fig. A.35 on page 592.

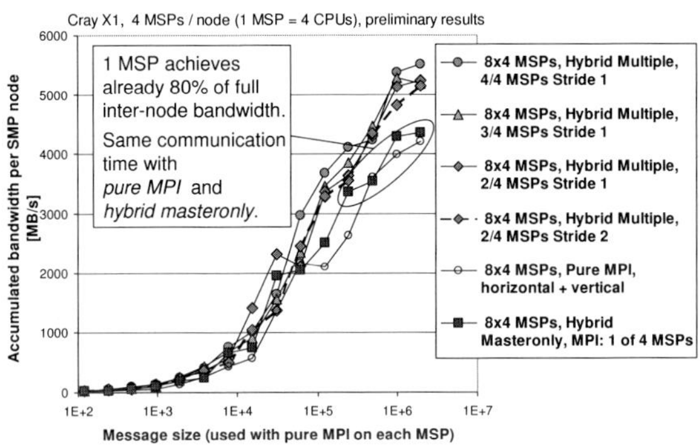

Fig. 10. Aggregated bandwidth per SMP node on Cray X1, MSP-based MPI-parallelization. The colour version of this figure can be found in Fig. A.36 on page 592.

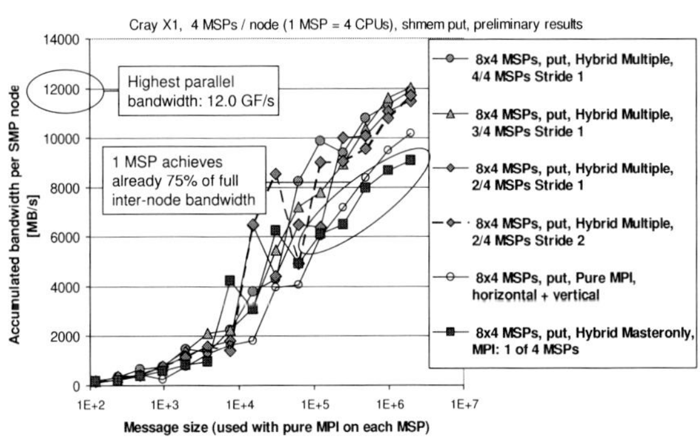

Fig. 11. Aggregated bandwidth per SMP node on Cray X1. MSP-based and MPI_Sendrecv is substituted by shmem_put. The colour version of this figure can be found in Fig. A.37 on page 592.

- If the topology problem can be solved, then we can compare hybrid masteronly with pure MPI: On some platforms, wasting inter-node bandwidth with masteronly style is the major problem; it causes more CPUs longer idling than with pure MPI. For example on an IBM SP system with 16 Power3+ CPUs on each SMP node, Fig. 6 shows the aggregated bandwidth per node with the experiment described in Section 3.1. The *pure MPI horizontal+vertical* bandwidth is defined in this diagram by dividing the amount of inter-node message bytes (without counting the intra-node messages)[4]) by the time needed for inter- and intra-node communication, i.e., the intra-node communication is treated as overhead. One can see, that more than 4 CPUs per node must communicate in parallel to achieve full inter-node bandwidth. At least 3 CPU per node must communicate in the hybrid model to beat the pure MPI model. Fig. 7 shows the ratio of the execution time in the hybrid models to the pure MPI model. A ratio greater than 1 shows that the hybrid model is slower than the pure MPI model.

On systems with 8 CPUs per node, the problem may be reduced, e.g., as one can see on a Hitachi SR 8000 in Fig. 8. On some vector type systems, one CPU may already be able to saturate the inter-node network, as shown in Fig. 9–11. Note: the aggregated inter-node bandwidth on the SX-6 is reduced, if more than one CPU per node tries to communicate at the same time over the IXS. Fig. 10 and 11 show preliminary results on a Cray X1 system with 16 nodes. Each SMP node consists of 4 MSPs (multi streaming processors). Each MSP itself consists of 4 SSPs (single streaming processors). With MSP-based programming, each MSP is treated as a CPU, i.e., each SMP node has 4 CPUs (=MSPs) that internally use an (automatic) thread-based parallelization (= *streaming*). With SSP-based programming, each SMP node has 16 CPUs (=SSPs). Preliminary results with the SSP-mode have shown, that the inter-node bandwidth is partially bound to the CPUs, i.e., that the behavior is similar to the 16-way IBM system.

Similar to the multi-threaded implementation of MPI on the Cray MSPs, it would be also possible on all other platforms to use multiple threads inside of the MPI communication routines if the application uses the hybrid masteronly scheme. The MPI library can easily detect whether the application is inside or outside of a parallel region. With this optimization (described in more detail in Section 5), the communication time of the hybrid masteronly model should always be shorter than the communication time in the pure MPI scheme.

On the other hand, looking on the Myrinet cluster with only 2 CPUs per SMP node, the hybrid communication model hasn't any drawback on such clusters because one CPU is already able to saturate the inter-node network.

[4] Because the intra-node messages must be treated as overhead if we compare pure MPI with hybrid communication strategies.

Programming with overlap of communication and computation

Although overlapping communication with computation is the chance to achieve fastest execution, this parallel programming style isn't widely used due to the lack of ease of use. It requires a coarse-grained and thread-rank-based OpenMP parallelization, the separation of halo-based computation from the computation that can be overlapped with communication, and the threads with different tasks must be load balanced. Advantages of the overlapping scheme are: (a) the problem that one CPU may not achieve the inter-node bandwidth is no longer relevant as long as there is enough computational work that can be overlapped with the communication; (b) the saturation problem is solved as long as not more CPUs communicate in parallel than necessary to achieve the inter-node bandwidth; (c) the sleeping threads problem is solved as long as all computation and communication is load balanced among the threads. A detailed analysis of the performance benefits of overlapping communication and computation can be found in [16].

5 Optimization Chance

On Cray X1 with MSP-based programming and on NEC SX-6, the *hybrid masteronly* communication pattern is faster than the *pure MPI*. Although both systems have vector-type CPUs, the reasons for these performance results are quite different: On the NEC SX-6, the hardware of one CPU is really able to saturate the inter-node network if the user data resides in global memory. On the Cray X1, each MSP consists of 4 SSPs (=CPUs). MPI communication issued by one MSP seems internally to be multi-streamed by all 4 SSPs. With this multi-threaded implementation of the communication, Cray can achieve 75–80 % of the full inter-node bandwidth, i.e., of the bandwidth that can be achieved if all MSPs (or all SSPs) communicate in parallel.

This approach can be generalized for the masteronly style. Depending on whether the application itself is translated for pure MPI approach, hybrid MPI + automatic SMP-parallelization, or hybrid MPI+OpenMP, the linked MPI library itself can also be parallelized with OpenMP directives or vendor-specific directives.

Often, the major basic capabilities of an MPI library are to put data into a shared memory region of the destination process (RDMA put), or to get data from the source process (RDMA get), or to locally calculate reduction operations on a vector, or to handle derived datatypes and data. All these operations (and not the envelop handling of the message passing interface) can be implemented multi-threaded, e.g., inside of a parallel region. In the case, that the application calls the MPI routines outside of parallel *application* regions, the parallel region inside of the MPI routines will allow a thread-parallel handling of these basic capabilities. In the case, the application overlaps communication and computation, the parallel region inside of the MPI library

Table 1. Inter- and Intra-node bandwidth for large messages compared with memory bandwidth and peak performance. All values are aggregated over one SMP node. Each message counts only once for the bandwidth calculation. Message size is 16 MB, except +) with 2 MB.

	Master-only, inter-node bandw. [GB/s]	pure MPI, inter-node bandw. [GB/s]	Master-only bw / max. inter-node bw [%]	pure MPI, intra-node bandw. [GB/s]	memory band-width bandw. [GB/s]	Peak and *Linpack* perfor-mance [GFLOP/s]	max. inter-node bw / peak *or Linpack* perf. [B/FLOP]	#nodes * #CPUs per SMP node
Cray X1, shmem_put preliminary results	9.27	**12.34**	**75 %**	33.0	136	**51.20** *45.03*	**0.241** *0.274*	8 * 4 MSPs
Cray X1, MPI preliminary results	4.52	**5.52**	**82 %**	19.5	136	**51.20** *45.03*	**0.108** *0.123*	8 * 4 MSPs
NEC SX-6, MPI with global memory	**7.56**	4.98	**100 %**	78.7 93.7+)	256	**64** *61.83*	**0.118** *0.122*	4 * 8 CPUs
NEC SX-5Be local memory	2.27	**2.50** a)	**91 %**	35.1	512	**64** *60.50*	**0.039** *0.041*	2 *16 CPUs a) only 8 CPUs
Hitachi SR8000	0.45	**0.91**	49 %	5.0	32+32	**8** *6.82*	**0.114** *0.133*	8 * 8 CPUs
IBM SP Power3+	0.16	**0.57+)**	28 %	2.0	16	**24** *14.27*	**0.023** *0.040*	8 *16 CPUs
SGI O3800 600MHz (2 MB messages)	0.427+)	**1.74+)**	25 %	1.73+)	3.2	**4.80** *3.64*	**0.363** *0.478*	16 *4 CPUs
SGI O3800 600MHz (16 MB messages)	0.156	**0.400**	39 %	0.580	3.2	**4.80** *3.64*	**0.083** *0.110*	16 *4 CPUs
SGI O3000 400MHz (preliminary results)	0.10	**0.30+)**	33 %	0.39+)	3.2	**3.20** *2.46*	**0.094** *0.122*	16 *4 CPUs
SUN Fire 6800[5] (preliminary results)	0.15	**0.85**	18 %	1.68		**43.1** *23.3*	**0.019** *0.036*	4 *24 CPUs
HELICS Dual-PC cluster with Myrinet	0.127+)	**0.129+)**	**98 %**	0.186+)		**2.80** *1.61*	**0.046** *0.080*	32 * 2 CPUs
HELICS Dual-PC cluster with Myrinet	**0.105**	0.091	**100 %**	0.192		**2.80** *1.61*	**0.038** *0.065*	32 * 2 CPUs
HELICS Dual-PC cluster with Myrinet	0.118+)	**0.119+)**	**99 %**	0.104+)		**2.80** *1.61*	**0.043** *0.074*	128 * 2 CPUs
HELICS Dual-PC cluster with Myrinet	**0.093**	0.082	**100 %**	0.101		**2.80** *1.61*	**0.033** *0.058*	128 * 2 CPUs
HELICS Dual-PC cluster with Myrinet	**0.087**	0.077	**100 %**	0.047		**2.80** *1.61*	**0.031** *0.054*	239 * 2 CPUs
Column[6]	1	2	3	4	5	6	7	8

is a nested region and will get only the (one) thread on which it is already running. Of course, the parallel region inside of MPI should only be launched, if the amount of data that must be transferred (or reduced) exceeds a given threshold.

This method optimizes the bandwidth without a significant penalty to the latency. On the Cray X1, currently only 4 SSPs are used to stream the communication in MSP mode achieving only 75–80 % of peak. It may be possible to achieve full inter-node bandwidth, if the SSPs of an additional MSP would also be applied. With such a multi-threaded implementation of the MPI communication for masteronly-style applications, there is no further need (with respect to the communication time) to split large SMP nodes into several MPI processes each with a reduced number of threads (as proposed in Section 3.2).

[5] A degradation may be caused by system processes because the benchmark used all processors of the SMP nodes.

[6] Columns 1, 2, 4 are benchmark results, Col. 3 is calculated from Col. 1 & 2, Col. 5 & 6 "peak" are theoretical values, Col. 6 "Linpack" is based on the TOP500 values for the total system [14], and Col. 7 is calculated from Col. 1, 2 & 6.

6 Conclusions

Different programming schemes on clusters of SMPs show different performance benefits or penalties on the hardware platforms benchmarked in this paper. Table 1 summarizes the results. Cray X1 with MSP-based programming and NEC SX-6 are well designed for the hybrid MPI+OpenMP masteronly scheme. On the other platforms, as well as on the Cray X1 with SSP-based programming, the master thread cannot saturate the inter-node network which is a significant performance bottleneck for the masteronly style.

To overcome this disadvantage, a multi-threaded implementation of the basic device capabilities in the MPI libraries is proposed in Section 5. Partially, this method is already implemented in the Cray X1 MSP-based MPI-library. Such MPI optimization would allow the saturation of the network bandwidth in the masteronly style. The implementation of this feature is important especially on platforms with more than 8 CPUs per SMP node.

This enhancement of current MPI implementations implies that the hybrid masteronly communication should be always faster than pure MPI communication. Both methods still include the sleeping threads or saturated network problem, i.e., that more CPUs are used for communicating than really needed to saturate the network. This drawback can be solved with overlapping of communication and computation, but this programming style needs extreme programming effort.

To achieve an optimal usage of the hardware, one can also try to use the idling CPUs for other applications, especially low-priority single-threaded or multi-threaded non-MPI applications if the parallel high-priority hybrid application does not use the total memory of the SMP nodes.

Acknowledgments

The authors would like to acknowledge their colleagues and all the people that supported this project with suggestions and helpful discussions. They would especially like to thank Dieter an Mey at RWTH Aachen, Thomas Ludwig, Stefan Friedel, Ana Kovatcheva, and Andreas Bogacki at IWR, Monika Wierse, Wilfried Oed, and Tom Goozen at CRAY, Holger Berger at NEC, Reiner Vogelsang at SGI, Gabriele Jost at NASA, and Horst Simon at NERSC for their assistance in executing the benchmark on their platforms. This research used resources of the HLRS Stuttgart, LRZ Munich, RWTH Aachen, University of Heidelberg, Cray Inc., NEC, SGI, NASA/AMES, and resources of the National Energy Research Scientific Computing Center, which is supported by the Office of Science of the U.S. Department of Energy. Part of this work was supported by KONWIHR project *cxHPC*.

References

[1] Rudolf Berrendorf, Michael Gerndt, Wolfgang E. Nagel and Joachim Prumerr, *SVM Fortran*, Technical Report IB-9322, KFA Jülich, Germany, 1993.
www.fz-juelich.de/zam/docs/printable/ib/ib-93/ib-9322.ps

[2] Frank Cappello and Daniel Etiemble, *MPI versus MPI+OpenMP on the IBM SP for the NAS benchmarks*, in Proc. Supercomputing'00, Dallas, TX, 2000. http://citeseer.nj.nec.com/cappello00mpi.html

[3] The Earth Simulator. www.es.jamstec.go.jp

[4] William Gropp, Ewing Lusk, Nathan Doss, and Anthony Skjellum, *A high-performance, portable implementation of the MPI message passing interface standard*, in Parallel Computing 22–6, Sep. 1996, pp 789–828.
http://citeseer.nj.nec.com/gropp96highperformance.html

[5] Shinichi Habataa, Mitsuo Yokokawa, and Shigemune Kitawaki, *The Earth Simulator System*, in NEC Research & Development, Vol. 44, No. 1, Jan. 2003, Special Issue on High Performance Computing.

[6] Jonathan Harris, *Extending OpenMP for NUMA Architectures*, in proceedings of the Second European Workshop on OpenMP, EWOMP 2000.
www.epcc.ed.ac.uk/ewomp2000/proceedings.html

[7] D. S. Henty, *Performance of hybrid message-passing and shared-memory parallelism for discrete element modeling*, in Proc. Supercomputing'00, Dallas, TX, 2000.
http://citeseer.nj.nec.com/henty00performance.html
www.sc2000.org/techpapr/papers/pap.pap154.pdf

[8] Matthias Hess, Gabriele Jost, Matthias Müller, and Roland Rühle, *Experiences using OpenMP based on Compiler Directed Software DSM on a PC Cluster*, in WOMPAT2002: Workshop on OpenMP Applications and Tools, Arctic Region Supercomputing Center, University of Alaska, Fairbanks, Aug. 5–7, 2002.
http://www.hlrs.de/people/mueller/papers/wompat2002/wompat2002.pdf

[9] Georg Karypis and Vipin Kumar. *A parallel algorithm for multilevel graph partitioning and sparse matrix ordering*, Journal of Parallel and Distributed Computing, 48(1): 71–95, 1998.
http://www-users.cs.umn.edu/~karypis/metis/
http://citeseer.nj.nec.com/karypis98parallel.html

[10] R. D. Loft, S. J. Thomas, and J. M. Dennis, *Terascale spectral element dynamical core for atmospheric general circulation models*, in proceedings, SC 2001, Nov. 2001, Denver, USA.
www.sc2001.org/papers/pap.pap189.pdf

[11] John Merlin, *Distributed OpenMP: Extensions to OpenMP for SMP Clusters*, in proceedings of the Second European Workshop on OpenMP, EWOMP 2000. www.epcc.ed.ac.uk/ewomp2000/proceedings.html

[12] Message Passing Interface Forum. *MPI: A Message-Passing Interface Standard*, Rel. 1.1, June 1995, www.mpi-forum.org.

[13] Message Passing Interface Forum. *MPI-2: Extensions to the Message-Passing Interface*, July 1997, www.mpi-forum.org.

[14] Hans Meuer, Erich Strohmaier, Jack Dongarra, Horst D. Simon, Universities of Mannheim and Tennessee, *TOP500 Supercomputer Sites*, www.top500.org.

[15] OpenMP Group, www.openmp.org.

[16] Rolf Rabenseifner and Gerhard Wellein, *Communication and Optimization Aspects of Parallel Programming Models on Hybrid Architectures*, International Journal of High Performance Computing Applications, Sage Science Press, Vol. 17, No. 1, 2003, pp 49–62.

[17] Rolf Rabenseifner, *Hybrid Parallel Programming: Performance Problems and Chances*, in proceedings of the 45th CUG Conference 2003, Columbus, Ohio, USA, May 12–16, 2003, www.cug.org.

[18] Mitsuhisa Sato, Shigehisa Satoh, Kazuhiro Kusano, and Yoshio Tanaka, *Design of OpenMP Compiler for an SMP Cluster*, in proceedings of the 1st European Workshop on OpenMP (EWOMP'99), Lund, Sweden, Sep. 1999, pp 32–39. http://citeseer.nj.nec.com/sato99design.html

[19] A. Scherer, H. Lu, T. Gross, and W. Zwaenepoel, *Transparent Adaptive Parallelism on NOWs using OpenMP*, in proc. of the Seventh Conference on Principles and Practice of Parallel Programming (PPoPP '99), May 1999, pp 96–106.

[20] Weisong Shi, Weiwu Hu, and Zhimin Tang, *Shared Virtual Memory: A Survey*, Technical report No. 980005, Center for High Performance Computing, Institute of Computing Technology, Chinese Academy of Sciences, 1998, www.ict.ac.cn/chpc/dsm/tr980005.ps.

[21] Lorna Smith and Mark Bull, *Development of Mixed Mode MPI / OpenMP Applications*, in proceedings of Workshop on OpenMP Applications and Tools (WOMPAT 2000), San Diego, July 2000. www.cs.uh.edu/wompat2000/

[22] Gerhard Wellein, Georg Hager, Achim Basermann, and Holger Fehske, *Fast sparse matrix-vector multiplication for TeraFlop/s computers*, in proceedings of VECPAR'2002, 5th Int'l Conference on High Performance Computing and Computational Science, Porto, Portugal, June 26–28, 2002, part I, pp 57–70. http://vecpar.fe.up.pt/

An Object-Oriented Approach to Specification and Composition of Web Services

Le Thanh Sach, Tru H. Cao, Le Nam Thang, and Le Thanh Son

Faculty of Information Technology, University of Technology, Ho Chi Minh City, Vietnam
ltsach@dit.hcmut.edu.vn
tru@dit.hcmut.edu.vn

Summary. The web service technology has emerged, facilitating interoperation between agents on the World Wide Web. When the number of services on the web becomes so large, it is necessary to have a way to reuse and compose new services from existing ones easily. Up until now, there are some proposals for this, such as DAML-S, XL, and WSFL. However, they do not allow one to manage and reuse services easily like object-oriented languages. This paper proposes an object-oriented web service framework and its language, in order to provide service creators with the ability to organize services into hierarchy of classes, and compose existing services into new ones. An operational architecture for object-oriented web services is also presented.

Key words: web service, WSFL, DAML-S, SOAP, UDDI, WSDL

1 An Introduction

Up until now, the World Wide Web has fundamentally been used via the interactions from human to application and mainly designed for human-readable interfaces; business services are also realized by this way. However, since web service technology emerged there have been more and more business services interacting with others on the Web.

Web services can be defined as modular programs, which have a collection of operations, each of which can be also seen as a service. Each service on the web is an independent and self-described module. Unlike services in the previous generation, web services can be requested by others or any Internet-enable programs by using the Internet protocols such as HTTP and FTP.

As web services play a dominant role in the web environment, an increasing number of online services are being published over the Internet. These make new opportunities for businesses, allowing new web services to be built by integration and composition of existing web services from other resources on the web.

Web services composition produces value-added services for its customers through composition of basic services, possibly offered by different companies. Building a new web service from existing services is not a trivial task, it involves a sequence of service invocations (control flows), and data flows between services [1] and manages execution of composition as a transaction unit.

Currently, there are many projects in both academic and industry research to develop a language to specify and compose web services. For examples, DAML-S [5] is an extended language of DAML, whose main goal is to automate some tasks in web service technology such as discovering, selecting, executing and monitoring web services. DAML-S does not concern about the implementation of services. XL [4] itself is an imperative programming language for implementation and also composition of web services. WSFL [1], an effort of IBM in web services composition, is a description language that specifies a composition of services in terms of control and data flows between activities. With each task in the business process being considered as an activity; WSFL describes a composition through two complementary models: (i) a flow model that specifies an execution sequence of the activities provided by the composed web services (ii) a global model that describes the interactions between the service providers. In the flow model each activity is linked with others through control links and data links. Each control link contains transaction conditions to specify whether the next activity will be processed or not, while each data link specifies data moving between activities. Mapping elements can be used with data links in order to transform data from the head to fit to the end [1].

The advantage of the afore-mentioned languages is that they provide a mechanism to utilize existing web services for composing new ones. But web services are visible, on requesters, as a collection of operations without hierarchy. Moreover, the listed languages do not allow services to inherit operations from existing ones in such the way that the traditional object-oriented approach does with its classes.

Therefore, in this paper we propose an object-oriented approach to web services, whereby each service is considered as a class consisting of its operations and all the services are organized into a class hierarchy. With this approach, not only web service operations can be reused, but also developed in a structured way without duplication in different services. The language we present here is named Object-Oriented Web Services Language (OOWSL) that inherits features from traditional object-oriented languages, besides features from the above web service languages.

The following sections of this paper introduce the above OOWSL model. Section 2 presents our object-oriented approach to modeling web services. Section 3 presents the architecture of service providers and its operation. OOWSL is presented in section 4. Some examples are provided in section 5. Finally, Section 6 gives concluding remarks and suggests future work for the OOWSL project.

2 An object-oriented approach to modeling web services

The current architecture for web services contains three main agents, namely, *Provider*, *Registry* and *Requester*. *Provider* contains web services and has the duty to execute demanded services. After finishing the execution, *Provider* returns the results to clients. *Registry* keeps yellow pages for web services, from which *Requester* can find available services. The final agent, *Requester*, comprises programs or services that need other services to finish its business, by sending requests to *Provider*.

Coming along with the emergence of web services technology, three technologies have also been created, namely, WSDL [2], SOAP [3] and UDDI. WSDL is a language to describe web services. Thereby, requesters can automate communications with providers. UDDI defines a standard way to publish and discover information about web services. SOAP can be seen as a way to encode messages between agents. Each SOAP message itself is encoded in XML and can be transported by using the current Internet protocols.

In the current web services technology, services are implemented by any programming languages such as Java and C#. When implementing web services, their low-level components such as classes and interfaces can be organized into hierarchies of classes. However, after services are published, requesters can only see services as collections of operations. These collections, which are described by WSDL documents, can also be used to generate stub classes in native languages that are used to develop requesting services. Programmers can follow WSDL documents to write codes for requesting services or use stub classes to automate communications with providers.

In our framework, we provide service creators with the ability to model and organize services into a hierarchy of classes. In this paper, we use object-oriented services, or OO-services for short, to mean web services that are developed using our proposed language OOWSL. Each OO-service can be seen as an object that can be located anywhere over the web and can talk to each other by sending/receiving messages.

In another aspect, the afore-mentioned languages such as XL and WSFL merge all information about services including their definitions and implementation, and thus the creation of services is carried out in one step. Meanwhile, the use of OO-analysis, OO-design and OO-programming in developing traditional programs has showed that it is better to separate the development process into several steps from the conceptual level to the implementation one. Therefore, in creating OO-services, we separate documents of OO-services into two layers, namely, logical layer and physical layer.

2.1 Logical Layer

This layer contains two kinds of diagrams, namely, service diagram and composition diagram. Service diagrams are used to define OO-services, including their names and operations. This kind of diagram has the same functionality as the class diagram in the unified modeling language. However, several concepts such as *association*, *dependency* and *interface* in the UML can not be used in our approach. Composition diagrams are used to define the composition for operations of OO-services. Speaking generally, several concepts in composing web services have the same meaning with the ones in specifying activity diagrams in the UML. However, some new concepts are recommended to be appropriate with the web service domain.

Service Diagram

At the first step in the creation process of OO-services, creators use service diagrams to describe OO-services. Each service diagram shows a hierarchy of services. Each service can contain a number of operations. Each operation can have some parameters and can return a result. Operations can also be supplemented with some modifiers such as public, private or protected. If the modifier is not specified, it is assumed to be public. However, when reusing OO-services on a requester side, the requester can see only public and protected operations. The private operations of an OO-service can be used only in other operations of that service.

Each parameter or result of an operation can have a data type. In our framework, all of the XML data types are supported and standard data types in Java language can be used.

For example, suppose that a certain company named ABC wants to publish a book service with three following operations:

- *BrowseBooks*: provides requesters with the ability to search for books using certain keywords.

- *OrderBooks*: provides requesters with the ability to order a list of books.

- *GetBookContent*: allows requesters to get the contents of certain books.

The service creators of ABC company can model ABC's service with the name *ABC_BookService*, as shown in figure 1.

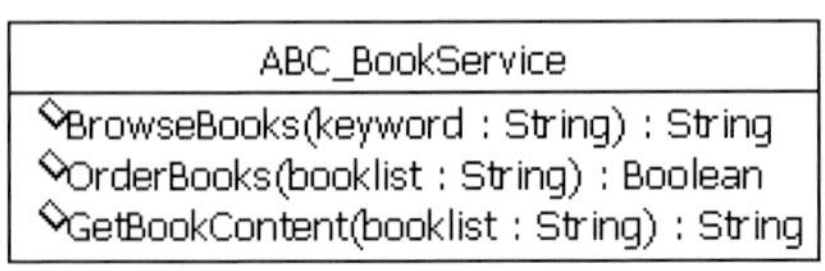

Fig. 1. A typical OO-service

An OO-service can inherit operations from others in the same way as an object does from others in object-oriented languages such as C++ or Java. As in C++, our framework accommodates multiple inheritance, but there is no more than one operation having the same signature in two parents of an OO-service.

For inheriting operations from a service, a service creator only needs to declare his/her service as a child of that parent service. A child OO-service can inherit protected and public operations from parent services.

For example, suppose that DEF company also wants to publish a book service, and the service creator realizes that it can inherit two operations from the *ABC_BookService*. Beside, the service creator of DEF realizes that it needs to override operation *OrderBooks* and support online sales. The *DEF_BookService* can be modeled as shown in figure 2.

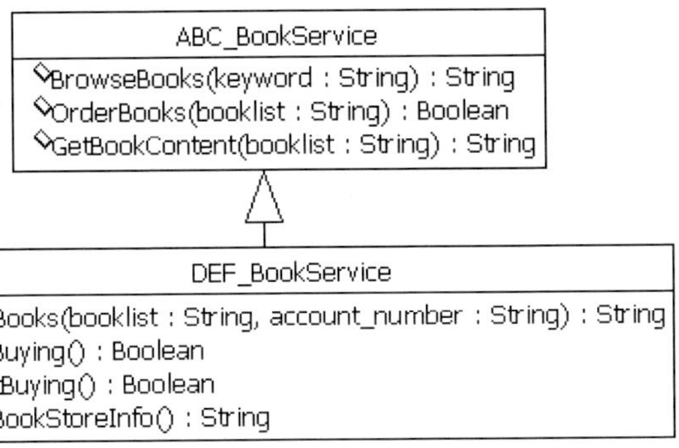

Fig. 2. Inheritance between OO-services

DEF_BookService supports two operations, *BrowseBooks* and *GetBookContent* as in *ABC_BookService* and have some supplement operations as follows:

- *BuyBooks*: provides requesters with the ability to buy books online. This operation can be seen from requesters, because it has the modifier public. The composition diagram section below will present a composition for this operation, which employs two private operations of *DEF_BookService* and two operations in other services.

- The service creator of DEF company also wants to provide requesters with ability to get general information from the company; therefore he/she adds to *DEF_BookService* a public operation named *GetBookStoreInfo*. This operation is also seen from requesters.

- *DEF_BookService* contains two private operations, namely, *PreBuying* and *PostBuying*, which are called at the beginning and the end of each call to *BuyBooks* respectively. These operations will not be seen from the requester side; they can only be used by operations of *DEF_BookService.*

- Finally, *DEF_BookService* overrides the operation *OrderBooks* from *ABC_BookService* in order to save order-invoices into the database of DEF company.

Composition Diagram

Today, there are many languages such as Java and C# that can be used for programming web services. However, when the number of web services on the Internet becomes so large, discovering and composing services from existing ones are the quest. Like WSFL [1] our proposed framework only concerns about description and composition for OO-services. Implementation of OO-services can be carried out in traditional languages.

Service diagrams just define OO-services by describing signatures of their operations, but not their bodies. The next step in creating OO-services is to assign the implementation to operations. There are two kinds of mapping signatures to executable documents. In many cases, creators just assign a composition diagram to each operation in OO-services. However, creators can also assign to an operation a function call to underlying component, e.g. a COM object or an EJB object. The assignment of an operation to a composition will be presented here and the mapping to a function call will be presented in the physical layer section.

Each composition diagram describes one operation in OO-services, showing connections between activities to be executed to finish that operation. The composition presented here is much the same as one in WSFL. However, we use two new concepts Input and Output instead of *Flow Source* and *Flow Sink* for the compatibility with the ones in defining OO-service operations.The concept *performedBy* in WSFL is represented by the dash-dot arrows that connect activities to operations on OO-services that are represented by left-right arrows, as shown in figure 3.

Suppose that, for buying books online, the service creators of DEF company employs the following goods transportation service and online paying service from other companies:

- *TransportService*: it has operation *TranferGoods* that can be called to request delivering goods.

- *BankingService*: it has operation *Paying* to transfer money between accounts.

Therefore, the creator can compose operation *BuyBooks* as shown in figure 3. with five steps as follows:

1. Receive invoices and account number from requesters.

2. Put those data to operation *PreBuying* to prepare something before carrying out trading.

3. If there is any error, e.g. the requested books are not available, *PreBuying* will send the error to the output and finish operation BuyBooks.

4. If *PreBuying* sees that the trading can be carried out, then *TransferGoods* will be called to demand the transportation company sending requested books to customers. *Paying* will also be called to request transferring money from customer accounts DEF company and transport company accounts.

5. After both the operations *TransferGoods* and *Paying*, *PostBuying* will be called to finish the trading.

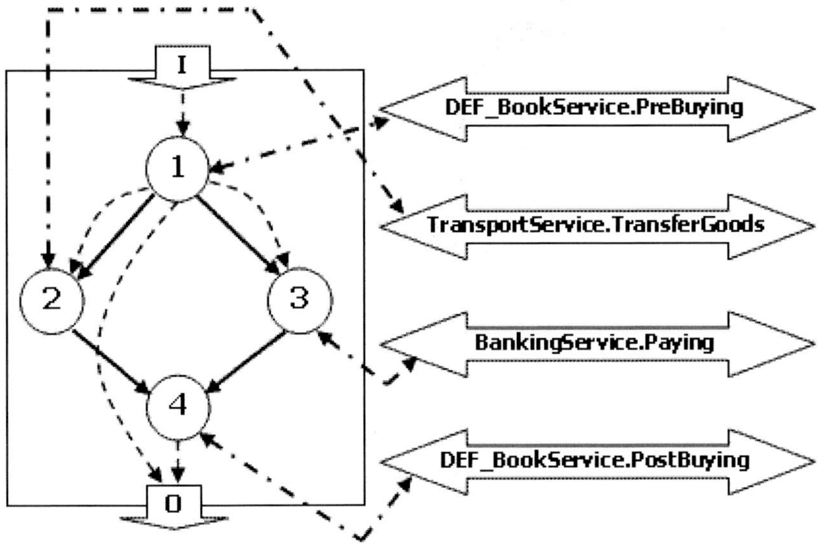

Fig. 3. An example of composition diagram

2.2 Physical Layer

After creating the service and composition diagrams in the logical layer and before deploying OO-services to applications servers, the creator has to describe the implementation and binding diagrams in the physical layer:

- Implementation diagram: This diagram maps operations of OO-services to functions in underlying components such as COM objects on Windows platform or EJB objects on Java platform. The service creator does not need to provide mappings for operations that have been assigned composition diagrams.

- Binding diagram: This diagram shows mappings between OO-services to locations where the services are deployed.

An OO-service can be described completely in the four diagrams above. However, to make them available to the World Wide Web, the service creator has to deploy them to application servers. The deployment process for OO-services is quite simple, whereby the creator carries out only two following steps:

1. Generate server and client documents for OO-services:

- Server documents are encoded by OOWSL language, which will be presented in Section 4. This document contains four diagrams as presented above.

- Client documents are used by requesters in order to call operations of providers. To make OO-services compatible with the current web service technology, WSDL documents can be generated from the description of OO-services. However, the requester can also use OOWSL documents directly.

In this case, OOWSL documents contains only service diagram and binding diagram.

- Beside, a hierarchy of OO-services can be mapped to classes in traditional languages such as Java, C++, C#, etc.

2. Deploy OO-services to application servers: application servers in our framework have been structured in such a way as to deploy OO-services easily, whereby service creators just add server documents to predefined folders of application servers.

3 A proposed architecture and its operation

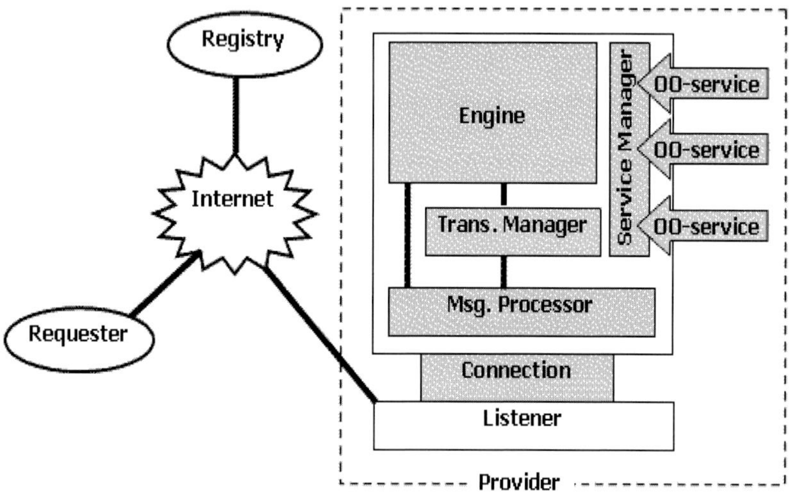

Fig. 4. The proposed OOWSL engine

3.1 Elements in the proposed architecture

Until today, there have been several commercial application servers such as the product of Microsoft -Internet Information Server - and some Java-based servers. They can be utilized to execute traditional web services. However, in order to realize OO-services we have to develop another server. This kind of server should be able to receive a call, to process the call and also to return the final result. In three jobs above, the first and the last can be interchanged with the components that have the same capability in the listed servers. For that reason, our proposed architecture for providers is separated into two modules, namely, *Listener* and *OOWSL-Engine*. *Listener* is designed to be able to replaced by others servers. We only focus on building the OOWSL-Engine. By this way, we can incorporate our engine to existing commercial application servers.

In this architecture, the responsibility of *Listener* and *OOWSL-Engine* are as follows:

Listener:

- Ready to send/receive messages to/from others. As such, listeners are like receptionists of organizations in real life.

- In the current version of our framework, SOAP is used as a language to format messages that are exchanged on the web.

OOWSL-Engine:

- First of all, OOWSL Engine should be able to talk with *Listener*. It means that the engine should be able to process messages to/from Listener. The sub-module *Msg. Processor* inside the engine bears this responsibility.

- At any time, OOWSL Engine should have to know what services that it currently contains, and what operations can be supported for each service. Therefore, the engine is provided with a sub-module called *Service Manager* to carry out this duty.

- Besides, OOWSL Engine should also be able to collect all messages for each client and manage transactions with each client. For this reason, the sub-module *Transaction Manger* is added to the engine.

- Finally, when OOWSL Engine has information about requested operations, it has to execute those operations and returns results to requesters. Therefore, it is equipped with the sub-module *Engine* inside, which has the same duty as Java Virtual Machine (JVM) in Java runtime environment (JRE). However, there is a difference that JVM executes byte-codes for Java classes while this engine calls a function in underlying component or executes a flow graph as described in composition diagrams.

3.2 The operation

This section presents the ways to use OO-services on the requester side. Suppose that a developer wants to create a general application or a web service, and knows that his application needs to call an operation on an existing OO-service. In that circumstance, the developer should write some codes to request that operation of the service. There are different ways to make a remote call, depending on which language the developer uses to code his/her application. As for calling a service in the current web service technology, the developer can follow directions in WSDL documents and code the communication with the service. Another way is that, the developer generates WSDL documents from

OOWSL documents for the service. After that, through using existing tools such as WSDL.EXE, the developer can generate stub-classes in the programming language that is used to code the application. With the stub-classes, the developer can call the service as calling to an operation in the programming language that he is using. The easiest way is that, the developer can use OOWSL documents directly when he creates OO-services in OOWSL itself.

In OOWSL framework, the implementation of an operation belongs to one of the three following situations, (i) the operation is assigned to a function in an underlying component such as a COM or an EJB object, (ii) the operation is inherited from parent service, and (iii) the operation is assigned to a composition document. Some tasks in calling an operation for each situation are listed below.

For example, suppose that the above-mentioned *DEF_BookService* is deployed into the OOWSL-Engine described in figure 4 and the other services *ABC_BookService*, *TransportService*, *BankingServicealso* are deployed somewhere on the web. Moreover, the operation *GetBookStoreInfo* of *DEF_BookService* is assigned to a function of a certain COM object, and the operation *BuyBooks* is assigned to a composition diagram as showed in figure 3 and figure 5 shows message flows in this example:

(i) If the requester calls the operation *GetBookStoreInfo* of *DEF_BookService*:

1. The requester makes a SOAP message and sends it to the listener. The message contains the information about the requested operation, including its arguments and the requested service name. The basic elements of that message looks like as below.

```
 SOAPAction:
"http://ltsach/DEF_BookService.GetBookStoreInfo"

<?xml version="1.0" encoding="utf-8"?> <soap:Envelope>
  <soap:Body>
    <GetBookStoreInfo/>
  </soap:Body>
</soap:Envelope>
```

The *SOAPAction* attribute contains the requested service name and the operation name.

2. The listener receives the message from the requester and forwards it to the *OOWSL-engine* to finish its work.

3. The sub-module *Msg. Processor* of the OOWSL Engine receives the message, processes and reports it to the sub-module *Engine* with the requested operations, services and its arguments.

4. The sub-module *Engine* uses *Service Manager*, which contains information about all services that have been deployed, to verify the correctness of the call.

5. The sub-module *Engine* uses *Transaction Manager* to create and manage the required transactions.

6. The sub-module *Engine* then uses the information returned from *Service Manager* to execute the operation.

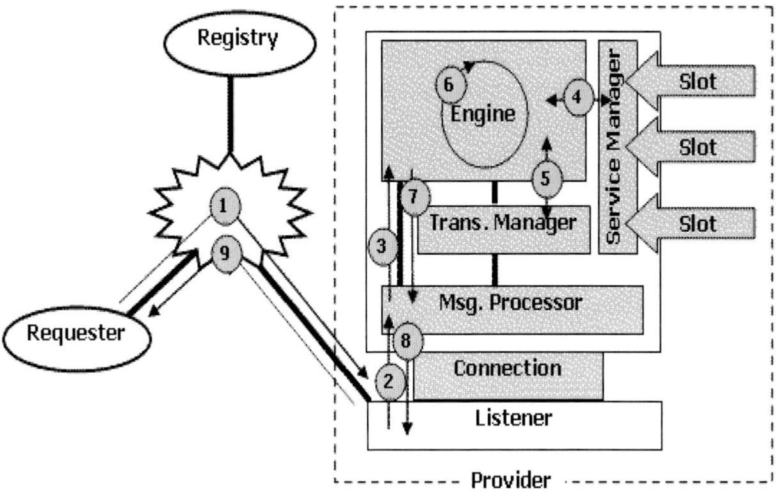

Fig. 5. Messages in calling a service

Since the requested service *DEF_BookService* is deployed locally and the requested operation has been mapped to a function of a COM object, *Engine* calls that function. The result from the function is transferred to *Msg. Processor* to be packed into a SOAP message returned to the requester.

7. The sub-module *Msg. Processor* receives the result from *Engine*. It packs the result into a SOAP message and forwards it to the listener.

8. The listener sends the message to the requester. The requester receives the message that contains the result for the call to the requested operation, displayed on an application form or a webpage.

(ii) If the requester calls the operation *BrowseBooks* on *DEF_BookService*:

This call still has the nine message flows as above. However, when carrying out the sixth, *Engine* has to do more tasks as follows:

The information from *Service Manager* shows that the requested operation should be called on service *ABC_BookService*, as shown in figure 2. Therefore, *Engine* has to play a role of a requester making another call to *BrowseBooks* on *ABC_BookService*, which is located on another site somewhere on the web.

- The result from *ABC_BookService* is transferred to *Msg. Processor* to be packed into a SOAP message returned to the requester.

(iii) If the requester calls the operation *BuyBooks* on *DEF_BookService*:

This call still has the nine message flows as above. However, when carrying out the sixth, *Engine* executes an algorithm to navigate activities on the composition diagram. This navigation algorithm is much the same as the one described in WSFL documents [1].

4 Object - Oriented Web Service Language

Like other programming languages, OOWSL is used to encode OO-services. However, there are some differences between OOWSL and others. While other languages are used to express instructions of programs, OOWSL is concerned with a high level of execution. Therefore, it provides users with the ability to create the description of an OO-service. OOWSL documents contain information conveyed by both WSDL and WSFL documents.

A typical OOWSL document, which is used for OO-services on the server side, contains the information that is described in four diagrams, namely, logic, composition, implementation and binding diagrams. However, OOWSL documents that are used on the client side contain only the information in logic and binding diagrams.

To be easily used on the Web, OOWSL has been specified using XML containing several elements. In this paper, we will present some basic elements along with their semantics. The definition of an OO-service in OOWSL documents is separated into four parts that correspond to the four kinds of diagrams in modeling OO-services. A sample document looks like as below:

```
<definitions
      version="1.0"
      xmlns="http://www.dit.hcmut.edu.vn/sw/oowsl/oowsl.xsd">
    <services>
    </services>
    <compositions>
    </compositions>
    <implementations>
    </implementations>
    <bindings>
    </bindings>
</definitions>
```

(i) The definition of an OO-service:

Creators use the element *services* of OOWSL to describe OO-services as shown in the above fragment. There is an only one element *services* for each OOWSL document. To define an OO-service, creators place element *service* inside element *services*. A definition for an OO-service includes a list of parents and operations of the service.

Inheritance in OOWSL is supported by using two elements, *extends* and *parent*. Information of operations for OO-services is described by using elements *operation* and *parameter*.

(ii) The description of composition:

The element compositions of OOWSL describes connections between existing operations in OO-services to make a new one. Basically, this composition is much the same as the one in WSFL [1]. However, this kind of information is described in the logical layer, where the service creator only concerns about OO-services and the operations for each one. Therefore, there are some differences between OOWSL and WSFL. For example, in WSFL when a creator wants to map a data item from a source activity to a target activity, he/she describes the mapping from source message to target message. Meanwhile, in OOWSL, each activity is actually an operation of an OO-service, and thus the creator can map the output of one operation to the input of another other.

(iii) The description of implementation:

In OOWSL, a service creator can assign each operation of an OO-service a composition diagram or a mapping to a function in underlying components such as a COM object or an EJB object. If the creator assigns the operation to a composition diagram then the *OOWSL-Engine* will look for the execution information for that operation in the composition part of OOWSL documents. However, if the creator assigns a mapping, the engine will look the information to execute the operation in the implementation part of OOWSL documents.

(iv) The description of binding:

Before deploying OO-services to application servers, the creator should have to assign a real address to each OO-service. This kind of information will be contained in the element bindings of OOWSL documents. The following is a typical part of the element bindings.

```
<bindings>
        <bind service="Service1" address="Address1">
        <bind service="Service2" address="Address2">
</bindings>
```

5 Conclusion and Future Work

In our project, we have developed a new framework based on the IBM's WSFL. This new framework extends the architecture, operational model and language of WSFL systems in order to enable inheritance between web services. As an extension of WSFL, OOWSL provides developers with the ability to compose new services from existing ones. Another modification has been made to WSFL is the presented transaction management mechanism.

This framework can be improved by building an automatic web service execution monitoring mechanism. Another improvement can be made is to fully support dynamic bindings. To realize these improvements, OOWSL should be provided with some elements to specify semantic constraints on OO-services.

Towards really complex applications we have built an *OOWSL-Engine* [6] for executing and monitoring. We are also investigating to realize the ability of dynamic binding.

References

[1] Laymann, F. IBM Web Services Flow Language (WSFL) version 1.0 (2001)
[2] Christensen, E. et al. Web Services Description Language (WSDL 1.2). W3C Recommendation. (2002)
[3] Box, D. et al. Simple Object Access Protocol (SOAP 1.1). W3C Recommendation. (2000)
[4] Florescu, D. and Grünhaghen. An XML Programming Language for the Service Specification and Composition. Bulletin of the IEEE Computer Society Technical Committee on Data Engineering.
[5] Anupriya Ankolekar et all. DAML-S: Web Service Description for the Semantic Web. (2002)
[6] Nam, L.D.L. Building OOWSL Engine. BEng Thesis, Ho Chi Minh City University of Technology (2003).

Applied Stochastic Integer Programming: Scheduling in the Processing Industries

Guido Sand[1], Sebastian Engell[1], A. Märkert[2], and Rüdiger Schultz[2]

[1] Department of Biochemical and Chemical Engineering, Universität Dortmund 44221 Dortmund, Germany
g.sand@bci.uni-dortmund.de
s.engell@bci.uni-dortmund.de

[2] Institute of Mathematics, Universität Duisburg-Essen, 47048 Duisburg, Germany
maerkert@math.uni-duisburg.de
schultz@math.uni-duisburg.de

Summary. In this contribution, we consider scheduling problems of flexible batch plants in the processing industries. Special emphasis is put on the aspect of uncertainty, which is undoubtedly relevant but was often neglected so far. Motivated by a real-world example process, we describe an "engineered" solution concept based upon *two-stage stochastic integer programming* along with a decomposition-based solution algorithm and numerical experiences.

1 Introduction

In the chemical processing industries, the concept of flexible batch plants is widespread especially for the production of small amounts of high valued products. In particular, the class of *multiproduct batch plants* is most popular and typically employed for similar products which require the same sequence of processing steps. Their high degree of flexibility in terms of structure and capacity makes rapid adaptations of the product supply to dynamically changing demands possible.

However, to operate such flexible plants in dynamic environments efficiently, logistic decisions have to be made in a high frequency. Thus, scheduling problems, i.e. assignments of processing steps to processing units over time, have to be solved online on moving horizons. According to [7], "engineered" mathematical programming (MP) approaches provide the most powerful techniques to model and solve these types of problems. Engineered MP-approaches differ from the classical ones by using highly customized models and applying problem-specific solution algorithms.

In fact, a considerable number of publications has demonstrated the suitability of mixed-integer (non)linear programming (MI(N)LP) approaches to tackle scheduling problems, for surveys see [1, 10, 6]. Nevertheless, most of them neglect the significant aspect of uncertainty: Typically, chemical production processes are subject to disturbances in the plant, in the processes and in the demands.

In this contribution, we present an engineered MP-approach to scheduling problems in the processing industries which is based upon uncertainty conscious models, namely two-stage stochastic integer programs. The focus is on structural aspects of the model, the generic solution approach, and numerical experiences for a real-world problem. For a detailed explanation and discussion of the model we refer the reader to [8].

2 A Real-World Example Process

As a real-world example, we consider a multi-product plant from the polymer industries which is used to produce expandable polystyrene (EPS). As can be seen from the flowsheet in Fig. 1, the plant consists of three stages and produces ten different products: two types of EPS in five grain size fractions each.

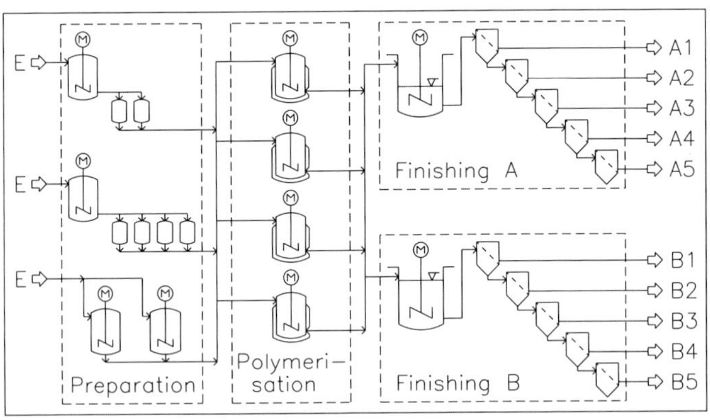

Fig. 1. Flowsheet of the EPS-Process

The preparation as well as the polymerization stage are operated in batch mode, whereas the finishing lines run continuously. The semi-continuously driven mixers are part of the finishing lines and serve as material buffers to couple the batch reactors with the continuous separation units. The mixing process introduces non-convex non-linearities into the problem.

According to the standard operating concept, the plant is recipe-driven. For the EPS-production, ten basic recipes exist which specify the EPS-type and the grain size distribution. The polymerization process is not selective, such that each batch contains one main grain size fraction but in addition significant amounts of the others as by-products.

This coupled production of different products in combination with the interconnection of the batchwise and continuously driven stages makes the scheduling problem a highly complex task. In order to meet specified demand profiles, four types of degrees of freedom must be fixed: 1. choice of the recipes, 2. timing of the polymerizations, 3. start-up/shut-down-times of the finishing lines, and 4. feedrates into the separation units. The decision making process is affected by four types of uncertainties: 1. processing times of the polymerizations, 2. grain size distributions, 3. capacity of the polymerization stage, and 4. demand profiles.

In total, the scheduling problem at hand constitutes a large-scale, stochastic, mixed-integer non-linear optimization problems which has to be solved in relatively short computing times.

3 Approximative Solution Concept

The engineered solution concept is on one hand based upon approximative models and on the other hand based on an efficient solution algorithm. Algorithmic issues are discussed in Sections 4.2 and 6. The approximation scheme is based on the following key ideas:

1. The non-linearities are approximatively linearized.
2. The monolithic scheduling problem is hierarchically decomposed into a master and a detailed scheduling (MS/DS) problem. Both are based on models for moving horizons and solved in intervals which are significantly smaller than the horizons, i.e the are solved multiple times within the horizon length.
3. The alternating sequence of receiving new information and making decisions constitutes a multi-stage information and decision structure, which is approximated by a two-stage structure.
4. The uncertain evolution of the future is represented by a finite number of scenarios with certain probabilities.

Formally, these ideas are reflected by the framework of *two-stage stochastic integer programs*, which are presented in the next section.

4 Two-Stage Stochastic Integer Programming

4.1 Modelling Framework

Two-stage stochastic integer programs (2-SSIP) reflect the tree structure as depicted in Fig. 2 for 10 intervals. The tree represents the time-dependency of the uncertainty: The evolution of the process is assumed to be certainly known for the first three intervals and branches out in the fourth interval. Note that the entire uncertainty is realized in the fourth interval, i.e. there is exactly one point where information is received. Accordingly, the decisions are divided into 1st stage or here&now decisions, which must be made under uncertainty, and 2nd stage or recourse decisions, which can be made based on certain information of the future evolution. A comprehensive introduction into stochastic programming can be found in [2].

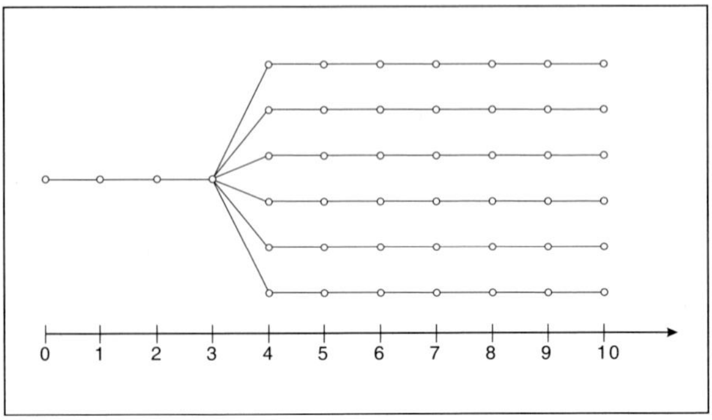

Fig. 2. Two-Stage Tree Structure

For a linearized 2-SSIP with a finite number of scenarios, a deterministic equivalent can be stated as the following mixed-integer linear program (MILP):

$$z = \min_{\mathbf{x},\mathbf{y}_\omega} \; \mathbf{c}^{\mathrm{T}}\mathbf{x} + \sum_{\omega=1}^{\Omega} \pi_\omega \mathbf{q}_\omega^{\mathrm{T}} \mathbf{y}_\omega \tag{1}$$
$$\text{s.t.} \quad \mathbf{T}_\omega \mathbf{x} + \mathbf{W}_\omega \mathbf{y}_\omega \le \mathbf{h}_\omega, \quad \mathbf{x} \in X, \mathbf{y}_\omega \in Y, \omega = 1 \ldots \Omega \,.$$

The 1st and 2nd stage variable-vectors $\mathbf{x}$ and $\mathbf{y}_\omega$ belong to polyhedral sets X and Y with integer requirements. The parameter Ω denotes the number of scenarios ω with corresponding probabilities π_ω. The constraints are formulated by means of the matrices $\mathbf{T}_\omega$ and $\mathbf{W}_\omega$ and the right hand side vector $\mathbf{h}_\omega$ of suitable dimensions. The objective is to minimize the expectation value over all scenarios computed as a weighted sum of $\mathbf{x}$ and $\mathbf{y}_\omega$ subject to the weighting-vectors $\mathbf{c}$ and $\mathbf{q}_\omega$. The transformation of a maximization problem into an equivalent minimization problem is straightforward.

4.2 Dual Decomposition Algorithm

The matrix structure of (1) exhibits a typical block-angular shape shown in Fig. 3, left. The columns correspond to the variables and the lines to the constraints. It is obvious that the various scenarios (each corresponding to a pair of matrices $\mathbf{T}_\omega$ and $\mathbf{W}_\omega$) are coupled by the matrices $\mathbf{T}_\omega$, i.e. the first stage variables, only.

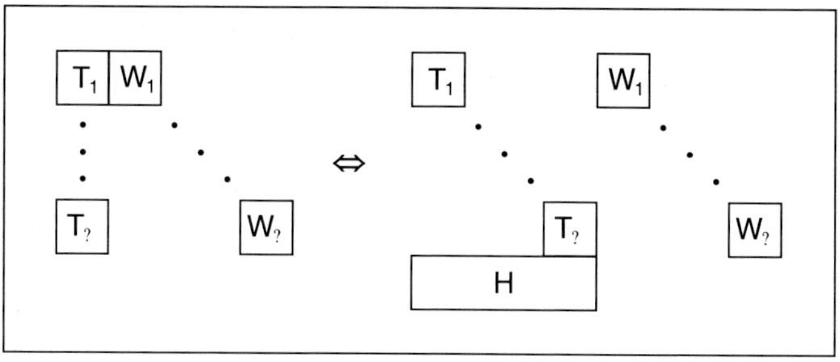

Fig. 3. Matrix Structure of Two-Stage Stochastic Programs

This specific structure is well-suited for the application of a solution approach based upon decomposition according to the scenarios. To this end, (1) is equivalently transformed into the following form:

$$z = \min_{\mathbf{x}_\omega, \mathbf{y}_\omega} \sum_{\omega=1}^{\Omega} \pi_\omega \left(\mathbf{c}^{\mathrm{T}} \mathbf{x}_\omega + \mathbf{q}_\omega^{\mathrm{T}} \mathbf{y}_\omega \right) \tag{2}$$
$$\text{s.t.} \quad \mathbf{T}_\omega \mathbf{x}_\omega + \mathbf{W}_\omega \mathbf{y}_\omega \le \mathbf{h}_\omega \, ,$$
$$\mathbf{x}_{\omega=1} = \ldots = \mathbf{x}_{\omega=\Omega}, \ \mathbf{x}_\omega \in X, \mathbf{y}_\omega \in Y, \omega = 1 \ldots \Omega \, .$$

As shown in Fig. 3, right, copies of the first stage variables are introduced for each scenario and the couplings between the scenarios are made explicit by means of the so-called *non-anticipativity constraints* (represented by the matrix $\mathbf{H}$). The dualization of the non-anticipativity constraints yields the following Lagrangian dual problem ($\boldsymbol{\lambda}$ denotes Lagrangian multipliers):

$$z^{\mathrm{LD}} = \max_{\boldsymbol{\lambda}} \min_{\mathbf{x}_\omega, \mathbf{y}_\omega} \sum_{\omega=1}^{\Omega} \pi_\omega \left(\mathbf{c}^{\mathrm{T}} \mathbf{x}_\omega + \mathbf{q}_\omega^{\mathrm{T}} \mathbf{y}_\omega \right) + \boldsymbol{\lambda}^{\mathrm{T}} \mathbf{H}_\omega \mathbf{x}_\omega \tag{3}$$
$$\text{s.t.} \quad \mathbf{T}_\omega \mathbf{x}_\omega + \mathbf{W}_\omega \mathbf{y}_\omega \le \mathbf{h}_\omega, \ \mathbf{x}_\omega \in X, \mathbf{y}_\omega \in Y, \omega = 1 \ldots \Omega \, .$$

The Lagrangian dual problem is a strict relaxation of the original 2-SSIP such that (3) provides a lower bound to (2). For a fixed $\boldsymbol{\lambda}$ (3) decomposes into Ω independent scenario-subproblems. However, a solution of (3) may yield Ω different vectors $\mathbf{x}_\omega$ such that the non-anticipativity constraints are infeasible. The feasibility can be re-established by the following branch&bound scheme proposed in [3]:

Step 1 – Initialization: Set the objective value of the best current solution z^* to ∞, and let the list of untreated problems $\mathcal{P}$ consist of problem (2).
Step 2 – Termination: If $\mathcal{P} = \emptyset$ then z^* is optimal.
Step 3 – Node Selection: Select and delete a problem $P \in \mathcal{P}$, and solve its Lagrangian dual. If P is infeasible, set $z^{\mathrm{LD}}(P)$ to ∞. If $z^{\mathrm{LD}}(P)$ is greater or equal z^* go to Step 2.

Step 4 – Bounding: Determine heuristically a tentative solution $\mathbf{x}^{\mathrm{R}}(P)$, and solve problem (2) with $\mathbf{x} = \mathbf{x}^{\mathrm{R}}$. If $z^*(P)$ is smaller than z^*, then set z^* to $z^*(P)$ and delete all $P' \in \mathcal{P}$ with $z^*(P')$ greater or equal z^*.

Step 5 – Branching: Select a component $x_{(k)}$ of $\mathbf{x}$, and add two new problems to $\mathcal{P}$ by extending P by the additional constraint $x_{(k)} \leq \lfloor \overline{x}_{(k)} \rfloor$ and $x_{(k)} \geq \lfloor \overline{x}_{(k)} \rfloor + 1$, respectively ($\overline{x}_{(k)}$ denotes the average over all ω). Got to step 3.

If $\mathbf{x}$ is purely integer, the algorithm terminates after finitely many steps yielding the global optimum. For continuous 1st stage variables see [3].

5 An Aggregated Scheduling Model

The modelling and solution concept is demonstrated here for the MS-layer (see Section 3). It requires a model which reflects the scheduling problem in a coarser manner then the DS-layer. To this end, a temporal aggregation approach is applied, which is based on a multi-period time representation. For each period the decisions are represented by aggregated variables and restricted by aggregated constraints. The periods of the model correspond to the intervals introduced in Section 3.

In the following, the aggregated model is presented to demonstrate its structure and complexity; for a detailed discussion see [9]. The optimization problem is to maximize

$$\sum_{\omega} \pi_\omega \Big(\textstyle\sum_{lif_p} \alpha_{lif_p\omega} M_{lif_p\omega} - \sum_{if_p} \alpha^+_{if_p\omega} M^+_{if_p\omega} - \sum_{if_p} \alpha^-_{if_p\omega} B^-_{if_p\omega} \qquad (4)$$
$$- \textstyle\sum_{ir_p} \beta_{ir_p\omega} N_{ir_p\omega} - \sum_{ip} \gamma_{ip\omega} w_{ip\omega} \Big)$$

subject to:

$$\sum_{j=i}^{k} \sum_{p,r_p} N_{jr_p\omega} \leq N^{\max}_{ik\omega} \quad \forall\, i \leq k, \omega\,, \qquad (5)$$

$$\begin{Bmatrix} z_{kp\omega} C^{\min}_{p\omega} & \text{if } k = I \\ y_{k+1,p\omega} C^{\min}_{p\omega} & \text{else} \end{Bmatrix} - \begin{Bmatrix} C^0_{p\omega} & \text{if } i = 1 \\ y_{ip\omega} C^{\max}_{p\omega} & \text{else} \end{Bmatrix} + \textstyle\sum_{j=i}^{k} z_{jp\omega} F^{\min}_{p\omega} \qquad (6)$$
$$\leq \textstyle\sum_{j=i}^{k} \sum_{r_p} N_{jr_p\omega} \leq$$
$$\begin{Bmatrix} z_{kp\omega} C^{\max}_{p\omega} & \text{if } k = I \\ y_{k+1,p\omega} C^{\max}_{p\omega} & \text{else} \end{Bmatrix} - \begin{Bmatrix} C^0_{p\omega} & \text{if } i = 1 \\ y_{ip\omega} C^{\min}_{p\omega} & \text{else} \end{Bmatrix} + \textstyle\sum_{j=i}^{k} z_{jp\omega} F^{\max}_{p\omega}$$
$$\forall\, i \leq k, p, \omega\,,$$

$$z_{i-1,p\omega} \geq y_{ip\omega},\; z_{ip\omega} \geq y_{ip\omega},\; z_{i-1,p\omega} + z_{ip\omega} - 1 \leq y_{ip\omega} \quad \forall\, i \geq 2, p, \omega\,, \qquad (7)$$

$$\begin{Bmatrix} z^0_{p\omega} & \text{if } i=1 \\ z_{i-1,p\omega} & \text{else} \end{Bmatrix} - z_{i,p\omega} \le 1 - z_{jp\omega} \quad \forall\, i+1 \le j \le i + I^{\text{off}}_p - 1 \le I, p, \omega \;, \tag{8}$$

$$-\begin{Bmatrix} z^0_{p\omega} & \text{if } i=1 \\ z_{i-1,p\omega} & \text{else} \end{Bmatrix} + z_{i,p\omega} \le z_{jp\omega} \quad \forall\, i+1 \le j \le i + I^{\text{on}}_p - 1 \le I, p, \omega \;, \tag{9}$$

$$\sum_{l|i+l-1\le I} M_{l(i+l-1)f_p\omega} = B^{\text{soll}}_{if_p\omega} - B^-_{if_p\omega} \quad \forall\, i, f_p, p, \omega \;, \tag{10}$$

$$\begin{aligned} &\textstyle\sum_{j=1}^{i} \sum_l M_{ljf_p\omega} + M^+_{if_p\omega} + \begin{Bmatrix} M^0_{f_p\omega} & \text{if } i=1 \\ 0 & \text{else} \end{Bmatrix} \\ &\quad = \textstyle\sum_{j=1}^{i} \sum_{r_p} \bar{\rho}_{f_p r_p \omega} N_{jr_p\omega} \quad \forall\, i, f_p, p, \omega \;, \end{aligned} \tag{11}$$

$$\begin{Bmatrix} z^0_{p\omega} & \text{if } i=1 \\ z_{i-1,p\omega} & \text{else} \end{Bmatrix} - z_{i,p\omega} \le w_{ip\omega} \quad \forall\, i, p, \omega \;, \tag{12}$$

$$-\begin{Bmatrix} z^0_{p\omega} & \text{if } i=1 \\ z_{i-1,p\omega} & \text{else} \end{Bmatrix} + z_{i,p\omega} \le w_{ip\omega} \quad \forall\, i, p, \omega \;, \tag{13}$$

$$N_{if_p,\omega=1} = N_{if_p,\omega=2} = \ldots = N_{if_p,\omega=\Omega} \quad \forall\, i, f_p, p \mid i \le 3 \;, \tag{14}$$

$$N_{if_p,\omega} \in \mathbb{N};\, z_{i,p,\omega} \in \{0;1\};\, M_{lif_p\omega}, M^+_{if_p\omega}, B^-_{if_p\omega}, w_{ip\omega}, y_{ip\omega} \in \mathbb{R}^+ \;.$$

The objective is to maximize the profit (4) computed from revenues for satisfied demands and costs for the production. Revenues are obtained subject to demand and supply constraints (10) and (11), respectively. The plant is modelled by capacity constraints for the polymerization and the finishing stage (5) and (6), in addition to (8) and (9) which ensure a smooth operation of the finishing lines. The non-anticipativity constraints are represented by (14). Equations (7), (12) and (13) result from exact linearizations.

On the MS-layer, uncertainties in the capacity of the polymerization stage and the demand profiles are considered, namely by the parameters $N^{\max}_{ik\omega}$ and $B^{\text{soll}}_{if_p\omega}$. The typical size of the aggregated model for *one scenario* is 538 continuous and 120 discrete variables and 605 constraints, and for *1,024 scenarios* it is 550,912 continuous and 122,880 discrete variables and 650,210 constraints. The problem contains 30 first stage variables which are all integer.

6 Numerical Experiences

In Section 4.2 the employed solution algorithm was presented in principle. However, its application requires to specify how to select nodes, how to branch, how to solve the Lagrangian dual problems, how to solve the scenario subproblems, etc. The aim in solving scheduling problems is typically not the optimal solution regardless of the computing time but a relatively good and quick feasible solution. Therefore, an efficient parameterization of the algorithm must be found by means of numerical experiments.

In the following, the major results of extensive numerical studies are presented. Since the employed branch&bound scheme necessitates multiple solutions of scenario subproblems, these are analyzed in Section 6.1 first, followed by an analysis of the master problems in Section 6.2.

6.1 Subproblems

The scenario subproblems are mixed-integer linear programs (MILPs) the matrix structure of which is schematically shown in Fig. 2. The variables are relatively strongly coupled and the matrix exhibits no obvious decomposition structure. The typical solution approach to that type and size of problems is based upon the relaxation of the integrality requirements: The relaxed linear programs (LPs) are solved within a branch&bound algorithm which searches for feasible integer solutions. We applied CPLEX [4] which is a commercial state-of-the-art implementation of branch&bound based MILP-algorithms.

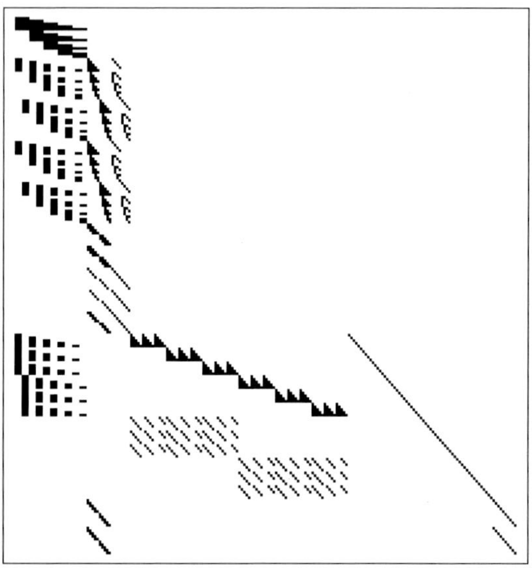

Fig. 4. Matrix Structure for One Scenario

The test of several modifications of the solution algorithm indicated that the standard parameterization leads to the best performance. The computations were performed on a SUN Ultra II workstation with CPLEX in version 7.5 for 40 randomly generated capacity and demand profiles. After 20 CPU-seconds feasible solutions were found with mean optimality gaps of 3.0%, i.e. the distance to the optimal solution is at most 3.0%. An average number of 1,081 nodes were visited; increasing the number of nodes to 2,000 and 10,000 reduces the to gaps to 2.7% and 2.4%, respectively. Due to the combinatorial complexity, the number of nodes has to be increased by orders of magnitude to reduce the optimality gap significantly.

An analysis of the capacity constraints shows that the difficulty of the example problem is mostly determined by the constraints for the finishing stage. These couple discrete variables corresponding to the polymerization stage with discrete variables corresponding to the finishing lines. Omitting these constraints reduces the optimality gaps to 1.2%, whereas omitting the capacity constraints of the polymerization does not affect the gaps significantly (but they do of course affect the solutions).

6.2 Master Problems

The 2-SSIPs were solved on a SUN Ultra Enterprise 450 workstation with CPLEX in version 8.0 for the scenario subproblems and an implementation of the conic bundle method described by [5] for the Lagrangian dual problems. Several modifications of the algorithm were tested on the basis of a bound of four hours CPU-time and randomly generated scenarios. With the best found algorithm parameters, the (mean) optimality gaps were 2.9% for a problem with eight scenarios of uncertain capacity, 1.7% for a problem with 128 scenarios of uncertain demands and 8.6% with 1,024 scenarios of uncertain capacity *and* demands. The number of visited nodes for the three settings were 81, 8 and 3, respectively.

The two most important observations are that 1. it is not efficient to solve the Lagrangian dual problems at all, and 2. the number of visited nodes is relatively small compared to the number of first stage variables. The iterative solution procedure of the Lagrangian dual requires the solution of all scenario subproblems (inner optimization problem in (3)) in each iteration step (outer optimization problem in (3)). It turned out to be more efficient to solve scenario subproblems to explore a higher number of nodes rather than to improve the bound in one particular node (which is done by solving the Lagrangian dual).

Furthermore, the relatively small number of nodes visited stresses the importance of efficient heuristics for suggestions $\mathbf{x}^{\mathrm{R}}$ of the 1st stages solutions. Specifically, the presented results were obtained by the following rule: "Choose the 1st stage vector of that scenario solution with the smallest difference (L_1-norm) to the weighted average of the 1st stage vectors over all scenarios (with the probabilities π_ω as weightings)".

7 Conclusions

In this contribution, we considered batch scheduling problems in the chemical industries with a special emphasis on the aspect of uncertainty. We presented an "engineered" approach based on two-stage stochastic integer programming along with a decomposition-based solution algorithm. The applicability of the approach was demonstrated by means of an aggregated model for a real-world example process. The resulting mixed-integer linear program with some 10^6 continuous variables and constraints and roughly 10^5 discrete variables was solved within four hours computing time with an optimality gap of less than 10%.

8 Acknowledgements

The financial support by the Deutsche Forschungsgemeinschaft under grants EN 152/17-1,2,3 is gratefully acknowledged.

References

[1] G. Applequist, O. Samikoglu, J. Pekny, and G. Reklaitis. Issues in the use, design and evolution of process scheduling and planning systems. *ISA Transactions*, 36:81–121, 1997.

[2] J. F. Birge and F. Louveaux. *Introduction to Stochastic Programming.* Springer, New York, 1997.

[3] C.C. Carøe and R. Schultz. Dual decomposition in stochastic integer programming. *Operations Research Letters*, 24:37–45, 1999.

[4] CPLEX. *Using the CPLEX Callable Library.* ILOG Inc., Mountain View, CA, 2002.

[5] C. Helmberg. *Semidefinite Programming for Combinatorial Optimization.* Habilitationsschrift, TU Berlin. ZIB-Report ZR-00-34, Konrad-Zuse-Zentrum, Berlin, 2000.

[6] S. J. Honkomp, S. Lombardo, O. Rosen, and J. F. Pekny. The curse of reality - why scheduling problems are so difficult in practice. *Computers and Chemical Engineering*, 24:323–328, 2000.

[7] J. Pekny and G. Reklaitis. Towards the convergence of theory and practice: a technology guide for scheduling/planning methodology. In *Proc. 3rd Conf. Foundations of Computer-Aided Process Operations*, pages 91–111, Michigan, 1998. CACHE Publications.

[8] G. Sand and S. Engell. Modelling and solving real-time scheduling problems by stochastic integer programming. *Computers and Chemical Engineering*, 2003. submitted.

[9] G. Sand and S. Engell. Risk conscious scheduling of batch processes. In *Computer Aided Process Engineering*, 8th Symposium on Process Systems Engineering. Elsevier, 2003. to appear.

[10] N. Shah. Single- and multisite planning and scheduling: current status and future challenges. In *Proc. 3rd Conf. Foundations of Computer-Aided Process Operations*, pages 75–90, Michigan, 1998. CACHE Publications.

Newton-Type Methods for Nonlinear Least Squares Using Restricted Second Order Information

Hubert Schwetlick

Institut für Numerische Mathematik, Technische Universität Dresden
D-01062 Dresden, Germany
schwetlick@math.tu-dresden.de

Summary. In the paper, a special approximated Newton method for minimizing a sum of squares $f(x) = \frac{1}{2}\,\|F(x)\|^2 = \frac{1}{2}\,\sum_{i=1}^{m}[F_i(x)]^2$ is introduced. In this *Restricted Newton method*, the Hessian $H = G + S$ of f where $G = (F')^T F'$, $S = F \circ F''$, is approximated by $A_{\mathrm{RN}} = G + B$ where $B = Z_2 Z_2^T S Z_2 Z_2^T$ is the restriction of the second order term S on the subspace im Z_2 spanned by the eigenvectors of the Gauss-Newton matrix G which belong to the q smallest eigenvalues of G. Some properties of this approximation are derived, and a related trust region method is tested on hand of some test functions from the literature.

Key words: nonlinear parameter estimation, nonlinear least squares, approximated Newton methods, Gauss-Newton methods, restricted second order derivatives

1 Introduction

Consider the *nonlinear least squares problem*

$$f(x) := \frac{1}{2}\,\|F(x)\|^2 = \frac{1}{2}\sum_{i=1}^{m}[F_i(x)]^2 \longrightarrow \operatorname*{Min}_{x\in\mathbb{R}^n} \qquad \text{(NLS)}$$

where $F : \mathbb{R}^n \rightarrow \mathbb{R}^m$ with $m \geq n$, typically $m \gg n$, is supposed to be sufficiently smooth. In this paper, the norm $\|\,.\,\|$ is always the Euclidean vector norm $\|y\| = \sqrt{\sum_{i=1}^{m} y_i^2}$.

Such problems arise from *parameter estimation* in a nonlinear explicit model equation

$$y = r(t,x), \quad r : \mathbb{R}^{\dim_t} \times \mathbb{R}^n \rightarrow \mathbb{R}^{\dim_y} \qquad (1)$$

with independent variables $t \in \mathbb{R}^{\dim_t}$, dependent variables (responses) $y \in \mathbb{R}^{\dim_y}$, and parameters $x \in \mathbb{R}^n$ via *nonlinear regression:* The unknown parameters x have to be estimated from observations

$$(t_l, y_l) \approx (t_l, y_l^*), \quad (l = 1, \ldots, L)$$

of 'true' states (t_l, y_l^*) with $y_l^* = r(t_l, x^*)$ using the least squares criterion, where x^* denotes the unknown 'true' parameters that characterize the real system under consideration. This means that the estimate to x^* is the solution of of the nonlinear least squares problem

$$f(x) := \frac{1}{2} \sum_{l=1}^{L} \|y_l - r(t_l, x)\|^2 \longrightarrow \operatorname*{Min}_{x \in \mathbb{R}^n}. \tag{2}$$

Here we have $m := L \times \dim_y$ scalar squares in the objective function, and the corresponding F has the special block structure

$$F(x) := \begin{pmatrix} y_1 - r(t_1, x) \\ \vdots \\ y_L - r(t_L, x) \end{pmatrix}.$$

Finally, to come to the topic of Minisymposium on Parameter Estimation and Experimental Design, the simplest approach for estimating parameters x in a differential equation, more precisely: in an initial value problem

$$\dot{y} = \Phi(t, y, x) \quad \text{for} \quad t \in [t_0, t_e], \quad y(t_0) = y_0(x) \tag{3}$$

with solution $y = y(t, x)$ that depends on the time t and also on the parameters x which occur in the right hand side $\Phi(t, y, x)$ of the ordinary differential equation and possibly in the initial values $y_0(x)$ consists in *least output estimation* or *simple shooting*: The response function $r(t, x) := y(t, x)$ of (2) is implicitly defined by the solution $y = y(t, x)$ of (3) at time t for given parameters x. Here we have $\dim_t = 1$, $t_0 \le t_1 < \cdots < t_i < t_{i+1} < \cdots < t_L \le t_e$, and

$$\begin{aligned} y &: [t_0, t_e] \times \mathbb{R}^n \to \mathbb{R}^{\dim_y}, \\ y_0 &: \mathbb{R}^n \to \mathbb{R}^{\dim_y}, \\ \Phi &: [t_0, t_e] \times \mathbb{R}^{\dim_y} \times \mathbb{R}^n \to \mathbb{R}^{\dim_y}. \end{aligned}$$

We assume Φ and y_0 to be sufficiently smooth so that the solution y has sufficiently smooth derivatives with respect to the parameters x.

Let us come back to the nonlinear least squares problem (NLS). The standard way for solving such a problem consists in using *descent methods*

$$x_+ = x + s, \quad s \text{ such that } \quad f(x_+) \le f(x) \quad \text{'sufficiently'}$$

where $x = x_k$ is the current iterate and $s = s_k$ is the *step* which leads to the subsequent iterate $x_+ = x_{k+1}$. The step s is determined using a local quadratic approximation, a so called *model*,

$$\varphi(s) := f + g^T s + \tfrac{1}{2} s^T A s \approx f(x + s) \tag{4}$$

to f around $x = x_k$ with $f := f(x)$, exact first order gradient information $g := g(x) := \nabla f(x) = F'(x)^T F(x)$ and a symmetric approximation $A = A^T$ to the Hessian $H := H(x) := \nabla^2 f(x)$ which has the additive structure

$$H = H(x) = \nabla^2 f(x) = F'(x)^T F'(x) + F(x) \circ F''(x) =: G + S \qquad (5)$$

with

$$G := G(x) := F'(x)^T F'(x),$$

$$S := S(x) := F(x) \circ F''(x) := \sum_{i=1}^{m} F_i(x) \cdot \nabla^2 F_i(x)$$

where $F'(x)$ and $F''(x)$ denote the first and second derivative of F at x, respectively.

In the following Section 2 we sketch how the step s can be obtained from the quadratic model (4) using line search or trust region techniques. Moreover, we review the two standard approximations, namely the Gauss-Newton approximation $A_{\mathrm{GN}} := G$ and the Newton approximation $A_{\mathrm{N}} := G + S = H$.

In Section 3 we propose an approximated Newton method that uses only restricted second order information $B \approx S$ with respect to the subspace spanned by the eigenvalues of G which belong to the q smallest eigenvalues of G. Since then nothing can be said about the definiteness of the resulting approximation $A_{\mathrm{RN}} = G + B$, in general, trust region techniques have to be used.

In Section 4, some first numerical tests will be summarized. The number $q = q_k$ is adaptively determined dependent on $\lambda_{\max}(G)$. It turns out that in most cases the adaptive strategy for choosing q_k gives $q_k = 0$, i. e., no additional second order information is computed. However, for some problems with nonzero residuals or strong nonlinearity, the adaptive strategy leads to $q_k > 0$, and the additional second order information used then improves the convergence compared to the Gauss-Newton approximation. So the extension can be considered as a tool that in the seldom case of very bad Gauss-Newton models can improve the convergence and robustness of Gauss-Newton methods. Moreover, some possible modifications will be discussed which do not require to compute the smallest eigenvalues and corresponding eigenvectors of G.

2 Computation of the Step and Quadratic Models

There are essentially two strategies for computing the step s from the quadratic model (4):

In *line search methods* a basic or *full* step s_{full} is computed as solution of the unconstrained quadratic minimization problem

$$\mathrm{Min}\{\, \varphi(s) : \; s \in \mathbb{R}^n \} \qquad (6)$$

which requires A to be positive definite since otherwise there is no or no unique solution. Then a *damping factor* $\alpha > 0$ is determined by an appropriate line search algorithm such that the step s defined by $s := \alpha \cdot s_{\text{full}}$ leads to a sufficient decrease in f.

Alternatively, in *trust region methods* the *trust region step* s is computed as solution of the constrained quadratic minimization problem

$$\text{Min}\{\, \varphi(s) : \; s \in \mathbb{R}^n, \; \|s\| \le \Delta\} \tag{7}$$

where the size of s is controlled by the *trust region radius* $\Delta > 0$. The so-called *trust region subproblem* (7) is always solvable, also if A is indefinite. See [10] and [3] for details and theoretical properties of both classes of methods.

When choosing an approximation A to H, there are also two basic possibilities:

The simplest approximation is the *Gauss-Newton* approximation $A_{\text{GN}} := G = (F')^T F'$ which depends only on first order derivatives of F and is always positive semidefinite. It defines the *Gauss-Newton model*

$$\varphi_{\text{GN}}(s) := f + g^T s + \tfrac{1}{2} s^T (F')^T F' s = \tfrac{1}{2} \left\|F + F's\right\|^2 \tag{8}$$

where $F := F(x)$ and $F' := F'(x)$. Hence, (6) becomes the linear least squares problem

$$\text{Min}\{\, \tfrac{1}{2} \left\|F + F's\right\|^2 : \; s \in \mathbb{R}^n\}. \tag{9}$$

which has always the solution $s_{\text{full}} = s_{\text{GN}} = -(F')^\dagger F$, the so-called Gauss-Newton step s_{GN}. Here $J^\dagger$ denotes the Moore-Penrose pseudo inverse to the matrix J. If F' has full rank n, i.e., if $G = (F')^T F'$ is positive definite, then $s_{\text{GN}} = -G^{-1} g$ is the unique (and stable) solution of (9). Otherwise the solutions are $s_{\text{GN}} + \ker F'$, and s_{GN} is the unique (but instable with respect to perturbations of F') solution of minimal norm.

It is well known that descent methods based on the Gauss-Newton model perform well, in most cases even perfectly, when $F' = F'(x_k)$ has full rank n and moderate condition number $\text{cond}(F') = \sqrt{(\text{cond}(G))}$ during the iteration and if, in addition, the residuals $F = F(x_k)$ are 'sufficiently small' in the limit.

Let us point out that the full step Gauss-Newton method $x_+ = x + s_{\text{GN}}$ converges locally and Q-linearly with convergence factor ϱ_{GN} toward a stationary point x_{opt} of f if $\text{rank}(F'(x_{\text{opt}})) = n$ and

$$\varrho_{\text{GN}} := \varrho(G(x_{\text{opt}})^{-1} S(x_{\text{opt}})) = \max_{h \ne 0} \frac{|h^T S(x_{\text{opt}}) h|}{h^T G(x_{\text{opt}}) h} < 1 \tag{10}$$

where $\varrho(M)$ denotes the spectral radius of M. If (10) is satisfied then $H(x_{\text{opt}})$ is even positive definite, hence, x_{opt} is a strong local minimizer.

However, when $F'(x_k)$ has small or vanishing singular values $\sigma_j(F')$ or equivalently, since $\lambda_j(G) = [\sigma_j(F')]^2$, if G has small or vanishing eigenvalues

$\lambda_j(G)$, then the convergence may be poor do to lacking information with respect to the subspace that is spanned by the right singular vectors of $F'(x_k)$ which belong to the smallest singular values. Note that these singular vectors are just the eigenvectors of G which belong to its smallest eigenvalues.

In this case or when the residual $F(x_{\text{opt}})$ is large, using the full second order information, i.e., taking the *Newton* approximation $A_{\text{N}} := H$ may help to overcome the difficulties described. Then (4) becomes the *Newton model*

$$\varphi_{\text{N}}(s) := f + g^T s + \tfrac{1}{2} s^T H s \tag{11}$$

which is the second order Taylor approximation to f. If H is positive definite, then the unconstrained problem (6) is uniquely solvable, and its solution $s_{\text{full}} = s_{\text{N}} = -H^{-1}g$ is obtained from the Newton equation $Hs = -g$ for the system $g(x) = \nabla f(x) = 0$ which motivates the name 'Newton' for this approach. If the Newton model is used in a trust region approach then the methods goes also through points x_k where H is indefinite and, under weak conditions, see again [3], converges to a stationary point x_∞ where H is positive semidefinite.

The main disadvantage of the Newton model is that it requires the term $S = F \circ F''$ of $H = G + S$ which depends on all second order derivatives of F. Its computation is expensive also when using state of the art techniques of automatic differentiation, cf. [6].

Therefore many authors have tried to approximate H in a cheaper way. Naive application of quasi-Newton updates as, e.g., BFGS, to f ignores the structure of f as sum of squares and the additive composition of H according to (5). Therefore, so called *structured* updates for the second order part S alone has been developed, see [9], [4], [2], [12], [5], [13] for such and other modifications of the Gauss-Newton method.

An alternative to using or approximating the whole second order term S which is the aim of this contribution consists in using S only on the subspace $\operatorname{im} Z_2$ spanned by the q eigenvectors $Z_2 = [z_{n-q+1}, \ldots, z_n]$ of G which belong to the q smallest eigenvalues

$$\underbrace{\lambda_1 \geq \cdots \geq \lambda_{n-q}}_{p=n-q} \geq \underbrace{\lambda_{n-q+1} \geq \cdots \geq \lambda_n}_{q} \geq 0 \tag{12}$$

of G. This will be discussed in the next Section.

3 Models with Restricted Second Order Information

Let

$$\hat{G} := Z^T G Z = [\,Z_1 \,|\, Z_2\,]^T G\, [\,Z_1 \,|\, Z_2\,] = \left[\begin{array}{c|c} \Lambda_1 & 0 \\ \hline 0 & \Lambda_2 \end{array}\right] \tag{13}$$

with orthogonal $Z = [\,Z_1 \,|\, Z_2\,]$ and diagonal

$$\begin{array}{ll} \Lambda_1 = \operatorname{diag}(\lambda_1, \ldots, \lambda_{n-q}) \in \mathbb{R}^{p\times p} & \text{`(large) good part'}, \\ \Lambda_2 = \operatorname{diag}(\lambda_{n-q+1}, \ldots, \lambda_n) \in \mathbb{R}^{q\times q} & \text{`(small) bad part'} \end{array}$$

be the spectral decomposition of G where the eigenvalues are ordered as in (12), and decompose

$$\hat{S} := Z^T S Z = [\,Z_1 \,|\, Z_2\,]^T S\, [\,Z_1 \,|\, Z_2\,] = \left[\begin{array}{c|c} \hat{S}_{11} & \hat{S}_{12} \\ \hline \hat{S}_{21} & \hat{S}_{22} \end{array}\right] \tag{14}$$

analogously. Then we have

$$H = G + S = Z(\hat{G} + \hat{S})Z^T$$

Now we approximate $S = Z\hat{S}Z^T$ by the matrix

$$B := Z \left[\begin{array}{c|c} 0 & 0 \\ \hline 0 & \hat{S}_{22} \end{array}\right] Z^T = Z_2 \hat{S}_{22} Z_2^T = Z_2 Z_2^T S Z_2 Z_2^T, \tag{15}$$

the restriction of S on $\operatorname{im} Z_2$, which leads to the restricted Newton approximation

$$A_{RN} := G + B = Z \left[\begin{array}{c|c} \Lambda_1 & 0 \\ \hline 0 & \Lambda_2 + \hat{S}_{22} \end{array}\right] Z^T \tag{16}$$

Since $\|H - A_{\mathrm{GN}}\|_F^2 = \|H - A_{\mathrm{RN}}\|_F^2 + \left\|\hat{S}_{22}\right\|_F^2$ we have

$$\|H - A_{\mathrm{RN}}\|_F < \|H - A_{\mathrm{GN}}\|_F \quad \text{if } \hat{S}_{22} \neq 0, \tag{17}$$

i. e., A_{RN} is a better approximation to H as A_{GN} in the Frobenius norm. Note that (17) does not necessarily hold in the spectral norm $\|\,.\,\|_2$ as the example $n = 2$, $q = 1$, $\hat{S} = \left[\begin{smallmatrix} 1 & 0.5 \\ 0.5 & -0.1 \end{smallmatrix}\right]$ shows. Here we have $\|H - A_{\mathrm{RN}}\|_2 = 1.2071$ but $\|H - A_{\mathrm{GN}}\|_2 = 1.1933$.

Let us remark that the matrix B defined in (15) solves

$$\operatorname{Min} \left\{ \|B\|_F : u^T B v = u^T S v \quad \forall u, v \in \operatorname{im} Z_2 \right\}.$$

Hence, B is the *minimum norm approximation* of S with respect to the 'bad' eigenspace $\operatorname{im} Z_2$ where G may have small eigenvalues.

For computing the correction term $B = Z_2\hat{S}_{22}Z_2^T$ in $A_{\mathrm{RN}} = G + B$ we use the representation

$$\hat{S}_{22} = \hat{S}_{22}^T = Z_2^T S Z_2 = F(x)^T F''(x)[Z_2, Z_2] \in \mathbb{R}^{q\times q}$$

for the projection $\hat{S}_{22}$ of S onto $\operatorname{im} Z_2$. It consists of the q^2 entries

$$(\hat{S}_{22})_{ij} = z_{p+i}^T S z_{p+j} = F(x)^T F''(x)[z_{p+i}, z_{p+j}] \quad (i, j = 1, \ldots, q)$$

$q(q-1)/2$ of which are different. Recall that the vectors

$$D_{ij}^2 := F''(x)[z_{p+i}, z_{p+j}] = \left.\frac{\partial^2 F(x + \alpha z_{p+i} + \beta z_{p+j})}{\partial\alpha\partial\beta}\right|_{\alpha=\beta=0}$$

are second order directional derivatives of F with respect to the one- or two-dimensional subspace $\operatorname{span}\{z_{p+i}, z_{p+j}\}$. Therefore, if q is small compared to n, they can efficiently be

1. computed by *automatic differentiation* with Z_2 as so-called *seed matrix*, see Griewank's plenary lecture at this Conference and his latest book [6] at SIAM, or
2. approximated by *second order divided differences* requiring $(q^2 + 3q)/2$ additional F-evaluations as

$$D_{ij}^2 = \begin{cases} \frac{1}{\delta^2}[F(x - \delta z_{p+i}) - 2F(x) + F(x + \delta z_{p+i})] + \mathcal{O}(\delta^2) & \text{if } i = j, \\ \frac{1}{\delta^2}[F(x + \delta z_{p+i} + \delta z_{p+j}) - F(x + \delta z_{p+i}) & \\ \qquad -F(x + \delta z_{p+j}) + F(x)] + \mathcal{O}(\delta) & \text{if } i < j \end{cases}$$

   with an appropriately chosen discretization stepsize $\delta > 0$ which may depend on $x = x_k$.

It is interesting that the sufficient conditions for the local convergence of the full step Gauss-Newton method around a stationary point x_{opt} of f (which guarantee $H(x_{\mathrm{opt}})$ to be positive definite) also imply the positive definiteness of A_{RN}.

Proposition 1. *Suppose that $g(x_{\mathrm{opt}}) = 0$, $G(x_{\mathrm{opt}})$ is positive definite, and condition (10) is satisfied. Then $A_{\mathrm{RN}}(x_{\mathrm{opt}}) = G(x_{\mathrm{opt}}) + B(x_{\mathrm{opt}})$ is positive definite.*

Proof. By using (13), (14) and representing h in the system of the eigenvectors $\{z_i\}$ of G as $h = Zu$ where $u = Z^T h = \left[\begin{smallmatrix} u_1 \\ u_2 \end{smallmatrix}\right]$ is splitted analogously, we obtain

$$\begin{aligned} h^T G h &= u^T \hat{G} u = u_1^T \Lambda_1 u_1 + u_2^T \Lambda_2 u_2, \\ h^T S h &= u^T \hat{S} u = u_1^T \hat{S}_{11} u_1 + 2u_1^T \hat{S}_{12} u_2 + u_2^T \hat{S}_{22} u_2, \\ h^T B h &= u^T \hat{B} u = u_2^T \hat{S}_{22} u_2. \end{aligned}$$

Thus, (10) means

$$|u_1^T \hat{S}_{11} u_1 + 2u_1^T \hat{S}_{12} u_2 + u_2^T \hat{S}_{22} u_2| = |h^T S h|$$
$$\leq \varrho_{\rm GN}\,(h^T G h) = \varrho_{\rm GN}\,(u_1^T \Lambda_1 u_1 + u_2^T \Lambda_2 u_2) \quad \forall\, u_1, u_2.$$

In particular, by setting $u_1 = 0$, we obtain

$$|h^T B h| = |u_2^T \hat{S}_{22} u_2| \leq \varrho_{\rm GN}\,(u_2^T \Lambda_2 u_2).$$

Then we have

$$\begin{aligned} h^T A_{\rm RN} h &= h^T (G+B) h = u_1^T \Lambda_1 u_1 + u_2^T \Lambda_2 u_2 + u_2^T \hat{S}_{22} u_2 \\ &\geq u_1^T \Lambda_1 u_1 + u_2^T \Lambda_2 u_2 - \varrho_{\rm GN}\,(u_2^T \Lambda_2 u_2) \\ &= u_1^T \Lambda_1 u_1 + (1-\varrho_{\rm GN})\, u_2^T \Lambda_2 u_2 \geq (1-\varrho_{\rm GN})(h^T G h) \end{aligned}$$

due to $0 \leq \varrho_{\rm GN} < 1$. This implies $\lambda_i(A_{\rm RN}) \geq (1-\varrho_{\rm GN})\lambda_{\min}(G) > 0$. □

Proposition 1 shows that, under the assumptions posed there, the full step $s_{\rm RN} = -A_{\rm RN}^{-1}\, g$ of the restricted Newton method is uniquely defined by (6) in a neighborhood of $x_{\rm opt}$.

Unfortunately, we were not able to prove that, under the assumptions of Prop. 1, the full step restricted Newton method converges asymptotically faster than the full step Gauss-Newton method, i. e., that

$$\varrho_{\rm RN} = \varrho(M_{\rm RN}) < \varrho(M_{\rm GN}) = \varrho_{\rm GN}$$

holds in case of $B \neq 0$ for the corresponding iteration matrices

$$M_{\rm RN} := -(G+B)^{-1}(S-B), \quad M_{\rm GN} := -G^{-1} S$$

These matrices are defined by

$$M := T'(x_{\rm opt}) = -A(x_{\rm opt})^{-1}(H(x_{\rm opt}) - A(x_{\rm opt}))$$

where T is the iteration operator

$$x_+ = T(x) := x - A(x)^{-1} g(x)$$

of the full step methods. Recall that the spectral radius $\varrho(M)$ characterizes the asymptotic rate of convergence of the underlying method, cf. [11]. In order to guarantee the required smoothness of $B(x)$ for fixed q we have to assume that the smallest q eigenvalues of $G(x_{\rm opt})$ are strictly separated from the others, cf. (12).

4 Numerical Tests and Conclusions

The restricted Newton method has been tested in [1] with adaptive choice of $q = q_k$ dependent on $\lambda_{\max}(G(x_k))$: an eigenvalue $\lambda = \lambda(G(x_k))$ is considered to be 'small', i.e., belongs to Λ_2, if $\lambda \leq tol \times \lambda_{\max}(G(x_k))$ with a small tolerance *tol*. Moreover, also hybrid methods along the lines of [4] which work with two models, typically the Gauss-Newton model and an approximated or exact Newton model, has been considered. The implementation is based on the trust region approach and follows the code NL2SOL of [4]. The test problems which are of low or moderate dimension n has been taken from the ('academic') Moré/Garbow/Hillstrom test set [8] and from [4].

The results can be summarized as follows:

1. In most cases the adaptive choice of q_k leads to $q_k = 0$, i.e., no second order correction B is taken, and the Gauss-Newton model is used.
2. In the average, adding $B(x_k)$ in case of ill-conditioned $G(x_k)$ slightly improves the convergence in case of nonzero residuals $F(x_{\text{opt}})$ and and compares to using full S. Sometimes, adding B may, however, also worsen the convergence.
3. On problems where H is indefinite during the iteration, adding B improves the convergence.

Further investigations will address

- Numerical tests with real life and/or large size problems. For large problems, only the smallest eigenvalues of G will be computed by appropriate methods as, e.g., linearly convergent inverse subspace iteration or cubically convergent block Rayleigh Quotient iteration, for the latter see [7].
- Using cheaper approximations not based on invariant subspaces which needs eigenvector computations but on pivoted Cholesky factorization of G: Cholesky factorization with diagonal (= complete) pivoting is performed as long as the current triangular factor L_p of dimension p is well conditioned which can easily be checked by using an condition estimator. Then the approximating subspace $\operatorname{im} Z_2$ is taken as $Z_2 = [e_{r(p+1)}, \ldots, e_{r(n)}]$ where e_j denotes the j-th coordinate vector and $r(p+1), \ldots, r(n)$ are the indices of the remaining $q = n - p$ variables in the Cholesky factorization terminated after the p-th step.

References

[1] K. Adam, *Quadratische Modelle zur Lösung nichtlinearer Quadratmittelprobleme*, Diplomarbeit, Institut für Numerische Mathematik, Techn. Univ. Dresden, Germany, 1996.

[2] M. Al-Baali and R. Fletcher, *An efficient line search for nonlinear least squares*, J. Optim. Theory Appl., 48 (1986), pp. 359–377.

[3] A. R. Conn, N. I. M. Gould, and P. L. Toint, *Trust-Region Methods*, SIAM, Philadelphia, 2000.

[4] J. E. Dennis, D. Gay, and R. Welsch, *An adaptive nonlinear least squares algorithm*, ACM Trans. Math. Software, 7 (1981), pp. 348–368.

[5] J. E. Dennis, H. J. Martínez, and R. A. Tapia, *A convergence theory for the structured BFGS secant method with an application to nonlinear least squares*, J. Optim. Theory Appl., 61 (1989), pp. 159–176.

[6] A. Griewank, *Evaluating Derivatives. Principles and Techniques of Algorithmic Differentiation*, SIAM, Philadelphia, 2000.

[7] R. Lösche, H. Schwetlick, and G. Timmermann, *A modified block Newton iteration for approximating an invariant subspace of a symmetric matrix*, Linear Algebra Appl., 275/276 (1998), pp. 381–400.

[8] J. J. Moré, B. S. Garbow, and K. E. Hillstrom, *Testing unconstrained optimization software*, ACM Trans. Math. Software, 7 (1981), pp. 17–41.

[9] L. Nazareth, *Some recent approaches to solving large residual nonlinear least squares problems*, SIAM Rev., 22 (1980), pp. 1–11.

[10] J. Nocedal and S. J. Wright, *Numerical Optimization*, Springer, New York, 1999.

[11] J. M. Ortega and W. C. Rheinboldt, *Iterative Solution of Nonlinear Equations in Several Variables*, Academic Press, New York, London, 1970. Republished 2000 by SIAM, Philadelphia.

[12] E. Spedicato and M. T. Vespucci, *Numerical experiments with variations of the Gauss-Newton algorithm for nonlinear least squares*, J. Optim. Theory Appl., 57 (1988), pp. 323–339.

[13] H. Yabe, *Variations of structured Broyden families for nonlinear least squares problems*, Optimization Methods and Software, 2 (1993), pp. 107–144.

Balance Algorithm - a New Approach to Solving the Mapping Problem on Heterogeneous Systems

Nguyen Thanh Son, Tran Nguyen Hoang Huy, and Nguyen Anh Kiet

Department of Information Technology, HCM University of Technology, Vietnam
sonsys@dit.hcmut.edu.vn
huy@dit.hcmut.edu.vn

Summary. A fundamental issue affecting the performance of a parallel program is the assignment of tasks to processors in order to achieve the minimum completion time. Most of state-of-the-art approaches consider homogeneous MIMD multiprocessor systems, in which all communication channels have the same bandwidth and all processors are equally powerful. These algorithms do not run efficiently on heterogeneous systems. In this paper, we present a new approach for the mapping problem on arbitrary systems. The main idea is based on the "global load balancing, local cut-size optimization" principle. This approach has achieved encouraged results that are verified by experiments for various random graphs and processor numbers.

Key words: partitioning, mapping, load balance, heterogeneous system

1 Introduction

One of the most important concerns in parallel computing is the proper distribution of work-load across processors, namely the mapping problem. The problem amounts to the balance of the computational weight on the processors and reduces the communication cost by keeping intensively intercommunicating processes on nearby processors.

A mapping is static if it is computed prior to the execution of the program and is never modified at runtime. Static mapping is NP-complete in general case. Therefore, numerous methods have been developed to solve heuristically the mapping problem for bus, ring, mesh and hypercube. If the communication network of the parallel system is a complete graph, then the mapping problem turns into the graph partitioning problem. There are many popular graph partitioning approaches such as Kernighan-Lin [11], Fiduccia-Mattheyses [2], Spectral Bisection [4], Genetic [1], Simulated Annealing [13], [6], etc. Most of them have investigated homogeneous systems, in which all communication channels have the same bandwidth and the processors are equally powerful.

Thus, it is not efficient to use these algorithms on heterogeneous systems without necessary improvements. Popular partitioning packages are METIS (by G. Karypis and V. Kumar at University of Minnesota) [10], and Chaco (by B. Hendrickson and R. Leland at Sandia National Lab.) [5]. Both of them use multilevel Kernighan-Lin method. One prominent feature of METIS is that it can handle the case of nonuniform partition sizes. However, it only consider the processor powers. The communication links have not yet been taken into account. Hence, traditional graph partitioning algorithms are not adequate to ensure the efficient mapping solutions for heterogeneous systems.

Based on the ideas of Kernighan-Lin and Fiduccia-Mattheyses algorithm, we combine their advantages to present an effective algorithm for heterogeneous systems. This algorithm, namely Balance Mapping, can be used for mapping a parallel program onto a parallel system. Furthermore, we proposed a flexible objective function that overcomes the disadvantages of many mapping packages to handle the case of heterogeneous systems.

The rest of the paper is organized as follows. In Section 2, we summarize recent mapping solutions. Section 3 focuses on our proposed objective function for heterogeneous systems. The main idea of new algorithm, namely Balance, is presented in Section 4. Experimental results, including the execution time, the edge-cut ratio and load imbalance ratio on several task graphs and system graphs are showed in Section 5. Then follows the conclusion.

2 Preliminaries

2.1 Overview and related works

A parallel program consists a number of tasks that run concurrently. These tasks might have to communicate with each other by sending or receiving messages. We model the parallel program as a weighted graph $S(V, E)$, namely task graph. Vertices v_S and edges e_S of S are assigned integer weight $w(v_S)$ and $w(e_S)$, which estimate the computation weight of the corresponding process and the amount of communication on the inter-process channels, respectively. In this paper, we consider the mapping problem for heterogeneous multicomputers. Message passing is used for communication between computers. The parallel system is modelled by a weighted graph T called system graph. Vertices v_T and edges e_T of T are assigned integer weight $w(v_T)$ and $w(e_T)$, which estimate the computational power of the corresponding processor and the cost of the inter-processor links, respectively.

Mapping a task graph onto a system graph is a function $M : S \rightarrow R$ such that $M(i)$ gives the processor onto which the task i is mapped. The main objective is to find an M that minimizes the overall execution time of the parallel program. Numerous methods have been developed to solve the mapping problem. The blind-search methods of artificial intelligence-breadth-first or depth-first-are exhaustive methods for finding an optimal mapping solution. With these methods, we can find the correct solution but it takes a large execution time. Thus, these methods are not suitable for large systems. The majority of the mapping methods used heuristics with reasonable running time to obtain a sub-optimal solution. There are several well-known heuristics such as Kernighan and Lin (KL) [2] and their improvements (by Fiduccia and Mattheyses (FM) [2]), Spectral Bisection [4], Genetic Algorithms [1], Simulated Annealing [6], [13], etc.

2.2 Kernighan-Lin and Fiduccia-Mattheyses algorithm

Kernighan-Line algorithm is one of the earliest local optimization algorithms proposed for graph partitioning. A graph model of the mapping problem is to partition the vertices of a task graph into k subsets to minimize the number of cross edges among subsets. It's often called k-way partitioning problem. The KL algorithm solves the k-way partitioning problem by recursively bisecting the task graph within $log_2 k$ steps.

Let (A, B) be a bisection of $G = (V, E)$, i.e, $A \cup B = V$ and $A \cap B = \varphi$. For vertices $a \in A$ and $b \in B$, denote by $g(a, b)$ the reduction in the edge-cut of the bisection when the two vertices a and b exchanged. Denote by g_v the reduction when vertex v is moved into the opposite side. It is easy to see that,

$$g(a, b) = g_a + g_b - 2\delta(a, b)$$

where

$$\delta(a, b) = \begin{cases} 1, \text{ if } (a, b) \in E \\ 0, \text{ otherwise.} \end{cases}$$

The pair (a, b) that maximizes $g(a, b)$ is selected. By that way, a sequence of pairs $(a_1, b_1), \ldots, (a_{n/2-1}, b_{n/2-1})$ are selected. The algorithm then chooses a pair (X, Y), with $X = a_1, \ldots, a_k$ and $Y = b_1, \ldots, b_k$, such that $\sum_{i=1}^{n} g(a_i, b_i)$ is maximized. The algorithm exchanges X and Y. This is defined as a pass of the KL. KL algorithm repeats the above pass until no improvement is possible.

Algorithm 1. Kernighan-Lin partitioning algorithm

```
while (exist at least an improvement) do
  Compute Ga, Gb where a ∈ A, b ∈ B;
  Q_A = ∅; Q_B = ∅;
  for i = 1 to n/2 − 1 do
    Choose a_i ∈ A − Q_A and b_i ∈ B − Q_B s. t. G(a_i, b_i) is maximal;
    Q_A = Q_A − {a_i}; Q_B = Q_B − {b_i};
    for each a ∈ A − Q_A do
      G_a = G_a + 2δ(a, a_i) − 2δ(a, b_i);
    end for
    for each b ∈ B − Q_B do
      Gb = Gb + 2d(b, b_i) − 2d(b, a_i);
    end for
```

Choose $k \in \{1, \dots, n/2 - 1\}$ to maximize $G = \sum_{i=1}^{k} g(a_i, b_i)$
Swap two subsets $\{a_1, a_2, \dots, a_k\}$ and $\{b_1, b_2, \dots, b_k\}$
end for
end while

Discussion

The worst case complexity of KL algorithm is $O(|E| \cdot |V|^3)$ [11]. Clearly, it is not efficient to use the algorithm with large graphs. The KL takes one subset from each of partition and exchanges them together. Originally, KL algorithm works fine on uniform task graphs (i.e., all vertices have the same weight). Hence, exchanging two elements that differ greatly in size might lead to an imbalance solution. Its improvement, FM algorithm, moves only one vertex once with the balance constraints on the partitions. The gains associated with the vertices are radix sorted by value, and the moves associated with each gain value are stored in a doubly-linked list. Choosing a move with the highest gain value involves in searching nonempty list with the highest gain, while updating the gain of a vertex is accomplished in constant time by removing it from the linked list and inserting it into another. This essential modification reduces the running time to $O(|E|)$.

The FM algorithm runs in a linear time. However, it is not effective if the input solution is nearly balanced. It is difficult (or unable) to find a legal move that does not violate load balance constraints. The key feature of both KL and FM is the generalization, i.e., it is a general purpose method which is applicable to any kind of graphs, regardless of its structure. Our study is based on this important characteristic to propose a new heuristic to solve the mapping problem on heterogeneous systems.

3 Adaptive objective functions

Solution quality of the mapping problem depends on the objective function. Graph partitioning algorithms such as KL, FM, Multilevel KL (in METIS, Chaco) have used an objective function that had no regard to the processor power, communication link bandwidth. In this paper, we propose a new function that covers such parameters in a flexible manner to adapt to heterogeneous systems. Let $P(P_1, P_2, \dots, P_n)$ be a k-way partition of task graph $S = (V, E)$, i.e., $P_1 \cup P_2 \cup \ldots \cup P_n = V$ and $P_i \cap P_j = \emptyset$, where $i \neq j$. The partition P_i will be mapped onto the ith processor. The work-load of processor i, denoted by WL_i, is the total computational weight of all tasks that mapped onto i:

$$WL_i = \sum_{j \in P_i} w(j) \qquad (1)$$

If we denote the power of the ith processor by PP_i, then the load imbalance WI_{ij} between the ith processor and the jth processor is given by:

$$WI_{ij} = \left| \frac{WL_i}{PP_i} - \frac{WL_j}{PP_j} \right| \tag{2}$$

The communication overhead between processors i and j, denoted by CC_{ij}, is the sum of total cross-edge weights, namely the edge-cut(i, j):

$$CC_{ij} = D_{ij} \times \text{edge-cut}(i, j) \tag{3}$$

where D_{ij} is the distance from processor i to j, which estimates the data transfer time on the communication link between two nodes i and j.
Objective function should take into account the load imbalance, and minimization of inter-processor communication. In general, all criteria are often combined into a unique aggregate function by means of weighted sums. However, the biggest drawback of aggregate functions lies in the setting of the coefficients. In particular, the trade-off between computation and communication cost is hard to tune, and must be evaluated for each different target systems.

Therefore, the total cost of a parallel program has been split into the computation cost and communication cost. The goal of our mapping algorithm is thus to minimize communication cost function, while keeping the load balance within a user-specified tolerance. The communication cost F_C is given by:

$$F_C = \sum_{i,j=1}^{N} CC_{ij} = \sum_{i,j=1}^{N} D_{ij} \times \text{edge-cut}(i, j) \tag{4}$$

For work-load balancing, we propose the load imbalance tolerance:

$$\varepsilon = max_{i,j \in N} \left\{ \left| \frac{WL_i}{PP_i} - \frac{WL_j}{PP_j} \right| \right\} \tag{5}$$

A mapping solution will be evaluated better than another if its communication cost F_C is smaller than the others, and its work-load imbalance does not exceed the pre-defined tolerance.

4 Balance Algorithm

The original KL and its improvement used in many packages such as METIS and Chaco, have worked fine on those uniform task graphs that have the same vertex weight. However, if the vertex weights of task graph are different, it might lead to an imbalance solution. FM algorithm can handle such task graphs effectively by moving vertex one by one. But, if the bisection is nearly balanced, FM algorithm will immediately converge without any improvement in communication cost.

In this study, we present a flexible approach that avoids the shortcoming of both KL and FM. Assume, without loss of generality, that we have an initial solution that has N partitions, denoted by $P_0, P_1, \ldots, P_n$ and ordered by the total vertices weight, i.e., $WL(P_0) < WL(P_1) < \cdots < WL(P_N)$, where $WL(P_i)$ is the total weight of the partition P_i. The parallel system has N processors with different processing powers. Balance algorithm (BA) works as presented in Algorithm 2.

Algorithm 2. Balance Mapping Algorithm

```
Initialize mapping solution with N partitions (P_1, P_2, ..., P_N);
Sort N partitions by the total vertices weight WL(P_0) < WL(P_1) < ··· <
WL(P_{N-1})
while (maximum load imbalance ≥ ε) do
  for i = 0 to N/2 do
    Exchange(P_{n-i-1}, P_i);
    Update N partitions, ordered by the total vertices weight;
  end for
end while
return mapping solution;

procedure Exchange(A,B)
stop = false;
while not stop do
  if (exist at least a ∈ A s.t. g_a is maximal) then
    Move a to partition B;
  else if (exist set of pair (a_i, b_i) s.t. g(a_i, b_i) ≥ 0) then
    swap-subset(a_i, b_i);
  else
    stop = true;
  end if
end while
```

Discussion

Trying to balance the work-load of P_i and P_{n-1-i}, Balance algorithm acts like FM algorithm. When the two partitions are nearly equal, it does the same as the KL does. The selection policy is to reduce communication cost F_C (Eq. 4) while preserving load balance between two partitions. The convergence of Balance algorithm is driven by the tolerant parameter in Eq. 5. A small ε means we like a strictly balanced solution with a large execution time, and vice versa.

We reduce the execution time of BA by using data structure of the FM to store the weight of vertices in order to quickly compute and update the gains in constant time. To avoid the rate of growth of $O(|V|^3)$ of the KL, we have implemented the KL variant of Bui et al. [1], which has the worst-case complexity $O(|E|)$.

5 Experimental results

The input task graphs belong two distinct classes. The first consists of sparse graphs, in which $|E|$ does not exceed $O(|V|)$. These task graphs represent the parallel applications with medium communication cost. The second comprises dense graphs that have $O(|V|^2)$-edges, represents intensive communication programs. There are 19 random task graphs that have 1000, 2000,..., 40000, 45000 and 50000 vertices, respectively.

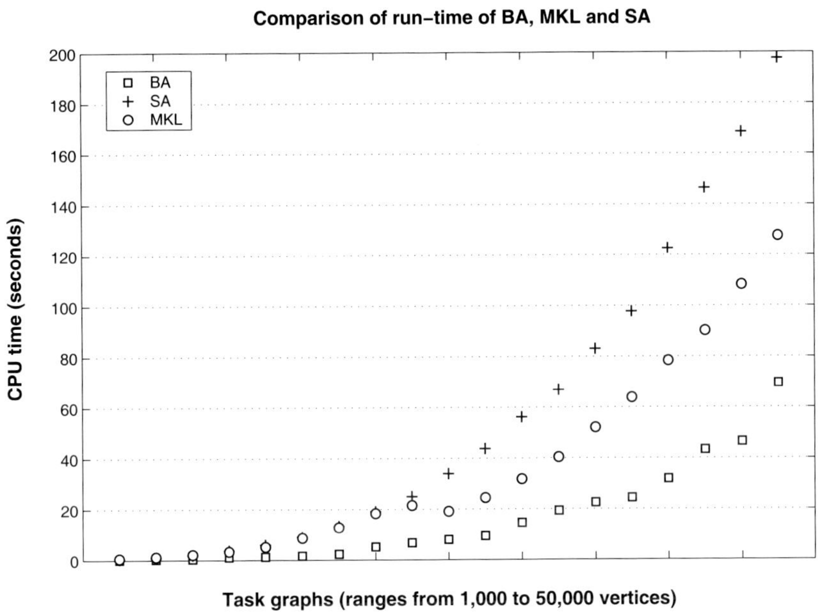

Fig. 1. The algorithm runtime of mapping task graphs onto eight-node system

We have chosen the load imbalance tolerance of $\varepsilon_0 = 0.01$ to obtain a sub-optimal solution within reasonable running time. The quality of mapping solution is measured by the CPU time, the edge-cut and average load imbalance ratio compared to our implementation of the KL algorithm with improvement of the FM. It is called the multiple-start KL (MKL). The MKL runs the KL algorithm 50 times with different initial solutions. The final solution of the MKL is the best one. Our mapping solutions are compared to other popular heuristic, namely Simulated Annealing (SA). We also have a comparison of quality of BA for mapping dense and sparse task graphs onto system graphs. All experiments are carried on an Intel Pentium 1.0 GHz processor with 128MB of memory running Linux Red Hat 7.2. For each case, we perform 100 trials.

We observe that BA consumes less CPU times than both MKL and SA. However, the execution time of BA is not linear because of the randomly initial solutions. SA again needs a lot of time to converge.

Figure 2a and 2b show the relative quality of our algorithm compared to

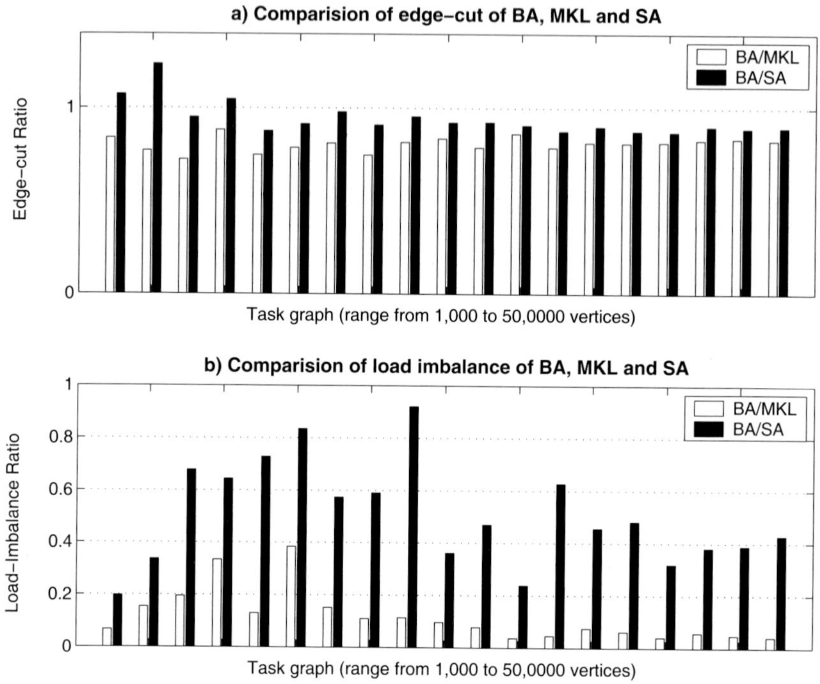

Fig. 2. Quality of BA compared to MKL and SA. For each task graph, the ratio of the edge-cuts (a) and load imbalance ratios (b) of BA to that of MKL, SA is plotted for mapping onto eight-node system graph

Multiple-Start KL (MKL) and Simulated Annealing (SA). For each task graph (ranges from 1,000 to 50,000 vertices), we plot the ratio of the edge-cuts of BA to that of MKL and SA. The values that are less than one indicate that BA produces better solutions than the others. We observe that in most task graphs, BA and SA provided solutions with similar quality. BA led MKL slightly of the edge-cuts and outperformed of load balancing. The KL algorithm is useful for reducing edge-cuts but it doesn't account for processor powers and communication channel bandwidths. Driven by our objective function, BA seems to be more effective to fit into the heterogeneous systems.

For dense task graphs, i.e., communication intensive applications, it should be noted that BA is not better than that of sparse graphs. To effectively use all system resources (e.g. processor power) to achieve maximum parallelism, we must pay for the increasing of communication overhead about 10%-20%.

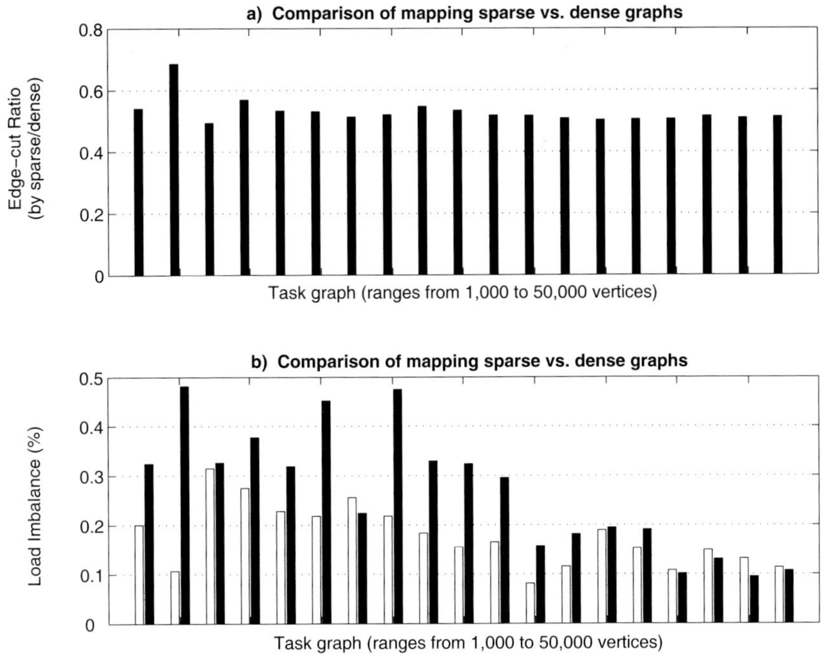

Fig. 3. The quality of mapping sparse vs. dense graphs onto eight-node system by BA

But, BA provides good load balance in both sparse and dense graphs. The work-load imbalance ratio is always less than pre-defined tolerance 0.01. We observe that the load imbalance ratios of mapping dense graphs are greater than that of mapping sparse graphs. It means that BA has tried to reduce as much communication costs as it can, and so it makes the load imbalance ratio slightly increase but does not exceed the threshold (Fig 3b).

6 Conclusions and further study

The paper presents a hybrid mapping algorithm that incorporates the advantages of both traditional algorithm, KL and FM, and the flexibility of the objective function. The proposed algorithm transforms the source graph into a set of partitions to map onto the parallel system. It takes into account for the processor powers and communication channel bandwidths to adjust the total work-load of each processor while reducing the communication costs. The new approach is suitable to the heterogeneous systems. The performance of BA is quite good and comparable to the well-known algorithms such as KL, FM and SA. However, BA still has a nonlinear complexity. It is observed that popular packages such as METIS [10], Chaco [5], use KL or its improvement to do partitioning and refining the results. Therefore, BA can supersede the KL algorithm in such packages to take the advantage of multilevel method and handle effectively the case of heterogeneous system.

References

[1] Bui, T.N., Moon, B.R.: Genetic Algorithm and Graph Partitioning. IEEE Transaction on Computer, **45**, 841–855 (1996)
[2] Fiduccia, C., Mattheyses, R.: A linear-time heuristic for improving network partitions. Technical Report 82CRD130, General Electric Co., Corporate Research and Development Center, Schenectady, NY (1982)
[3] Hui, C.C., Chanson, S.T.: Allocating Task Interaction Graph to processors in heterogeneous networks. IEEE Transactions on Parallel and Distributed Systems, **8**, 908–925 (1997)
[4] Hendrickson, B., Leland, R.:A multilevel graph partitioning. Technical Report SAND93-1301, Sandia National Laboratories (1993)
[5] Hendrickson, B., Leland, R.: The Chaco user's guide, version 1.0. Technical Report SAND93-2339, Sandia National Laboratories (1993)
[6] Kirkpatrick, S., Gelatt, C.D., Vecchi,M.P.: Optimization by Simulated Annealing. Science, Vol. 220, No. 4598, 671-680 (1983)
[7] Karypis, G., Kumar, V.: A Fast and High Quality Multilevel Scheme for Partitioning Irregular Graphs, Tech. Rep. TR 95-035,Department of Computer Science, University of Minnesota, Minneapolis, MN (1995)
[8] Karypis, G., Kumar, V.: A fast and high quality multilevel scheme for partitioning irregular graphs. Technical Report TR 95-035, Department of Computer Science, University of Minnesota (1995)
[9] Karypis, G., Kumar, V.: Parallel Multilevel K-way Partitioning Scheme for Irregular Graphs, Report 96036, University of Minnesota (1996)
[10] Karypis, G., Kumar, V.: METIS 4.0: Unstructured graph partitioning and sparse matrix ordering system. Technical report, Dept. of Computer Science and Engineering, Univ. of Minnesota (1998)
[11] Kernighan, B., Lin, S.: An effective heuristic procedure for partitioning graphs. The Bell System Technical Journal, 291–308 (1970)
[12] Lee, C.H., Shin, K.G: Optimal task assignment in homogeneous networks. IEEE Transactions on Parallel and Distributed System, **8**, 119–129 (1997)
[13] Van Laarhoven, P.J.M., Aarts, E. H. L.: Simulated Annealing: Theory and Applications, D. Reidel Publishing Company, Dordrecht, Holland (1987)

SMBOpt: A Software Package for Optimal Operation of Chromatographic Simulated Moving Bed Processes

Abdelaziz Toumi and Sebastian Engell

Process Control Laboratory, Biochemical and Chemical Engineering
Univeristät Dortmund
a.toumi@bci.uni-dortmund.de
s.engell@bci.uni-dortmund.de

Summary. In the last years, the Simulated Moving Bed (SMB) technique has increasingly been applied to isolate and purify high-valued pharmaceutical substances or fine chemicals and biotechnology products [13, 12, 9]. SMB-Processes are governed by mixed continuous and discrete dynamics and exhibit a complex behavior due to non-linear multi-component adsorption. Such processes cannot be easily designed and operated in the optimal way based solely on experience. In order to reduce separation costs and to shorten the overall design and development stages, a new integrated software package for the optimal operation of SMB-Processes is presented in this contribution.

1 Introduction

Preparative chromatography is attracting more and more interest from the fine chemicals and pharmaceutical industry, both for product development and commercial production. In the development of Life Science products, an important step is the choice and the design of cost-efficient unit operations for purification. Pharmaceuticals often have to be nearly 100 % pure due to regulatory demands. Sometimes their physico-chemical properties differ little from those of the byproducts, and they may be thermally unstable. In these cases, standard separations such as distillation are not applicable. Therefore in recent years chromatographic separation processes which can be operated at low temperatures gained a lot of attention not only for analytical applications (HPLC, GC) but also for preparative separations of products in the food and pharmaceutical industries.

In this area, mostly liquid chromatography is used where the substances to be separated are dissolved in a desorbent. The chromatographic separation is based on the different adsorptivities of the components to a specific adsorbent which is fixed in a chromatographic column. The most simple process, batch chromatography, involves a single column which is charged with pulses of the feed solution. These feed injections are carried through the column by pure desorbent. While traveling through the column, the more adsorptive specie is retained longer by the adsorbent thus leaving the column after the less adsorptive specie. The separated peaks can be withdrawn as different fractions at the end of the column with the desired purity. While the batch or elution chromatography mode has found the largest number of applications up to now, Simulated Moving Bed (SMB) chromatography as a continuous process is gaining more and more attention due to its advantages in terms of productivity and eluent consumption [1].

2 Principle of the SMB process

SMB is a practical way of implementing a counter-current chromatographic process. A simplified description of the process is given in Fig. 1. It consists of several chromatographic columns connected in series which constitute a closed-loop system. A counter-current motion of the solid phase with respect to the liquid phase is simulated by periodically and simultaneously moving the inlet and outlet lines by one column in the direction of the liquid flow [2].

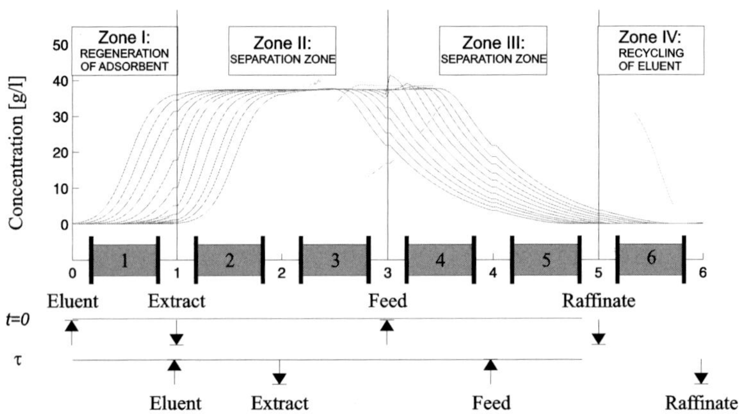

Fig. 1. Principle of the SMB process

After a startup phase, SMB-Processes reach a Cyclic Steady State (CSS). Fig. 1 shows the CSS-evolution along the columns plotted for different time instants within a switching period. At every axial position, the concentrations vary over time, the values reached at the end of each switching period are equal to the initial one. Thus the concentration profile is mainly shifted one column forward in direction of the liquid flow (compare the bold lines in Fig.1).

2.1 Process Intensification

The integration of unit operations, e. g. of a chromatographic separator, with a biochemical or a chemical reaction in one single apparatus may allow significant improvements in process performance. In such combined operations, the reaction can benefit from the separation due to the removal of reactants from the reaction zone thus overcoming equilibrium constraints, or the separation efficiency can be enhanced by a chemical reaction [17].

In this contribution, we focus on a Reactive SMB (RSMB) process for glucose isomerization composed of 6 reactive chromatographic fixed beds [5]. A pure glucose solution is injected to the system at the feed line. At the extract line, a mixture of glucose and fructose called High Fructose Corn Syrup (HFCS) is withdrawn. Water is used as solvent and it is fed continuously to the system at the desorbent-line. The raffinate line is not used in this system so that only a 3-zones SMB process results. It has to be remarked that the software package is of course written for general SMB-Processes with 4 or even more zones. The special 3-zones RSMB process considered here is a case study to demonstrate the numerical tools and their usage.

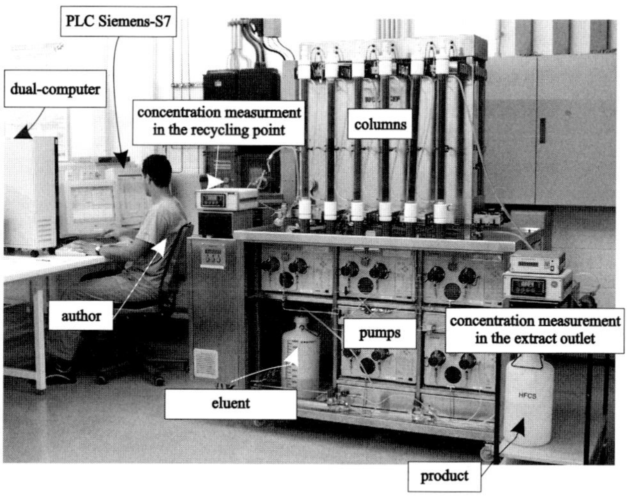

Fig. 2. RSMB plant for glucose isomerization

Fig. 2 shows a photograph of the pilot plant. It consists of a LicoSep12-50 plant (NovaSEP, France) which can be used with up to 12 preparative chromatographic columns with a maximal diameter of 50 millimeters. The columns are packed homogeneously with the ion exchange resin Amberlite CR-1320Ca and the immobilized enzyme Sweetzyme IT (supplied by Novo Nordisk Bioindustrial). The complete process is controlled by a modern programmable logic controller (PLC) PC-S7 of the SIEMENS S7-400 series (CPU S7-414-

2DP). Since the algorithms implemented to control the process require a lot of computing power, they can not be realized on the CPU of the PLC. They are performed on a PC (Dual-Athlon, 2xCPU's with 1.5 Ghz each) which communicates via an industrial Ethernet with the WinCC tool using the C-script interface Global Script.

The Reactive Simulated Moving Bed (RSMB) plant is equipped with several sensors: the temperatures of the columns and the incoming streams are kept constant by a closed heating water circuit which is controlled by a thermostat. The concentrations in the product line and in the recycling loop are measured online using a combination of a density measurement unit and a polarimeter [16, 11, 10]. The pressure in the recycling loop (i. e. at the inlet of the recycling pump (P_3)) is maintained constant using a P-controller which manipulates the extract flow rate.

3 Integrated Approach for Process Design and Control

Fig. 3 shows the different steps required to design and to operate a chromatographic SMB-Process. First a mathematical model is built which takes into account the main effects in the chromatographic columns: multi-component adsorption, mass transfer, diffusion, dispersion as well as the discrete dynamics. The non-idealities resulting from the volumes between the columns are also considered.

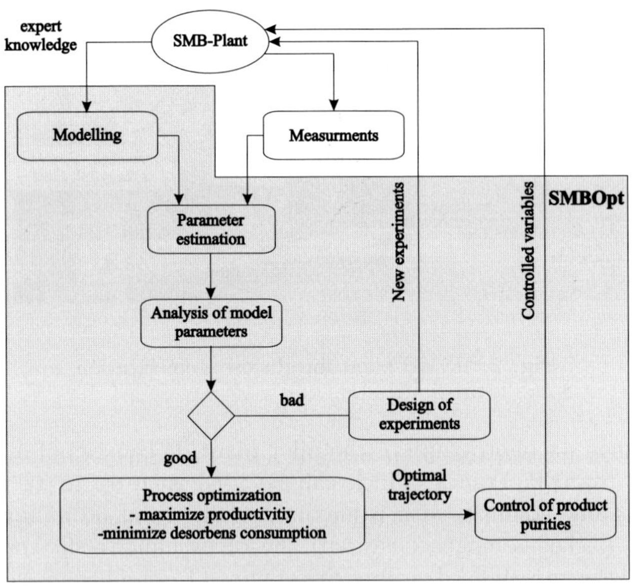

Fig. 3. Integrated approach for process design and control

The next important issue is parameter estimation. The model parameters can be either directly measured or derived from experimental data by least-squares-fitting. However, under some operating conditions it is sometimes hard or even impossible to estimate certain model parameters. The reason is the low sensitivity of the measured data to variations in the model parameters. In this case, it is necessary to design new experiments to estimate the model parameters reliably. Even during plant operation, some physical parameters may change due to degradation. Therefore they have to be re-estimated on-line.

The goal of process optimization is to calculate the operating conditions which lead to minimal separation costs while satisfying the product purity requirements and the plant constraints. In this context, the validated process model is used to evaluate the process behavior.

However, the inevitable model/plant mismatch and process disturbances may lead to off-spec products. In this case, a feedback control must be used to stabilize the plant at the desired conditions. This is a challenging task since the SMB process exhibits a strongly non-linear behavior and only few state variables can be directly measured. For this purpose a nonlinear model-predictive controller is proposed where a sequence of optimization problems have to be solved on-line. The nonlinear model-predictive controller (NMPC) minimizes the separation costs and thus corrects the off-line calculated trajectory in an optimal way. Another advantage is the possibility to include the rigorous non-linear model as well as physical constraints on the input and state variables. This results in a non-linear optimization problem which must be solved on-line during a prescribed sampling period.

All tools for simulation, optimization and control have been developed completely in the programming language Fortran^{90}. They can be accessed directly from the graphical user interface which has been developed in the more flexible programming language JAVA. Several separation tasks can be investigated in parallel. The model parameters and the numerical settings of one separation task are held consistent for all tools using a common data-buffer. The results are plotted graphically using the compaq array visualizer [3] which offers an interface to Fortran code and provides high-quality plots in 2- and 3-dimensions.

4 Dynamic Process Model

A lot of work has already been published on modeling of chromatographic processes [6, 7]. Only the relevant aspects will be briefly discussed here. Accurate dynamic models of multi-column continuous chromatographic processes consist of dynamic models of each column and take into account the periodic port switching.

We assume that the solid phase consists of porous, uniform and spherical particles (radius R_p, radial coordinate r) with void fraction ϵ_p, and that a local equilibrium is established within the pores. The concentration of component i is denoted by $c_{b,i}$ in the fluid phase and by q_i in the solid phase. ϵ_b is the void fraction of the bulk or liquid phase, D_{ax} the axial dispersion coefficient, c_i^{eq} the equilibrium concentration, $k_{l,i}$ the film mass transfer resistance and $D_{p,i}$ the diffusion coefficient within the particle pores. The concentration within the pore is denoted by $c_{p,i}$. The interstitial liquid velocity is represented by u. Depending on the reaction system considered, the reaction rate expression $r_{\mathrm{kin},i}$ can be introduced in the fluid-phase balance equations (e. g. to describe a homogeneous reaction catalyzed by an enzyme) or in the solid-phase balance equations (e. g. to describe a heterogeneously catalyzed esterification) [4]. X denotes the ratio of the volume of the adsorptive resin to the volume of the reactive enzyme. The following set of partial differential equations can be derived from a mass balance around an infinitely small cross-section area of the column, if a constant radial distribution of u and c_i is assumed:

$$\begin{aligned}\frac{\partial c_{b,i}}{\partial t} + \frac{(1-\epsilon_b)\,3\,k_{l,i}}{\epsilon_b R_p} X\,(c_{b,i} - c_{p,i}|_{r=R_p}) + (1-X) r_{\mathrm{kin},i}^{\mathrm{liq}} \\ = D_{\mathrm{ax}} \frac{\partial^2 c_{b,i}}{\partial x^2} + u \frac{\partial c_{b,i}}{\partial x},\end{aligned} \tag{1}$$

$$(1-\epsilon_p)\frac{\partial q_i}{\partial t} + \epsilon_p \frac{\partial c_{p,i}}{\partial t} - \epsilon_p D_{p,i}\left[\frac{1}{r^2}\frac{\partial}{\partial r}\left(r^2 \frac{\partial c_{p,i}}{\partial r}\right)\right] - r_{\mathrm{kin},i}^{\mathrm{sol}} = 0, \tag{2}$$

with appropriate initial and boundary conditions [18]. The adsorption equilibrium and the reaction kinetics are expressed by additional algebraic relationships between q_i and $c_{p,i}$ which are summarized in Appendix B. The parameters of the Glucose/Fructose RSMB process have been determined experimentally and are listed also in Appendix B.

The resulting system of coupled partial differential equations (1-2) can be solved efficiently by using the numerical approach proposed in [6], where a finite element discretization of the bulk phase is combined with an orthogonal collocation of the solid phase.

Accurate values of the model parameters are needed to use the process model for optimization and control, as the process is very sensitive to some of the parameters. These can be obtained by mathematical fitting of simulation runs to experimental data using the model parameters as optimization variables. But one must pay attention to the fact that certain model parameters cannot be estimated well from given experiments.

The mathematical approach used to investigate which parameters can be estimated reliably is based on the Fisher Information Matrix which is described in more detail in [14].

5 Optimal operation

The goal is to minimize the specific separation cost for a given plant while meeting the required product purities after the process has reached the cyclic steady state (CSS). For the description of the CSS, the operator Φ is introduced which represents the process dynamics $\mathbf{f}(\mathbf{x}, \mathbf{u})$ and the switching operations between two switching intervals:

$$\mathbf{x}_{k+1} = \Phi(x_k) \Leftrightarrow \begin{cases} \mathbf{x}^*_{k+1} = \mathbf{x}_k + \int_{t=0}^{\tau} \mathbf{f}(\mathbf{x}(t), \mathbf{u}(t))dt, \\ \mathbf{x}_{k+1} = \mathbf{P}\mathbf{x}^*_{k+1}. \end{cases} \tag{3}$$

The switching operation causes a re-initialization of the state vector of the dynamic simulation and is represented by the permutation matrix $\mathbf{P}$ (see Appendix A). In the CSS, the axial concentration profile $\mathbf{x}_k$ at the end of a period k does not change from period to period, which can be checked numerically as:

$$\|\Phi(\mathbf{x}_k) - \mathbf{x}_k\| \leq \epsilon_{\text{steady}}. \tag{4}$$

Then, given a SMB process with a *fixed* column partition, the optimization problem can be stated as follows:

$$\begin{aligned} \min_{Q_{\text{De}}, Q_{\text{Fe}}, Q_{\text{III}}, \tau} \quad & \text{Cost}_{\text{spec}}(k) \\ \text{s.t.} \quad \|\Phi(\mathbf{x}_k) - \mathbf{x}_k\| &\leq \epsilon_{\text{steady}}, \\ \text{Pur}_{\text{Ex,k}} &\geq \text{Pur}_{\text{Ex,min}}, \\ Q_{\text{I}} &\leq Q_{\text{max}}. \end{aligned} \tag{5}$$

An inequality constraint is imposed on the product purity:

$$\text{Pur}_{\text{Ex,k}} = \frac{\int_0^{\tau} c_A^{Ex,k}(t) Q_{Ex}^k(t)\, dt}{\int_0^{\tau} (c_A^{Ex,k}(t) + c_B^{Ex,k}(t)) Q_{Ex}^k(t)\, dt}, \tag{6}$$

where A denotes the more retained specie Fructose and B represents Glucose.

Since the flow rate in zone I is the highest in the plant, it is constrained in order to avoid violation of the maximal pressure which can be delivered by the pumps. Finally the objective function must be specified based on the available data of the operating cost.

The natural degrees of freedom are the flow rates of desorbent Q_{De}, feed Q_{Fe}, recycle $Q_{\text{Re}} = Q_{III}$, and the switching period τ. In the framework of optimization, they are transformed to the so-called β-factors [4], where the apparent solid flow rate Q_{S} is introduced:

$$Q_{\text{S}} = \frac{(1 - \epsilon_b) A\, L}{\tau}, \quad \beta_1 = \frac{1}{H_A} \left(\frac{Q_1}{Q_{\text{S}}} - \frac{1}{F} \right)$$

$$\beta_2 = \frac{1}{H_B} \left(\frac{Q_2}{Q_{\text{S}}} - \frac{1}{F} \right), \quad \frac{1}{\beta_3} = \frac{1}{H_A} \left(\frac{Q_3}{Q_{\text{S}}} - \frac{1}{F} \right). \tag{7}$$

These transformations reflect the fact that in SMB processes the absolute flow rates are less important than their values relative to the apparent solid flow rate. H_A and H_B denote the linear terms of the isotherm which are dominating the adsorption behavior.

We use a direct sequential algorithm for the solution of the optimization problem (5). The process is simulated until the cyclic steady state is reached. The constraints and the objective value are then evaluated and given back to the non-linear SQP-based optimizer FFSQP [20].

6 Control of the RSMB Process

Toumi and Engell [19] recently presented a nonlinear model predictive scheme for the control of the reactive SMB process for glucose isomerization. This approach and its experimental implementation will be discussed in more detail in the next subsections. The key feature of our approach is that the production cost is minimized on-line while the product purities are considered as constraints.

6.1 Formulation of the control problem

The essence of model predictive control (MPC) is to optimize, over the future values of the inputs, the future process evolution. The future process behavior is predicted by a process model over a finite time interval which is called the *prediction horizon*. The first input of the optimal input sequence, which spans the *control horizon*, is applied to the plant, and the problem is solved again at the next time interval using updated process measurements and a shifted horizon. In the framework of MPC control, it is simple to include hard constraints on the state and the input variables. The process behavior can be predicted using a linear model or a nonlinear model. In the latter case, computing the *global* solution of a non-convex optimization problem may require formidable effort and may be impossible within a fixed sampling time. Therefore, we modify the nonlinear model predictive algorithm: the emphasis here is on the calculation of a *suboptimal* but *feasible* solution which can be performed under real-time constraints.

We formulate the following optimal control problem over the finite *control horizon* H_r:

$$\min_{[\beta_k,\dots,\beta_{k+H_r}]} \quad \Gamma = \sum_{j=k}^{k+H_p} \left(\mathrm{Cost}(j) + \Delta\beta_j^T \mathbf{R}_j \Delta\beta_j\right)$$

$$\text{s.t.} \quad \begin{cases} \dot{\mathbf{x}}_j = \mathbf{f}(\mathbf{x}_j, \beta_j), \\ \mathbf{x_{j+1,0}} = \mathbf{P}\mathbf{x}_j(\tau(j)), \\ j = k, \cdots, k+H_p. \end{cases} \tag{8}$$

$$\mathrm{Pur}_{\mathrm{Ex,H_r},k} + \Delta\mathrm{Pur}_{\mathrm{Ex},k} \geq \mathrm{Pur}_{\mathrm{Ex,min},k},$$
$$\mathrm{Pur}_{\mathrm{Ex,H_p},k} + \Delta\mathrm{Pur}_{\mathrm{Ex},k} \geq \mathrm{Pur}_{\mathrm{Ex,min},k},$$
$$Q_{\mathrm{I},j} \leq Q_{\max},$$
$$\mathbf{g}(\beta_j) \geq \mathbf{0}, j = k, \cdots, k+H_p.$$

We discretize the *prediction horizon* in switching periods. The optimization problem (8) constitutes a dynamic optimization problem with the transient behavior of the process as a constraint. The objective function Γ is the sum of costs incurred for each cycle (e. g. desorbent consumption) and a regularizing term added in order to smooth the input sequence to avoid high fluctuations in the input sequence from cycle to cycle. The first equality constraint represents the plant model evaluated over the finite prediction horizon H_p. The switching dynamics are introduced via the permutation matrix $\mathbf{P}$. Since the maximal attainable pressure drop by the pumps must not be exceeded, constraints are imposed on the flow rates in zone I. Further inequality constraints $\mathbf{g}(\beta_j)$ are added in order to avoid negative flow rates during the optimization.
The control objective is reflected by the purity constraint over the control horizon:

$$\mathrm{Pur}_{\mathrm{Ex,H_r},k} = \frac{1}{H_r} \sum_{j=k}^{k+H_r} \mathrm{Pur}_{\mathrm{Ex,j}}, \tag{9}$$

which is corrected by a bias term $\Delta\mathrm{Pur}_{\mathrm{Ex}}$ resulting from the difference between the last simulated and the last measured process output to compensate unmodelled effects:

$$\Delta\mathrm{Pur}_{\mathrm{Ex},k} = \mathrm{Pur}_{\mathrm{Ex},(k-1),\mathrm{meas}} - \mathrm{Pur}_{\mathrm{Ex},(k-1)}. \tag{10}$$

A second purity constraint over the whole prediction horizon acts as a terminal (stability) constraint forcing the process to converge towards the optimal cyclic steady state. It has to be pointed out that the control goal (i. e. to fulfill the extract purity) is introduced as a constraint. We are using a feasible path SQP algorithm for the optimization [20] which generates a feasible point before it starts to minimize the objective function.

6.2 On-line parameter adaptation

The concentration profiles in the recycling line are measured and collected during a cycle. Since this measurement point is fixed in the closed-loop arrangement, the sampled signal includes information of all three zones. During the start-up phase, an on-line estimation of the actual model parameters is started in every cycle. The quadratic cost functional $J_{\text{est}}(\mathbf{p})$

$$J_{\text{est}}(\mathbf{p}) = \sum_{i=1}^{n_{\text{species}}} \left(\int_0^{N_{\text{col}}} (c_{i,\text{meas}}(t) - c_{i,\text{Re}}(t))^2 \, dt \right) \tag{11}$$

is minimized with respect to the parameters $\mathbf{p}$. For this purpose, the least squares solver E04UNF from the NAG-library is used [15]. n_{species} denotes the number of species and N_{col} the number of chromatographic columns.

6.3 Experimental study

The analysis using the Fisher information matrix showed that the process is highly sensitive to the values of the Henry coefficients H_i, the mass transfer resistances $k_{l,i}$ and the reaction rate k_m. These are therefore key-parameters of the RSMB process. These parameter are re-estimated online at every cycle (a cycle is equal to switching time multiplied by the number of columns). In Fig. 4, the concentration profiles collected in the recycling line are compared to the simulated ones. The parameters are initialized with the values given in Appendix B. At the end of the experiment all system parameters have converged towards stationary values. The developed mathematical model describes the behavior of the RSMB process well.

The formulation of the optimization problem of the NMPC-controller was slightly modified for the experimental investigation. The switching time was still a control variable, but modified only from cycle to cycle. This is due to the asymmetry of the RSMB process that results from the volume of the recycling pump in the closed loop. It disturbs the overall performance of the process and is corrected by adding a delay for the switching of the inlet/outlet line passing the recycling pump [8]. Therefore the shift of the valves is not synchronous to compensate for the technical imperfection of the real system and to get closer to the ideal symmetrical SMB system. To avoid port overlapping, the switching time must be held constant during a cycle.

In the real process, the enzyme concentration changes from column to column. The geometrical lengths of the columns also differ slightly. Moreover, the temperature is not constant over the columns due to the inevitable gradient of the heating-circuit. These problems cause an asymmetry of the concentration profiles at the product outlet. Even at the cyclic steady state, the product purity changes from period to period. Using the bias term given by Eq. (10) causes large variations of the controlled inputs from period to period. This can be damped by using the minimal value over the last cycle

$$\Delta \text{Pur}_{\text{Ex}} = \min_{j=(k-1,\cdots,k-1-N_{\text{col}})} (\text{Pur}_{\text{Ex},j} - \text{Pur}_{\text{Ex},j,\text{meas}}). \tag{12}$$

The desired purity for the experiment reported below was set to 55.0% and the controller was started at the 60th period. A diagonal matrix $R_j = 0.02\, I_{(3,3)}$

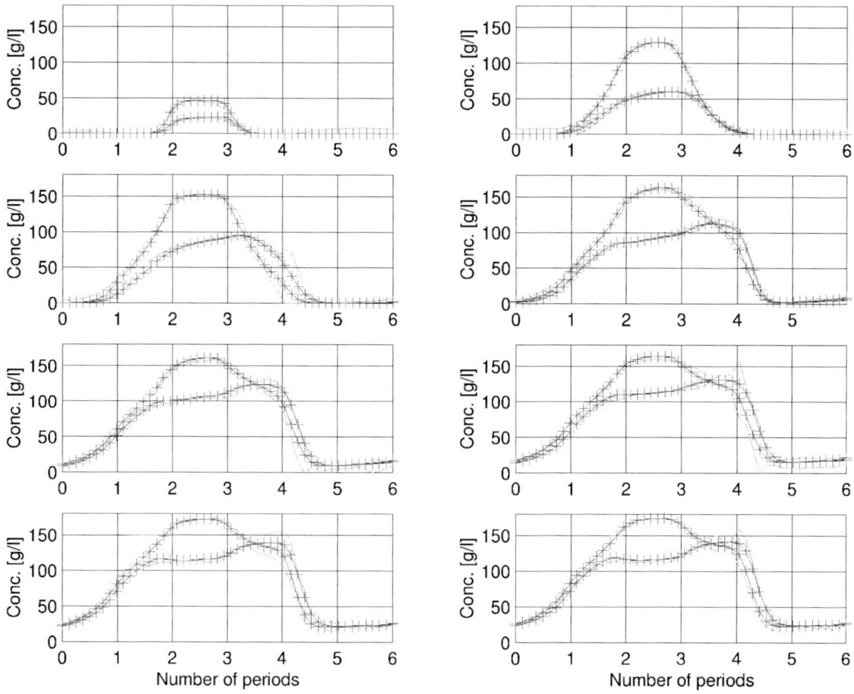

Fig. 4. Comparison of experimental and simulated concentration profiles collected in the recycle line. The colour version of this figure can be found in Fig. A.12 on page 583.

was chosen for regularization. The control horizon is set to $H_r = 1$ and the prediction horizon is $H_p = 60$ periods. Fig. 5 shows the evolution of the product purity as well as of the controlled variables. In the open-loop mode where the operating point was calculated based on the initial model, the product purity was violated at periods number 48 and 54. After one cycle the controller increased the purity above 55.0% and kept it there. The controller first reduces the desorbent consumption. This action seems to be in contradiction to the intuitive idea that more desorbent injection should enhance the separation. In the presence of a reaction this is not true, as shown by this experiment. The controlled variables converge towards a steady state, but they still change from period to period, due to the non ideality of the plant.

7 Conclusions and future work

In this work, a toolset for the model-based optimization of SMB processes and the application to a reactive simulated moving bed process for glucose isomerization was described. In order to maintain the product purity despite process

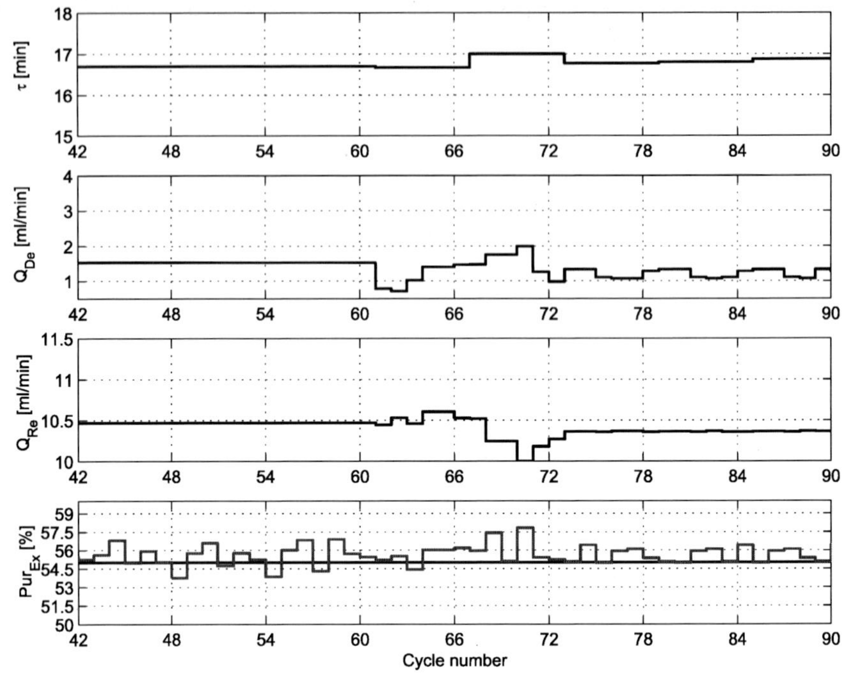

Fig. 5. Control Experiment for a target purity of 55 %

disturbances, a nonlinear model predictive controller was implemented. A very good controller performance was obtained in simulations and the controller was implemented at a real plant of small production scale. The experiments confirmed the excellent properties of the control scheme.

The paper presented a fully optimization-based integrated approach to the parameter estimation, the computation of optimal operating parameters and the on-line control of SMB processes. User friendly software has been developed to support all these steps.

A Permutation Matrix

The permutation matrix $\mathbf{P}$ is given as:

$$\mathbf{P} = \begin{pmatrix} \mathbf{0} & \cdots & \cdots & \mathbf{0} & \mathbf{I} \\ \mathbf{I} & \ddots & \mathbf{0} & \cdots & \mathbf{0} \\ \mathbf{0} & \mathbf{I} & \ddots & \ddots & \vdots \\ \vdots & \ddots & \mathbf{I} & \ddots & \mathbf{0} \\ \mathbf{0} & \cdots & \mathbf{0} & \mathbf{I} & \mathbf{0} \end{pmatrix}$$

$\mathbf{I}$ denotes the (n, n)-identity matrix and n the number of state variables of a single chromatographic column. N_{col} denotes the total number of chromatographic columns.

B RSMB Process for Glucose Isomerization

The model parameters of the reactive SMB process for glucose isomerization are summarized in the following table [18]:

L	57.0 cm	k_i	[1.46e-1,1.33e-1] cm^3/g
D	2.6 cm	k_{ij}	[2.90e-1,9.30e-2] cm^3/g
ϵ_p	0.01 [-]	X	0.9 [-]
ϵ_b	0.4 [-]	$Q_{\max}$	18.0 ml/min
d_p	16.25 μm	k_m	4.70e-3 1/s
$k_{l,i}$	[3.80e-5,2.05e-5] 1/s	k_{eq}	1.0798 [-]
ρ	1.0 g/cm^3	Q_f	1.3 ml/min
η	5.8e-3 g/(cm s)	ϵ_{steady}	1.0e-5 [-]
D_p	1.0e-3 cm^2/s	N_i	[2 2 2] or [1 2 3]
ν_i	[+1,-1] [-]	c_{gl}	300 g/l
H_i	[0.2545,0.1958] [-]	c_{fr}	0 g/l

$$q_i = H_i c_{b,i} + k_i c_{b,i}^2 + k_{ij} c_{b,i} c_{b,j}, \quad i,j = 1,2 \text{ and } i \neq j.$$
$$r_{\text{kin},i}^{\text{liq}} = \nu_i k_m \left(\frac{c_{b,i}}{k_{eq}} - c_{b,j}\right), \quad i,j = 1,2 \text{ and } i \neq j$$

References

[1] Bléhaut, J. and R. M. Nicoud: Recents aspects in simulated moving bed. Analusis Magazine **26(7)** (1998) 60–69.

[2] Broughton, D.B.: Continuous simulated countercurrent sorption process employing desorbent made in said process. US Patent 3.291.726 (1966).

[3] Compaq, HP: Compaq Array Visualizer Version 1.6. HP Compaq (2003)

[4] Dünnebier, G., J. Fricke and K.-U. Klatt: Optimal design and operation of simulated moving bed chromatographic reactors. Ind. Eng. Chem. Res. **39** 2290–2304 (2000).

[5] Fricke, J. and H.Schmidt-Traub: Design of chromatographic SMB-reactors. In: *International Symposium on Preparative and Industrial Chromatography and Allied Techniques*. Heidelberg, Germany, 03-09.02.

[6] Gu, T.: Mathematical Modelling and Scale Up of Liquid Chromatography. Springer, New York (1995)

[7] Guiochon, G.: Preparative liquid chromatography. Journal of Chromatography A **965** (2002) 129–161.

[8] Hotier, G.: Physically meaningful modeling of the 3-zone and 4-zone simulated moving bed processes. AIChE Journal **42** (1996) 154–160.

[9] Imamoglu, S.: Simulated moving bed chromatography (SMB) for applications in bioseparation. *Advances in Biochemical Engineering/ Biotechnology* **76**, 211–231 (2002).

[10] Jupke, A.: Experimentelle Modellvalidierung und Modellbasierte Auslegung von Simulated Moving Bed (SMB) Chromatographieverfahren. Dr.-Ing. Dissertation. Universität Dortmund, Fachbereich Bio- und Chemieingenieurwesen (in preparation) (2002).

[11] Jupke, A., A. Epping and H. Schmidt-Traub: Optimal design of batch and simulated moving bed chromatographic separation processes. *Journal of Chromatography A* **944**, 93–117 (2002).

[12] Juza, M., M. Mazzotti and M. Morbidelli: Simulated moving-bed chromatography and its application to chirotechnology. *Trends in Biotechnology* **18**, 108–118 (2000).

[13] Juza, M., O. Di Giovanni, G. Biressi, V. Schurig, M. Mazzotti and M. Morbidelli: Continuous enantiomer separation of the volatile inhalation anesthetic enfluran with gas chromatographic simulated moving bed unit. *Journal of Chromatography A* **813**, 333–347 (1998).

[14] Majer, M. C.: Parameterschätzung, Versuchsplanung und Trajektorienoptimierung für verfahrenstechnische Prozesse. Dr.-Ing. Dissertation. Universität Stuttgart, Institut für Systemdynamik und Regelungstechnik, VDI Reihe 3, Nr. 538, VDI Verlag, ISBN 3-18-353803-2, Düsseldorf (1998).

[15] NAG The NAG fortran library mark 15. Technical Report 1-10. NAG Ltd, Oxford (1991).

[16] Schloß, J. Vom: Auslegung integrierter Reaktions- und Trennprozesse am Beispiel der enzymatisch katalysierten Glucoseisomerisierung. Master's thesis. Department of Chemical Engineering, Universität Dortmund (2001).

[17] Stankiewicz, A.: Reactive separations for process intensification: An industrial perspective. *Chemical Engineering and Processing* **42**, 137–144 (2003).

[18] Toumi, A. and S. Engell: Optimization-based control of a reactive simulated moving bed process for glucose isomerization. *Chemical Engineering Science (submitted)* (2003).

[19] Toumi, A. and S. Engell (2004). Optimal operation and control of a reactive simulated moving bed process. *Accepted for IFAC Symposium on Advanced Control of Chemical Processes, Hongkong 2004.* (2004)

[20] Zhou, J. L., A. L. Tits and C. T. Lawrence: *User's Guide for FFSQP Version 3.7: A FORTRAN code for solving constrained nonlinear (Minimax) optimization problems, generating iterates satisfaying all inequality and linear constraints.* University of Maryland (1997).

Partly Convex and Convex-Monotonic Optimization Problems

Hoang Tuy

Institute of Mathematics, 18 Hoang Quoc Viet Road, 10307 Hanoi, Vietnam
htuy@thevinh.ncst.ac.vn

Summary. A class of nonconvex optimization problems is studied that exhibits partial convexity combined with partial monotonicity. To exploit this particular hybrid structure a natural approach is to use a branch and bound scheme with branching performed on the nonconvex variables and bounds computed by lagrangian or convex relaxation. We discuss conditions that guarantee convergence of such branch and bound algorithms. Incidentally, several incorrect results in the recent literature on related subjects are reviewed.

Key words: Nonconvex optimization, hybrid convex-monotonic optimization, partly convex programming, branch and bound method, lagrangian relaxation, dual bound, consistent bound.

1 Introduction

Most optimization methods developed over the last five decades have been based on exploiting convexity in various forms: ordinary convexity, reverse convexity, or d.c. (difference of convex) structure. To cope with the complexity of nonlinear phenomena diverse generalizations of the concept of convexity have also been introduced (quasiconvexity, pseudoconvexity, etc.). Aside from convexity, monotonicity is another important structure underlying a wide variety of problems arising from engineering and economics. To be specific, a function $f : \mathbb{R}^n_+ \to \mathbb{R}$ is said to be monotonic if it is either *increasing*: $f(x') \geq f(x)$ whenever $x'_i \geq x_i$ $(i = 1, \ldots, n)$, or *decreasing* : $f(x') \leq f(x)$ whenever $x'_i \geq x_i$ $(i = 1, \ldots, n)$; it is said to be *d.m.* if it can be represented as a difference of two monotonic functions. Recently a theory of monotonic optimization has emerged with the objective to study optimization problems described by means of monotonic, or more generally, d.m. functions [19], [18], [9], [22], [7].

Although d.c. and d.m. functions form a very wide class of functions, many problems encountered in various fields do not exhibit a pure d.c. or d.m. structure. Instead, a hybrid structure frequently occurs, in which partial convexity is combined with partial monotonicity. We will refer to such hybrid problems as *convex-monotonic* optimization problems. The interplay between convexity and monotonicity often generates new difficulties for handling this class of problems. At the same time this interplay creates new properties which, if properly exploited, may give rise to quite efficient solution methods. The aim of the present paper is to demonstrate this possibility by focussing on branch and bound approaches.

The general convex-monotonic problem can be formulated as follows:

$$\min\{F(x,y)|\ G_i(x,y) \le 0\ (i=1,\ldots,m), x \in C,\ y \in D\} \qquad \text{(P)}$$

where C is a nonempty compact convex subset of a convex set $X \subset \mathbb{R}^n_+, D$ is a nonempty closed convex subset of a convex set $Y \subset \mathbb{R}^p_+$, and $F(x,y)$: $X \times Y \to \mathbb{R},\ G_i(x,y) : X \times Y \to \mathbb{R}$ are functions, convex with respect to y for every fixed x and increasing or decreasing or even d.m. with respect to x for every fixed y.

Clearly every polynomial $P(x),\ x \in \mathbb{R}^n_+$, is a d.m. function on $\mathbb{R}^n_+$, since it can be represented as $P_+(x) + P_-(x)$, where $P_+(x)$ ($P_-(x)$, resp.) is the sum of all terms of $P(x)$ with positive (negative, resp.) coefficients. Using this fact, polynomial and fractional programming problems can easily be transformed into special convex-monotonic problems of the form (P). This class of problems also contains a variety of other problems of interest of the form

$$\min\{\langle c, y\rangle :\ A(x)y \le b,\ y \ge 0, x \in X\}$$

which have been studied e.g. in Ben-Tal et al. [1].

Due to their hybrid convex-monotonic structure, problems of the class (P) are generally very difficult and cannot be efficiently solved without understanding this structure and properly taking advantage of it.

When the functions $F(x,y), G_i(x,y),\ i=1,\ldots,m$, are only assumed to be convex in y for every fixed x but otherwise arbitrary, the problem (P) can be referred to as a *partly convex program.* To exploit this partial convexity, a classical approach is Benders-Geoffrion decomposition by viewing the nonconvex variables as "complicating variables". However, this method is often hindered by several restrictive conditions required for the structure of the functions $F(x,y)$ and $G_i(x,y),\ i=1,\ldots,m$. Therefore, in many cases a more practical approach for overcoming the difficulty is decomposition via branch and bound methods in which branching is performed upon the nonconvex variables $x \in \mathbb{R}^n_+$ and the x-space is partitioned into hyperrectangles (boxes), simplices or cones. In such branch and bound algorithms, a fundamental operation is *bounding*: for each partition set (rectangle, simplex or cone, respectively) $M \subset \mathbb{R}^n_+$ compute a lower bound $\beta(M)$ for the optimal value of the subproblem

$$\gamma(M) := \inf\{F(x,y)|\ G_i(x,y) \le 0\ (i=1,\ldots,m,\ x \in M,\ y \in D\}. \quad \text{(P(M))}$$

In most cases an exhaustive partitioning (subdivision) scheme is used, so that any filter $\{M_k\}$ of partition sets — i.e., any infinite nested sequence of partition sets $\{M_k\}$ in which M_{k+1} is a child of M_k) — satisfies $\lim_{k\to+\infty}\text{diam}M_k = 0$, and hence, collapses to a single point, e.g. $\cap_{k=1}^{\infty} M_k = \{x^*\}$. The convergence as well as the efficiency of the procedure then critically depend upon the bounding method. The most important property to demand of a bounding method is its ability to guarantee convergence of a BB procedurre using an exhaustive partitioning scheme. We shall refer to this property as "*consistency*" of the bounds. It turns out that for partly convex and convex-monotonic problems consistent bounding methods can be developed that are reasonably efficient.

The paper is organized as follows. First, in Section 2 a generic BB (branch and bound) algorithm for partly convex programming will be described. Then in Section 3 a general theorem on the consistency of lagrangian relaxation for partly convex programming will be established under fairly general assumptions. In Section 4 various special cases will be discussed and an example will be provided to show how the generic BB algorithm with lagrangian bounds can be applied successfully to solve the bilinear matrix inequality problem from control theory.

A major obstacle to the use of lagrangian relaxation is that the lagrangian dual problem may often be itself a difficult nonconvex problem. It is therefore of interest to investigate important classes of partly convex problems for which the lagrangian dual problem can be solved without too much effort, or some relaxed lagrangian bound can be efficiently computed which is still consistent.

Fortunately enough, for linear-monotonic and certain convex-monotonic problems, linear or convex relaxations equivalent to lagrangian relaxations can often be derived in a straightforward manner. Specifically, in Section 5 we will show that under mild conditions the lagrangian dual of a linear-monotonic optimization problem reduces to a mere linear relaxation. The class of such linear-monotonic problems (P) includes a variety of optimization problems of practical and theoretical interest such as polynomial fractional programs, bilinear programs (e.g. pooling/blending problems in oil refinery), etc.

For general convex-monotonic optimization problems, lagrangian dual bounds may be very hard to compute. However, as will be shown in Section 6, under rather natural assumptions, a simple convex relaxation can provide a consistent bound. Therefore, this class of problems, too, can be solved by the generic BB algorithm. An example of applications is furnished by the optimal design of water distribution networks.

Finally, Section 7 is devoted to the relation of our results to those previously published. It turns out that quite a few incorrect results on this subject have appeared in the recent literature that require a critical review.

Thus, virtually any problem of the class (P) can be solved by a convergent branch and bound algorithm with branching performed upon the nonconvex variables according to an exhaustive subdivision scheme and bounds computed by lagrangian or convex relaxation. Though such a branch and bound algorithm may not be always the most efficient way to handle a given problem (P), it offers a convenient framework within which a range of improvements is possible by exploiting additional structure present in the problem.

2 A generic BB algorithm for partly convex programming

Following a general principle which has become commonplace in global optimization, an efficient BB method for solving a nonconvex optimization problem like (P) should not branch upon the total set of variables x, y but rather, only upon the nonconvex variables x (see e.g. [16], [15]). The rationale is simply that this allows the partitioning procedure to be carried out in the x-space (of dimension n) rather than in the (x, y)-space (of dimension $n+p$) (see e.g. [17]).

Without much restriction, we can assume that X is a rectangle or a simplex if simplicial subdivisions are to be used. Define $G(x,y) = (G_1(x,y), \ldots, G_m(x,y))$, and write $G(x,y) \leq 0$ to mean $G_i(x,y) \leq 0$, $i = 1, \ldots, m$.

Algorithm GBB(α)

Select an exhaustive subdidivion rule (e.g. the standard bisection). Take a number $\alpha \geq \sup\{F(x,y)|\ x \in C,\ y \in D\}$. If no such $\alpha \in \mathbb{R}$ is readily available let $\alpha = +\infty$.

Initialization. If some feasible solutions are available, let CBS$=(\bar{x}^0, \bar{y}^0)$ be the best among them . Set $M_1 = X$, $\mathcal{P}_1 = \mathcal{S}_1 = \{M_1\},\ k = 1$.

Step 1. For each rectangle $M \in \mathcal{P}_k$ compute a lower bound $\beta(M)$ for

$$\gamma(M) := \inf\{F(x,y)|\ G(x,y) \leq 0,\ x \in M \cap C,\ y \in D\}. \tag{1}$$

Step 2. Update the incumbent CBS by setting $(\bar{x}^k, \bar{y}^k)$ equal to the best among all feasible solutions available at the completion of the previous step.

Step 3. Delete every $M \in \mathcal{S}_k$ such that $\beta(M) \geq \min\{\alpha, F(\bar{x}^k, \bar{y}^k)\}$ (with the convention $F(\bar{x}^k, \bar{y}^k) = +\infty$ if $(\bar{x}^k, \bar{y}^k)$ is not defined). Let $\mathcal{R}_k$ be the set of remaining members of $\mathcal{S}_k$.

Step 4. If $\mathcal{R}_k = \emptyset$ then terminate: CBS $= (\bar{x}^k, \bar{y}^k)$ yields a global optimal solution, or else the problem is infeasible (if CBS$= \emptyset$).

Step 5. Choose $M_k \in \operatorname{argmin}\{\beta(M)|M \in \mathcal{R}_k\}$. Subdivide M_k according to the chosen subdivision rule. Let $\mathcal{P}_{k+1}$ be the partition of M_k.

Step 6. Let $\mathcal{S}_{k+1} = (\mathcal{R}_k \setminus \{M_k\}) \cup \mathcal{P}_{k+1}$. Set $k \leftarrow k+1$ and go back to Step 1.

The basic operation in this algorithm is bounding. Below it is assumed that the bounds $\beta(M)$ satisfy the following natural conditions:

$$\text{(a)}\quad M' \subset M \Rightarrow \beta(M') \geq \beta(M); \qquad \text{(b)}\quad \beta(M) < +\infty \Rightarrow M \cap C \neq \emptyset. \tag{2}$$

Usually $\beta(M)$ is the optimal value of some relaxation of the subproblem (1). The most popular relaxations are

1) *linear relaxation*: $F(x,y)$ and $G_i(x,y)$, $i = 1,\ldots,$ are replaced by their respective linear minorants, and C, D by outer approximating polyhedrons;

2) *convex relaxation,* in particular SDP (semidefinite programming) relaxation: a suitable convex program (in particular a SDP) is derived whose optimal value is an underestimator of $\gamma(M)$;

3) *lagrangian relaxation*: the subproblem (1) is replaced by its lagrangian dual

$$\beta(M) = \sup_{\lambda\in\mathbb{R}^m_+} \inf\{F(x,y) + \langle\lambda, G(x,y)\rangle|\ x \in M\cap C,\ y\in D\}. \tag{3}$$

Note that condition (a) in (2) holds or can easily be made to hold for most bounds, whereas condition (b) holds for lagrangian bounds (3) but not necessarily for bounds via linear or convex relaxation. A stronger condition not satisfied even by lagrangian bounds would be to require nonemptiness of the set $\{(x,y)|\ G_i(x,y) \le 0\ (i = 1,\ldots,m),\ x \in M\cap C,\ y\in D\}$ when $\beta(M) < +\infty$. In fact, this set may well be empty while $\beta(M) < +\infty$ (see an example in [2]).

Theorem 1. *Whenever Algorithm GBB(α) is infinite, it generates a filter of partition sets $\{M_{k_\nu}\} \subset \{M_k\}$ collapsing to a point $x^* \in C$. If*

$$\lim_{k\to\infty} \beta(M_k) = \inf\{F(x^*,y)|\ G(x^*,y) \le 0,\ y\in D\}, \tag{4}$$

then $\beta^* := \lim_{k\to\infty}\beta(M_k)$ is the optimal value of (P) and any optimal solution y^* of the problem in (4) yields an optimal solution (x^*,y^*) of (P). If, in addition, $\alpha < +\infty$ then the problem (4) has an optimal solution.

Proof. The existence of a filter $\{M_{k_\nu}\} \subset \{M_k\}$ follows from the general theory of branch and bound algorithms (see e.g. [17]). By exhaustiveness, such a filter collapses to a point x^*. Since every partition set M with $\beta(M) \ge \alpha$ is pruned, one must have $\beta(M_k) < +\infty$ and hence, by virtue of (2), $M_k\cap C \ne \emptyset\ \forall k$. The sets $M_k \cap C$ then form a nested sequence of nonempty compact sets, so by Cantor theorem, $\cap_{k=1}^{+\infty}(M_k\cap C) = (\cap_{k=1}^{+\infty}(M_k)\cap C \ne \emptyset$. Therefore, $x^* \in C$. In view of (2) the sequence $\beta(M_k)$ is nondecreasing, so $\beta^* = \lim_{k\to+\infty}\beta(M_k)$ exists and $\beta^* \le \alpha$. Let

$$\gamma := \inf\{F(x,y)|\ G(x,y) \le 0,\ x\in C,\ y\in D\}. \tag{5}$$

Then, obviously, $\beta^* \le \gamma$. But in view of (4),

$$\begin{aligned}\beta^* &= \inf\{F(x^*,y)|\ G(x^*,y)\le 0,\ y\in D\} \ge \\ &\ge \inf\{F(x,y)|\ G(x,y)\le 0,\ x\in C,\ y\in D\} \\ &= \gamma,\end{aligned}$$

hence $\beta^* = \gamma$. If $\alpha < +\infty$, then $\beta^* \le \alpha < +\infty$ and the conclusion follows. □

Condition (4) simply means that the bound must be *exact at the limit* (the gap must be eventually closed). This condition alone does not ensure the solvability of the problem (4), because there still is the possibility that $\beta^* = +\infty$. However this possibility is excluded if $\alpha < +\infty$.

3 Consistency of lagrangian bounds for partly convex programming

In this section we study the convergence of Algorithm GBB(α) when lagrangian bounds are used throughout, i.e. when $\beta(M)$ in Step 1 is given by formula (3). As already mentioned, these bounds satisfy conditions (2). It turns out that the convergence of Algorithm GBB(α) is guaranteed under mild conditions.

Theorem 2. *Assume that C is a compact convex subset of $\mathbb{R}^n$, D is a compact convex subset of $\mathbb{R}^p$, and $F(x,y)$, $G_i(x,y)$, $i = 1,\ldots,m$, are lower semi-continuous real-valued functions, convex in y for every fixed x.*

Then Algorithm GBB(α) using lagrangian bounds (3) applied to problem (P) either terminates after finitely many iterations (yielding an optimal solution or proving that (P) is infeasible), or else it generates a filter of partition sets $M_{k_\nu}, \nu = 1,2,\ldots,$ collapsing to a point $x^* \in C$ such that the convex problem

$$\min\{F(x^*,y)|\ G(x^*,y) \le 0, y \in D\} \tag{6}$$

is solvable and any optimal solution $y^* \in D$ of (6) yields an optimal solution (x^*,y^*) of the problem (P).

Proof. According to Theorem 1 it suffices to prove that

$$\lim_{k\to\infty} \beta(M_k) = \inf\{F(x^*,y)|\ G(x^*,y) \le 0,\ y \in D\} \tag{7}$$

when $\sup\{F(x,y)|\ x \in C, y \in D\} \le \alpha < +\infty$. From the obvious inequalities

$$\begin{aligned}\beta(M_k) &\le \inf\{F(x,y)|\ G(x,y) \le 0,\ x \in M_k \cap C, y \in D\} \\ &\le \inf\{F(x^*,y)|\ G(x^*,y) \le 0,\ y \in D\}\end{aligned}$$

we have

$$\beta(M_k) \nearrow \beta^* \le \inf\{F(x^*,y)|\ G(x^*,y) \le 0,\ y \in D\} \le \alpha < +\infty.$$

Suppose that (7) does not hold, i.e.

$$\inf\{F(x^*,y)|\ G(x^*,y) \le 0,\ y \in D\} > \beta^*. \tag{8}$$

We show that this leads to a contradiction. Since

$$\sup_{\lambda\in\mathbb{R}^m_+} \{F(x^*,y) + \langle\lambda, G(x^*,y)\rangle\} = \begin{cases} F(x*,y) \text{ if } & G(x^*,y) \le 0 \\ +\infty & \text{otherwise}\end{cases}$$

we have

$$\inf\{F(x^*,y)|\ G(x^*,y) \le 0,\ y \in D\} = \inf_{y\in D}\sup_{\lambda\in\mathbb{R}^m_+} \{F(x^*,y) + \langle\lambda, G(x^*,y)\rangle\}. \tag{9}$$

But, in view of the compactness of D, the minimax theorem (see Appendix) yields

$$\inf_{y\in D}\sup_{\lambda\in\mathbb{R}^m_+}\{F(x^*,y)+\langle\lambda,G(x^*,y)\rangle\} = \sup_{\lambda\in\mathbb{R}^m_+}\inf_{y\in D}\{F(x^*,y)+\langle\lambda,G(x^*,y)\rangle\}. \tag{10}$$

Therefore, (8) implies that

$$\sup_{\lambda\in\mathbb{R}^m_+}\inf_{y\in D}\{F(x^*,y)+\langle\lambda,G(x^*,y)\rangle\} > \beta^*. \tag{11}$$

Hence, there is $\tilde{\lambda}$ such that

$$\min_{y\in D}\{F(x^*,y)+\langle\tilde{\lambda},G(x^*,y)\rangle > \beta^*.$$

Using the lower semi-continuity of the function $(x,y)\mapsto\{F(x,y)+\langle\tilde{\lambda},G(x,y)\rangle\}$ we can then find, for every fixed $y\in D$, an open ball U_y in $\mathbb{R}^n$ around x^* and an open ball V_y in $\mathbb{R}^p$ around y such that

$$F(x',y')+\langle\tilde{\lambda},G(x',y')\rangle > \beta^* \quad \forall x'\in U_y\cap C, \forall y'\in V_y.$$

Since the balls $V_y, y\in D$ form a covering of the compact set D there is a finite set $S\subset D$ such that the balls V_y, $y\in S$, still form a covering of D. If $U=\cap_{y\in S}U_y$ then for every $y\in D$ we have $y\in V_{y'}$ for some $y'\in S$, hence for all $x\in U\subset U_{y'}$ we will have

$$F(x,y)+\langle\tilde{\lambda},G(x,y)\rangle > \beta^* \quad \forall x\in U\cap C,\ \forall y\in D.$$

But $M_k\subset U$ for all sufficiently large k, because $\cap_k M_k=\{x^*\}$. Then the just established inequality implies that

$$\sup_{\lambda\in\mathbb{R}^m_+}\inf\{F(x,y)+\langle\lambda,G(x,y)|\ x\in M_k\cap C, y\in D\} > \beta^*.$$

Hence, $\beta(M_k)>\beta^*$, which is a contradiction.

We have thus proved (7). The last assertion of the theorem is obvious, since if $\alpha<+\infty$ then

$$\inf\{F(x^*,y)|\ G(x^*,y)\le 0,\ y\in D\}\le\max\{F(x,y)|\ x\in C,\ y\in D\}<+\infty.\square$$

In many cases the condition that set D be bounded (hence compact) may be too restrictive. The following Theorem may then be useful

Theorem 3. *Assume that all conditions specified in Theorem 2 are fulfilled, except that the condition that D be compact convex is replaced by the following*

(R) *There exists a compact convex set $\overline{D}$ satisfying*

$$\{y\in D|\ (\exists x\in C)\ G(x,y)\le 0\}\subset\overline{D}\subset D.$$

Then the conclusions of Theorem 2 hold, provided D is replaced by $\overline{D}$ in the formula (3) for computing lagrangian bounds, i.e.

$$\beta(M) = \sup_{\lambda\in\mathbb{R}^m_+} \inf\{F(x,y) + \langle\lambda, G(x,y)\rangle|\ x \in M\cap C, y\in\overline{D}\}. \qquad (12)$$

Proof. In fact the problem (P) is equivalent to

$$\min\{F(x,y)|\ G_i(x,y)\le 0\ (i=1,\dots,m), x\in C,\ y\in\overline{D}\} \qquad (\overline{\text{P}})$$

which satisfies all conditions required in Theorem 2. □

An important case of condition (R) is the following

Lemma 1. *Assume that* $\sup\{F(x,y)|\ x\in C, y\in D\}\le\alpha$. *Then condition* (R) *is implied by*

(S) *There exists* $\lambda^*\in\mathbb{R}^m_+$ *such that* $\inf_{x\in C}\{F(x,y)+\langle\lambda^*, G(x,y)\rangle\}\to+\infty$ *as* $y\in D, \|y\|\to+\infty$.

More specifically, if (S) *holds then* (R) *holds with* $\overline{D}$ *being the closed convex hull of the set*

$$D^* := \{y\in D|\ \varphi(y)\le\alpha\}, \text{ where } \varphi(y) := \inf_{x\in C}\{F(x,y)+\langle\lambda^*, G(x,y)\rangle\}$$

Proof. Clearly the set D^* is bounded, because any sequence $\{y^\nu\}\subset D^*$ such that $\|y^\nu\|\to+\infty$ would satisfy $\varphi(y)\to+\infty$ by Assumption (S). But for any $y\in D$ such that $G(x,y)\le 0$ for some $x\in C$ we have $\inf_{x\in C}\{F(x,y)+\langle\lambda^*, G(x,y)\rangle\}\le\inf_{x\in C}\{F(x,y)|\ G(x,y)\le 0, y\in D\}\le\alpha$, i.e. $\varphi(y)\le\alpha$, and hence, $y\in D^*$. Thus, $\overline{D}$ is compact convex and satisfies (R). □

Theorem 4. *Let C be a compact convex subset of R^n, D a polyhedron in $\mathbb{R}^p$ having at least one vertex, and let functions $F(x,y), G_i(x,y), i=1,\dots,m$ be continuous in x and affine in y for every fixed x. If either condition (S) in Lemma 1 or the following condition (T) is satisfied then the conclusions of Theorem 2 hold.*

(T) *For every $x^*\in C$ there exists $y^*\in D$ such that $G_i(x^*,y^*)<0,\ i=1,\dots,m$.*

Proof. As previously it suffices to prove (7) for x^* such that $\cap_k M_k = \{x^*\}$.

If (S) is satisfied then $\{F(x^*,y)+\langle\lambda^*, G(x^*,y)\ \ge \inf_{x\in C}\{F(x,y)+\langle\lambda^*, G(x,y)\rangle\}\to+\infty$ as $y\in D, \|y\|\to+\infty$. On the other hand, if (T) is satisfied then $F(x^*,y^*)+\langle\lambda, G(x^*,y^*)\rangle\to-\infty$ as $\lambda\to+\infty$. In both cases, the minimax theorem holds for the function $(y,\lambda)\mapsto F(x^*,y)+\langle\lambda, G(x^*,y)\rangle$ (see Appendix), i.e.

$$\inf_{y\in D}\sup_{\lambda\in\mathbb{R}^m_+}\{F(x^*,y)+\langle\lambda, G(x^*,y)\rangle\}$$
$$= \sup_{\lambda\in\mathbb{R}^m_+}\inf_{y\in D}\{F(x^*,y)+\langle\lambda, G(x^*,y)\rangle\}.$$

Therefore, if (7) is not true then there exists $\tilde\lambda\in\mathbb{R}^m_+$ such that

$$\inf_{y\in D}\{F(x^*,y)+\langle\tilde{\lambda},G(x^*,y)\rangle\}>\beta^*. \tag{13}$$

Denote by vert(D) the vertex set of D. Since the function $y\mapsto F(x^*,y)+\langle\tilde{\lambda},G(x^*,y)$ is affine this inequality holds if and only if

$$\min_{y\in\text{vert}(D)}\{F(x^*,y)+\langle\tilde{\lambda},G(x^*,y)\rangle\}>\beta^*.$$

Now for every $y\in\text{vert}(D)$, in view of the lower semi-continuity of the function $y\mapsto F(x^*,y)+\langle\tilde{\lambda},G(x^*,y)\rangle$ at x^* there exists an open ball $U(y)$ in $\mathbb{R}^n$ around x^* such that

$$F(x,y)+\langle\tilde{\lambda},G(x,y)\rangle>\beta^*\quad\forall x\in U(y).$$

Letting $U=\cap_{y\in\text{vert}(D)}U(y)$ we then have

$$F(x,y)+\langle\tilde{\lambda},G(x,y)\rangle>\beta^*\quad\forall x\in U,\forall y\in D.$$

Therefore for all k so large that $M_k\subset U$:

$$F(x,y)+\langle\tilde{\lambda},G(x,y)\rangle>\beta^*\quad\forall x\in M_k\cap C,\forall y\in D,$$

and hence

$$\beta(M_k)=\sup_{\lambda\in\mathbb{R}^m_+}\inf\{F(x,y)+\langle\tilde{\lambda},G(x,y)\rangle|\ x\in M_k\cap C,y\in D\}>\beta^*,$$

conflicting with $\beta(M_k)\nearrow\beta^*$. □

4 Special cases

From the above theorems we can derive the convergence of Algorithm GBB(α) in most important special cases.

4.1 Nonconvex programming.

Consider the nonconvex optimization problem

$$\min\{f(x)|\ g_i(x)\le 0\ (i=1,\dots,m),\ x\in C\} \tag{SP}$$

where C is a nonempty compact convex subset of $\mathbb{R}^n$, and $f,g_i:\mathbb{R}^n\to\mathbb{R}$, $i=1,\dots,m$. Setting $D=\{y^*\}\subset\mathbb{R}^m$ and $F(x,y)\equiv f(x)$, $G_i(x,y)\equiv g_i(x)\ \forall y\in\mathbb{R}^m$ we can view (SP) as a special problem (P) with a bounded set D. Suppose Algorithm GBB(α) is applied to this problem, starting from an initial rectangle (or simplex) $M_1\supset C$, and using for every partition set $M\subset M_1$ the lagrangian bound

$$\beta(M)=\sup_{\lambda\in\mathbb{R}^m_+}\inf\{f(x)+\sum_{i=1}^m\lambda_i g_i(x)|\ x\in M\cap C\}.$$

As a consequence of Theorem 2 we can state

Corollary 1. *Assume that the functions $f(x)$, $g_i(x)$, $i = 1, \ldots, m$, are lower semi-continuous. Then either the algorithm stops after finitely many iterations (yielding the incumbent as an optimal solution or indicating infeasibility of (SP)), or else it generates an infinite nested sequence $M_{k_\nu} \subset M_k$ collapsing to an optimal solution x^* of (SP).*

Proof. If the algorithm generates an infinite nested sequence M_{k_ν} collapsing to a point x^* then by Theorem 2, $x^* \in C$ and $\lim \beta(M_k) = \inf\{f(x)|\ g_i(x) \leq 0\ (i = 1, \ldots, m),\ x \in C\}$. We contend that $\lim \beta(M_k) < +\infty$, i.e. that x^* is an optimal solution. Since $x^* \in C$ it suffices to show that x^* satisfies $g_i(x^*) \leq 0$, $i = 1, \ldots, m$. But if it were not so, then $g_{i_0}(x^*) > 0$ for some $i_0 \in \{1, \ldots, m\}$, and by lower semi-continuity of $g_{i_0}(x)$, there would be a ball W around x^* such that $g_{i_0}(x) > \rho := \frac{1}{2} g_{i_0}(x^*)\ \forall x \in W$. Then for k so large that $M_k \subset W$ we would have $\beta(M_k) = \sup_{\lambda \in \mathbb{R}^m_+} \inf_{x \in M_k \cap C}\{f(x) + \sum_{i=1}^m \lambda_i g_i(x)\} \geq \sup_{\lambda_{i_0} \geq 0} \inf_{x \in M_k \cap C}\{f(x) + \lambda_{i_0}\rho\} = +\infty$, conflicting with $\beta(M_k) < +\infty$. □

Corollary 1 is stronger and more general than the results in [2], while the proof is simpler, especially regarding the capability of the algorithm to detect infeasibility.

4.2 Partly linear programming.

A special case of problem (P) when $F(x, y), G_i(x, y)$ are linear in y is the problem enountered in several applications (see e.g. [1]):

$$f^*_{[r,s]} = \min\{\langle c(x), y\rangle|\ A(x)y - b(x) \leq 0, r \leq x \leq s, y \geq 0\} \tag{PL}$$

where $x \in \mathbb{R}^n$, $y \in \mathbb{R}^p$, $c : \mathbb{R}^n \to \mathbb{R}^p$, $A : \mathbb{R}^n \to \mathbb{R}^{m \times p}$, $b : \mathbb{R}^m \to \mathbb{R}^p$, $[r, s] := \{x|\ r \leq x \leq s\} \subset \mathbb{R}^n$. Clearly this is a problem (P) with

$$F(x, y) = \langle c(x), y\rangle, \quad G(x, y) = A(x)y - b(x), \quad C = [r, s],\ D = R^p_+.$$

The lagrangian bound over a partition set $M \subset [r, s]$ is given by

$$\begin{aligned} \beta(M) &= \sup_{u \geq 0} \inf_{x \in M} \inf_{y \geq 0}\{\langle c(x), y\rangle + \langle u, A(x)y - b(x)\rangle\} \\ &= \sup_{u \geq 0}\{-\langle b(x), u\rangle + \inf_{x \in M} \inf_{y \geq 0}\{\langle c(x) + (A(x))^T u, y\rangle\}\} \\ &= \sup_{u \geq 0}\{-\langle b(x), u\rangle + h(u)\} \end{aligned} \tag{14}$$

where

$$h(u) = \begin{cases} 0 & \text{if } (A(x))^T u + c(x) \geq 0\ \forall x \in M \\ -\infty & \text{otherwise} \end{cases}$$

Hence

$$\beta(M) = \sup_{u \geq 0}\{-\langle b(x), u\rangle|\ (A(x))^T u + c(x) \geq 0\ \forall x \in M\}. \tag{15}$$

Corollary 2. *Let* $A(x) = [a_{ij}(x)] \in \mathbb{R}^{m\times p}$, *and assume* $a_{ij}(x), c_j(x), b_i(x)$ *are lower semi-continuous functions. If* $\alpha := \sup\{\langle c(x), y\rangle |\ x \in [r,s], y \geq 0\} < +\infty$ *and either of the following conditions holds:*

(S1) $(\forall x^* \in [r,s])\ (\exists u^* \in \mathbb{R}^m_+)\quad (A(x^*))^T u^* + c(x^*) > 0;$

(T1) $(\forall x^* \in [r,s])\ (\exists y^* \in \mathbb{R}^p_+)\quad A(x^*)y^* - b(x^*) < 0;$

then Algorithm GBB(α) using lagrangian bounds either terminates after finitely many steps with an optimal solution, or else it generates a filter $\{M_{k_\nu}\}$ *collapsing to a point* $x^* \in C$ *such that the linear program*

$$\inf\{\langle c, y\rangle |\ A(x^*)y \leq b(x^*),\ y \geq 0\}$$

is solvable and any optimal solution $y^* \in \mathbb{R}^p_+$ *of it yields an optimal solution* (x^*, y^*) *of (PL).*

Proof. If (T1) holds then this follows from Theorem 4. If (S1) holds then by lower semi-continuity of the functions, there exists a ball U around x^* such that $\min_{x\in U\cap C}(A(x))^T u^* + c(x) > 0$. Hence

$$\begin{aligned} &\inf_{x\in U\cap C}[\langle c(x), y\rangle + \langle u^*, A(x)y - b(x)\rangle] \\ &= \inf_{x\in U\cap C}[\langle c(x) + (A(x))^T u^*, y\rangle - \langle u^*, b(x)\rangle] \\ &\to +\infty \text{ as } y \geq 0, \|y\| \to +\infty, \end{aligned}$$

and the conclusion follows from Theorem 3. □

Remark 1 In the particular case when the functions $a_{ij}(x)$, are continuous, $c(x) \equiv c, b(x) \equiv b$, and (S1) holds, the above result was established earlier in [1] by quite an elaborate argument that heavily depends on parametric optimization and set-valued mappings theory. A simpler direct proof was later given in [17].

It is also easily seen that (S1) implies (R), so that when (S1) holds Corollary 2 is a mere consequence of Theorem 2. In fact, if $a^1(x), \ldots, a^m(x)$ are the rows of $A(x)$, then (S1) implies that there exists a ball U around x^* such that $0 \in \text{intconv}\{a^1(x), \ldots, a^m(x), c(x)\}\ \forall x \in U$ (because by a slight perturbation of u^* in the inequality $(A(x^*))^T u^* + c(x^*) > 0$ one can assume $u^* > 0$, and $(A(x))^T u^* + c(x) > 0$ for all x sufficiently close to x^*). Hence there is a ball W, say of radius $r > 0$, around 0 such that $W \subset \text{conv}\{a^1(x), \ldots, a^m(x), c(x)\}\ \forall x \in U$. Since for every fixed x the set $\text{conv}\{a^1(x), \ldots, a^m(x), c(x)\}$ is the polar of the set $\{y |\ A(x)y \leq e, \langle c(x), y\rangle \leq 1\}$ (where $e = (1, \ldots, 1) \in \mathbb{R}^m$), it follows that for every $x \in U$ the latter set is contained in the ball of radius $1/r$ around 0. This implies in turn that for all $x \in U$ the set $\{y |\ A(x) \leq b(x), \langle c(x), y\rangle \leq \alpha\}$ is contained in a ball; in other words the set $\{y |\ A(x) \leq b(x), \langle c(x), y\rangle \leq \alpha, x \in U\}$ is bounded.

Example 1. The general bilinear matrix inequalities (BMI) problem in control theory can be formulated as follows (see e.g. [15]):

$$\min\langle c, x\rangle + \langle d, y\rangle \tag{16}$$

$$\text{s.t. } G_0 + \sum_{j=1}^{m} y_j G_j \preceq 0 \tag{17}$$

$$L_0 + \sum_{i=1}^{n} x_i L_{i0} + \sum_{j=1}^{m} y_j L_{0j} + \sum_{i=1}^{n}\sum_{j=1}^{m} x_j y_j L_{ij} \prec 0 \tag{18}$$

$$x \in X = [p, q] \subset \mathbb{R}^n, \ y \in \mathbb{R}^m_+ \tag{19}$$

where x, y are the decision variables, $G_0, G_j, L_0, L_{0i}, L_{j0}, L_{ij}$ are symmetric matrices of appropriate sizes, and the notation $G \preceq 0, L \prec 0$ means that the matrix G is semidefinite negative, L is definite negative.

For ease of notation we write

$$\begin{bmatrix} A \\ B \end{bmatrix}_d \text{ for } \begin{bmatrix} A & 0 \\ 0 & B \end{bmatrix}$$

and define

$$A_{00}(x) = \begin{bmatrix} G_0 \\ L_0 + \sum_{i=1}^{n} x_i L_{i0} \\ \langle x, c\rangle \end{bmatrix}_d, A_{j0}(x) = \begin{bmatrix} G_j \\ L_{0j} + \sum_{i=1}^{n} x_i L_{ij} \\ d_j \end{bmatrix}_d,$$

$$Q_{00} = \begin{bmatrix} 0 \\ 0 \\ 1 \end{bmatrix}_d.$$

Then, as was shown in [15], this problem can be converted to the form

$$\min\{t |\ A_0(x, p, q) + \sum_{j=1}^{m} y_j A_j(x, p, q) \preceq tQ, \ y \geq 0, \ x \in X\}$$

where

$$A_j(x, p, q) = \begin{bmatrix} A_{j0}(x) \\ A_{j1}(x, p, q) \end{bmatrix}_d, \ Q = \begin{bmatrix} Q_{00} \\ Q_{01} \end{bmatrix}_d, \ Q_{01} = 0$$

$$A_{j1}(x, p, q) = \begin{bmatrix} (x_1 - p_1)G_j \\ (q_1 - x_1)G_j \\ \cdots \\ (x_n - p_n)G_j \\ (q_n - x_n)G_j \end{bmatrix}_d, \ j = 0, 1, \ldots, n.$$

In this form the (BMI) problem appears to be a partly linear problem for which all conditions of Theorem 2 (or Corollary 2) are fulfilled. The branch and bound algorithm proposed in [15] is just a version of Algorithm GBB(α) using lagrange dual bounds. The convergence of the algorithm, guaranteed by Corollary 2, was established in that paper under the assumption

$$(\forall x \in X)(\exists Z_1 \succeq 0) \qquad \mathrm{Tr}(Z_1 Q_{00}) = 1,\ \mathrm{Tr}(Z_1 A_{j0}(x)) > 0,\ j = 1, \ldots, m$$

which can easily be recognized as (S1). Note that the Lagrange dual of the problem

$$\max_{Z \succeq 0}\ \min_{t \in \mathbb{R}, y \geq 0, x \in M} \left\{ t + \mathrm{Tr}\left[Z(A_0(x,p,q) + \sum_{j=1}^{m} y_j A_j(x,p,q) - tQ) \right] \right\}$$

has been shown in [15] to be equivalent to the LMI program

$$\begin{aligned} \max\ \{t|\ & \mathrm{Tr}(ZA_0(x,p,q)) \geq t, \\ & \mathrm{Tr}(ZA_j(x,p,q)) \geq 0\ \forall x \in \mathrm{vert}X,\ j = 1, \ldots, m, \\ & \mathrm{Tr}(ZQ) = 1,\ Z \succeq 0\} \end{aligned}$$

where vert$X$ denotes the vertex set of X. This is a special case of a general result to be established in the next section.

5 Lagrangian bound for linear-monotonic problems

The main difficulty with the lagrangian approach is how to solve the lagrangian dual problem (3). One way to get round this difficulty is to further relax the lagrangian relaxation to obtain a still consistent bound, or to find equivalent reformulations of the problem for which the lagrangian dual is easier to solve. Alternatively, convex or other computationally easy relaxations can be used.

In this and the next section we will show that for a large majority of convex-monotonic problems either the lagrangian relaxation is easy to solve or a convex relaxation is readily available.

Consider the following partly linear programming problem which is slightly more general than the problem (PL) considered in the preceding section:

$$f^*_{[r,s]} = \min\{\langle c(x), y\rangle + c_0(x)|\ A(x)y + B(x) \leq b, r \leq x \leq s, y \geq 0\} \qquad . \tag{GPL}$$

where $x \in \mathbb{R}^n$, $y \in \mathbb{R}^p$, $c : \mathbb{R}^n \to \mathbb{R}^p$, $c_0 : \mathbb{R}^n \to \mathbb{R}$, $A := \mathbb{R}^n \to \mathbb{R}^{m \times p}$, $B : \mathbb{R}^n \to \mathbb{R}^{m \times n}$, $b \in \mathbb{R}^m$, $[r, s] \subset \mathbb{R}^n_+$. Setting $A(x) = [a_{ij}(x)]$, and denoting the i-th row of $B(x)$ by $B_i(x)$, we can write the problem in the expanded form as

$$\begin{aligned} \min\ & \sum_{j=1}^{p} y_j c_j(x) + c_0(x) \\ \text{s.t.}\ & \sum_{j=1}^{p} y_j a_{ij}(x) + B_i(x) \leq b_i\ (i = 1, \ldots, m) \\ & y \geq 0, \quad r \leq x \leq s. \end{aligned} \tag{GPL}$$

Clearly this is a special partly convex problem, with

$F(x,y) = \langle c(x), y\rangle + c_0(x), G(x,y) = A(x)y + B(x) - b,\ C = [r,s], D = \{y \geq 0\}.$

In this section we will show that under a monotonicity condition the lagrangian dual of problem (PL) reduces to a linear program.

Specifically, assume that

(i) *$c_0(x)$ and $B(x)$ are linear, i.e. $c_0(x) = \langle c_0, x\rangle$, with $c_0 \in \mathbb{R}^n$, $B(x) = Bx$ with $B = [b_{ik}] \in \mathbb{R}^{m\times n}$.*

(ii) *For every j, the functions $c_j(x), a_{ij}(x), i = 1,\ldots,m$, are lower semi-continuous and either all increasing on $[r,s]$, or all decreasing on $[r,s]$.*

Assumption (ii) is satisfied in particular if for every j, each of the functions $c_j(x), a_{ij}(x), i = 1,\ldots,m$ or its negative is a polynomial in x with positive coefficients and integral or rational positive exponents.

The lagrangian dual of (GPL) with respect to the nonlinear constraints is

$$\varphi^*_{[r,s]} = \sup_{\lambda\geq 0} \inf\{\langle y, c(x)\rangle + \langle c_0, x\rangle + \langle \lambda, A(x)y + Bx - b\rangle |\ x \in [r,s],\ y \geq 0\}. \quad (20)$$

For fixed $\lambda \geq 0$ we have

$$\begin{aligned}
&\inf\{\langle y, c(x)\rangle + \langle c_0, x\rangle + \langle \lambda, A(x)y + Bx - b\rangle |\ x \in [r,s],\ y \geq 0\}\\
&\quad = -\langle b, \lambda\rangle + \inf_{x\in[r,s]} \inf_{y\geq 0} \{\langle Bx, \lambda\rangle + \langle c_0, x\rangle + \langle c(x) + (A(x))^T\lambda, y\rangle\}\\
&\quad = -\langle b, \lambda\rangle + h(\lambda)
\end{aligned}$$

where

$$h(\lambda) = \begin{cases} \inf_{x\in[r,s]}[\langle Bx, \lambda\rangle + \langle c_0, x\rangle] & \text{if} \quad c(x) + (A(x))^T\lambda \geq 0 \quad \forall x \in [r,s],\\ -\infty & \text{otherwise.}\end{cases} \quad (21)$$

Lemma 2. *For any $q \in \mathbb{R}^n$:*

$$\inf_{r\leq x\leq s} \langle q, x\rangle = \langle q, r\rangle + \max\{\langle r - s, t\rangle |\ t \geq 0, t \geq -q\}.$$

Proof. Clearly $\min\{\langle q, x\rangle |\ r \leq x \leq s\} = \min\{\langle q, r\rangle + \langle q, x - r\rangle |\ 0 \leq x - r \leq s - r\} = \langle q, r\rangle + \sum_{q_i<0} q_i(s_i - r_i) = \langle q, r\rangle + \max\{\langle r - s, t\rangle |\ t \geq 0, t \geq -q\}$. □

Therefore, under assumption (i):

$$\begin{aligned}
&\inf_{r\leq x\leq s}\{\langle Bx, \lambda\rangle + \langle c_0, x\rangle\}\\
&\quad = \langle B^T\lambda + c_0, r\rangle + \max\{\langle r - s, t\rangle |\ t \geq 0, t \geq -B^T\lambda - c_0\}.
\end{aligned} \quad (22)$$

Denote the j-th column of A by A_j, so that the condition $\langle A(x), \lambda\rangle + c(x) \geq 0$ $\forall x \in [r,s]$ means

$$\langle A_j(x), \lambda\rangle + c_j(x) \geq 0 \quad \forall x \in [r,s],\ \forall j = 1,\ldots,p.$$

Let J_+ be the set of all j such that all $c_j(x), a_{ij}(x), i = 1,\ldots,m$ are increasing and let J_- be the set of all j such that all $c_j(x), a_{ij}(x), i = 1,\ldots,m$ are decreasing. By assumption (ii) $J^+ \cup J^- = \{1, \ldots, p\}$. Then for every $j = 1, \ldots, p$, we have

$$\begin{aligned} &\langle A_j(x), \lambda \rangle + c_j(x) \geq 0 \quad \forall x \in [r, s] \\ \Leftrightarrow &\sum_{i=1}^{m} \lambda_i a_{ij}(x) + c_j(x) \geq 0 \quad \forall x \in [r, s] \\ \Leftrightarrow &\begin{cases} \sum_{i=1}^{m} \lambda_i a_{ij}(r) + c_j(r) \geq 0 & \text{if } j \in J_+ \\ \sum_{i=1}^{m} \lambda_i a_{ij}(s) + c_j(s) \geq 0 & \text{if } j \in J_- \end{cases} \end{aligned} \tag{23}$$

From (20), (21) and (23) we derive

$$\begin{aligned} \varphi^*_{[r,s]} = \sup\{-\langle b, \lambda \rangle + &\inf_{x \in [r,s]} \{\langle Bx, \lambda \rangle + \langle c_0, x \rangle | \\ &\langle A_j(r), \lambda \rangle + c_j(r) \geq 0 \ (j \in J_+); \\ &\langle A_j(s), \lambda \rangle + c_j(s) \geq 0 \ (j \in J_-), \ \lambda \geq 0\} \end{aligned}$$

Denote the i-th row of B by B_i. In view of (22) the lagrangian dual problem (20) thus reduces to the linear program

$$\langle c_0, r \rangle + \max\{\langle r - s, t \rangle + \sum_{i=1}^{m} \lambda_i [\langle B_i, r \rangle - b_i]\}$$

$$\text{s.t.} \left| \begin{array}{l} -\sum_{i=1}^{m} \lambda_i B_i - t \leq c_0 \\ -\sum_{i=1}^{m} \lambda_i a_{ij}(r) \leq c_j(r) \ (j \in J_+) \\ -\sum_{i=1}^{m} \lambda_i a_{ij}(s) \leq c_j(s) \ (j \in J^-) \\ \lambda \geq 0, \ t \geq 0 \end{array} \right.$$

whose dual is

$$\begin{aligned} \langle c_0, r \rangle + \min\{&\sum_{j \in J_+} c_j(r) y_j + \sum_{j \in J_-} c_j(s) y_j + \langle c_0, z \rangle\} | \\ \text{s.t.} \quad &\sum_{j \in J_+} a_{ij}(r) y_j + \sum_{j \in J_-} a_{ij}(s) y_j + \langle B_i, r + z \rangle \leq b_i \ (i = 1, \ldots, m), \\ &y \geq 0, \ 0 \leq z \leq s - r. \end{aligned}$$

By setting $x = r + z$, this dual becomes

$$\min \{\sum_{j \in J^+} c_j(r) y_j + \sum_{j \in J} c_j(s) y_j + \langle c_0, x \rangle\} \tag{24}$$

$$\text{s.t.} \sum_{j \in J_+} a_{ij}(r) y_j + \sum_{j \in J_-} a_{ij}(s) y_j + \langle B_i, x \rangle \leq b_i \ (i = 1, \ldots, m), \tag{25}$$

$$y \geq 0, \ r \leq x \leq s. \tag{26}$$

which is a linear program obtained from (GPL) by the substitution

$$a_{ij}(x) \leftarrow a_{ij}(r), \quad c_j(x) \leftarrow c_j(r) \ \text{ for } \ j \in J_+)$$
$$a_{ij}(x) \leftarrow a_{ij}(s), \quad c_j(x) \leftarrow c_j(s) \ \text{ (for } \ j \in J_-).$$

Noting that under condition (ii) $\sum_{j\in J_+} c_j(r)y_j + \sum_{j\in J_-} c_j(s)y_j$ is a linear underestimator of $\langle y, c(x)\rangle$ and similarly, $\sum_{j\in J_+} a_{ij}(r)y_j + \sum_{j\in J_-} a_{ij}(s)y_j$ is a linear underestimator of $\langle A(x), y\rangle$, we can thus state:

Theorem 5. *Under assumptions* (i) *and* (ii) *the lagrangian dual problem of* (GPL) *reduces to a linear program, whose dual is a linear relaxation of* (GPL).

Of course, the optimal value of the linear program (24)(25)(26) gives a lower bound for the optimal value of (GPL). Although this is often a good bound, it is not true, even in the special case considered in [12], that this bound must be at least as good as any lower bound obtained by convex relaxation. In fact, since the problem (24)-(25)- (26) is a mere linear relaxation of (GPL), it is quite conceivable that for certain problems tigher relaxation than the lagrangian one may exist. Examples can easily be constructed to show that, contrary to assertions in [12], lagrangian dual bounds can sometimes be rather poor, if the primal problem is not suitably reformulated (see e.g. [20]).

Remark 2 The lagrangian dual is still a linear program if assumption (i) is replaced by the condition

($\star$) $\langle \lambda, B(x)\rangle + c_0(x)$ *is quasiconcave in* x,

or if asssumption (ii) is replaced by the condition

(#) *For each* $j = 1, \ldots, n$ *the function* $\langle A_j(x), \lambda\rangle + c_j(x)$ *is quasiconcave in* x.

In fact, if condition ($\star$) holds instead of (i), then, since a quasiconcave function attains its minimum over a box $[r, s]$ at one corner of it, we have

$$\inf_{x\in[r,s]} \{\langle B(x), \lambda\rangle + c_0(x)\} = \min\{\langle \lambda, B(w^i)\rangle + c_0(w^i)|\ i = 1, \ldots, 2^n\}$$

where $w^1, \ldots, w^{2^n}$ are the extreme points of $[r, s]$, so the lagrangian dual (20) reduces to the linear program

$$\max\{-\langle b, \lambda\rangle + t$$
$$\text{s.t.} \left| \begin{array}{l} -\langle B(w^i), \lambda\rangle + t \le c_0(w^i), \ i = 1, \ldots, 2^n \\ -\sum_{i=1}^m \lambda_i a_{ij}(r) \le c_j(r) \ (j \in J_+) \\ -\sum_{i=1}^m \lambda_i a_{ij}(s) \le c_j(s) \ (j \in J^-) \\ \lambda \ge 0 \end{array} \right.$$

Likewise, if condition (#) holds instead of (ii) then for each $j = 1, \ldots, n$:

$$\langle A_j(x), \lambda\rangle + c_j(x) \ge 0 \Leftrightarrow \langle A_j(w^k), \lambda\rangle + c_j(w^k) \ge 0 \quad k = 1, \ldots, 2^n,$$

so the lagrangian dual reduces to the linear program

$$
\begin{aligned}
&\langle c_0, r\rangle + \max\{\langle r-s, t\rangle + \langle Br-b, \lambda\rangle\} \\
&\text{s.t. } \langle A_j(w^k), \lambda\rangle + c_j(w^k) \geq 0 \quad j=1,\ldots,n, k=1,\ldots,2^n \\
&\qquad \lambda \geq 0,\ t \geq 0
\end{aligned}
$$

Thus, Theorem 5 includes the main result in [1] which corresponds to the special case $c(x) \equiv c \in \mathbb{R}^p,\ c_0(x) \equiv 0,\ B(x) \equiv 0$ with the assumption (#).

Example 2. The general bilinear matrix inequality (BMI) problem as discussed at the end of Section 2 is an example of linear-monotonic optimization problem. Another example of interest is the pooling/blending problem in oil refineries (see e.g. [6]) which can be formulated as follows. Let x_{il} denote the amount of component i allocated to pool l, y_{lj} the amount going from pool l to product j, z_{ij} the amount of component i going directly to product j, p_{lk} the level of quality k (relative sulfur content level) in pool l, and C_{ik} the level of quality k in component i. Then, for given the unit prices c_i, d_j of component i and product j, the problem is to determine the variables $x_{il}, y_{lj}, z_{ij}, p_{lk}$ so as to maximize the profit

$$
-\sum_i \sum_\nu c_i x_{i\nu} + \sum_\nu \sum_j d_j y_{\nu j} + \sum_i \sum_j (d_j - c_i) z_{ij}.
$$

under the constraints

$$
\begin{aligned}
&\left.\begin{array}{l}\sum_\nu x_{il} + \sum_j z_{ij} \leq A_i \\ \sum_i x_{il} - \sum_j y_{lj} = 0\end{array}\right\} \quad \text{(component balance)} \\
&\sum_i x_{i\nu} \leq S_\nu \qquad \text{(pool balance)} \\
&\sum_\nu (p_{\nu k} - P_{jk}) y_{\nu j} + \sum_i (C_{ij} - P_{jk}) z_{ij} \leq 0 \quad \text{(pool quality)} \\
&\sum_\nu y_{\nu j} + \sum_i z_{ij} \leq D_j \quad \text{(product demands constraints)} \\
&x_{i\nu}, y_{\nu j}, z_{ij}, p_{\nu k} \geq 0
\end{aligned}
$$

where A_i, S_ν, P_{jk}, D_j, are the upper bounds for component availabilities, pool sizes, product qualities and product demands, respectively, and C_{ik} the level of quality k in component i. Clearly this is a linear-monotonic optimization problem, with nonconvex variables p_{lk}. It is also easy to verify that the conditions of Corollary 2 are satisfied, so that this problem can be solved by Algorithm GBB(α) with bounds by linear relaxation.

6 Consistent bound for convex-monotonic problems

We now turn to the general convex-monotonic problem

$$
\begin{aligned}
&\min f_1(x,y) + f_2(x,y) + u(x) \\
&\text{s.t. } g_i(x,y) + h_i(x,y) + v_i(x) \leq 0 \ (i=1,\ldots,m) \\
&\qquad x \in C, \quad y \in D.
\end{aligned}
\tag{PC}
$$

where $C \subset [a,b] \subset \mathbb{R}^n_+$, D is a compact convex subset of $\mathbb{R}^p, u(x), v_i(x)$ are affine functions, $f_1(x,y), f_2(x,y), g_i(x,y),\ h_i(x,y)\ (i = 1,\ldots,m)$ are lower semi-continuous functions, convex in y for every fixed x, and monotonic in x for every fixed y. Without loss of generality one can assume that $f_1(x,y), g_i(x,y)$ are decreasing in x and $f_2(x,y), h_i(x,y)$ are increasing. For any subrectangle $M = [r,s] \subset [a,b]$ consider the subproblem (PC(M)) obtained from (PC) by replacing the constraint $x \in C$ with $x \in C \cap [r,s]$.

Theorem 6. *A lower bound for the optimal value of the subproblem* (PC(M)) *is furnished by*

$$\begin{aligned} \beta(M) := \min\ & f_1(s,y) + f_2(r,y) + u(x) \\ \text{s.t. } & g_i(s,y) + h_i(r,y) + v_i(x) \le 0\ (i = 1,\ldots,m) \\ & r \le x \le s, \quad y \in D. \end{aligned} \tag{RC(M)}$$

This bound is consistent for Algorithm GBB(α).

Proof. Clearly $f_1(s,y), f_2(r,y), g_i(s,y), h_i(r,y)$ are convex underestimators of the functions $f_1(x,y), f_2(x,y), g_i(x,y), h_i(x,y)$, so (RC($M$)) is a convex relaxation of (PC(M)). The bound $\beta(M)$ computed via this relaxation obviously satisfies (2). Therefore, according to Theorem 1, to prove the consistency of the bound it suffices to show that for any nested sequence of boxes $\{M_k = [r_k, s_k]\}$ collapsing to a point x^* we have

$$\lim_{k\to\infty} \beta(M_k) = \beta^*, \quad \text{where}$$

$$\begin{aligned} \beta^* = \min\{ & f_1(x^*,y) + f_2(x^*,y) + u(x^*)| \\ & g_i(x^*,y) + h_i(x^*,y) + v_i(x^*) \le 0\ (i = 1,\ldots,m),\ y \in D\}. \end{aligned} \tag{27}$$

For every k let (x^k, y^k) be an optimal solution of the problem (RC(M_k)), so that

$$\begin{gathered} f_1(s^k,y^k) + f_2(r^k,y^k) + u(x^k) = \beta(M_k), \\ g_i(s^k,y^k) + h_i(r^k,y^k) + v_i(x^k) \le 0\ (i = 1,\ldots,m), \\ r^k \le x^k \le s^k,\ y^k \in D. \end{gathered}$$

Since $\cap_{k=1}^{+\infty} M_k = \{x^*\}$, we must have $r^k \to x^*, s^k \to x^*$ as $k \to +\infty$. In view of the compactness of the set D we may assume, by passing to a subsequence if necessary, that $y^k \to y^* \in D$. Then, by the lower semi-continuity of the functions involved,

$$\begin{aligned} & f_1(x^*,y^*) + f_2(x^*,y^*) + u(x^*) \le \lim_{k\to+\infty} \beta(M_k), \\ & g_i(x^*,y^*) + h_i(x^*,y^*) + v_i(x^*) \le 0\ (i = 1,\ldots,m). \end{aligned} \tag{28}$$

But, since RC(M_k)) is a relaxation of the problem (PC(M_k)),

$$\begin{aligned} \beta(M_k) \le \min\{ & f_1(x,y) + f_2(x,y) + u(x)| \\ & g_i(x,y) + h_i(x,y) + v_i(x) \le 0\ (i = 1,\ldots,m), \\ & r^k \le x \le s^k, y \in D\}, \end{aligned}$$

and since $x^* \in \cap_{k=1}^{+\infty} M_k$ it follows that for every k :

$$\beta(M_k) \le \min\{f_1(x^*, y) + f_2(x^*, y) + u(x^*)| \\ g_i(x^*, y) + h_i(x^*, y) + v_i(x^*) \le 0 \ (i = 1, \ldots, m), \ y \in D\} = \beta^*.$$

Hence, for every k :

$$\beta(M_k) \le \beta^* \le f_1(x^*, y^*) + f_2(x^*, y^*) + u(x^*). \tag{29}$$

From (28) and (29) we finally deduce

$$\lim_{k \to +\infty} \beta(M_k) = f_1(x^*, y^*) + f_2(x^*, y^*) + u(x^*) = \beta^*,$$

completing the proof. □

Of course, (RC(M)) is a mere convex relaxation of (PC(M)) which can be derived immediately from the monotonicity assumptions, without appealing to lagrangian relaxation.

Example 3. The water distribution network problem (see e.g. [5]) provides an example of convex-monotonic problem. In the case of a single source, a single demand pattern, and a new pumping facility at the source node, this problem consists in determining the pump pressure H, the flow rate q_i and the head losses J_i along the arcs $i = 1, \ldots, s$, so as to minimize the cost :

$$f(q, H, J) := b_1 \sum_i L_i (KL_i/J_i)^{\beta/\alpha} (q_i/c)^{\beta\lambda/\alpha} + b_2 |a_1|^{e_1} H^{e_2} + b_3 |a_1| H$$

under the constraints

$$\begin{array}{ll}
\sum_{i \in \text{in}(k)} q_i - \sum_{i \in \text{out}(k)} q_i = a_k \quad k = 1, \ldots, n & \text{(flow balance)} \\
\sum_{i \in \text{loop } p} \pm J_i = 0 \quad p = 1, \ldots, m & \text{(head loss balance)} \\
\sum_{i \in r(k)} \pm J_i \le H + h_1 - h_k^{\min} \quad p = 1, \ldots, m & \text{(hydraulic requirement)} \\
\left.\begin{array}{l} q_i^{\max} \ge q_i \ge q_i^{\min} \ge 0 \quad i = 1, \ldots, s \\ H^{\max} \ge H \ge H^{\min} \ge 0 \end{array}\right\} & \text{(bounds)} \\
\left.\begin{array}{l} KL_i (q_i/c)^\lambda / d_{\max}^\alpha \le J_i \\ J_i \le KL_i (q_i/c)^\lambda / d_{\min}^\alpha \end{array}\right\} i = 1, \ldots, s & \text{(Hazen-Williams equations)}
\end{array}$$

(L_i is the length of arc i, K and c are the Hazen-Williams coefficients, $-a_1$ is the supply water rate at the source and $b_1, b_2, b_3, e_1, e_2, \beta$ are constants such that $0 < e_1, e_2 < 1$ and $1 < \beta < 2.5, \lambda = 1.85$ and $\alpha = 4.87$).

It is easily seen that the objective function $f(q, H, J)$ in this problem is convex in J for fixed (q, H), whereas the function $KL_i(q_i/c)^\lambda$, is monotonic in q so this is a convex-monotonic program. For any given rectangle $M = \{(q, H)|\ \underline{q}^M \le q \le \overline{q}^M,\ \underline{H}^M \le H \le \overline{H}^M\}$ we have

$$f(\underline{q}^M, \underline{H}^M, J) \le f(q, H, J) \tag{30}$$

$$KL_i(\underline{q}_i^M/c)^\lambda \le KL_i(q_i/c)^\lambda \le KL_i(\overline{q}_i^M/c)^\lambda, \ i = 1, \ldots, s \tag{31}$$

for all $(q, H) \in M$. Therefore, a lower bound for the objective function value over all feasible solutions (q, H, J) with $(q, H) \in M$ can be computed by solving a convex relaxation obtained by replacing the objective function with its convex underestimator and the expression $KL_i(q_i/c)^\lambda$ in the Hazen-Williams equations with $KL_i(\underline{q}_i^M/c)^\lambda$ or $KL_i(\overline{q}_i^M/c)^\lambda$ accordingly.

Algorithm GBB(α) method can thus be applied to solve this problem, with branching performed upon (q, H) and bounds computed by convex relaxation based on (30)-(31) as just indicated.

7 Relation to previously published results

Before closing the paper some remarks are in order about the relation of the above presented results to those previously published. This is worthwhile in view of the peculiar situation that has resulted from the appearance of quite a few incorrect results in the recent literature on this topic.

The use of lagrangian dual bounds in a branch and bound algorithm for solving nonconvex optimization problems originated from Falk [4], where linearly constrained nonconvex minimization problems were considered. Shor [11] also extensively investigated lagrangian dual bounds for use in nonconvex quadratic programming. On the other hand, the idea of exploiting partial convexity in global optimization by branch and bound procedures based on an exhaustive partitioning scheme of the space of nonconvex variables appeared first in the paper [16], where nonconvex optimization problems with separated convexity were solved in that way. It should be noted that in many cases the linear relaxation proposed in [16] coincides with the lagrangian relaxation. Later partial convexity in nonseparated form such as in problem (P), was exploited in Ben-Tal et al. [1] also by a branch and bound scheme operating in the space of nonconvex variables but using lagrangian dual bounds. Convergence results (in particular, conditions for closing the duality gap) were obtained in [1] only for the partly linear case (problem (PL)) and under rather restricted conditions (e.g. continuity of $c(x), a_{ij}(x)$, and feasibility of the problem), though using quite elaborate arguments. A much simpler proof of these results can be found in [17] which, however, cannot be extended to the general case considered in the present paper.

More recently, properties of the dual bound method for partly convex problems have also been investigated in Thoai [13]. However, most results in [13] turned out to be incorrect, as can be demonstrated by easily constructed counter-examples [20].

The decomposition approach using a branch and bound scheme can be extended to problems of the general form

$$\min\{F(x,y)|\ G_i(x,y) \le 0\ (i=1,\ldots,m),\ x \in C,\ y \in D\} \qquad (GP)$$

where $C,\ D$ are closed subsets of $\mathbb{R}^n,\ \mathbb{R}^p$, respectively, and $F, G_i : C \times D \to \mathbb{R}$ are functions satisfying some general conditions to be stated below.

These problems attract interest when for every fixed $x \in C$ the functions $F(x,y), G_i(x,y)$ belong to some nice class of functions in y, so that the variables x can be regarded as "complicating". Theorem 1 applies obviously to this problem. On the other hand, close scrutiny of the proof of Theorem 2 shows that the essential role of partial convexity in this proof is to allow the use of the minimax theorem to derive the equality (10):

$$\inf_{y \in \overline{D}} \sup_{\lambda \geq 0} \{F(x^*,y) + \sum_{i=1}^{m} \lambda_i G_i(x^*,y)\} = \sup_{\lambda \geq 0} \inf_{y \in \overline{D}} \{F(x^*,y) + \sum_{i=1}^{m} \lambda_i G_i(x^*,y)\},$$

where $\overline{D}$ is some compact convex subset of D. Therefore, if we now replace the partial convexity assumption by the requirement that the above equality should hold, then this proof carries over to the general case of problem (GP). Specifically, the following theorem holds.

Theorem 7. *Assume that the set C is compact, the set D is closed, and $F(x,y)$, $G_i(x,y), i = 1,\ldots,m$, are lower semi-continuous functions. Furthermore, assume that:*

(*) For every $x^* \in C$ there exist a ball W around x^* and a compact set $\overline{D} \subset D$ such that

$$\inf\{F(x,y) |\ G(x,y) \leq 0,\ x \in W \cap C,\ y \in D\}$$
$$= \inf\{F(x,y) |\ G(x,y) \leq 0,\ x \in W \cap C,\ y \in \overline{D}\}. \tag{32}$$
$$\inf_{y \in \overline{D}} \sup_{\lambda \geq 0} \{F(x^*,y) + \langle \lambda, G(x^*,y)\rangle\} = \sup_{\lambda \geq 0} \inf_{y \in \overline{D}} \{F(x^*,y) + \langle \lambda, G(x^*,y)\rangle\} \tag{33}$$

Then Algorithm GBB(α) using lagrangian bounds (3) applied to problem (GP) either terminates after finitely many iterations (yielding an optimal solution or proving that (P) is infeasible), or else it generates an infinite nested sequence of partition sets $M_{k_\nu}, \nu = 1,2,\ldots,$ collapsing to a point $x^* \in C$ such that the problem

$$\min\{F(x^*,y) |\ G(x^*,y) \leq 0,\ y \in D\}$$

is solvable and any optimal solution $y^* \in D$ of (33) yields an optimal solution (x^*,y^*) of (GP).

In [14], where the problem (GP) has also been investigated, a branch and bound algorithm (called Algorithm BB) is proposed which differs from Algorithm GBB(α) presented in Section 2 only in that the lower bound $\beta(M)$ over a partition set M is defined in GBB(α) by (3) while in BB it is given by

$$\beta(M) = \sup_{\lambda \geq 0} \inf\{F(x,y) + \langle \lambda, G(x,y)\rangle |\ x \in M, y \in D\}.$$

With this definition, the lower bound in BB does not satisfy condition (b) in (2), so to cope with irregular situations that may arise from this, Algorithm BB requires checking the nonemptiness of the set $Q(M) := M \cap C$ for each newly generated M in order to delete partition sets M with $M \cap C = \emptyset$. Thus, the lower bound is actually

$$\beta(M) = \begin{cases} \sup_{\lambda \geq 0} \inf\{F(x,y) + \langle \lambda, G(x,y) \rangle | \ x \in M, y \in D\} & \text{if } M \cap C \neq \emptyset \\ +\infty & \text{otherwise} \end{cases} \tag{34}$$

Clearly computing $\beta(M)$ according to (34) is no easier than solving (3), because checking the nonemptiness of a nonconvex set like $M \cap C$ may be in itself as difficult a problem as (GP).

In any case it can easily be verified that Theorem 7 and its proof, as given above, are still valid when Algorithm BB is used instead of GBB(α). Therefore, Theorem 7 includes both Theorems 3.1 and 3.2 in [14] as special cases when problem (GP) has an optimal solution and all functions F, G_i are continuous. The assumptions used in [14] are rather restrictive, while difficult to check, as is apparent from the examples of applications. Furthermore, the proof of the main Theorem 3.1 in [14] is not quite valid when certain problems mentioned in condition (iv) have no optimal solution. The algorithms for the problems treated as examples of applications (concave minimization, optimization over the efficient set, general quadratic programming), are not quite new, since the lagrangian duals for these problems are merely the duals of standard linear relaxations by convex envelope (see e.g. [20], [21]). A new but incorrect algorithm is derived for the linear multiplicative programming problem

$$\min\{\prod_{i=1}^{n} c_i(y) | \ y \in D\} \tag{35}$$

where D is a polytope in $\mathbb{R}^p$, $c_i(y)$ are affine functions satisfying $c_i(y) > 0 \, \forall D$. In fact, the lagrangian dual for (35) as given in [14] is false, since the function $\prod_{i=1}^{n} x_i - \langle \lambda, x \rangle$ may not be quasiconcave as supposed by the author (counter-example: $f(x) = x_1 x_2 - (x_1 + x_2)$ is not quasiconcave on $\mathbb{R}^2_+$). To obtain a correct lagrangian relaxation for problem (35) one can rewrite it in the form

$$\min\{\sum_{i=1}^{n} \log x_i | \ Cy - x \leq c, \ Ay \leq b, \ x \in [r,s], y \geq 0\}$$

which appears as a (GPL) satisfying conditions ($\star$) and (i) (Remark 2). Then the associated lagrangian dual is the linear program

$$\max\{-\langle c, \lambda^1 \rangle - \langle b, \lambda^2 \rangle + t$$

$$\text{s.t.} \left| \begin{array}{l} \langle w, \lambda^1 \rangle + t \leq \sum_{i=1}^{n} \log(w_i), \ \forall w \in W \\ -C^T \lambda^1 - A^T \lambda^2 \leq 0 \\ \quad (\lambda^1, \lambda^2) \geq 0 \end{array} \right.$$

where W is the vertex set of the rectangle $[r, s]$.

8 Conclusion

We have presented the theoretical foundation of a generic branch and bound algorithm for a class of global optimization problems with a hybrid convex-monotonic structure. In this algorithm branching is performed upon the non-convex variables, via an exhaustive subdivision process, while bounds over partition sets are computed by solving linear or convex programs which can often be identified as lagrangian relaxations. Convergence of such a branch and bound method is guaranteed under mild conditions.

9 Appendix

A general minimax theorem has been established in [21] which yields the following propositions as corollaries:

Theorem 8. *Let C, D be two closed convex sets in $\mathbb{R}^n, \mathbb{R}^m$ respectively, and let $f(x, y) : C \times D \to \mathbb{R}$ be a function quasiconvex and lower semi-continuous in x, quasiconcave and upper semi-continuous in y, and satisfying, furthermore, either of the following conditions:*

(i) *C is compact;*

(ii) *there exists $\bar{y} \in D$ such that $f(x, \bar{y}) \to +\infty$ as $x \in D, \|x\| \to +\infty$.*

Then $\min_{x\in\overline{C}} \sup_{y\in D} f(x, y) = \sup_{y\in D} \inf_{x\in C} f(x, y)$, *where $\overline{C}$ is the compact set*

$$\overline{C} = \begin{cases} C & \text{if (i) holds} \\ \{x \in C |\ f(x, \bar{y}) \leq \gamma\} & \text{if (ii) holds.} \end{cases}$$

and $\gamma := \inf_{x\in C} \sup_{y\in D} f(x, y)$.

Theorem 9. *Let C, D be two closed convex sets in $\mathbb{R}^n, \mathbb{R}^m$ respectively, and let $f(x, y) : C \times D \to \mathbb{R}$ be a function quasiconvex and lower semi-continuous in x, quasiconcave and upper semi-continuous in y, and satisfying, furthermore, either of the following conditions:*

(i) *D is compact;*

(ii) *there exists $\bar{x} \in C$ such that $f(\bar{x}, y) \to -\infty$ as $y \in D, \|y\| \to +\infty$.*

Then $\inf_{x\in C} \sup_{y\in D} f(x, y) = \max_{y\in\overline{D}} \inf_{x\in C} f(x, y)$, *where $\overline{D}$ is the compact set*

$$\overline{D} = \begin{cases} D & \text{if (i) holds} \\ \{y \in D |\ f(\bar{x}, y) \geq \beta\} & \text{if (ii) holds.} \end{cases}$$

and $\beta := \sup_{y\in D} \inf_{x\in C} f(x, y)$.

References

[1] A. Ben-Tal et al. :'Global minimization by reducing the duality gap'. *Math. Prog.* 63(1994)193-212.

[2] M. Dür : 'Dual bounding procedures lead to convergent branch-and-bound algorithms', *Math. Program.* Ser A 91(2001)117-2001.

[3] I. Ekeland and R. Temam: *Convex Analysis and Variational Problems.* North Holland, Amsterdam, 1976.
[4] J. Falk: 'Lagrange multipliers and nonconvex programs' *SIAM J. Control* 7(1969), 534-545.
[5] O. Fujiwara and D.B. Khang: 'A two-phase decomposition method for optimal design of looped water distribution networks' *Water Resources Research* 23(1990)977-982.
[6] C. Floudas and A. Aggarwal: 'A decomposition strategy for global optimum search in the pooling problem', *ORSA Journal on Computing*, 2(3), 1990.
[7] Ng.T. Hoai Phuong and H. Tuy: 'A unified monotonic approach to generalized linear fractional programming', *Journal of Global Optimization*, 2002, to appear.
[8] R. Horst and H. Tuy: *Global Optimization – deterministic approaches*, 3rd edition, Springer 1996.
[9] A. Rubinov, *Abstract Convexity and Global Optimization*, Kluwer 2000.
[10] H.D. Sherali and E.P. Smith: 'A global optimization approach to a water distribution network problem' *Journal of Global Optimization* 11(1997), 107-132.
[11] N.Z. Shor and S.I. Stetsenko: *Quadratic extremal problems and nondifferentiable optimization*, Naukova Dumka, Kiev, 1989 (Russian)
[12] N. V. Thoai: 'Duality Bound Method for the General Quadratic Programming Problem with Quadratic Constraints', *Journal of Optimization Theory and Applications* 107 (2000), 331-354.
[13] N.V. Thoai: 'On duality bound method in partly convex programming', *Journal of Global Optimization* 22(2002), 263-270.
[14] N.V. Thoai:'Convergence and Applications of a Decomposition Method Using Duality Bounds for Nonconvex Global Optimization'*Journal of Optimization Theory and Applications* 113(2002), 165-193.
[15] H.D. Tuan, P. Apkarian and Y. Nakashima: 'A new Lagrangian dual global optimization algorithm for solving bilinear matrix inequalities', *Int. J. Robust Nonlinear Control*, (2000); **10**: 561-578.
[16] H. Tuy: 'On Nonconvex Optimization Problems with Separated Nonconvex Variables', *Journal of Global Optimization,* 2(1992), 133-144.
[17] H. Tuy: *Convex Analysis and Global Optimization*, Kluwer 1998.
[18] H. Tuy: 'Normal Sets, Polyblocks, and Monotonic Optimization' *Vietnam Journal of Mathematics*, Springer Verlag, 27:4(1999), 277-300.
[19] H. Tuy: 'Monotonic optimization: Problems and solution approaches', *SIAM Journal on Optimization,* **11**(2000), 464-494.
[20] H. Tuy: 'Counter-Examples to Some Results on D.C. Optimization', Preprint, Institute of Mathematics, Hanoi, Vietnam, 2002. Submitted.
[21] H. Tuy: 'A New General Minimax Theorem', Preprint, Institute of Mathematics, Hanoi, Vietnam, 2003. Submitted.
[22] H. Tuy and L.T. Luc, 'A new approach to optimization under monotonic constraint', *Journal of Global Optimization*, 18(2000), 1-15.

Efficient 1-Bit-Communication Cellular Algorithms

Hiroshi Umeo[1], Koshi Michisaka[2], Naoki Kamikawa[3], and Yuichi Kinugasa[1]

[1] Univ. of Osaka Eletcro-Communication, Japan
umeo@umeolab.osakac.ac.jp
[2] Internet Initiative Japan
koshi@umeolab.osakac.ac.jp
[3] Noristu Koki
naoki@umeolab.osakac.ac.jp

Summary. We propose several efficient algorithms for a large scale of cellular automata having 1-bit inter-cell communications ($CA_{1-\mathrm{bit}}$). A 1-bit inter-cell communication model studied in this paper is a new class of cellular automata (CA) whose inter-cell communication is restricted to 1-bit. We call the model 1-bit CA in short. The number of internal states of the 1-bit CA is assumed to be finite in a usual way. The next state of each cell is determined by the present state of itself and two binary 1-bit inputs from its left and right neighbor cells. Thus the 1-bit CA can be thought to be one of the most powerless and simplest models in a variety of CAs.

We study a sequence generation problem, a firing squad synchronization problem and an early bird problem, all of which are known as the classical and fundamental problems in cellular automata.

First we consider the sequence generation problem. It is shown that there exists a 1-state $CA_{1-\mathrm{bit}}$ that can generate in real-time a context-sensitive sequence such that $\{2^n \,|\, n = 1, 2, 3, ...\}$. Prime sequence can also be generated in real-time by $CA_{1-\mathrm{bit}}$ with 34 states. Secondary, we study the firing squad synchronization problem on two-dimensional $CA_{1-\mathrm{bit}}$. We give a two-dimensional $CA_{1-\mathrm{bit}}$ which can synchronize any $n \times n$ square and $m \times n$ rectangular arrays in $2n - 1$ and $m + n + \max(m, n)$ steps, respectively. In addition, we propose a generalized synchronization algorithm that operates in linear steps on two-dimensional rectangular arrays with the general located at an arbitrary position of the array. The time complexities for the first two algorithms developed are one to two steps larger than optimum ones proposed for O(1)-bit communication model. In the last, we give a 1-bit implementation for an early bird problem. It is shown that there exists a 12-state $CA_{1-\mathrm{bit}}$ that solves the early bird problem in linear time.

1 Introduction

In recent years cellular automata (CA) have been establishing increasing interests in the study of modeling real phenomena occurring in biology, chemistry, ecology, economy, geology, mechanical engineering, medicine, physics, sociology, public traffic, etc. Cellular automata are considered to be a good model of complex systems in which an infinite one-dimensional array of finite state machines (cells) updates itself in synchronous manner according to a uniform local rule.

In this paper, we study a sequence generation problem [1, 4, 7, 19, 20], a firing squad synchronization problem [2, 5, 6, 10-13, 15-17, 21-25] and an early bird problem [3, 8, 9, 14, 23], all of which are known as the classical and fundamental problems studied extensively on O(1)-bit communication models of cellular automata. An O(1)-bit communication model is a conventional CA where the amount of communication bits exchanged at one step between neighboring cells is assumed to be O(1)-bit, however, such bit-information exchanged between inter-cells has been hidden behind the definition of conventional automata-theoretic finite state descriptions. On the other hand, a 1-bit inter-cell communication model studied in this paper is a new CA whose inter-cell communication is restricted to 1-bit. We call the model 1-bit CA in short, and it is denoted as $\mathrm{CA}_{1-\mathrm{bit}}$. The number of internal states of the 1-bit CA is assumed to be finite in a usual way. The next state of each cell is determined by the present state of itself and two binary 1-bit inputs from its left and right neighbor cells. Thus the 1-bit CA can be thought to be one of the most powerless and simplest models in a variety of CAs.

In the next section 2, we define formally a 1-bit communication cellular automaton ($\mathrm{CA}_{1-\mathrm{bit}}$) and gives a computational relation between the conventional CA and the $\mathrm{CA}_{1-\mathrm{bit}}$. In section 3, we consider a sequence generation problem on $\mathrm{CA}_{1-\mathrm{bit}}$ and give several non-regular sequences that can be generated in real-time by $\mathrm{CA}_{1-\mathrm{bit}}$. In section 4, a synchronization problem is studied and three 1-bit implementations of synchronization algorithms for two-dimensional square and rectangular arrays will be given. In the last section, an early bird problem is considered and an efficient 12-state implementation will be given. Due to the space available, we omit the details of the proofs of theorems given below.

2 One-Bit Communication Cellular Automata

A one-dimensional 1-bit inter-cell communication cellular automaton [13, 18-20] consists of an infinite array of identical finite state automata, each located at positive integer point. Each automaton is referred to as a cell. A cell at point i is denoted by C_i where $i \geq 1$. Each C_i, except C_1, is connected with its left and right neighbor cells via a left or right one-way communication link, where those communication links are indicated by right- and left-going arrows, as is shown in Fig. 1, respectively. Each one-way communication link can transmit only one bit at each step in each direction. One distinguished leftmost cell C_1, the communication cell, is connected to outside world. A cellular automaton with 1-bit inter-cell communication (abbreviated by $\mathrm{CA}_{1-\mathrm{bit}}$) consists of an infinite array of finite state automaton $A = (Q, \delta, F)$, where

1. Q is a finite set of internal states.
2. δ is a function, defining the next state of any cell and its binary outputs to its left and right neighbor cells, such that δ: $Q \times \{0,1\} \times \{0,1\} \rightarrow Q \times \{0,1\} \times \{0,1\}$, where $\delta(p, x, y) = (q, x', y')$, p, $q \in Q$, $x, x', y, y' \in \{0,1\}$, has the following meaning: We assume that at step t the cell C_i is in state p and receiving binary inputs x and y from its left and right communication links, respectively. Then, at the next step $t+1$, C_i assumes state q and outputs x' and y' to its left and right communication links, respectively. Note that binary inputs to C_i at step t are also outputs of C_{i-1} and C_{i+1} at step t. A quiescent state $q \in Q$ has a property such that $\delta(q, 0, 0) = (q, 0, 0)$.
3. $F(\subseteq Q)$ is a special subset of Q. The set F is used to specify designated state of C_1 in the definition of sequence generation.

Thus the $\mathrm{CA}_{1-\mathrm{bit}}$ is a special subclass of *normal* (i.e., *conventional*) cellular automata studied so far. Let N be any normal cellular automaton with a set of states Q and a transition function $\delta : Q^3 \rightarrow Q$. The state of each cell on N depends on previous states of itself and its nearest neighbor cells. This means that the total information exchanged per one step between neighboring cells is O(1)-bit. By encoding each state in Q with a binary sequence of length $\lceil \log_2 |Q| \rceil$, sending the sequences sequentially bit by bit in each direction via each one-way communication link, receiving them bit by bit again, and decoding them into their corresponding states in Q, the $\mathrm{CA}_{1-\mathrm{bit}}$ can simulate one step of N in $\lceil \log_2 |Q| \rceil$ steps. This observation gives the following computational relation between the normal CA and $\mathrm{CA}_{1-\mathrm{bit}}$.

Lemma 1. *Let N be any normal cellular automaton with time complexity $T(n)$. Then, there exists a $\mathrm{CA}_{1-\mathrm{bit}}$ which can simulate N in $kT(n)$ steps, where k is a positive constant integer such that $k = \lceil \log_2 |Q| \rceil$ and Q is the set of N's states.*

3 Real-time Sequence Generation Problem on $\mathrm{CA}_{1-\mathrm{bit}}$

Now we define the sequence generation problem on $\mathrm{CA}_{1-\mathrm{bit}}$. Let M be a $\mathrm{CA}_{1-\mathrm{bit}}$ and $\{t_n | n = 1, 2, 3, ...\}$ be an infinite monotonically increasing positive integer sequence defined on natural numbers such that $t_n \geq n$ for any $n \geq 1$. We have a semi-infinite array of cells, shown in Fig. 1, and all cells, except C_1, are in quiescent state and output 0 to their left and right communication links at time $t = 0$. The communication cell C_1 assumes a special state in Q and outputs 1 to its right communication link at time $t = 0$ for an initiation of the sequence generator. We say M generates a sequence $\{t_n | n = 1, 2, 3...\}$ in k *linear-time* if and only if the left end cell of M falls into

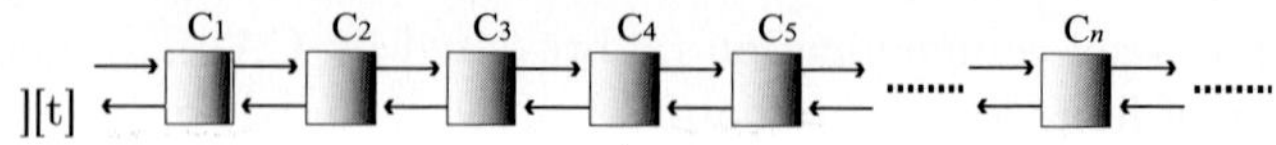

Fig. 1. One-dimensional cellular automaton having 1-bit inter-cell communication links.

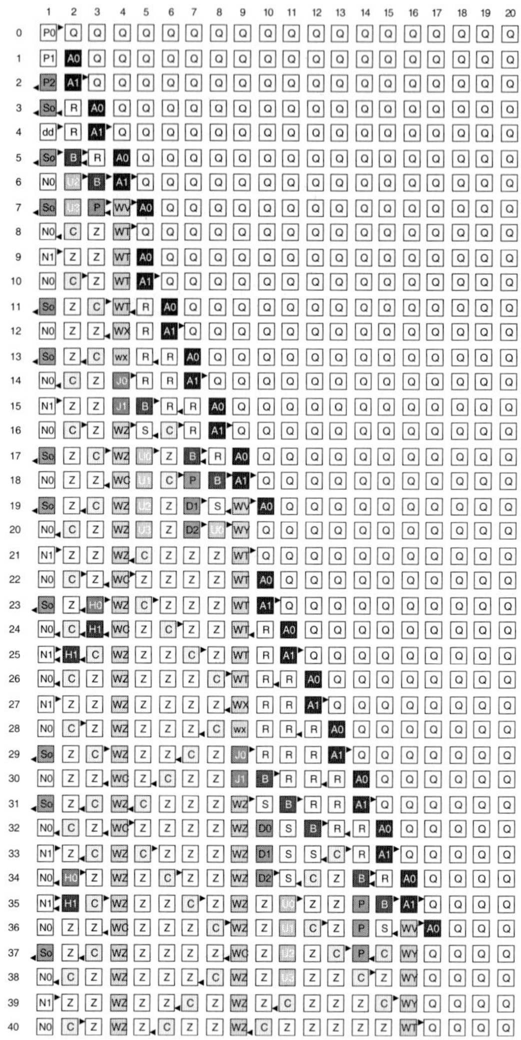

Fig. 2. A configuration of real-time generation of primes on $CA_{1-\text{bit}}$ with 34 states. The colour version of this figure can be found in Fig. A.38 on page 593.

a special state in $F(\subseteq Q)$ and outputs 1 to the outside world via its left communication link at time $t = kt_n$, where k is a fixed positive integer. We call M *real-time* generator, when $k = 1$.

The 1-bit CA can be thought to be one of the most powerless and simplest models in a variety of CAs. In spite of its simplicity, the $\mathrm{CA}_{1-\mathrm{bit}}$ can generate a variety of context-sensitive sequences given below. In Fig. 2, we show some snapshots for real-time prime sequence generation on 34-state $\mathrm{CA}_{1-\mathrm{bit}}$.

Theorem 1. [19] *There exists a 3-state $CA_{1-\mathrm{bit}}$ that can generate $\{n^2 | n = 1, 2, 3, ...\}$ in real-time.*

Theorem 2. [19] *There exists a 9-state $CA_{1-\mathrm{bit}}$ that can generate Fibonacci sequence in real-time.*

Theorem 3. [20] *Prime sequence can be generated in real-time by a 34-state $CA_{1-\mathrm{bit}}$.*

A class of 1-state $\mathrm{CA}_{1-\mathrm{bit}}$ is the simplest one in $\mathrm{CA}_{1-\mathrm{bit}}$. We show that there exists a context-sensitive sequence that can be generated in real-time by a 1-state $\mathrm{CA}_{1-\mathrm{bit}}$. The context-sensitive sequence is as follows:$\{2^n | n = 1, 2, 3, ...\}$. Figure 3 is the transition rule set for the $\mathrm{CA}_{1-\mathrm{bit}}$ that generates $\{2^n | n = 1, 2, 3, ...\}$ in real-time. The leftmost cell C_1 always assumes a state a and $\mathrm{C}_i (i \geq 2)$ takes a state q at any step.

Current state	Input from right link
Input from left link	(next state, left output, right output)

1

a	R=0	R=1
L=0	(a,0,0)	(a,1,0)
L=1	--	--

2

q	R=0	R=1
L=0	(q,0,0)	(q,1,1)
L=1	(q,1,1)	(q,0,0)

Fig. 3. A transition rule for real-time generation of $\{2^n | n = 1, 2, 3, ...\}$ on a 1-state $\mathrm{CA}_{1-\mathrm{bit}}$.

In Fig. 4, we show some snapshots for the real-time generation of the sequence. Small right and left black triangles, ▶ and ◀, shown in the figure, indicate a 1-bit signal transfer in the right or left direction between neighbor cells. A symbol in a cell shows its internal state.

Theorem 4. *An infinite sequence $\{2^n | n = 1, 2, 3, ...\}$ can be generated in real-time by a 1-state $CA_{1-\mathrm{bit}}$.*

4 Firing Squad Synchronization Problem on $\mathrm{CA}_{1-\mathrm{bit}}$

In this section, we study a famous firing squad synchronization problem on the newly introduced 1-bit CA model for which solution gives a finite-state protocol for synchronizing a large scale of cellular automata. The problem was originally proposed by J. Myhill to synchronize all parts of self-reproducing

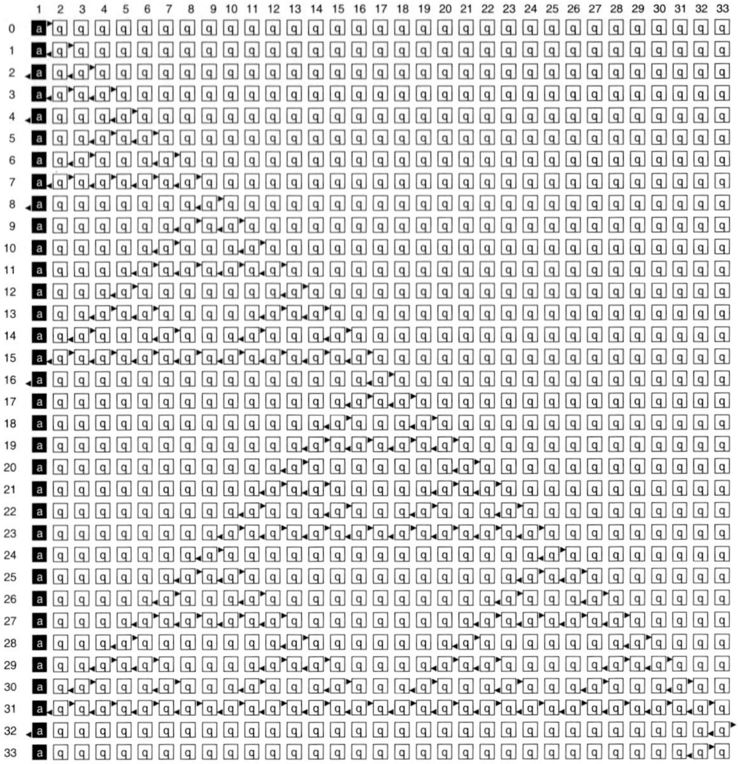

Fig. 4. Snapshots for real-time generation of $\{2^n | n = 1, 2, 3, ...\}$ on a 1-state $CA_{1-\mathrm{bit}}$.

cellular automata [12]. The firing squad synchronization problem has been studied extensively in more than 40 years [2, 5, 6, 10-13, 15-17, 21-25].

We develop some synchronization algorithms for 2-D 1-bit inter-cell communication CA models. Fig. 5 shows a finite two-dimensional cellular array consisting of $m \times n$ cells. A cell on (i, j) is denoted by $C_{i,j}$. Each cell is an identical (except the border cells) finite state automaton. The array operates in lock-step mode in such a way that the next state of each cell (except border cells) is determined by both its own present state and the present binary inputs from its north, south, east and west neighbors. All cells, except the general cell, are initially in the quiescent state with the property that the next state of a quiescent cell with four 0 inputs is the quiescent state again and outputs 0 to its four neighbors.

Given an array of $m \times n$ identical cellular automata, including a *general* on the $C_{1,1}$ cell which is activated at time $t = 0$, we want to give the description (state set and next-state function) of the automata so that, *at some future time*, all the cells will *simultaneously* and, *for the first time*, enter a special *firing* state. The set of states must be independent of m and n. The tricky

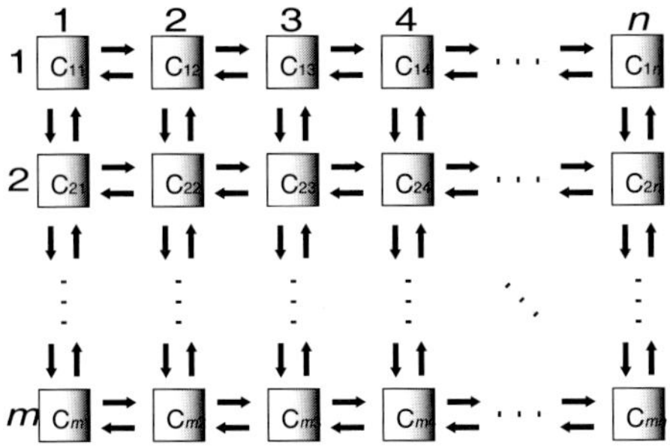

Fig. 5. Two-dimensional cellular automaton.

part of the problem is that the same kind of soldier with a fixed number of states is required to synchronize, regardless of the size m and n of the array.

Several 2-D synchronization algorithms and their implementations have been presented in Shinar [15] and Szwerinski [16] for O(1)-bit communication models. Before presenting our synchronization algorithms on 2-D $\mathrm{CA}_{1-\mathrm{bit}}$, we review two algorithms for synchronizing 1-D $\mathrm{CA}_{1-\mathrm{bit}}$ with the general at the left end or at an arbitrary position of the array. Nishimura, Sogabe and Umeo[13] designed an optimum-step firing squad synchronization algorithm on $\mathrm{CA}_{1-\mathrm{bit}}$, where $2n-2$ steps are required for synchronizing n cells on 1-D array and the general is located at the left end of the array. The algorithm, that is referred to as NSU algorithm, is stated as follows:

Theorem 5. [13] *There exists a $CA_{1-\mathrm{bit}}$ which can synchronize n cells with the general on the left end in $2n-2$ steps. The $CA_{1-\mathrm{bit}}$ constructed has 78 internal states and 208 transition rules.*

Theorem 6 given below is a generalized version of Theorem 5.

Theorem 6. [22] *There exists a $CA_{1-\mathrm{bit}}$ which can synchronize n cells in $n+max(k, n-k+1)$ steps, where k is any integer such that $1 \le k \le n$ and a general is located on the kth cell from the left end of the array. The total number of internal states and transition rules of the $CA_{1-\mathrm{bit}}$ realized on a computer is 282 and 721, respectively.*

4.1 Synchronization Algorithm on Square Arrays

We present a new synchronization algorithm that runs in $(2n-1)$ steps on $n \times n$ square arrays. Our algorithm is one step slower than that of Shinahr [15] for O(1)-bit communication model and operates as follows. By dividing the entire square array into n L-shaped 1-D arrays such that the length of the ith L is $2n-2i+1$ $(1 \le i \le n)$, we treat the square firing as n independent 1-D firings with the general located at the center cell. On the ith L, a general is generated at $C_{i,i}$ at time $t = 2i-1$, and the general initiates the horizontal

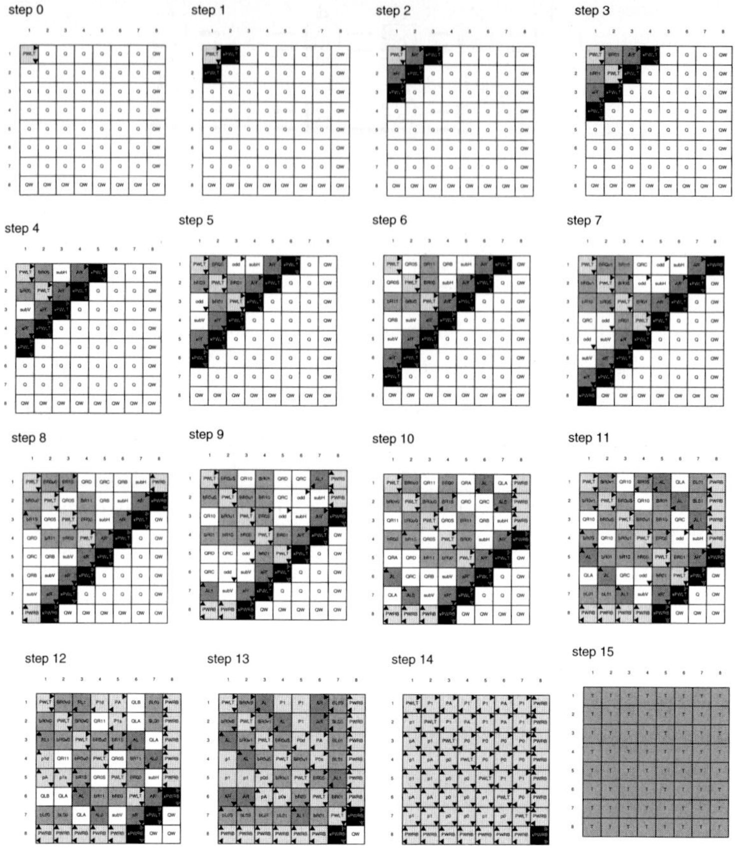

Fig. 6. Snapshots of the $(2n-1)$-step square firing squad synchronization algorithm with the general on the north west corner. The colour version of this figure can be found in Fig. A.39 on page 594.

and vertical firings on the row and column arrays. In our construction, we apply the previous NSU algorithm [13] for each row and column firing. The array fires in optimum time $t = 2i - 1 + 2(n - i + 1) - 2 = 2n - 1$.

We have tested our transition rule set on squares of size 2×2 to 1000×1000. The total number of internal states and transition rules of the $\mathrm{CA}_{1-\mathrm{bit}}$ realized on a computer is 127 and 405, respectively. Figure 6 shows snapshots of configurations of our 127-state synchronization algorithm running on a square of size 8×8. Thus we have:

Theorem 7. *There exists a 2-D $\mathrm{CA}_{1-\mathrm{bit}}$ which can synchronize $n \times n$ cells in $2n - 1$ steps.*

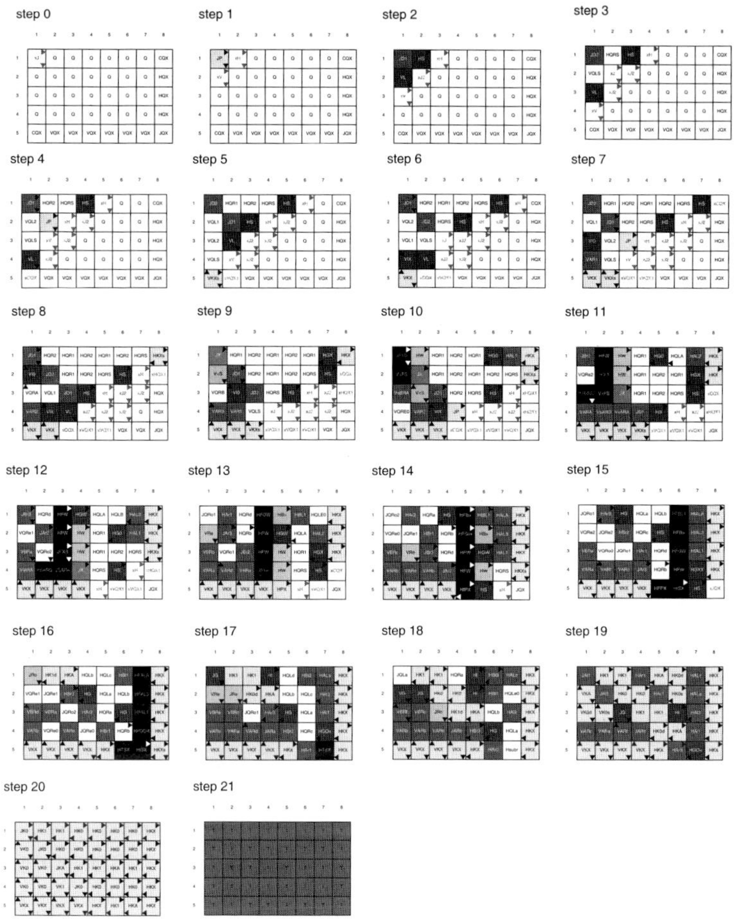

Fig. 7. Snapshots of our rectangular firing squad synchronization algorithm with the general at the north-west corner. The colour version of this figure can be found in Fig. A.40 on page 595.

4.2 Synchronization Algorithm on Rectangular Arrays

The generalized firing squad synchronization algorithm presented in [Theorem 6] can be applied to the problem of synchronizing rectangular arrays with the general at the north-west corner. The configuration of the generalized firing on 1-D arrays can be mapped on 2-D array.

The rectangular array is regarded as $\min(m, n)$ L-shaped 1-D arrays, where they are synchronized independently using the generalized firing squad synchronization algorithm. We have implemented the algorithm on a computer. In Fig. 7, we show snapshots of the synchronization process on 5×8 rectangular array. The total number of internal states and transition rules of the $\mathrm{CA}_{1-\mathrm{bit}}$ realized on a computer are 862 and 2217, respectively. Thus we have:

Theorem 8. *There exists a 2-D $CA_{1-\mathrm{bit}}$ which can synchronize $m \times n$ rectangular arrays in $m+n+$ $max(m,n)$ steps.*

4.3 Generalized Synchronization Algorithm on 2-D Rectangular Arrays

In this subsection, we study the generalized synchronization algorithm on rectangular arrays. Let r, s be any integer such that $1 \le r \le m$, $1 \le s \le n$. At time $t = 0$ the general cell $\mathrm{C}_{r,s}$ is in *fire-when-ready* state that is an initiation signal to the array. Before presenting the 1-bit algorithm, we show a simple and efficient mapping scheme developed for O(1)-bit CA model that embeds any generalized one-dimensional synchronization algorithms onto two-dimensional arrays [21].

Now we consider a 2-D array of size $m \times n$. We divide mn cells into $m+n-1$ groups g_k, $1 \le k \le m+n-1$, defined as follows;

$$g_k = \{C_{i,j} | (i-1)+(j-1) = k-1\}.$$

That is

$$g_1 = \{C_{1,1}\}, g_2 = \{C_{1,2}, C_{2,1}\}, g_3 = \{C_{1,3}, C_{2,2}, C_{3,1}\}, \ldots, g_{m+n-1} = \{C_{m,n}\}.$$

Let M be any one-dimensional $\mathrm{CA}_{1-\mathrm{bit}}$ that fires ℓ cells in $T(\ell, k)$ steps, where the general is on C_k and k be any integer such that $1 \le k \le \ell$. We assume that M has $m+n-1$ cells. We consider the one-to-one correspondence between the ith group g_i and the ith cell C_i on M such that $g_i \leftrightarrow C_i$, where $1 \le i \le m+n-1$. We can construct a 2-D $\mathrm{CA}_{1-\mathrm{bit}}$ N so that all cells in g_i simulates the ith cell C_i in real-time and N can fire any $m \times n$ arrays with the general $\mathrm{C}_{r,s}$ at time $t = T(m+n-1, r+s-1)$ if and only if M fires any 1-D arrays of length $m+n-1$ with the general on C_{r+s-1} at time $t = T(m+n-1, r+s-1)$.

Based on the generalized 1-D algorithm given in [Theorem 6], we get the following 2-D generalized synchronization algorithm that fires in $T(m,n,r,s)$ steps given below. The total number of internal states and transition rules of the $\mathrm{CA}_{1-\mathrm{bit}}$ realized on a computer is 300 and 2333, respectively. In Fig. 8 we show snapshots of the 300-state generalized synchronization algorithm running on rectangular array of size 5×8 with the general on $\mathrm{C}_{3,4}$. Thus we have:

Theorem 9. *There exists a 2-D 1-bit communication $CA_{1-\mathrm{bit}}$ that can synchronize any $m \times n$ rectangular arrays in $T(m,n,r,s)$ steps, where (r,s) is an arbitrary initial position of the general and $T(m,n,r,s)$ is defined as follows: $T(m,n,r,s) = m+n-2+\max(r+s, m+n-r-s+2) \pm O(1)$.*

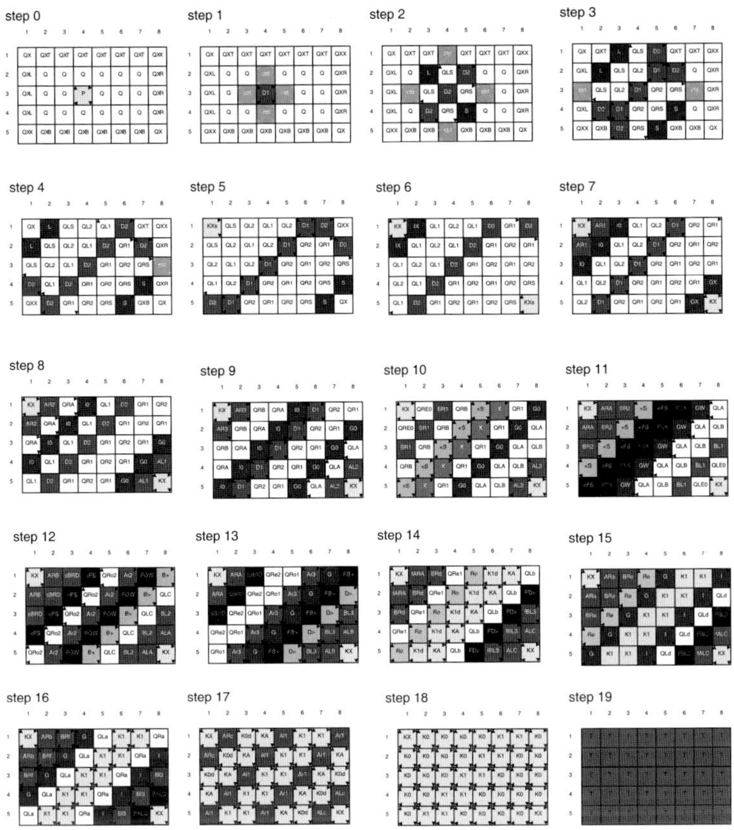

Fig. 8. Snapshots of our generalized rectangular firing squad synchronization algorithm operating on an array of size 5×8 with the general on $C_{3,4}$. The colour version of this figure can be found in Fig. A.41 on page 596.

Szwerinski [16] proposed an optimum-time generalized 2-D firing algorithm with 25600 internal states that fires any $m \times n$ array in $m + n + \max(m, n) - \min(r, m - r + 1) - \min(s, n - s + 1) - 1$ steps. Our 2-D generalized synchronization algorithm is relatively larger than the optimum one proposed by Szwerinski [16], however, the number of internal states required for the firing is the smallest known at present.

5 Early Bird Problem on $CA_{1-\text{bit}}$

In this section we study an early bird problem on $CA_{1-\text{bit}}$. Consider a one-dimensional CA consisting of n cells in which any cell initially in quiescent state may be excited from the outside world. The problem is to give a description (state set and next state function) of the automata so that the first excitation(s) can be distinguished from the later ones. The problem was originally devised by Rosenstiehl, Fiksel and Holliger [14] to design some graph-theoretic algorithms operating on networks of finite state automata. Büning [3] showed that a 5-state solution developed by Legendi and Katona [8] is an optimum one in the number of states on O(1)-bit communication model. We have got a 12-state implementation on $CA_{1-\mathrm{bit}}$ that operates in $3n+O(1)$ steps for one-dimensional $CA_{1-\mathrm{bit}}$ of size n. In Fig. 9, we show some snapshots of the 12-state implementation.

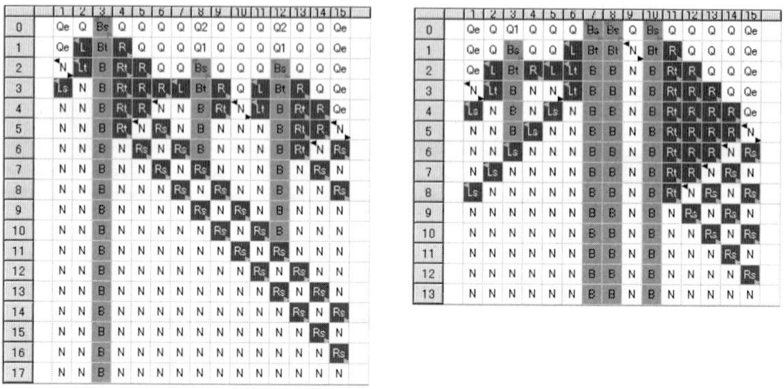

Fig. 9. Snapshots of a 12-state implementation of the early bird problem on $CA_{1-\mathrm{bit}}$. The colour version of this figure can be found in Fig. A.42 on page 596.

Theorem 10. *Ther exists a 12-state $CA_{1-\mathrm{bit}}$ that can solve the early bird problem in $3n+O(1)$ steps.*

6 Conclusion

A sequence generation problem, a firing squad synchronization problem and an early bird problem are known as the classical and fundamental problems which have been studied extensively on O(1)-bit communication models of cellular automata. In this paper, we have developed several optimum algorithms for those problems and given their efficient implementations on $CA_{1-\mathrm{bit}}$. It has been shown that there exists a context-sensitive sequence that can be generated in real-time by a 1-state $CA_{1-\mathrm{bit}}$. We have proposed several new synchronization algorithms for two-dimensional $CA_{1-\mathrm{bit}}$and implemented them on a computer. Most of the algorithms proposed are one to four steps larger than optimum ones proposed for O(1)-bit communication model. We are convinced that there still exist interesting new synchronization algorithms, although more than 40 years have passed since the development of the problem. A 12-state implementation on $CA_{1-\mathrm{bit}}$ is also given for the early bird problem.

References

[1] M. Arisawa: On the generation of integer series by the one-dimensional iterative arrays of finite state machines (in Japanese). *The Trans. of IECE* 71/8, Vol. 54-C, No.8, pp. 759-766, (1971).

[2] R. Balzer: An 8-state minimal time solution to the firing squad synchronization problem. *Information and Control*, vol. 10(1967), pp. 22-42.

[3] H. K. Büning: The early bird problem is unsolvable in a one-dimensional cellular space with 4 states. *Acta Cybernetica*, vol. 6(1983), pp.23-31.

[4] P. C. Fischer: Generation of primes by a one-dimensional real-time iterative array. *J. of ACM*, Vol., 12, No.3, pp. 388-394, (1965).

[5] E. Goto: A minimal time solution of the firing squad problem. Dittoed course notes for Applied Mathematics 298, Harvard University, (1962), pp. 52-59.

[6] M. Hisaoka, H. Yamada, M. Maeda, T. Worsch, and H. Umeo: A design of firing squad synchronization algorithms for a multi-general problem and their implementations (in Japanese). *Technical Report of IEICE NLP2002-133*(2003), 103-108.

[7] I. Korec: Real-time generation of primes by a one-dimensional cellular automaton with 11 states. *Proc. of 22nd Intern. Symp. on MFCS '97, Lecture Notes in Computer Science*, 1295, pp. 358-367, (1997).

[8] T. Legendi and E. Katona: A 5-state solution of the early bird problem in a one-dimensional cellular space. *Acta Cybernetica*, Vol.5, No.2, pp. 173-179, (1981).

[9] T. Legendi and E. Katona: A solution of the early bird problem in an *n*-dimensional cellular space. *Acta Cybernetica*, Vol.7, No.1, pp. 81-87, (1984).

[10] J. Mazoyer: A six-state minimal time solution to the firing squad synchronization problem. *Theoretical Computer Science*, vol. 50(1987), pp. 183-238.

[11] J. Mazoyer: On optimal solutions to the firing squad synchronization problem. *Theoretical Computer Science*, vol. 168(1996), pp. 367-404.

[12] E. F. Moore: The firing squad synchronization problem. in *Sequential Machines, Selected Papers* (E. F. Moore ed.), Addison-Wesley, Reading MA., (1964), pp. 213-214.

[13] J. Nishimura, T. Sogabe and H. Umeo: A Design of Optimum-Time Firing Squad Synchronization Algorithm on 1-Bit Cellular Automaton. *Proc. of The 8th International Symposium on Artificial Life and Robotics*, pp. 381-386, (2003).

[14] P. Rosenstiehl, J. R. Fiksel, and A. Holliger: Intelligent graphs: Networks of finite automata capable of solving graph problems. in *Graph Theory and Computing* (R. C. Read ed.), (1972), Academic Press, New York, pp.219-265.

[15] I. Shinahr: Two- and three-dimensional firing squad synchronization problems. *Information and Control*, vol. 24(1974), pp. 163-180.

[16] H. Szwerinski: Time-optimum solution of the firing-squadsynchronization-problem for n-dimensional rectangles with the general at an arbitrary position. *Theoretical Computer Science*, vol. 19(1982), pp. 305-320.

[17] S. L. Torre, M. Napoli and M. Parente: A compositional approach to synchronize two dimensional networks of processors. *Theoretical Informatics and Applications*, 34 (2000) pp. 549-564.

[18] H. Umeo: Linear-time recognition of connectivity of binary images on 1-bit inter-cell communication cellular automaton. *Parallel Computing*, 27, pp. 587-599, (2001).

[19] H. Umeo and N. Kamikawa: A design of real-time non-regular sequence generation algorithms and their implementations on cellular automata with 1-bit inter-cell communications. *Fundamenta Informaticae*, 52 (2002) 255-275.

[20] H. Umeo and N. Kamikawa: An infinite prime sequence can be generated in real-time by 1-bit inter-cell communication cellular automaton. *Preproc. of The 6th International Conference on Developments in Language Theory*, Univ. of Kyoto Sangyo,(2002), pp.372-382.

[21] H. Umeo, M. Maeda and N. Fujiwara: An efficient mapping scheme for embedding any one-dimensional firing squad synchronization algorithm onto two-dimensional arrays. *Proc. of the 5th International Conference on Cellular Automata for Research and Industry*, LNCS 2493, Springer-Verlag, pp.69-81(2000).

[22] H. Umeo, M. Hisaoka, K. Michisaka, K. Nishioka and Masashi Maeda: Some new generalized synchronization algorithms and their implementations for large scale cellular automata. *Proc. of The Third International Conference on Unconventional Models of Computation*(C. S. Calude, M. J. Dinneen and F. Pepper Eds.), 2002, pp.276-286.

[23] R. Vollmar: On two modified problems of synchronization in cellular automata. *Acta Cybernetica*, Vol.3, No.4, pp. 293-300, (1978).

[24] R. Vollmar: Algorithmen in Zellelarautomaten. Teubner, p. 192, Stuttgart, (1979).

[25] A. Waksman: An optimum solution to the firing squad synchronization problem. *Information and Control*, vol. 9(1966), pp. 66-78.

Adaptive Finite Elements for Output-Oriented Model Calibration

Boris Vexler

Institut für Angewandte Mathematik, Universität Heidelberg
Im Neuenheimer Feld 294/293, D-69120 Heidelberg, Germany
boris.vexler@iwr.uni-heidelberg.de

Summary. We consider the problem of model calibration involving partial differential equations. The problem is formulated as a parameter identification problem with finite number of unknown parameters, which can occur in the differential operator as well as in the boundary conditions.

The finite element discretization on locally refined meshes is adaptively chosen with regard to a goal functional, which corresponds to a quantity of interest (output) of the state equation. An a posteriori error estimator for the error in the goal functional is derived. It is used for the quantitative error control and successive improvement of the accuracy by appropriate mesh refinement.

Numerical examples illustrate the behavior of the method.

1 Introduction

We consider the problem of model calibration involving partial differential equations of the following form: We are given a mathematical model of a physical process, described by a system of partial differential equations (*state equation*). This model depends on a finite number of unknown (or just imprecisely known) model parameters, which can not be measured directly. We are interested in the value of some quantity of the process (*quantity of interest* or *output*), which can not be measured directly either (or whose measurement would require too much effort). In the simplest case this quantity corresponds to the value of one of the model parameters, but in general, the quantity of interest depends on the *state variable* of the process as well as on the parameters. Further, we are given a set of measurements and our aim is to compute this quantity for the calibrated model with parameters, which fit the measurements.

One possible approach to this kind of problems is to split this problem in a pure parameter identification problem and an output-oriented simulation of the process. By this approach, first the unknown parameters are estimated from the given measurements using some parameter identification techniques, and then the quantity of interest is computed by an appropriate simulation. But this approach has some drawbacks: First, the appropriate discretizations for the parameter identification problem and for the output-oriented simulation can be completely different; and second, the required accuracy of the parameters for achieving a given tolerance for the error in the quantity of interest is a priori unknown due to the (in general unknown) derivatives of the quantity of interest with respect to the parameters.

We suggest another approach for more effective solution of this kind of problems. We consider this problem as a parameter identification problem and chose the finite element discretizations adaptively according to the quantity of interest. To this end, we derive an a posteriori error estimator, which aims at controlling the error in the quantity of interest. This error estimator guides an adaptive mesh refinement algorithm with purpose to improve the accuracy of the computed output in an efficient way.

Our a posteriori error estimator is based on the a posteriori error estimation for parameter identification problems derived in Becker & Vexler [5] and on the optimal control approach to error estimation developed in Becker & Rannacher [2, 3]. The idea of our approach is to combine the error estimator for the error in parameter (see Becker & Vexler [5]) and the output-oriented error estimator for solution of uncontrolled equations (see, e.g., Becker & Rannacher [2]). This allows to estimate the error directly in the quantity of interest.

Our approach can be successfully used for very different types of parameters and goal functionals. The parameters can occur in the state equation as well as in the boundary conditions. This is important for some application, where the precise boundary conditions are unknown.

The problem of output-oriented model calibration described above can be formulated as follows: We are given a partial differential equation (*state equation*) with some unknown parameters in a weak form:

$$a(u,q)(\varphi) = f(\varphi) \quad \forall \varphi \in V. \tag{1}$$

Here, $q \in Q = \mathbb{R}^{n_p}$ denotes the unknown parameters, $u \in g(q) + V$ is the *state variable*, where $g : Q \to \hat{V}$ describes the the Dirichlet boundary conditions, and V and $\hat{V}$ denote appropriate Hilbert spaces with $V \subset \hat{V}$. The semi-linear form $a(\cdot,\cdot)(\cdot)$ is defined on the Hilbert space $\hat{V} \times Q \times V$. Semi-linear forms are written with two parentheses, the first one refers to the nonlinear arguments, whereas the second one embraces all linear arguments. The partial derivatives of the semi-linear form $a(\cdot,\cdot)(\cdot)$ are denoted by $a'_u(\cdot,\cdot)(\cdot,\cdot)$, $a'_q(\cdot,\cdot)(\cdot,\cdot)$ etc.

Further, we are given an observation operator $C : \hat{V} \to Z$, which maps the *state variable* u to the space of measurements $Z = \mathbb{R}^{n_m}$, assuming $n_m \geq n_p$. We denote by $\langle \cdot,\cdot \rangle_Z$ the scalar product of Z and by $\|\cdot\|_Z$ the corresponding norm. Similar notation are used for the scalar product and norm in the space Q.

The model parameters are calibrated from a given set of measurements $\bar{C} \in Z$ using a least squares approach. In this way we obtain the constrained optimization problem:

$$\text{Minimize} \quad \frac{1}{2}\left\|C(u)-\bar{C}\right\|_Z^2 \tag{2}$$

under the constraint (1). The cost functional (2) here is the squared norm of the so called *model residual* defined by

$$R_m(u) := \bar{C} - C(u). \tag{3}$$

The state equation is discretized by conforming finite elements on a regular mesh $\mathcal{T}_h$, resulting in finite element spaces $V_h \subset V$ and $\hat{V}_h \subset \hat{V}$ with $V_h \subset \hat{V}_h$ (for precise definitions see Section 2). The corresponding discrete state $u_h \in g_h(q_h) + V_h$ and parameter $q_h \in Q$ are determined by:

$$\text{Minimize} \quad \frac{1}{2}\left\|C(u_h)-\bar{C}\right\|_Z^2 \tag{4}$$

under the constraint

$$a(u_h, q_h)(\varphi_h) = f(\varphi_h) \quad \forall \varphi_h \in V_h, \tag{5}$$

where $g_h : Q \to \hat{V}_h$ is an approximation of the operator g, for example $g_h = i_h \circ g$ with an appropriate interpolation operator $i_h : \hat{V} \to \hat{V}_h$ (see, e.g., Clement [8]).

The quantity of interest is given by a goal functional $E : \hat{V} \times Q \to \mathbb{R}$ and we are interested in its value $E(u, q)$ in the optimum. We prove the following error representation:

$$E(u, q) - E(u_h, q_h) = \eta_h + R, \tag{6}$$

where η_h denotes the a posteriori error estimator and R is a remainder term, which may usually be neglected; see the discussion in Section 4.

In order to illustrate the typical use of the error estimator η_h, we sketch a generic adaptive mesh refinement algorithm. Such an algorithm generates a sequence of locally refined meshes and corresponding finite element spaces until the estimated error with respect to E is below a given tolerance TOL. For the following iteration, we have a mesh refinement procedure that adaptively refines a given regular mesh to obtain a new regular mesh for the next iteration. The refinement procedure is guided by information based on the cell-wise contributions of the estimator η_h.

Remark 1. In step 3, the least squares problem is solved on a fixed mesh. As initial data, we use the values from the computation on the previous mesh. This allows us to avoid unnecessary iterations of the optimization loop on fine meshes.

The outline of this article is as follows: In the next section we reformulate the problem under consideration as an unconstrained optimization problem and

Adaptive Mesh Refinement Algorithm

1. Choose an initial mesh $\mathcal{T}_{h_0}$ and set $k = 0$
2. Construct the finite element space V_{h_k}
3. Compute $u_{h_k} \in V_{h_k}, q_{h_k} \in Q$ by solving (4,5)
4. Evaluate the a posteriori error estimator η_{h_k}
5. If $\eta_{h_k} \le TOL$ quit
6. Refine $\mathcal{T}_{h_k} \to \mathcal{T}_{h_{k+1}}$ using information from η_{h_k}
7. Increment k and go to 2.

make some assumptions on the problem (1,2), which we need throughout the paper. In Section 3 we define the finite element discretization of the parameter identification problem on a locally refined mesh and describe a typical optimization loop, which acts on a fixed mesh. In Section 4 we derive an a posteriori error estimator for the error in the goal functional $E(u, q)$ and prove the corresponding error representation. In the last section we demonstrate the behavior of the method for some numerical examples.

2 Unconstrained formulation

In this section we make some assumptions on the problem (1,2), which we need throughout the paper. Furthermore we introduce a reduced observation operator in order to reformulate the problem under consideration as an unconstrained optimization problem and express its derivative with the help of the solutions of some tangent problems.

Throughout this paper, we assume that the parameter identification problem described so far admits a (locally) unique solution. Moreover we assume the semi-linear form $a(\cdot,\cdot)(\cdot)$, the boundary condition operator $g(\cdot)$ and the observation operator $C(\cdot)$ to be three times continuously differentiable and make the following assumption on the derivative $a'_u(\cdot,\cdot)(\cdot,\cdot)$:

Assumption 1. *In a neighborhood $B(u,q) \subset \hat{V} \times Q$ of the solution (u,q) to problem (1,2) the derivative $a'_u(\cdot,\cdot)(\cdot,\cdot)$ is coercive, i.e. there exists a constant $\gamma > 0$ with*

$$a'_u(v,p)(w,w) \ge \gamma \, \|w\|_V^2 \qquad \forall (v,p) \in B(u,q), \quad \forall w \in V. \tag{7}$$

Due to the implicit function theorem in Banach spaces (see, e.g., Dieudonné [11]) this assumption implies the existence of a continuously differentiable solution operator S for the state equation in a neighborhood $Q_0 \subset Q$ of the solution to the problem (1,2). For all $q \in Q_0$ we have: $S(q) \in g(q) + V$ and

$$a(S(q), q)(\varphi) = f(\varphi) \quad \forall \varphi \in V. \tag{8}$$

Using this solution operator S we define the reduced observation operator $c : Q_0 \to Z$ by:

$$c(q) := C(S(q)) \tag{9}$$

in order to reformulate the problem under consideration as an unconstrained optimization problem:

$$\text{Minimize} \quad \frac{1}{2} \left\| c(q) - \bar{C} \right\|_Z^2 . \tag{10}$$

Denoting by $J = c'(q)$ the Jacobian of the reduced observation operator c, the first-order necessary condition for (10) reads:

$$J^* c(q) = J^* \bar{C}. \tag{11}$$

In the following proposition we compute the Jacobian J.

Proposition 1. *Let the reduced observation operator c be defined as in (9). Then its partial derivatives can be computed as follows:*

$$\frac{\partial c_i}{\partial q_j}(q) = J_{ij} = C_i'(u)(w_j), \quad i = 1 \ldots n_m,\ j = 1 \ldots n_p, \tag{12}$$

with $u = S(q)$, C_i and c_i denote the components of the observation and the reduced observation operators respectively; J_{ij} denotes the entries of the Jacobian matrix $J = c'(q)$ and $w_j \in g'_{q_j}(q) + V$ is the solution to the following tangent (sensitivity) problem:

$$a_u'(u, q)(w_j, \varphi) = -a_{q_j}'(u, q)(1, \varphi) \quad \forall \varphi \in V. \tag{13}$$

Proof. The proof is given by using the implicit function theorem and the chain rule.

We will throughout suppose that the problem is non-degenerate in the following sense:

Assumption 2. *The Jacobian matrix J of the reduced observation operator c has full rank n_p in a neighborhood of the solution to problem (1,2).*

3 Discretization and optimization loop

We consider two- and three dimensional meshes consisting of *cells* K which are either triangles, tetrahedra, quadrilaterals, or hexahedra and constitute a non-overlapping covering of the computational domain:

$$\Omega = \bigcup K.$$

Note that this implies that the boundary $\partial\Omega$ of the domain is polygonal. The general case requires the treatment of cells with curved boundaries and is neglected here.

The corresponding mesh is denoted by $\mathcal{T}_h$, where the mesh parameter h is defined as a cell-wise constant function by setting $h|_K = h_K$ and h_K is the diameter of K. The straight parts which make up the boundary ∂K of a cell K are called *faces*.

A mesh $\mathcal{T}_h$ is called regular, if it fulfills the standard conditions for shape-regular finite element mesh, see e.g. Ciarlet [7]. However, the cells are allowed to have nodes, which lie on midpoints of faces of neighboring cells. But at most one such *hanging node* is permitted for each face. On a regular mesh, we construct continuous finite element spaces V_h in the standard way, see e.g. Ciarlet [7]. Only the case of hanging nodes requires some additional remark. There are no degrees of freedom corresponding to these irregular nodes and the value of the finite element function is determined by point-wise interpolation. This implies continuity and therefore global conformity. For implementation details see e.g. Carey & Oden [6].

For a given finite element space V_h the corresponding discrete solution $(u_h, q_h) \in V_h \times Q$ is determined by the constrained least squares problem (4,5). We assume the discrete analog of Assumption 1, in order to guarantee the existence of a continuously differentiable discrete solution operator S_h in a neighborhood $Q_{0,h} \subset Q$ of the solution to the discrete problem, i.e, there holds for all $q \in Q_{0,h}$: $S_h(q) \in g_h(q) + V_h$ and

$$a(S_h(q), q)(\varphi_h) = f(\varphi_h) \quad \forall \varphi \in V_h. \tag{14}$$

As before, we turn the discrete problem (4,5) into an unconstrained minimization problem:

$$\text{Minimize} \quad \frac{1}{2} \left\| c_h(q_h) - \bar{C} \right\|_Z^2. \tag{15}$$

We denote by $J_h = c_h'(q_h)$ the Jacobian of the discrete reduced function and assume again that it has full rank. The first-order necessary condition for (15) reads:

$$J_h^* c_h(q_h) = J_h^* \bar{C}. \tag{16}$$

Therefore, the discrete solution (u_h, q_h) is determined by the system of equations constituted by the state equation (5) and the optimality condition (16).

Remark 2. The Jacobian J_h of the discrete observation operator c_h can be computed in the same way as in Proposition 1.

The iterative solution of (16) on a fixed mesh $\mathcal{T}_h$ is organized as follows: Let q_h^0 be an initial guess (which will be the solution on the previous mesh in the adaptive algorithm). Then we iterate

$$q_h^{k+1} = q_h^k + \delta q_h, \tag{17}$$

where δq_h is the solution of the linear problem

$$(J_h^* J_h)\delta q_h = J_h^*(\bar{C} - c_h(q_h^k)). \tag{18}$$

This is the Gauß-Newton algorithm which can be interpreted as the solution to the linearized minimization problem

$$\text{Minimize} \quad \frac{1}{2}\left\|c_h(q_h^k) + J_h \delta q_h - \bar{C}\right\|^2. \tag{19}$$

Remark 3. In our practical realization, we use trust-region techniques in order to improve global convergence, see e.g. [9, 10, 12].

4 A posteriori error estimation

In this section we derive an a posteriori error estimator for the error in the quantity of interest, i.e. in the goal-functional $E(u, q)$. We show, that this error can be written as:

$$E(u, q) - E(u_h, q_h) = \eta_h + R, \tag{20}$$

where η_h denotes the a posteriori error estimator and R is a remainder term, which my be usually neglected. This error estimator is used in the adaptive mesh refinement algorithm, described in Section 1.

In order to derive this error estimator we rewrite the error in the following way:

$$E(u, q) - E(u_h, q_h) = E(u, q) - E(S(q_h), q_h) + E(S(q_h), q_h) - E(u_h, q_h). \tag{21}$$

We denote by $\tilde{u} = S(q_h) \in g(q_h) + V$ the solution of the state equation for the parameter q_h, i.e.

$$a(\tilde{u}, q_h)(\varphi) = f(\varphi) \quad \forall \varphi \in V. \tag{22}$$

Moreover, we introduce two error functional, $E^{(1)} : Q \to \mathbb{R}$ and $E^{(2)} : \hat{V} \to \mathbb{R}$ by

$$E^{(1)}(r) = E(S(r), r) \tag{23}$$

and

$$E^{(2)}(v) = E(v, q_h), \tag{24}$$

and obtain the following error representation:

$$E(u,q) - E(u_h, q_h) = E^{(1)}(q) - E^{(1)}(q_h) + E^{(2)}(\widetilde{u}) - E^{(2)}(u_h). \tag{25}$$

This allows us to split the error in two parts: the error with respect to the parameters characterized by $E^{(1)}$ and the error in the state characterized by $E^{(2)}$.

The next step is to derive separately a posteriori error estimations for the error in both functionals. For the error in the first functional we can use the result from Becker & Vexler [5]:

Theorem 1.

$$E^{(1)}(q) - E^{(1)}(q_h) = \frac{1}{2}\rho(u_h)(y - i_h y) + \frac{1}{2}\rho^*(u_h, y_h)(u - i_h u) + P + R_1, \tag{26}$$

where $y \in V$ is the solution of the adjoint problem:

$$a'_u(u,q)(\varphi, y) = -\langle J(J^*J)^{-1}\nabla E^{(1)}(q), C'(u)(\varphi)\rangle \quad \forall \varphi \in V \tag{27}$$

and $\rho(\cdot)(\cdot)$ and $\rho^*(\cdot)(\cdot)$ are the residuals of the state and adjoint equation defined by:

$$\begin{aligned} \rho(u_h)(\varphi) &:= f(\varphi) - a(u_h, q_h)(\varphi) \\ \rho^*(u_h, y_h)(\varphi) &:= -\langle J_h(J_h^* J_h)^{-1}\nabla E^{(1)}(q_h), C'(u_h)(\varphi)\rangle - a'_u(u_h, q_h)(\varphi, y_h). \end{aligned} \tag{28}$$

The remainder term R_1 is quadratic in the error and the additional remainder term P admits the estimate:

$$|P| \leq \widetilde{C}\left(\|e_u\|_V + \|e_q\|_Q + \|\delta_h v\|_V + \|\delta_h \bar{z}\|_V\right)\|R_m(u)\|_Z, \tag{29}$$

where $e_u := u - u_h$, $e_q := q - q_h$, $\delta_h\varphi := \varphi - i_h\varphi$ is an interpolation error operator on $\hat{V}$ and $R_m(u)$ is the model residual defined in (3). The mean tangent solution $v \in \hat{V}$ is given by

$$v = -\sum_{j=1}^{n_p}\left((J^*J)^{-1}\nabla E^{(1)}(q)\right)_j w_j \tag{30}$$

and the normalized adjoint solution $\bar{z} \in V$ is determined by:

$$a'_u(u,q)(\varphi, \bar{z}) = \langle -\frac{R_m(u)}{\|R_m(u)\|}, C'(u)(\varphi)\rangle_Z \quad \forall \varphi \in V, \tag{31}$$

if the model residual $R_m(u)$ does not vanish; otherwise we set $\bar{z} = 0$. The constant $\widetilde{C}$ does not depend on the mesh parameter h nor on the measurements $\bar{C}$.

Remark 4. For this error representation we need information about $\nabla E^{(1)}(q)$. It can be obtained analogously to Proposition 1 in the following way:

$$\frac{\partial}{\partial q_j} E^{(1)}(q) = E'_u(u,q)(w_j) + E'_{q_j}(u,q)(1), \tag{32}$$

where w_j is defined in (13). The term $\nabla E^{(1)}(q_h)$ is obtained analogously.

For the second part of (25) we use the error representation for uncontrolled equations from Becker & Rannacher [3]:

Theorem 2.

$$E^{(2)}(\widetilde{u}) - E^{(2)}(u_h) = \frac{1}{2}\rho(u_h)(\widetilde{y} - i_h\widetilde{y}) + \frac{1}{2}\widetilde{\rho}^*(u_h, y_h)(\widetilde{u} - i_h\widetilde{u}) + \widetilde{R}, \tag{33}$$

where $\widetilde{y} \in V$ is the solution of the adjoint problem:

$$a'_u(\widetilde{u}, q_h)(\varphi, \widetilde{y}) = E'_u(\widetilde{u}, q_h)(\varphi) \quad \forall \varphi \in V, \tag{34}$$

$\rho(\cdot)(\cdot)$ and $\widetilde{\rho}^*(\cdot)(\cdot)$ are the residuals of the state and adjoint equation defined by:

$$\begin{aligned} \rho(u_h)(\varphi) &:= f(\varphi) - a(u_h, q_h)(\varphi) \\ \widetilde{\rho}^*(u_h, \widetilde{y}_h)(\varphi) &:= E'_u(u_h, q_h)(\varphi) - a'_u(u_h, q_h)(\varphi, \widetilde{y}_h). \end{aligned} \tag{35}$$

The remainder term $\widetilde{R}$ is cubic in the error.

For evaluation of error estimators the local (interpolation) errors $y - i_h y$, $u - i_h u$, $\widetilde{y} - i_h\widetilde{y}$ and $\widetilde{u} - i_h\widetilde{u}$ can be approximated by using second order difference quotients or higher order reconstruction. In our numerical examples, we use interpolation of the computed bilinear finite element solutions on the space of biquadratic finite elements on patches of cells. For analysis of different procedures in this direction see Becker & Rannacher [2].

We note, that by all these procedures the error $u - i_h u$ is approximated only by using information obtained by u_h. For this reason we assume (for evaluation of error estimators):

$$\delta u := u - i_h u \approx \widetilde{u} - i_h\widetilde{u}. \tag{36}$$

We use this assumption in order to avoid the solution of two different adjoint problems (27) and (34). Instead of it, we introduce another adjoint problem, whose right hand side is given by the sum of the right hand sides of the adjoint problems (27) and (34). This new adjoint solution $\overline{y} \in V$ is determined by

$$a'_u(u,q)(\varphi, \overline{y}) = -\langle J(J^*J)^{-1}\nabla E^{(1)}(q), C'(u)(\varphi)\rangle + E'_u(\widetilde{u}, q_h)(\varphi) \quad \forall \varphi \in V. \tag{37}$$

The resulting a posteriori error estimator has the following form:

$$E(u,q) - E(u_h, q_h) \approx \eta_h = \frac{1}{2}\rho(u_h)(\overline{y} - i_h\overline{y}) + \frac{1}{2}\overline{\rho}^*(u_h, \overline{y}_h)(\delta u), \tag{38}$$

where $\rho(\cdot)(\cdot)$ and $\overline{\rho}^*(\cdot)(\cdot)$ are the residuals of the state and adjoint equation defined by:

$$\begin{aligned} \rho(u_h)(\varphi) &:= f(\varphi) - a(u_h, q_h)(\varphi) \\ \overline{\rho}^*(u_h, \overline{y}_h)(\varphi) &:= -\langle J_h (J_h^* J_h)^{-1} \nabla E^{(1)}(q_h), C'(u_h)(\varphi)\rangle \\ &+ \ E'_u(u_h, q_h)(\varphi) - a'_u(u_h, q_h)(\varphi, \overline{y}_h). \end{aligned} \tag{39}$$

Remark 5. For the adaptive mesh refinement algorithm described in Section 1 we need a cell-wise representation of the error estimator discussed above. It can be simply obtained by cell-wise integration by parts of (38). For more details about the localization of error estimators see e.g. Becker & Rannacher [2].

5 Numerical Examples

In this section we demonstrate the behavior of our method on two model problems. We compare our error estimator with some other ones and test the robustness and the efficiency of it. Throughout, the discretization of the state equation uses piecewise bilinear finite elements on locally refined meshes consisting of quadrilaterals. The resulting nonlinear state equations are solved by Newton's method and the solution of the linear subproblems are computed using multigrid. With these ingredients, the total numerical cost for solution on a given mesh behaves like $O(N)$, where N is the number of nodes. All computation are done with the finite element library Gascoigne3D, see [1].

5.1 Example 1

We consider a convection-diffusion equation with unknown constant transport direction (q_1, q_2) in the unit square $\Omega = (0,1)^2$:

$$\begin{aligned} -\Delta u + q_1 u_x + q_2 u_y &= 2 \qquad \text{in } \Omega, \\ u &= 0 \qquad \text{on } \partial\Omega. \end{aligned} \tag{40}$$

The quantity of interest is here the mean value

$$E(u) = \int_{\Omega_0} u \, dx \tag{41}$$

on a subdomain Ω_0 defined by

$$\Omega_0 = (\frac{15}{16}, 1) \times (0, \frac{1}{16}).$$

The measurements are given by the values of the state variable at five different points:

$$\begin{aligned} &\xi_1 = (0.25, 0.5), \quad \xi_2 = (0.5, 0.25), \\ &\xi_3 = (0.75, 0.5), \quad \xi_4 = (0.5, 0.75), \quad \xi_5 = (0.5, 0.5). \end{aligned} \tag{42}$$

The components of the corresponding observation operator C have the following form:

$$C_i(v) = v(\xi_i), \tag{43}$$

and the parameter identification problem is formulated as follows:
For $(u, q) \in V \times Q$ with $V = H_0^1(\Omega)$ and $Q = \mathbb{R}^2$

$$\text{Minimize} \quad \frac{1}{2} \sum_{i=1}^{5} (u(\xi_i) - \bar{C}_i)^2 \tag{44}$$

under the constraint (40), where $\bar{C}_i$ denote the components of the measurement vector $\bar{C} \in Z = \mathbb{R}^5$ and are given by the values of the state variable u for the exact parameter $q = (8, 8)$, i.e. $\bar{C}_i = u(\xi_i)$.

For this problem we present some numerical results using the a posteriori error estimation described above. For comparison we also consider some other types of mesh refinement: global refinement, refinement corresponding to the error estimator derived for the error in parameter ($E_s(q) = q_1 + q_2$) and the refinement based on the estimating of the error in the cost functional ($E_c(u) = \frac{1}{2} \|C(u)\|^2$). The latter corresponds to the a posteriori error estimator for optimization problems, see Becker & Rannacher [3] and Becker, Kapp & Rannacher [4].

The comparison of the accuracy achieved on the meshes resulting from this four types of mesh refinement is made in Figure 1.

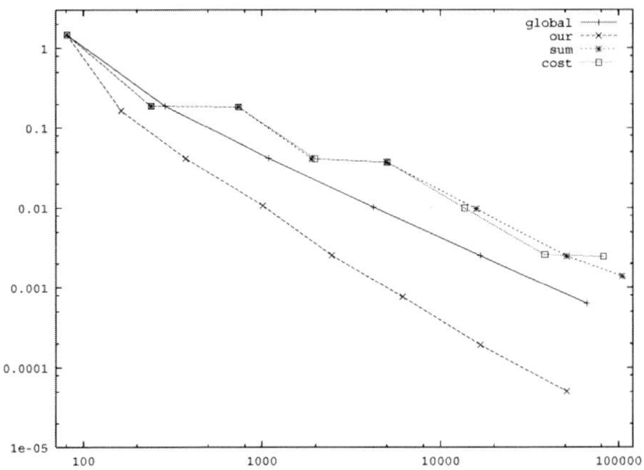

Fig. 1. Errors in $E(u)$ for different refinement strategies vs. number of nodes

As seen from Figure 1 the strategy of the a posteriori error estimation derived above is the most efficient one. The strategies, which do not regard

the quantity of interest, i.e. the error functional $E(u)$, are even worse than the global mesh refinement.
The a posteriori error estimator is also used for quantitative error control. The comparison between the error in the quantity of interest and the estimation of it is shown in Table 1. The effectivity index of the error estimator defined by

$$I_{eff} := (E(q) - E(q_h))/\eta \tag{45}$$

closes to 1. This computation is done on the sequence of locally refined meshes produced by the error estimator for $E(u)$.

Table 1. Efficiency of the error estimator

N	$E(u) - E(u_h)$	η	I_{eff}
81	1.46e-0	4.07e-0	0.36
163	1.65e-1	8.10e-2	2.04
375	4.13e-2	4.24e-2	0.97
1009	1.07e-2	1.07e-2	1.00
2463	2.55e-3	2.56e-3	0.99
6135	7.65e-4	7.79e-4	0.98
16713	1.92e-4	1.95e-4	0.98

The comparison of typical meshes resulting from the a posteriori error estimator for $E(u)$ and the estimator for the cost functional is made in Figure 2.

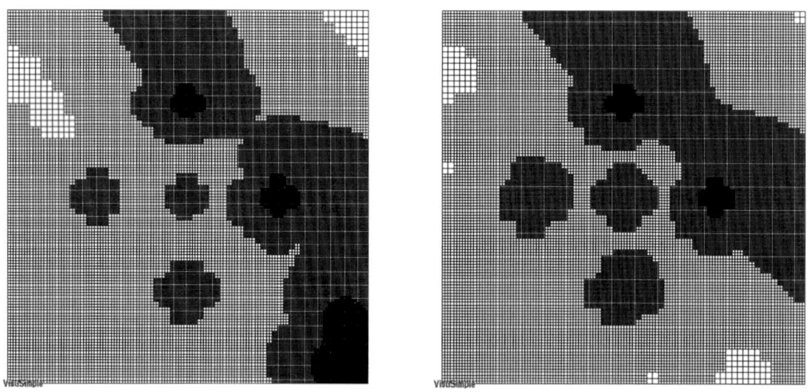

Fig. 2. Typical meshes produced by a posteriori error estimator for $E(u)$ (left) and for the cost functional $E_c(u)$ (right)

5.2 Example 2

We consider an example, where the parameters are involved in the differential operator as well as in the boundary condition. The quantity of interest depends here also on the state variable and on the parameters. We state equation is considered on the unit square $\Omega = (0,1)^2$. The boundary of Ω consists of three parts $\partial\Omega = \Gamma_1 \cup \Gamma_2 \cup \Gamma_3$, see Figure 3.

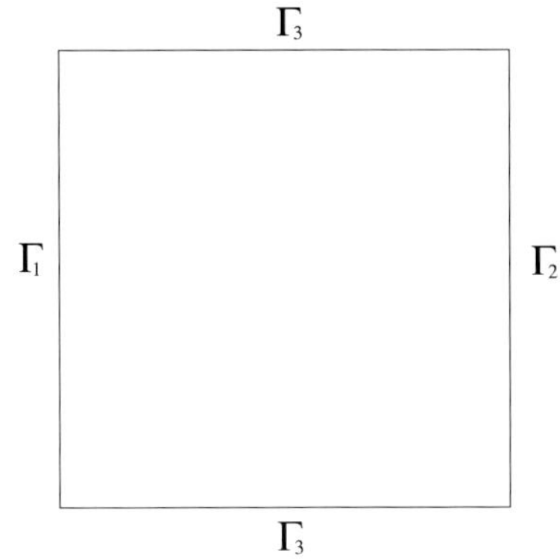

Fig. 3. The computational domain

The state equation has the following form:

$$\begin{aligned} -\Delta u + q_1 y u &= 2 \quad \text{in } \Omega, \\ u &= 0 \quad \text{on } \Gamma_1, \\ u &= q_2 \text{ on } \Gamma_2, \\ \partial u / \partial n &= 0 \quad \text{on } \Gamma_3. \end{aligned} \tag{46}$$

The quantity of interest is here defined by:

$$E(u,q) = q_2 \int_{\Gamma_1} \partial u / \partial n \, ds. \tag{47}$$

We are given a set of measurements (values of the state variable in three different points):

$$\xi_1 = (0.25, 0.75), \quad \xi_2 = (0.5, 0.75), \quad \xi_3 = (0.75, 0.75). \tag{48}$$

The components of the corresponding observation operator C have the following form:

$$C_i(v) = v(\xi_i), \tag{49}$$

and the parameter identification problem is formulated as follows:
For $(u,q) \in (g(q) + V) \times Q$ with $V = \{v \in H^1(\Omega) \,|\, v = 0 \text{ on } \Gamma_1 \cup \Gamma_2\}$ and $Q = \mathbb{R}^2$

$$\text{Minimize} \quad \frac{1}{2}\sum_{i=1}^{3}(u(\xi_i) - \bar{C}_i)^2 \tag{50}$$

under the constraint (46), where the Dirichlet operator g is given by

$$g(q) = q_2\, x, \tag{51}$$

$\bar{C}_i$ denote the components of the measurement vector $\bar{C} \in Z = \mathbb{R}^3$ and are given by the values of the state variable u for the exact parameter $q = (50, 1)$, i.e. $\bar{C}_i = u(\xi_i)$.
For this problem we present some numerical results using the a posteriori error estimation described above. As in Example 1 we compare it with some other types of mesh refinement: global refinement, refinement corresponding to the error estimator derived for the error in parameter ($E_s(q) = q_1 + q_2$) and the "uncontrolled" or "state" refinement, i.e the refinement using the error estimator for the uncontrolled equation. The comparison of the accuracy achieved on the meshes resulting from this four types of mesh refinement is made in Figure 4.

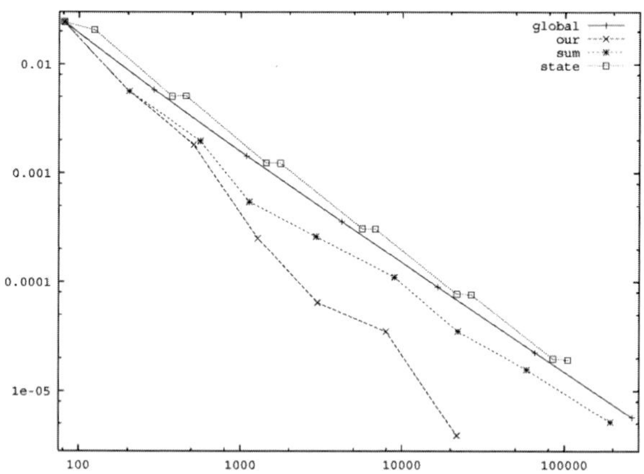

Fig. 4. Errors in $E(u, q)$ for different refinement strategies vs. number of nodes

The comparison between the error in $E(u, q)$ and the estimation of it is shown in Table 2. The effectivity index of the error estimator is defined as in Example 1 by (45).

The comparison of typical meshes resulting from these types of mesh refinement is made in Figure 5.

Table 2. Efficiency of the error estimator for the error functional $E(u,q)$

N	$E(u,q) - E(u_h, q_h)$	η	I_{eff}
81	2.43e-2	2.73e-2	0.89
203	5.62e-3	5.56e-3	1.01
513	1.82e-3	1.86e-3	0.98
1275	2.53e-4	2.79e-4	0.91
2993	6.44e-5	6.25e-5	1.03
7931	3.50e-5	3.57e-5	0.98
21817	3.89e-6	3.74e-6	1.04

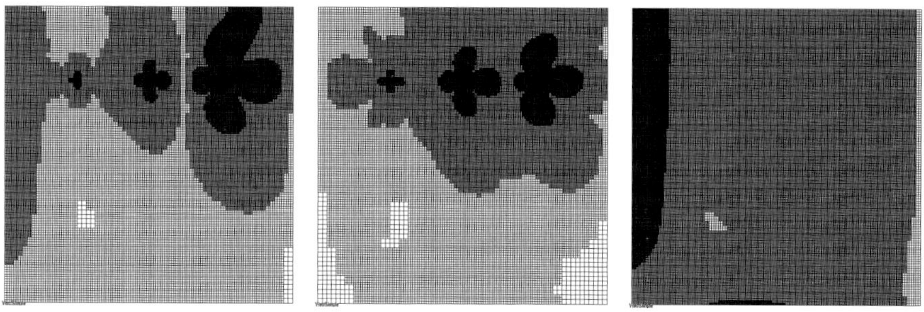

Fig. 5. Typical meshes produced by a posteriori error estimator for $E(u,q)$ (left), for the parameters (middle) and for the "uncontrolled" equation (right)

References

[1] R. Becker and M. Braack: Gascoigne3D. A finite element toolkit for flow problems, SFB-preprint, Heidelberg (2003)

[2] R. Becker and R. Rannacher: A feed-back approach to error control in finite element methods: Basic analysis and examples. East-West J. Numer. Math. **4**(4), 237–264 (1996)

[3] R. Becker and R. Rannacher: An optimal control approach to a posteriori error estimation in finite element methods. In Acta Numerica 2001 (A. Iserles, ed.), Cambridge University Press, Cambridge, 1–102 (2001)

[4] R. Becker, H. Kapp, and R. Rannacher: Adaptive finite element methods for optimal control of partial differential equations: Basic concept. SIAM J. Cont. Opt.**39**(1), 113–132 (2000)

[5] R. Becker and B. Vexler: A posteriori error estimation for finite element discretization of parameter identification problems. Numerische Mathematik, published online (2003)

[6] C.F. Carey and J.T. Oden: Finite Elements, Computational Aspects. Vol. III., New Jersey: Prentice-Hall (1984)

[7] P.G. Ciarlet: The Finite Element Method for Elliptic Problems. North-Holland Publishing Company, Amsterdam (1978)

[8] Ph. Clement: Approximation by finite element functions using local regularization. Revue Franc. Automat. Inform. Rech. Operat. **9** (R-2), 77–84 (1975)

[9] A.R. Conn, N. Gould and Ph.L. Toint. Trust-region methods, SIAM, MPS, Philadelphia, 2000.

[10] J.E. Dennis and R.B. Schnabel: Numerical methods for unconstrained optimization and nonlinear equations. Number 16 in Classics in Applied Mathematics. SIAM, Society for Industrial and Applied Mathematics, Philadelphia (1996)

[11] J. Dieudonné: Fondation of Modern Analysis. Academic Press, New York (1960)

[12] J. Nocedal and S.J. Wright: Numerical Optimization. Springer Series in Operations Research, Springer New York (1999)

Simulation Study of Vehicle Platooning Maneuvers with Full-State Tracking Control

Danwei Wang[1], Minhtuan Pham[1], and Cat T. Pham[2]

[1] Nanyang Technological University, School of Electrical and Electronic Engineering
Block S2, Nanyang Avenue, 639798 Singapore
edwwang@ntu.edu.sg

[2] Department of Automation Technology, Institute of Information Technology
National Center for Sciences and Technology, 18 Hoang Quoc Viet Street, Hanoi, Vietnam
ptcat@ioit.ncst.ac.vn

Summary. The paper presents a co-simulation platform to study the motion and control of a vehicle platooning system. In the study of a vehicle platooning system, the key factors include dynamic performance of all vehicles, the relative positions and oriantations of between vehicles, and the real-time interactions between an advanced controller and an autonomous vehicle in the platooning formation. This co-simulation platform features all these key components by integrating two softwares: ADAMS and MATLAB/SIMULINK. The advantages of ADAMS in mechanical system prototyping and those of MATLAB in advanced controller designing are exploited and combined to offer an integrated, visual and pseudo-practical platform for the investigation of control systems of mobile robots. The platooning dynamics is modelled in ADAMS with multiple vehicles with proper driving, steering mechanisms and sensing system to detect the distances and relative orientations between vehicles. An advanced nonlinear tracking controller is modelled in SIMULINK. The co-simulation of the platooning dynamics and the advanced controller is made possible by proper definition of inputs/outputs variables. This co-simulation platform is illustrated by a pair of car-like vehicles modelled in ADAMS with a nonlinear controller built in SIMULINK.

1 Introduction

Vehicle platooning in the recent years has gained much attention in research. Most existing car-following control algorithms are based on longitudinal control or lateral control. The former is actually the velocity or acceleration control of a vehicle to maintain desired spacing between two vehicles [1], [2]. On the other hand, the latter keeps the following vehicle on the track either of a planned road or the leading vehicle's trajectory [3]-[5]. Recently, a method integrating both lateral and longitudinal controllers in one was introduced in [6] and [8]. The method is a nonlinear controller based on the theory of output feedback and isomorphic transformation.

Though well developed, theory needs to be extensively tested to prove its value and to ensure it work properly. However, experiments normally require expensive equipments and lengthy process of part design and system integration. Extensive, comprehensive and near-reality simulations are helpful and sometimes necessary. One widely used and powerful simulation software is MATLAB [3], which provides an general-purpose environment to study system motions and control algorithms. SIMULINK, a diagram programming method of MATLAB, simplifies much more the programming procedure with function blocks. SIMULINK is an excellent platform for extensive and comprehensive simulation of control algorithms.

However, in MATLAB, vehicle dynamics must be represented in differential equations, which have limitations in modelling a complex mechanical system. In most cases, the vehicle is modelled as a simplified representation which could bypass some characteristics of the vehicle. CAD (*Computer-Aided Design*) softwares, however, include several powerful mechanical design and simulation softwares providing tools to built up a mechanical structure and do 3D visual simulations such as ANSYS, SolidWorks, and particularly, ADAMS (*Automatic Dynamic Analysis of Mechanical Systems*). Many have used ADAMS in design and simulation of robotic manipulators and stationary machines [9]-[10]. One of its advantages is that it provides an ability to integrate ADAMS models with other simulation softwares such as MATRIX, EASY and MATLAB so that its capacities are not limited.

This paper introduces a new method to simulate vehicle platooning maneuvers in particular, and for autonomous vehicle research, in general, by using ADAMS and MATLAB/SIMULINK. A full-state tracking algorithm is used as an example to drive the following vehicle in the platoon.

2 Vehicle platooning and vehicle modelling

A vehicle platooning system consists of at least two vehicles. Without loss of generality, we consider a pair of vehicles forming a convoy. A leading vehicle is moving in front and a following vehicle is to track autonomously the leading vehicle with a predetermined spacing. Distance and orientation differences between the two vehicles are measured and fed back to a tracking control algorithm in the following vehicle for driving and steering.

This section presents the modelling of a vehicle platooning system of two vehicles. The two vehicles are identical in mechanical design to simplify the process of modelling by using ADAMS, which is a software platform consisting of many software packages. Different packages provide various specific-purposed functionalities.

2.1 Vehicle modelling

ADAMS/View is a prototyping module that allows users to build a mechanical system using some provided common basic parts. A GUI (Graphic User Interface) exhibits all these common basic parts and a user can *drag and drop* any required parts to build a complex mechanical system. The parts can be made of different materials and the user can choose various properties and parameter values, such as density, elasticity, stiffness and friction at contacts.

A vehicle platooning system consists of two vehicles moving on a ground plane. The ground is created using a 2D plane and placed at height level 0. The gravity is chosen with the normal value of $9.8m/s^2$. Both vehicles are car-like with four-wheel-driving and front-wheel-steering. The major mechanisms of a vehicle include body, tyres, drive systems, and steering systems. Their models in ADAMS as well as the integration are described below

- **Vehicle body**: The vehicle body is a metal plate to support and hold all other parts. It weights about $250kg$, including batteries, and has a shape as shown later in Fig. 3 with dimensions of $1.85m * 1.21m * 0.06m$.
- **Tyres**: Each tyre is modelled with a torus as shown in Fig. 1. Its major and minor radii are $14cm$ and $6cm$, respectively, and its weight is $15kg$. The mass of a tyre is defined using a mass inertia tensor diagonal matrix. The contact with ground is defined as a contact between a circle and a plane. Contacting forces between the wheels and the ground are modelled. The normal force at a contact point is computed based on the IMPACT function from the library as a nonlinear spring-damper and dependent on parameters: stiffness ($10kN/mm$), damping ($200Ns/mm$) and penetration depth ($2cm$). The friction forces are also defined using the Coulomb friction model. The friction coefficients vary depending on the slip velocity of the tyre and changing from 0.6 to 0.75.
- **Drive systems**: Each wheel is equipped with a drive system as shown in Fig. 1. A revolute joint connects each of the four wheels to a cylindrical axis which in turn links to the vehicular body by either another revolute joint (for front wheels) or a fixed joint (for rear wheels).

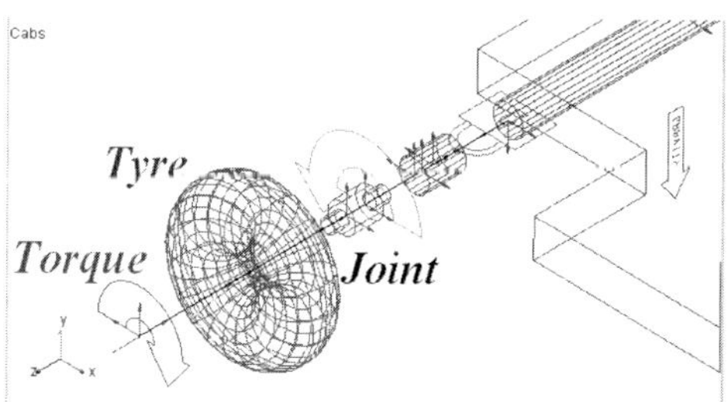

Fig. 1. Driving wheel model. The colour version of this figure can be found in Fig. A.43 on page 597.

A torque is also applied at each wheel to simulate the driving motor. The joint torque of the leading vehicle is given by user to simulate its moving in a manner unknown to the following vehicle. For the autonomous following vehicle, the joint torque is a function of the desired acceleration provided by the tracking controller.

$$\tau_{driving} = \frac{Iu_d + F_{friction}}{r} \tag{1}$$

where I is the moment of inertia of the wheel ($0.227kgm^2$); u_d is the desired acceleration; $F_{friction}$ is the friction force at the contact point as described above; and r is the radius of the wheel ($20cm$). The four desired accelerations are calculated based on the desired acceleration at the reference point given through *ADAMS/Control*'s variable named *DesAcc*.

- **Steering systems**: The steering system is located at the front axle of the vehicle. A main steering joint in the middle links with a steel jack (Fig. 2.a), which in turn joins to 2 other links. Each of these two links is connected to the wheel holder (i.e. the cylindrical axis mentioned above) of a wheel using a cylindric joint (Fig. 2.b). When the main steering joint rotates, the two wheels linked to the main steering joint are accordingly steered. The steering angle is limited due to mechanical constraints. It is easy to simulate that by limiting the angular position of the main steering joint (e.g. ±20 degrees). The angular position of the front steering main joint is controlled by a variable created in *ADAMS/Control* as an input to the model, namely *DesFSteer*.

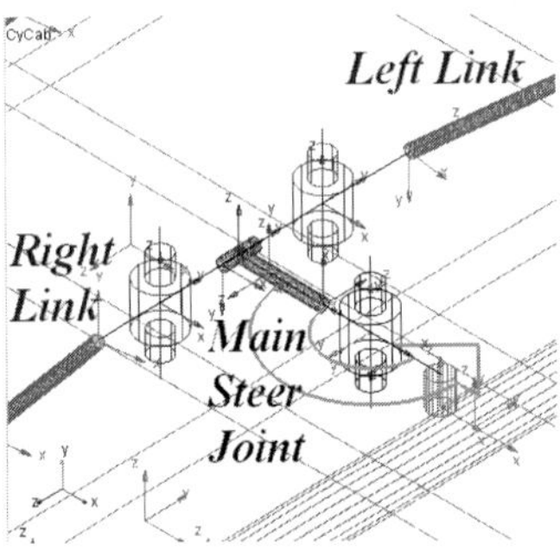

(a) Main Steer Joint

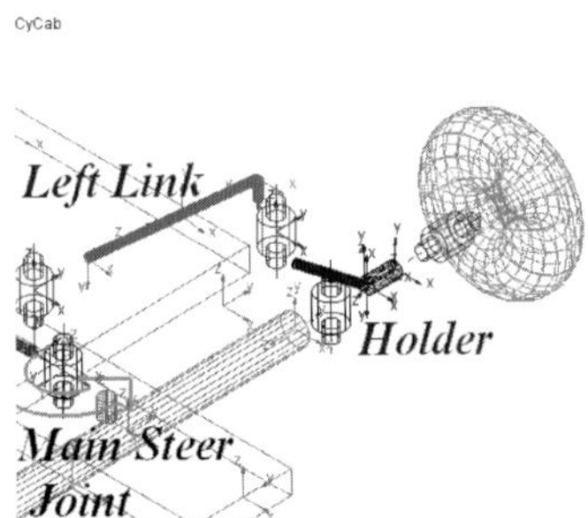

(b) Links to a wheel

Fig. 2. Steering mechanism. The colour version of this figure can be found in Fig. A.44 on page 597.

With all the mechanisms built, a vehicle can be assembled and a platooning system with two vehicles is shown in Fig. 3. In this simulation study, the following vehicle is to keep a distance of l meters from the leading vehicle and control algorithm will be discussed in the next section.

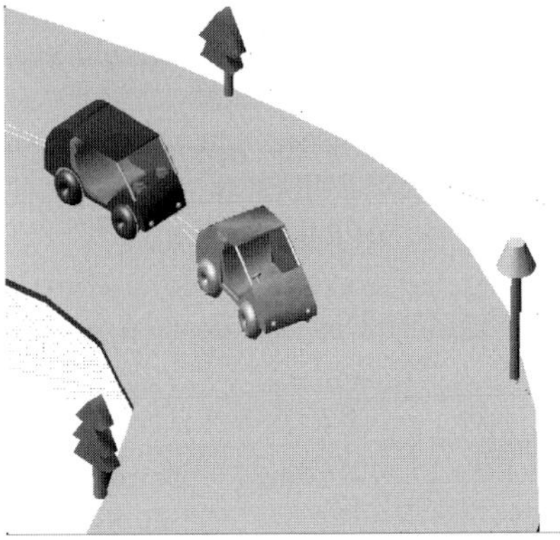

Fig. 3. ADAMS model of vehicle platooning. The colour version of this figure can be found in Fig. A.45 on page 597.

3 A tracking controller and its modelling

3.1 Tracking control

The algorithm, named ***Full-State Tracking Control***, is briefly described as below for the *Look-Ahead Tracking* case only (see full details in [7]), as shown in Fig. 4.

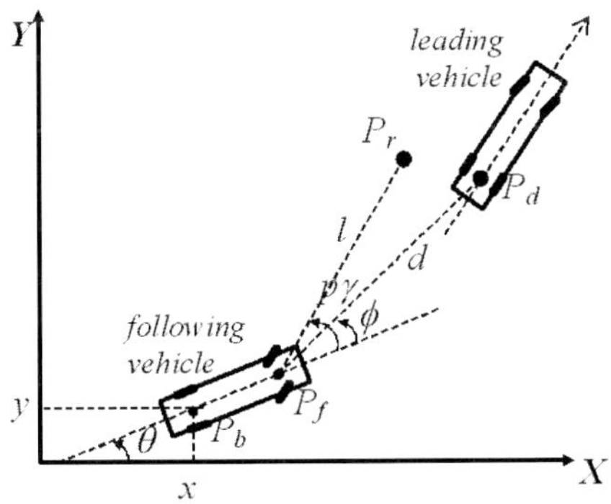

Fig. 4. Look-Ahead situation of platooning

A focus point (P_r) is defined in front of the following vehicle which is l meters away from the front point P_f of the vehicle. And the angle, formed by the longitudinal axis of the vehicle and the line P_fP_r, is p times of the steering angle γ.

$$P_r = z = \begin{bmatrix} x + a\cos\theta + l\cos(\theta + p\gamma) \\ y + a\sin\theta + l\sin(\theta + p\gamma) \end{bmatrix} \tag{2}$$

where
x, y: the reference point of the following vehicle
θ: the heading angle of the following vehicle
a: the length of the longitudinal axis
l, p: two design parameters of tracking controller

The coordinates of the leading vehicle, considered as the desired point or the tracked point, P_d, are

$$P_d = z_d = \begin{bmatrix} x + a\cos\theta + d\cos(\theta + \varphi) \\ y + a\sin\theta + d\sin(\theta + \varphi) \end{bmatrix} \tag{3}$$

where
d: the spacing distance $P_f P_d$
φ: the angle, formed by the longitudinal axis and $P_f P_d$.

The basic idea of the control scheme is to drive the focus point, P_r, to track the desired point, P_d. If this can be done, the following vehicle will follow the leading vehicle. With appropriately chosen parameters l and p, [7] has revealed a controller as follows:

$$u = \bar{E}^{-1}(\gamma)F\left(v, u_m, \gamma, \omega, d, \dot{d}, \ddot{d}, \varphi, \dot{\varphi}, \ddot{\varphi}\right) \tag{4}$$

where $E(\gamma)$ is a nonlinear 2x2 matrix of steering angle γ only and F is a highly nonlinear column vector (see [7] for details). Control input u , the product of the inverted matrix of $E(\gamma)$ and F, is such a highly nonlinear function vector.

3.2 Controller modelling

ADAMS can model the platooning system and also provides tools to build linear controllers. However, our controller is highly nonlinear and can not be modelled in ADAMS. We use MATLAB, instead, to model the controller. MATLAB's library of functions covers almost all the needs for developing and simulating highly nonlinear and sophisticated control algorithms. SIMULINK, a part of MATLAB, simplifies further the process of programming the controller and organizes it in terms of dataflow diagrams. A SIMULINK model includes blocks whose inputs and/or outputs are directionally defined. The output of one block connects to the input of another block so on and so forth.

Using SIMULINK, the closed-loop system presented in the previous subsection is modelled and composed of the ADAMS block to represent the ADAMS model, the feedbacks of measurements, the controller itself and the input commands to the ADAMS block (Fig. 5).

- **ADAMS block**: to make use of the ADAMS model from SIMULINK, the model needs to be imported as follows
  - *Export model from ADAMS*: *ADAMS/Control* plays an important role in connecting ADAMS models to other control applications. The key is

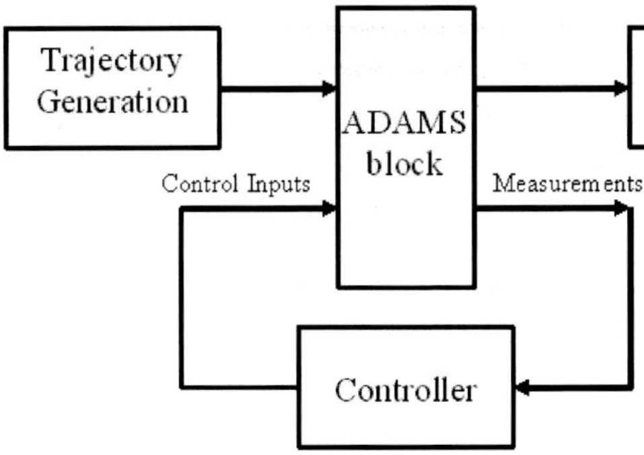

Fig. 5. The closed-loop system.

the capability of creating state variables and using them to exchange data with other softwares. Utilizing this feature, several variables are created in the ADAMS model of the platooning system. These variables can be catalogued into 3 groups:

1. *Input Commands to the model*: They are the desired steering angle and velocity set by high-level controller in SIMULINK for both vehicles of the platoon.
2. *Feedbacks from the model*: current states of the following vehicle, for example current acceleration, velocity and steering angles etc., as well as all the required measurements such as $d, \dot{d}, \ddot{d}, \varphi, \dot{\varphi}$ and $\ddot{\varphi}$ are acquired and sent back to the controller.
3. *Variables for display*: these are for display and comparison purpose only. They include the position and heading angle of the two vehicles in the generalized coordinates as well as the current states of the leading vehicle.

ADAMS model is then exported to MATLAB with *Plant Inputs* referring to *Input Commands* variables and *Plant Outputs* referring to the rest of variables.

– *Import model to MATLAB*: The ADAMS model is imported to MATLAB with the call to the exported file name, e.g. **ad_2_csd**, followed by a call: **adams_sys**. A new SIMULINK block, namely *adams_sub*, representing the ADAMS model is created and ready to be added to a SIMULINK model.

- **Input commands**: There are two sets of command inputs consisting of steering angle and acceleration: one is for the following vehicle and the other for the leading vehicle. The former is generated by the controller with the feedbacks from the ADAMS model. The latter is composed of several blocks which generate a desired steering angle and a desired acceleration feasible for the leading vehicle.

- **Feedbacks**: The outputs from ADAMS block consist of required measurements for the controller.
- **Controller**: Mathematical blocks are used to form controller (4).

By assembling all the parts altogether, a SIMULINK model representing the designed closed-loop system is created and ready for simulation (Fig. 6).

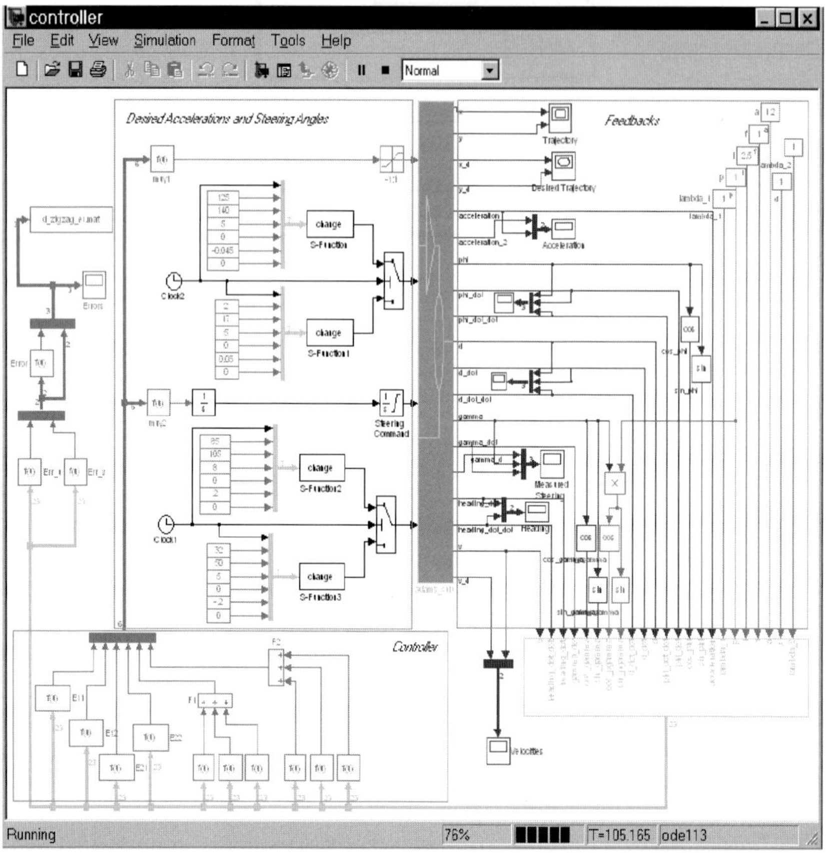

Fig. 6. SIMULINK model.

4 Co-simulation and results

4.1 Co-simulation

When the simulation written in SIMULINK is started, it will call ADAMS/-Control and ADAMS/Solver to initialize and to exchange data with the ADAMS model. The two models will run simultaneously and co-operatively.

There are two modes for ADAMS model: *interactive* or *batch*. They are different in terms of user-interface. In the *interactive* mode, the ADAMS model will be actually activated and run. The visible graphic simulation is shown. On the other hand, the *batch* mode turns off the graphic interface and run the simulation purely numerically (i.e. only calculations). Apparently, the simulation running in the *batch* mode is much faster.

All the intermediate data is stored both in MATLAB workspace and in ADAMS database. One can generate a video clip showing the whole process of motion of the ADAMS mode. This feature is really helpful to understand the behaviors of the system before doing practical experiments.

4.2 Simulation results and discussions

Simulations with different values of l, p, λ and ξ have been done using the proposed method. One of the results is presented here with $(l, p, \lambda, \xi) = (2.5, 1, 1, 1)$. The initial tracking error is assumed to be zero, i.e. the two vehicles are initially l meters away. The trajectory of the leading vehicle is planned and composed of a few portions: speeding up from zero velocity, turning left, turning right, and slowing down to a complete stop.

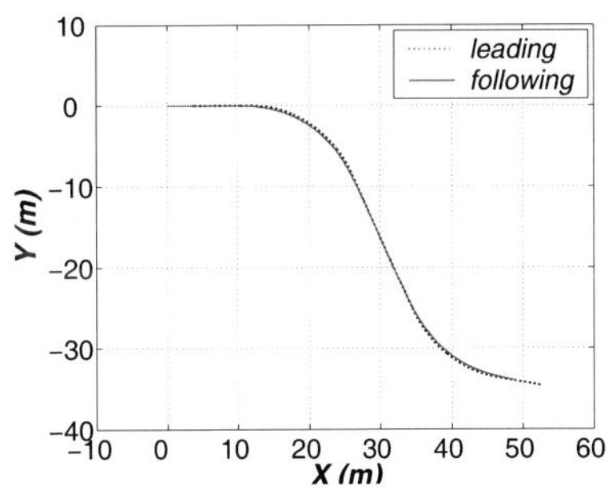

Fig. 7. Trajectories for $l = 2.5m$.

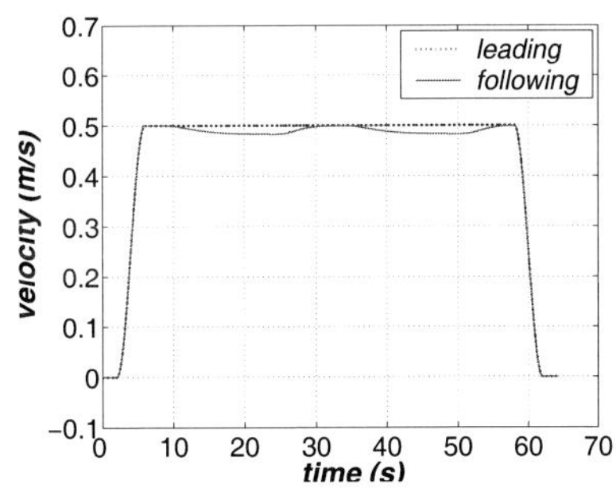

Fig. 8. Velocity.

The tracking result, shown in Fig. 7-8, reveals that the following vehicle manages to follow the leading vehicle. The tracking error is so small that the distance between two vehicles is almost maintained about $2.5m$ all the time.

In addition, a video clip showing 3D animation is also generated and it provides us a better view of simulation. Such an animation clip also helps us to study the system performance before an experimental set-up is built and/or an advanced feedback controller is implemented.

5 Conclusions

This paper exploits the advantages of two modelling and simulation softwares: ADAMS and MATLAB, in order to verify and investigate the performance of a platooning system based on full-state tracking control, including two car-like vehicles. Development in each of the two environments as well as the integration have been discussed. The simulation results conclude that the proposed control scheme work properly to drive autonomously the following vehicle to follow the leading vehicle. The methodology has revealed a powerful platform to do simulations not only for vehicle platooning but also for developments of autonomous mobile vehicles.

References

[1] J.K. Hedrick, D. McMahon, V.K. Narendran, and D. Swaroop, *Longitudinal Vehicle Controller Design for IVHS systems*", Proc. of the 1991 American Control Conference, vol. 3, June 1991, pp. 3107-3112.

[2] D. Yanakiev, and I. Kanellakopoulos, *Longitudinal Control of Heavy-Duty Vehicles for Autonomated Highway Systems*", Proc. of the 1995 American Control Conference, June 1995, pp. 3096-3100.

[3] G. Lee, S. Kim, Y. Yim, J. Jung, S. Oh, and B. Kim, *Longitudinal and Lateral Control System Development for a Platoon of Vehicles*, Proc. of the IEEE/IEEJ/JSAI International Conference on Intelligent Transportation Systems, 1999, pp. 605-610.

[4] S.K. Gehrig, and F.J. Stein, *A Trajectory-Based Approach for the Lateral Control of a Car Following System*, Proc. of the 1998 IEEE International Conference on Systems, Man, and Cybernetics, Vol. 4 , 1998, pp 3596 -3601.

[5] R. White, and M. Tomizuka, *Autonomous Following Lateral Control of Heavy Vehicles Using Laser Scanning Radar*, Proc. of the American Control Conference, Vol. 3, June 2001, pp. 2333-2338.

[6] D. Wang , and G. Xu, *Full State Tracking and Internal Dynamics of Nonholonomic Wheeled Mobile Robots*, Proc. of the American Control Conference, June 2000, pp 3274-3277.

[7] D. Wang, G. Xu, and M. Pham, *Look-Ahead/behind Control of a Car-Like Vehicle*, Proc. of the 2001 Robotics and Mechatronics Congress, Singapore, June 2001, pp. 101-106.

[8] M. Pham, and D. Wang, *Dynamics-Based Full-State Tracking for a Car-Like Mobile Robot*, Proc. of the 7th International Conference on Conference on Control, Automation, Robotics and Vision (ICARCV) 2002, Singapore, December 2002, pp. 752-756.

[9] B. Ozdalyan and M.V. Blundell, *Anti-lock Braking System Simulation and Modelling in ADAMS*, Proc. of the International Conference on Simulation '98., 1998, pp. 140 -144.

[10] B. Hale-Heighway, C. Murray, S. Douglas, and M. Gilmartin, *Multibody Dynamic Modelling of Commercial Vechicles*, Computing & Control Engineering Journal, Vol. 13, No. 1, Feb 2002, pp. 11-15.

The Modeling of Spectral Lines

Rainer Wehrse

Institute f. Theoret. Astrophysics, University of Heidelberg
Tiergartenstr. 15, D 69121 Heidelberg, Germany,
and
Interdisciplinary Center for Scientific Computing, University of Heidelberg,
Im Neuenheimer Feld 368, D 69120 Heidelberg, Germany
wehrse@ita.uni-heidelberg.de

1 Introduction

Light is of crucial importance for astronomers since first by far most of the information they gather from celestial objects and processes is obtained by means of light and second the momentum and energy balances of many stars and gaseous nebulae are strongly influenced by the photon field. Whereas in celestial mechanics the direction of the light and in radiation-hydrodynamical modeling usually only the time evolution of frequency integrated quantities as that of the radiation energy density or of the total momentum density play a role, in spectral diagnostics –in addition to the direction and its temporal behavior– the spectral composition of the light is of highest interest, since it consists of slowly varying components ("continua"), abrupt changes ("edges"),and a very large number of narrow resonances ("lines") and therefore has a very large information content. In many cases it allows the simultaneous determination of the pressures, temperatures, velocity fields and of the the chemical composition of the matter along the ray of propagation.

In the previous paper by Meinköhn the detailed modeling of one single line has been described. In this paper we will address methods to treat several (i.e. about 10 to 10^3) and many (about 10^3 to more than 10^9) spectral lines and investigate briefly the role of spatial inhomogeneities. The examples that will be given refer to star forming regions since they are often well observed but the corresponding theory is still very fragmentary. The latter is mainly due to the very complex geometry and velocity structure as well as large density and temperature contrasts. Among others, these lead to non-equilibrium level occupations of the atoms and molecules in such regions and very complex radiation fields. On the other hand, this makes the use of models valuable that involve only simple hydrodynamics but sophisticated radiation fields –as described below– so that the results can be compared in detail with observed spectra and important empirical parameters be derived.

In the following section 2 we introduce the rate equations that govern the time evolution of the level populations, which allow to calculate the optical properties (in particular the extinction coefficient and the source function) of the matter. They enter the radiative transfer equation (Sect. 3) which allows to derive the global radiation field from the local state of the matter. In Sect 4 the calculation of a few lines in moving 3D configurations and of the corresponding level occupations is described. When there are many lines this procedure becomes infeasible and one has to assume that the levels are occupied according to a Boltzmann distribution and that their positions, strengths and shapes follow stochastic distributions. In Sect. 5 a corresponding point process model is presented. In all these calculations it is assumed that the density and pressure distributions are smooth; however, observations indicate that the matter in star forming regions is highly turbulent and that therefore density fluctuations could be important. First results of an attempt to derive the temperature structure and to estimate the consequences for the emergent radiation field are given in Sect. 6. We close with a discussion and an outlook.

2 The thermodynamical state of the matter

We assume that in addition to the chemical composition the spatial distributions of the total density, the kinetic temperature, and of the velocity field are time-independent and prescribed. In the matter we consider here atoms (or ions or molecules) of a species with energy levels E_i ($i = 0 \ldots N_l$). Transitions from level i to level j are possible to collisions (coefficients $\mathcal{C}_{i\to j}$) and due to radiative processes (coefficients $\mathcal{R}_{i\to j}$), for details see [2], [5]. We just note here that $\mathcal{R}_{i\to j}$ is a functional of the specific intensities of the radiation field and that therefore one has to deal with a non-local coupling of the transitions. Furthermore, this makes the problem non-linear, cf. Sect. 3. If the number of particles per unit volume in state i at position $\mathbf{x}$ and at time t is $n_i(\mathbf{x}, t)$, then the rate of change is given by the *rate equation*

$$\frac{dn_i(\mathbf{x},t)}{dt} = - \sum_{j=1, i\neq j}^{N_l} n_i(\mathbf{x},t)\,(\mathcal{C}_{i\to j} + \mathcal{R}_{i\to j}) + \sum_{j=1, i\neq j}^{N_l} n_j(\mathbf{x},t)\,(\mathcal{C}_{j\to i} + \mathcal{R}_{j\to i})\,. \tag{1}$$

In many cases only the equilibrium state $dn_i(\mathbf{x}, t)/dt = 0$ is of interest so that the system 1 of ordinary differential equations could be replaced by algebraic equations which seem much easier to solve. However, it is presently not known how many solutions exist for the corresponding system and how they are distributed.

3 The radiative transfer equation

We describe the radiation field by the *specific intensity* as measured in the comoving (i.e. Lagrangian) frame. It is a function of frequency ν (or the dimensionless variable $\xi = -\ln\nu + const.$), the time t, position variable $\mathbf{x}$ and the direction vector $\mathbf{n}$. It can either be regarded as a photon distribution function normalized in a particular way (see [5]) or as the radiative energy flowing through a unit area, for details see e.g. [2]. It obeys the *radiative transfer equation* which reads in our case for a 3D configuration with non-relativistic velocities $\boldsymbol{\beta} = \mathbf{v}/c$ (cf. [6] and Meinköhn's contribution in this volume)

$$\begin{aligned}\mathbf{n}\cdot\nabla_{\mathbf{x}} I(\xi,t,\mathbf{x},\mathbf{n}) + w(\mathbf{x},\mathbf{n})\frac{\partial I(\xi,t,\mathbf{x},\mathbf{n})}{\partial\xi} \\ = -\chi(t,\mathbf{x},\xi)(I(\xi,t,\mathbf{x},\mathbf{n}) - S(t,\mathbf{x},\xi))\end{aligned} \tag{2}$$

where

$$w(\mathbf{x},\mathbf{n}) = \mathbf{n}\left(\frac{\partial\boldsymbol{\beta}(\mathbf{x})}{\partial\mathbf{x}}\right)\mathbf{n} \tag{3}$$

takes account of the wavelength shift due to the motions. χ is the *extinction coefficient* which reads for non-overlapping lines in our case in terms of the Einstein coefficients $B_{j\to i}$ and $B_{i\to j}$, and the profile function $\varphi(\hat{\xi},\vartheta,\xi)$ ($\hat{\xi}$ is the linecenter, ϑ a line shape parameter)

$$\chi(t,\mathbf{x},\xi) = \frac{h\,\exp(-\hat{\xi})}{4\pi}\left(n_i(\mathbf{x},t)B_{i\to j} - n_j(\mathbf{x},t)B_{j\to i}\right)\varphi(\hat{\xi},\vartheta,\xi). \tag{4}$$

The source function

$$S(t,\mathbf{x},\xi) = \frac{n_i(\mathbf{x},t)A_{i\to j}}{n_j(\mathbf{x},t)B_{j\to i} - n_i(\mathbf{x},t)B_{i\to j}} \tag{5}$$

describes the photons entering the direction $\mathbf{n}$ after a scattering process or after creation from the thermal pool. $A_{i\to j}$ is the Einstein coefficient for spontaneous emission). In Eq. 5 it is assumed that in the scattering process absorption and re-emmission are not correlated ("complete redistribution", for details see [5]). For boundary conditions the specific intensities of all rays impinging from the outside on the domain of interest have to be given. In this paper we assume them to be zero.

4 The treatment of a few NLTE lines in 3D media

For 1D media with the level populations deviating from the Boltzmann population of the local temperature line profiles and strengths have been calculated for many years, cf. [2]. Since usually here the stationary state is of interest only, $dn_i/dt = 0$ is assumed and Eqs. 1 and 2 are solved as an DAE system. In order to avoid multiple solutions we do not use this approach in our multidimensional calculations but instead employ a combination of a line and a (pseudo-)timestep method, i.e. we discretize the geometric and frequency space as well as the directions but keep the time coordinate continuous. Using up-wind discretisations on tensor-product grids we solve the transfer equation 2 by means of finite differences. The resulting ODE system for the occupation numbers $n_i(\mathbf{x}_k, t)$ is then solved by means of a standard Adams-Bashford or Bulirsch-Stoer extrapolation algorithm. The integration is stopped when the time derivatives get sufficiently small. Unfortunately, the system is usually very large so that algorithms for stiff systems cannot be employed. In addition, the CPU times and memory requirements are high so that only a moderate number of atomic levels (and therefore lines) can be considered. However, these disadvantages are –according to our experience– by far overcompensated by the high flexibility of the approach (e.g. if necessary the time evolution can be followed without additional effort) and by the unique connection between the initial and final states.

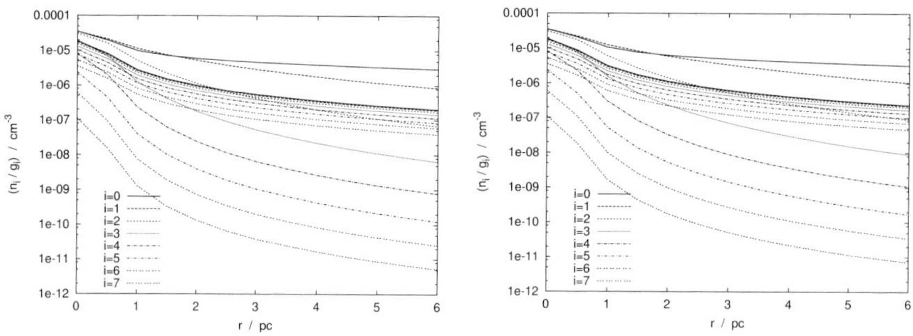

Fig. 1. Examples for the Occupation numbers of the lowest rotational levels of the CS molecule in an axially symmetric protostellar core (a molecular cloud in which a group of stars is about to be formed) in statistical equilibrium (NLTE) and in local thermodynamical equilibrium (LTE) as a function of distance from the center in the equatorial plane (left) and in the polar direction (right). Note the strong underpopulation of the higher states in NLTE which is caused by the very low density in these objects. From [3]

Figures 1 and 2 show examples of application of this method to star forming clouds (see Ph.D. thesis of P. Müller, [3]). The clouds contract and rotate and therefore are axially symmetric (this property is not exploited in the calcuations, however). They emit rotational lines of the CS molecule which can be observed with high resolution in frequency and and space (as projected on the celestial sphere). In Fig. 1 the NLTE occupation numbers are compared with the corresponding Boltzmann values. It is seen that there are strong dif-

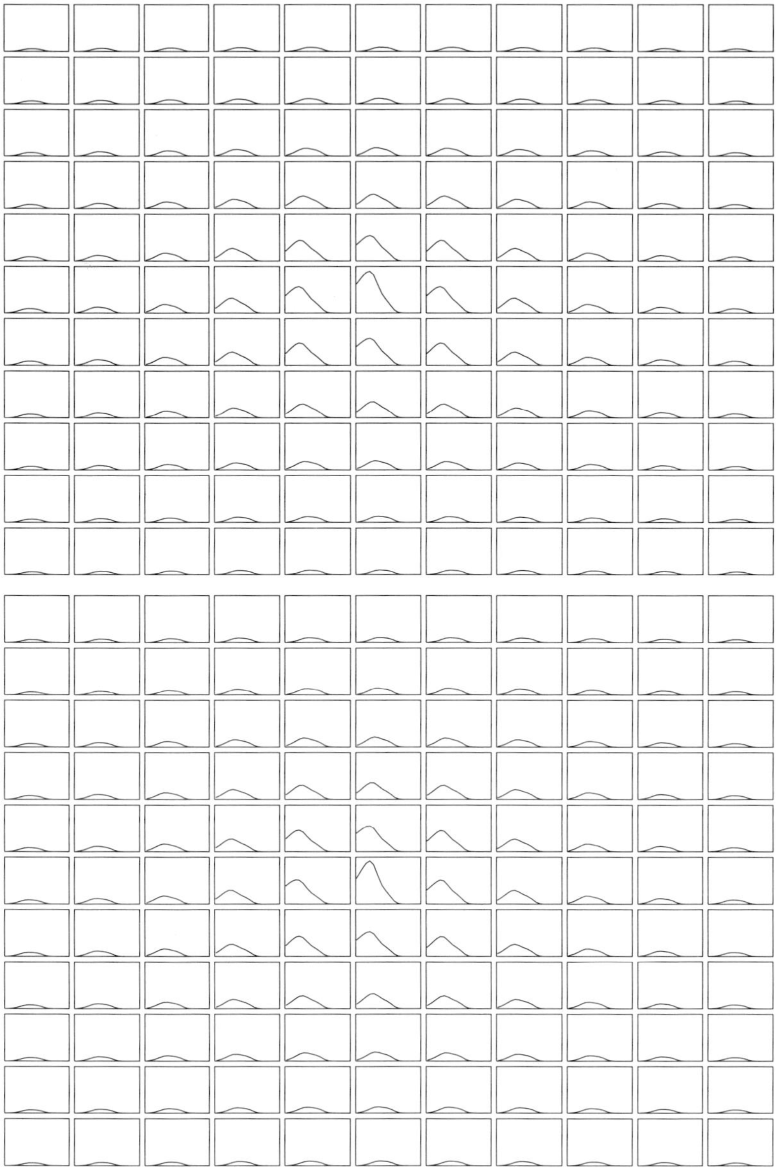

Fig. 2. Examples of spatially resolved line profiles for the $J = 0 \rightarrow 1$ rotational transition of the CS molecule emitted from an axially symmetric protostellar core as seen essentially pole-on (upper panel) and equator-on (lower panel). The shapes are calculated according to the algorithms described in Sections 2 – 4 and reflect the density distribution, the expansion velocities as well as the rotational velocities. They can be compared directly with modern observations from [3]

ferences, in particular for the excited levels in the outer regions, where the optical depths and the densities are low. Spatially resolved line profiles calculated with the LTE asumption would therefore look quite different from those derived from the NLTE level populations shown in Fig. 2 and would by no means agree with observations.

5 Stochastic treatment of many lines

If one has to deal with many lines (cf. [11]) the treatment described above is no longer feasible and one has to assume that the excitation temperature and the kinetic temperature are identical. The assumption implies that the extinction coefficient and the source function become independent of the radiation field. S becomes equal to the Planck function and can therefore be calculated from the kinetic temperature and the frequency. χ depends in addition on the density and may comprise contributions from various lines and from the continuum. However, for convenience we neglect subsequently the spatial and temporal variations of S and χ and keep w depth independent. Equation 2 can then easily be solved along a ray (or characteristic) s

$$I(s,\xi;w) = S(s,\xi) - S(0,\xi - ws)\exp\left(-\tfrac{1}{w}\int_{\xi-ws}^{\xi}\chi(\zeta)d\zeta\right) \tag{6}$$
$$-\int_{\xi-ws}^{\xi}\exp\left(-\tfrac{1}{w}\int_{\eta}^{\xi}\chi(\zeta)d\zeta\right)\frac{dS\left(s-\frac{\xi-\eta}{w},\eta\right)}{d\eta}\,d\eta.$$

In many cases (e.g. if one has to deal with astronomical observations obtained with a spectrograph of low resolution) one is not interested in the specific intensity $I(s,\xi;w)$ at frequency ξ exactly but in the average specific intensity $\bar{I}(s,\xi;w)$ over a frequency range $\xi\ldots\xi+\Delta\xi$. If the ergodic hypothesis is valid this expectation value is equal to the average value $\left\langle I(s,\xi;w)\right\rangle$ in which the extinction coefficients at ξ are taken as stochastic quantities. According to Eq. 6 this average value is given by

$$\left\langle I(s,\xi;w)\right\rangle = S(s,\xi) - S(0,\xi - ws)\left\langle\exp\left(-\tfrac{1}{w}\int_{\xi-ws}^{\xi}\chi(\zeta)d\zeta\right)\right\rangle \tag{7}$$
$$-\int_{\xi-ws}^{\xi}\left\langle\exp\left(-\tfrac{1}{w}\int_{\eta}^{\xi}\chi(\zeta)d\zeta\right)\right\rangle\frac{dS\left(s-\frac{\xi-\eta}{w},\eta\right)}{d\eta}\,d\eta.$$

With the assumption that the extinction coefficient is the sum of contributions from lines at positions $\hat{\xi}_l$ $(l = 1\ldots)$ and that the number of lines in a frequency interval and their positions within this interval follow a *Poisson point process* the expectation values on the rhs can be evaluated, see [10], [8],

$$\left\langle \exp\left(-\frac{1}{w}\int_{\xi-ws}^{\xi}\chi(\zeta)d\zeta\right)\right\rangle =$$
$$\exp\left(\int_{\Theta}\int_{-\infty}^{\infty}\varrho(\hat{\xi},\vartheta)\left\{\exp\left(-\frac{1}{|w|}\int_{\xi-|w|s/2}^{\xi+|w|s/2}\chi(\hat{\xi},\vartheta,\zeta-\hat{\xi})d\zeta\right)-1\right\}d\hat{\xi}d\vartheta\right) \tag{8}$$

with Θ being the set of possible ϑ-values. Note that this expectation value is essentially the characteristic function of the opacity distribution function $P(\chi)$. Levy's theorem then states that $P(\chi)$ can be calculated essentially by means of a Fourier transform (see [1], [8]). This approach is of great astronomical relevance since it generalizes the opacity distribution functions for static media which have been known for long time and since provides an elegant new way to calculate $P(\chi)$.

Several examples for $P(\chi)$ are given in [1], [11]. Since the formalism can be applied to the case of radiative diffusion we present here (Fig. 3) two results that solve long-standing open problems in the theory of (super-)nova and accretion disk modelling and that seem to be solvable only in the framework of our approach.

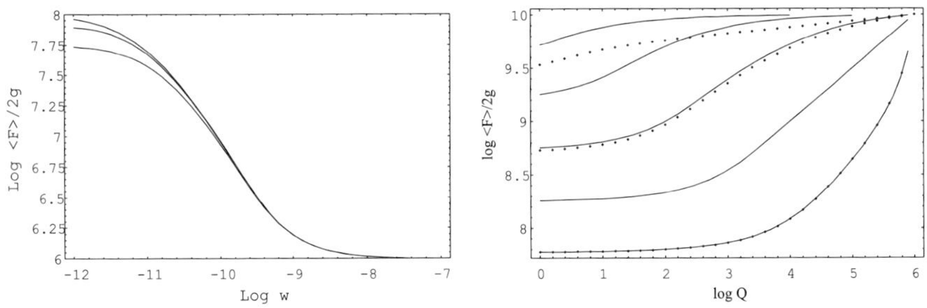

Fig. 3. The normalized radiative flux in the deep interior of a medium (diffusion regime) as a function of the velocity gradient w (left) and the line completeness factor (right). In the left graph the curves refer to different intrinsic line widths. It is seen that the flux is monotonically decreasing function of the velocity gradient and that the line shapes are important for small w only. The full curves in the right graphs refer to infinitely narrow lines whereas the dotted curves refer to lines with Lorentzian shapes of finite widths. It follows from the graph that an incomplete line list has quite different consequences in static and differentially moving media, for details see [8], [9].

6 Stochastic heating

As in many other astrophysical objects the Reynolds number is very high in star forming regions. The resulting turbulence is seen as an additional broad-

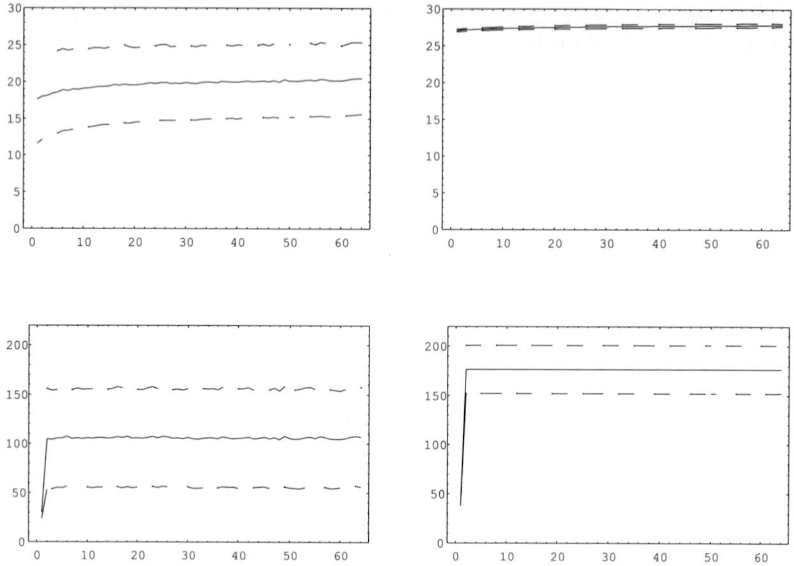

Fig. 4. Temperatures averaged over xy planes in the medium as a function of the z coordinate, i.e. full curves indicate the mean temperature of the xy surface plane for configurations with stochastic heating (left) and smooth distributions of the heating sources (right). The upper panel refers to a temperature dependence of the absorption coefficient $\kappa \propto T^5$, in the lower panel $\kappa \propto T^{-5}$ is assumed. The broken curves indicate the corresponding standard deviations.

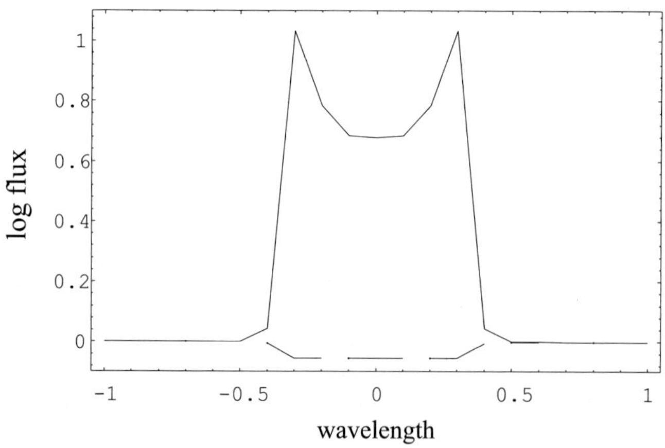

Fig. 5. Profiles of a synthetic spectral line emitted from a smooth configuration (broken curve) and from a corresponding stochastically heated configuration (full curve). Note that the doubly peaked shape is not caused by large scale velocities or photon diffusion in frequency but a consequence of the temperature fluctuations.

ening in the shapes of the emitted lines (cf. [2]). Unfortunately, there is neither a proper description of the turbulence nor a homogenization scheme for the radiative transfer equation to model reliably the spectroscopic consequences. We present here first results of an attempt to study heuristically the influence of the dissipation of turbulent energy on the temperature stratification and the consequences for emitted line profiles.

For this purpose we consider a box in which the matter is assumed to be grey and its temperature to be determined by the energy equation

$$\kappa(T(\mathbf{x}))\,(B(T(\mathbf{x})) - J(\mathbf{x})) = H(T(\mathbf{x})) \tag{9}$$

(κ absorption coefficient, T the local temperature, B Planck function, J mean intensity, H heating rate). H is assumed to vary from grid point to grid point according to

$$H(T(\mathbf{x})) = q * T(\mathbf{x}) \times \left(1 + (\mathcal{R}(\mathbf{x}) - 1/2)^5\right) \tag{10}$$

($0 < \mathcal{R} < 1$ uniformly distributed random variable with mean value $1/2$, q constant). Equation 10 implies that we allow variations in the heating law on the smallest possible scale and that high values of the heating rate are quite rare. Evidently, the results are a strong function of the temperature dependence of the absorption coefficient. In order to obtain an estimate of possible variations we assume simple power law dependencies $\kappa \propto T^{\pm 5}$.

The mean intensity J is to be determined by the radiative transfer equation so that we insert B from Eq. 9 into Eq. 2 which we solve by means of a finite difference scheme with up-wind discretisation on a tensor product grid (cf. [4]) and a fixed point iteration for the mean intensity. The resulting temperature distribution is then used to compute the the strength and shape of a (synthetic) resonance line emitted from the medium (cf. Meinköhn's contribution in this volume).

Some results are shown in Figs. 4 and 5 . Whereas –as expected– the temperature variations in the random configurations are larger than the non-random ones, it comes as a surprise that the mean temperatures are considerably lower in the random case. In a similar way, the calculations confirm that absorption profiles may change to emission profiles when the medium becomes randomly heated. However, it was not expected that the profiles may show doubly-peaked shapes even without large scale velocity fields and frequency redistribution. For spectral diagnostics this poses a severe problem.

7 Discussion and outlook

In the previous sections we have described algorithms recently developed by us (i) to model a moderate number of spectral lines emerging from 3D media outside LTE, (ii) to treat a very large number of lines by means of a stochastic model, and (iii) to study the effects of stochastic heating on the temperature distribution of a 3D medium and the resulting consequences for emergent line profiles. The algorithms certainly have a large range of applicability and have been run successfully even on single-processor modern PCs. However, it also became clear that much work is still ahead, e.g. an NLTE algorithm that uses locally refined spatial grids is urgently needed and a homogenization scheme to treat radiation fields in highly inhomogeneous media would facilitate very much the discussion of spectral lines and continua emitted (or reflected) by such configurations. Last but not least a proper algorithm for the solution of the corresponding inverse problems has to be developed.

Acknowledgement

I thank B. Baschek, G.V. Efimov, E. Meinköhn, G. Shaviv, W. von Waldenfels, and D.T. Wickramasinghe for their collaboration and many very helpful comments. The project was supported by the Deutsche Forschungsgemeinschaft (Sonderforschungsbereich 359, Teilprojekt C2).

References

[1] Baschek, B., von Waldenfels, W., Wehrse, R., Opacity distribution in static and moving media, A&A **371**, 1084 (2001)
[2] Mihalas, D., Stellar Atmospheres, Freeman, San Francisco (1978)
[3] Müller, P., Ph.D. Thesis, 3D-NLTE Linienbildung in differentiell bewegten Molekülwolken und analytische Untersuchungen zur hydrodynamischen Struktur axialsymmetrischer Systeme, Heidelberg (2001)
http://www.ita.uni-heidelberg.de/publications/thesis/index.html#Dissertations
[4] Richling, S., Meinköhn, E., Kryzhevoi, N.V., Kanschat, G., A&A **370**, 707, 2001
[5] Oxenius, J., Kinetic Theory of Particles and Photons, Springer (1986)
[6] Wehrse, R., Baschek, B., von Waldenfels, W., The diffusion of radiation in moving media, I. Basic assumptions and formulae, A&A, **359**, 780 (2000)
[7] Wehrse, R., Baschek, B., von Waldenfels, W., The diffusion of radiation in moving media, II. Limits for large and small velocity gradients for deterministic lines, A&A, **359**, 788 (2000)
[8] Wehrse, R., Baschek, B., von Waldenfels, W., The diffusion of radiation in moving media, III. Stochastic representation of spectral lines, A&A, **390**, 1141 (2002)
[9] Wehrse, R., Baschek, B., von Waldenfels, W., The diffusion of radiation in moving media, IV. Flux vector, effective opacity, and expansion opacity, A&A, **401**, 43 (2003)
[10] Wehrse, R., von Waldenfels, W., Baschek, B., Differentially moving media with many spectral lines: stochastic approach, J.Q.S.R.T. **60**, 963 (1998)
[11] Wehrse, R., in: Multiscale Problems in Science and Technology, Antonić, N., van Duijn, C.J., Jäger, W., Mikelić, A., eds., Springer, p. 291 (2002)

Divergence Free High Order Filter Methods for the Compressible MHD Equations*

H. C. Yee[1] and Björn Sjögreen[2]

[1] NASA Ames Research Center, USA
yee@nas.nasa.gov
[2] Royal Institute of Technology, Sweden
bjorns@nada.kth.se

Summary. The generalization of a class of low-dissipative high order filter finite difference methods for long time wave propagation of shock/turbulence/combustion compressible viscous gas dynamic flows to compressible MHD equations for structured curvilinear grids has been achieved. The new scheme is shown to provide a natural and efficient way for the minimization of the divergence of the magnetic field numerical error. Standard divergence cleaning is not required by the present filter approach. For certain MHD test cases, divergence free preservation of the magnetic fields has been achieved.

1 Introduction

An integrated approach for the control of numerical dissipation in high order finite difference schemes in structured curvilinear grids for the compressible Euler and Navier-Stokes equations has been developed and verified by the authors and collaborators [27, 28, 18, 29, 21, 25]. These schemes are suitable for complex multiscale compressible viscous flows, especially for high speed turbulence combustion and acoustics problems. Standard high-resolution shock-capturing schemes are too dissipative for these types of flow problems. For the performance of these schemes on the aforementioned flows, see [27, 28, 18, 29, 21, 19, 20] and references cited therein. Basically, the scheme consists of sixth-order or higher non-dissipative spatial difference operators as the base scheme. To control the amount and types of numerical dissipation, an artificial compression method (ACM) indicator or multiresolution wavelets are used as sensors to adaptively limit the amount and to aid in the selection and/or blending of the appropriate types of numerical dissipation to be used. This adaptive control of numerical dissipation is accomplished by a filter step after the completion of each full time step integration of the base

* Part of this work was performed while the second author was a RIACS visiting scientist at NASA Ames Research Center.

scheme. Hereafter, we refer to these schemes as the high order ACM-filter and WAV-filter methods.

The type of base schemes used in the high order ACM-filter and WAV-filter methods is divergence free preserving for the magnetohydrodynamics (MHD) equations. However, straightforward application of the filter step to the MHD equations will not automatically preserve the divergence free magnetic field condition. With careful modification of the gas dynamic scheme, the filter mechanism offers several natural and efficient alternatives (without the standard divergence cleaning procedures) for minimizing the $\nabla\cdot\mathbf{B}$ numerical error which are not easily attainable without additional work in the standard high-resolution shock-capturing schemes. The focus of this paper is to present a filter approach that exhibits divergence free preservation for certain test cases. Extensive grid convergence comparisons with standard high-resolution shock-capturing schemes will be shown.

2 Relevance

This paper is concerned with the compressible MHD equations, henceforth, for ease of reference, referred to simply as MHD equations. Throughout the paper, the term "$\nabla\cdot\mathbf{B}$ numerical error" refers to the "amount of non-zero value of the discretized form of $\nabla\cdot\mathbf{B}$ of the underlying scheme". The following discussion pertains to schemes involving the use of Riemann solvers or the eigen-structure of the MHD equations. In addition, our discussion is restricted to the finite difference formulation for structured grids.

An important ingredient in our method is the use of the dissipative portion of high-resolution shock-capturing schemes as part of the nonlinear filters. These nonlinear filters involve the use of approximate Riemann solvers. We will therefore first present a new form of high-resolution shock-capturing schemes for the conservative MHD equations using the non-conservative eigensystem.

Consider the 3-D conservative and non-conservative forms of the ideal compressible MHD equations in Cartesian grids,

$$\begin{pmatrix} \rho \\ \rho u \\ \rho v \\ \rho w \\ e \\ B_x \\ B_y \\ B_z \end{pmatrix}_t + \operatorname{div}\begin{pmatrix} \rho\mathbf{u} \\ \rho\mathbf{u}\mathbf{u}^T + (p + B^2/2)I - \mathbf{B}\mathbf{B}^T \\ \mathbf{u}(e + p + B^2/2) - \mathbf{B}(\mathbf{u}^T\mathbf{B}) \\ \mathbf{u}\mathbf{B}^T - \mathbf{B}\mathbf{u}^T \end{pmatrix} = 0 \qquad (1)$$

and

$$\begin{pmatrix} \rho \\ \rho u \\ \rho v \\ \rho w \\ e \\ B_x \\ B_y \\ B_z \end{pmatrix}_t + \text{div} \begin{pmatrix} \rho \mathbf{u} \\ \rho \mathbf{u}\mathbf{u}^T + (p + B^2/2)I - \mathbf{B}\mathbf{B}^T \\ \mathbf{u}(e + p + B^2/2) - \mathbf{B}(\mathbf{u}^T\mathbf{B}) \\ \mathbf{u}\mathbf{B}^T - \mathbf{B}\mathbf{u}^T \end{pmatrix} = -(\nabla\cdot \mathbf{B}) \begin{pmatrix} 0 \\ B_x \\ B_y \\ B_z \\ \mathbf{u}^T\mathbf{B} \\ u \\ v \\ w \end{pmatrix} \tag{2}$$

where the velocity vector $\mathbf{u} = (u, v, w)^T$, the magnetic field vector $\mathbf{B} = (B_x, B_y, B_z)^T$, ρ is the density, and e is the total energy. The notation $B^2 = B_x^2 + B_y^2 + B_z^2$ is used. The pressure is related to the other variables by

$$p = (\gamma - 1)(e - \frac{1}{2}\rho(u^2 + v^2 + w^2) - \frac{1}{2}(B_x^2 + B_y^2 + B_z^2)).$$

For plasmas, γ is usually equal to 5/3 (for monatomic gases). The vector on the right hand side of (2) is the non-conservative portion of the MHD equations [16, 17, 24]. The non-conservative term is proportional to $(\nabla\cdot \mathbf{B})$. Physically, it is zero if $\nabla\cdot \mathbf{B} = 0$ initially. In symbolic form, the conservative and non-conservative forms can be written as

$$U_t + \nabla \cdot \mathbf{F} = 0, \; U_t + \nabla \cdot \mathbf{F} = S,$$

where U is the corresponding state vector, $\mathbf{F}$ is the conservative inviscid flux vector tensor and S is the non-conservative portion of the equations in (2). The non-conservative term S can also be written as $S = \sum_{i=1}^{3} N_i(U)U_{x_i}$, where the notation $(x_1, x_2, x_3) = (x, y, z)$ is used. The curvilinear grid formulation follows the same methodology as in [25].

2.1 Divergence Condition

The $\nabla\cdot \mathbf{B} = 0$ condition is an initial constraint for the MHD equations, and it is **not** part of the MHD differential system. This is unlike the divergence-free condition of the velocities for the incompressible Euler or Navier-Stokes equations which is part of the differential system and is needed to close the system and must be explicitly enforced. For the MHD equations with $\nabla\cdot \mathbf{B} = 0$ as the initial data, all one needs is to construct schemes with the discretized form of $\nabla\cdot \mathbf{B}$ on the order of the truncation error, which goes to zero when the grid is refined. Unfortunately, straightforward extension of existing gas dynamic schemes to the MHD equations does not necessarily preserve the divergence-free condition.

Presently, there are basically two camps in solving the multi-dimensional MHD equations; namely, that which solves the conservative form, and the one which solves the non-conservative symmetrizable form [10, 16, 17]. For both forms of the MHD equations, high-resolution shock-capturing methods suffer from the need to perform extra work to drive the $\nabla \cdot \mathbf{B}$ numerical error down to machine zero. The popular approaches for minimizing the $\nabla \cdot \mathbf{B}$ numerical error include augmenting an extra PDE to the system [2], using variants of the staggered approach of K.S. Yee [31, 6, 7, 4] and using a projection method [32]. There is a key advantage to solving the conservative equations over the non-conservative equations, since the conservative form guarantees correct propagation speeds and locations of discontinuities. The disadvantage is that the conservative form is a non-strictly hyperbolic system with non-convex inviscid fluxes. There exist states (e.g., triple umbilic points for 1-D) for which the Jacobian of the flux of the conservative form does not have a complete set of eigenvectors. In this paper and companion papers [22, 30], both the non-conservative and conservative forms of the multidimensional compressible MHD system are considered.

2.2 Conservative and Non-Conservative Formulations Involving the Use of Approximate Riemann Solvers

For convenience of presentation we will describe our numerical methods for the x-flux on a uniform grid. A more detailed discussion can be found in [22]. Let $A(U)$ denote the Jacobian $\partial F/\partial U$ with the understanding that the present F and S are the x-component of the 3-D description above. For later discussion we write the non-conservative S term in the x-direction as $N(U)U_x$.

Gallice [9] and Cargo and Gallice [1] observed that seven of the eigenvalues and eigenvectors are identical for the "conservative" Jacobian matrix A and the "non-conservative" Jacobian matrix $(A - N)$. The eighth eigenvector of A of the conservative system (which is distinct from the non-conservative system) associated with the degenerate zero eigenvalue can sometimes coincide with one of the other eigenvectors, thereby, making it impossible to define the MHD Roe's approximate Riemann solver in the standard way. The eigenvectors of the non-conservative Jacobian $(A - N)$ always form a complete basis, and can be obtained from analytical formulas [9, 1]. A Roe type average state was developed in [9, 1].

We formulate our scheme together with the Gallice form of the MHD Roe's approximate Riemann solver in curvilinear grids for both the conservative and non-conservative MHD equations. We propose to use the non-conservative form of the eigen-decomposition but with the degenerate eigenvalue replaced by an entropy correction [11, 26] of what was supposed to be the zero eigenvalue for the conservative form (e.g., a small parameter ϵ that is scaled by the largest eigenvalue of $A(U)$). Our rationale for doing this is that only the eighth eigenvector of the non-conservative form is not the same as the eighth eigenvector for the conservative form. The incorrect eigenvector for the conservative form will be multiplied by an eigenvalue, which is close to zero (the eigenvalue will not be exactly zero when an entropy correction is used). Thus the effect of a "false" eigenvector will be small. By using the eighth eigenvector of the non-conservative system instead, the difficulty of dealing with an incomplete set of eigenvectors for the conservative system can be avoided.

The conservative filter approach, the conservative Harten-Yee, MUSCL and the fifth-order WENO [12] schemes used in this paper are formed by using the non-conservative eigen-decomposition described above in solving the conservative MHD equation set (1). The non-conservative filter approach, and the non-conservative Harten-Yee, MUSCL and WENO schemes are just the non-conservative eigen-decomposition in solving the non-conservative MHD equation set (2).

3 Description of High Order Filter Methods

Our high order ACM-filter and WAV-filter methods consist of two stages, a divergence- preserving base scheme stage (not involve the use of approximate Riemann solvers) and a filter stage (involve the use of approximate Riemann solvers). The filter stage can be divergence-free preserving depending on the type of filter operator being used and the method of applying the filter step. In order to have a good shock-capturing capability and improved nonlinear stability related to spurious high frequency oscillations, the blending of a high order nonlinear filter and a high order linear filter were proposed in our gas dynamic schemes. The nonlinear filter consists of the product of an ACM or wavelet sensor and the nonlinear dissipative portion of a high-resolution shock-capturing scheme. The high order linear filter is just the centered linear dissipative operator that is compatible with the order of the base scheme being used.

3.1 Divergence-Free Preserving Base Scheme Step

The first stage of the numerical method consists of a time step having a non-dissipative high order spatial and high order temporal base scheme operator L (e.g., a divergence-free preserving sixth-order central in space and fourth-order Runge-Kutta in time),

$$U^* = L(U^n), \tag{3}$$

where U^n is the numerical solution vector at time level n. When necessary, a high order linear numerical dissipation operator can be used. For example, a divergence-free preserving eighth-order linear dissipation with the sixth-order centered base scheme to approximate $F(U)_x$ is written as

$$\frac{\partial F}{\partial x} \approx D_{06} F_j + d\Delta x^7 (D_+ D_-)^4 U_j, \tag{4}$$

where D_{06} is the standard sixth-order accurate centered difference operator, and D_+D_- is the standard second-order accurate centered approximation of the second derivative. The small parameter d is a scaled value in the range of 0.00001 to 0.01, depending on the flow problem, and has the sign which gives dissipation in the forward time direction. The operators are modified at boundaries in a stable way [29].

This highly accurate base scheme is employed to numerically preserve the divergence-free condition of the magnetic field (to the level of round-off error) for curvilinear grids. When the solution is smooth, the filter step might not be needed. Thus the use of a high order centered difference operator as the base scheme will perfectly preserve the divergence-free condition. In this case the result will be the same, whether we solve the conservative system (1) or non-conservative system (2). Under a shock/shear and turbulence/combustion environment, the use of a dissipative portion of the shock-capturing scheme as part of the filter is necessary. In this case, a possible source of violation of the divergence-free condition can be from the filter step.

3.2 Adaptive Numerical Dissipation Filter Step

After the completion of a full time step of the divergence-free preserving base scheme stage, the second stage is to adaptively filter the solution by the **product** of "an ACM indicator or wavelet sensor" and the "**nonlinear dissipative portion** of a high-resolution shock-capturing scheme". The final update step after the filter can be written as (assume 1-D for ease of illustration)

$$U_j^{n+1} = U_j^* - \frac{\Delta t}{\Delta x}[H_{j+1/2}^f - H_{j-1/2}^f]. \tag{5}$$

The filter numerical flux vector is $H_{j+1/2}^f = R_{j+1/2}\overline{H}_{j+1/2}$. Here $R_{j+1/2}$ is the matrix of right eigenvectors of the Jacobian of the non-conservative MHD flux vector $(A_{j+1/2} - N_{j+1/2})$ evaluated at the Gallice average state $U_{j+1/2}^*$ as discussed in the previous subsection. The $\overline{H}_{j+1/2}$ are also evaluated from the same characteristic quantities derived from these eigenvectors using the Gallice average state based on the U^* values of (4). Due to the fact that the base scheme step is divergence free preserving and does not involve the use of approximate Riemann solvers, there is no difference in solving the conservative or non-conservative system for the filter approach. To reduce un-necessary computations (the non-conservative portion), the non-conservative filter approach only solves the conservative system on the base scheme step. Thus, the conservative and non-conservative filter approaches differ merely by the eighth eigenvalue.

Denote the elements of the vector $\overline{H}_{j+1/2}$ by $\overline{h}_{j+1/2}^l, l = 1, 2, ..., 8$. They have the form

$$\overline{h}_{j+1/2}^l = (\omega)_{j+1/2}^l(\varphi_{j+1/2}^l). \tag{6}$$

Here $(\omega)_{j+1/2}^l$ is a sensor to activate the shock-capturing nonlinear filter. For example, $(\omega)_{j+1/2}^l$ is designed to be zero in regions of smooth flow and near one in regions with discontinuities. It varies from one grid point to another and is obtained either from a wavelet analysis of the solution (WAV-filter scheme), or from a gradient-based detector (ACM-filter scheme) [27, 28, 18, 29, 21]. The blending of nonlinear filters with high order linear filter is discussed in [29].

The dissipative portion of the nonlinear filter $\varphi^l_{j+1/2} = g^l_{j+1/2} - b^l_{j+1/2}$ is the dissipative portion of a high order high-resolution shock-capturing scheme for the lth-characteristic wave. Here $g^l_{j+1/2}$ and $b^l_{j+1/2}$ are numerical fluxes of the uniformly high order high-resolution scheme and a high order central scheme for the lth characteristic, respectively. It is noted that $b^l_{j+1/2}$ might not be unique since there is more than one way of obtaining $\varphi^l_{j+1/2}$. For the forms of the $\varphi^l_{j+1/2}$ used in the numerical experiment section, see [27, 28, 18, 29, 21]. For example, the form of Harten and Yee and symmetric TVD schemes are already in the proper form in the sense that they are written in a central differencing portion $b^l_{j+1/2}$ and a nonlinear dissipation portion $\varphi^l_{j+1/2}$. No work is required to obtain $\varphi^l_{j+1/2}$ in this case.

With the exception of some smooth flows using the WAV-filter scheme, the filter given by (6), if applied to the entire MHD system (denoted by "filter all") normally will not preserve the divergence free magnetic field condition. In order to minimize the numerical error of the divergence-free magnetic condition, the nonlinear filter step only acts on the gas dynamic portion of the equations (denoted by "no filter on **B**"). With the divergence free spatial base scheme and the manner that we update the solution on the filter step, the divergence free property should be preserved by the "no filter on **B**" option. There are additional variants of the filter approach that from a theoretical standpoint, are divergence free. See [30] for more details.

4 2-D Compressible MHD Numerical Examples

For illustrative purposes, numerical experiments using sixth-order central spatial discretization as the base scheme is chosen for the ACM-filter and WAV-filter schemes. The sixth-order base scheme together with the nonlinear filter with wavelet sensor will be denoted WAV66. When a more conventional gradient based sensor ACM is used, the scheme is denoted ACM66. If high order linear numerical dissipation is also used in the base scheme, the methods will be denoted WAV66+AD8 and ACM66+AD8 respectively. The strength of the eighth-order dissipation will be denoted by a tunable coefficient d, as in (4). In all of the filter scheme computations, the nonlinear dissipative portion of Harten-Yee is used as part of the nonlinear filter term. For all test cases, the entropy fix parameter is 0.25 for the ACM and WAV-filter schemes.

The fifth-order weighted ENO scheme [12] (WENO5), and second-order Harten-Yee and MUSCL schemes are used for comparison. Classical fourth-order Runge-Kutta time stepping is used for all sixth-order schemes, as well as for the WENO5 scheme. The second-order Harten-Yee and MUSCL are integrated in time by the second-order TVD Runge-Kutta method. Except for WENO5, the minmod limiter, the van Leer version of the van Albada limiter and the Colella-Woodward limiter are considered.

The $\nabla \cdot \mathbf{B}$ numerical error is obtained by approximating the spatial derivatives by sixth-order centered differences for WAV66, ACM66 and WENO5, whereas the corresponding $\nabla \cdot \mathbf{B}$ numerical error is obtained by second-order centered differences for the second-order TVD schemes (MUSCL and Harten-Yee). The L^2-norm of $\nabla \cdot \mathbf{B}$ of a particular scheme is computed by taking the square root of the sum over the square of all three spatial directions of the discretized form of $\nabla \cdot \mathbf{B}$ at all grid points.

4.1 MHD Kelvin-Helmholtz Instabilities ($\gamma = 1.4$, Periodic BC)

The magnetohydrodynamic Kelvin-Helmholtz instabilities have been studied by many previous investigators [2, 13, 8]. We have used the set up in [2] which is shown in Fig. 1. Snapshots of the time evolution of the x-velocity is also shown in Fig. 1 by CEN66+AD8 (sixth-order central with an eighth-order linear dissipative added to the base scheme ($d = 0.001$). The solution is obtained without the filter step. At stopping time $T = 0.5$, the problem is smooth enough that it can be solved by the base scheme alone. Density contours at time $T = 0.5$ with 30 equidistant contour levels between 0.4 and 1.2 are used. Five levels of grid refinement are considered, namely, 51×101, 101×201, 201×401, 401×801 and 801×1601. Grids of increasing refinements by the eighth-order central difference with a tenth-order linear dissipation added (CEN88+AD10, $d = 0.001$) are used as the reference solution. Computations using $d = 0$ (CEN88) are not stable for the five grids.

At time 0.5 the problem is smooth enough and there is no need for the more CPU intensive shock-capturing schemes. However, as the flow evolves at a later time, shock-capturing methods are required. Here, the purpose is to examine the $\nabla \cdot \mathbf{B}$ numerical error when the flow is still smooth using shock-capturing methods. Figure 2 shows the density (left) and $\nabla \cdot \mathbf{B}$ (middle) contours at $T = 0.5$, and L^2-norm of $\nabla \cdot \mathbf{B}$ as a function of time (right) by MUSCL (top row) and WENO5 (bottom row). The same computations by ACM66 and WAV66 are shown in Fig. 3. The $\nabla \cdot \mathbf{B}$ contours with 30 equidistant contour levels between -150 and 150 are used. The CPU time used was considerably larger (around a factor 2.5) for the WENO5 scheme. MUSCL, Harten-Yee and WENO5 exhibit small oscillations at the outer edges of the vortices as the grid is refined. It is possible to decrease these oscillations by increasing the multi-dimensional entropy fix parameters of the Harten-Yee scheme [26].

Density contours using ACM66, ACM66+AD8, WAV66 and WAV66+AD8 for all limiters exhibit an accuracy similar to CEN88+AD8 (figures not shown). There is no gain in solving the conservative over the non-conservative system for these two filter schemes. However, their $\nabla \cdot \mathbf{B}$ numerical errors are very different when using the "no filter on **B**" option verses the "filter all"

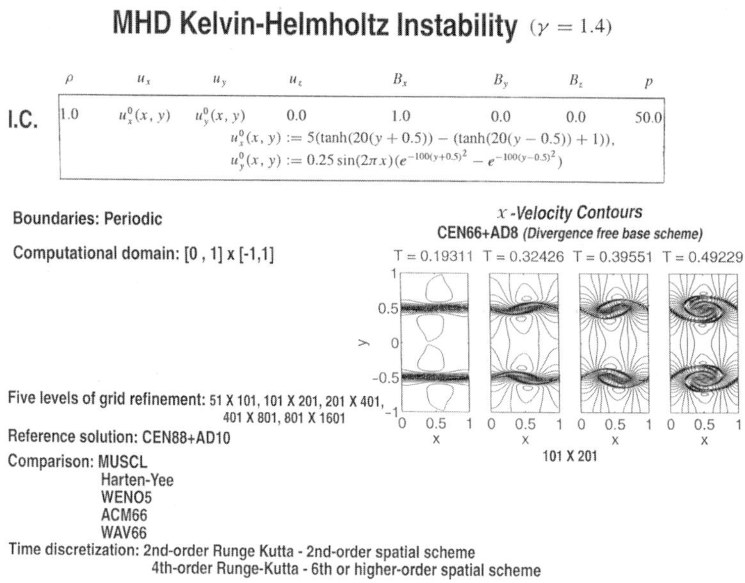

Fig. 1. Problem setup and time evolution of the Kelvin-Helmholtz problem. x-velocity contours by CEN66+AD8 on 101×201 grid points.

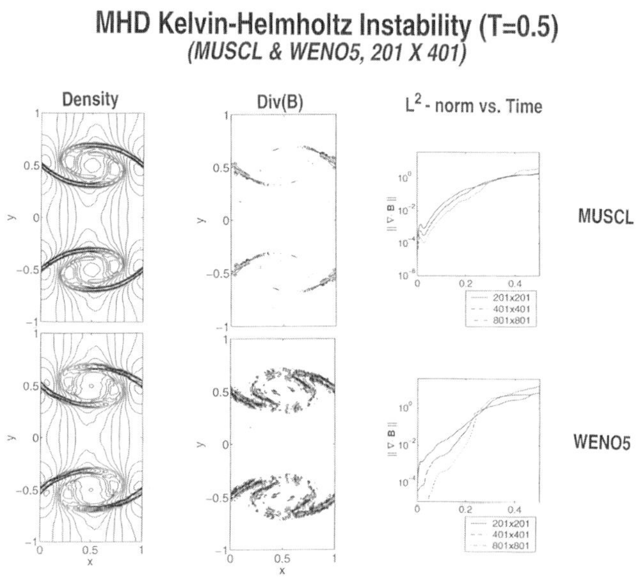

Fig. 2. Density (left) and $\nabla\cdot\mathbf{B}$ (middle) contours at $T = 0.5$, and L^2-norm of $\nabla\cdot\mathbf{B}$ as a function of time (right) by MUSCL (top row) and WENO5 (bottom row).

option. They are also very different from the standard MUSCL, Harten-Yee and WENO5 schemes. For the no filter on B equations option, divergence free preservation is achieved by the ACM66 and WAV66. The three standard shock-capturing methods exhibit similar $\nabla\cdot\mathbf{B}$ numerical errors.

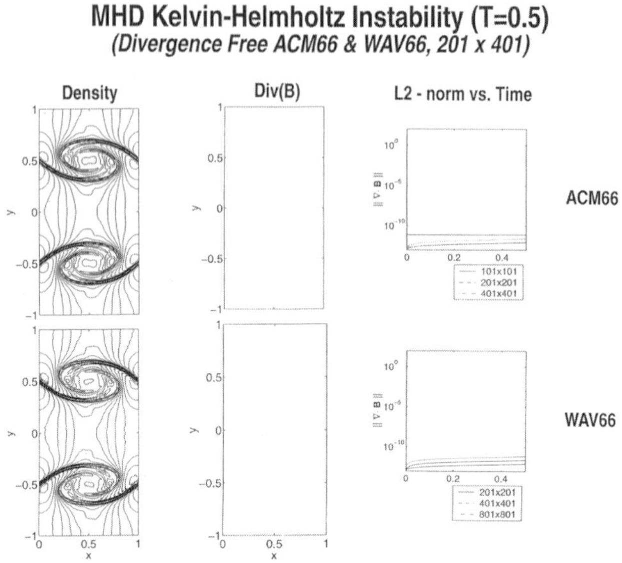

Fig. 3. Density (left) and $\nabla\cdot\mathbf{B}$ (middle) contours at $T = 0.5$, and L^2-norm of $\nabla\cdot\mathbf{B}$ as a function of time (right) by ACM66 (top row) and WAV66 (bottom row).

4.2 A 2-D Compressible MHD Riemann Problem ($\gamma = 5/3$)

We examine the same 2-D Riemann problem as in [2]. It consists of four constant states at time zero, as shown in Fig. 4. Grid convergence studies solving conservative (top) and non-conservative (bottom) system by WENO5 are shown in Fig. 5 for density contours at $T = 0.2$ with 40 contours equally spaced between 0.75 and 2.1. The accuracy in a solution of a Riemann problem away from discontinuities is difficult to improve by increasing the order of the scheme. A large part of the solution is constant, and the structure that develops is affected by low order errors from the discontinuity in the initial data. Since all five methods can capture shocks within 2-4 grid cells, their density contours look very similar even-though the $\nabla\cdot\mathbf{B}$ contours or the L_2 norm of the $\nabla\cdot\mathbf{B}$ numerical errors are all very different.

The effect on $\nabla\cdot\mathbf{B}$ when switching from a non-conservative system to a conservative system is less significant for the Harten-Yee and WENO5 than for MUSCL. The ACM66, ACM66+AD8 WAV66 and WAV66+AD8 methods all exhibit divergence free preservation when no nonlinear filter is applied on the **B** equations. Figures 6 show a comparison.

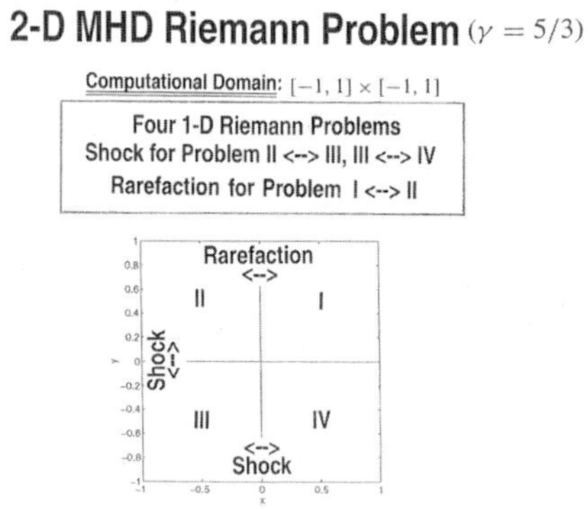

I.C.

Qdr.	ρ	ρu_x	ρu_y	ρu_z	B_x	B_y	B_z	e
I	0.9308	1.4557	−0.4633	0.0575	0.3501	0.9830	0.3050	5.0838
II	1.0304	1.5774	−1.0455	−0.1016	0.3501	0.5078	0.1576	5.7813
III	1.0000	1.7500	−1.0000	0.0000	0.5642	0.5078	0.2539	6.0000
IV	1.8887	0.2334	−1.7422	0.0733	0.5642	0.9830	0.4915	12.999

Fig. 4. Schematic of the initial data for the 2-D Riemann problem.

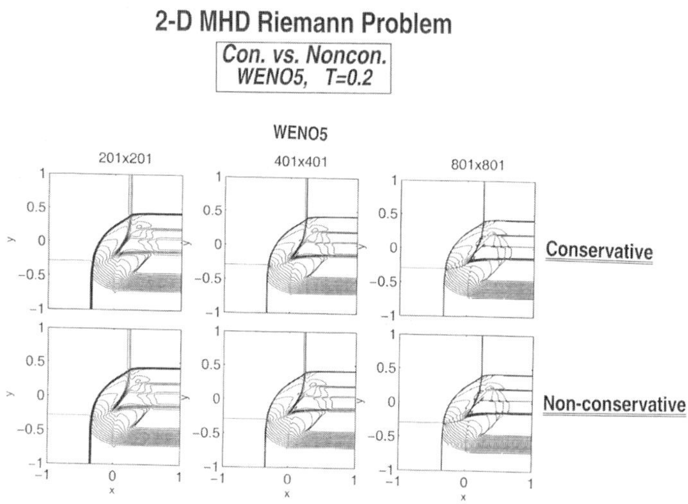

Fig. 5. Grid refinement by WENO5. Density contours solving the conservative (upper row) and non-conservative (lower row) form of the equations.

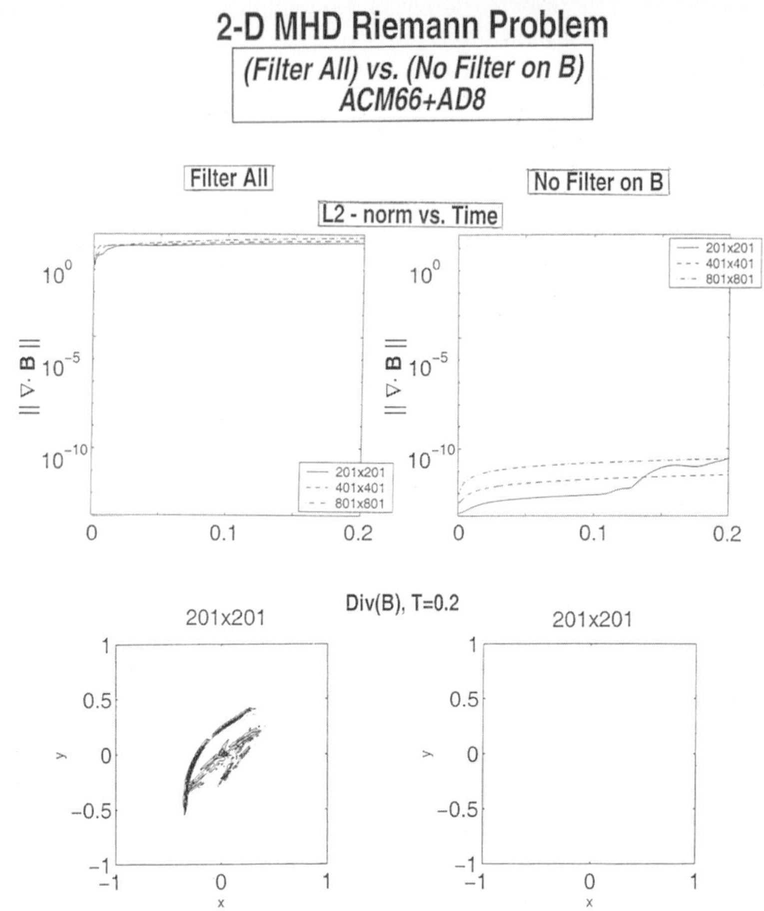

Fig. 6. L^2 norm of $\nabla \cdot \mathbf{B}$ vs. time and $\nabla \cdot \mathbf{B}$ contours at $T = 0.2$ by ACM66+AD8 when the non-linear filter is not applied on the magnetic field (left) and when it is applied to all components (right). 201×201, 401×401, and 801×801 grid points.

4.3 Compressible MHD Orszag-Tang Vortex ($\gamma = 5/3$, Periodic BC)

The 2-D Compressible MHD Orszag-Tang vortex problem [14, 3, 15, 4, 5] consists of periodic boundary conditions with smooth initial data is shown in Fig. 7. Density contours by WAV66+AD8 at $T = 3.14$ using "filter all" and "no filter on **B**" are shown in the same figure. The density contours are almost identical.

The initial sine waves break into discontinuities at a later time with complicated flow interactions. The computation stops at time T= 3.14 ($\approx \pi$), when discontinuities have formed and interacted. The solution has both complicated structure and discontinuities. It is a problem well suited for demonstrating our approach with highly accurate methods for solutions with discontinuities. Density contours with 30 equally spaced contours between 0.9 and 6.1 are used for illustration. Again, the same five levels of grid refinement study were performed on all five methods.

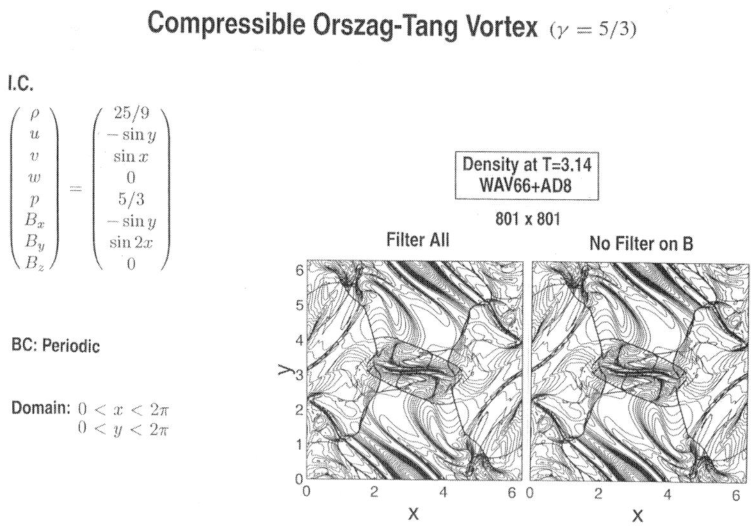

Fig. 7. Schematic, problem setup and density contours by WAV66+AD8 for the Orszag-Tang problem using a 801×801 grid at time $T = 3.14$.

Figures 8 and 9 show the comparison of WENO5 (solving both systems) with WAV66+AD8. Divergence free preservation is achieved by WAV66+AD8 using the "no filter on **B**" option. ACM66+AD8 exhibits a similar behavior as WAV66+AD8 with the exception that divergence free is also possible for the "filter all" option for WAV66+AD8 for $T < 0.7$, whereas the ACM66+AD8 losses its divergence free preservation at a much earlier time. The behavior of WAV66 ($d = 0$) and ACM66 ($d = 0$) is similar.

The resolution of the global structure of the density contours is well captured by all five methods. However, small fine structures were captured by the ACM-filter and WAV-filter schemes on a 101×101 grid, and not by MUSCL, Harten-Yee and WENO5.

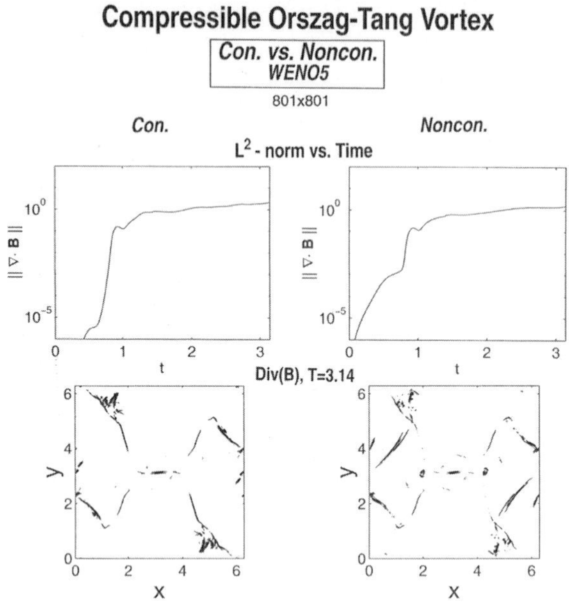

Fig. 8. L^2 norm of $\nabla \cdot \mathbf{B}$ in time (top) and $\nabla \cdot \mathbf{B}$ contours (bottom) by WENO5 for the conservative (left) and the non-conservative (right) systems.

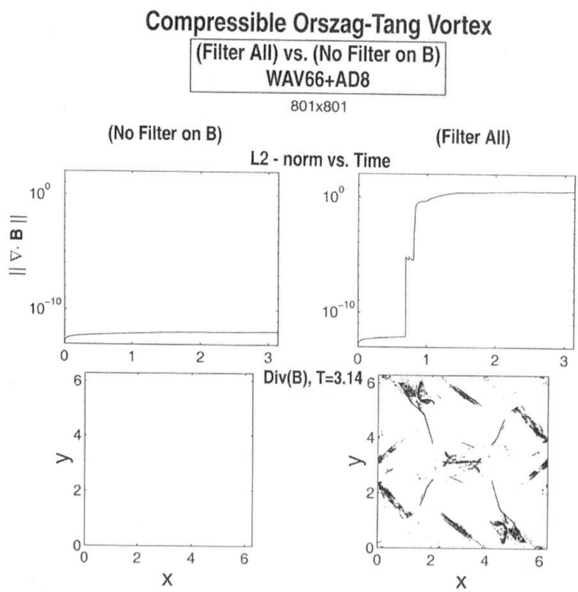

Fig. 9. WAV66+AD8 L^2 norm of $\nabla \cdot \mathbf{B}$ in time (top row), $\nabla \cdot \mathbf{B}$ contours at $T = 3.14$ (bottom row). No non-linear filter on $\mathbf{B}$ (left) and non-linear filter on all components (right).

5 Concluding Remarks

A natural and efficient high order filter approach in the sense of not needing traditional divergence cleaning for the minimization of the $\nabla \cdot \mathbf{B}$ numerical error was proposed and validated using three 2-D compressible MHD test cases. Five levels of grid refinement on three different flow types were compared with three standard high-resolution shock-capturing schemes, namely, a second-order MUSCL and Harten-Yee upwind TVD schemes, and the fifth-order WENO scheme. The role that the proper treatment of the corresponding numerical boundary conditions can play on the effect of reducing the $\nabla \cdot \mathbf{B}$ numerical error was studied in [30]. Among the three test cases we can safely conclude that divergence free high order filter schemes for the compressible MHD equations are possible without the need of standard divergence cleaning. These schemes are applicable to a wide variety of flow physics problems. Application of these schemes to viscous MHD flows with resistivity and multiscale structure is forthcoming.

References

[1] P. Cargo and G. Gallice, *Roe Matrices for Ideal MHD and Systematic Construction of Roe Matrices for Systems of Conservation Laws*, J. Comput. Phys., **136**(1997), 446-466.

[2] A. Dedner, F. Kemm, D. Kröner, C.-D. Munz, T. Schnitzer, and M. Wesenberg, *Hyperbolic Divergence Cleaning for the MHD Equations*, J. Comput. Phys., **175**(2002), 645-673.

[3] R.B. Dahlburg and J.M. Picone, *Evolution of the Orszag-Tang Vortex System in a Compressible Medium. I. Initial Average Subsonic Flow,* Phys. fluid B, **1**(1989), 2153-2171.

[4] W. Dai and P. R. Woodward, *A Simple Finite Difference Scheme for Multidimensional Magnetohydrodynamical Equations* J. Comput. Phys., **142**(1998), 331-369.

[5] W. Dai and P.R. Woodward, *On the Divergence-Free Condition and Conservation Laws in Numerical Simulations for Supersonic Magnetohydrodynamic Flows,* Astrophys. J., **494**(1998), 317-335.

[6] H. De Sterck, *Multi-Dimensional Upwind Constrained Transport on Unstructured Grids for Shallow Water Magnetohydrodynamics*, AIAA Paper 2001-2623, (2001).

[7] C.R. Evans and J.F. Hawley, *Simulation of Magnetohydrodynamic Flows: A Constrained Transport Method*, Astrophys. J. **332**(1988), 659-677.

[8] A. Frank, T.W. Jone, D. Ryu and J.B. Gaalaas, *The Magnetohydrodynamic Kelvin-Helmholtz Instability: A Two- Dimensional Study,* Astrophys. J. **460**(1996), 777-793.

[9] G. Gallice, *Systéme DÉuler-Poisson, Magnétohydrodynamique et Schemeas de Roe*, PhD Thesis, L'Université Bordeaux I, 1997.

[10] S.K. Godunov *Symmetric Form of the Equations of Magnetohydrodynamics,* Numerical Methods for Mechanics of Continuum Medium, **13**(1972), 26-34.

[11] A. Harten and J.M. Hyman, *A Self-Adjusting Grid for the Computation of Weak Solutions of Hyperbolic Conservation Laws,*, J. Comput. Phys., **50**(1983), 235-269.

[12] G.-S. Jiang and C.-W. Shu, *Efficient Implementation of Weighted ENO schemes*, J. Comput. Phys., **126** (1996), 202-228.

[13] A. Malagoli, G. Bodo and R. Rosner, *On the Nonlinear Evolution of Magnetohydrodynamic Kelvin-Helmholtz Instabilities'* Astrophys. J. **456**(1996), 708-716.

[14] S.A. Orszag and C.M. Tang *Small Scale Structure of Two-Dimensional Magnetohydrodynamic* Turbulence, J. Fluid Mech., **90**(1979), 129-143.

[15] J.M. Picone and R.B. Dahlburg, *Evolution of the Orszag-Tang Vortex System in a Compressible Medium. II. Supersonic Flow,* Phys. Fluid B, **3**(1991), 29-44.

[16] K.G. Powell, *An Approximate Riemann Solver for Magnetohydrodynamics (That works in More than One Dimension),* ICASE-Report 94-24, NASA Langley Research Center, April 1994.

[17] K.G. Powell, P.L. Roe, T.J. Linde, T.I. Gombosi and D.L. De Zeeuw, *A Solution-Adaptive Upwind Scheme for Ideal Magnetohydrodynamics*, J. Comput. Phys., **154**(1999), 284-309.

[18] B. Sjögreen and H. C. Yee, *Multiresolution Wavelet Based Adaptive Numerical Dissipation Control for Shock-Turbulence Computation*, RIACS Technical Report TR01.01, NASA Ames research center (Oct 2000), to appear, J. Scient. Computing.

[19] B. Sjögreen and H. C. Yee, *Grid Convergence of High Order Methods for Multiscale Complex Unsteady Viscous Compressible Flows*, RIACS Technical Report TR01.06, April, 2001, NASA Ames research center; AIAA 2001-2599, Proceedings of the 15th AIAA CFD Conference, June 11-14, 2001, Anaheim, CA., also, J. Comput. Phys., **185**(2003), 1-26.

[20] B. Sjögreen and H. C. Yee, *Low Dissipative High Order Numerical Simulations of Supersonic Reactive Flows*, RIACS Report TR01-017, NASA Ames Research Center (May 2001); Proceedings of the ECCOMAS Computational Fluid Dynamics Conference 2001, Swansea, Wales, UK, September 4-7, 2001.

[21] B. Sjögreen and H. C. Yee, *Analysis of High Order Difference Methods for Multiscale Complex Compressible Flows*, Proceedings of the 9th International Conference on Hyperbolic Problems, March 25-29, 2002, Pasadena, CA.

[22] B. Sjögreen and H.C. Yee, *Efficient Low Dissipative High Order Schemes for Multiscale MHD Flows, I: Basic Theory*, AIAA 2003-4118, Proceedings of the 16th AIAA/CFD Conference, June 23-26, 2003, Orlando, Fl.

[23] G. Tóth, *The div B=0 Constraint in Shock-Capturing Magnetohydrodynamic Codes*, J. Comput. Phys., **161**(2000), 605-652.

[24] M. Vinokur, *A rigorous derivation of the MHD Equations Based only on Faraday's and Ampére's Laws,* Presentation at LANL MHD Workshop on $\nabla \cdot \mathbf{B}$ Cleaning, August, 1996.

[25] M. Vinokur and H.C. Yee, *Extension of Efficient Low Dissipative High Order Schemes for 3-D Curvilinear Moving Grids,* NASA TM 209598, June 2000.

[26] H.C. Yee, *A Class of High-Resolution Explicit and Implicit Shock-Capturing Methods,* VKI Lecture Series 1989-04, March 6-10, 1989, also NASA TM-101088, Feb. 1989.

[27] H.C. Yee, N.D. Sandham, N.D., and M.J. Djomehri, *Low Dissipative High Order Shock-Capturing Methods Using Characteristic-Based Filters,* J. Comput. Phys., **150**(1999), 199-238.

[28] H.C. Yee, M. Vinokur, M., and M.J. Djomehri, *Entropy Splitting and Numerical Dissipation,* J. Comput. Phys., **162**(2000), 33-81.

[29] H.C. Yee and B. Sjögreen, *Designing Adaptive Low Dissipative High Order Schemes for Long-Time Integrations,* Turbulent Flow Computation, (Eds. D. Drikakis & B. Geurts), Kluwer Academic Publisher (2002); also RIACS Technical Report TR01-28, Dec. 2001.

[30] H.C. Yee and B. Sjögreen, *Efficient Low Dissipative High Order Schemes for Multiscale MHD Flows, II: Minimization of $\nabla \cdot \mathbf{B}$ Numerical Error* RIACS Report TR03.10, July 2003.

[31] K.S.Yee, *Numerical solution of initial boundary value problems involving Maxwell's equations in isotropic media,* IEEE Trans. Antennas Propagat., **14**(1966), 302–307.

[32] A.L. Zachary, A. Malagoli and P. Colella, *A Higher-Order Godunov Method for Multidimensional Ideal Magnetohydrodynamics,* SIAM J. Sci. Comput., **15**(1994), 263-284.

A

Colour Figures

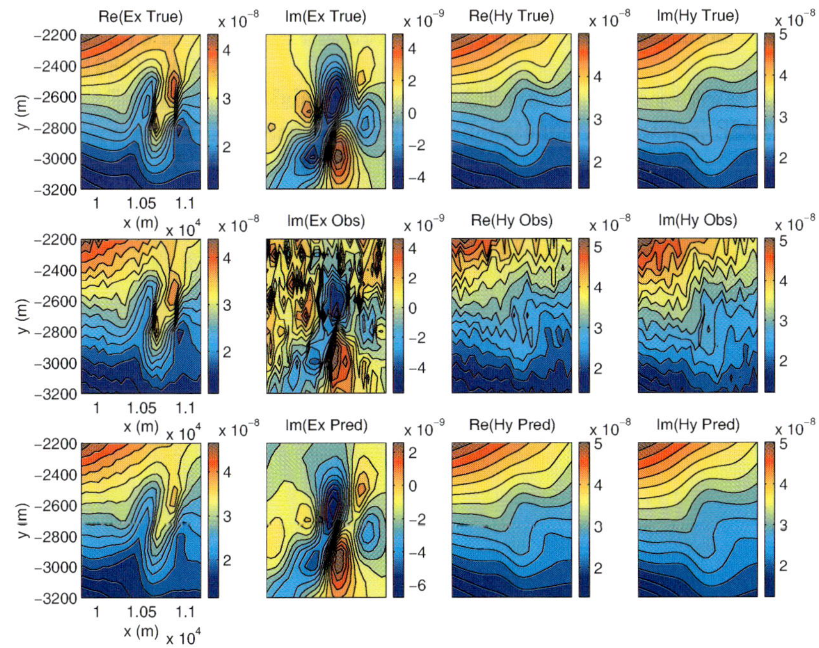

Fig. A.1. [U. Ascher et al] The accurate E_x, H_y data for 512 Hz are shown in the top row. The error contaminated data are shown in the middle row and the bottom row displays the data predicted from the inverted model.

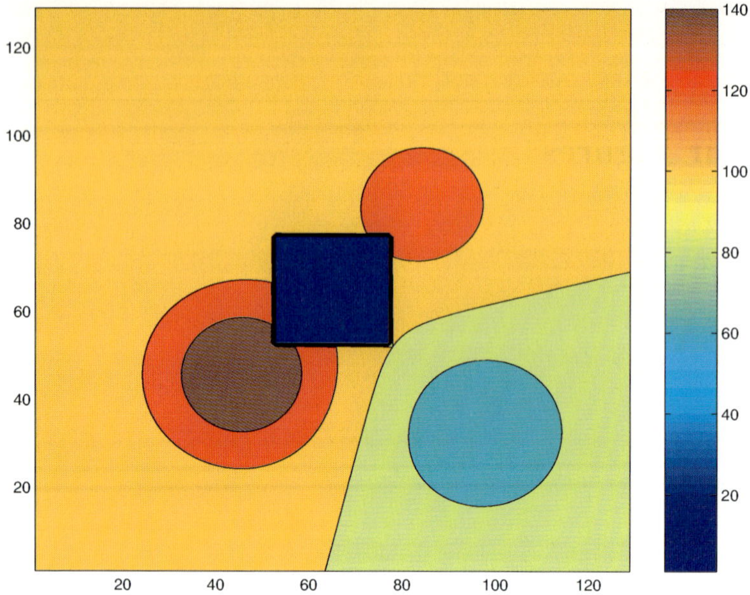

Fig. A.2. [U. Ascher et al] Contour plot of the "true model" for Example 3.

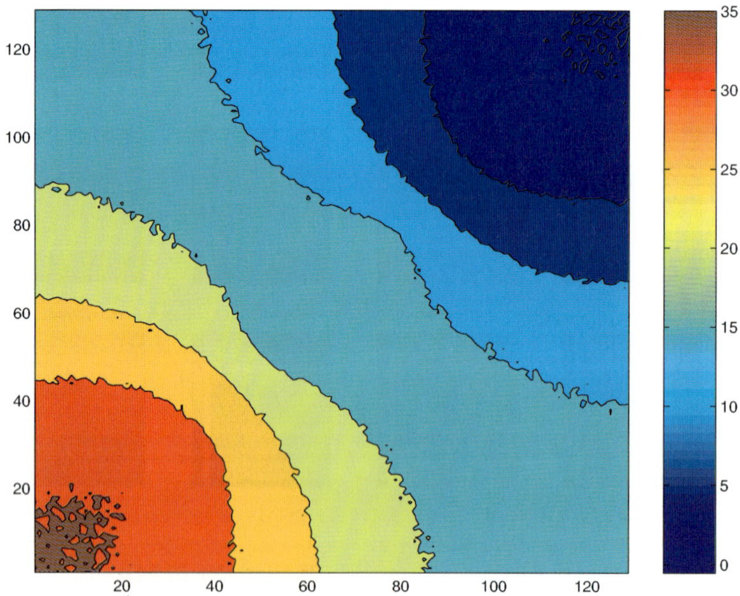

Fig. A.3. [U. Ascher et al] Data for Example 3.

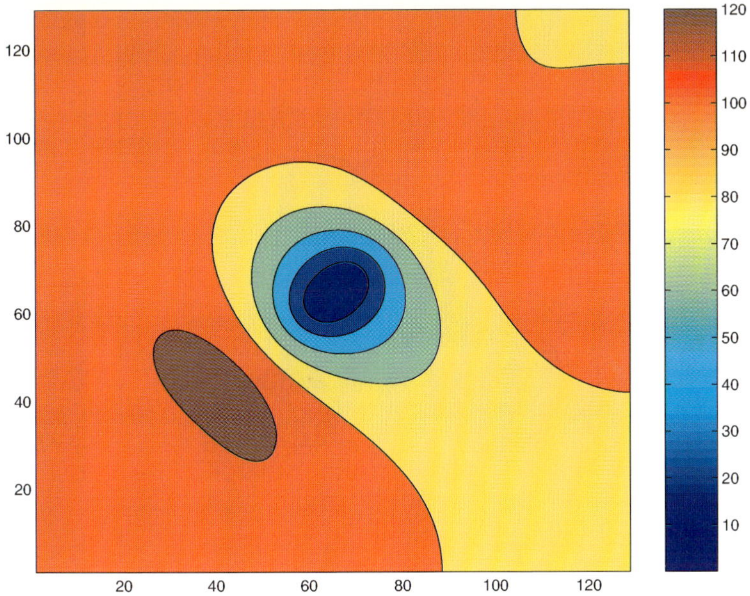

Fig. A.4. [U. Ascher et al] Recovered model using least squares with $\beta = 10^{-5}$.

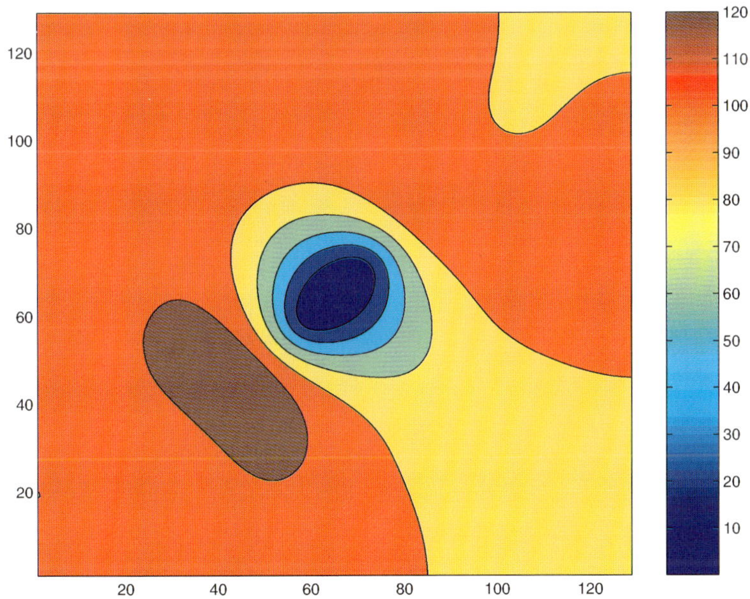

Fig. A.5. [U. Ascher et al] Recovered model using least squares with $\beta = 3 \times 10^{-6}$.

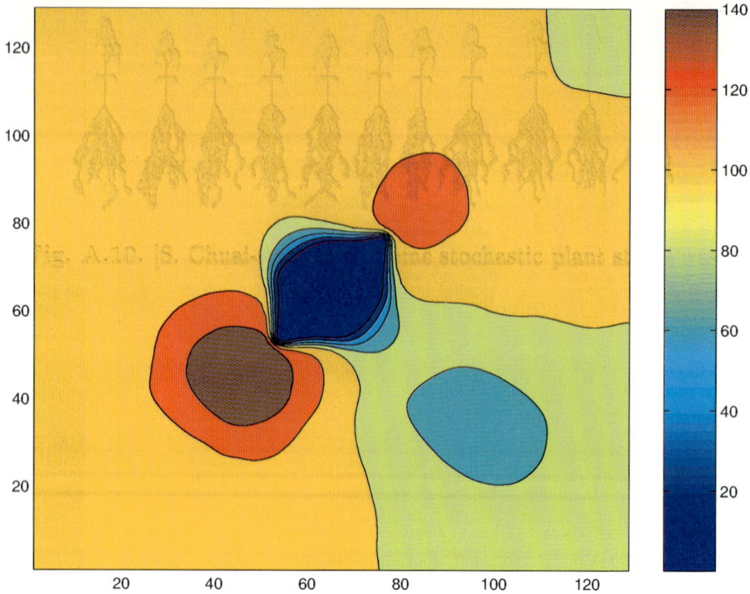

Fig. A.6. [U. Ascher et al] Recovered model using Huber's norm with $\beta = 10^{-5}$.

Fig. A.7. [U. Ascher et al] The noiseless data, viz., the field corresponding to the "true model" using the applied discretization.

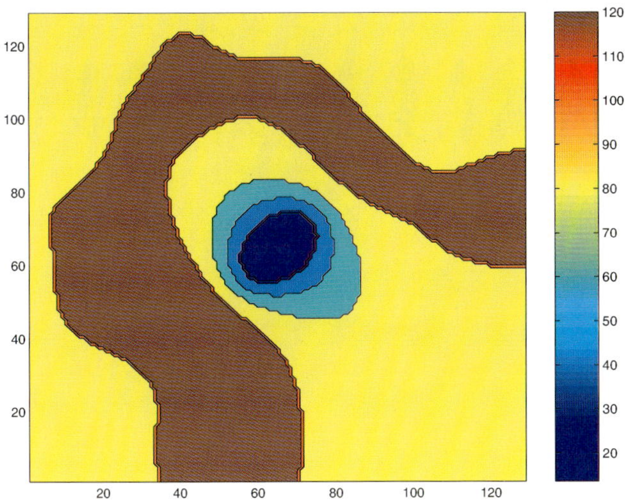

Fig. A.8. [U. Ascher et al] The model of Figure 8 replaced by a piecewise constant approximation with 5 constant values.

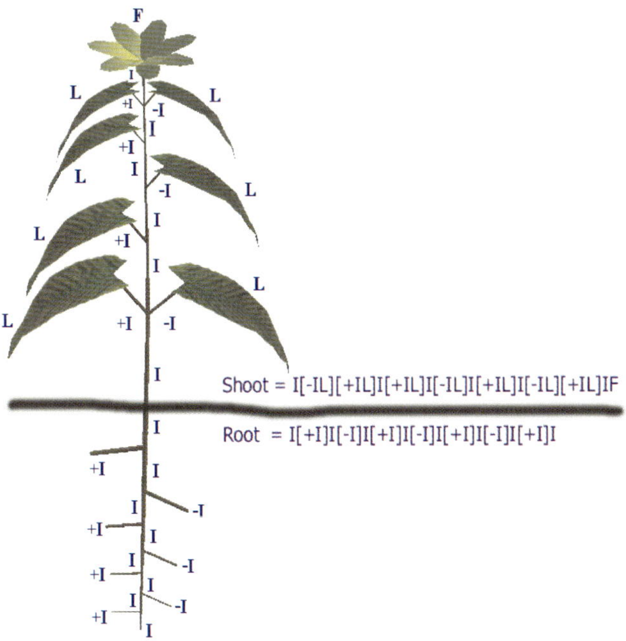

Fig. A.9. [S. Chuai-Aree et al] A simple L-System interpretation.

Fig. A.10. [S. Chuai-Aree et al] Some stochastic plant structures.

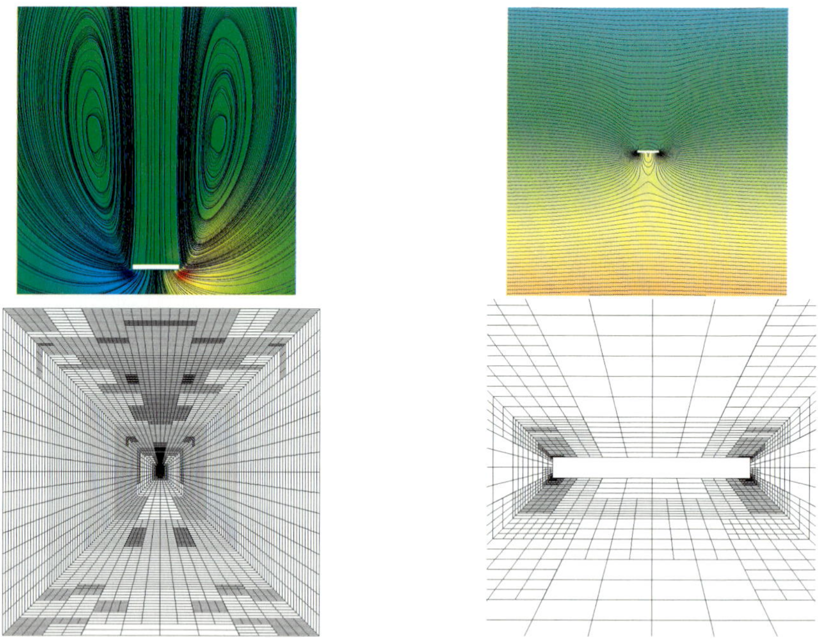

Fig. A.11. [S. Bönisch et al] (top left) Streamlines around the falling body for $\nu = 0.1$; (top right) pressure isolines; (bottom left and right) adaptive mesh obtains by means of (31) on a domain with diameter $D = 800$ and corresponding zoom around the body.

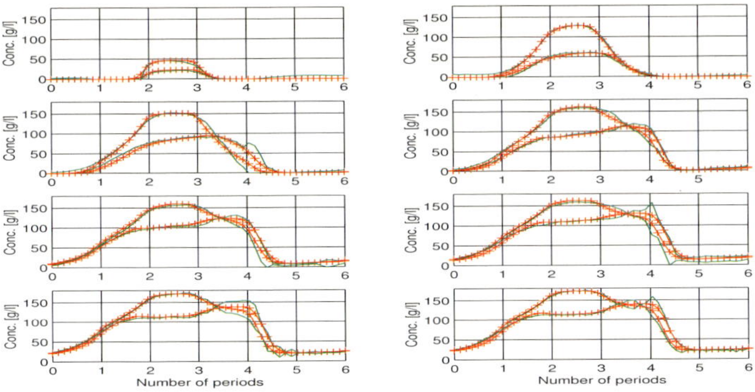

Fig. A.12. [A. Toumi et al] Comparison of experimental and simulated concentration profiles collected in the recycle line

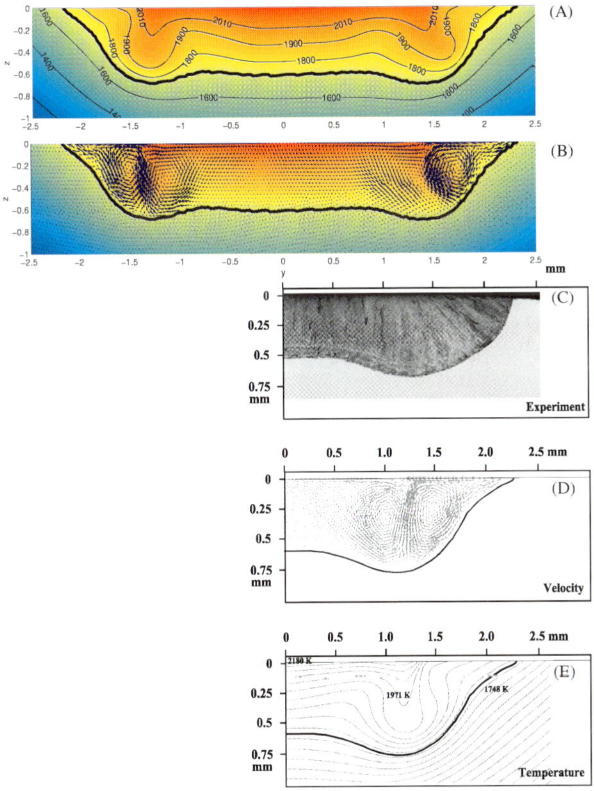

Fig. A.13. [M. Do-Quang et al] AP1-100A: Comparison of experimental and 2D, 3D numerical results after 1 second, (C)-(E) from [12]

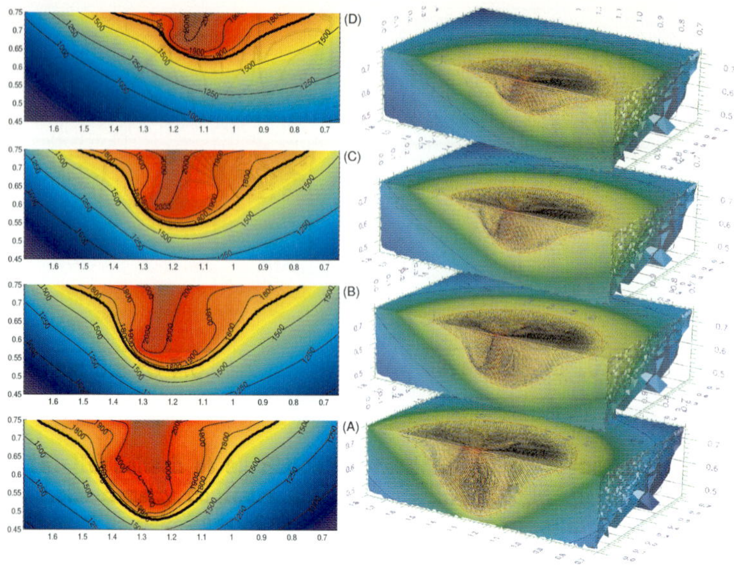

Fig. A.14. [M. Do-Quang et al] AP5-100A: Temperature and velocities fields of welding pool. $U_s = 0mm \cdot s^{-1}$ (A); $3mm \cdot s^{-1}$ (B); $6mm \cdot s^{-1}$ (C) and $9mm \cdot s^{-1}$ (D).

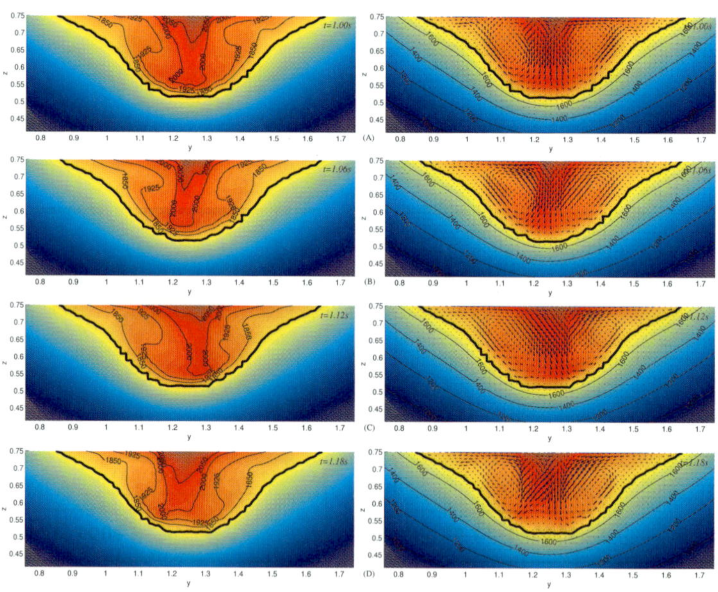

Fig. A.15. [M. Do-Quang et al] 3D simulation, temperature distributation and velocity field time dependent.

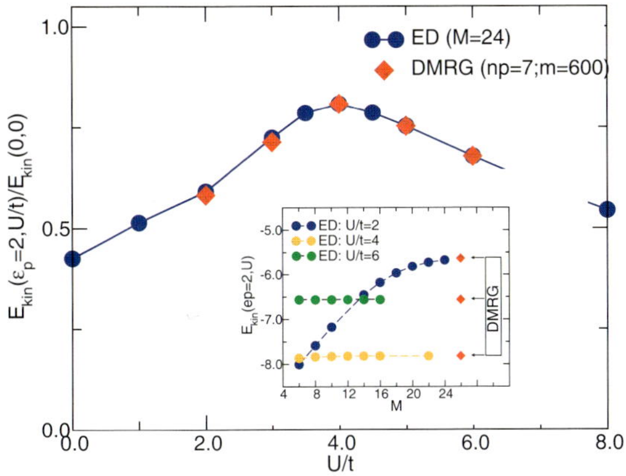

Fig. A.16. [G. Hager et al] Kinetic energy as a function of Hubbard interaction U/t with $g^2 = 2$ and $\omega_0/t = 1$. Inset: Comparison of ED and DMRG results for $E_{\rm kin}$ at different U with $g^2 = 2$ and $\omega_0/t = 1$; Convergence of ED results as a function of the cut-off parameter M is demonstrated. The corresponding results from DMRG calculations using $np = 6$ pseudo-sites and $m = 1000$ are represented by stars.

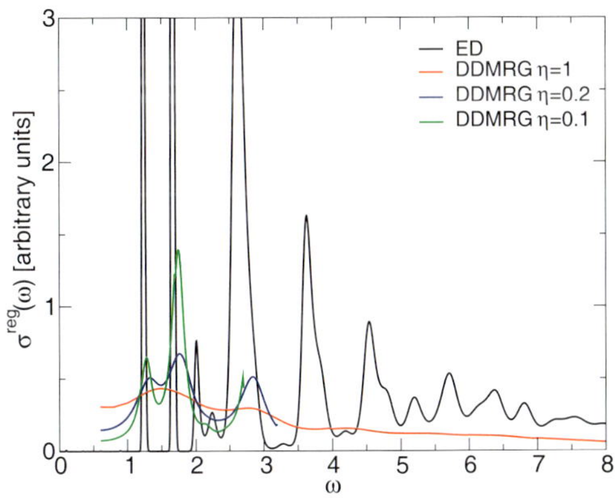

Fig. A.17. [G. Hager et al] Optical conductivity for the $N = 8$ site HHM at $U/t = 6$, $g^2 = 2$ and $\omega_0 = 1$. The DDMRG data was calculated at $m = 200$ using three different broadenings $\eta = 1$, $\eta = 0.2$ and $\eta = 0.1$, respectively. Curves for each η were drawn to a point where the CPU-time for the different runs is comparable.

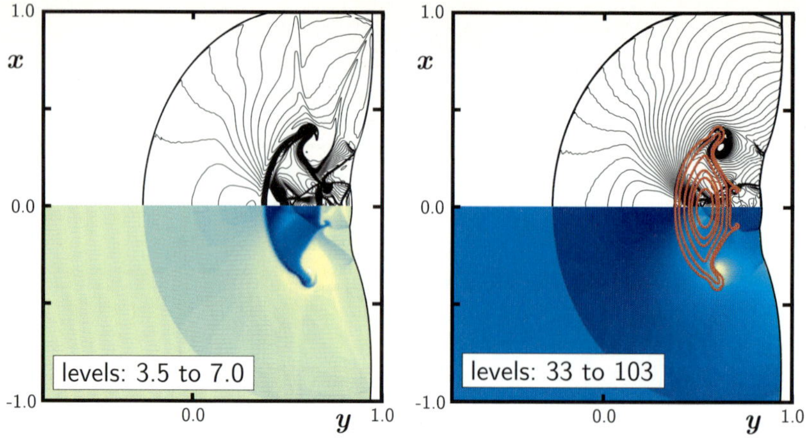

Fig. A.18. [R. Jeltsch et al] Shock interaction with a magnetized cloud at time $t = 0.3$. Left: density contours and contour lines. Right: pressure contours and contour lines. The field lines of the magnetic flux are superimposed in the pressure plot.

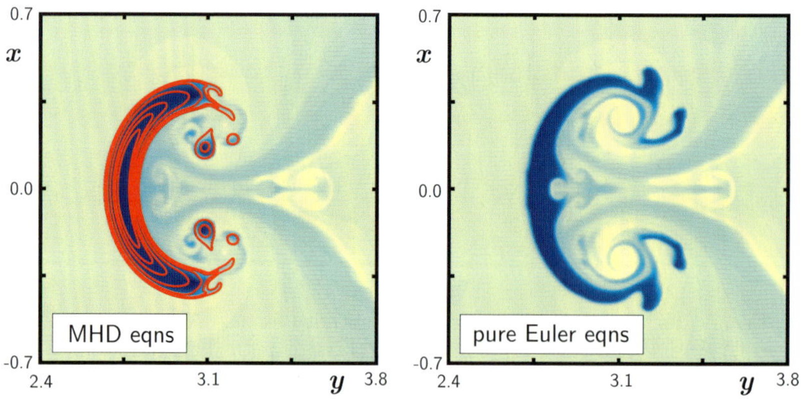

Fig. A.19. [R. Jeltsch et al] Simulation results for the shock interaction with a cloud at time $t = 0.5$. Left: magnetized cloud. Density contours with magnetic field lines superimposed. Right: non-magnetized cloud. Density contours. In both plots the contour levels have the range 3.5 - 6.5.

Belief control: Direct control:

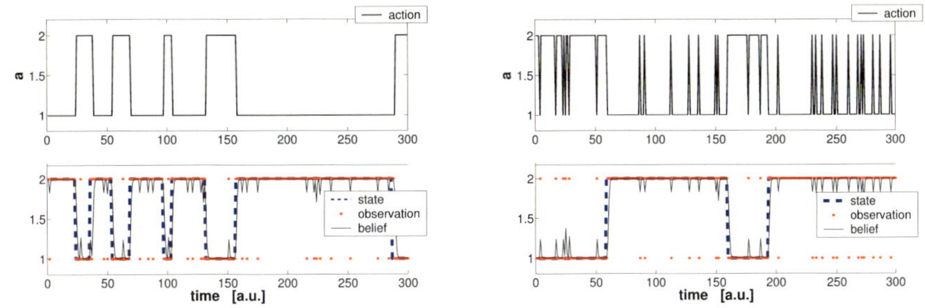

Fig. A.20. [C. Kreutz et al] The figures show typical realizations of belief control and direct control. Belief control (upper left plot) shows a clearly smoother policy than direct control (upper right plot), because of optimal smoothing properties of the belief.

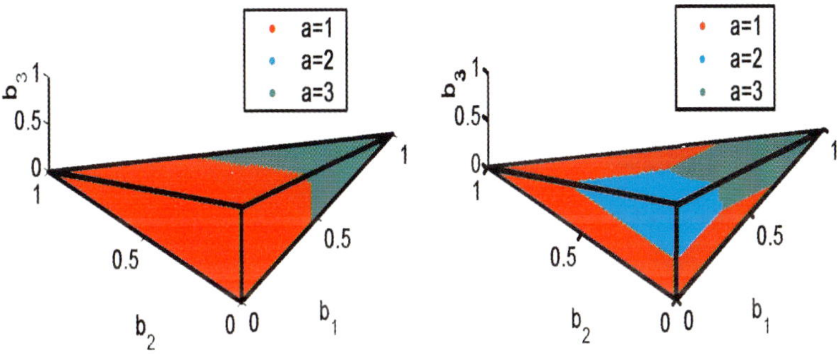

Fig. A.21. [C. Kreutz et al] Optimal belief control for CPAP in the case of slight affection (left) and heavy affection (right).

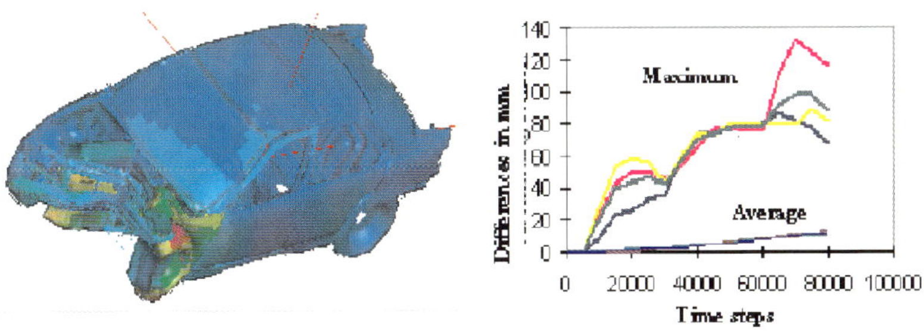

Fig. A.22. [L. Mei et al] BMW model after a 40% offset crash using PAM-CRASH and the differences between simulation runs on a 32 node IBM SP2.

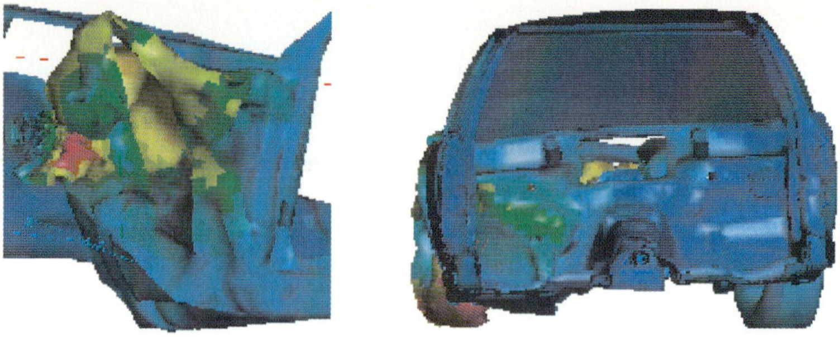

Fig. A.23. [L. Mei et al] Scatter of simulation results on the motor carrier and in the interior.

Fig. A.24. [L. Mei et al] Simcluster as color for BMW testcase of whole car and the interior at time 80ms.

Fig. A.25. [L. Mei et al] Development of the clusters with time (70ms, 60ms, 50ms, 40ms).

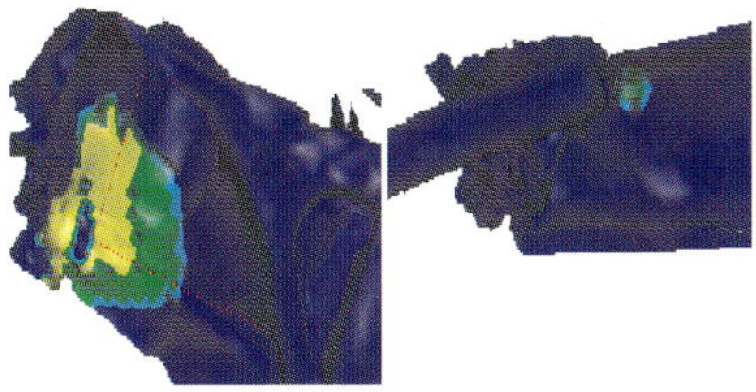

Fig. A.26. [L. Mei et al] Simcluster for the motor carrier at time 35ms (left) and at 28ms (right) for its inner part.

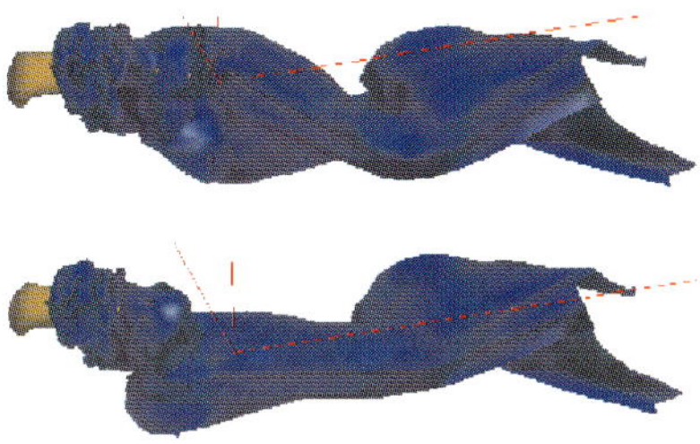

Fig. A.27. [L. Mei et al] Top view on the motor carrier at 80ms for two extremely different simulation runs.

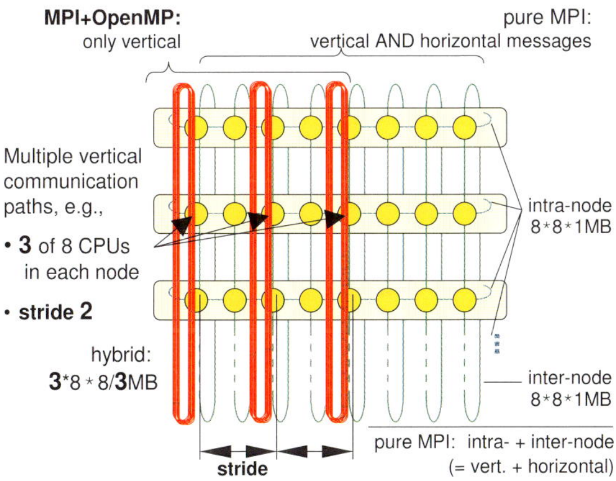

Fig. A.28. [R. Rabenseifner et al] Communication pattern with *hybrid MPI+OpenMP* style and with *pure MPI* style.

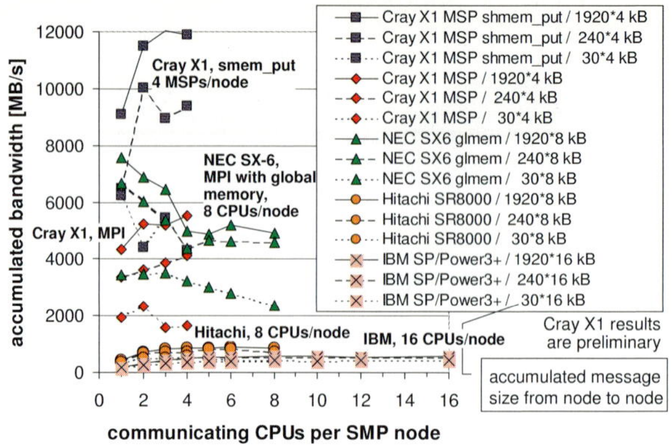

Fig. A.29. [R. Rabenseifner et al] Aggregated bandwidth per SMP node.

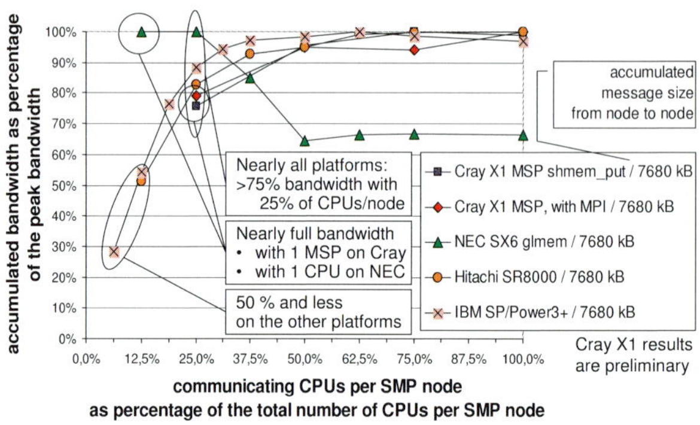

Fig. A.30. [R. Rabenseifner et al] Aggregated bandwidth per SMP node.

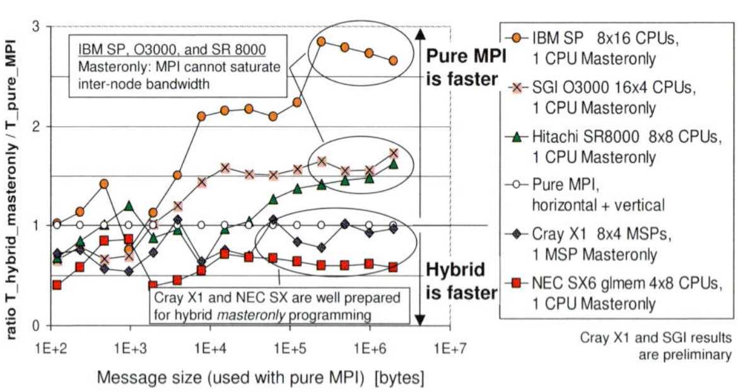

Fig. A.31. [R. Rabenseifner et al] Ratio of hybrid communication time to pure MPI communication time.

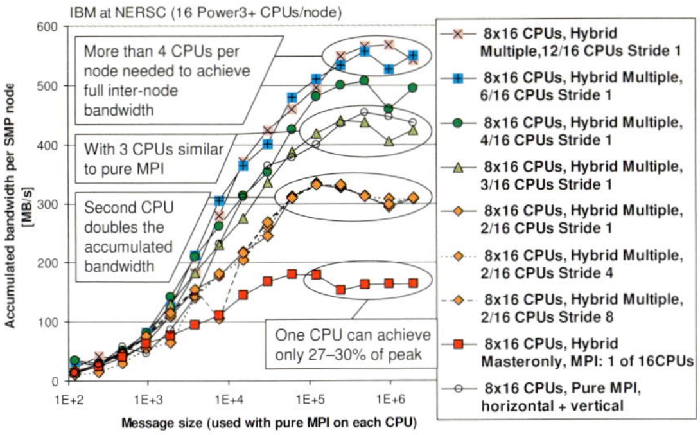

Fig. A.32. [R. Rabenseifner et al] Aggregated bandwidth per SMP node on IBM SP with 16 Power3+ CPUs per node.

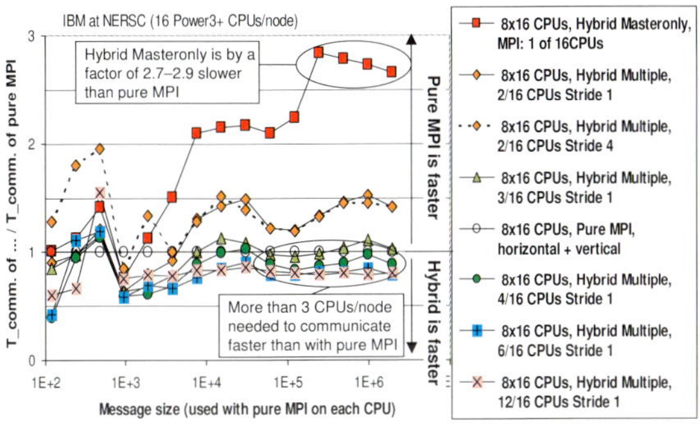

Fig. A.33. [R. Rabenseifner et al] Ratio of communication time in hybrid models to pure MPI programming on IBM SP.

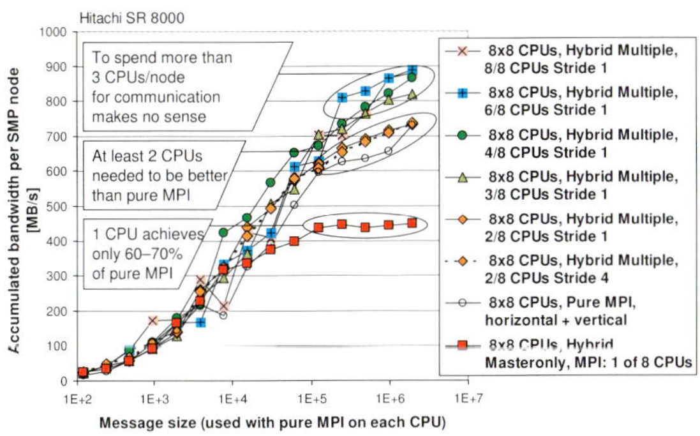

Fig. A.34. [R. Rabenseifner et al] Aggregated bandwidth per SMP node on Hitachi SR 8000.

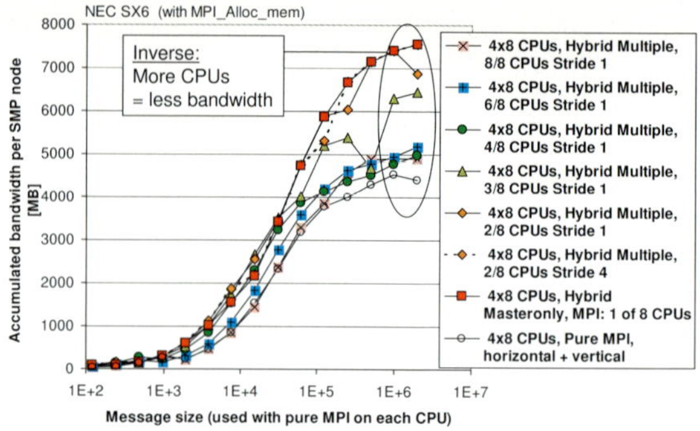

Fig. A.35. [R. Rabenseifner et al] Aggregated bandwidth per SMP node on NEC SX-6.

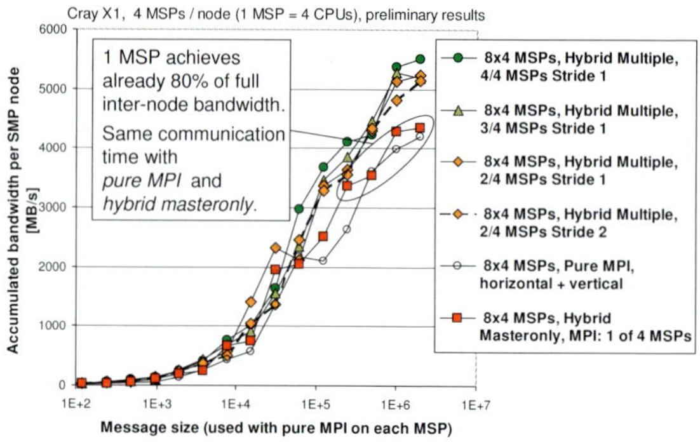

Fig. A.36. [R. Rabenseifner et al] Aggregated bandwidth per SMP node on Cray X1, MSP-based MPI-parallelization.

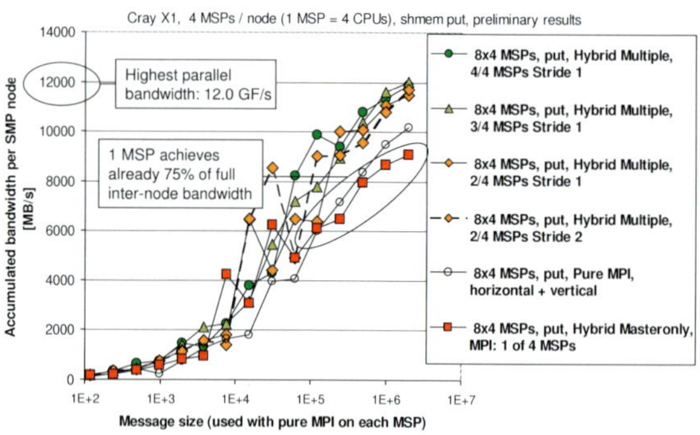

Fig. A.37. [R. Rabenseifner et al] Aggregated bandwidth per SMP node on Cray X1. MSP-based and MPI_Sendrecv is substituted by shmem_put.

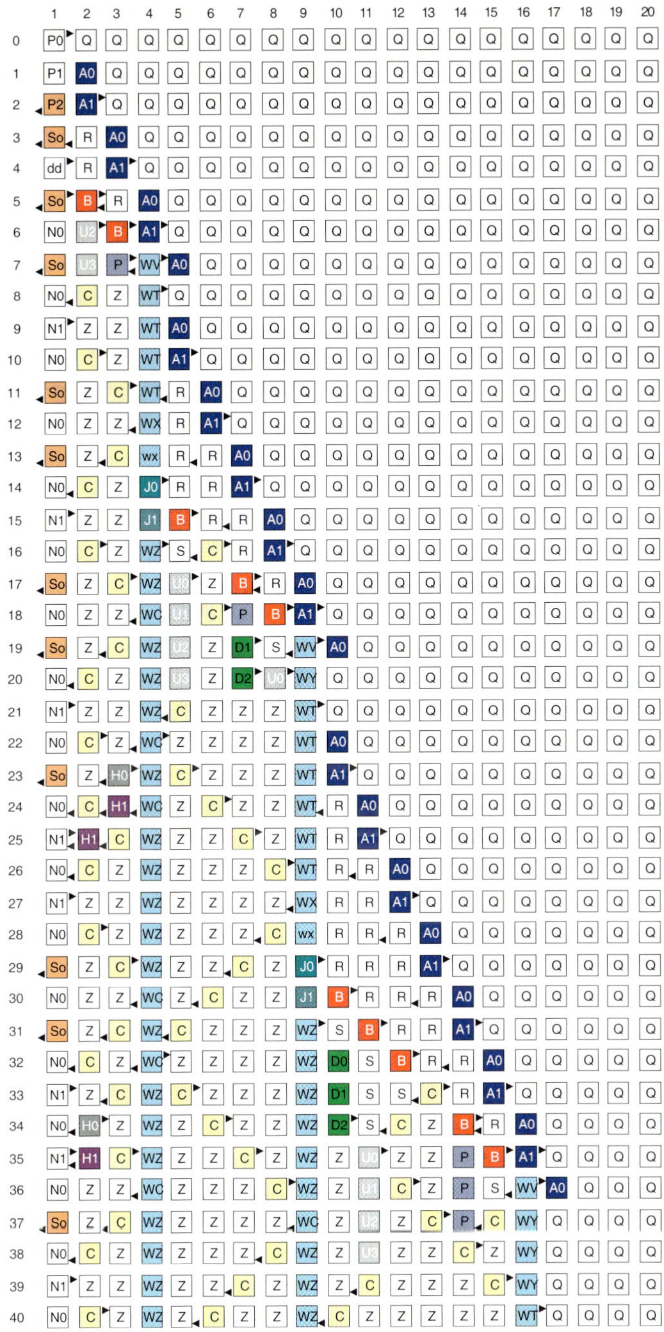

Fig. A.38. [H. Umeo et al] A configuration of real-time generation of primes on CA_{1-bit} with 34 states.

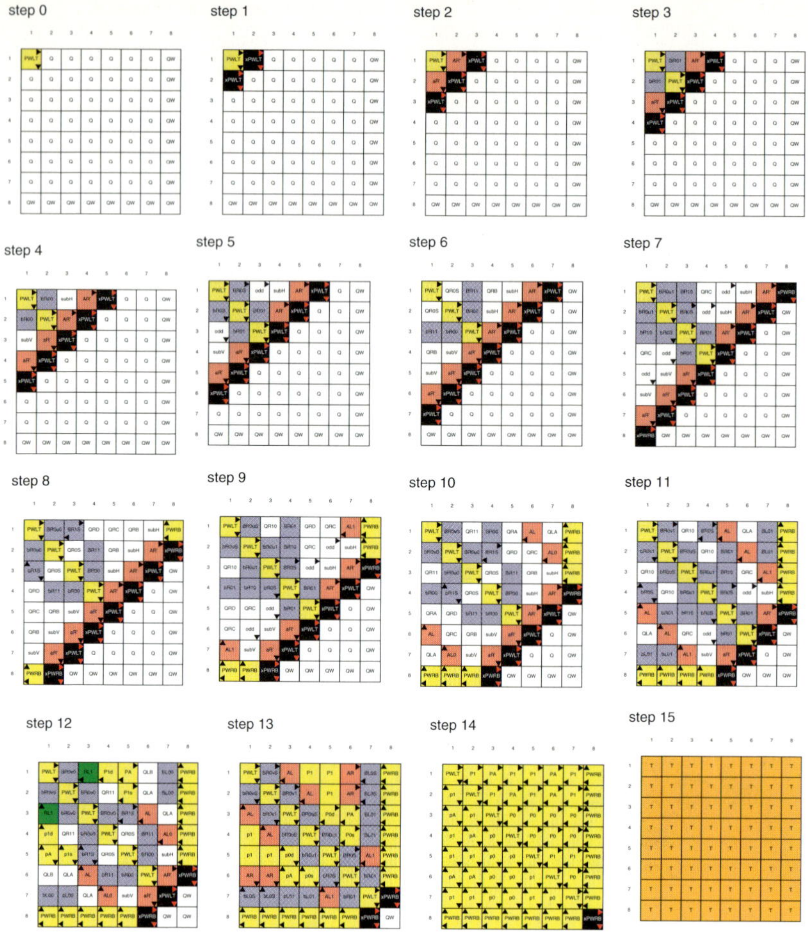

Fig. A.39. [H. Umeo et al] Snapshots of the $(2n-1)$-step square firing squad synchronization algorithm with the general on the north west corner.

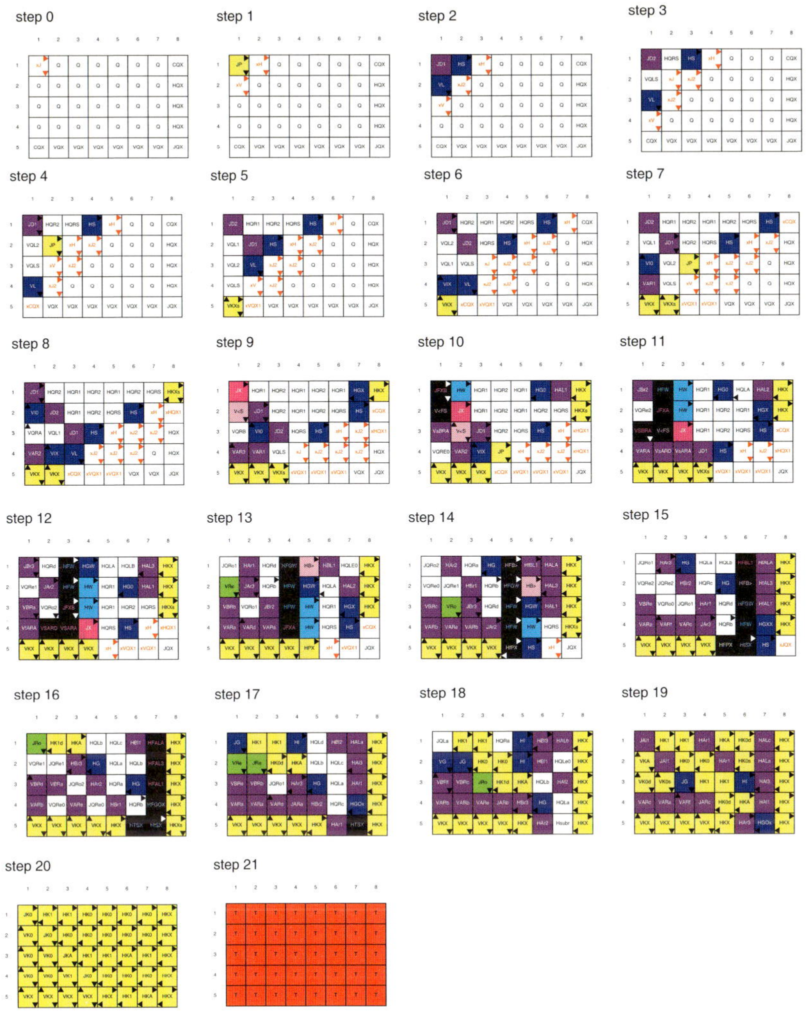

Fig. A.40. [H. Umeo et al] Snapshots of our rectangular firing squad synchronization algorithm with the general at the north-west corner.

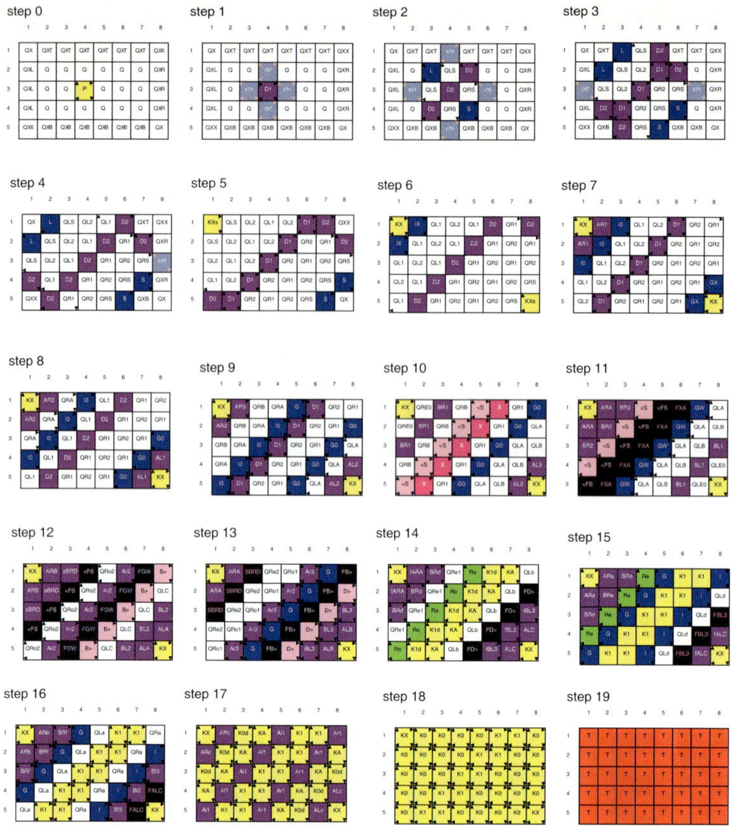

Fig. A.41. [H. Umeo et al] Snapshots of our generalized rectangular firing squad synchronization algorithm operating on an array of size 5×8 with the general on $C_{3,4}$.

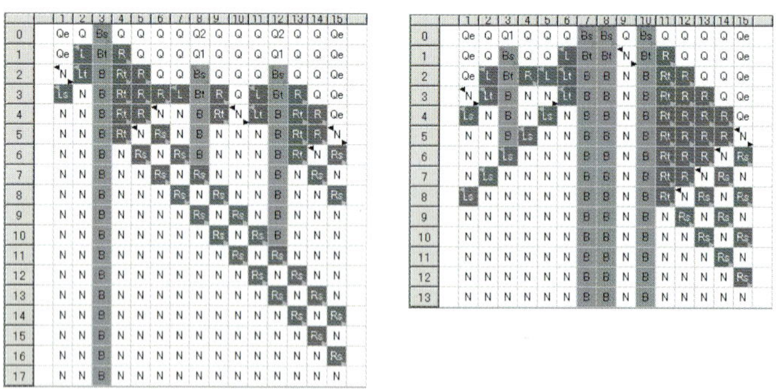

Fig. A.42. [H. Umeo et al] Snapshots of a 12-state implementation of the early bird problem on $CA_{1-\mathrm{bit}}$.

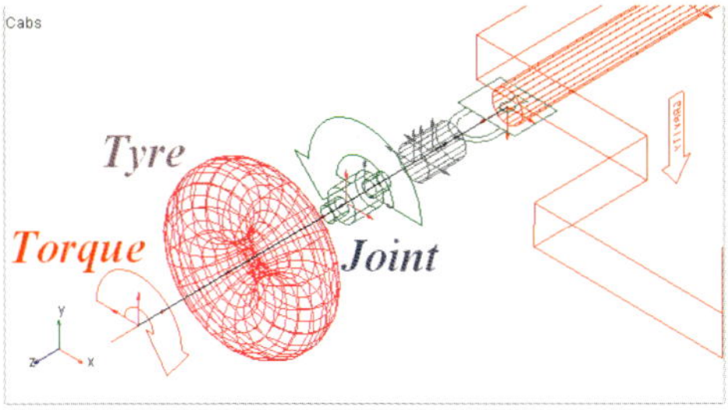

Fig. A.43. [D. Wang et al] Driving wheel model.

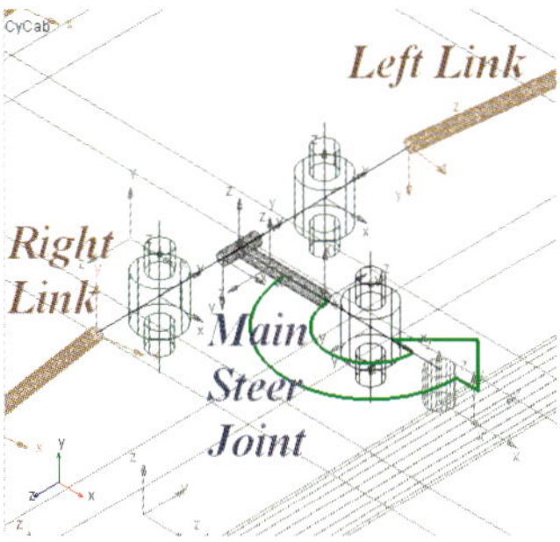

(a) Main Steer Joint

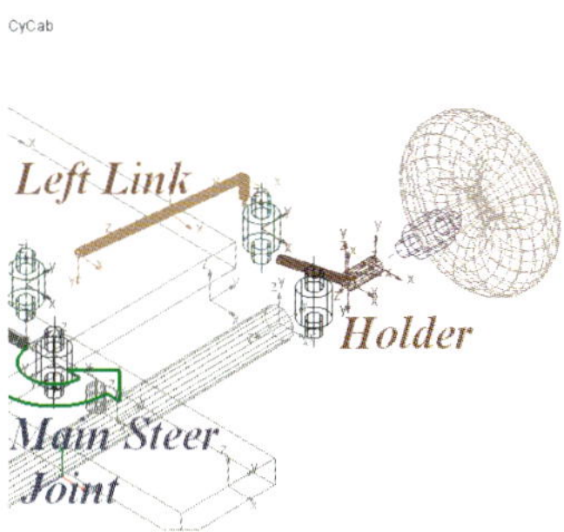

(b) Links to a wheel

Fig. A.44. [D. Wang et al] Steering mechanism.

Fig. A.45. [D. Wang et al] ADAMS model of vehicle platooning.

Printing: Mercedes-Druck, Berlin
Binding: Stein+Lehmann, Berlin